PRINCIPLES OF HUMAN ANATOMY

Gerard J. Tortora
Bergen Community College

Canfield Press San Francisco
A Department of Harper & Row, Publishers, Inc.
New York Hagerstown London

Illustrators: Line illustrations are by *Nelva B. Richardson,* with the exception of the following figures, which were prepared by *Russell Peterson* (1–1, 1–2, 1–6, 1–7, 1–8, 1–9, 2–1, 2–3, 2–4, 2–5, 2–6, 2–7, 2–8, 2–9, 2–10, 2–11, 2–12, 2–13, 2–14, 2–15, 2–16, 2–17, 4–5, 5–5, 6–1, 6–8, 7–1, 7–3, 7–4, 7–6, 7–8, 7–9, 7–10, 7–12, 7–13, 7–14, 7–15, 7–16, 7–17, 8–1, 8–2, 8–4, 8–5, 8–6, 8–7, 8–8, 8–9, 9–1, 9–2, 9–3, 9–5, 11–2, 11–9, 11–10, 11–11, 11–12, 11–14, 11–15, 11–16, 11–18, 11–19, 11–20, 11–23, 11–24, 11–25a, 14–6, 16–1, 16–3, 16–4, 16–6a, 16–8, 16–9, 16–10, 16–11, 16–12, 16–13, 16–14, 16–15, 16–16, 17–1, 17–8, 17–12, 19–4a, 19–6, 19–10, 19–11, 19–12, and Exhibits 1–1 and 12–1), and the following figures, which were prepared by *Marsha J. Dohrmann* (12–1, 12–2, 12–3a, 12–4, 13–1, 13–2, 13–3, 13–4, 13–5, 13–6, 13–7, 13–8, 13–9, 13–10, 13–11, 13–12, 13–13, 13–14, 14–1, 14–2, 14–3a, 14–4, 14–5, 14–7, 14–8, 14–9, 14–11, 15–1, 15–2, 15–3, and Exhibits 3–1 and 3–2).

Designer: Rita Naughton

Cover photograph: © Elaine Faris Keenan, 1976, NEST, S.F.

Principles of Human Anatomy

Library of Congress Cataloging in Publication Data

Tortora, Gerard J
Principles of human anatomy.

Bibliography: p.
Includes index.
1. Human physiology. 2. Anatomy, Human.
I. Title.
QP34.5.T68 612 76–51306
ISBN 0–06–388775–4

78 79 80 10 9 8 7 6 5 4 3

CONTENTS

PREFACE

AUDIENCE

Designed for the introductory course in human anatomy, *Principles of Human Anatomy* assumes no previous study of the human body. The text has been geared to students in biological, medical, and health-oriented programs. Among the students specifically served by this volume are those aiming for a career as a nurse, medical assistant, physician's assistant, medical laboratory technologist, radiologic technologist, respiratory therapist, dental hygienist, physical therapist, mortician, and medical record keeper. However, because of the scope of the text, *Principles of Human Anatomy* may also be useful for students in the liberal arts and physical education.

OBJECTIVES

Because the subject matter of human anatomy is an exceedingly large and complex body of knowledge to present in an introductory course, I have attempted to present unified concepts and data considered useful to a basic understanding of the structure of the human body. I have minimized data unessential to this objective. My second principal objective has been to present the material at a reading level that can be handled by the average student. In accordance with this objective, I have not avoided technical vocabulary or vital, but difficult, concepts. Rather, I have attempted to develop step-by-step, easy-to-comprehend explanations of each concept.

THEMES

This textbook departs from the approaches of other anatomy texts in that somewhat more emphasis is given to physiology and disorders. I feel that it is motivationally and conceptually more effective to present basic anatomy with some reference to function. This framework is especially true of certain systems of the body, such as the nervous system. It makes little sense to present students with anatomical detail without applying the anatomy to function because structure determines function. By including function, this text will give students a better understanding of anatomical concepts. Moreover, I feel that a study of anatomy should include frequent reference to clinical situations. For this reason, I have included disorders that can be understood once the normal anatomy is learned. The disorders are treated under a special heading, called "Applications to Health," in selected chapters. The basic content of this book is anatomy, but by presenting it within the context of physiology and disorders, it should be more readily assimilated by the student.

ORGANIZATION

The book is organized by systems rather than by regions. The first chapter provides students with an introduction to the levels of structural organization, body cavities, anatomical terms, directional terms, planes of the body, surface anatomy, and units of measurement. Then, a generalized cell is used to demonstrate the basic features of cells. Tissue organization is presented through descriptions of the structure, functions, and locations of the principal kinds of epithelium and connective tissue. The histology of bone, muscle, nervous tissue, and blood is dealt with under the relevant organ systems. The skin and its accessory structures are utilized to acquaint the student with the organ and system level of organization.

The first body system studied in detail is the skeletal system. This is accomplished by examining the principal features of osseous tissue, the axial skeleton, the appendicular skeleton, and articulations. Then the muscular system is analyzed through a study of muscle tissue and the locations and actions of the principal muscles of the body.

The student is next introduced to the blood vascular and lymphatic systems. The chapter includes a study of blood, interstitial fluid, lymph, the heart, blood vessels, circulatory routes, and lymphatic structures.

The next major area of emphasis is the nervous system. Students are introduced to the structure and physiology of nervous tissue, the spinal cord and spinal nerves, the brain and cranial nerves, and the autonomic system. The structure and function of the visual, auditory, gustatory, and olfactory receptors are also considered.

Attention is then turned to a discussion of the anatomy and physiology of the endocrine system and reproductive systems. Here the interrelationships between hormones and reproduction are established.

The text concludes with a description of the respiratory system, digestive system, and urinary system.

Discussions of all major body systems include some reference to physiology, disorders, and a listing of pertinent key terms.

SPECIAL FEATURES

The text contains a number of special learning aids for students, including the following:

1. "Student Objectives" appear at the beginning of each chapter. Each objective describes a knowledge or skill the student should acquire while studying the chapter. (See "Note to the Student" for an explanation of how the objectives can be utilized.) End-of-chapter "Review Questions and Problems" are designed specifically to help meet the stated objectives. In addition, each "Chapter Summary in Outline" provides a checklist of major topics the student should learn. The end-of-chapter material concludes with a list of "Selected Readings."
2. The health-science student is generally expected to learn a great deal about the anatomy of certain organ systems—specifically, bones, skeletal muscles, blood vessels, and nerves. In these high anatomy areas, I have pulled many anatomical details out of the narrative and placed them in Exhibits, most of which are closely tied to illustrations. This method organizes the data and deemphasizes rote learning of concepts presented in the narrative.
3. An unusually large number of disorders is described in the section entitled "Applications to Health" in appropriate chapters. The topics provide review of normal anatomy and physiology and allow the student to see why the study of anatomy is fundamental to a career in any of the health fields.
4. Glossaries of selected "Key Terms" appear at the end of every chapter that deals with a major body system.
5. Photomicrographs and electron micrographs are frequently accompanied by adjacent labeled diagrams that amplify and aid observation.
6. There are numerous roentgenograms, especially of bones. These are labeled and designed to provide students with an opportunity to transfer anatomical knowledge to clinical situations.
7. Many students find muscle identification an onerous task. To help the student, I have provided the following learning aids: The illustrations of muscles are shown with duplicates of the drawings that are used for bone identification. In this way, the student is given consistent points of reference. I have also presented a brief section on the criteria for naming skeletal muscles. Each Exhibit dealing with muscles also contains a listing of prefixes, suffixes, roots, and definitions for each muscle discussed. These will help the student understand why a particular skeletal muscle is so named.
8. Another distinctive feature of this textbook is the inclusion of a very large number of photomicrographs of various tissues of the body. These are designed to help the student understand anatomy at the microscopic level.
9. A comprehensive "Glossary" of important terms used in the textbook, with a pronunciation key, concludes the book.

SUPPLEMENTARY MATERIAL

A complimentary Instructor's Manual accompanies *Principles of Human Anatomy.* For each chapter of the textbook, the Manual contains a listing of key instructional concepts, selected audiovisual materials, and twenty multiple-choice questions. A directory of the distributors of the audiovisual materials is also provided. The questions have been carefully designed to evaluate student understanding of data, concepts, clinical situations, and their applications.

ACKNOWLEDGMENTS

Since the inception of this textbook, Canfield Press has provided me with the services of several individuals who reviewed various portions of the manuscript at different times. These reviewers provided me with invaluable assistance in preparing the manuscript. Among those to whom I wish to express my deepest gratitude for reviewing the outline of the manuscript are George W. Bond, Jr., of Fitchburg State College; John W. Douthit, of Mt. San Antonio College; Ralph P. Eckerlin, of Northern Virginia Community College; John P. Harley, of Eastern Kentucky University; Louise B. Katz, of Sinclair Community College; Francis C. Monette, of Boston University; Harry S. Reasor, Jr., of Miami-Dade Junior College; and Alfreda G. Suskie, of Mohawk Valley Community College. The following individuals reviewed the entire first draft of the manuscript: Shirley Davis, of Montgomery College; Bruce Grayson, of the University of Miami at Coral Gables; Ronald Plakke, of the University of Northern Colorado; and Martha Van Bolt, of Charles Stewart Mott Community College. Each has made an outstanding contribution to the textbook. The difficult task of reviewing the artwork was undertaken by Richard Sugerman, of Wichita State University, and Ernest Gardner, of the University of California at Davis. Both art reviewers provided me with many excellent suggestions for improving the illustrations. The final manuscript review was done by Richard Mast, of Miami-Dade Junior College, North. His helpful suggestions and pointed comments are greatly appreciated. All of the reviewers have helped me to develop an accurate, logical, and pedgogically sound presentation of human anatomy.

I am especially pleased with the superb quality of the line drawings. The bulk of the artwork was prepared by Nelva B. Richardson. Her talent, inspiration, and cooperation have made the art program a distinctive feature of the textbook. Selected pieces of art were ably and imaginatively prepared by Marsha J. Dohrmann.

I wish to particularly acknowledge Victor B. Eichler, of Wichita State University, and Steve Harper, for providing me with many excellent photographs and photomicrographs and John C. Bennett, of St. Mary's Hospital in San Francisco, for supplying many high-quality roentgenograms. Gratitude is also extended to the many other individuals, publishers, and companies that provided photographs, photomicrographs, and electron micrographs.

The editorial assistance provided by Canfield Press for the development of this project has been outstanding. I wish to express my appreciation to R. Wayne Oler, Editor-in-Chief and Associate Publisher, who offered me all the resources of Canfield Press to successfully complete the project; Howard Boyer, Senior Editor, who personally supervised and became involved in all phases of the project, providing continuous guidance and encouragement; Pearl C. Vapnek, Production Editor, who coordinated the various facets of the project; and Alice S. Goehring, who brought copyeditorial consistency and accuracy to the final manuscript.

All drafts of the manuscript were typed by Geraldine C. Tortora, my wife. She also handled all secretarial duties related to the project, for which I am deeply grateful.

I would like to invite readers of this book to send their reactions and suggestions concerning the book to me at the address given below. These responses will be helpful to me in formulating plans for subsequent editions.

Gerard J. Tortora
Biology Department
Bergen Community College
400 Paramus Road
Paramus, NJ 07652

NOTE TO THE STUDENT

At the beginning of each chapter is a listing of "Student Objectives." Before you read the chapter, please read the objectives carefully. Each objective is a statement of a skill or knowledge that you should acquire. To meet these objectives, you will have to perform several activities. Obviously, you must read the chapter very carefully and if there are sections of the chapter that you do not understand after one reading, you should reread those sections before continuing. In conjunction with your reading, pay particular attention to the Figures and Exhibits; they have been carefully coordinated to the textual narrative. At the each of each chapter are three other guides that you may find useful. The first, "Chapter Summary in Outline," is a concise summary of important topics discussed in the chapter. This section is designed to consolidate the essential points covered in the chapter, so that you may recall and relate them to one another. The second guide, "Review Questions and Problems," is a series of questions designed specifically to help you meet your objectives. After you have answered the questions, you should return to the beginning of the chapter and reread the objectives to determine whether or not you have achieved the goals. The final guide, "Selected Readings," lists pertinent materials on the chapter topics for supplementary examination.

CHAPTER 1

AN INTRODUCTION TO THE HUMAN BODY

STUDENT OBJECTIVES

After you have read this chapter, you should be able to:

1. Define anatomy, with its subdivisions, and physiology
2. Determine the relationship between structure and function
3. Compare the levels of structural organization that comprise the human body
4. Define a cell, a tissue, an organ, a system, and an organism
5. List by name and location the principal body cavities and their major organs
6. Describe the subdivisions of the abdominopelvic cavity into nine regions and four quadrants
7. Define the anatomical position
8. Compare common and anatomical terms used to describe the external features of the body
9. Define directional terms used in association with the body
10. Describe the common anatomical planes of the body
11. Identify by visual inspection or palpation various surface features of the head, neck, trunk, upper extremity, and lower extremity
12. Define the common metric units of length, mass, and volume, and their English equivalents, that are used in measuring the human body

You are about to begin a study of the human body in order to learn not only how your body is organized but also, in many instances, how it functions. The study of the human body involves many branches of science, each of which contributes to a more comprehensive understanding of the parts of your body and how they work. Once you learn how your body normally works, you can understand what happens to your body when it is injured, diseased, or placed under stress.

Two branches of science that will help you understand your body parts and functions are anatomy and physiology. **Anatomy** refers to the study of *structure* and the relationships among structures. Anatomy is a very broad science, and the study of structure becomes more meaningful when specific aspects of the science are considered. For example, **gross anatomy** deals with structures that can be studied without using a microscope. Another kind of anatomy, **systemic anatomy,** covers particular systems of the body, such as the system of nerves, spinal cord, and brain, or the system of heart, blood vessels, and blood. **Regional anatomy** is a division of anatomy dealing with a specific region of the body, such as the head, neck, chest, or abdomen. **Developmental anatomy** is the study of development from the fertilized egg to the adult form. **Embryology** is generally restricted to the study of development from the fertilized egg to the end of the eighth week in utero. Other branches of anatomy are **pathological anatomy,** the study of structural changes caused by disease, and **histology,** the microscopic study of the structure of tissues and cells.

Whereas anatomy and its branches deal with structures of the body, **physiology** deals with the *functions* of the body parts. In other words, physiology is concerned with how a part of the body actually works. As you will see in later chapters, physiology cannot be completely separated from anatomy. It is for this reason that limited reference will be made to physiology, even though the primary concern of this text is anatomy. Each structure of the body is custom-modeled to carry out a particular set of functions. For instance, the interior of the nose is lined with hairs that allow the nose to perform the function of filtering dust from inhaled air. Bones are able to function as rigid supports for the body because they are constructed of hard minerals. In a sense, then, the structure of a part determines what functions it will perform. In turn, body functions often influence the size, shape, and health of the structures. For example, glands perform the function of manufacturing chemicals. Some of these chemicals stimulate bones to build up minerals so they become hard and strong. Other chemicals cause the bones to give up some of their minerals so they do not become too thick or too heavy.

HOW ARE YOU PUT TOGETHER?

The human body consists of several levels of structural organization that are associated with one another in several ways. The lowest level of structural organization, the *chemical level,* includes all chemical substances essential for maintaining life. All these chemicals are made up of atoms joined together in various ways (Figure 1–1). The chemicals, in turn, are put together to form the next higher level of organization, the *cellular level.* **Cells,** as you probably know, are the basic structural and functional units of the organism. Among the many kinds of cells found in your body are muscle cells, nerve cells, and blood cells. Figure 1–1 shows several isolated cells from the lining of the stomach. Each of these cells has a different structure, and each performs a different job.

From the cellular level, the next higher level of structural organization is the *tissue level.* **Tissues** are made up of groups of similar cells that perform a specific function. For example, when the isolated cells shown in Figure 1–1 are joined together, they form a tissue called *epithelium,* which lines the stomach. Each kind of cell in the tissue has a specific function. Mucous cells produce mucus, a slime that lubricates food as it passes through the stomach. Parietal cells produce acid in the stomach, and chief cells produce enzymes needed to digest proteins. Other examples of tissues in your body are muscle tissue, connective tissue, and nervous tissue.

In many places in the body, different kinds of tissues are joined together to form an even higher level of organization, the *organ level.* **Organs** are groups of two or more different tissues that perform a particular function. Examples of organs are the heart, liver, lungs, brain, and stomach. Referring to Figure 1–1, you will see that the stomach is an organ since it consists of two or more kinds of tissues. Three of the tissues that make up the stomach are shown here. The serous layer (also called the serosa) protects the stomach and reduces friction when the stomach moves and rubs against other organs. The muscle tissue layers of the stomach contract to mix food and pass the food on to the next digestive organ. The epithelial tissue layer produces mucus, acid, and enzymes.

The next higher level of structural organization in the body is the *system level.* A **system** consists of

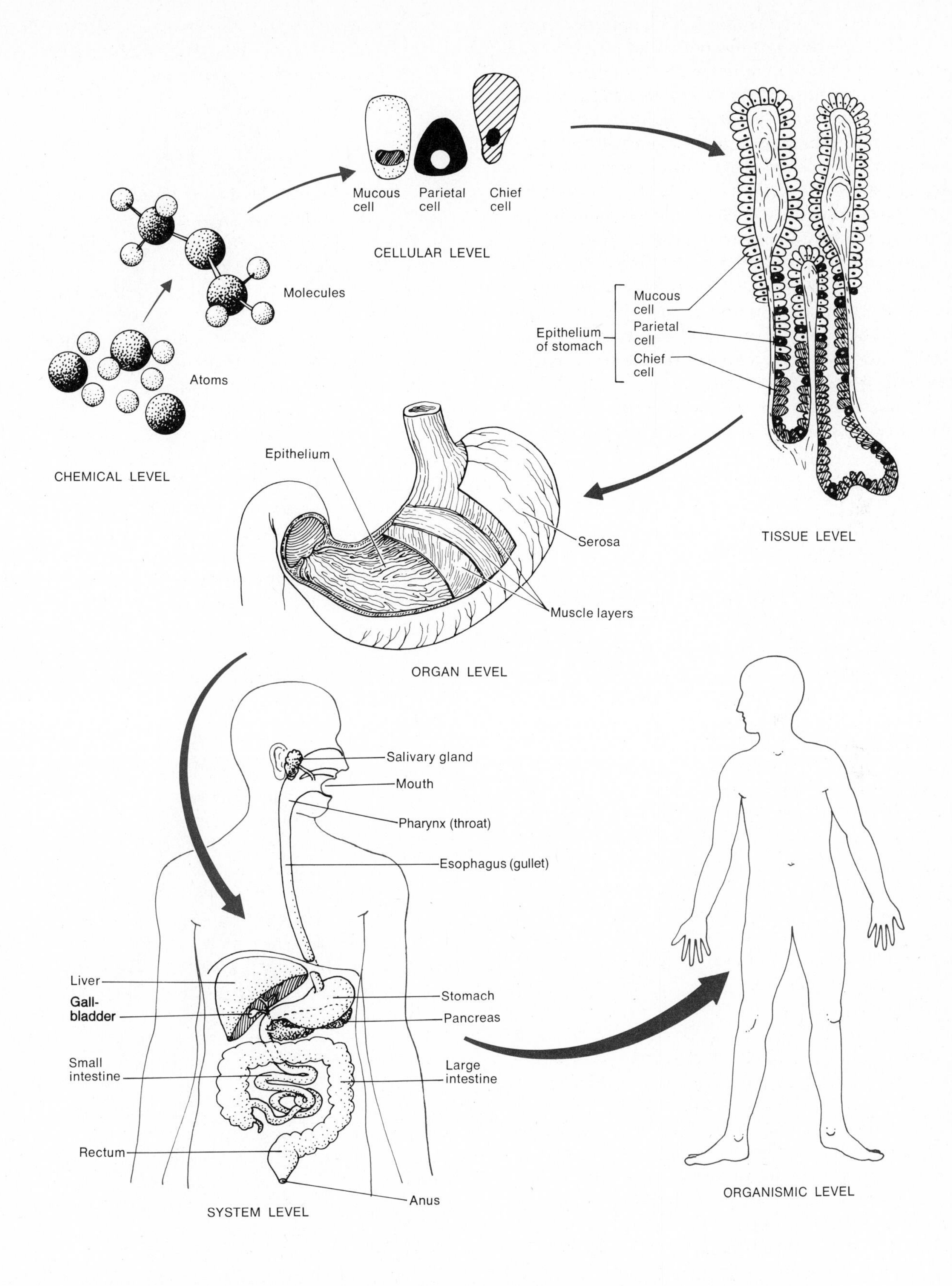

Figure 1–1. Levels of structural organization that compose the human body.

an association of organs that have a common function. The digestive system, which functions in the breakdown of food, comprises the mouth, saliva-producing glands, throat, esophagus, stomach, small intestine, large intestine, rectum, liver, gallbladder, and pancreas. All the parts of the body functioning with each other constitute the total **organism.**

In the chapters that follow, you will examine the anatomy and physiology of the following body systems: integumentary (pertaining to the skin), skeletal, muscular, nervous, endocrine (pertaining to hormones), circulatory, respiratory, digestive, urinary, and reproductive. Exhibit 1–1 contains a listing of these systems, their representative organs, and their general functions.

The systems are presented in the Exhibit in the order in which they will be studied in later chapters.

Exhibit 1–1. PRINCIPAL SYSTEMS OF THE HUMAN BODY, REPRESENTATIVE ORGANS, AND FUNCTIONS

1. INTEGUMENTARY

Hair

Nipple

Finger nails

The skin and its associated structures such as hair, nails, and glands

FUNCTION: Protects the body, regulates body temperature, and eliminates wastes

2. SKELETAL

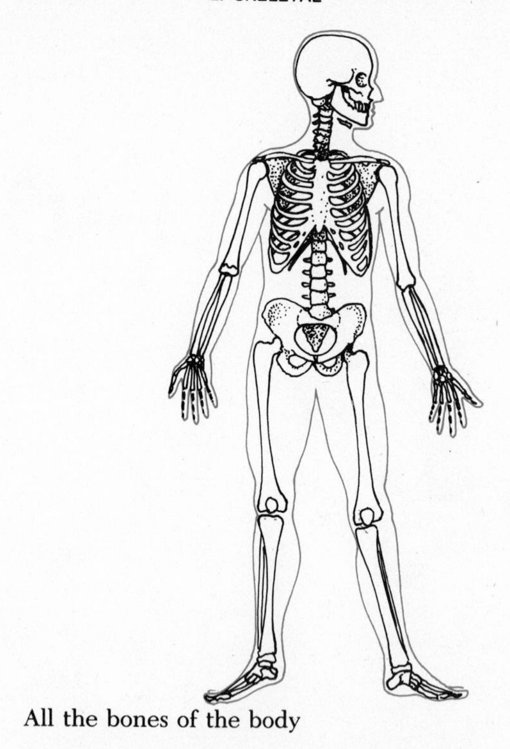

All the bones of the body

FUNCTION: Supports and protects the body, gives leverage, produces blood cells, and stores minerals

3. MUSCULAR

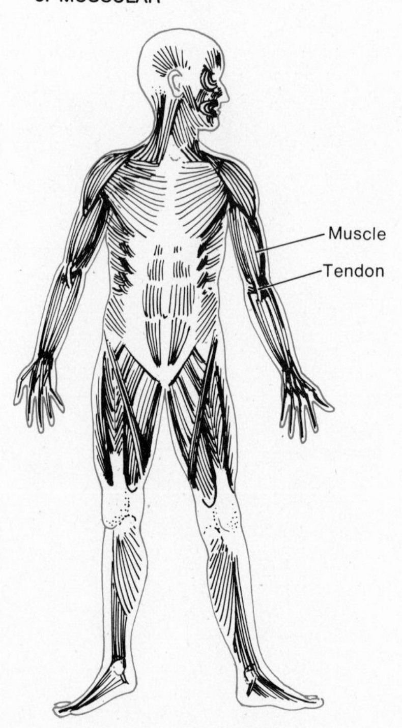

All the muscle tissue of the body, including skeletal, visceral, and cardiac

FUNCTION: Allows movement, maintains posture, and produces heat

4. BLOOD VASCULAR

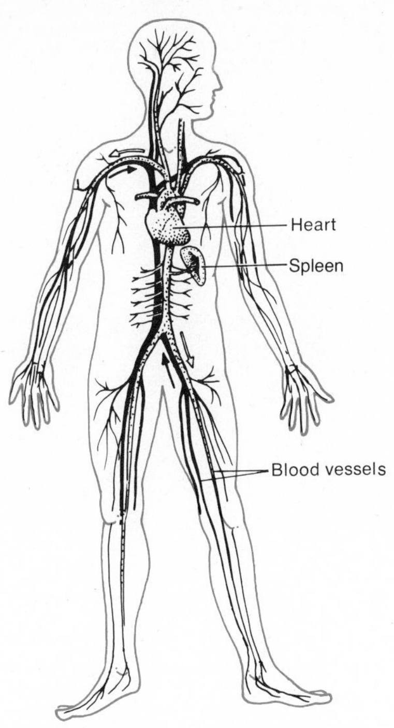

Blood, heart, and blood vessels

FUNCTION: Distributes nutrients to cells, eliminates wastes from cells, carries oxygen and carbon dioxide to and from cells, maintains the acid-base balance of the body, and protects against disease

5. LYMPH VASCULAR

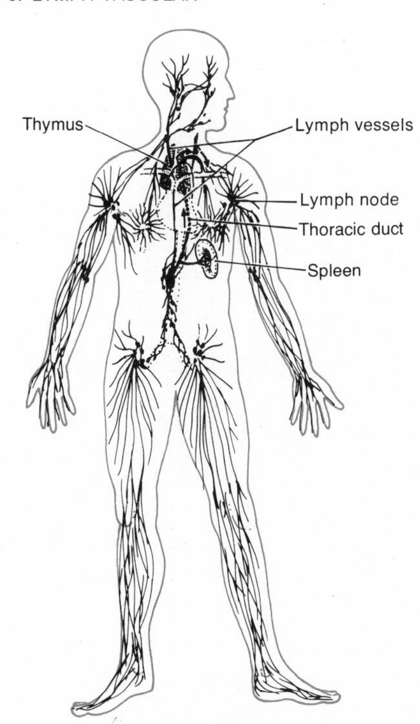

Lymph, lymph nodes, lymphatic organs, and lymph vessels

FUNCTION: Filters the blood and protects the body against disease

6. NERVOUS

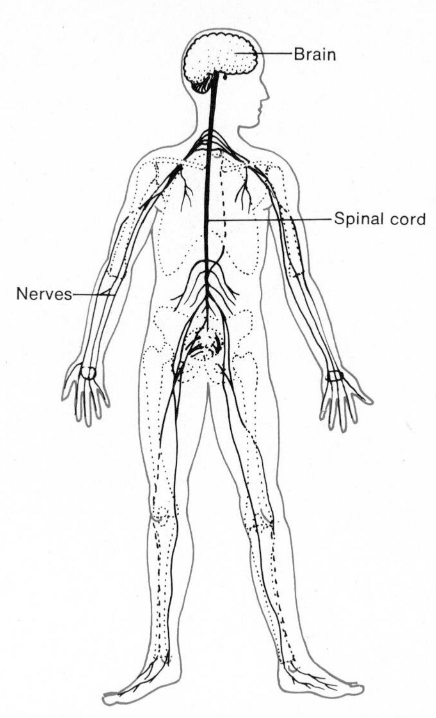

Brain, spinal cord, nerves, sense organs, e.g., eye and ear

FUNCTION: Regulates body activities through nerve impulses

7. ENDOCRINE

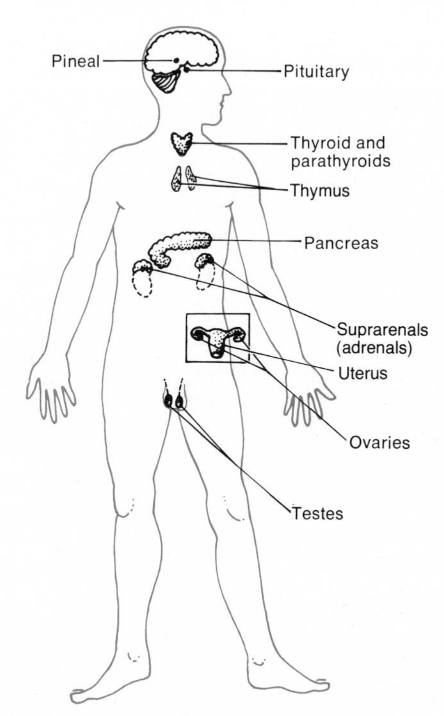

All glands that produce hormones

FUNCTION: Regulates body activities through hormones transported by blood

Exhibit 1–1 (cont.)

8. REPRODUCTIVE

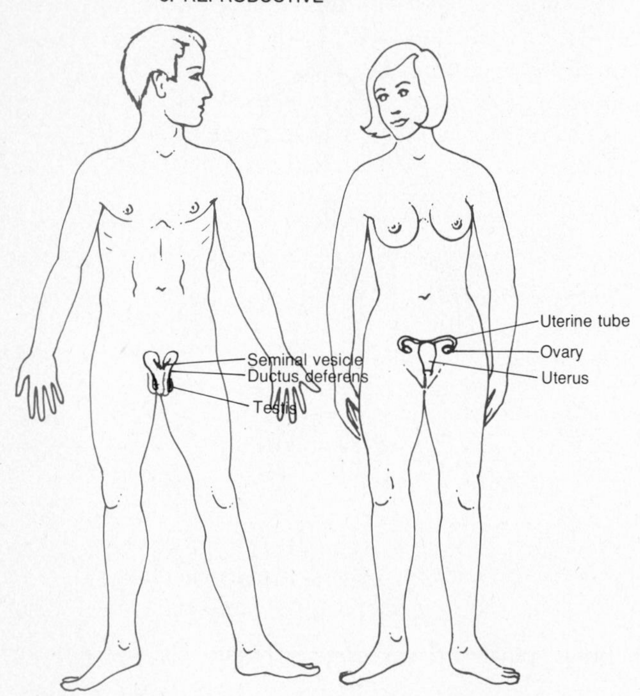

Organs (testes and ovaries) that produce reproductive cells (sperms and ova) and organs that deliver and store reproductive cells

FUNCTION: Reproduces the species

9. RESPIRATORY

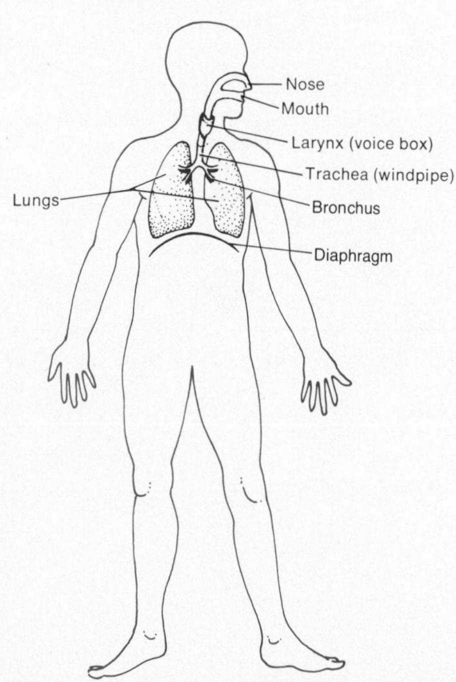

Lungs, tubes connecting them with the atmosphere, and vasculature

FUNCTION: Supplies oxygen, eliminates carbon dioxide, regulates the acid-base balance of the body

10. DIGESTIVE

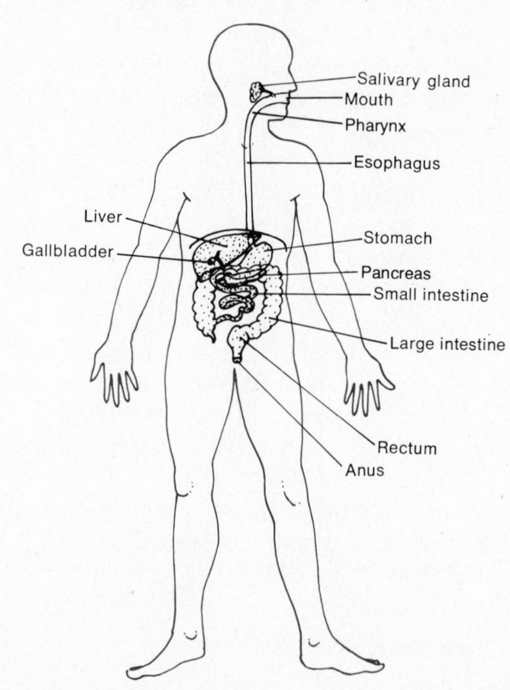

A long tube and associated organs such as the liver and gallbladder

FUNCTION: Performs the physical and chemical breakdown of food for use by the body, eliminates solid wastes

11. URINARY

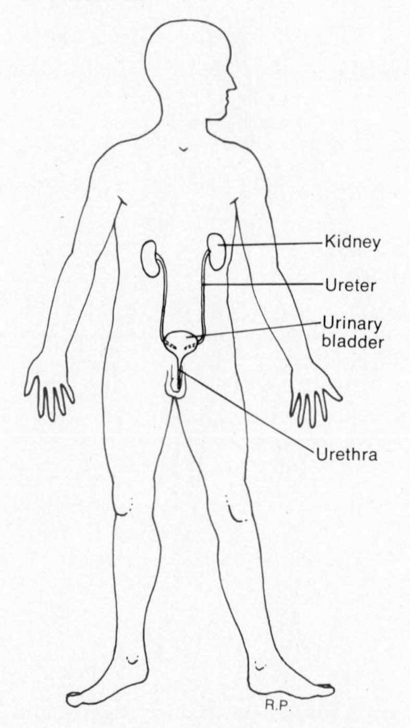

Organs that produce and eliminate urine

FUNCTION: Eliminates wastes, regulates fluid balance, maintains the acid-base balance of the body

STRUCTURAL PLAN OF THE HUMAN BODY

Now that you have a general idea as to how you are put together, we will take a closer look at the structural plan of your body.

The human organism possesses certain anatomical characteristics that are easily identifiable and can, therefore, serve as reference landmarks. For example, humans have a *backbone*, a characteristic that places them in a large group of organisms called vertebrates. In terms of body form, humans are said to be bilaterally symmetrical. *Bilateral symmetry* means that the left and right sides of the body are mirror images. Another characteristic of the body's organization is that it resembles a *tube within a tube*. The outer tube is formed by the body wall, whereas the inner tube is the digestive tract. In the chapters that follow, you will need to know such general characteristics as well as the terms used to describe positions and directions in the body. Cavities of the body will be considered first.

Body cavities

Spaces within the body that contain various internal organs are called **body cavities.** Specific cavities may be distinguished if the body is viewed after making a *median,* or *midsagittal, section*—that is, after cutting it into right and left halves. Figure 1–2a shows the two principal body cavities. The first of these, which may be called the *dorsal body cavity,* is located near the dorsal (posterior), or back, surface of the body. It is a bony cavity formed by the skull bones and the bones that enclose the spinal cord. These are referred to as the cranial bones and the vertebrae, respectively. The dorsal body cavity is further subdivided into a *cranial cavity,* which contains the brain, and a *vertebral* or *spinal canal,* which contains the spinal cord and the beginnings of spinal nerves.

The second principal body cavity may be called the *ventral body cavity.* This cavity, also called the *coelom,* is located inside the ventral (anterior), or

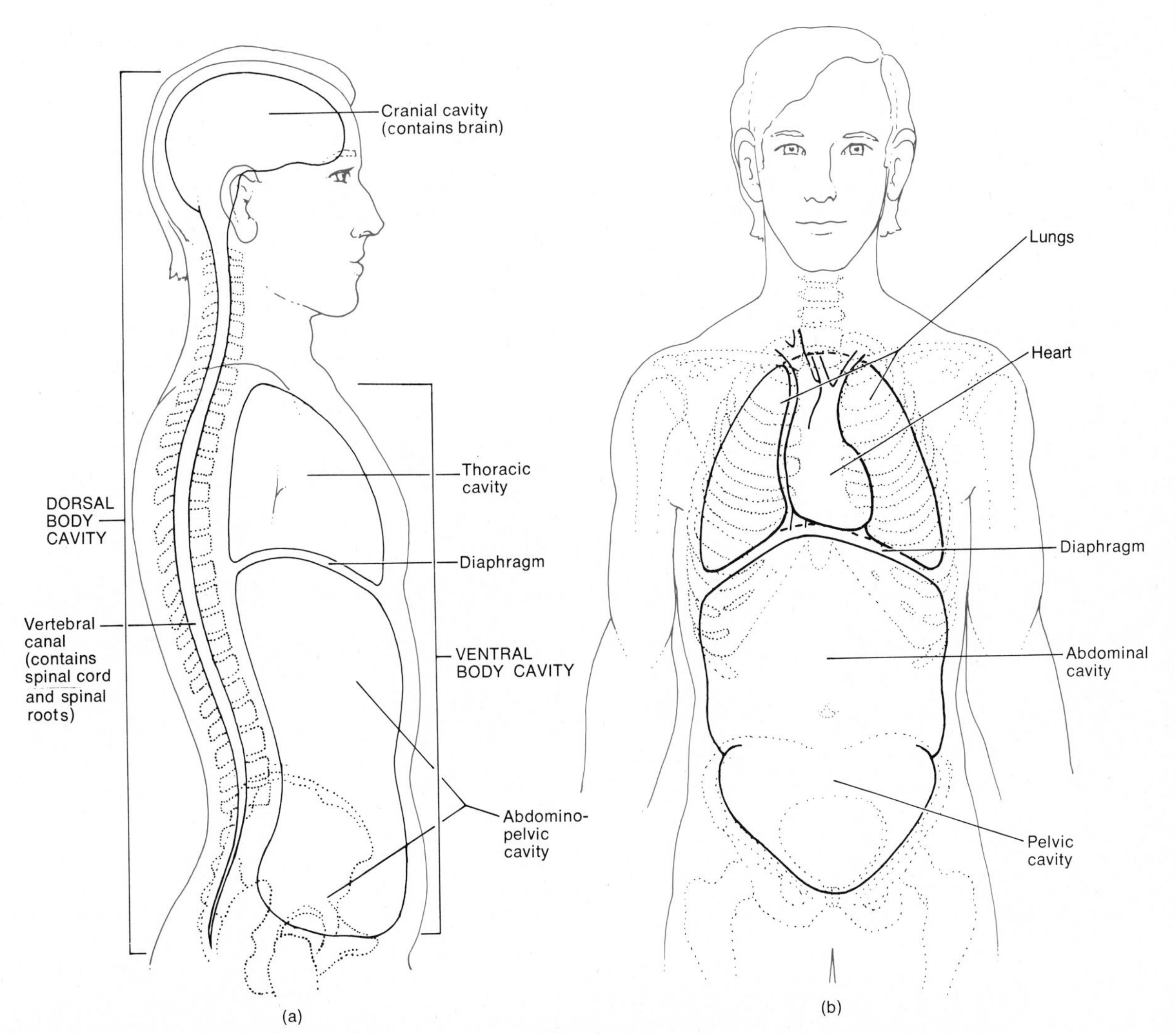

Figure 1–2. Body cavities. (a) Median section through the body to indicate the location of the dorsal and ventral body cavities. (b) Subdivisions of the ventral body cavity. Broken lines indicate borders of the mediastinum.

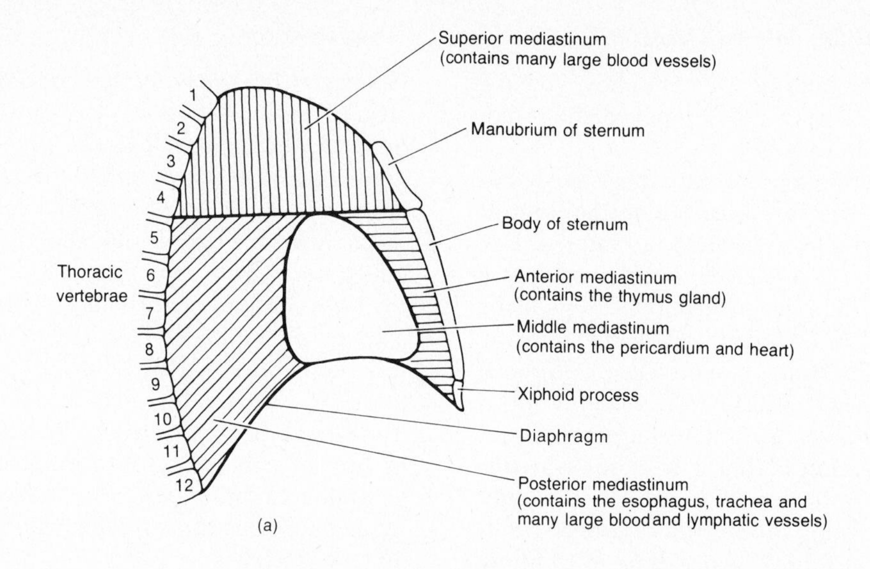

(a)

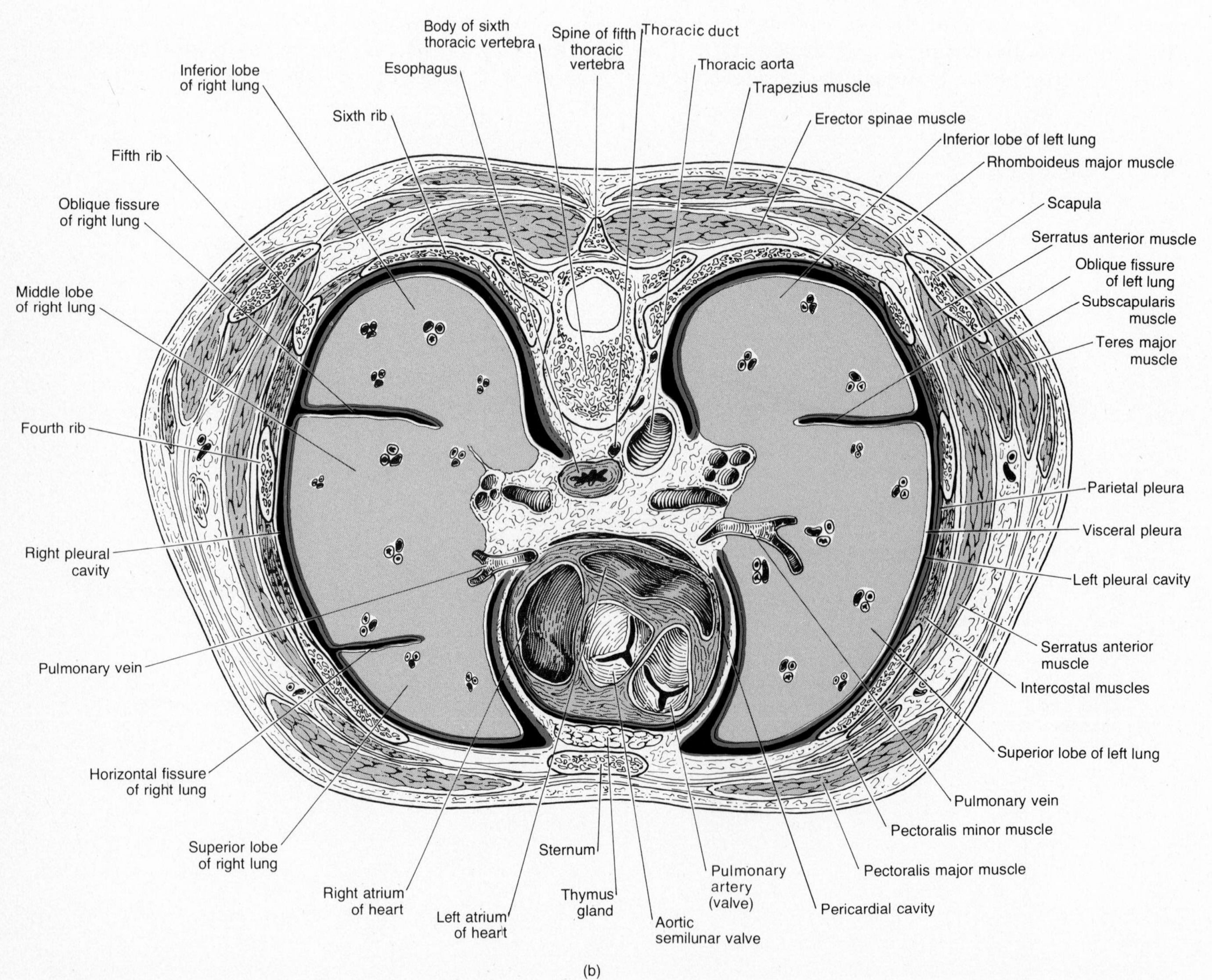

(b)

Figure 1–3. The mediastinum, the space between the pleurae of the lungs that extends from the sternum to the vertebral column. (a) Subdivisions of the mediastinum seen in right lateral view. (b) Mediastinum seen in a cross section of the thorax. Although many unfamiliar structures are shown and labeled in the cross section, do not become concerned. They will be referred to in later chapters, at which points they will have more meaning.

front, surface of the body. The organs inside the ventral body cavity are collectively called the **viscera.** Its walls are composed of skin, connective tissue, bone, muscles, and serous membrane. Like the dorsal body cavity, the ventral body cavity has two principal subdivisions. These are the upper portion, called the *thoracic cavity,* or chest cavity, and the lower portion, called the *abdominopelvic cavity.* The anatomical landmark that divides the ventral body cavity into the thoracic and abdominopelvic cavities is the muscular diaphragm. The thoracic cavity, in turn, contains several divisions.

There are two *pleural cavities* (Figure 1–3b), one around each lung, and a **mediastinum,** a mass of tissue between the pleurae of the lungs that extends from the sternum (breastbone) to the vertebral column (Figure 1–3). The *pericardial cavity* is located around the heart (Figure 1–3b). The abdominopelvic cavity, as the name suggests, is divided into two portions, although no wall lies between them (Figure 1–2b). The upper portion, called the *abdominal cavity,* contains the stomach, spleen, liver, gallbladder, pancreas, small intestine, most of the large intestine, the kidneys, and the ureters. The lower portion, called the *pelvic cavity,* contains the urinary bladder, sigmoid colon, rectum, the female reproductive organs, and some of the male reproductive organs. The abdominal cavity is arbitrarily separated from the pelvic cavity by drawing an imaginary line across the tops of the hipbones.

In order to describe the location of organs more easily, the abdominopelvic cavity may be divided into the nine regions shown in Figure 1–4. The *epigastric region* contains the left lobe and medial

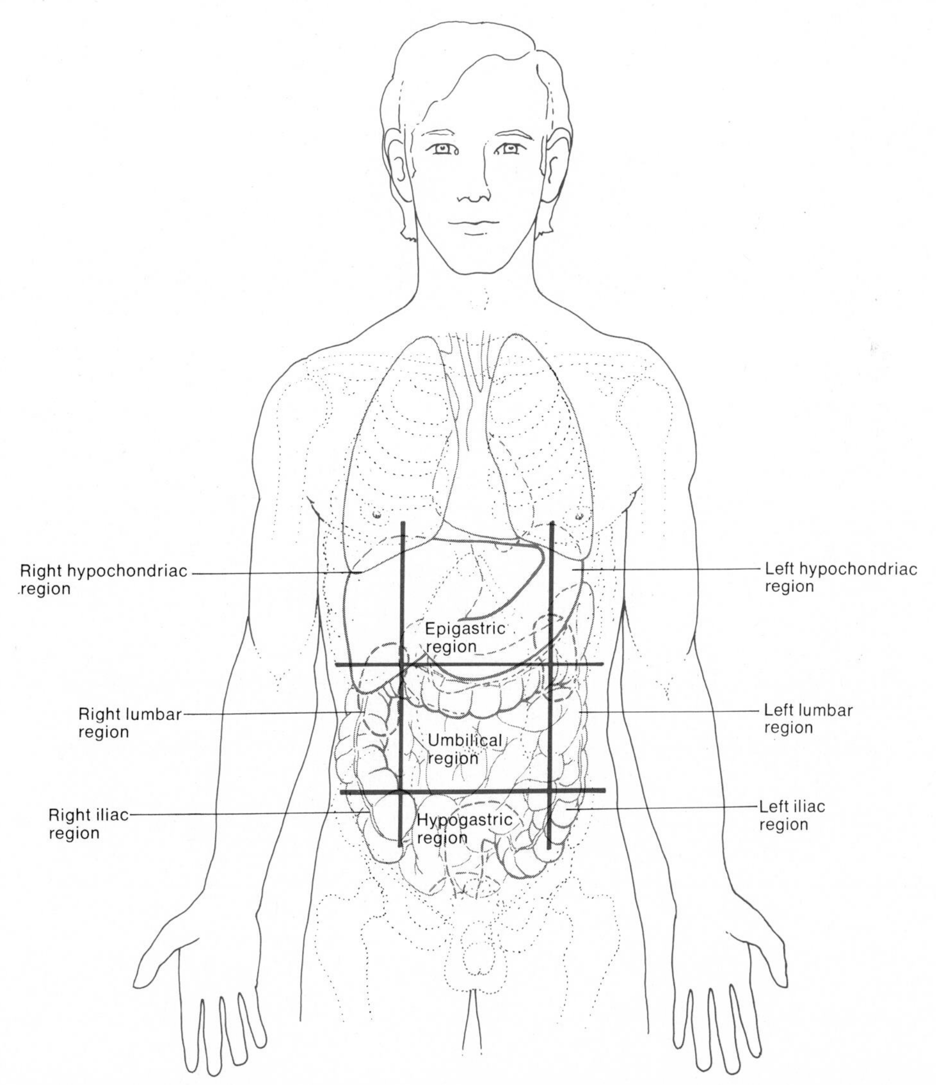

Figure 1–4. Nine regions of the abdomen and pelvis. The top horizontal plane is drawn near the bottom of the rib cage. The bottom horizontal plane is drawn below the tops of the hipbones. The two vertical planes are drawn just to the inside of the nipples. These planes divide the area into a larger middle section and smaller left and right sections. The names of the regions are labeled.

part of the right lobe of the liver, the pyloric part and lesser curvature of the stomach, the superior and descending portions of the duodenum, the body and upper part of the head of the pancreas, and the two adrenal glands. The *right hypochondriac region* contains the right lobe of the liver, the gallbladder, and the upper third of the right kidney. The *left hypochondriac region* contains the body and fundus of the stomach, the spleen, the left colic (splenic) flexure, the upper two-thirds of the left kidney, and the tail of the pancreas. The *umbilical region* contains the middle of the transverse colon, the inferior part of the duodenum, the jejunum, the ileum, the hilar regions of the two kidneys, and the bifurcations (divisions into two like a fork) of the abdominal aorta and inferior vena cava. The *right lumbar region* contains the upper part of the cecum, the ascending colon, the right colic (hepatic) flexure, the lower lateral portion of the right kidney, and the small intestine. The *left lumbar region* contains the descending colon, the lower third of the left kidney, and the small intestine. The *hypogastric region* contains the urinary bladder when full, the small intestine, and part of the sigmoid colon. The *right iliac region* contains the inferior end of the cecum, the appendix, and the small intestine. The *left iliac region* contains the junction of the descending and sigmoid parts of the colon and the small intestine.

A simpler method also used to designate subdivisions of the abdominopelvic cavity is to divide it into four quadrants (Figure 1–5). This method is frequently used by clinicians. According to this method, a horizontal plane is passed through the

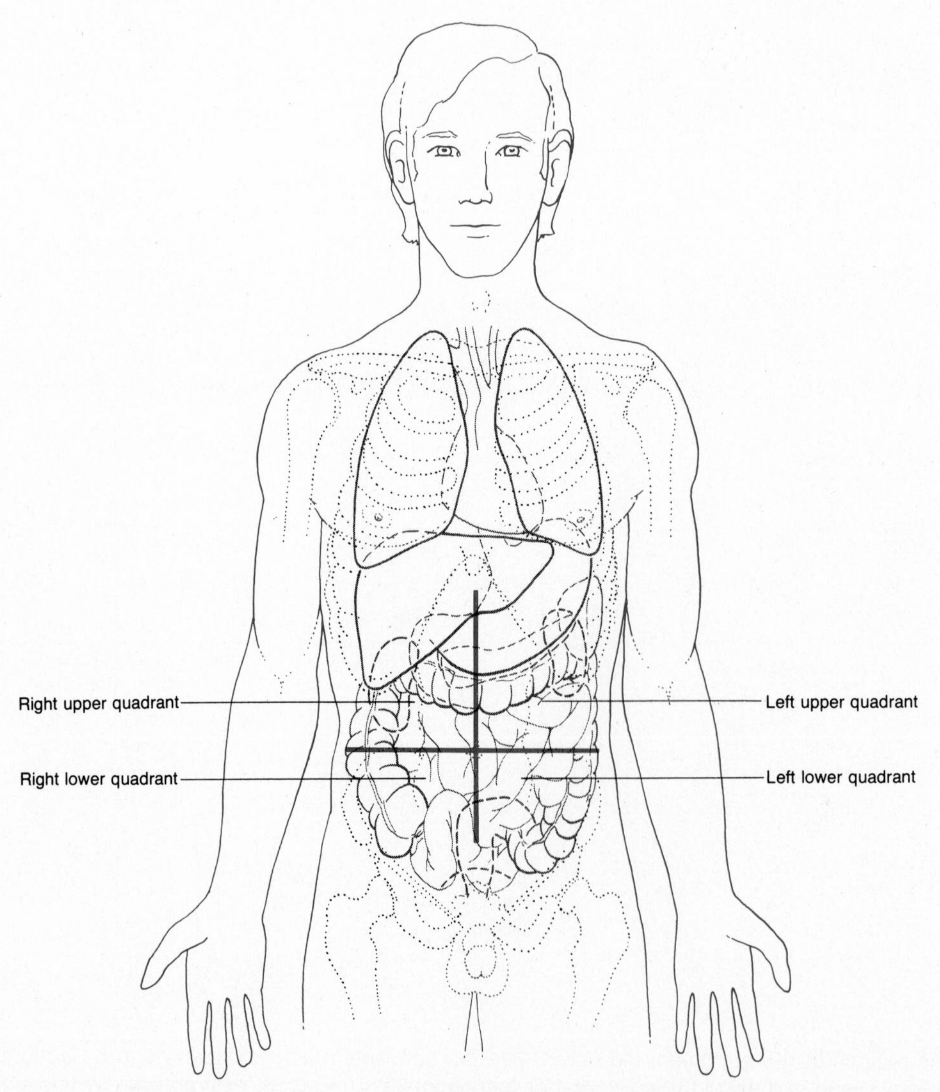

Figure 1–5. Subdivisions of the abdominal wall and abdominopelvic cavity into four quadrants.

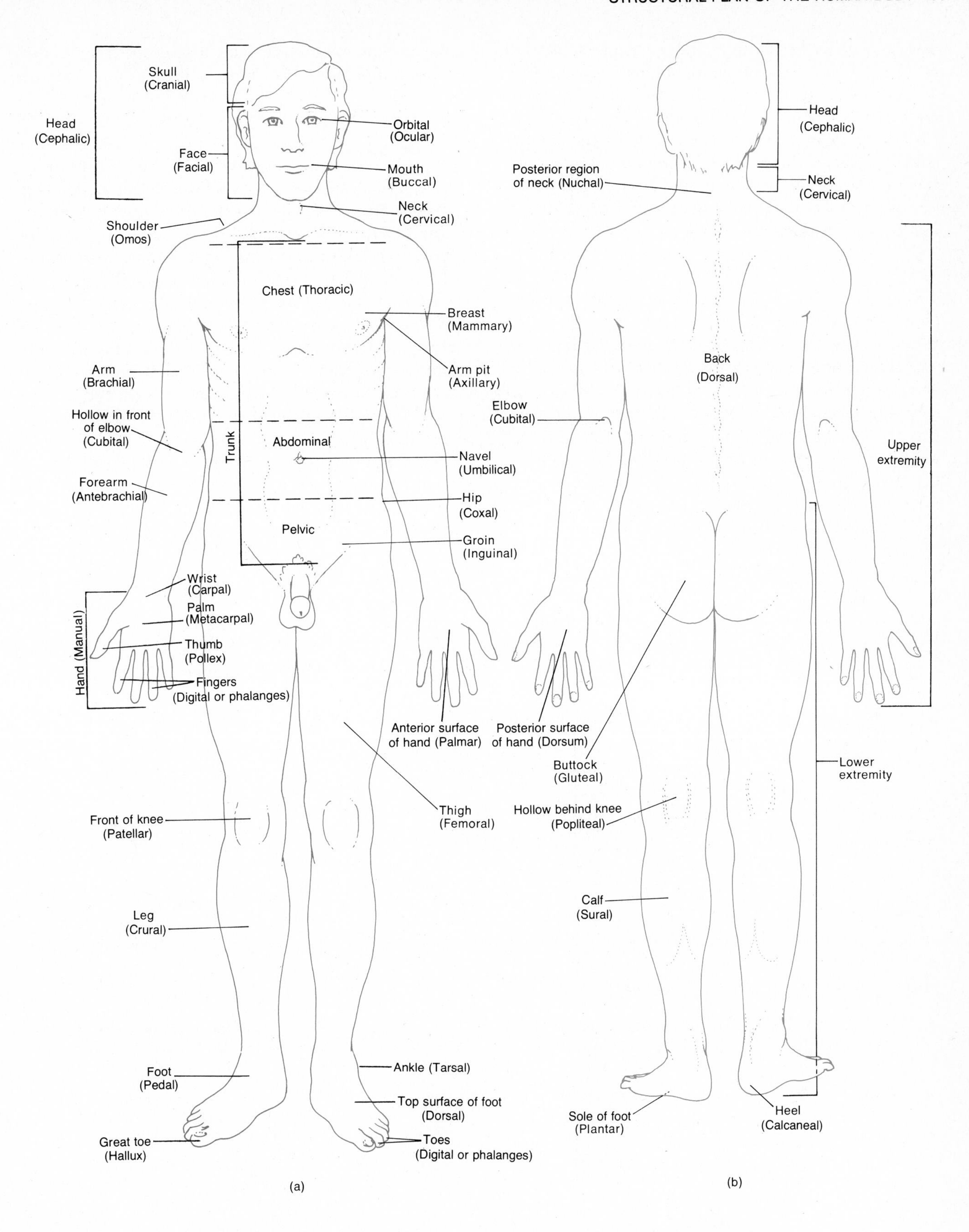

Figure 1–6. The anatomical position. (a) Anterior view. (b) Posterior view. Both common terms and anatomical terms, in parentheses, are indicated for many of the various regions of the body.

umbilicus together with a midsagittal plane. These two planes divide the abdomen into a *right superior (upper) quadrant, left superior (upper) quadrant, right inferior (lower) quadrant,* and a *left inferior (lower) quadrant.* Whereas the nine-region designation is more widely used for anatomical studies, the four-quadrant designation is better suited for locating the site of an abdominal pain or an abdominal tumor or other abnormality.

The anatomical position and regional names

When a region of the body is described in an anatomical text or chart, we assume that the body is in the anatomical position. The body in the **anatomical position** is erect and facing the observer. The arms are at the sides, and the palms of the hands are turned forward, as illustrated in Figure 1–6. The common terms and the anatomical terms, in parentheses, of certain body regions have been included in Figure 1–6. The anatomical names may not mean much to you at this point, but gaining some familiarity with them will assist you in later chapters.

Directional terms

In order to explain exactly where a structure of the body is located, anatomists must use certain **directional terms.** For instance, if you want to point out the sternum to someone who knows where the clavicle (collarbone) is, you can say that the sternum is inferior (farther away from the head) and to the medial (toward the middle of the body) part of the clavicle. As you can see, using the terms *inferior* and *medial* avoids a great deal of complicated description. Since the directional terms listed in Exhibit 1–2 will be used throughout this book and in your laboratory work, we suggest that you learn them. Many of the terms defined in the Exhibit may be understood by referring to Figure 1–7. Essential parts of the figure are labeled so that you can see the directional relationships among parts. Refer to

Exhibit 1–2. SUMMARY OF DIRECTIONAL TERMS FOR THE HUMAN BODY

TERM	DEFINITION	EXAMPLE
1. Superior (cephalic or cranial)	Toward the head; toward the upper part of a structure.	The heart is superior to the liver.
2. Inferior (caudal)	Away from the head; toward the lower part of a structure.	The rectum is inferior to the liver.
3. Anterior (ventral)	Nearer to or at the front of the body.	The urinary bladder is anterior to the rectum.
4. Posterior (dorsal)	Nearer to or at the back of the body.	The lungs are posterior to the sternum.
5. Medial	Nearer the midline of the body.	The ulna is on the medial side of the forearm.
6. Lateral	Farther from the midline of the body.	The kidneys are lateral to the backbone.
7. Proximal	Nearer the attachment of a limb to the trunk.	The humerus is proximal to the radius.
8. Distal	Away from the attachment of a limb to the trunk.	The phalanges are distal to the wrist.
9. External (superficial)	Toward the surface of the body.	The muscles of the abdominal wall are external to the viscera in the abdominal cavity.
10. Internal (deep)	Away from the surface of the body.	The muscles of the upper extremity are internal to the skin of the upper extremity.
11. Parietal	The walls of a cavity.	The parietal layer of the serous pericardium helps form the wall of the pericardial cavity (see Figure 11–4).
12. Visceral	The covering of an organ.	The visceral layer of the serous pericardium covers the external aspect of the heart (see Figure 11–4).

the figure after you read the definition for the directional term.

Planes of the body

The structural plan of the human body may also be discussed with respect to **planes,** that is, flat surfaces, that pass through it. Several of these planes are used commonly. Figure 1–8 illustrates the planes described below. A *midsagittal plane* through the midline of the body runs vertically and divides the body into equal right and left sides. A *sagittal plane* or *parasagittal plane* also runs vertically but it divides the body into unequal left and right portions. A *frontal* or *coronal plane* runs vertically and divides the body into anterior and posterior portions. Finally, a *horizontal* or *transverse plane* runs parallel and divides the body into superior and inferior portions.

In subsequent chapters, you will view parts of the body that are sectioned in various ways in order to illustrate their most important features. When you examine sections of organs, it is important to understand how the section is made so that you can understand the anatomical relationship of one part to another. Figure 1–9a presents a series of diagrams that indicate how three different sections are made through a simple tube. The sections shown are a *cross section,* an *oblique section,* and *longitudinal sections.* Figure 1–9b is a series of diagrams showing how the same three kinds of sections are made through the spinal cord.

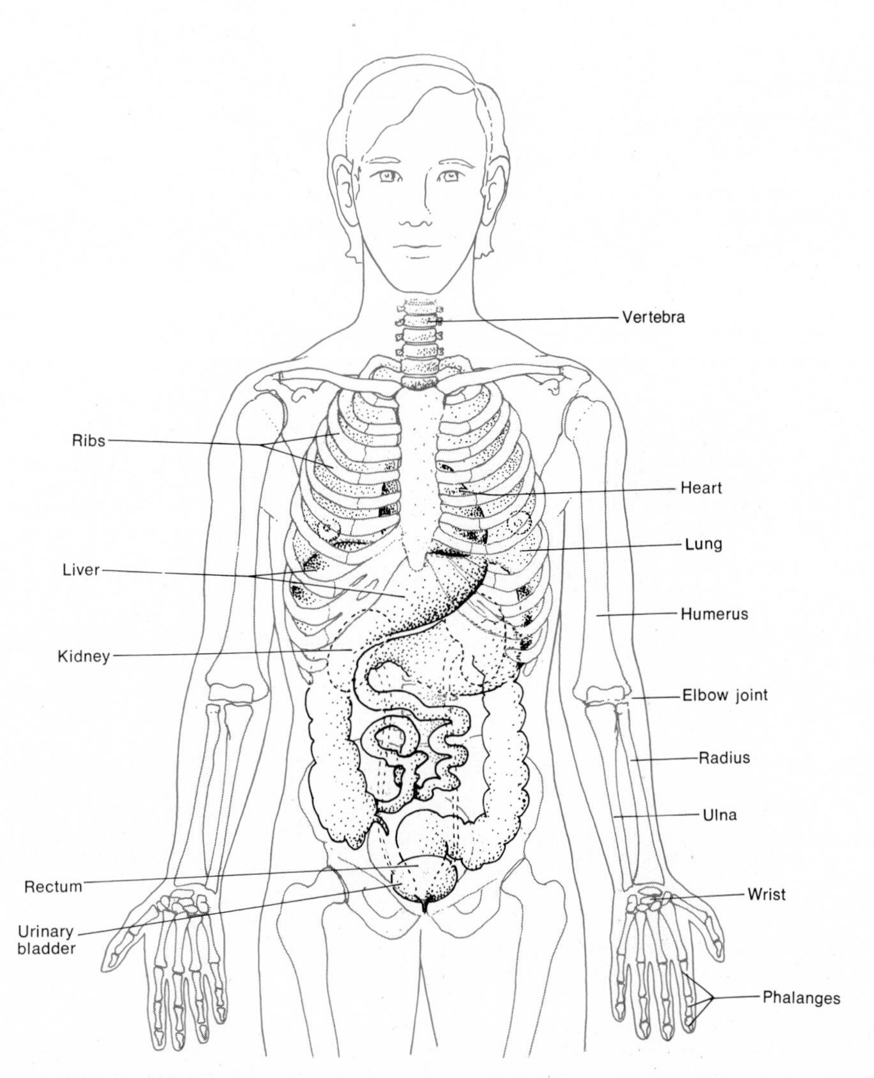

Figure 1–7. Relating anatomical and directional terms. By studying Exhibit 1–2 with this figure, you should gain an understanding of the meanings of the terms superior, inferior, anterior, posterior, medial, lateral, proximal, and distal.

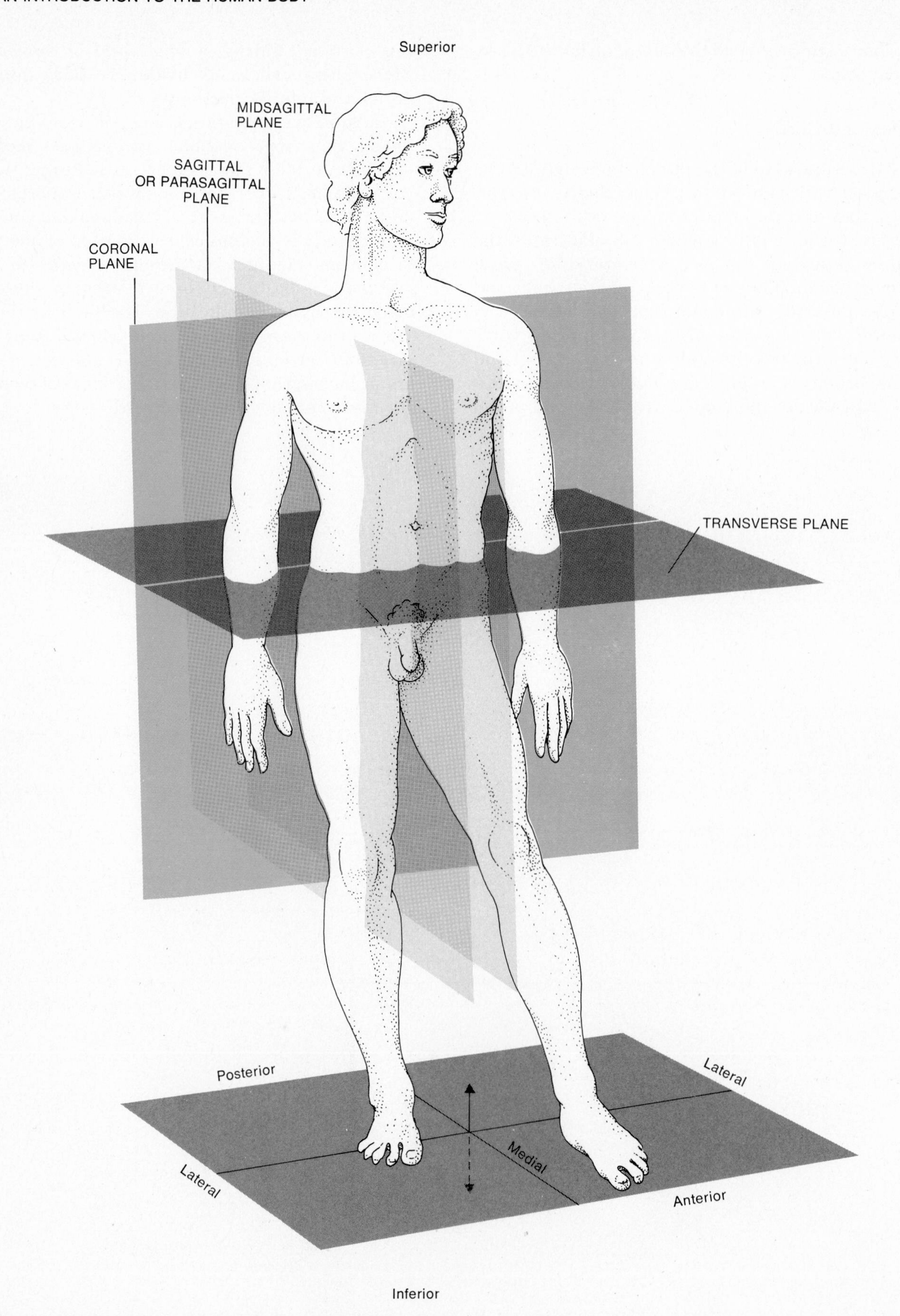

Figure 1–8. Planes of the body.

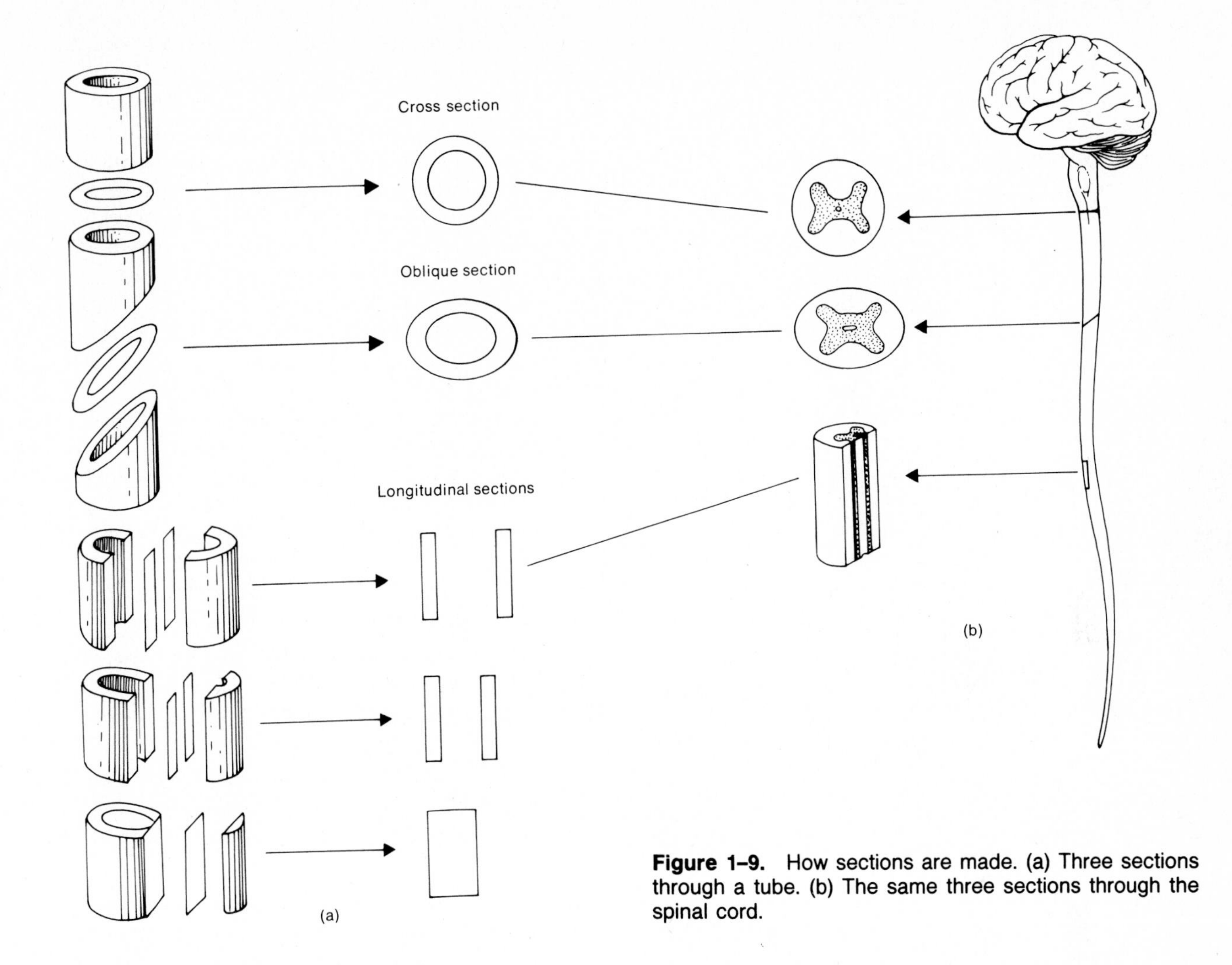

Figure 1–9. How sections are made. (a) Three sections through a tube. (b) The same three sections through the spinal cord.

INTRODUCTION TO SURFACE ANATOMY

In the beginning of this chapter, several branches of anatomy were defined and their importance to an understanding of the structure of the body was noted. Now that you have a fairly good idea as to how the body is organized, we can take a look at another branch of anatomy called surface anatomy.

Very simply, **surface anatomy** is the study of the form and markings of the surface of the body. A knowledge of surface anatomy will help you to identify certain superficial structures through visual inspection or palpation through the skin. *Palpation* means to feel with the hand.

To introduce you to surface anatomy, the body will first be divided into its primary divisions. These divisions are outlined as follows and may be reviewed in Figure 1–6:

- I. Head
  - A. Cranium
  - B. Face
- II. Neck
- III. Trunk
  - A. Back
  - B. Thorax
  - C. Abdomen
  - D. Pelvis
- IV. Extremities
  - A. Upper extremity
    1. Axilla
    2. Shoulder
    3. Arm
    4. Elbow
    5. Forearm
    6. Hand
       - *(a)* Wrist
       - *(b)* Palm
       - *(c)* Dorsum of hand
  - B. Lower extremity
    1. Buttocks
    2. Thigh
    3. Knee
    4. Leg
    5. Foot
       - *(a)* Heel
       - *(b)* Dorsum of foot
       - *(c)* Sole of foot

The head

Our discussion of surface anatomy will first consider the head. The **head** *(caput)* is divided into the cranium and face. The **cranium** *(brain case)* consists of a crown *(vertex)*; front *(sinciput)*, including the forehead *(frons)*; back *(occiput)*; sides *(tempora)*; and ears *(aures)*. The **face** *(facies)* is subdivided into the ocular region which includes the eyes *(oculi)*, eyebrows *(supercilia)*, and eyeballs *(bulbi oculorum)*; nose *(nasus)*; mouth *(os)*, including the lips *(labia)* and cavity *(cavum oris)*; cheek *(bucca* or *mala)*; and chin *(mentum)*.

Examine Exhibit 1–3. It contains illustrations of various surface features of the head with an accompanying description of each of the features. Although a large number of surface features are included in Exhibit 1–3 and other Exhibits dealing with surface anatomy, do not become alarmed. The Exhibits are presented to give you an understanding of the importance of surface anatomy. In subsequent chapters, when you have a better under-

Exhibit 1–3. SURFACE ANATOMY OF THE HEAD

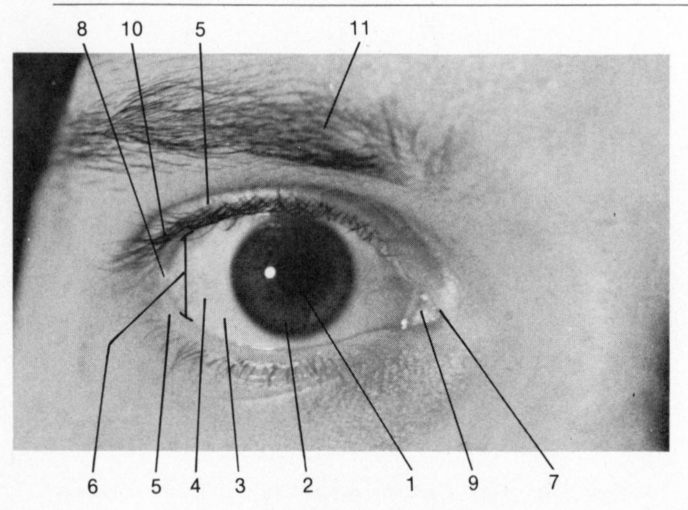

Anterior view of right eye

EYE

1. *Pupil.* Opening of center of iris of eyeball for light transmission.
2. *Iris.* Circular pigmented muscular membrane behind cornea.
3. *Sclera.* "White" of eye, a coat of fibrous tissue that covers entire eyeball except for cornea.
4. *Conjunctiva.* Membrane that covers exposed surface of eyeball and lines eyelids.
5. *Eyelids.* Folds of skin and muscle lined by conjunctiva.
6. *Palpebral fissure.* Space between eyelids when they are open.
7. *Medial commissure.* Site of union of upper and lower eyelids near nose.
8. *Lateral commissure.* Site of union of upper and lower eyelids away from nose.
9. *Lacrimal caruncle.* Fleshy, yellowish projection of medial commissure that contains modified sweat and sebaceous glands.
10. *Eyelashes.* Hairs on margins of eyelids, usually arranged in two or three rows.
11. *Eyebrows.* Several rows of hairs superior to upper eyelids.

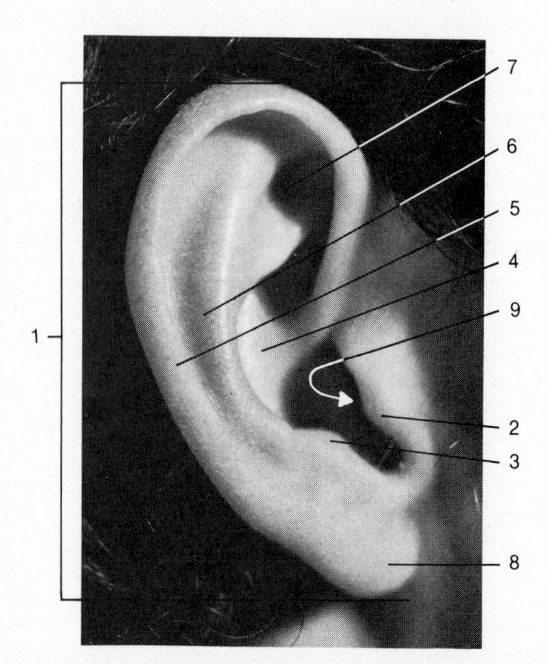

Lateral view of right ear

EAR

1. *Auricle.* Portion of external ear not contained in head, also called pinna or trumpet.
2. *Tragus.* Cartilaginous projection anterior to external opening to ear.
3. *Antitragus.* Cartilaginous projection opposite tragus.
4. *Concha.* Hollow of auricle.
5. *Helix.* Superior and posterior free margin of auricle.
6. *Antihelix.* Semicircular ridge posterior and superior to concha.
7. *Triangular fossa.* Depression in superior portion of antihelix.
8. *Lobule.* Interior portion of auricle devoid of cartilage.
9. *External acoustic (auditory) meatus.* Canal extending from external ear to eardrum.

standing of anatomical terms, we will refer back to the Exhibits. They will have even more meaning then.

The neck

The **neck** *(collum)* is divided into an anterior region called the *cervix,* two lateral regions, and a posterior region referred to as the *nucha.* The most prominent structure in the midline of the anterior region is the thyroid cartilage or Adam's apple. Just above it, the hyoid bone can be palpated. Below the Adam's apple, the cricoid cartilage of the larynx can be felt. A good portion of the lateral regions of the neck is formed by the sternocleidomastoid muscles. Each muscle extends from the mastoid process of the temporal bone, felt as a bump behind the auricle of the ear, to the sternum and clavicle. A muscle that extends downward and outward from the base of the skull and occupying a portion of the lateral region of the neck is the trapezius muscle. You might be interested to know that a "stiff neck"

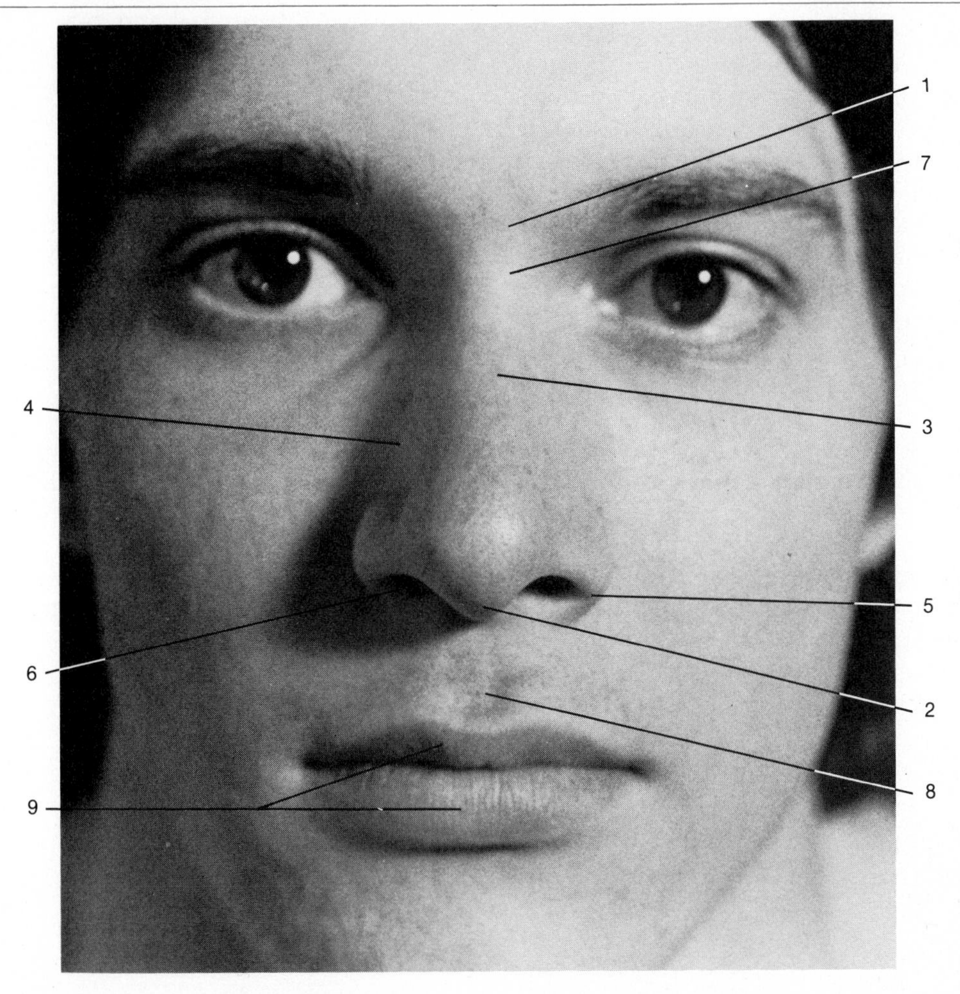

Anterior view of nose and lips

NOSE AND LIPS

1. *Root.* Superior attachment of nose at forehead.
2. *Apex.* Tip of nose.
3. *Dorsum nasi.* Rounded ridge connecting root and apex.
4. *Nasofacial angle.* Blending of ridge with tissues of face.
5. *Ala.* Inferior expansion of side uniting with upper lip.
6. *External nares.* External openings into nose.
7. *Bridge.* Superior part of external nose formed by union of nasal bones.
8. *Philtrum.* Vertical groove in medial portion of upper lip.
9. *Lips.* Upper and lower fleshy portion of mouth.

Photographs courtesy of Victor B. Eichler, of Wichita State University, and Steve Harper.

is frequently associated with an inflammation of this muscle. A very prominent vein that runs along the lateral surface of the neck is the external jugular vein. It is readily seen if you are angry or if your collar is too tight. The locations of a few surface features of the neck are shown in Exhibit 1–4.

The trunk

The **trunk** is divided into the back *(dorsum),* chest *(thorax),* abdomen *(venter),* and pelvis. One of the most striking surface features of the back is the vertebral spines, the dorsally pointed projections of the vertebrae. A very prominent vertebral spine is the vertebra prominens of the seventh cervical vertebra. When the head is bent forward, it is easily seen. Another easily identifiable surface landmark of the back is the scapula. In fact, several parts of the scapula may also be seen or palpated. In lean individuals, the ribs may also be seen. Among the superficial muscles of the back that can be seen are the latissimus dorsi, erector spinae, infraspinatus, trapezius, and teres major. These, as well as the other surface features of the back, are described in Exhibit 1–5.

The chest presents a number of anatomical landmarks. At its superior region are the clavicles. The sternum lies in the midline of the chest. Its superior end attaches to the clavicles. Between the medial ends of the clavicles, there is a depression on the superior surface of the sternum called the jugular notch. The sternal angle is formed by a junction line between the manubrium and body of the sternum and is palpable under the skin. It locates the sternal end of the second rib and is the most reliable surface landmark of the chest. The inferior portion of the sternum, the xiphoid process, may be palpated. Also visible or palpable are the ribs. Among the more prominent superficial chest muscles are the pectoralis major and serratus anterior. These, and other surface features of the chest, are described in Exhibit 1–5.

The abdomen and pelvis have already been discussed in terms of their nine regions or four quadrants (see Figures 1–4 and 1–5). External features of the abdomen include the umbilicus, serratus anterior muscle, external oblique muscle, and rectus abdominis muscle, all described in Exhibit 1–5.

The upper extremity

The **upper extremity** consists of the armpit *(axilla),* shoulder *(omos),* arm *(brachium),* elbow *(cubitus),* forearm *(antebrachium),* and hand *(manus).* The hand, in turn, is subdivided into the wrist *(carpus),* palm *(metacarpus),* and fingers (*digits* or *phalanges*). See Exhibit 1–6.

Moving laterally along the top of the clavicle, it is possible to palpate a slight elevation at the lateral end of the clavicle, the acromioclavicular joint. Less than 2½ cm (1 in.) from this joint, one can also feel the acromion of the scapula which forms the tip of the shoulder. The rounded prominence of the shoulder is formed by the deltoid muscle, a frequent site for intramuscular injections.

Exhibit 1–4. SURFACE ANATOMY OF THE NECK

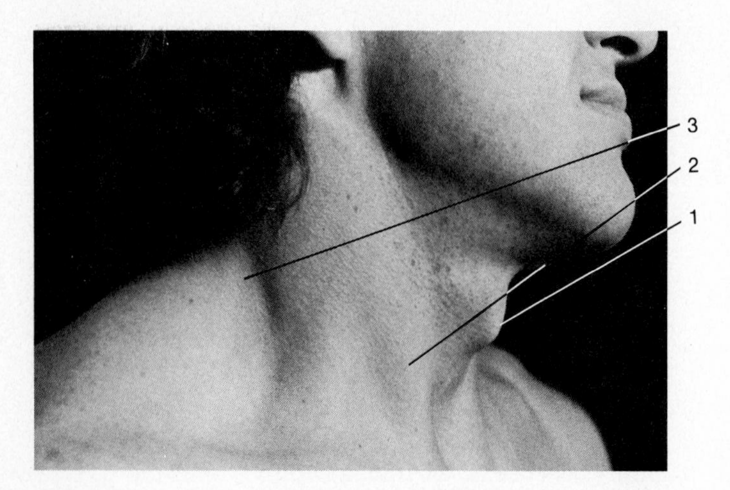

Lateral view of neck

NECK

1. *Thyroid cartilage.* Anterior portion of larynx (voicebox).
2. *Sternocleidomastoid muscle.* Forms major portion of lateral surface of neck; bends head on neck and turns it to opposite side.
3. *Trapezius muscle.* Occupies a portion of lateral surfaces of neck; used in shrugging shoulders.

Photographs courtesy of Victor B. Eichler, of Wichita State University, and Steve Harper.

Exhibit 1–5. SURFACE ANATOMY OF THE TRUNK

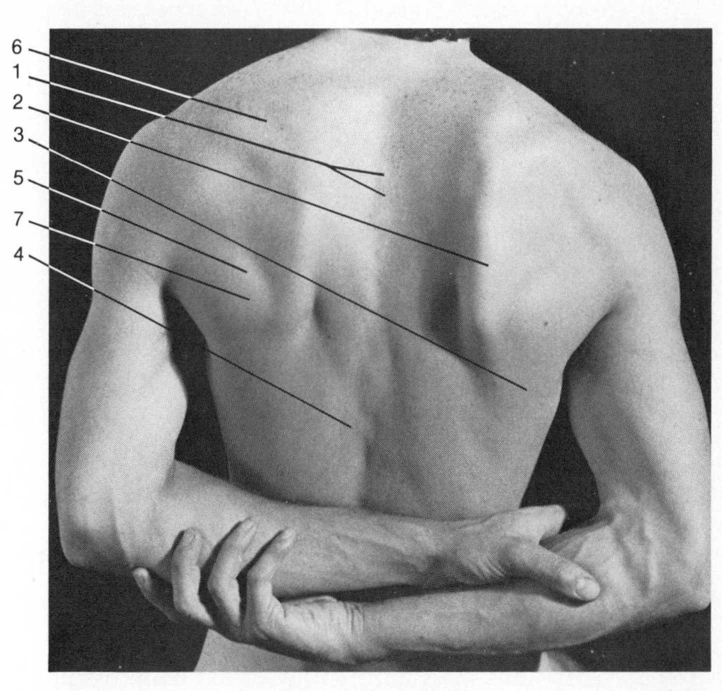

Posterior view of back

BACK

1. *Vertebral spines.* Dorsally pointed projections of vertebrae (backbones).
2. *Scapula.* Shoulder blade.
3. *Latissimus dorsi muscle.* Broad muscle of back that helps draw shoulders backward.
4. *Erector spinae muscle.* Parallel to vertebral column; moves backbone.
5. *Infraspinatus muscle.* Located inferior to spine of scapula; helps rotate humerus (armbone) laterally.
6. *Trapezius.* See Exhibit 1–4 for description.
7. *Teres major.* Located inferior to infraspinatus; helps move humerus.

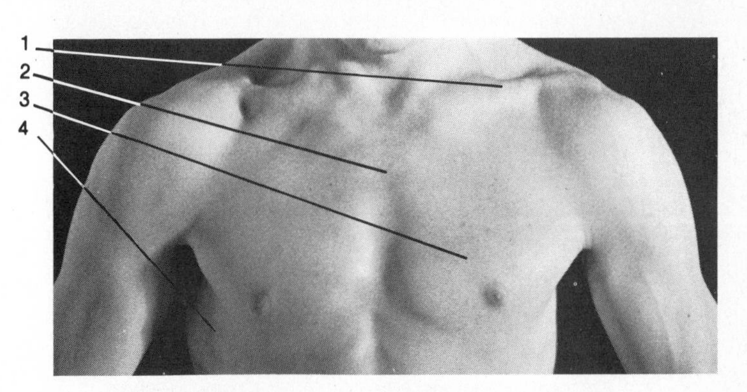

Anterior view of chest

CHEST

1. *Clavicle.* Collarbone.
2. *Sternum.* Breastbone.
3. *Pectoralis major muscle.* Principal upper chest muscle; helps move humerus.
4. *Serratus anterior muscle.* Below and lateral to pectoralis major; helps move scapula. See also anterior view of the abdomen.

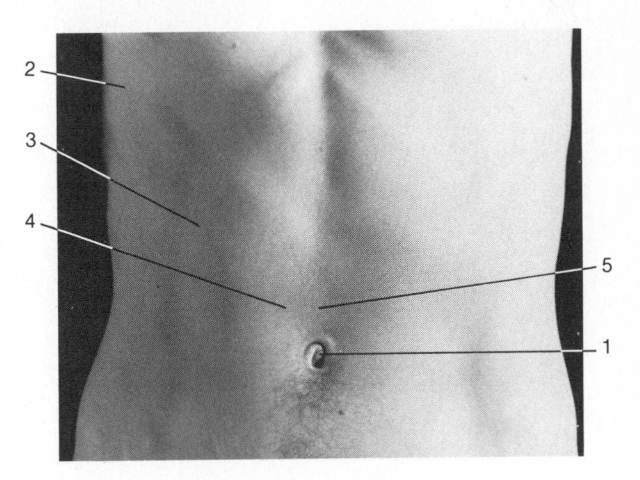

Anterior view of abdomen

ABDOMEN

1. *Umbilicus.* Also called navel; previous site of attachment of umbilical cord in fetus.
2. *Serratus anterior.* See chest above for description.
3. *External oblique.* Located inferior to serratus anterior; helps compress abdomen.
4. *Rectus abdominis.* Located just lateral to midline of abdomen; helps compress abdomen.
5. *Linea alba.* Flat, tendinous raphe between rectus abdominis muscles.

Photographs courtesy of Victor B. Eichler, of Wichita State University, and Steve Harper.

The bulk of the anterior surface of the upper extremity is occupied by the biceps muscle, whereas the bulk of the posterior surface is occupied by the triceps muscle. At the elbow it is possible to locate three bony protuberances. The medial and lateral epicondyles of the humerus form visible eminences on the dorsum of the elbow. The olecranon of the ulna forms the large eminence in the middle of the dorsum of the elbow and lies between and slightly superior to the epicondyles when the forearm is extended. The ulnar nerve can be palpated in a groove behind the medial epicondyle. The triangular space of the anterior region of the elbow is the cubital fossa. The median cubital vein crosses the cubital fossa transversely. This vein is the one frequently selected for removal of blood for diagnosis, transfusions, and intravenous therapy.

The most prominent landmarks of the forearm are the olecranon and the styloid process of the ulna. The styloid process of the ulna may be seen as a protuberance on the medial (little finger) side of the wrist. The ulna is the medial bone of the forearm, and the radius is the lateral bone of the forearm. On the lateral side of the upper forearm is the brachioradialis muscle. Next to it is the flexor carpi radialis muscle. On the medial side is the flexor carpi ulnaris muscle.

At the wrist, several structures may be felt. On the anterior surface of the wrist, it is possible to see the tendon of the palmaris longus muscle by making a fist. Next to this tendon, as you move toward

Exhibit 1–6. SURFACE ANATOMY OF THE UPPER EXTREMITY

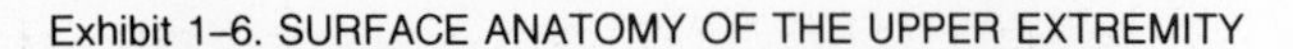

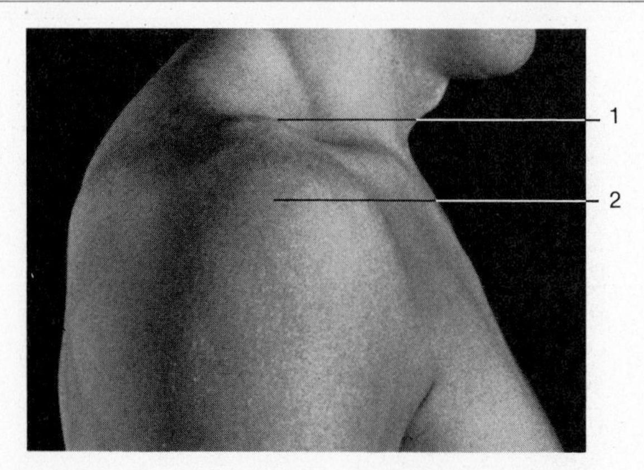

Lateral view of shoulder

SHOULDER

1. *Acromion.* Expanded end of spine of scapula; forms tip of shoulder.
2. *Deltoid muscle.* Triangular-shaped muscle that forms rounded prominence of shoulder.

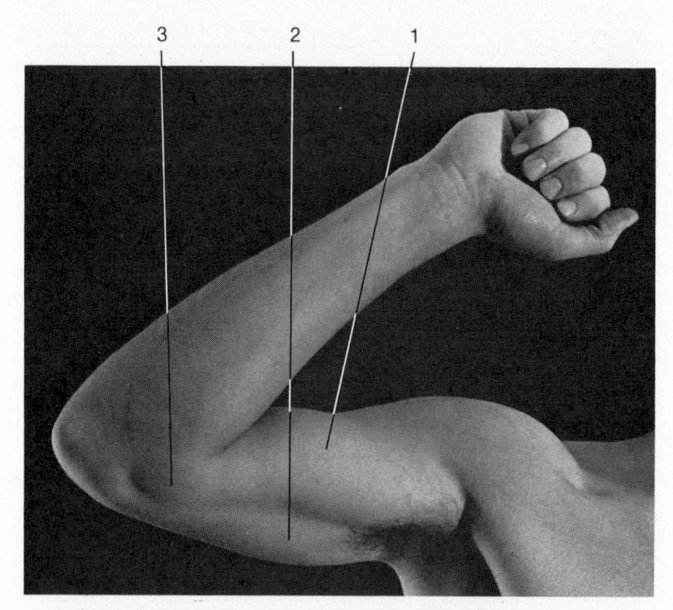

Anterior view of upper extremity

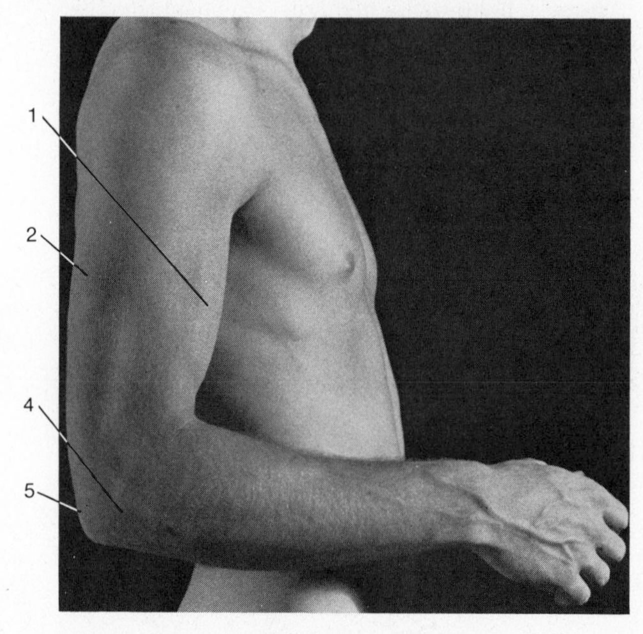

Lateral view of upper extremity

UPPER EXTREMITY

1. *Biceps brachii muscle.* Forms bulk of anterior surface of arm; helps move forearm.
2. *Triceps brachii muscle.* Forms bulk of posterior surface of arm; helps move forearm.
3. *Medial epicondyle.* Medial projection at distal end of humerus.
4. *Lateral epicondyle.* Lateral projection at distal end of humerus.
5. *Olecranon.* Projection of proximal end of ulna (medial bone of forearm); forms elbow.

Photographs courtesy of Victor B. Eichler, of Wichita State University, and Steve Harper.

FOREARM

1. *Styloid process of ulna.* Projection of distal end of ulna at medial side of wrist (little-finger side).
2. *Brachioradialis muscle.* Located at superior and lateral aspect of forearm; helps move forearm.
3. *Flexor carpi radialis muscle.* Located along midportion of forearm; helps move wrist.
4. *Flexor carpi ulnaris muscle.* Located at medial aspect of forearm; helps move wrist.

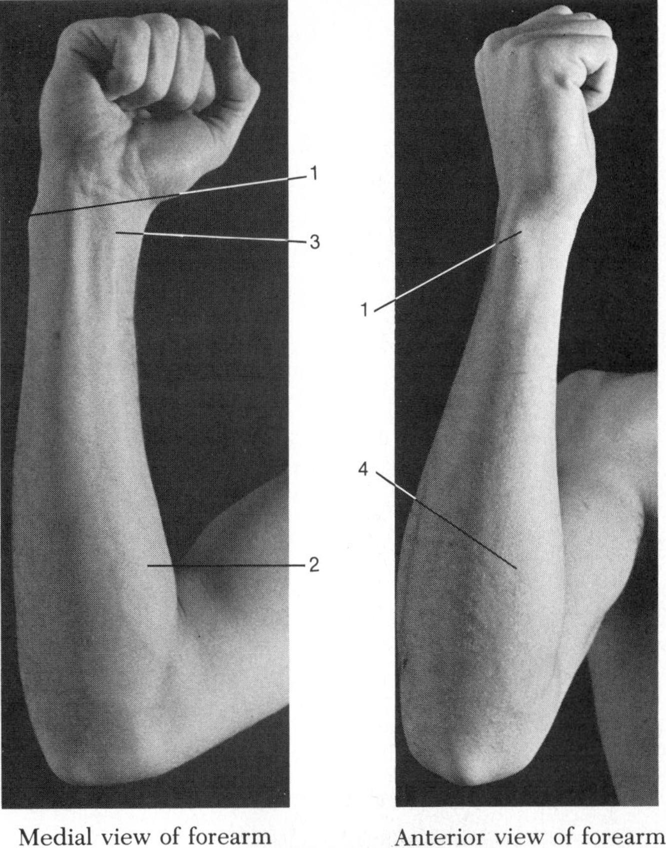

Medial view of forearm Anterior view of forearm

WRIST AND HAND

1. *Tendon of palmaris longus muscle.* Tendon on anterior surface of wrist nearer ulna.
2. *Tendon of flexor carpi radialis.* Tendon on anterior surface of wrist lateral to palmaris longus tendon.
3. *"Knuckles."* Distal ends of second through fifth metacarpal (palm) bones.
4. *Bracelet flexure lines.* Creases in wrist.
5. *Proximal transverse flexure line.* Crease in palm running in an oblique transverse direction; closer to wrist.
6. *Distal transverse flexure line.* Crease in palm running in an oblique transverse direction; farther from wrist.
7. *Phalanges.* Bones of fingers.

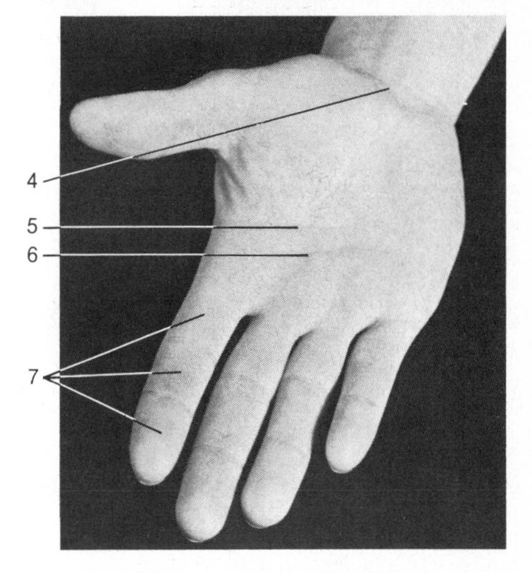

Anterior view of hand

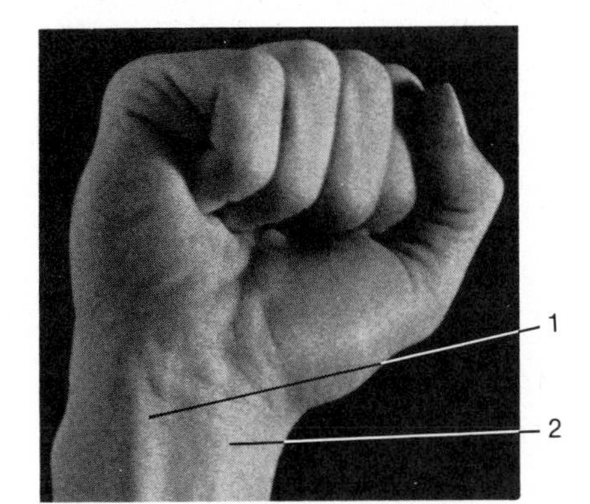

Anterior view of wrist

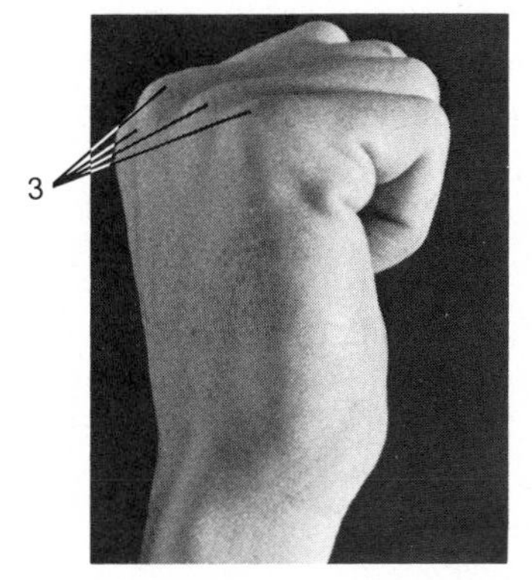

Medial view of wrist

the thumb, you can feel the tendon of the flexor carpi radialis muscle. If you continue toward the thumb, you can palpate the radial artery just before the styloid process of the radius. This artery is frequently used to take the pulse. By bending the thumb backward, two prominent tendons may be located along the posterior surface of the wrist. The one closer to the styloid process of the radius is the tendon of the extensor pollicus brevis muscle. The one closer to the styloid process of the ulna is the tendon of the extensor pollicus longus muscle. The depression between these two tendons is known as the "anatomical snuffbox." By palpating the depression, you can feel the radial artery. It might be of interest to note that the "knuckles" are the distal ends of the second through fifth metacarpal (palm) bones. By examining the palmar surface of the hand, it is possible to see a number of creases in the skin, as well as the location of the phalanges between joints of the fingers.

The lower extremity

The **lower extremity** consists of the buttocks *(gluteal region)*, thigh *(femoral region)*, knee *(genu)*, leg *(crus)*, and foot *(pes)*. The foot includes the ankle *(tarsus)* and toes *(digits* or *phalanges)*. See Exhibit 1–7.

The iliac crest forms the outline of the superior border of the buttock. About 20 cm (8 in.) below the highest portion of the iliac crest, the greater trochanter of the femur can be felt on the lateral side of the thigh. As you will see later, the iliac crest and greater trochanter are useful landmarks when giving an intramuscular injection in the gluteal muscle. Most of the prominence of the buttocks is formed by the gluteal muscles. The bony prominence in each buttock is the ischial tuberosity of the hipbone. This structure bears the weight of the body when you are seated. Among the prominent superficial muscles on the anterior surface of the thigh are the adductor longus, vastus lateralis, vastus medialis, and rectus femoris. The vastus lateralis is frequently used as an injection site by diabetics when administering insulin.

On the anterior surface of the knee, the patella, or kneecap, is observable. Below it is the patellar ligament. On the posterior surface of the knee is a diamond-shaped space, the popliteal fossa.

Just below the patella, on either side of the patellar ligament, the medial and lateral condyles of the femur and tibia can be felt. The bony prominence below the patella in the middle of the leg is the tibial tuberosity. The tibia is the medial bone of the leg, and the fibula is the lateral bone of the leg. Prominent superficial muscles of the leg include the tibialis anterior, gastrocnemius, and soleus.

At the ankle, the medial malleolus of the tibia and the lateral malleolus of the fibula can be noted as two prominent eminences. Arising from the heel bone (calcaneus) is the calcaneal (Achilles) tendon.

Exhibit 1–7. SURFACE ANATOMY OF THE LOWER EXTREMITY

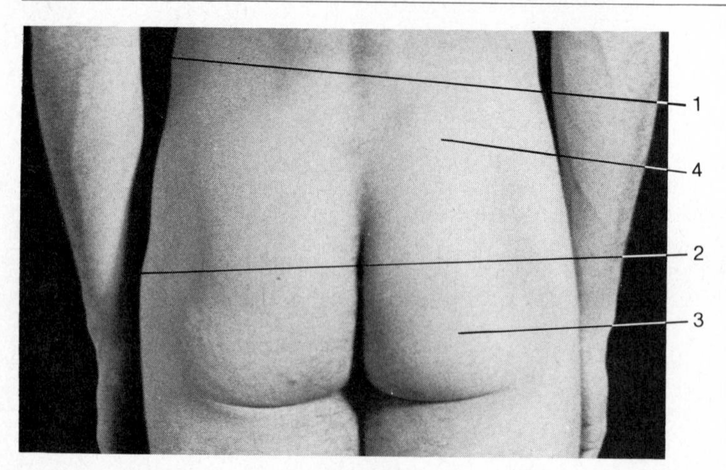

Posterior view of buttocks and thigh

Anterior view of thigh →

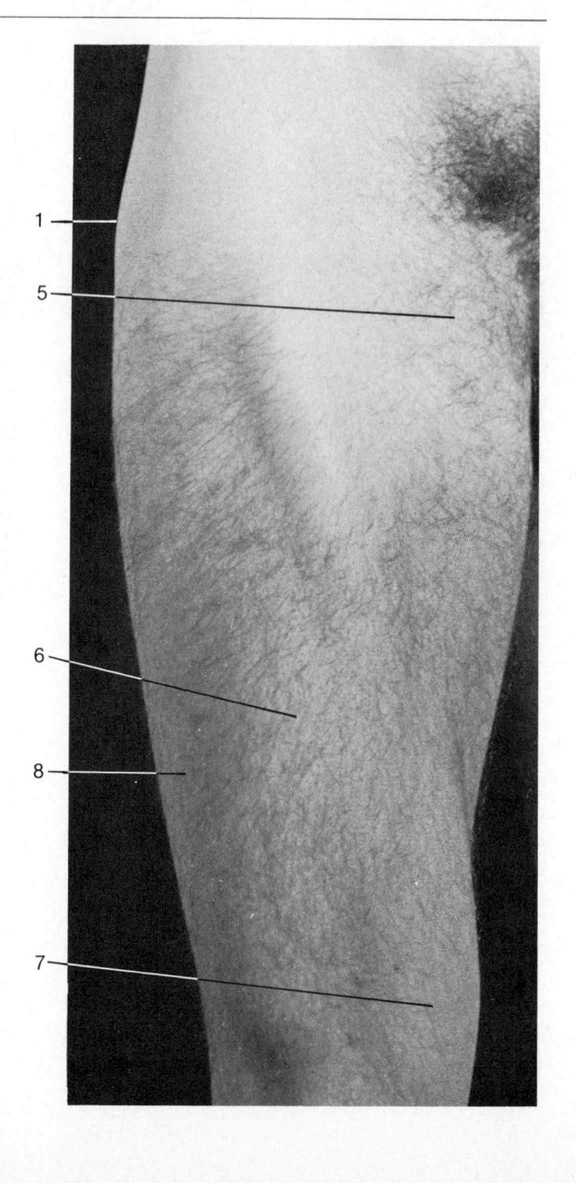

BUTTOCKS AND THIGH

1. *Iliac crest.* Superior portion of hipbone.
2. *Greater trochanter.* Projection of proximal end of femur (thighbone).
3. *Gluteus maximus muscle.* Forms major portion of prominence of buttock; helps move femur.
4. *Gluteus medius muscle.* Located above gluteus maximus; helps move femur.
5. *Adductor longus muscle.* Located on medial side of thigh; helps move femur.
6. *Rectus femoris muscle.* Located along midportion of thigh; helps move tibia (shinbone).
7. *Vastus medialis muscle.* Located at medial, inferior portion of thigh; helps move tibia.
8. *Vastus lateralis muscle.* Located along anterolateral surface of thigh; helps move tibia.

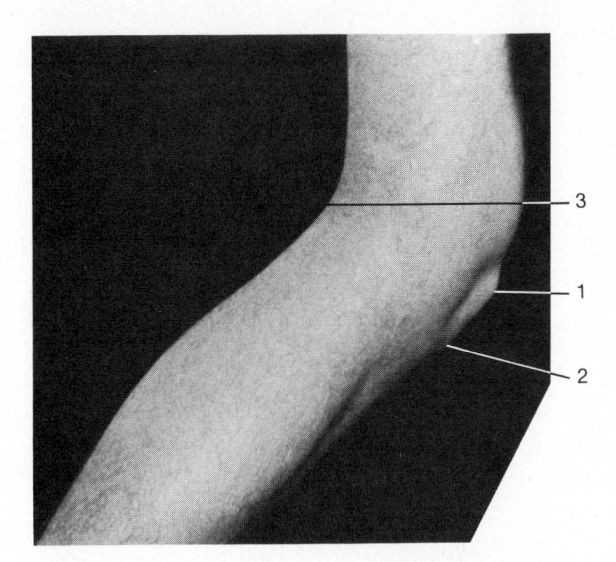

Posteromedial view of knee

KNEE

1. *Patella.* Also called kneecap; located on anterior surface of knee along midline.
2. *Patellar ligament.* Located below patella.
3. *Popliteal fossa.* Diamond-shaped space on posterior surface of knee.

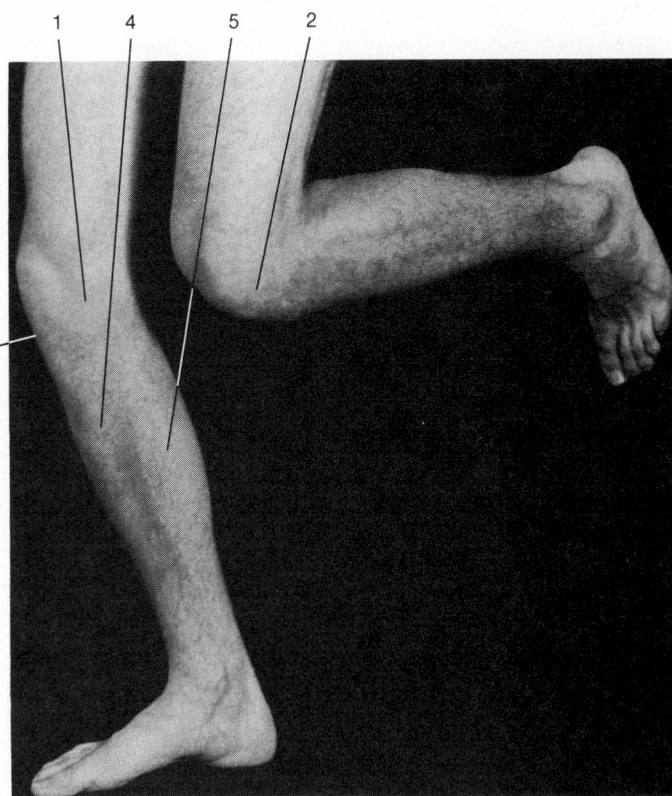

Medial view of right leg and lateral view of left leg

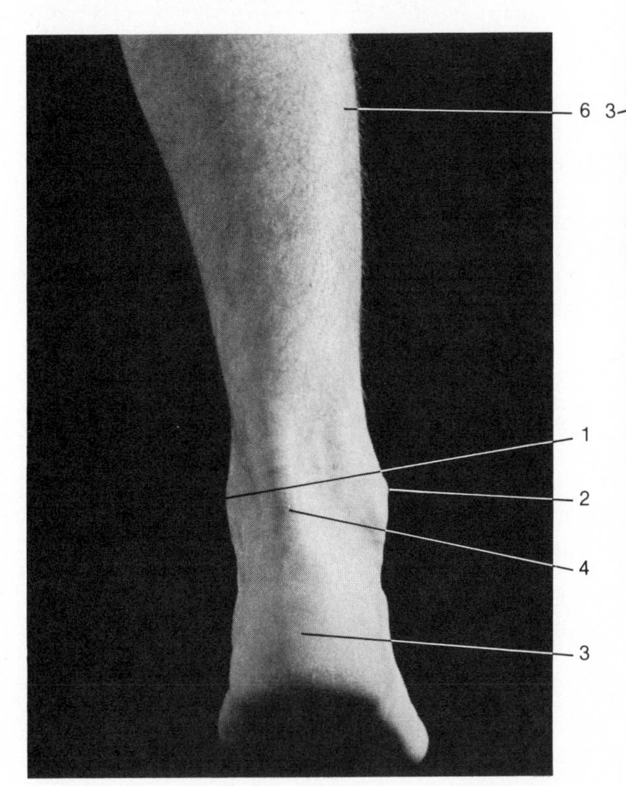

Posterior view of foot

FOOT

1. *Medial malleolus.* Projection of distal end of tibia that forms medial prominence of ankle.
2. *Lateral malleolus.* Projection of distal end of fibula (lateral bone of leg) that forms lateral prominence of ankle.
3. *Calcaneus.* Heel bone.
4. *Calcaneal tendon.* Also called Achilles tendon; conspicuous tendon attached to calcaneus.

LEG

1. *Medial condyles of femur and tibia.* Medial projections just below patella; upper part of projection belongs to distal end of femur; lower part of projection belongs to proximal end of tibia.
2. *Lateral condyles of femur and tibia.* Lateral projections just below patella; upper part of projection belongs to distal end of femur; lower part of projection belongs to proximal end of tibia.
3. *Tibial tuberosity.* Bony prominence below patella.
4. *Tibialis anterior muscle.* Located on anterior surface of leg along midportion; helps move foot.
5. *Gastrocnemius muscle.* Forms bulk of mid and upper portion of posterior surface of leg; helps move foot.
6. *Soleus muscle.* Located deep to the gastrocnemius; helps move foot. See also posterior view of foot.

Photographs courtesy of Victor B. Eichler, of Wichita State University, and Steve Harper.

Exhibit 1–8. METRIC UNITS OF LENGTH AND SOME ENGLISH EQUIVALENTS

METRIC UNIT (ABBREVIATION, SYMBOL)	MEANING OF PREFIX	METRIC EQUIVALENT	ENGLISH EQUIVALENT
1 kilometer (km)	kilo = 1,000	1,000 m	3,280.84 ft or 0.62 mi; 1 mi = 1.61 km
1 hectometer (hm)	hecto = 100	100 m	328. ft
1 dekameter (dam)	deka = 10	10 m	32.8 ft
1 meter (m)	Standard unit of length		39.37 in. or 3.28 ft or 1.09 yd
1 decimeter (dm)	deci = $\frac{1}{10}$	0.1 m	3.94 in.
1 centimeter (cm)	centi = $\frac{1}{100}$	0.01 m	0.394 in.; 1 in. = 2.54 cm
1 millimeter (mm)	milli = $\frac{1}{1,000}$	0.001 m	0.0394 in.
1 micrometer (μm) [formerly micron (μ)]	micro = $\frac{1}{1,000,000}$	0.000,001 m	3.94×10^{-5} in.
1 nanometer (nm) [formerly millimicron (mμ)]	nano = $\frac{1}{1,000,000,000}$	0.000,000,001 m	3.94×10^{-8} in.
1 angstrom (Å)		0.000,000,000,1 m	3.94×10^{-9} in.

MEASURING THE HUMAN BODY

One of the most important concepts you should understand about your body is the concept of measurement. You will constantly be exposed to various kinds of measurement in order to understand how your body works, how big various organs are, how much various organs weigh, and how much of a given medication should be administered, to mention a few. Measurements involving time, weight, temperature, size, length, and volume are a routine part of your studies in a medical science program.

Whenever you come across a measurement in the text, the measurement will be given in metric units. To help you compare the metric unit to a familiar unit, the approximate English equivalent will also be given in parentheses directly after the metric unit. For example, you might be told that the length of a particular part of the body is 2.54 cm (1 in.).

As a first step in helping you understand the relationships of the metric to English systems of measurement, three Exhibits have been prepared. Exhibit 1–8 contains metric units of length and some English equivalents. Exhibit 1–9 contains metric units of mass and some English equivalents. Exhibit 1–10 contains metric units of volume and some English equivalents. Before you begin reading Chapter 2, carefully examine the Exhibits. Even if you do not learn all the metric units and their equivalents at this point, you can still refer back to the Exhibits later.

We will now turn our attention to the structural and functional unit of the body, the cell.

Exhibit 1–9. METRIC UNITS OF MASS AND SOME ENGLISH EQUIVALENTS

METRIC UNIT (ABBREVIATION, SYMBOL)	METRIC EQUIVALENT	ENGLISH EQUIVALENT
1 kilogram (kg)	1,000 g	2.205 lb
1 hectogram (hg)	100 g	
1 dekagram (dag)	10 g	
1 gram (g)	1 g	1 lb = 453.6 g; 1 oz = 28.35 g
1 decigram (dg)	0.1 g	
1 centigram (cg)	0.01 g	
1 milligram (mg)	0.001 g	
1 microgram (μg)	0.000,001 g	

Exhibit 1–10. METRIC UNITS OF VOLUME AND SOME ENGLISH EQUIVALENTS

METRIC UNIT (ABBREVIATION, SYMBOL)	METRIC EQUIVALENT	ENGLISH EQUIVALENT
1 liter (l)	1,000 ml	33.81 fl oz or 1.057 qt; 946 ml = 1 qt
1 milliliter (ml)	0.001 liter	0.0338 fl oz; 30 ml = 1 fl oz; 5 ml = 1 teaspoon
1 cubic centimeter (cm^3)	0.999972 ml	0.0338 fl oz

Chapter summary in outline

ANATOMY AND PHYSIOLOGY

1. Anatomy is the study of structure and how structures are related to each other.
2. Subdivisions of anatomy include gross anatomy (macroscopic), systemic anatomy (systems), regional anatomy (regions), embryology (development prior to eighth week), pathological anatomy (disease), and histology (microscopic study of tissues and cells).
3. Physiology is the study of how structures function.

HOW ARE YOU PUT TOGETHER?

1. The human body consists of levels of structural organization from the chemical level to the organismic level.
2. The chemical level is represented by all the atoms and molecules in the body. The cellular level consists of cells. The tissue level is represented by tissues. The organ level consists of body organs, and the system level is represented by organs that work together to perform a more general function.
3. The human organism is a collection of structurally and functionally integrated systems.

STRUCTURAL PLAN OF THE HUMAN BODY

Body cavities

1. Spaces in the body that contain internal organs are called cavities.
2. Dorsal and ventral cavities are the two principal body cavities. The dorsal cavity contains the brain and spinal cord. The organs of the ventral cavity (coelom) are collectively called the viscera.
3. Dorsal cavity is subdivided into the cranial cavity and vertebral canal.
4. Ventral body cavity is subdivided by the diaphragm into an upper thoracic cavity and a lower abdominopelvic cavity.
5. Thoracic cavity contains two pleural cavities, pericardial cavity, and the mediastinum.
6. Abdominopelvic cavity, which is actually an upper abdominal cavity and a lower pelvic cavity, is divided into nine anatomical regions. It may also be divided into four quadrants.

The anatomical position and regional names

1. The position in which the body is studied and described is the anatomical position. The subject stands erect and faces the observer with arms at sides and palms turned forward.
2. Regional names are terms given to specific regions of the body for reference purposes. Examples of regional names include cranium (skull), thorax (chest), brachium (arm), patellar (knee), caput (head), and gluteal (buttock).

Directional terms

1. These are names given to indicate the relationship of one part of the body to another.
2. Some examples of directional terms are superior (toward the head), anterior (near the front), medial (nearer the midline), distal (farther from the attachment of a limb), and external (toward the surface).

Planes of the body

1. Planes of the body are flat surfaces that divide the body into definite areas. The midsagittal plane divides the body into equal right and left sides; the sagittal (parasagittal) plane, into unequal right and left sides; the frontal (coronal) plane, into anterior and posterior portions; the horizontal (transverse) plane, into superior and inferior portions.
2. Sections of organs include cross sections, oblique sections, and longitudinal sections.

INTRODUCTION TO SURFACE ANATOMY

1. Surface anatomy is the study of the form and markings of the surface of the body.
2. Surface-anatomy features may be noted by visual inspection or palpation.
3. The primary divisions of the body used to study surface anatomy are the head, neck, trunk, upper extremity, and lower extremity.
4. A review of surface anatomy is presented in Exhibits 1–3 through 1–7.

MEASURING THE HUMAN BODY

1. Measurements involving length, mass, and volume are integral components of a knowledge of the human body.
2. Metric units of length may be reviewed in Exhibit 1–8, metric units of mass may be reviewed in Exhibit 1–9, and metric units of volume may be reviewed in Exhibit 1–10.

Review questions and problems

1. Define anatomy. How does each subdivision of anatomy help you understand the structure of the human body? Define physiology.
2. Construct a diagram to illustrate the levels of structural organization that characterize the body. Be sure to define each level.
3. Outline the function of each system of the body, and list several organs that compose each system.
4. What does bilateral symmetry mean? Why is the body considered to be a tube within a tube?
5. Define a body cavity. List the body cavities discussed, and tell which major organs are located in each. What landmarks separate the various body cavities from each other?
6. Discuss how the abdominopelvic area is subdivided into nine regions. Name and locate each region and list the organs, or parts of organs in each. Describe how the abdominopelvic cavity is divided into four quadrants and list the name of each quadrant.
7. When is the body in the anatomical position? Why is the anatomical position used?
8. Review Figure 1–6. See if you can locate each region on your own body, and name each by its common and anatomical term.
9. What is a directional term? Why are these terms important? Can you use each of the directional terms listed in Exhibit 1–2 in a complete sentence?
10. Construct a series of figures of the human body, and indicate each directional term used in Exhibit 1–2 by labeling the relationship of organs to each other.
11. Describe the various planes that may be passed through the body. Explain how each plane divides the body.
12. What is meant by the phrase, "A part of the body has been sectioned"? Given an orange, can you make a cross section, oblique section, and longitudinal section with a knife?
13. Using Exhibits 1–3 through 1–7 as an outline, locate as many of the surface features as you can on your partner's body, wall charts, models, photographs, and skeletons.
14. Convert the following lengths:
    (a) If a bacterial cell measures 100 μm in length, how many nanometers is this?
    (b) How many meters are in 1 mi?
    (c) If a road sign reads 35 km/hour, what would your speedometer have to read to obey the sign?
    (d) How many millimeters are in 1 km? In 1 in.?
    (e) A person's arm measures 2 ft in length. How many centimeters is this?

(f) Convert 0.40 m to millimeters.
(g) How many millimeters are in 5 in.?
(h) If you ran 295.2 ft, how many meters would you have run?
(i) If the distance to the moon is 239,000 mi, what is it in meters?

15. Solve the following conversions of mass:
(a) Calculate the milligrams in 0.4 kg and in 1 lb.
(b) If a bottle contains 1.42 g, how many centigrams does it contain?
(c) The indicated dosage of a certain drug is 50 μg. How many milligrams is this?
(d) If you weigh 110 lb, how many kilograms do you weigh?
(e) How many centigrams are in 1 g?

16. Convert the following volumes:
(a) If you excrete 1,200 ml of urine in a day, how many liters is this?
(b) How many milliliters are in 2 liters?
(c) Convert 2 pt to milliliters.
(d) If you remove 15 cm^3 of blood from a patient, how many milliliters have you removed?

Selected readings

Basmajian, John V. *Primary Anatomy.* 6th ed. Baltimore: Williams and Wilkins, 1970. Pp. xiff., 11–18.

Gray, Henry. *Anatomy of the Human Body.* 29th ed., edited by Charles Mayo Goss. Philadelphia: Lea and Febiger, 1973. Chap. 3.

Hollinshead, W. H. *Textbook of Anatomy.* 3d ed. New York: Harper and Row, 1974.

Royce, Joseph. *Surface Anatomy.* Philadelphia: F. A. Davis, 1965.

CHAPTER 2

CELLS: STRUCTURAL UNITS OF THE BODY

STUDENT OBJECTIVES

After you have read this chapter, you should be able to:

1. Define and list a cell's generalized parts
2. Describe the molecular organization of the plasma membrane
3. List the factors related to semipermeability of the plasma membrane
4. Define diffusion, facilitated diffusion, osmosis, filtration, dialysis, active transport, phagocytosis, and pinocytosis
5. Describe the chemical composition and list the functions of cytoplasm
6. Describe two general functions of a cell nucleus
7. Distinguish between agranular and granular endoplasmic reticulum
8. Define the function of ribosomes
9. Describe the role of the Golgi complex in the synthesis, storage, and secretion of glycoproteins
10. Describe the function of mitochondria as "powerhouses of the cell"
11. Explain why a lysosome in a cell is called a "suicide packet"
12. Describe the structure and function of centrioles in cellular reproduction
13. Distinguish between the structural and functional differences of cilia and flagella
14. Define and list several examples of a cell inclusion
15. Define and list several examples of an extracellular material
16. Describe the sequence of events involved in cell division
17. Describe the significance of cell division

A study of the body at the cellular level of organization is important to a total understanding of the structure and function of the body because many activities essential to life occur in cells and many disease processes originate in cells. A **cell** may be defined as the basic, living, structural and functional unit of the body and, in fact, of all organisms. It is the smallest structure capable of performing all the activities vital to life. **Cytology** is the specialized branch of science concerned with the study of cells. In this chapter, we shall concentrate on the structure of cells, the functions of cells, and the reproduction of cells. A series of illustrations accompanies each cell structure that you study. The first illustration shows the location of the structure within the cell. The second is an *electron micrograph,* which is a photograph taken with a high-powered electron microscope.* The third illustration that accompanies each structure is a drawing of the electron micrograph. The drawing will clarify some of the small details by exaggerating their outlines.

* The electron microscope can magnify an object up to 200,000 times its size. In comparison, the light microscope that you probably use in your laboratory magnifies objects 1,000 to 2,000 times their size (1,000–2,000X).

THE GENERALIZED ANIMAL CELL

In the discussions that follow, we shall speak about a **generalized animal cell,** which is a composite of many different kinds of cells found in the body. The characteristics of specific cells will be discussed in later chapters as parts of the systems to which they belong. Bone cells, for example, will be discussed in the chapter on the skeleton. Examine the generalized cell illustrated in Figure 2–1, and keep in mind that no such single cell actually exists.

For convenience, we shall divide the generalized cell into four principal parts:

1. The *cell surface,* or the outer, limiting membrane separating the internal parts of the cell from the extracellular fluid and external environment
2. *Cytoplasm,* or the ground substance of the cell in which organelles are embedded
3. *Organelles,* the cellular components that are highly specialized for particular cellular activities
4. *Inclusions,* or the secretions and storage areas of cells

Extracellular materials, which are substances external to the cell surface, will also be examined in connection with cells.

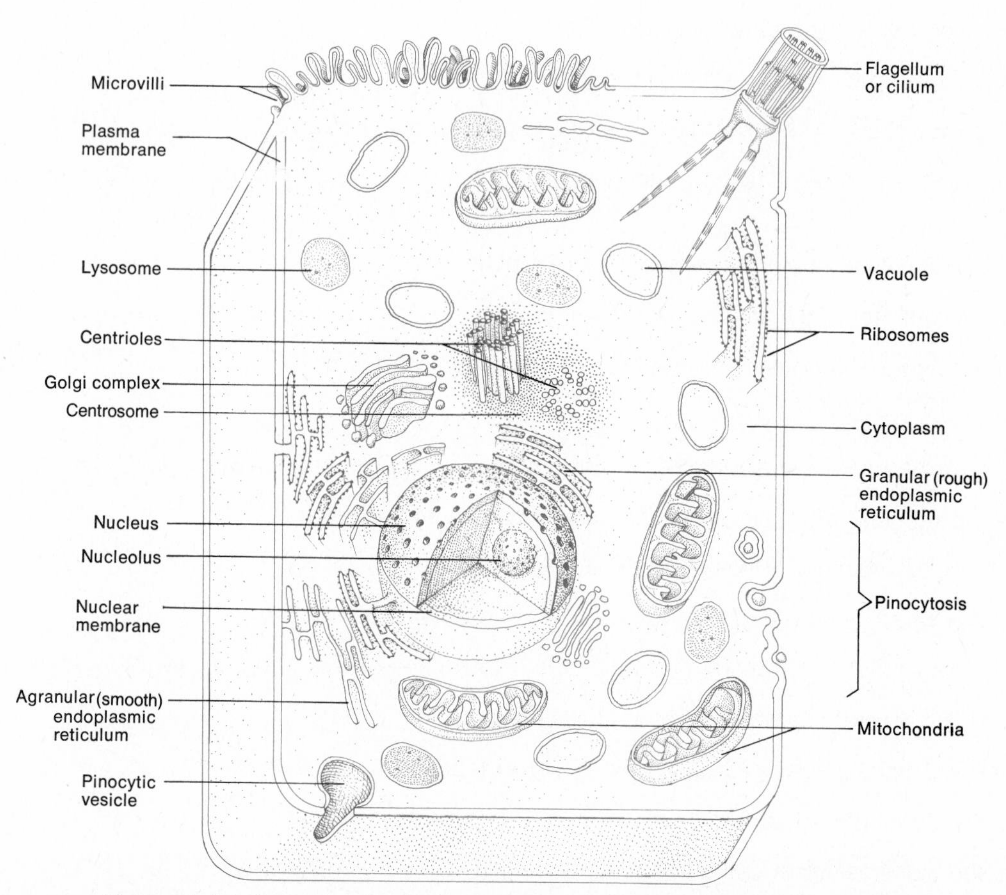

Figure 2–1. Generalized animal cell based on electron microscope studies.

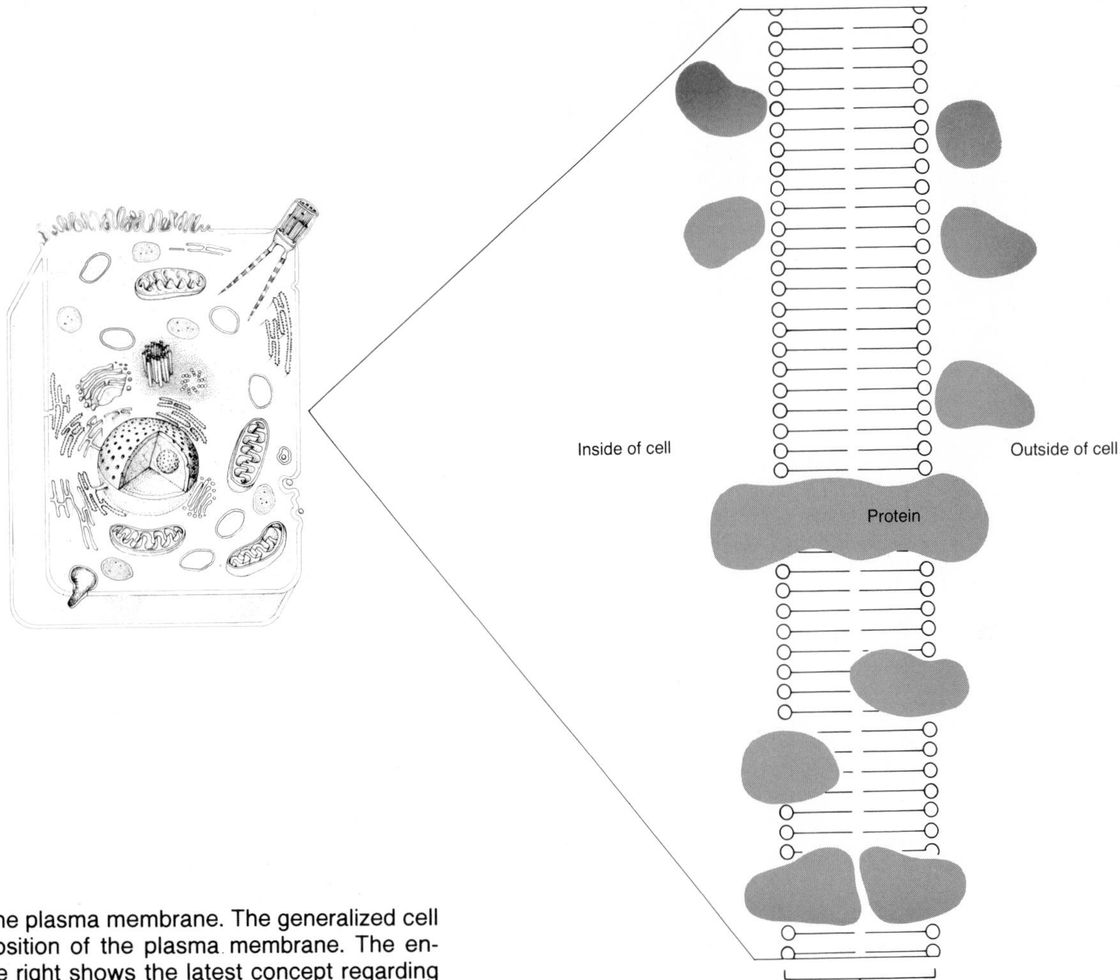

Figure 2–2. The plasma membrane. The generalized cell indicates the position of the plasma membrane. The enlargement to the right shows the latest concept regarding the relationship of the phospholipid and protein molecules.

THE CELL SURFACE

The exceedingly thin structure that separates one cell from other cells and from the external environment is called the **plasma membrane,** or **cell membrane.** Electron microscopy studies have shown that the plasma membrane ranges from 65 to 100 angstroms in thickness, a dimension below the limits of the light microscope.* Scientists have known for a long time that plasma membranes are composed of phospholipid and protein molecules. Very recent studies of membrane structure suggest a new concept regarding the possible arrangement of phospholipid and protein molecules. This new concept of membrane structure is called the *fluid mosaic hypothesis.* These studies seem to indicate that the phospholipid molecules, which account for about one-half the mass of the membrane, are arranged in two parallel rows. This double row of phospholipid molecules is termed a bilayer (Figure 2–2). The protein molecules are arranged somewhat differently. Some lie at or near the inner and outer surface of the membrane. Others penetrate the membrane partway, completely, singly, or in pairs. Such an arrangement suggests that membranes are not static but that the proteins and phospholipids have a considerable degree of movement. It is quite possible that many key functions of membranes may be fully explained once the reasons for protein and phospholipid movements are understood. For example, the fluid mosaic hypothesis may explain how specific receptor sites on membranes may attach to hormones (for example, insulin), to transmitter substances produced by axons, and to antigens on the surfaces of red blood cells. Other features of the membrane structure are areas that look like breaks along the membrane surface. These breaks appear at intervals and range in size from 7 to 10 Å in diameter. Researchers suspect they may be pores. (See Figure 2–8b.)

The basic functions of the plasma membrane are to enclose the components of the cell and to serve as a boundary through which substances must pass

* One **angstrom,** usually written as 1 Å, = 0.000,000,000,1 m ($^1/_{250,000,000}$ in.). Another microscopic unit of measurement is the **micrometer** (μm), which is equal to 0.000,001 m ($^1/_{25,000}$ in.).

to enter or exit the cell. One important characteristic of the plasma membrane is that it permits certain ions and molecules to enter or exit the cell but restricts the passage of others. For this reason, plasma membranes are described as **semipermeable.** In general, plasma membranes are freely permeable to water. In other words, they let water into and out of the cell. However, they act as barriers to the movement of almost all other substances. The ease with which a substance passes through a membrane is called the membrane's *permeability* to that substance. The permeability of a plasma membrane appears to be a function of several factors. We can list four of these factors:

1. *Size of the entering molecules.* Large-sized molecules cannot pass through the plasma membrane. Water and amino acids are relatively small molecules and can enter and exit the cell easily. However, most proteins, which consist of many amino acids linked together, seem to be too large to pass through the membrane. Many scientists believe that the giant-sized molecules do not enter the cell because they are larger than the diameters of the suspected membrane pores.
2. *Solubility in lipids.* Substances that dissolve easily in lipids pass through the membrane more readily than other substances since a major part of the plasma membrane consists of lipid molecules.
3. *Charge on ions.* The charge of an ion attempting to cross the plasma membrane can determine how easily the ion can enter or leave the cell. This is true because the protein portion of the membrane is capable of ionization. If an ion has a charge opposite that of the membrane, it is attracted to the membrane and passes through more readily. Conversely, if the ion attempting to cross the membrane has the same charge as the membrane, it is repulsed by the membrane, and its passage is restricted. This phenomenon conforms to the rule of physics that opposite charges attract, whereas like charges repel each other.
4. *Presence of carrier molecules.* Plasma membranes contain special molecules called carriers that are capable of attracting and transporting substances across the membrane regardless of size, ability to dissolve in lipids, or membrane charge. This mechanism will be considered later in the chapter.

These four aspects of the plasma membrane work together to determine whether the membrane will be permeable to the various substances attempting to enter or leave the cell. Before discussing further aspects of the membrane, we need to consider some of the processes involved when substances move from one region to another.

Moving materials across plasma membranes

The mechanisms whereby substances move across the plasma membrane are important to the life of the cell. Certain substances, for example, must move into the cell to support life, whereas waste materials or substances that may be harmful must be moved out of the cell. The processes involved in these movements may be divided into two broad categories, depending upon whether the cell participates in the process by expending energy. Accordingly, the process may be classed as either passive or active. In *passive processes,* substances move across plasma membranes without any help from the cell. The substances move, on their own, from an area where their concentration is greater to an area where their concentration is less. The substances could also be pushed through the plasma membrane by pressure from an area where the pressure is greater to an area where it is less. These phenomena will be explained shortly. In *active processes,* by contrast, the cell contributes energy and assumes a role in moving the substance across the membrane. Both active and passive processes should be understood in the study of cell physiology.

Passive processes

DIFFUSION. A passive process called **diffusion** occurs when there is a *net* or greater movement of molecules or ions from a region of high concentration to a region of low concentration. The movement from high to low concentration continues until the molecules are evenly distributed. At this point, they move in both directions at an equal rate. This point of even distribution is called *equilibrium.* The difference between higher and lower concentrations is called the *concentration gradient.* Molecules moving from the high-concentration area to the low-concentration area are said to move *down* or *with* the concentration gradient. Consider the following example. If a dye pellet is placed in a beaker filled with water, the color of the dye is seen immediately around the pellet. At increasing distances from the pellet, the color becomes lighter (Figure 2–3). If the beaker is observed some time later, however, the water solution will be a uniform color. This happens because the dye molecules possess kinetic energy, which causes them to move about at random, dispersing them throughout the

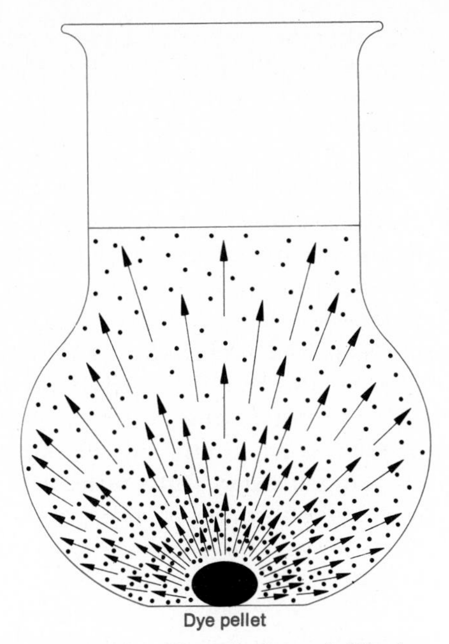

Figure 2–3. The principle of diffusion.

entire area. The dye molecules move down the concentration gradient from an area of high concentration to an area of low concentration. The water molecules also move from a high-concentration to a low-concentration area. When dye molecules and water molecules are evenly distributed among themselves, equilibrium is reached and diffusion ceases, even though molecular movements continue. As another example of diffusion, consider what would happen if you opened a bottle of perfume in a room. The perfume molecules would diffuse until an equilibrium was reached between the perfume molecules and the air molecules in the room.

In the examples cited, no membranes were involved. Diffusion may occur, however, through semipermeable membranes in the body. One of the best examples of this kind of diffusion in the human body is the movement of oxygen from the blood into the cells and the movement of carbon dioxide from the cells back into the blood.

Another example of diffusion through a semipermeable membrane occurs by a process called **facilitated diffusion.** Although some chemical substances are insoluble in lipids, they can still pass through the plasma membrane. Among these are different sugars, especially glucose. In the process of facilitated diffusion, glucose combines with a carrier substance. The combined glucose-carrier is soluble in the lipid layer of the membrane, and the carrier transports the glucose to the inside of the membrane. At this point, the glucose separates from the carrier and the glucose passes to the inside of the cell. The carrier then returns to the outside of the membrane to pick up more glucose and transport it to the inside. The purpose of the carrier is to make the glucose soluble in the lipid portion of the membrane so that it can pass through the membrane. By itself, glucose is insoluble and cannot penetrate the membrane. You can get a fairly good idea as to how a carrier works by taking a look at Figure 2–6a. In the process of facilitated diffusion, the cell does not have to expend any energy, and the movement of the substance is from a region of higher to lower concentration.

The rate at which facilitated diffusion occurs depends on several factors. Among these are (1) the difference in concentration of the substance on either side of the membrane, (2) the amount of carrier available to transport the substance, and (3) how fast the carrier and substance combine. With regard to the facilitated diffusion of glucose, the process is greatly accelerated by the presence of insulin, a hormone produced by the pancreas. One of the functions of insulin is to lower the blood-glucose level by accelerating the transportation of glucose from the blood into body cells. This, as we have just seen, is by facilitated diffusion.

OSMOSIS. Another passive process by which materials move across membranes is **osmosis.** Unlike diffusion, this process specifically refers to the net movement of water molecules through a semipermeable membrane from an area of high water concentration to an area of lower water concentration. Once again, a simple apparatus may be used to demonstrate the process. The apparatus shown in Figure 2–4 consists of a tube constructed from cellophane, a semipermeable membrane. The cellophane tube is filled with a colored 20 percent sugar (sucrose) solution. The upper portion of the cellophane tube is plugged with a rubber stopper through which a glass tubing is fitted. The cellophane tube is placed into a beaker containing pure water. Initially, the relative concentrations of water on either side of the semipermeable membrane are different. There is a lower concentration of water inside the cellophane tube than there is outside of it. As a result of this difference, water moves from the beaker into the cellophane tube. The force with which the water moves is called osmotic pressure. Very simply, **osmotic pressure** is the force under which a solvent moves from a solution of lower solute concentration to a solution of higher solute concentration when the solutions are separated by a semipermeable membrane. There is no movement of sugar from the cellophane tube into the beaker, however, since the cellophane is impermeable to molecules of

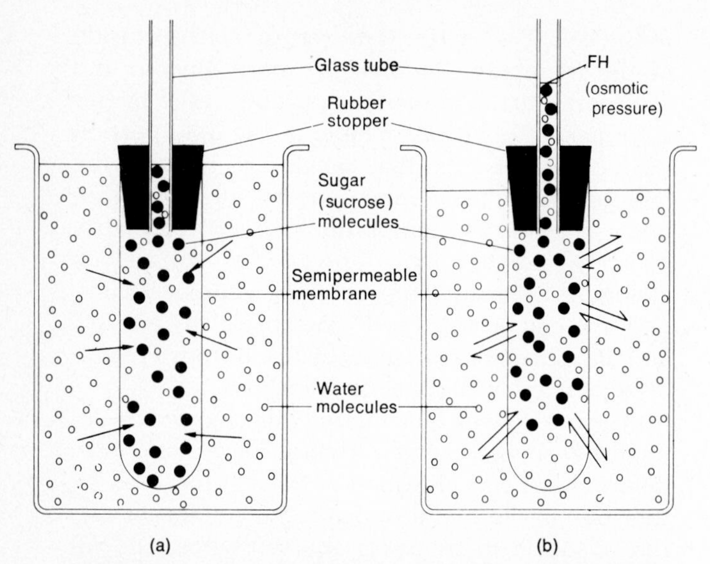

Figure 2–4. The principle of osmosis. (a) Apparatus at the start of the experiment. (b) Apparatus at equilibrium. In (a), the cellophane tube contains a 20 percent sugar (sucrose) solution and is immersed in a beaker of distilled water. The arrows indicate that water molecules can pass freely into the cellophane tube, but that sugar molecules are held back by the semipermeable membrane. As water moves into the cellophane tube by osmosis, the sugar solution is diluted, and the volume of the solution in the cellophane tube increases. This increased volume is shown in (b), with the sugar solution moving up the glass tubing. The final height reached (FH) occurs at equilibrium and represents the osmotic pressure. At this point, the number of water molecules leaving the cellophane tube is equal to the number of water molecules entering the tube.

sugar. This is because sugar molecules are too large to go through the pores of the membrane. As water movement into the cellophane tube continues, the sugar solution becomes increasingly diluted and begins to move up the glass tubing. After a period of time, the water that has accumulated in the cellophane tube and the glass tubing exerts a downward pressure that forces water molecules back out of the cellophane tube and into the beaker. When water molecules leave the cellophane tube and enter the tube at the same rate, equilibrium is reached.

Osmosis may also be understood by considering the effects of different water concentrations on red blood cells. If the normal shape of a red blood cell is to be maintained, the cell must be placed in an **isotonic solution** (Figure 2–5a). This is a solution in which the concentrations of water molecules and solute molecules are the same on both sides of the semipermeable membrane. In the case of red blood cells, the concentrations of water and solute in the extracellular fluid outside the red blood cell must be the same as the concentration of the intracellular fluid. Under ordinary circumstances, a 0.85 percent NaCl solution is isotonic for red blood cells. In this condition, water molecules enter and exit the cell at the same rate, allowing the cell to maintain its normal shape. A different situation results if red blood cells are placed in a solution that has a lower concentration of solutes and, therefore, a higher concentration of water. This is called a **hypotonic solution.** In this condition, water molecules enter the cells faster than they can leave. This causes the red blood cells to swell and eventually burst (Figure 2–5b). The rupture of red blood cells in this manner is called *hemolysis* or *laking.* A good hypotonic solution is distilled water. On the other hand, a **hypertonic solution** has a higher concentration of solutes and a lower concentration of water than the red blood cells. One example of a hypertonic solution is a 10 percent NaCl solution. In such a solution, water molecules move out of the cells faster than they can enter. This causes the cells to shrink (Figure 2–5c). The shrinkage of red blood cells in this manner is referred to as *crenation.* Quite obviously, red blood cells may be greatly impaired or destroyed if placed in solutions that deviate significantly from the isotonic state.

FILTRATION. A third passive process involved in moving materials in and out of cells is **filtration.** This process involves the movement of solvents such as water and dissolved substances such as sugar across a semipermeable membrane by mechanical pressure. Such a movement is always from an area of higher pressure to an area of lower pressure and continues as long as a pressure difference exists. Most small- to medium-sized molecules can be forced through a cell membrane by pressure. An example of filtration occurs in the kidneys, where the blood pressure supplied by the heart forces water and urea through thin cell membranes of tiny blood vessels and into the kidney cells. In this basic

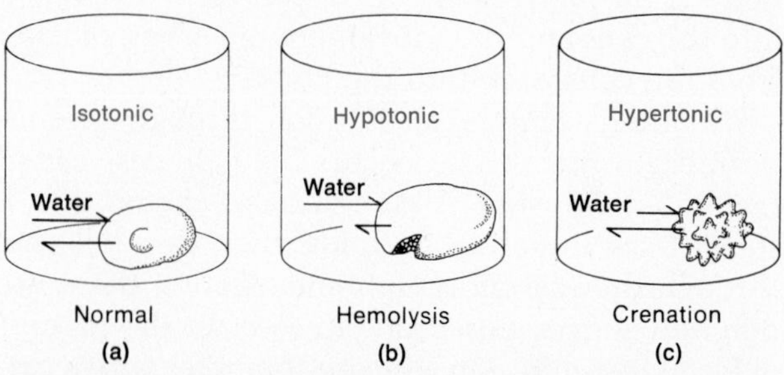

Figure 2–5. The principle of osmosis applied to red blood cells. Shown here are the effects on red blood cells when placed in (a) an isotonic solution, in which case they maintain normal shape; (b) a hypotonic solution, in which case they undergo hemolysis; and (c) a hypertonic solution, in which case they undergo crenation.

process, protein molecules are retained by the body since they are too large to be forced through the cell membranes of the kidney cells. Harmful substances, such as urea, however, are small enough to be forced through and eliminated.

DIALYSIS. The final passive process to be considered is **dialysis,** the process by which the artificial kidney works. Dialysis involves the separation of small molecules from large molecules by diffusion of the smaller molecules through a semipermeable membrane. For example, assume that a solution containing molecules of various sizes is placed in a tube that is permeable only to the smaller molecules. The tube is then placed in a beaker of distilled water. Eventually, the smaller molecules will move from the tube into the water in the beaker, and the larger molecules will be left behind. This principle of dialysis is employed in artificial kidneys. In the operation of an artificial kidney, the blood of the patient is passed into a dialysis tube outside the patient's body. The dialysis tube takes the place of the patient's kidneys. As the blood moves through the tube, waste products pass from the blood into a solution surrounding the dialysis tube. At the same time, certain nutrients are passed from the solution into the blood. The blood is then returned to the body.

Active processes

We shall now turn our attention to processes in which cells actively participate in moving substances across membranes. In these processes, the cell must expend energy. By participating in the transport of substances, the cell can even move them against a concentration gradient. Among the active processes we shall consider are active transport, phagocytosis, and pinocytosis.

ACTIVE TRANSPORT. The process by which substances, usually ions, are transported across plasma membranes from an area of lower concentration to an area of higher concentration is called **active transport** (Figure 2–6a). Although the exact mechanism is not known, the following sequence is believed to occur:

1. An ion outside the plasma membrane is attached to an enzymelike carrier molecule located in or on the plasma membrane.
2. The ion-carrier complex forms a compound that is soluble in the lipid portion of the membrane.
3. The compound moves toward the interior portion of the membrane where it is split by enzymes.
4. The ion is then transported into the cell, and the carrier returns to the surface of the membrane to pick up another ion.

The energy for the attachment and release of the carrier molecule is supplied by ATP.

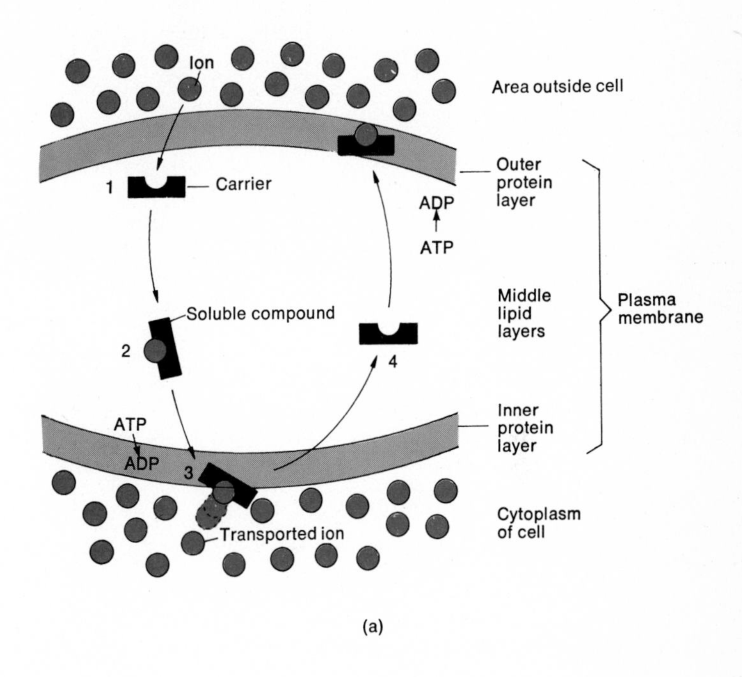

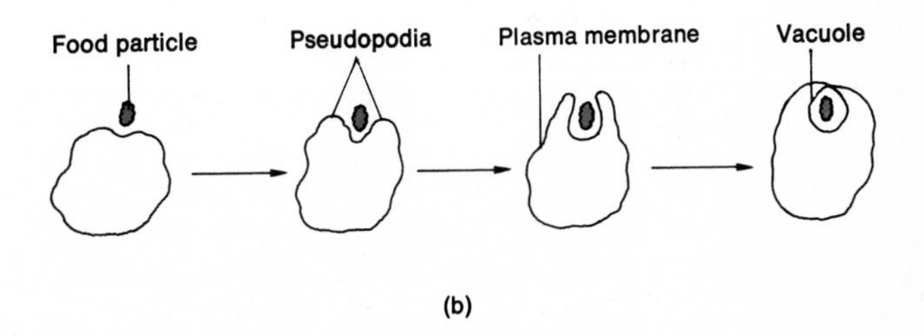

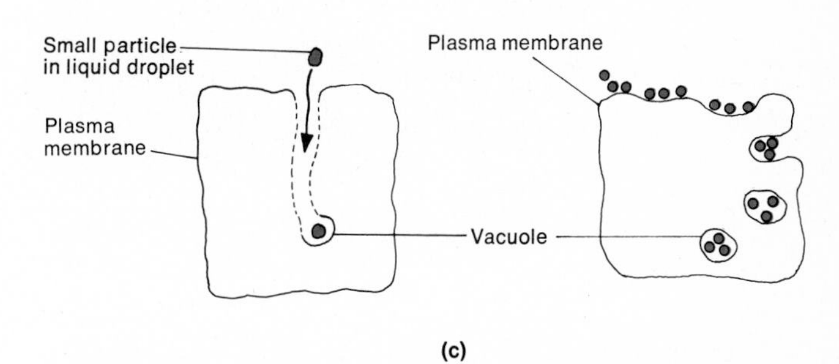

Figure 2–6. Active processes. (a) Mechanism of active transport. (b) Phagocytosis. (c) Two variations of pinocytosis. In the variation on the left, the ingested substance enters a channel formed by the plasma membrane and becomes enclosed in a vacuole at the base of the channel. In the variation on the right, the ingested substance becomes enclosed in a vacuole that forms and detaches at the surface of the cell.

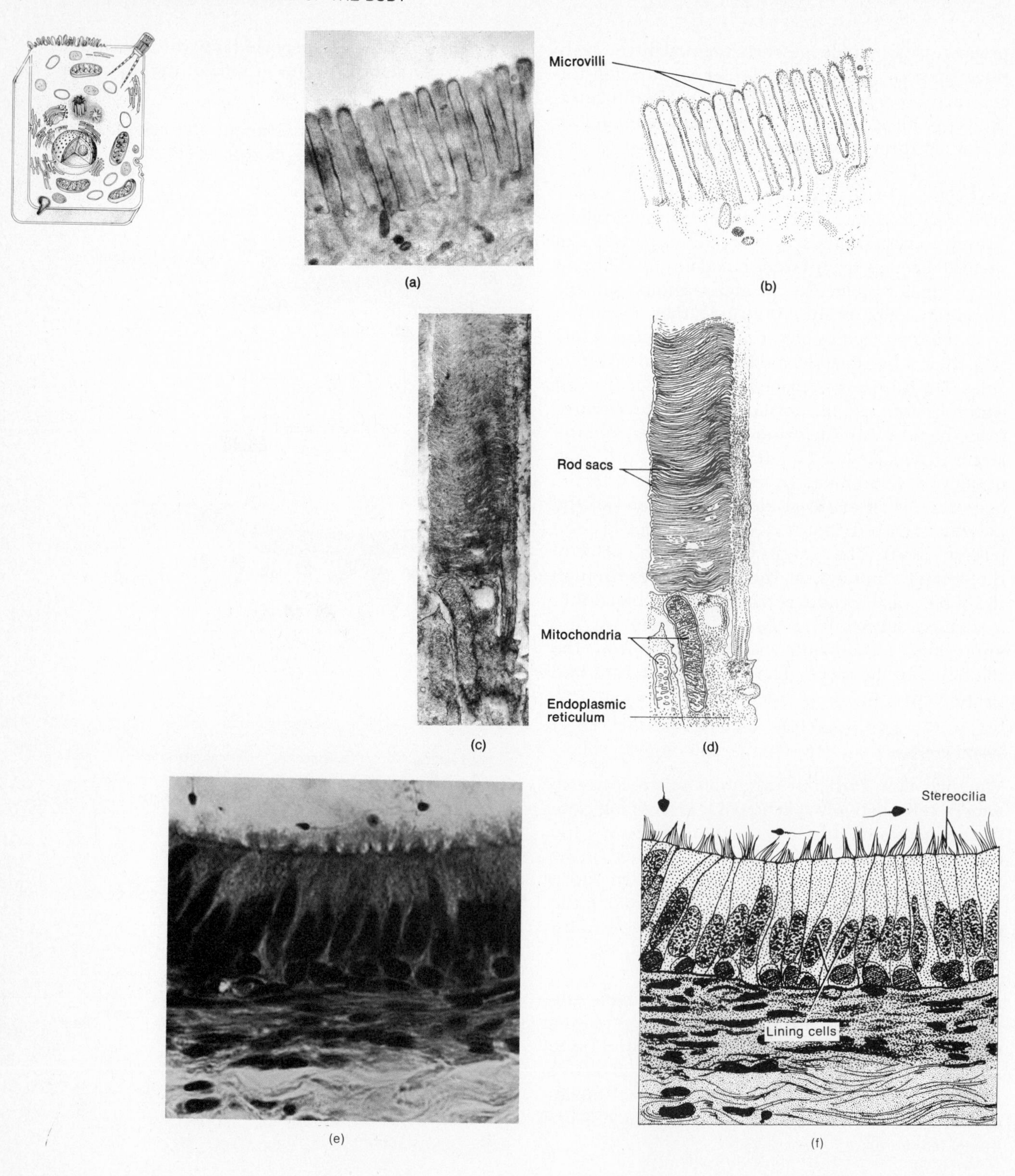

Figure 2–7. Modified plasma membranes. (a) Microvilli: an electron micrograph of a portion of small intestine at a magnification of 20,000X. (b) Microvilli: a labeled diagram of the electron micrograph. (c) Rod cell sacs: an electron micrograph of a portion of a rod cell of the eye at a magnification of 2,000X. (d) Rod cell sacs: a labeled diagram of the electron micrograph. (e) Stereocilia: a photomicrograph of stereocilia projecting from the lining cells of the epididymis at a magnification of 200X. (f) Stereocilia: a labeled diagram of the photomicrograph. (Electron micrographs courtesy of E. B. Sandborn, M.D., University of Montreal. Photomicrograph courtesy of Victor B. Eichler, Wichita State University.)

PHAGOCYTOSIS. Another active process by which cells take in substances across the plasma membrane is called **phagocytosis,** or "cell eating" (Figure 2–6b). In this process, projections of cytoplasm, called *pseudopodia,* engulf solid particles exterior to the cell. Once the particle is surrounded, the membrane folds inwardly, forming a membrane sac around the particle. This newly formed sac, called a *digestive vacuole,* breaks off from the outer cell membrane, and the solid material inside the vacuole is digested. Indigestible particles are removed from the cell by a reverse phagocytosis. This process is important because molecules and particles of material that would normally be restricted from crossing the plasma membrane can be brought into the cell. The phagocytic white blood cells of the body make up an important defense mechanism. Through phagocytosis, the white blood cells destroy bacteria and other foreign substances.

PINOCYTOSIS. The final active process to be discussed is called **pinocytosis,** or "cell drinking." In this process, the engulfed material consists of a liquid rather than a solid (Figure 2–6c). Moreover, no cytoplasmic projections are formed. Instead, the liquid is attracted to the surface of the membrane. The membrane folds inwardly, surrounds the liquid, and detaches from the rest of the intact membrane. Whereas relatively few cells are capable of phagocytosis, many cells may carry on pinocytosis. Examples include cells in the kidneys and urinary bladder.

Modified plasma membranes

Electron microscope studies have revealed that plasma membranes of certain cells contain a number of modifications. That is, they have different structures for very specific purposes. For example, the membranes of some of the cells lining the small intestine have small, cylindrical projections called **microvilli** (Figure 2–7a, b).These fingerlike projections enormously increase the absorbing area of the cell surface. A single cell may have as many as 3,000 microvilli, and 1 sq mm (0.0394 sq in.) of intestine may contain as many as 200 million microvilli.

Another membrane modification is found in the rod and cone cells of the eye. These cells serve as photoreceptors, or light-receiving cells. The upper portion of each rod cell contains two-layered, disc-shaped membranes called **sacs** that contain the pigments involved in vision (Figure 2–7c, d). A final example of a membrane modification is known as **stereocilia.** They are found only in cells lining a duct (epididymis) of the male reproductive system. They appear by light microscopy as long, slender, branching processes at the free surfaces of the lining cells (Figure 2–7e, f). Electron micrographs show stereocilia to be long, branching microvilli.

CYTOPLASM

The living matter inside the cell's plasma membrane and external to the nucleus is called **cytoplasm** (Figure 2–8a, b). It is the matrix or ground substance of the cell in which a variety of organelles and inclusions are found. Physically, cytoplasm may be described as a thick, semitransparent, elastic fluid containing suspended particles. Chemically,

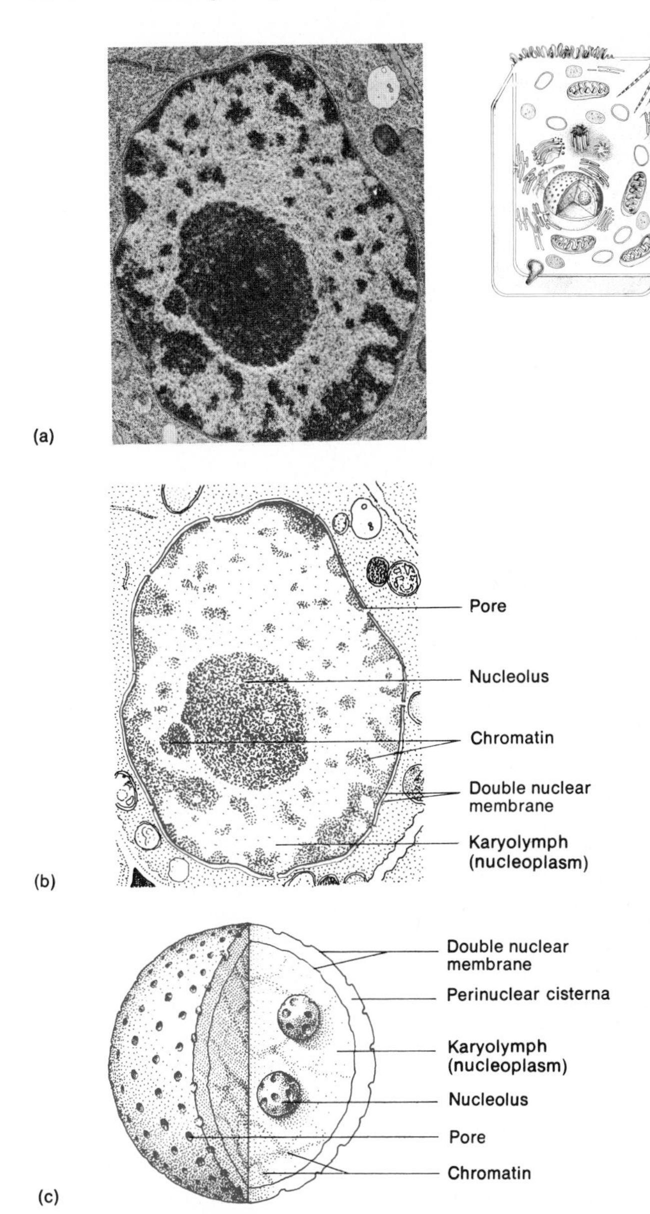

Figure 2–8. Cytoplasm and the nucleus. (a) Electron micrograph of cytoplasm and the nucleus at a magnification of 31,600X. (Courtesy of Myron C. Ledbetter, Brookhaven National Laboratory.) (b) Labeled diagram of the electron micrograph. (c) Diagrammatic representation of a nucleus with two nucleoli.

cytoplasm is 75 to 90 percent water plus solid components. Proteins, carbohydrates, lipids, and inorganic substances comprise the bulk of the solid components. The inorganic substances and most carbohydrates are soluble in water and are present as a true solution. The majority of organic compounds, however, are found as colloids, or particles that remain suspended in the surrounding ground substance. Since the particles of a colloid bear electrical charges that repel each other, they remain suspended and separated from each other.

Functionally, cytoplasm is the substance in which chemical reactions occur. The cytoplasm receives raw materials from the external environment and converts the raw materials into usable energy by decomposition reactions. Cytoplasm is also the site where new substances are synthesized for cellular use. It packages chemicals for transport to other parts of the cell or other cells of the body and facilitates the excretion of waste materials.

ORGANELLES

Despite the myriad chemical activities occurring simultaneously in the cell, there is little interference of one reaction with another. This is so because the cell has a system of compartmentalization that is provided by structures collectively called **organelles.** These structures are specialized portions of the cell that assume various roles in growth, maintenance, repair, and control. An understanding of the structure and function of representative organelles will aid you in understanding subsequent discussions of systems in the body.

The nucleus

Generally a spherical- or oval-shaped organelle, the **nucleus** contains a gel-like fluid that is thicker than the fluid of the surrounding cytoplasm (Figure 2–8a, b). In addition to being the largest structure in the cell, the nucleus controls cellular structure, directs many cellular activities, and contains the hereditary factors of the cell, called genes. Certain cells, such as mature red blood cells and cells in the center of the lens of the eye, do not have nuclei. These cells carry on only limited chemical activity and are not capable of growth or reproduction.

Structurally, the nucleus is separated from the cytoplasm by a double membrane called the *nuclear membrane* (Figure 2–8b, c). Between the two layers of the nuclear membrane is a space referred to as the *perinuclear cisterna.* This arrangement of the nuclear membrane resembles the structure of the plasma membrane. Minute pores are present in the nuclear membrane, allowing the nucleus to communicate with a membranous network in the cytoplasm called the endoplasmic reticulum. This network will be described shortly. Substances entering and exiting the nucleus are believed to pass through the tiny pores. Inside the nucleus, three prominent structures are visible. The first of these is a gel-like fluid called *karyolymph (nucleoplasm).* One or more spherical bodies called the *nucleoli* are also present. These structures are composed primarily of RNA and assume a role in directing protein synthesis. Finally, there is the *genetic material* consisting principally of DNA. When the cell is not reproducing, the genetic material appears as a threadlike mass and is called *chromatin.* Prior to cellular reproduction the chromatin shortens and thickens into rod-shaped bodies called *chromosomes.* These bodies also will be discussed subsequently.

Endoplasmic reticulum and ribosomes

Within the cytoplasm, there is a system consisting of pairs of parallel membranes enclosing narrow cavities of varying shapes. This system is known as the **endoplasmic reticulum,** or **ER** (Figure 2–9). The ER, in other words, is a network of canals running through the entire cytoplasm. These canals are continuous with both the plasma membrane and nuclear membrane. It is believed that the ER provides a surface area for chemical reactions, a pathway for transporting molecules within the cell, a storage area for synthesized molecules, and, together with the Golgi complex, secretes certain chemicals. This last function will be discussed in connection with the Golgi complex. Attached to the outer surfaces of the ER are exceedingly small, dense, spherical bodies called **ribosomes.** In these areas, the ER is referred to as *granular,* or *rough,* reticulum. Portions of the ER that have no ribosomes are called *agranular,* or *smooth,* reticulum. Ribosomes are thought to serve as the sites of protein synthesis in the cell.

Golgi complex

Another structure found in the cytoplasm is the **Golgi complex.** This structure usually consists of four to eight flattened channels, stacked upon each other with expanded areas at their ends. The stacked elements are called *cisternae,* and the expanded, terminal areas are referred to as *Golgi vacuoles* (Figure 2–10). Generally, the Golgi complex is located near the nucleus and is directly connected, in parts, to the ER. One of the functions of

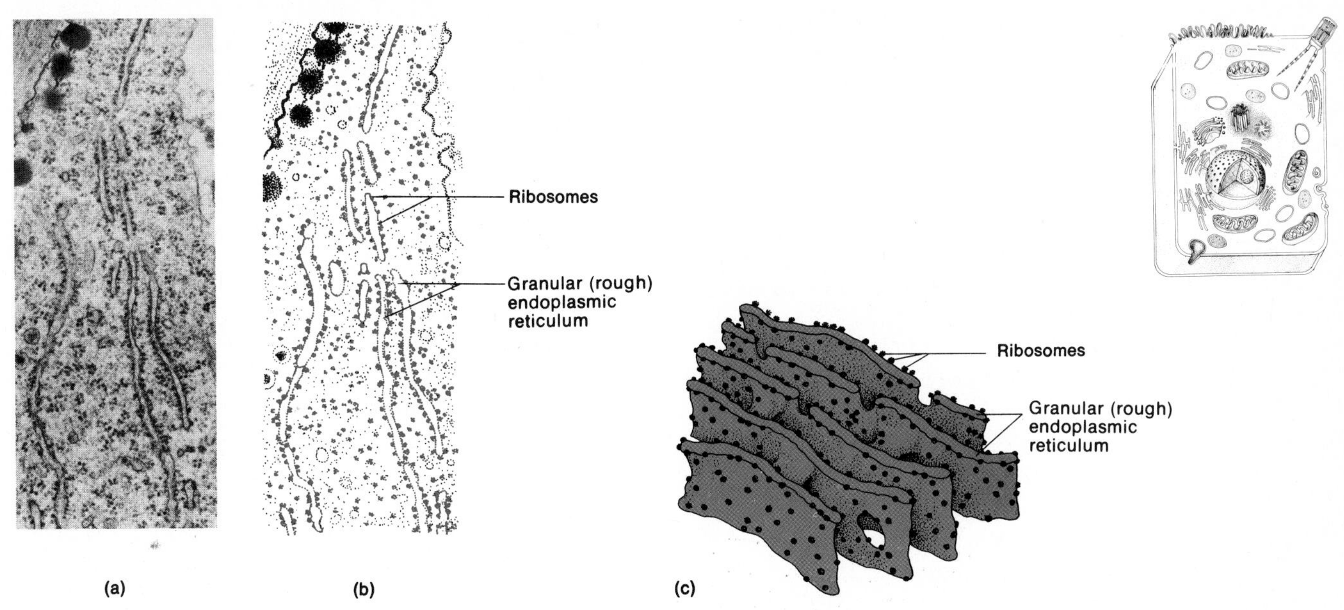

Figure 2–9. Endoplasmic reticulum and ribosomes. (a) Electron micrograph of the endoplasmic reticulum and ribosomes at a magnification of 76,000X. (Courtesy of Myron C. Ledbetter, Brookhaven National Laboratory.) (b) Labeled diagram of the electron micrograph. (c) Diagrammatic representation of the endoplasmic reticulum and ribosomes. See if you can find the agranular (smooth) endoplasmic reticulum in Figure 2–10.

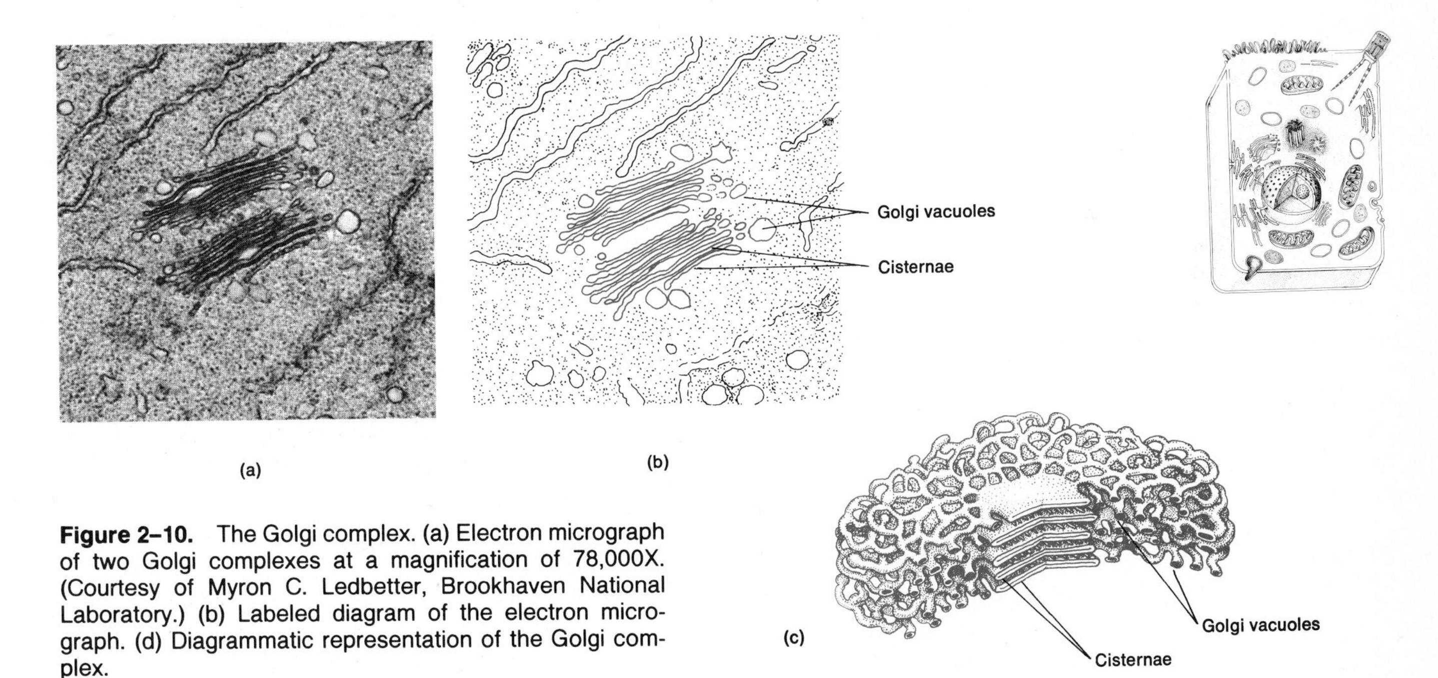

Figure 2–10. The Golgi complex. (a) Electron micrograph of two Golgi complexes at a magnification of 78,000X. (Courtesy of Myron C. Ledbetter, Brookhaven National Laboratory.) (b) Labeled diagram of the electron micrograph. (d) Diagrammatic representation of the Golgi complex.

the Golgi complex is the secretion of proteins. Proteins synthesized by the ribosomes associated with granular ER are transported into the ER tubules. They migrate along the ER tubules until they reach the Golgi complex. As the proteins accumulate in the cisternae of the Golgi complex, the cisternae expand to form Golgi vacuoles. After a certain critical size is reached, the vacuoles pinch off from the cisternae. The protein and its associated vacuole is referred to as a *secretory granule.* The secretory granule then moves toward the surface of the cell where the protein is secreted from the granule. Cells of the digestive tract that secrete protein enzymes utilize this mechanism. The vacuole prevents "digestion" of the cytoplasm of the cell as it moves toward the cell surface.

Another function of the Golgi complex is associated with lipid secretion. It occurs in essentially the same way as protein secretion, except the lipids are synthesized by the agranular ER. The lipids pass through the ER into the Golgi complex. As in the mechanism just described, the lipids migrate into the cisternae and vacuoles and are discharged at the surface of the cell. In the course of moving through the cytoplasm, the vacuole may release some of the lipids into the cytoplasm before discharge from the cell. These appear in the cytoplasm as lipid droplets. Among the lipids secreted in this manner are the steroids.

The Golgi complex also functions in the synthesis of carbohydrates. Recent evidence indicates that the carbohydrates synthesized by the Golgi complex are combined with proteins synthesized by the ribosomes associated with granular ER to form carbohydrate-protein complexes. These complexes of carbohydrate and protein are called *glycoproteins.* As the glycoproteins are assembled, they accumulate in the flattened channels of the Golgi complex. The channels expand and form Golgi vacuoles. After a certain critical size is reached, the vacuoles pinch off from the channel, migrate through the cytoplasm, and pass out of the cell through the plasma membrane. Outside the plasma membrane, the vacuoles rupture and release their contents. Essentially, the Golgi complex synthesizes carbohydrates and combines them with proteins. It then packages the resulting glycoprotein and secretes it from the cell. The Golgi complex is well developed and highly active in secretory cells such as those found in the pancreas and salivary glands.

Mitochondria

Throughout the cytoplasm appear small, spherical, rod-shaped, or filamentous structures called **mitochondria.** When sectioned and viewed under an electron microscope, each reveals an elaborate internal organization (Figure 2–11). A mitochondrion consists of a double membrane similar in structure to the plasma membrane. The outer mitochondrial membrane is smooth, but the inner membrane is thrown into a series of folds called *cristae.* The center of the mitochondrion is referred to as the *matrix.* Because of the nature and arrangement of the cristae, the inner membrane provides an enormous surface area for chemical reactions. Enzymes involved in energy-releasing reactions which form ATP are arranged on the cristae. Mitochondria are frequently called the "powerhouses of the cell" because of the central role they play in the production of ATP. Very active cells have a large number of mitochondria because of their high energy expenditure.

Lysosomes

When viewed under the electron microscope, **lysosomes** appear as membrane-enclosed spheres somewhat smaller than mitochondria (Figure 2–12). Unlike mitochondria, however, lysosomes have only a single membrane and lack detailed structure. Moreover, they contain powerful digestive enzymes capable of breaking down many kinds of

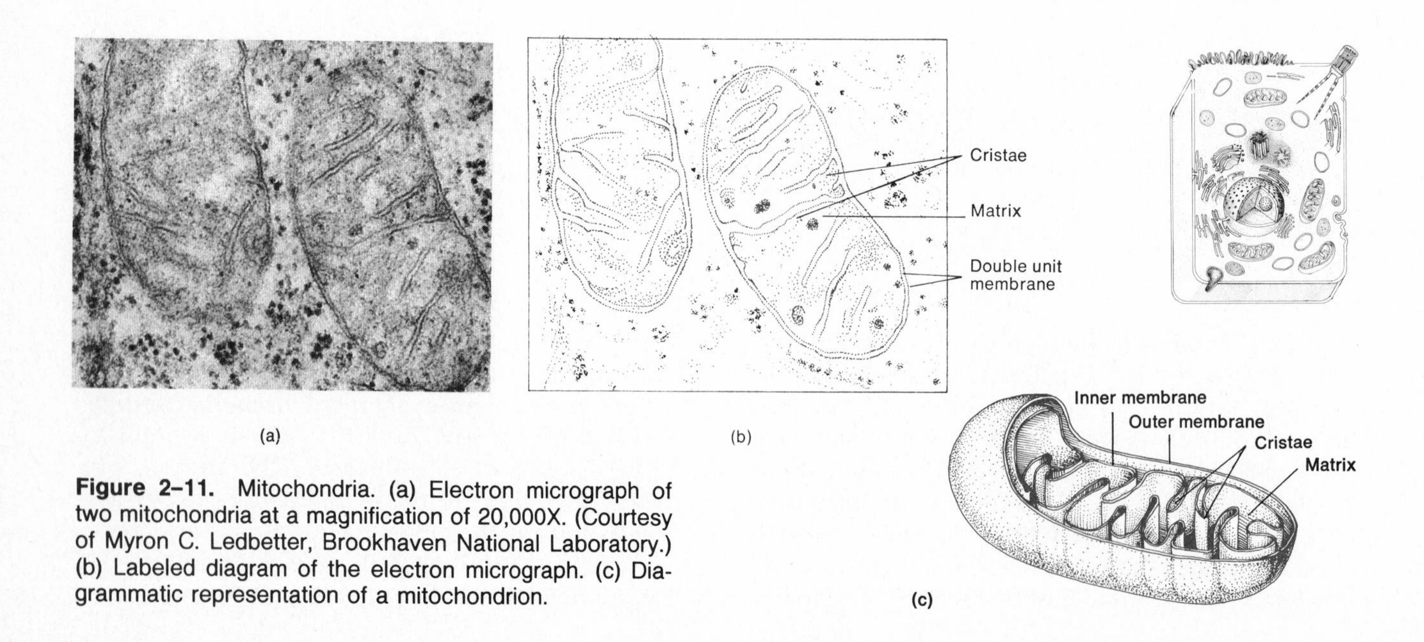

Figure 2–11. Mitochondria. (a) Electron micrograph of two mitochondria at a magnification of 20,000X. (Courtesy of Myron C. Ledbetter, Brookhaven National Laboratory.) (b) Labeled diagram of the electron micrograph. (c) Diagrammatic representation of a mitochondrion.

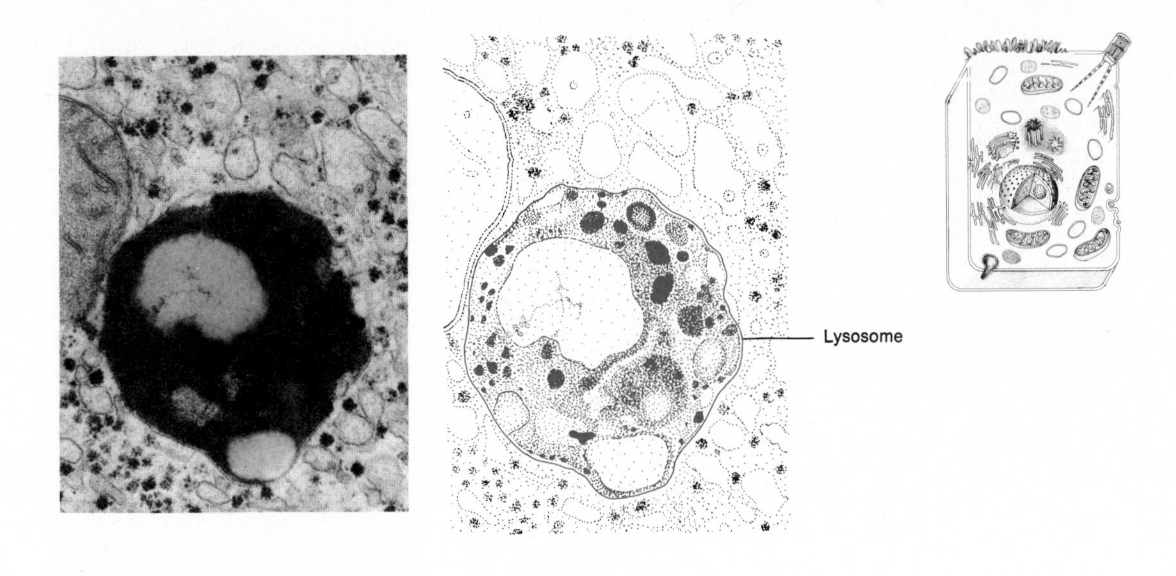

(a) (b)

Figure 2–12. The lysosome. (a) Electron micrograph of a lysosome at a magnification of 55,000X. (Courtesy of F. Van Hoof, Université Catholique de Louvain.) (b) Labeled diagram of the electron micrograph.

molecules. These enzymes are also capable of digesting bacteria that may enter the cell. White blood cells, which ingest bacteria by phagocytosis, contain large numbers of lysosomes. Scientists have wondered why these powerful enzymes do not also destroy their own cells. The suspected reason is that the lysosome membrane in a healthy cell is impermeable to enzymes so they cannot move out into the cytoplasm. When a cell is injured, though, the lysosomes release their enzymes. The enzymes then promote reactions that break the cell down into its chemical constituents. The chemical remains are either reused by the body or excreted. Because of this function, lysosomes have been called "suicide packets."

Centrosome

A rather dense area of cytoplasm, generally spherical in shape and located near the nucleus, is called the **centrosome** or **centrosphere.** Within the centrosome is a pair of cylinder-shaped structures, the **centrioles** (Figure 2–13). Each centriole is composed of a ring of nine evenly spaced bundles. Each bundle, in turn, consists of three hollow tubules. The two centrioles are situated so that the long axis of one is at right angles to the long axis of the other. Centrioles assume a role in cell reproduction—a process that will be described later in this chapter. Certain cells, such as mature nerve cells, do not

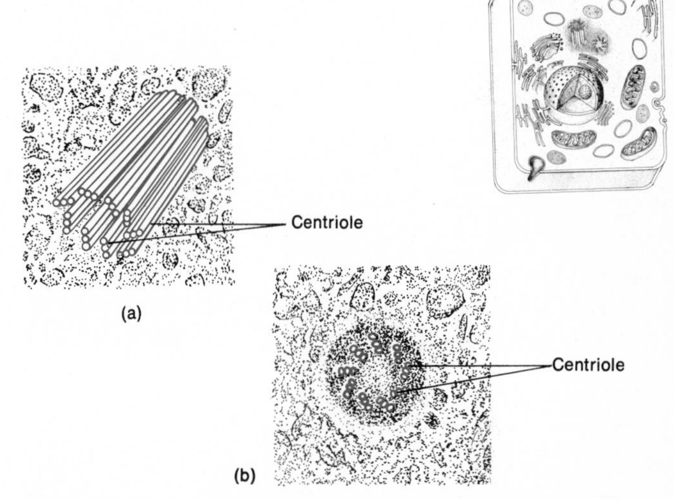

Figure 2–13. Centrioles. (a) Longitudinal section. (b) Cross section.

possess a centrosome. As a result, these cells do not reproduce. This is why nerve cells cannot be replaced if they are destroyed.

Flagella and cilia

Some cells of the body possess projections that are utilized for moving the entire cell or for moving substances along the surface of the cell. These projections contain cytoplasm and are bounded by the plasma membrane. If the projections are few and long in proportion to the size of the cell, they are called **flagella.** An example of a flagellum is the tail of a sperm cell. The tail is used for the locomotion of the sperm. If, on the other hand, the projections are numerous and short, resembling many hairs, they are called **cilia.** In humans, ciliated cells of the respiratory tract move lubricating fluids over the

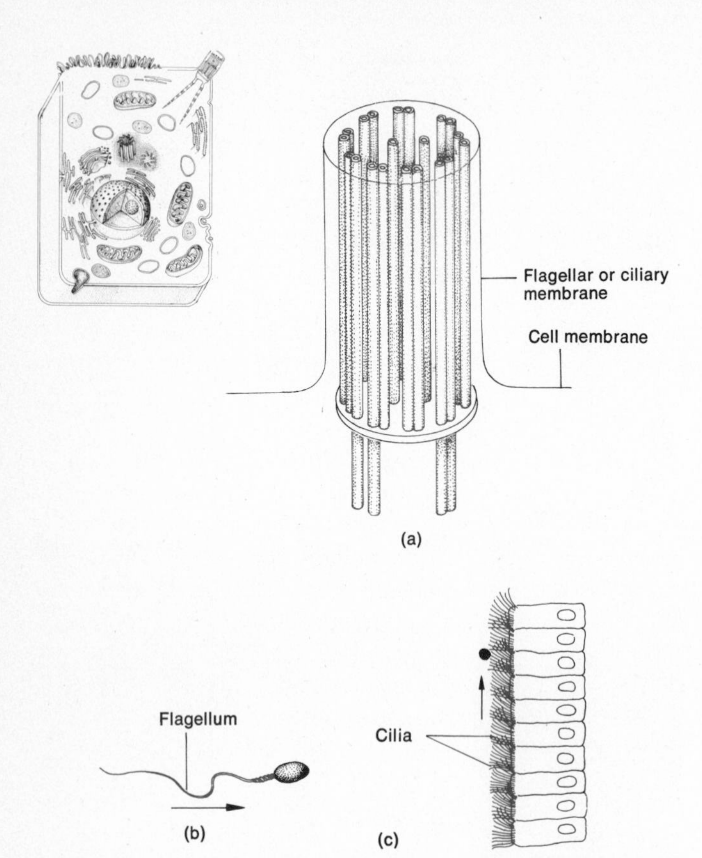

Figure 2–14. Flagella and cilia. (a) Diagrammatic representation of a flagellum or cilium. (b) Flagellum of a sperm cell. (c) Cilia of the respiratory tract moving a particle upward toward the mouth.

surface of the tissue and trap foreign particles. Electron microscopy has revealed that there is no fundamental structural difference between cilia and flagella (Figure 2–14).

Organelles, whether they are nuclei or flagella, work continuously to maintain the life of the cell. As part of their activities, some of the organelles manufacture products that are stored in the cells or secreted. The stored products are called inclusions, whereas the secreted products become part of the extracellular materials that lie outside the cells.

CELL INCLUSIONS

The **cell inclusions** are a rather large and diverse group of chemical substances. These products are principally organic in nature and may appear or disappear at various times in the life of the cell. Some inclusions, such as hemoglobin crystals and melanin, are kept within the cell. *Hemoglobin crystals* lie inside red blood cells. They perform the function of attaching to oxygen molecules and carrying the oxygen to other cells. *Melanin* is a pigment stored in the cells of the skin, hair, and eyes. It protects the body by screening out harmful ultraviolet rays from the sun. Other inclusions are temporarily stored in the cell. When the body needs them, they are released. One example is *glycogen*, a polysaccharide that is stored in liver and skeletal muscle cells. When the body requires quick energy, liver cells can break down the glycogen into glucose and release the glucose. *Lipids*, which are stored in fat cells, may be decomposed when the body runs out of carbohydrates for producing energy. Still other inclusions are secreted fairly continuously by the cell. An example is mucus, which is produced by cells that line organs. Its function is to provide lubrication. The major parts of the cell and their functions are summarized in Exhibit 2–1.

Exhibit 2–1. CELL PARTS AND THEIR FUNCTIONS

PART	FUNCTIONS
I. Plasma or cell membrane	Protects and allows substances to enter or exit the cell through diffusion, osmosis, filtration, dialysis, active transport, phagocytosis, and pinocytosis
II. Cytoplasm	Serves as the ground substance in which chemical reactions occur
III. Organelles	
A. Nucleus	Controls cellular activities and contains genes
B. Endoplasmic reticulum	Provides a surface area for chemical reactions; provides a pathway for transporting chemicals; serves as a storage area
C. Ribosomes	Act as sites of protein synthesis
D. Golgi complex	Synthesizes carbohydrates, combines carbohydrates with proteins, packages materials for secretion, secretes lipids
E. Mitochondria	Produce ATP
F. Lysosomes	Digest chemicals and foreign microbes
G. Centrioles	Form spindles during cell division
H. Flagella and cilia	Afford movement of cell or movement of particles along surface of cell
IV. Inclusions	Involved in overall body functions; include materials retained in cell (hemoglobin), reserve materials (glycogen, fats), and secretions (mucus)

EXTRACELLULAR MATERIALS

The substances that lie outside cells are collectively called **extracellular materials.** They include the body fluids, which provide a medium for dissolving, mixing, and transporting substances. They include secreted inclusions like mucus. And they also include some special substances that form the matrix, or mold, in which some cells are embedded.

The matrix materials are produced by certain cells and are deposited outside their plasma membranes. The matrix supports the cells, binds them together, and gives strength and elasticity to the tissue. Some of the matrix materials are *amorphous,* which means they have no specific shape. These include hyaluronic acid and chondroitin sulfate. *Hyaluronic acid* is a viscous, fluidlike substance that binds cells together, lubricates joints, and maintains the shape of the eyeballs. *Chondroitin sulfate* is a jellylike substance that provides support and adhesiveness in cartilage, bone, heart valves, the cornea of the eye, and the umbilical cord. Other matrix materials are *fibrous,* or threadlike. Fibrous materials provide strength and support for tissues. Among these are **collagen,** or *collagenous fibers.* Collagen is found in all kinds of connective tissue, especially in bones, tendons, and ligaments. **Reticulin,** also called *reticular fibers,* is a matrix material that forms a network around fat cells, nerve fibers, muscle cells, and blood vessels. **Elastin,** found in *elastic fibers,* is a substance that gives elasticity to the skin and to the tissues which form the walls of blood vessels.

CELL DIVISION

Most of the cell activities mentioned thus far maintain the life of the cell on a day-to-day basis. However, cells become damaged, diseased, or wear out and die. Moreover, new cells must be produced for growth. The vital process by which cells are replaced is called cell division.

Cell division is the process by which cells reproduce themselves. For our purposes, assume that cell division or, more appropriately, nuclear division, may be one of two kinds. The first kind of division is the process by which a single parent cell duplicates itself. This process is known as mitosis (nuclear division) and cytokinesis (cytoplasmic division). It is the process by which body cells replace themselves. The second kind of division is a mechanism by which sperm and egg cells are produced. This process is called meiosis, which is the mechanism that enables the reproduction of an entirely new organism. Meiosis will be discussed in the chapter on reproduction (Chapter 18).

Mitosis

The function of mitosis and cytokinesis is to replace cells in the body. The process ensures that each new daughter cell has the same *number* and *kind* of chromosomes as the original parent cell. After the process is complete, the two daughter cells have the same hereditary material and genetic potential as the parent cell. This kind of cell division results in an increase in the number of body cells. Mitosis and cytokinesis are, therefore, the means by which dead or injured cells are replaced and also the means by which cells are added for body growth. In a 24-hour period, the average human adult loses about 500 million cells from different parts of the body. Obviously, these cells must be replaced. Cells that have a short life span, such as the cells of the outer layer of skin, the cornea of the eye, and the digestive tract, are continually being replaced. The succession of events that takes place during mitosis and cytokinesis is plainly visible under a microscope after the cells have been stained in the laboratory.

When a cell reproduces itself, it must replicate its chromosomes so that its heredity may be passed

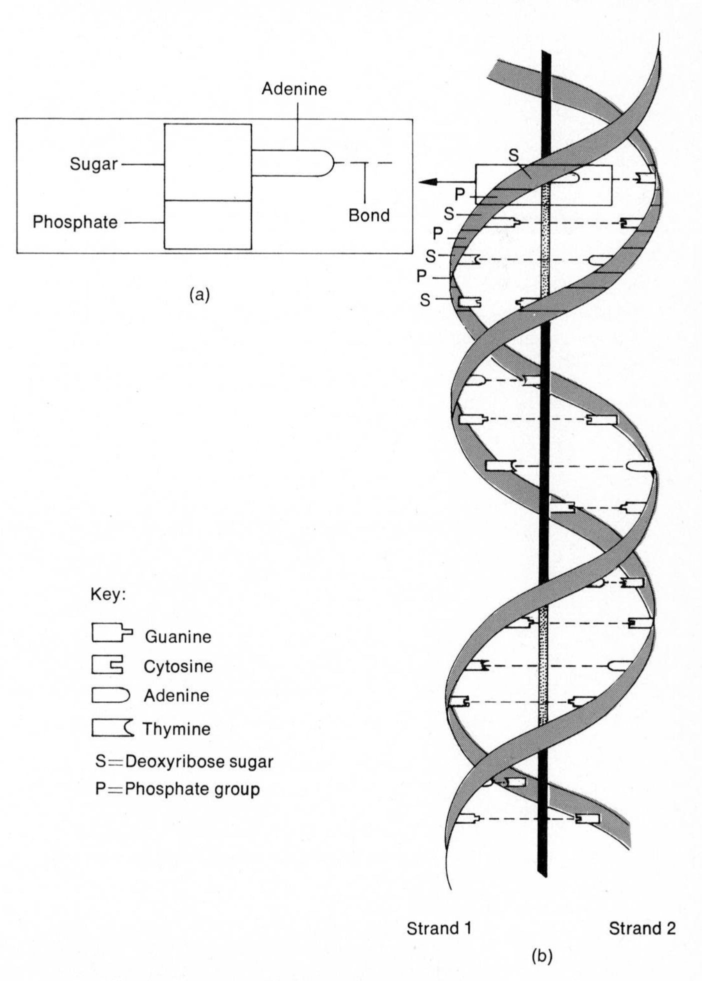

Figure 2–15. The DNA molecule. (a) Adenine nucleotide. (b) Portion of an assembled DNA molecule.

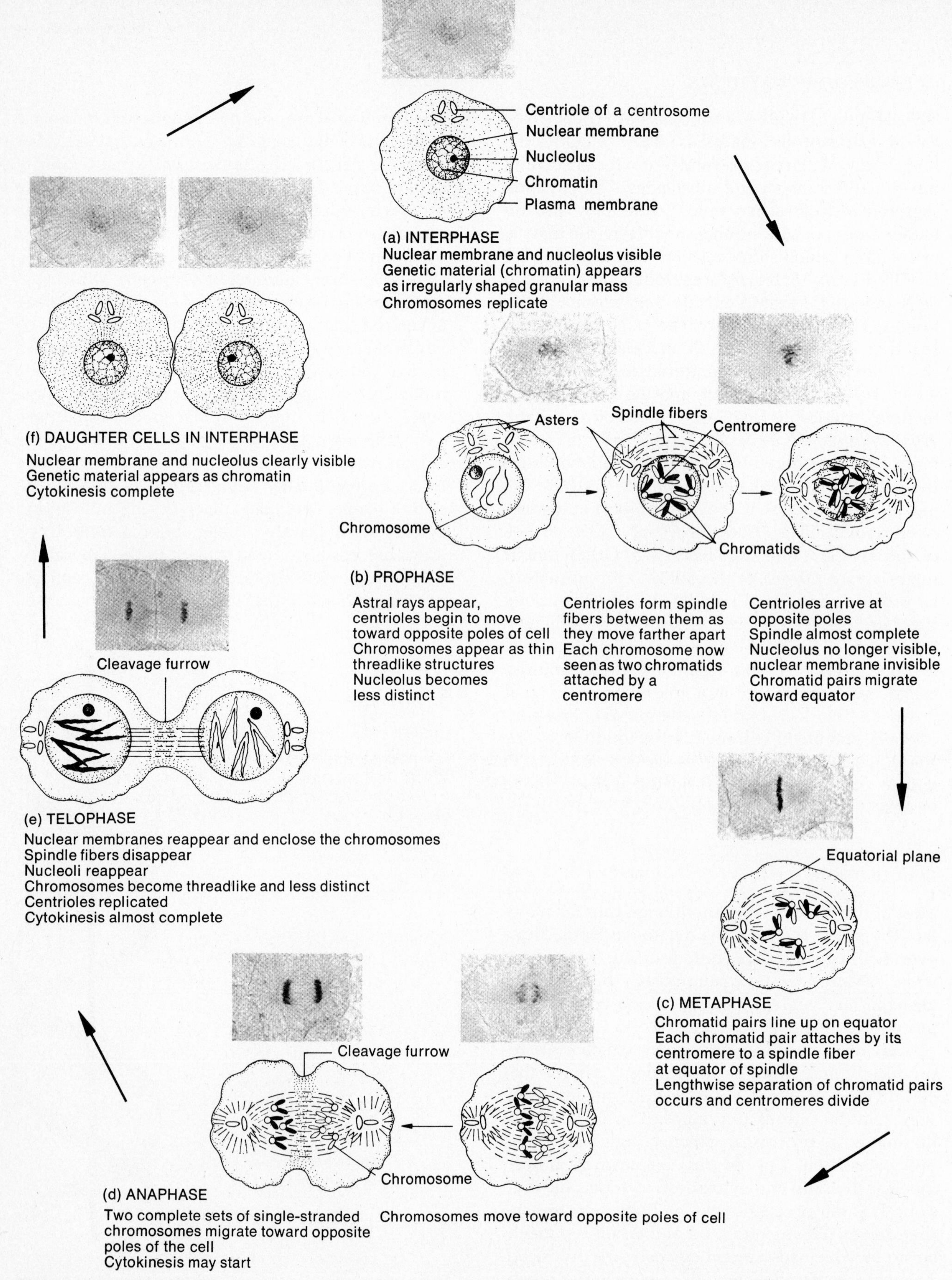

Figure 2–16. Cell division: mitosis and cytokinesis. Photomicrographs and diagrammatic representations of the various stages of division in whitefish eggs. Read the sequence starting at (a), and move clockwise until you complete the cycle. (Photomicrographs courtesy of Carolina Biological Supply Company.)

on to succeeding generations of cells. A **chromosome** is a highly coiled DNA molecule. This is the molecule that contains your hereditary information. A chromosome, in turn, consists of genes. Each human chromosome consists of about 20,000 genes.

Before taking a look at the relationship between chromosomes and cell division, it will first be necessary to briefly examine the structure of DNA, the basic component of chromosomes.

A molecule of DNA is a chain composed of repeating units called **nucleotides.** Each nucleotide of DNA consists of three basic parts (Figure 2–15a): (1) It contains one of four possible *nitrogen bases,* which are ring-shaped structures containing atoms of C, H, O, and N. The nitrogen bases found in DNA are named adenine, thymine, cytosine, and guanine. (2) It contains a sugar called *deoxyribose.* (3) It has a phosphoric acid called the *phosphate group.* The nucleotides are named according to the nitrogen base that is present. Thus, a nucleotide containing thymine is called a *thymine nucleotide.* One containing adenine is called an *adenine nucleotide,* and so on.

The chemical composition of the DNA molecule was known before 1900, but it was not until 1953 that a model of the organization of the chemicals was constructed. This model was proposed by J. D. Watson and F. H. C. Crick on the basis of data from many investigations. Figure 2–15b shows the following structural characteristics of the DNA molecule: (1) The molecule consists of two strands with crossbars. The strands twist about each other in the form of a *double helix* so that the shape resembles a twisted ladder. (2) The uprights of the DNA "ladder" consist of alternating phosphate groups and the deoxyribose portions of the nucleotides. (3) The rungs of the ladder contain paired nitrogen bases. As shown, adenine always pairs off with thymine, and cytosine always pairs off with guanine.

The process called **mitosis** is the replication of chromosomes and the distribution of the two sets of chromosomes into two separate and equal nuclei. For convenience, biologists break down the process into four stages: prophase, metaphase, anaphase, and telophase. These are arbitrary classifications, and mitosis is actually a continuous process, one stage merging imperceptibly into the next. Interphase is the stage that occurs between consecutive cell divisions.

When a cell is carrying on every life process except for division, it is said to be in **interphase** (Figure 2–16a). One of the principal events of interphase is the replication of DNA. When DNA replicates, its helical structure partially uncoils (Figure 2–17). Those portions of DNA that remain coiled

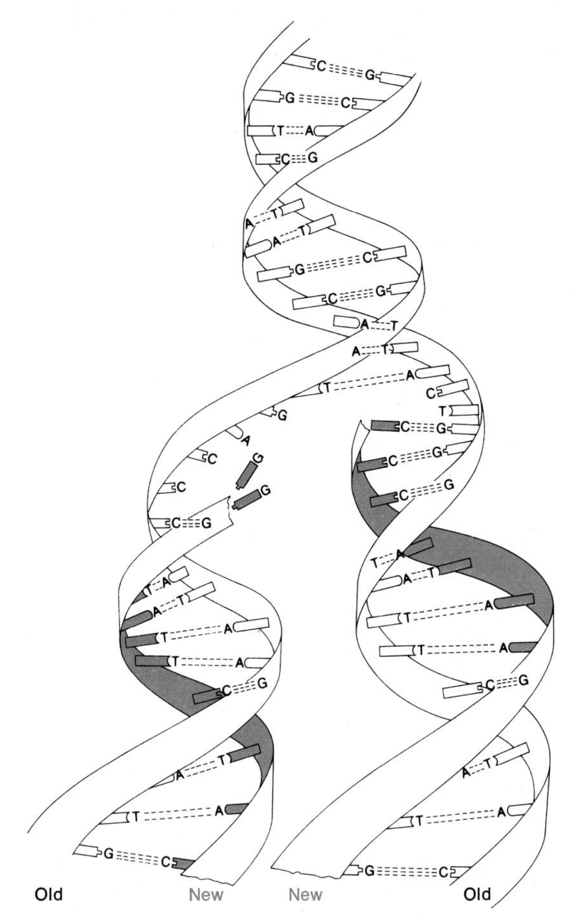

Figure 2–17. Replication of DNA. Each strand of the double helix separates by breaking the bonds between nucleotides. New nucleotides attach at the proper sites, and two new strands of DNA are paired off with the two old strands. After replication, the two DNA molecules, each consisting of a new and an old strand, return to their helical structure.

stain darker than the uncoiled portions. This unequal distribution of stain causes the DNA to appear as a granular mass called **chromatin.** (See Figure 2–16a). During uncoiling, DNA separates at the points where the nitrogen bases are connected. Each exposed nitrogen base then picks up a complementary nitrogen base (with associated sugar and phosphate group) from the cytoplasm of the cell. This uncoiling and complementary base pairing continues until each of the two original DNA strands is matched and joined with two newly formed DNA strands. The net effect is that the original DNA molecule has become two DNA molecules.

During interphase the cell is also synthesizing most of its RNA and proteins. It is producing chemicals so that all cellular components can be doubled during division. When you look at an interphase cell under a microscope, you will notice that the nucleus has a clearly defined membrane, nucleoli, karyolymph, and chromatin. As interphase progresses, you will also see two pairs of centrioles. The centrioles divide and the resulting two pairs of centrioles separate. Once a cell completes its activities during interphase, mitosis begins.

During **prophase** (Figure 2–16b), the centrioles move apart and project a series of radiating fibers called *asters.* The centrioles move to opposite poles of the cell and become connected by another system of fibers called *spindle fibers.* Simultaneously, the chromatin has been shortening and thickening into chromosomes. The nucleoli have become less distinct, and the nuclear membrane has disappeared. Each prophase "chromosome" is actually composed of a pair of separate structures called *chromatids.* Each chromatid is a complete chromosome, made of a double-stranded DNA molecule. Each chromatid is attached to its chromatid pair by a small spherical body called the *centromere.* During prophase, the chromatid pairs move toward the equatorial plane region of the cell.

During **metaphase** (Figure 2–16c), the second stage of mitosis, the chromatid pairs line up on the equatorial plane of the spindle fibers. The centromere of each chromatid pair attaches itself to a spindle fiber. The lengthwise separation of the chromatids now takes place. Each centromere divides, and the independent chromatids are moved to opposite poles of the cell.

Anaphase (Figure 2–16d), the third stage of mitosis, is characterized by the continued movement of complete sets of chromatids, or chromosomes, to opposite poles of the cell. During this movement, the centromeres that are attached to the spindle fibers seem to drag the trailing parts of the chromosomes toward opposite poles.

Telophase (Figure 2–16e), the final stage of mitosis, consists of a series of events approximately opposite that of prophase. By now, two identical sets of chromosomes have reached opposite poles. New nuclear membranes begin to enclose them. The chromosomes start to assume their chromatin form. Finally, nucleoli reappear, and the spindle fibers disappear. The formation of two nuclei identical to those in cells of interphase terminates telophase. A mitotic cycle has been completed (Figure 2–16f).

The time required for a complete mitotic cycle varies with the kind of cell, its location, and the influence of certain factors such as temperature. Furthermore, the different stages of mitosis are not equal in duration. Prophase is usually the longest stage, lasting from one to several hours. Metaphase is considerably shorter, ranging from 5 to 15 minutes. Anaphase is the shortest stage, lasting from 2 to 10 minutes. Telophase lasts from 10 to 30 minutes. These lengths of time are only relative, however, and should not be taken as exact limits.

Cytokinesis

The division of the cytoplasm, called **cytokinesis,** often begins in late anaphase and terminates in telophase. Cytokinesis begins with the formation of a cleavage furrow that runs around the cell at its equator. The furrow progresses inward, resembling a constricting ring, and cuts completely through the cell to form two separate portions of cytoplasm (Figure 2–16d to f).

Chapter summary in outline

THE GENERALIZED ANIMAL CELL

1. A cell is the basic, living, structural and functional unit of the body.
2. A generalized cell is a composite that represents various cells of the body.
3. Cytology is the science concerned with the study of cells.
4. The principal parts of a cell are the plasma membrane, cytoplasm, organelles, and inclusions. Extracellular materials are manufactured by the cell and deposited outside the plasma membrane.

THE CELL SURFACE

1. The plasma membrane, or cell membrane, surrounds the cell and separates it from other cells and the external environment.
2. The plasma membrane is composed of proteins and a bilayer of lipids. It is believed that the membrane contains pores.
3. The semipermeable nature of the membrane restricts the passage of certain substances. Substances can pass through the membrane depending on their molecular weight, lipid solubility, charges, and carriers.

Moving materials across plasma membranes

1. Passive processes involve the kinetic energy of individual molecules.
2. Diffusion is the net movement of molecules or ions from an area of higher concentration to an area of lower concentration until an equilibrium is reached.
3. In facilitated diffusion, certain molecules, like glucose, combine with a carrier to become soluble in the lipid portion of the membrane.
4. Osmosis is the movement of water through a semipermeable membrane from an area of higher water concentration to an area of lower water concentration.
5. Osmotic pressure is the force under which a solvent moves from a solution of lower solute concentration to a solution of higher solute concentration when the solutions are separated by a semipermeable membrane.
6. Filtration is the movement of water and dissolved substances across a semipermeable membrane by pressure.
7. Dialysis is the separation of small molecules from large molecules by diffusion through a semipermeable membrane.
8. Active processes involve the use of ATP by the cell.
9. Active transport is the movement of ions across a cell membrane from lower to higher concentration. This process relies upon the participation of carriers.
10. Phagocytosis is the ingestion of solid particles by pseudopodia. It is an important process used by white blood cells to destroy bacteria that enter the body.
11. Pinocytosis is the ingestion of a liquid by the plasma membrane. In this process, the liquid becomes surrounded by a vacuole.

Modified plasma membranes

1. The membranes of certain cells are structured for specific functions.
2. Microvilli are fingerlike projections of the plasma membrane that increase the surface area for absorption.
3. Rod cells of the eye contain sacs that pick up light for vision.
4. Stereocilia are long, slender, branching processes of the lining cells of the epididymis.

CYTOPLASM

1. The cytoplasm is the living matter inside the cell that contains organelles and inclusions.
2. It is composed mostly of water plus proteins, carbohydrates, lipids, and inorganic substances. The chemicals in cytoplasm are either in solution or in a colloid, or suspended, form.
3. Functionally, cytoplasm is the medium in which chemical reactions occur.

ORGANELLES

1. These are specialized structures that carry on specific activities.

The nucleus

1. Usually the largest organelle, the nucleus controls cellular activities.
2. Cells without nuclei, such as mature red blood cells, do not grow or reproduce.
3. The parts of the nucleus include the nuclear membrane, nucleoplasm, nucleoli, and genetic material (DNA).

Endoplasmic reticulum and ribosomes

1. The ER is a network of parallel membranes, continuous with the plasma membrane and nuclear membrane.
2. It functions in chemical reactions, transportation, and storage.
3. Granular or rough ER has ribosomes attached to it. Agranular or smooth ER does not contain ribosomes. Ribosomes are small spherical bodies that serve as sites of protein synthesis.

Golgi complex

1. This structure consists of four to eight flattened channels stacked on each other.
2. In conjunction with the ER, the Golgi complex synthesizes glycoproteins and secretes lipids.
3. It is particularly prominent in secreting cells such as those in the pancreas or salivary glands.

Mitochondria

1. These structures consist of a smooth outer membrane and a folded inner membrane. The inner folds are called cristae.
2. The mitochondria are called "powerhouses of the cell" because ATP is produced within them.

Lysosomes

1. Lysosomes are spherical structures containing digestive enzymes.
2. They are found in large numbers in white blood cells, which carry on phagocytosis.
3. If the cell is injured, lysosomes release enzymes and digest the cell. For this reason, they are called "suicide packets."

Centrosome

1. The dense area of cytoplasm containing the centrioles is called a centrosome.
2. Centrioles are paired cylinders arranged at right angles to one another. They assume an important role in cell reproduction.

Flagella and cilia

1. These cell projections have the same basic structure and are used in movement.
2. If projections are few and long, they are called flagella. If they are numerous and hairlike, they are called cilia.
3. The flagellum on a sperm cell serves to move the entire cell. The cilia on cells of the respiratory tract move foreign matter along the cell surfaces toward the throat for elimination.

CELL INCLUSIONS

1. These chemical substances are produced by cells. They may be stored, may participate in chemical reactions, and may have recognizable shapes.
2. Examples of cell inclusions are glycogen, hemoglobin crystals, mucus, and melanin.

EXTRACELLULAR MATERIALS

1. These are all the substances that lie outside the cell membrane.
2. They provide support and a medium for the diffusion of nutrients and wastes.
3. Some, like hyaluronic acid, are amorphous, or have no shape. Others, like collagen, are fibrous, or threadlike.

CELL DIVISION

1. The kind of cell division that results in the formation of new cells is called mitosis and cytokinesis. Nuclear division that results in the production of sperm and egg cells is termed meiosis.
2. Mitosis, division of the nucleus, consists of prophase, metaphase, anaphase, and telophase.
3. Cytokinesis, division of the cytoplasm, occurs in late anaphase and terminates in telophase.
4. Mitosis and cytokinesis replace and add body cells. Prior to mitosis and cytokinesis, the DNA molecules, or chromosomes, replicate themselves so that the same chromosomal complement can be passed on to future generations of cells.
5. A cell carrying on every life process except division is said to be in interphase.

Review questions and problems

1. Define a cell. What are the four principal portions of a cell? What is meant by a generalized cell?
2. Discuss the structure of the plasma membrane. What factors determine the permeability of the plasma membrane? How are plasma membranes modified for various functions?
3. What are the major differences between active processes and passive processes in moving substances across plasma membranes?
4. Define and give an example of each of the following: diffusion, facilitated diffusion, osmosis, filtration, active transport, phagocytosis, and pinocytosis.
5. Compare the effect on red blood cells of an isotonic, hypertonic, and hypotonic solution. What is osmotic pressure?
6. Discuss the chemical composition and physical nature of cytoplasm. What is its function?
7. What is an organelle? By means of a diagram, indicate the structure and describe the function of the following organelles: nucleus, endoplasmic reticulum, ribosomes, Golgi complex, mitochondria, lysosomes, centrosome, cilia, and flagella.
8. Define a cell inclusion. Provide examples and indicate their functions.
9. What is an extracellular material? Give examples and the functions of each.
10. How does DNA replicate itself?
11. Discuss mitosis and cytokinesis with regard to stages. What are the characteristics of each stage, the relative duration, and the importance?
12. To increase your knowledge of the parts of a cell, label the cell shown below. In addition, summarize the function of each part. When you have finished, refer to Figure 2–1 to see how well you labeled the cell.

Selected readings

DeRobertis, E. D. P., F. A. Saez, and E. M. F. DeRobertis, Jr. *Cell Biology.* 6th ed. Philadelphia: W. B. Saunders, 1975.

"Exploring the New Biology." *National Geographic,* Vol. 150, No. 3 (September 1976), 355–408.

Novikoff, Alex B., and Eric Holtzman. *Cells and Organelles.* New York: Holt, Rinehart and Winston, 1970.

Pfeiffer, John, and the Editors of *Life. The Cell.* New York: Time Inc., 1964.

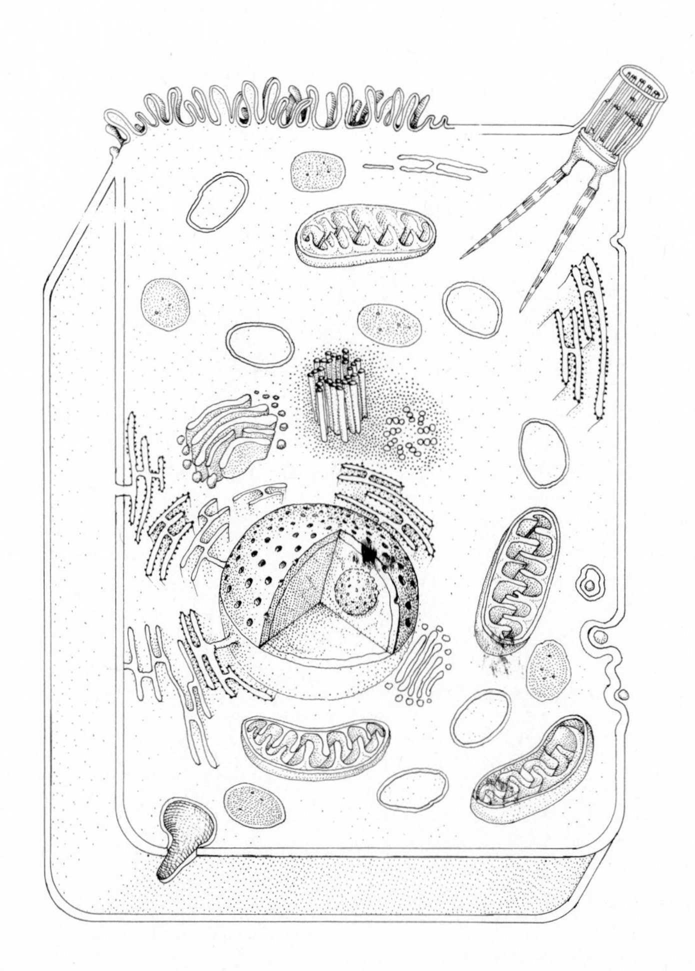

CHAPTER 3

TISSUES

STUDENT OBJECTIVES

After you have read this chapter, you should be able to:

1. Define a tissue
2. Classify the tissues of the body into four major types
3. Describe the distinguishing characteristics of epithelial tissue
4. Contrast the structural and functional differences of covering, lining, and glandular epithelium
5. Compare the shape of cells and the layering arrangements of covering and lining epithelium
6. List the structure, function, and location of simple, stratified, and pseudostratified epithelium
7. Define a gland
8. Distinguish between exocrine and endocrine glands
9. Classify exocrine glands according to structural complexity and physiology
10. Describe the distinguishing characteristics of connective tissue
11. Contrast the structural and functional differences between embryonal and adult connective tissues
12. Describe the ground substance, fibers, and cells that constitute connective tissue
13. List the structure, function, and location of loose connective tissue, adipose tissue, collagenous, elastic, and reticular connective tissue
14. List the structure, function, and location of the three types of cartilage
15. Define an epithelial membrane
16. List the location and function of mucous, serous, synovial, and cutaneous membranes

The cells discussed in the preceding chapter are highly organized units, but they do not function as isolated units. Instead, they work together in a group of somewhat similarly constructed cells, called a tissue. We shall now examine how the body operates at the tissue level of organization.

A **tissue** is an aggregration of similar cells working together to perform a specialized activity. For example, some tissues of the body function in moving body parts. Others move food through body organs. Some tissues protect and support the body. And still others function to produce chemicals such as enzymes and hormones. Depending on their functions and structure, the various tissues of the body are classified into four principal types: (1) epithelial tissue, which covers body surfaces, lines body cavities, and forms ducted or exocrine glands; (2) connective tissue, which protects and supports the body and its organs and binds organs together; (3) muscular tissue, which is responsible for movement; and (4) nervous tissue, which initiates and transmits nerve impulses that coordinate body activities. Some of these tissues will be discussed in later chapters as parts of a particular system. Others, such as epithelial tissue and most connective tissues, will be treated in detail in this chapter.

EPITHELIAL TISSUE

The tissues falling into this main category carry out many activities in the body, ranging from protection to secretion. **Epithelial tissue,** or more simply **epithelium,** may be divided into two subtypes: (1) *covering and lining epithelium* and (2) *glandular epithelium.* (See Exhibit 3–1.) Covering and lining epithelium forms the outer covering of external body surfaces and the outer covering of some internal organs. It lines the body cavities and the interiors of the respiratory and digestive tracts, blood vessels, and ducts. Along with nervous tissue, it constitutes the parts of the sense organs that are sensitive to stimuli such as light and sound. Glandular epithelium constitutes the secreting portion of glands.

Both types of epithelium consist largely or entirely of closely packed cells with little or no intercellular material such as fluid or fibers between the cells. In other words, there is little fluid or few fibers between cells. In addition, the epithelial cells are arranged in continuous sheets that may be either single or multilayered. Nerves may run through these sheets, but blood vessels do not. Thus, they are referred to as *avascular.* The vessels that supply nutrients and remove wastes are located in underlying connective tissue. Epithelium overlies and adheres firmly to connective tissue, which holds the epithelium in position and prevents it from being torn. The surface of attachment between epithelium and connective tissue is a thin layer of modified connective tissue called the **basement membrane.** Since all epithelium is subjected to a certain degree of wear, tear, and injury, its cells can divide and produce new cells to replace those that are destroyed. These general characteristics are found in both types of epithelial tissue. We can now see how the two types differ, by first looking at covering and lining epithelium.

Covering and lining epithelium

Covering and lining epithelium is arranged in several different ways, and the arrangement is related to its location and function. If the epithelium is specialized for absorption or filtration and is located in an area that has minimal wear and tear, the cells of the tissue are arranged in a single layer. Such an arrangement is called **simple epithelium.** If the epithelium is not specialized for absorption or filtration and is found in an area with a high degree of wear and tear, then the cells are stacked in several layers. This tissue is referred to as **stratified epithelium.** A third, less common, arrangement of epithelium is called **pseudostratified.** Like simple epithelium, pseudostratified epithelium has only one layer of cells. However, some of the cells do not reach the surface—an arrangement that gives the tissue a multilayered, or stratified, appearance. (See Exhibit 3–1.) The pseudostratified cells that do reach the surface either secrete mucus or contain cilia that move mucus and foreign particles for eventual elimination from the body.

In addition to classifying covering epithelium according to the number of its layers, it may also be categorized by cell shape. The cells may be flat, cubelike, columnar or may resemble a cross between shapes. **Squamous** cells are flattened, scalelike, and fitted together to form a mosaic. **Cuboidal** cells are usually cube-shaped when viewed in cross section. They sometimes appear as hexagons. **Columnar** cells are long and cylindrical, appearing as rectangles set on their ends. **Transitional** cells look like a combination of shapes and are found where there is a great degree of distention or expansion in the body. Transitional cells on the bottom layer of an epithelial tissue may range in shape from cuboidal to columnar. In the intermediate layer, they may be cuboidal or polyhedral. Transitional cells in the superficial layer may range from cuboidal to squamous, depending on how much they are pulled out of shape during certain body functions.

Considering layers and cell type in combination, we may classify covering and lining epithelium as follows:

1. Simple
   (a) Squamous
   (b) Cuboidal
   (c) Columnar
2. Stratified
   (a) Squamous
   (b) Cuboidal
   (c) Columnar
   (d) Transitional
3. Pseudostratified

Simple epithelium

SIMPLE SQUAMOUS EPITHELIUM. This type of simple epithelium consists of a single layer of flat, scalelike cells. When viewed from the surface, this epithelium resembles a tiled floor (Exhibit 3–1). The nucleus of each cell is centrally located and is oval or spherical in shape. Since simple squamous epithelium has only one layer of cells, it is highly adapted to the functions of diffusion, osmosis, and filtration. Thus, we find simple squamous epithelium lining the air sacs of the lungs, where oxygen is exchanged with carbon dioxide. It is present in the part of the kidney that filters the blood. It is also found in very delicate structures such as the crystalline lens of the eye and the lining of the eardrum. Simple squamous epithelium is found in parts of the body that have little wear or tear. A tissue that is very similar to simple squamous epithelium is called endothelium. **Endothelium** lines the heart and the blood and lymph vessels and forms the walls of capillaries. The term **mesothelium** is applied to another simple squamous epitheliumlike tissue that lines the ventral body cavity and covers the viscera.

Exhibit 3–1. SUMMARY OF EPITHELIAL TISSUES

I. COVERING AND LINING EPITHELIUM

SIMPLE SQUAMOUS (130X)

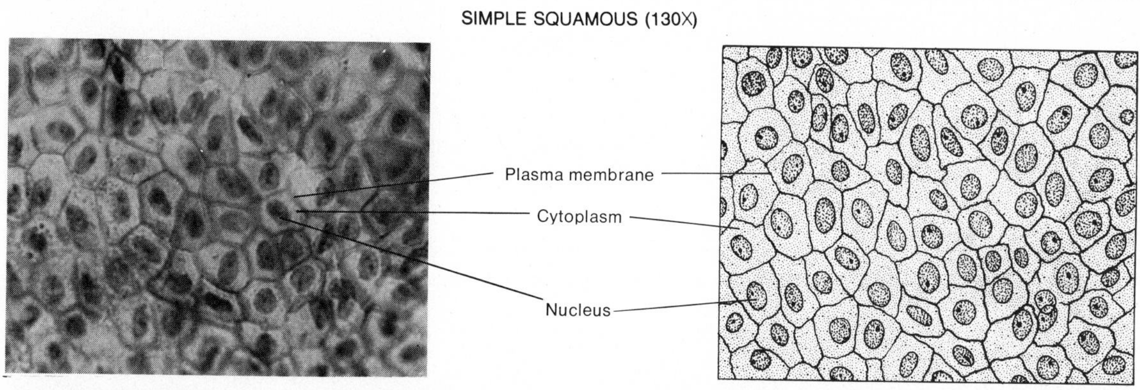

Description: Single layer of flat, scalelike cells; large, centrally located nucleus.

Location: Lines air sacs of lungs, glomerular capsule of kidneys, crystalline lens of eyes, and eardrum. Called endothelium when it lines heart, blood, and lymphatic vessels, and forms capillaries. Called mesothelium when it lines visceral organs.

Function: Filtration, absorption, and secretion in serous membranes.

SIMPLE CUBOIDAL (130X)

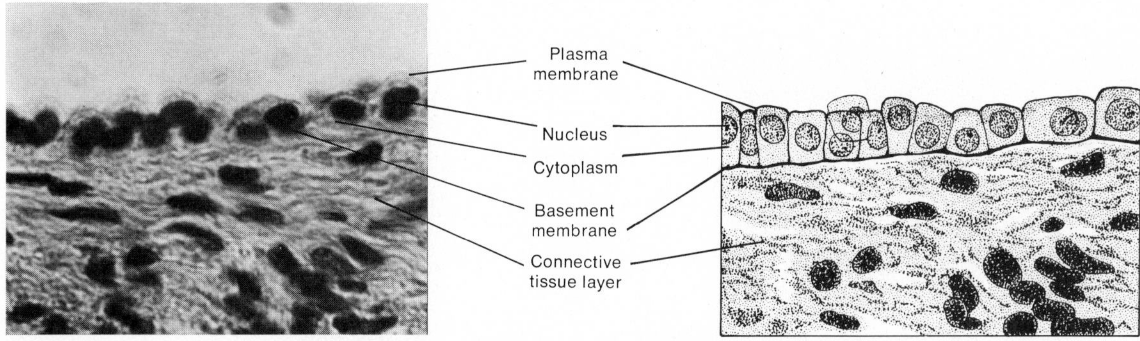

Description: Single layer of cube-shaped cells; centrally located nucleus.

Location: Covers surface of ovary; lines inner surface of cornea and lens of eye, kidney tubules, and smaller ducts of many glands.

Function: Secretion and absorption.

SIMPLE CUBOIDAL EPITHELIUM. When viewed from the top, the cells of simple cuboidal epithelium appear as polygons fitted closely together. The cuboidal nature of the cells is obvious only when the tissue is sectioned at right angles (Exhibit 3–1). Like simple squamous epithelium, these cells possess a centrally located nucleus. Simple cuboidal epithelium is found covering the surfaces of the ovaries and lining the inner surfaces of the cornea. In the kidneys, where it forms the kidney tubules and contains microvilli, it functions in water reabsorption. It also lines the smaller ducts and secreting units of many glands, such as the thyroid. This tissue performs the functions of secretion and absorption.

SIMPLE COLUMNAR EPITHELIUM. The surface view of simple columnar epithelium is similar to that of simple cuboidal tissue. When sectioned at right angles to the surface, however, the cells appear somewhat rectangular. The nuclei are located near the bases of the cells (Exhibit 3–1). Simple columnar epithelium is modified in several ways, depending on its location and function in the body. Simple columnar epithelium lines the stomach, the small and large intestines, the digestive glands, and the gallbladder. In such sites, the cells protect the underlying tissues. Many of them are also modified so that they can aid in food-related activities. In the small intestine especially, the plasma membranes of the cells are folded into many fingerlike projections called **microvilli.** (See Figure 2–7a.) The microvilli arrangement increases the surface area of the plasma membrane and thereby allows digested nutrients and fluids to diffuse into the body at a faster rate. Interspersed among the typical columnar cells of the intestine are other modified columnar cells called **goblet cells.** These cells, which secrete mucus, are so named because the mucus accumulates in the upper half of the cell, causing

Exhibit 3–1 (cont.)

SIMPLE COLUMNAR (NONCILIATED) (130X)

Plasma membrane
Goblet cell
Absorptive cell
Cytoplasm
Nucleus
Basement membrane
Connective tissue layer

Description: Single layer of nonciliated rectangular cells; contains goblet cells; nuclei at bases of cells.

Location: Lines stomach, small and large intestines, digestive glands, and gallbladder.

Function: Secretion and absorption.

SIMPLE COLUMNAR (CILIATED) (130X)

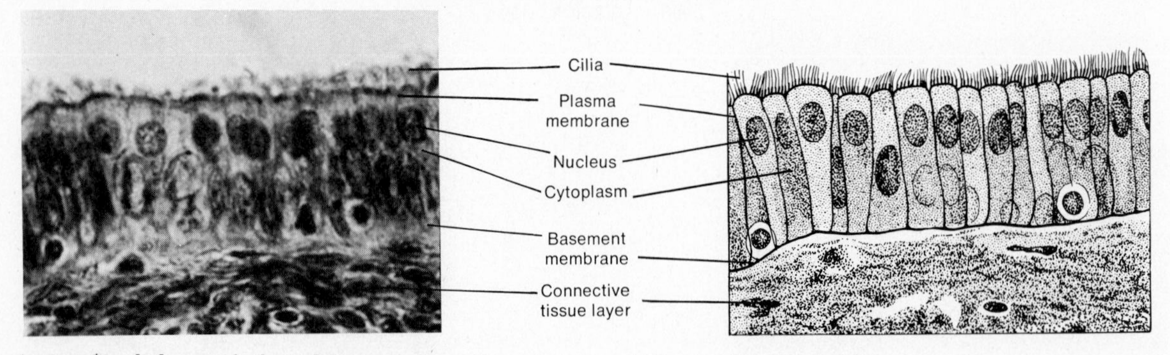

Description: Single layer of ciliated rectangular cells; contains goblet cells; nuclei at bases of cells.

Location: Lines upper respiratory tract, uterine (fallopian) tubes, and uterus.

Function: Moves mucus by ciliary action.

the area to bulge out. The whole cell resembles a goblet or wine glass. Mucus in this tissue serves as a lubricant that prevents friction between the food and the walls of the digestive tract. A third modification of columnar epithelium is found in cells with hairlike processes called **cilia.** In portions of the upper respiratory tract, ciliated columnar cells are interspersed with goblet cells. Mucus secreted by the goblet cells forms a film over the respiratory surface. This film traps foreign particles that are breathed in. The cilia, which wave in unison, move the mucus and foreign particles toward the throat, where it can be swallowed or eliminated. In this way, only air is allowed to enter the lungs. Ciliated columnar epithelium, combined with goblet cells, is also found in the uterus and uterine tubes of the female reproductive system.

Stratified epithelium

In contrast to simple epithelium, stratified epithelium consists of at least two layers of cells. This means that stratified epithelium is relatively durable and can protect underlying tissues from the outside environment and from wear and tear. Some stratified epithelium cells are also involved in secretion. The name of the specific kind of stratified epithelium is dependent upon the shape of the surface cells.

STRATIFIED SQUAMOUS EPITHELIUM. In the layers of this type of epithelium, the more superficial cells are flat, whereas the deeper cells vary in shape from cuboidal to columnar (Exhibit 3–1). The basal, or bottom, cells are continually multiplying by cell division. As the newly produced cells grow in size,

Exhibit 3–1 (cont.)

STRATIFIED SQUAMOUS (65X)

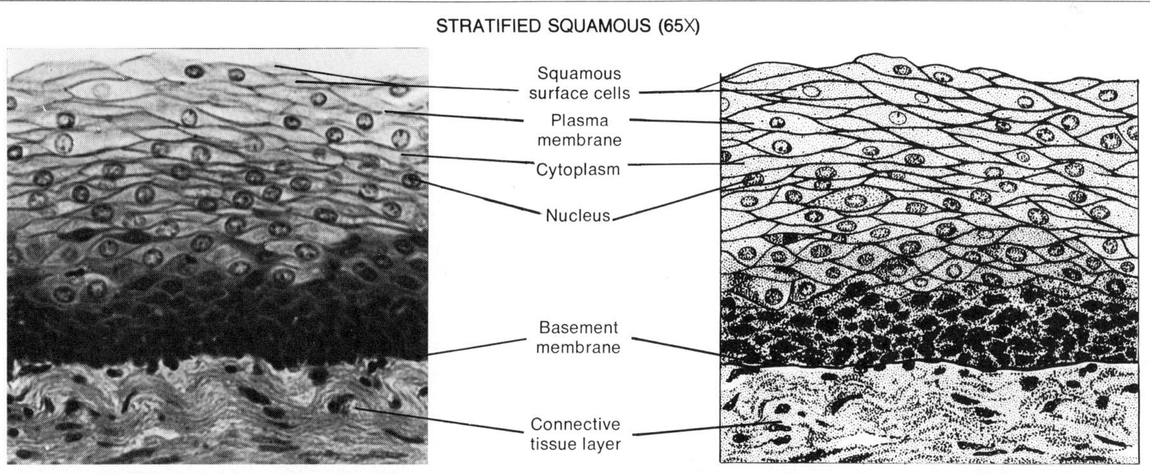

Description: Several layers of cells; deeper layers are cuboidal to columnar; superficial layers are flat and scale-like; basal cells replace surface cells as they are lost.

Location: Nonkeratinizing variety lines wet surfaces such as mouth, esophagus, part of epiglottis, and vagina. Keratinizing variety forms outer layer of skin.

Function: Protection.

STRATIFIED CUBOIDAL (80X)

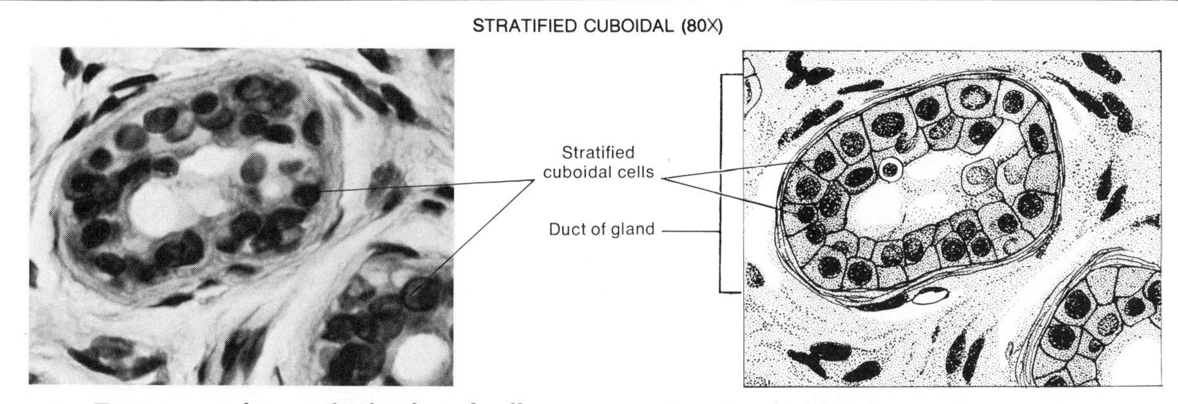

Description: Two or more layers of cube-shaped cells.

Location: Ducts of adult sweat glands.

Function: Protection.

they compress the cells on the surface and push them outward. According to this growth pattern, basal cells continually shift upward and outward. As they move farther away from the deep layer and their source of blood, they become dehydrated, shrink, and grow harder. Once at the surface, the cells are rubbed off. New cells continually emerge, are sloughed off, and replaced.

One form of stratified squamous is called *stratified squamous nonkeratinizing epithelium.* This type of tissue is found on wet surfaces that are subjected to considerable wear and tear and that typically do not perform the function of absorption. Such surfaces include the insides of the mouth, the gullet, and the vagina. Another form of stratified squamous is called *stratified squamous keratinizing epithelium.* The surface cells of this type of epithelium are modified into a tough, resistant layer of material containing keratin. **Keratin** is a protein that is waterproof, resistant to friction, and relatively impervious to bacterial invasion. The outer layer of skin consists of this tissue.

STRATIFIED CUBOIDAL EPITHELIUM. This relatively rare type of epithelium is found primarily in the ducts of sweat glands of adults (Exhibit 3–1). It sometimes consists of more than two layers of cells. Its function is mainly protective.

STRATIFIED COLUMNAR EPITHELIUM. Like stratified cuboidal, this type of tissue is also relatively infrequent in the body. Usually the basal layer or layers consist of shortened, irregularly polyhedral cells. Only the superficial cells are columnar in form (Exhibit 3–1). This kind of epithelium lines part of the male urethra and some larger excretory ducts such as lactiferous ducts in the mammary glands. It functions in protection and secretion.

TRANSITIONAL EPITHELIUM. This kind of epithelium is very much like stratified squamous nonkeratinizing epithelium. The distinction is that the outer layer of cells in transitional epithelium tend to be large and rounded rather than flat (Exhibit 3–1). This feature allows the tissue to be stretched

Exhibit 3–1 (cont.)

STRATIFIED COLUMNAR (130X)

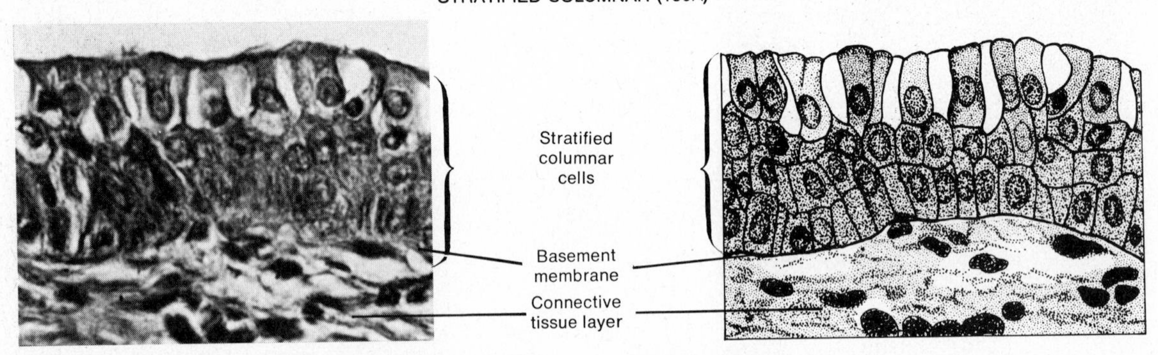

Description: Several layers of polyhedral cells; only superficial layer is columnar.

Location: Lines part of male urethra and some larger excretory ducts.

Function: Protection and secretion.

STRATIFIED TRANSITIONAL (110X)

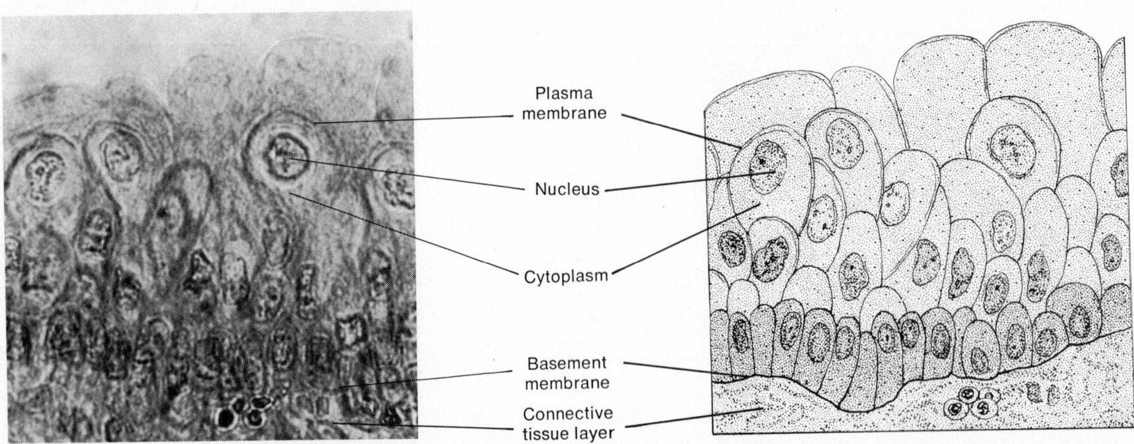

Description: Resembles stratified squamous nonkeratinizing tissue, except superficial cells are larger and more rounded.

Location: Lines urinary tract.

Function: Permits distention.

without the danger of the outer cells breaking apart from one another. When stretched, they are drawn out into squamouslike cells. Because of this arrangement, transitional epithelium lines hollow structures that are subjected to expansion from within, such as the urinary bladder. Its obvious function is to prevent a rupture of the epithelium.

Pseudostratified epithelium

The third category of covering and lining epithelium is called *pseudostratified epithelium.* The nuclei of the cells in this kind of tissue are at varying positions, some toward the surface and some in the basal region. Even though all the cells are attached to the basement membrane in a single layer, some of the cells do not reach the surface (Exhibit 3–1). This gives the impression of a multilayered tissue, which is the reason for the designation *pseudo*-stratified epithelium. It lines the larger excretory ducts of many glands and parts of the male urethra. In addition, pseudostratified epithelium may be ciliated and found with goblet cells. In this form, it lines the larger respiratory passages and some of the ducts of the male reproductive system.

The various kinds of epithelium that we have discussed so far are arranged in sheets which cover or line a body surface. Even though some of the covering and lining cells release products such as mucus, secretion is just a secondary function of this kind of epithelium. We shall now discuss a type of epithelium whose primary function is secretion.

Glandular epithelium

The function of glandular epithelium is secretion, which is accomplished by glandular cells that lie in clusters deep to the covering and lining epithelium. A **gland** can be one cell or a group of highly specialized epithelial cells that secrete substances either into ducts or into the blood. The production of such substances always requires active work by the glandular cells and results in an expenditure of energy. On the basis of this distinction, all glands of the body are classified as exocrine or endocrine glands (Exhibit 3–1). **Exocrine glands** secrete their products into ducts or tubes that empty out at the surface of the covering and lining epithelium. The product of an exocrine gland may be released at the surface of the skin or released into the lumen of a hollow organ. The secretions of exocrine glands include enzymes, oil, and sweat. But these glands never secrete hormones. Examples of exocrine glands are sweat glands, which eliminate perspiration to cool the skin, and salivary glands, which secrete a digestive enzyme. **Endocrine glands,** by contrast, are ductless and, consequently, must secrete their products directly into the blood. The secretions of endocrine glands are always hormones. Examples of endocrine glands include the pituitary, thyroid, and adrenal glands.

The various exocrine glands of the body may be classified according to their structural complexity and physiology. We will consider structural complexity first.

Exhibit 3–1 (cont.)

PSEUDOSTRATIFIED (170X)

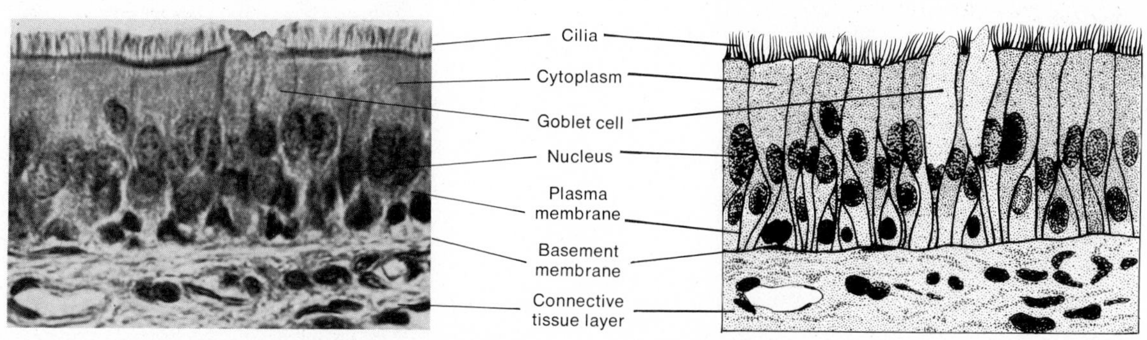

Description: Not a true stratified tissue; nuclei of cells are present at different levels; some cells do not reach surface, but all sit on the basement membrane.

Location: Lines larger excretory ducts of many large glands and male urethra; ciliated variety with goblet cells lines larger respiratory passageways and some ducts of male reproductive system.

Function: Secretion and movement of mucus by ciliary action.

According to structural complexity, exocrine glands are classified into two broad types: (1) unicellular and (2) multicellular. **Unicellular glands** are single-celled glands. A good example of a unicellular gland is a goblet cell (Exhibit 3–1). Goblet cells are found in the epithelial lining of the digestive, respiratory, urinary, and reproductive systems. They produce mucus to lubricate the free surfaces of the membranes. **Multicellular glands** are many-celled glands and occur in several different forms (Figure 3–1). For example, if the secretory portions of the gland are tubular in shape, they are referred to as *tubular glands.* If the secretory portions of the gland are flasklike, they are referred to as *acinar glands.* If the gland contains both tubular and flask-like secretory portions, it is called a *tubuloacinar* gland. Further, if the duct of the gland does not branch, it is referred to as a *simple gland,* and if the duct does branch, it is called a *compound gland.* If we now combine the shape of the secretory portion with the degree of branching of the duct, we arrive at the structural classification for exocrine glands outlined on the next page.

Exhibit 3–1 (cont.)

II. GLANDULAR EPITHELIUM

EXOCRINE GLAND (110X)

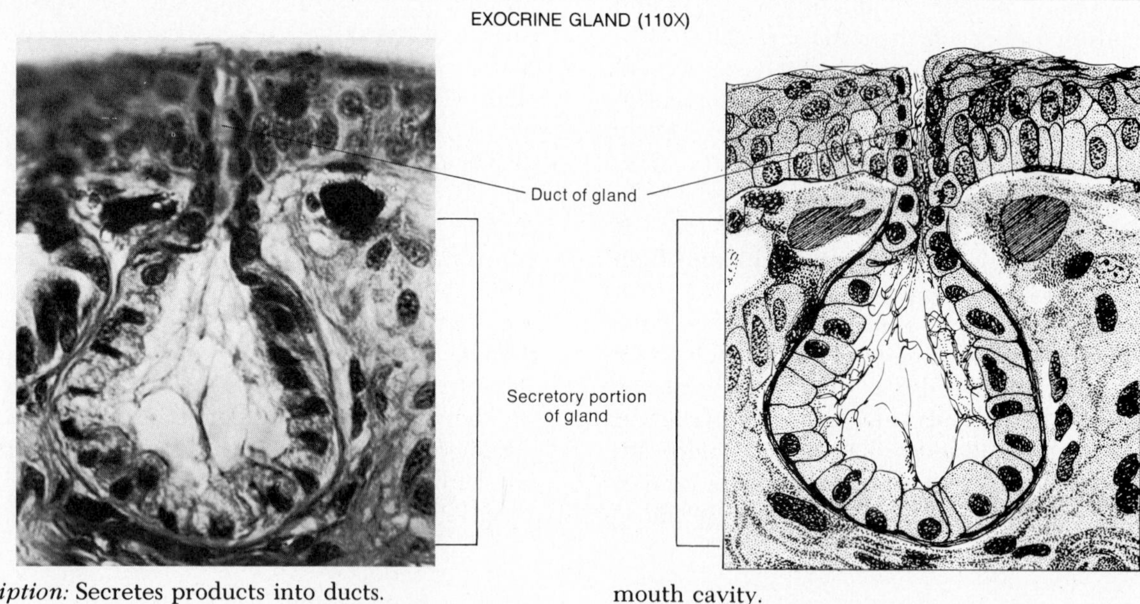

Description: Secretes products into ducts.

Location: Sweat, oil, and wax glands of the skin; digestive glands such as salivary glands which secrete into mouth cavity.

Function: Produce perspiration, oil, wax, and digestive enzymes.

ENDOCRINE GLAND (45X)

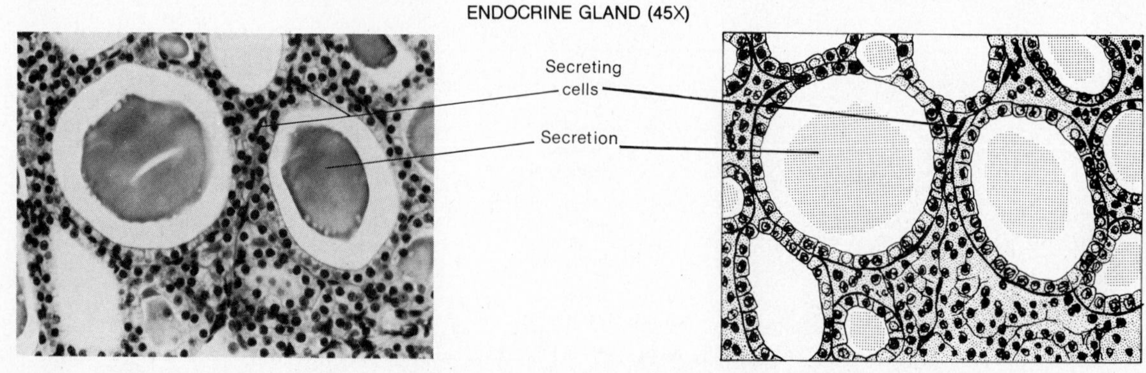

Description: Secretes hormones directly into blood.

Location: Pituitary gland at base of brain; thyroid gland near larynx; adrenal (suprarenal) glands above kidneys.

Function: Produce hormones that regulate various body activities.

Photomicrographs courtesy of Victor B. Eichler, Wichita State University, except for pseudostratified tissue, which is courtesy of Donald I. Patt, from *Comparative Vertebrate Histology,* by Donald I. Patt and Gail R. Patt, Harper & Row, Publishers, Inc., New York, 1969.

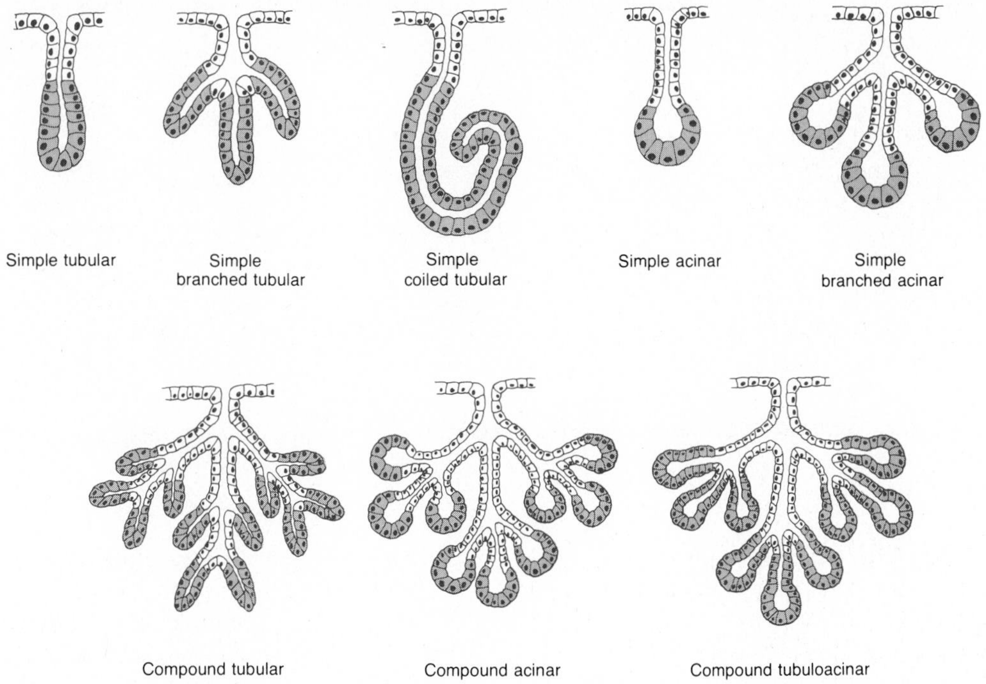

Simple tubular
Simple branched tubular
Simple coiled tubular
Simple acinar
Simple branched acinar

Compound tubular
Compound acinar
Compound tubuloacinar

Figure 3–1. Structural types of multicellular exocrine glands. The secretory portions of the glands are indicated in color. The uncolored areas represent the ducts of the glands.

I. *Unicellular.* One-celled gland that secretes mucus. Example: Goblet cell of the digestive system.

II. *Multicellular.* Many-celled glands.

A. *Simple.* Single, nonbranched duct.
1. *Tubular.* The secretory portion is straight and tubular-shaped. Example: Crypts of Lieberkuhn of intestines (see Figure 20–19).
2. *Branched tubular.* The secretory portion is branched and tubular-shaped. Example: Gastric and uterine glands (see Figure 20–13b).
3. *Coiled tubular.* The secretory portion is coiled. Example: Sweat glands (see Figure 4–1).
4. *Acinar.* The secretory portion is flasklike. Example: Seminal vesicle glands (see Figure 18–7).
5. *Branched acinar.* The secretory portion is branched and flasklike. Example: Sebaceous glands (see Figure 4–3).

B. *Compound.* Branched duct.
1. *Tubular.* The secretory portion is tubular. Example: Bulbourethral glands (see Figure 18–7), testes, liver.
2. *Acinar.* The secretory portion is flasklike. Example: Mammary gland and salivary glands (sublingual and submandibular) (see Figure 20–7a).
3. *Tubuloacinar.* The secretory portion is both tubular and flasklike. Example: Salivary glands (parotid) and pancreas (see Figure 20–7b).

The functional classification of exocrine glands is based upon how the gland releases its secretion. The three recognized categories are holocrine, merocrine, and apocrine glands. **Holocrine glands** first accumulate a secretory product in their cytoplasm. Then, the secreting cell dies, and the cell and its contents are discharged as the secretion of the gland (Figure 3–2a). The discharged cell is replaced by a new cell. One example of a holocrine gland is a sebaceous gland of the skin. **Merocrine glands** are those that secrete without any part of the cell being lost (Figure 3–2b). The secretion is simply formed and discharged by the cell. An example of a merocrine gland is the pancreas. Another is the salivary glands. **Apocrine glands** are those whose secretory product accumulates at the apical (outer)

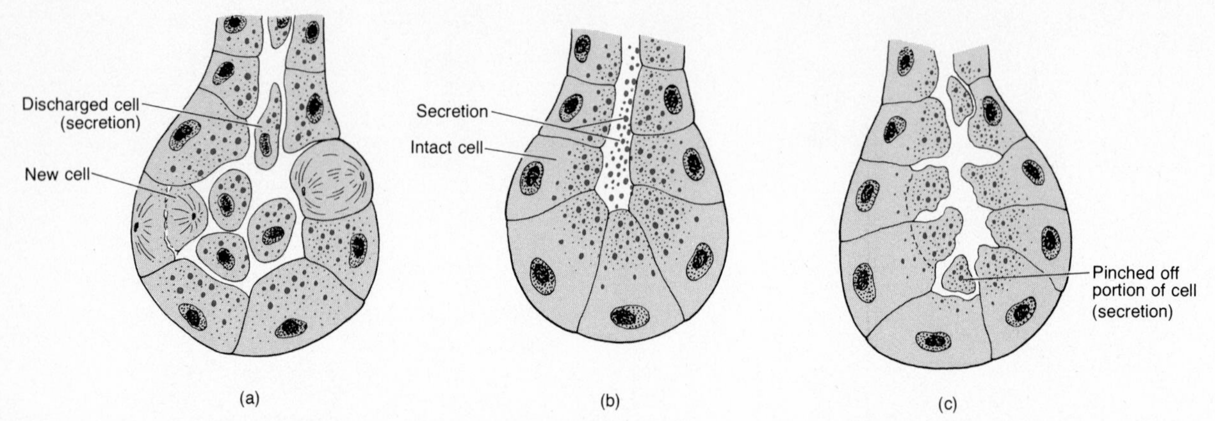

Figure 3–2. Functional classification of multicellular exocrine glands. (a) Holocrine gland. (b) Merocrine gland. (c) Apocrine gland.

margin of the secreting cell. The apical region of the cell and its secretory contents pinch off from the rest of the cell to form the secretion (Figure 3–2c). The remaining part of the cell repairs itself and repeats the process. An example of an apocrine gland is the mammary gland.

We shall now examine some of the structural characteristics, locations, and functions of connective tissue.

CONNECTIVE TISSUE

The most abundant tissue in the body is **connective tissue.** This binding and supporting tissue usually has a rich blood supply. Thus, it is referred to as *vascular.* The cells are widely scattered, rather than closely packed. And the tissue has a great deal of intercellular material, called the *matrix.* In contrast to epithelium, connective tissues do not occur on free surfaces, such as the surfaces of a body cavity or the external surface of the body. The general functions of connective tissues are protection, support, and the binding together of various organs.

The intercellular substance in a given connective tissue largely determines the qualities of the tissue. These substances are nonliving and may consist of fluid, semifluid, or mucoid (mucuslike) material. In cartilage, the intercellular material is firm. In bone, the substance is quite rigid. The living parts of connective tissue are the cells, which produce the intercellular substances. The cells may also store fats, ingest bacteria and cell debris, form anticoagulants, or give rise to antibodies, which protect against diseases.

Before classifying and studying the various kinds of connective tissue, it will be helpful to distinguish between loose and dense connective tissues. **Loose connective tissue** refers to the arrangement of intercellular substance. That is, the fibers in the intercellular substance are neither abundant nor so arranged as to prevent stretching. In addition, the intercellular substance is soft or jellylike in consistency. By contrast, **dense connective tissue** is characterized by the close packing of fibers, and there is less intercellular substance. In areas of the body where tensions are exerted in all directions, the fiber bundles are interwoven and without regular orientation. Such a dense connective tissue is referred to as *irregularly arranged.* Dense, irregularly arranged connective tissue occurs in sheets and forms most fasciae, the dermis of the skin, the fibrous capsules of some organs (testes, liver, lymph nodes), the periosteum of bone, and the perichondrium of cartilage. In other areas of the body, dense connective tissue is adapted for tension in one direction and the fibers have an orderly, parallel arrangement. Such a connective tissue is known as *regularly arranged.* Dense, regularly arranged connective tissue comprises tendons, ligaments, and aponeuroses.

Although the fibers composing connective tissues will be described shortly, it should be noted that loose connective tissue contains all three kinds of fibers. These are collagenous, elastic, and reticular. Dense, irregularly arranged connective tissue also contains all three kinds of fibers, but the collagenous fibers predominate. The dense, regularly arranged connective tissue of many ligaments and aponeuroses is composed of collagenous fibers. Such ligaments are known as collagenous ligaments. However, some ligaments are composed of elastic fibers and are known as yellow elastic ligaments. Examples of these include the ligamenta flava of the vertebrae (Figure 8–26), the suspensory ligament of the penis, and the true vocal cords.

The various kinds of connective tissue may be classified in several ways depending upon the criteria employed. We shall classify them as follows:

I. Embryonal connective tissue
II. Adult connective tissue
  A. Connective tissue proper
    1. Loose connective or areolar
    2. Adipose tissue
    3. Collagenous connective
    4. Elastic connective
    5. Reticular connective
  B. Cartilage
    1. Hyaline cartilage
    2. Fibrocartilage
    3. Elastic cartilage
  C. Bone
  D. Vascular tissue

Embryonal connective tissue

Connective tissue that is present primarily in the embryo or fetus is called *embryonal connective tissue* (Exhibit 3–2). Whereas the term *embryo* refers to a developing human from fertilization through the first 2 months of pregnancy, a *fetus* is regarded as a developing human from the third month of pregnancy to birth. One example of connective tissue found almost exclusively in the embryo is *mesenchyme.* This is the tissue from which all other connective tissues eventually arise. Mesenchyme may be observed under the skin and along the developing bones of the embryo. Some mesenchymal cells are scattered irregularly throughout adult connective tissue, most frequently around blood vessels. Here, mesenchymal cells differentiate into fibroblasts (described shortly) that assist in wound healing. Another kind of embryonal connective tissue is *mucous connective tissue,* which is found only in the fetus. This tissue, also called *Wharton's jelly,* is located in the umbilical cord of the fetus where it provides support for the wall of the cord.

Exhibit 3–2. SUMMARY OF CONNECTIVE TISSUES

I. EMBRYONAL

MESENCHYMAL (130X)

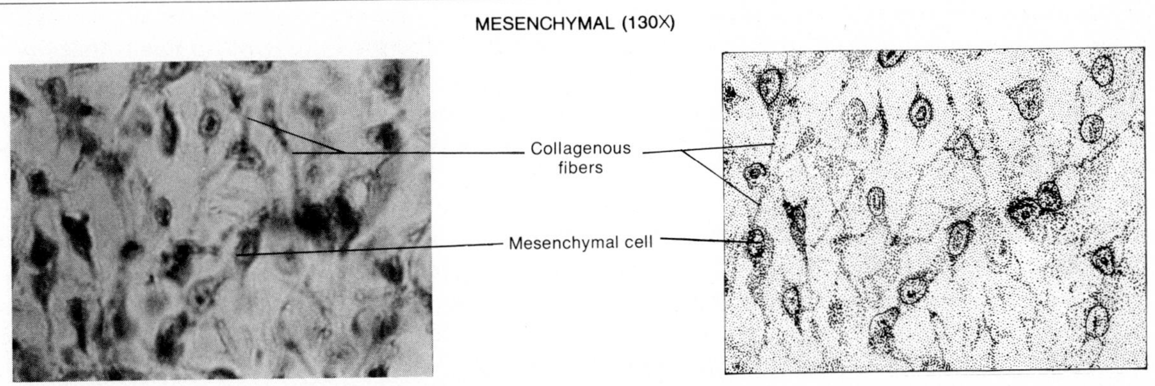

Description: Consists of highly branched mesenchymal cells embedded in a fluid substance with scattered collagenous fibers.

Location: Under skin and along developing bones of embryo; some mesenchymal cells are found in adult connective tissue, especially along blood vessels.

Function: Forms all other kinds of connective tissue.

MUCOUS (130X)

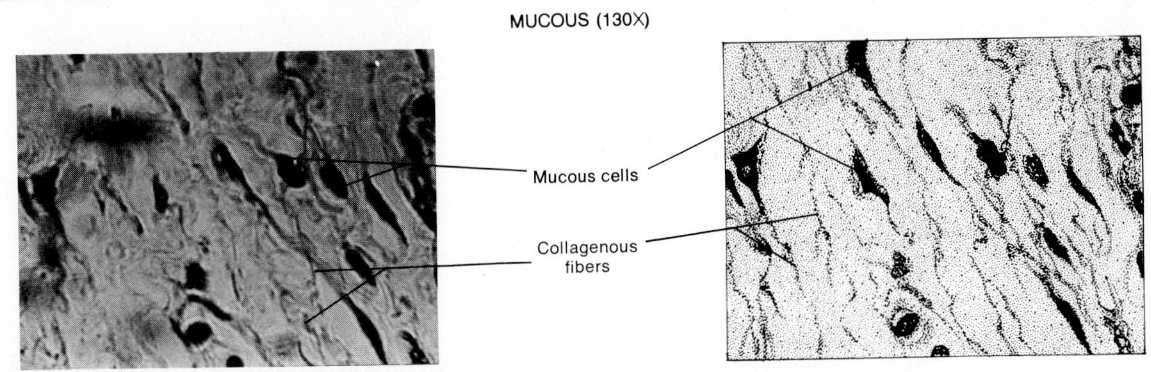

Description: Consists of flattened or spindle-shaped cells embedded in a mucouslike substance containing collagenous fibers.

Location: Umbilical cord of fetus.

Function: Support.

Adult connective tissue

Adult connective tissue is connective tissue that exists in the newborn and that does not change after birth. On the basis of kinds of cells and the nature of intercellular substance present, adult connective tissue is subdivided into several kinds.

Connective tissue proper

Connective tissue that has a more or less fluid intercellular material and a fibroblast (described below) as the typical cell is termed *connective tissue proper.* Five examples of such tissues may be distinguished.

LOOSE CONNECTIVE, OR AREOLAR. This type of tissue is one of the most widely distributed connective tissues in the body. Structurally, it consists of fibers and several kinds of cells embedded in a semifluid ground substance (Exhibit 3–2). This ground substance consists of a viscous material called hyaluronic acid. Normally, the thick consistency of this material impedes the movement of substances through the tissue. However, if an enzyme called hyaluronidase is injected into the tissue, hyaluronic acid changes to a watery consistency. This is of great clinical importance because the reduced viscosity hastens the absorption and diffusion of injected drugs and fluids through the tissue and thus can lessen tension and pain.

The three types of fibers embedded between the cells of loose connective tissue are collagenous fibers, elastic fibers, and reticular fibers. **Collagenous,** or **white, fibers** are very tough and resistant to a pulling force, yet are somewhat flexible because they are usually wavy. These fibers often occur in bundles. They are composed of many minute, wavy fibers called fibrils lying parallel to one another. The bundle arrangement affords a great deal of strength. Chemically, collagenous fibers consist of the protein collagen. **Elastic,** or **yellow, fibers,** by contrast, are smaller than collagenous fibers and freely branch and rejoin one another. Elastic fibers consist of a protein called elastin. These fibers also provide some strength and have a high degree of elasticity, up to 50 percent of their length. **Reticular fibers** are protein-polysaccharide complexes. They are very fine fibers that branch extensively. Some authorities believe that reticular fibers are immature collagenous fibers. Like collagenous fibers, reticular fibers provide support and strength and form the *stroma (framework)* of many soft organs.

The cells in loose connective tissue are numerous and varied. The majority of the cells are **fibroblasts,** which are large, flat cells with branching processes. If the tissue is injured, the fibroblasts are believed to form collagenous fibers. Some evidence suggests that fibroblasts also form elastic fibers and the viscous ground substance. In actuality, mature fibroblasts may be referred to as **fibrocytes.** The basic distinction between the two is that fibroblasts (or *"blasts"* of any form of cell) are involved in the formation of the immature tissue or repair of the mature tissue, and fibrocytes (or *"cytes"* of any form of cell) are involved in the maintenance of the health of the mature tissue. Other cells found in loose connective tissue are called **macrophages,** or **histiocytes.** They are irregular in form with short branching processes. These cells are capable of engulfing bacteria and cellular debris by the process of phagocytosis. Thus, they provide an important defense for the body. A third kind of cell in loose

Exhibit 3–2 (cont.)

II. ADULT

LOOSE or AREOLAR (65X)

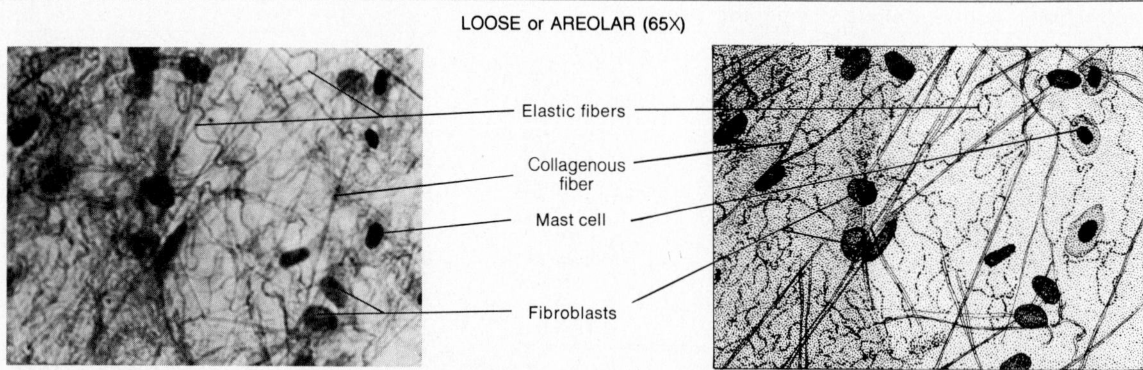

Description: Consists of fibers (collagenous and elastic) and several kinds of cells (fibroblasts, macrophages, plasma cells, and mast cells) embedded in a semifluid ground substance.

Location: Subcutaneous layer of skin, mucous membranes, blood vessels, nerves, and body organs.

Function: Strength, elasticity, support, phagocytosis, produces antibodies, and produces an anticoagulant.

connective tissue is the **plasma cell.** These cells are small and either round or irregularly shaped. They give rise to antibodies and, accordingly, provide a defensive mechanism through immunity. Plasma cells are found in many places of the body, but the majority are found in connective tissue, especially in that of the digestive tract and the mammary glands. Another kind of cell in loose connective tissue is called the **mast cell.** This cell is found in abundance along blood vessels. It forms an anticoagulant, a substance that prevents blood from clotting within the vessels, called heparin. Mast cells are also believed to produce histamine, a chemical that dilates, or enlarges, small blood vessels. Other cells in loose connective tissue include **melanocytes,** or pigment cells, fat cells, and white blood cells.

Loose connective tissue is continuous throughout the body. It is present in all mucous membranes and around all blood vessels and nerves. And it occurs around body organs. Combined with adipose tissue, it forms the *subcutaneous layer,* or the layer of tissue that attaches the skin to underlying tissues and organs. The subcutaneous layer is also referred to as the *superficial fascia.*

ADIPOSE TISSUE. This kind of tissue is basically a form of loose connective tissue in which the fibroblasts are specialized for fat storage. Adipose cells have the shape of a "signet ring" because the cytoplasm and nucleus are pushed to the edge of the cell by a large droplet of fat (Exhibit 3–2). In general, adipose tissue is found throughout the body wherever loose connective tissue is located. Specifically, it is found in the subcutaneous layer below the skin, around the kidneys, at the base and on the surface of the heart, in the marrow of long bones, and as a padding around joints. Adipose tissue is a poor conductor of heat and therefore reduces heat loss through the skin. It is also an important food reserve and generally supports and protects various organs.

COLLAGENOUS CONNECTIVE. This kind of tissue has a predominance of collagenous fibers, or white fibers, arranged in bundles (Exhibit 3–2). The cells found in collagenous connective tissue are fibroblasts, which are placed in rows between the bundles. The tissue is silvery white in appearance. It is tough, yet somewhat pliable. Because of the great strength of this tissue, dense, regularly arranged collagenous connective tissue is the principal component of tendons, which attach muscles to bones; many ligaments, which hold bones together at joints; and aponeuroses, which are flat bands connecting one muscle with another or with a bone. Dense, irregularly arranged collagenous connective tissue forms membrane capsules around various organs such as the kidney, heart, testes, liver, and lymph nodes. It also forms the *deep fasciae,* which are sheets of connective tissue wrapped around muscles to hold them in place.

ELASTIC CONNECTIVE. Unlike collagenous connective tissue, elastic connective tissue has a predominance of elastic fibers that branch freely (Exhibit 3–2). These fibers give the tissue a yellowish color. Fibroblasts are present only in the spaces between the fibers. Elastic connective tissue can be stretched. As the name implies, it is elastic and will snap back into shape. It is a component of the cartilages of the larynx, the walls of elastic arteries, the trachea, bronchial tubes to the lungs, and the lungs

Exhibit 3–2 (cont.)

ADIPOSE (90X)

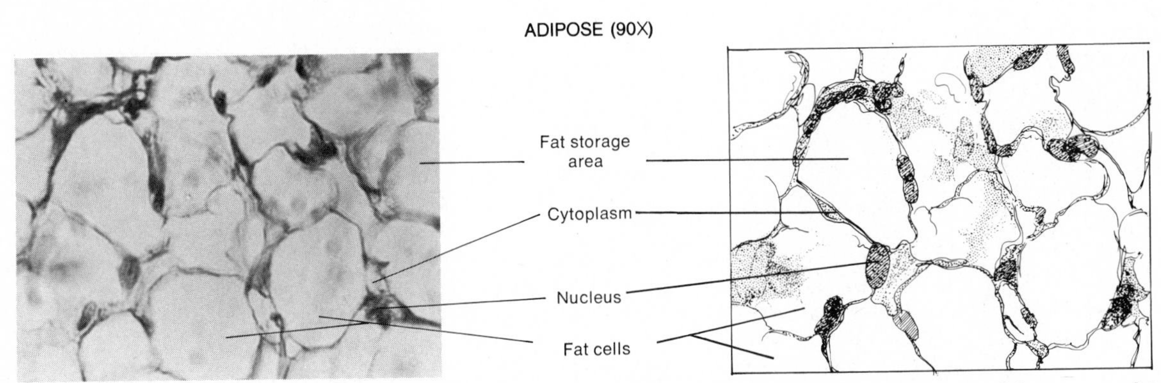

Description: Contains fibroblasts specialized for fat storage; cells have a "signet-ring shape."

Location: Subcutaneous layer of skin, around heart and kidneys, marrow of long bones, padding around joints.

Function: Reduces heat loss through skin, serves as a food reserve, supports, and protects.

Exhibit 3–2 (cont.)

COLLAGENOUS (65X)

Fibroblasts

Bundles of collagenous fibers

Description: Collagenous, or white, fibers predominate and are arranged in bundles; fibroblasts are in rows between bundles.

Location: Forms tendons, ligaments, aponeuroses, membranes around various organs, and fasciae.

Function: Provides strong attachment between various structures.

ELASTIC (65X)

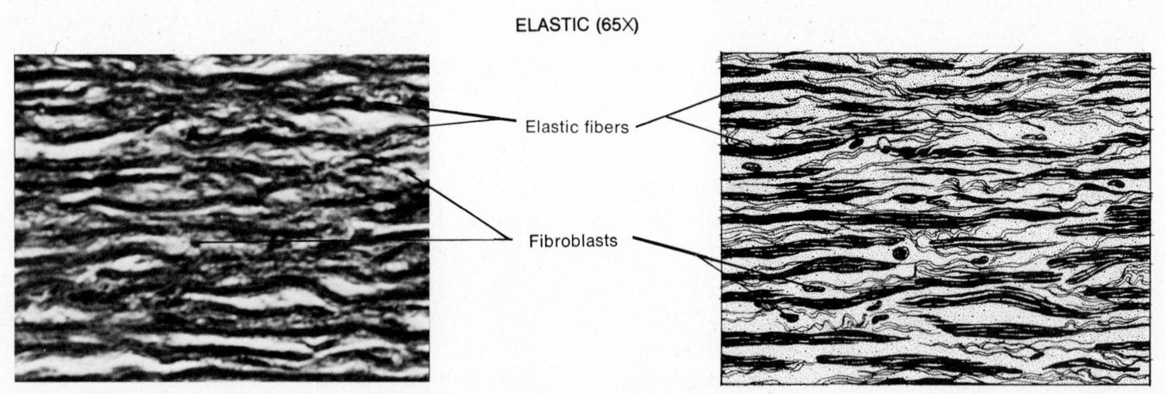

Description: Elastic, or yellow, fibers predominate and branch freely; fibroblasts present in spaces between fibers.

Location: Lung tissue, cartilage of larynx, walls of arteries, trachea, bronchial tubes, true vocal cords, and between vertebrae.

Function: Allows stretching of various organs.

RETICULAR (65X)

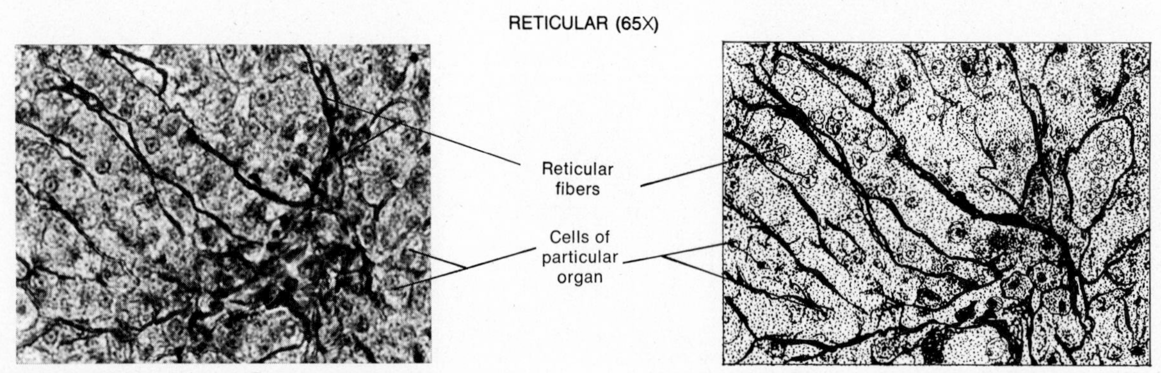

Description: Consists of a network of interlacing fibers; cells are thin and flat and wrapped around fibers.

Location: Liver, spleen, and lymph nodes.

Function: Forms stroma, or framework, of organs; binds together smooth muscle tissue cells.

themselves. It is also found between the vertebral spaces of the backbone. Elastic connective tissue provides stretch and strength, allowing the structures in which it is found to perform their functions more efficiently. Yellow elastic ligaments, composed mostly of elastic fibers, comprise the ligamenta flava of the vertebrae and the true vocal cords.

RETICULAR CONNECTIVE. This kind of connective tissue consists of interlacing reticular fibers (Exhibit 3–2). It helps to form the framework, or body, of many organs, including the liver, spleen, and lymph nodes. Reticular connective tissue also helps to bind together the cells of smooth muscle tissue. It is especially adapted to providing strength and support in the structures in which it is found.

As a whole, connective tissue proper is characterized by a more or less fluid intercellular material. Moreover, the typical cell is a fibroblast. Cartilage, the next type of adult connective tissue to be studied, has a more solid intercellular material and a different type of cell.

Cartilage

This type of connective tissue is capable of enduring considerably more stress than the connective tissues just discussed. *Cartilage* consists of a fairly dense network of collagenous fibers and some elastic fibers firmly embedded in a gel-like substance. The cells of mature cartilage, called *chondrocytes,* occur singly or in groups in spaces called *lacunae* in the intercellular substance. The surface of cartilage is surrounded by a connective tissue covering called the *perichondrium.* The combining form *chondro,* which you will see often, means cartilage, and the form *peri* means around. On the basis of the texture or fiber type of the intercellular substance, three kinds of cartilage are recognized. These are, in the order of discussion, hyaline cartilage, fibrocartilage, and elastic cartilage.

HYALINE CARTILAGE. This cartilage, also called gristle, appears as a bluish white, glossy, homogenous mass. The collagenous fibers, although present, are not visible, and the prominent chondrocytes are found in lacunae (Exhibit 3–2). Hyaline cartilage is the most abundant kind of cartilage in the body. It is found at joints, where it is called *articular cartilage.* And it forms the ventral ends of the ribs, where it is referred to as *costal cartilage.* Hyaline cartilage also helps to form the nose, larynx, trachea, and the bronchi and bronchial tubes leading to the lungs. Most of the embryonic skeleton consists of hyaline cartilage. This kind of cartilage affords flexibility and support.

Exhibit 3–2 (cont.)

HYALINE CARTILAGE (30X)

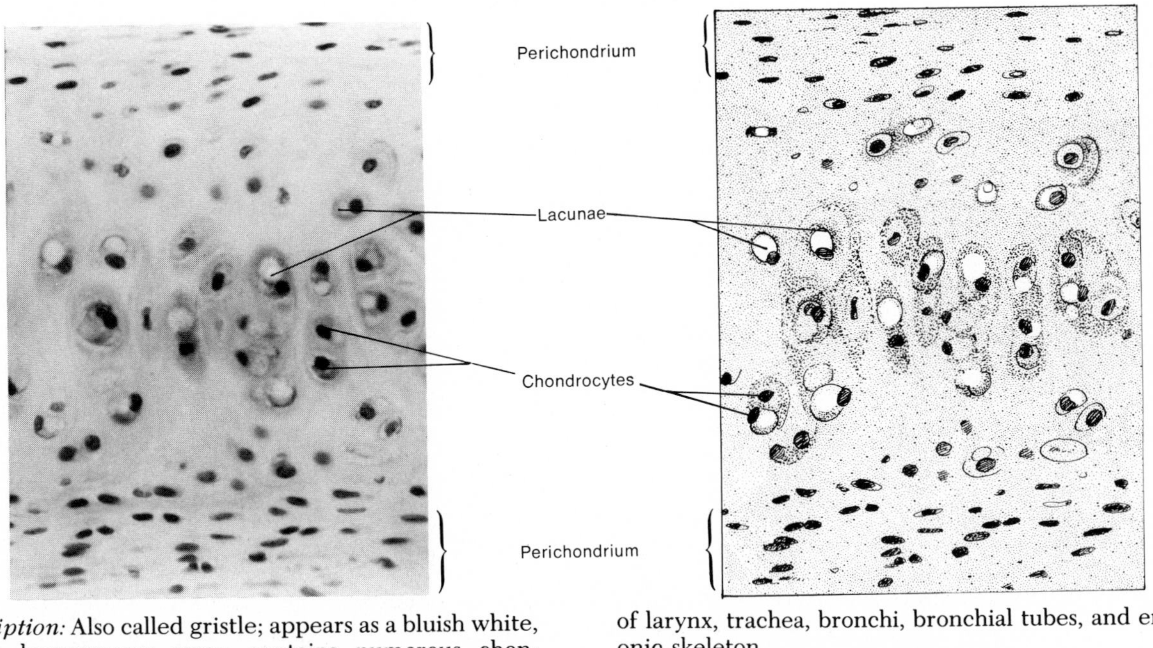

Perichondrium

Lacunae

Chondrocytes

Perichondrium

Description: Also called gristle; appears as a bluish white, glossy, homogenous mass; contains numerous chondrocytes; is the most abundant type of cartilage.

Location: Ends of long bones, ends of ribs, nose, parts of larynx, trachea, bronchi, bronchial tubes, and embryonic skeleton.

Function: Provides movement at joints, flexibility, and support.

FIBROCARTILAGE. Chondrocytes that are scattered through many bundles of visible collagenous fibers are found in this type of cartilage (Exhibit 3–2). Fibrocartilage is found at the symphysis pubis, or the point where the coxal (hip) bones fuse at the midline anteriorly. It is also in the discs that lie between each of the vertebrae. This tissue combines the properties of strength and rigidity.

ELASTIC CARTILAGE In this tissue, chondrocytes are located in a threadlike network of elastic fibers (Exhibit 3–2). Elastic cartilage provides strength and maintains the shape of certain organs. Among these are the larynx, the external part of the ear or pinna, and the internal tubes that connect the nose to the ears, the Eustachian tubes.

Bone

The details of bone or osseous tissue, another kind of connective tissue, will be discussed in Chapter 5 as part of the skeletal system.

Vascular tissue

This kind of connective tissue, also known as blood, will be treated in Chapter 11 as a component of the circulatory system.

Exhibit 3–2 (cont.)

FIBROCARTILAGE (65X)

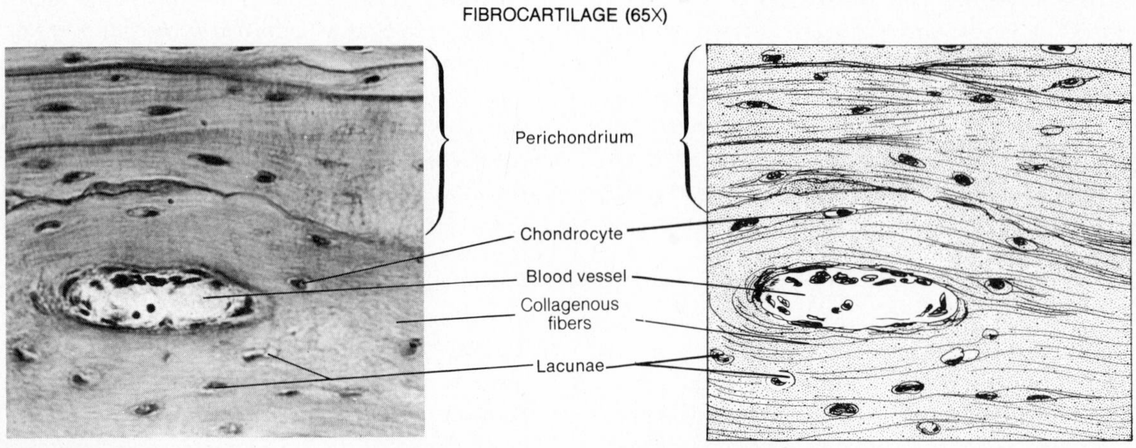

Description: Consists of chondrocytes scattered among bundles of collagenous fibers.

Location: Symphysis pubis and intervertebral discs.

Function: Support and fusion.

ELASTIC CARTILAGE (65X)

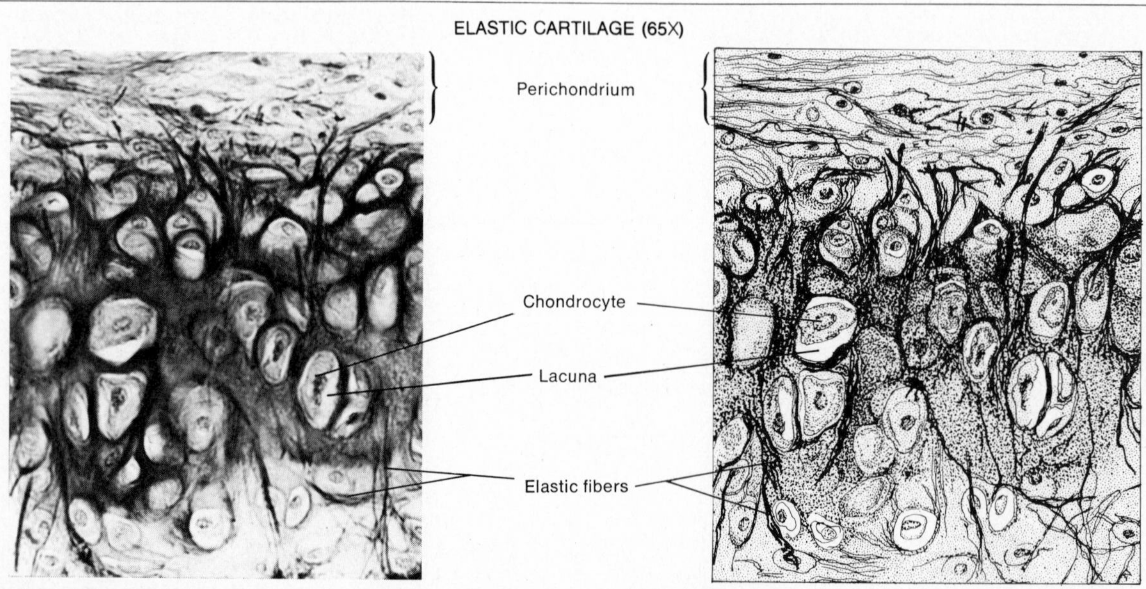

Description: Consists of chondrocytes located in a threadlike network of elastic fibers.

Location: Epiglottis, parts of larynx, external ear, and Eustachian tubes.

Function: Gives support and maintains shape.

Photomicrographs courtesy of Victor B. Eichler, Wichita State University.

MUSCLE TISSUE AND NERVOUS TISSUE

Epithelial and connective tissue can take a variety of forms to provide a variety of services for the body. In a sense, they are all-purpose tissues. By contrast, the third major type of tissue, called muscle tissue, consists of highly modified cells that perform one basic function. This function is to contract. We shall look at muscle tissue in Chapter 9, when we discuss how the body moves. The fourth major type, called nervous tissue, is specialized to perform the function of sending electrical impulses. Nerve cells and their tissue will be examined in Chapter 12. Meanwhile, let us look at some structures that are composed of epithelial and connective tissue. We shall start here with the membranes of the body. In Chapter 4, we shall study the skin.

EPITHELIAL MEMBRANES

The combination of an epithelial layer and an underlying connective tissue layer constitutes an **epithelial membrane.** The principal epithelial membranes of the body are: (1) mucous membranes, (2) serous membranes, and (3) the cutaneous membrane or skin.

1. **Mucous membranes.** These membranes line the body cavities that open to the exterior. Examples include the membranes lining the mouth and the entire digestive tract, the respiratory passages, the reproductive system, and the urinary system. The surface membrane of a mucous membrane may vary in type. For example, it is stratified squamous in the esophagus and simple columnar in the intestine. The epithelial layer of a mucous membrane secretes mucus, which keeps the cavities from being dried out by air. It also traps dust in the respiratory passageways and lubricates food as it moves through the digestive tract. In addition, the epithelial layer is responsible for functions such as the secretion of digestive enzymes and the absorption of food. The connective tissue layer of a mucous membrane binds the epithelium to the underlying structures. It is referred to as the *lamina propria* and allows some flexibility of the membrane. It holds the blood vessels in place, provides a covering that protects underlying muscles from abrasion or puncture, and provides the epithelium covering it with oxygen and nutrients and removes wastes.

2. **Serous membranes.** These membranes are found lining body cavities that do not open to the exterior, and they cover the organs that lie within those cavities. Serous membranes consist of two portions. The part that is attached to the wall of a cavity is called the *parietal* portion. The part that covers the organs lying inside these cavities is called the *visceral* portion. The serous membrane lining the thoracic cavity and covering the lungs is called the pleura. The membrane lining the heart cavity and covering the heart is referred to as the pericardium. The root word *cardio* means heart. Last, the serous membrane lining the abdominal cavity and covering the abdominal organs and some pelvic organs is called the peritoneum.

The epithelial layer of a serous membrane is mesothelium and secretes a lubricating fluid, which allows the organs to glide easily over each other or along the walls of the cavities. The connective tissue layer ties the organs to the cavity walls and keeps them from falling to the bottom of the chest or abdomen.

3. **The cutaneous membrane.** The cutaneous membrane, or skin, constitutes an organ of the integumentary system and will be discussed in the next chapter.

Synovial membranes line the cavities of the joints. Like serous membranes, they line structures that do not open to the exterior. Unlike mucous and serous membranes, synovial membranes do not contain epithelium. They are composed of loose connective tissue with elastic fibers and varying amounts of fat. Synovial membranes, therefore, are not epithelial membranes. Synovial membranes secrete *synovial fluid,* a fluid that lubricates the ends of bones as they move at joints.

Chapter summary in outline

EPITHELIAL TISSUE

1. A tissue is a group of cells specialized for a particular function.
2. Epithelium covers and lines body surfaces and forms glands.
3. Epithelium has many cells, little intercellular material, and no blood vessels. It is attached to connective tissue by a basement membrane. It is also capable of replacing itself.

Covering and lining epithelium

1. Simple epithelium is a single layer of cells adapted for absorption or filtration.
2. Stratified epithelium is several layers of cells adapted for protection.
3. Epithelial cell shapes include squamous (flat), cuboidal (cube-like), columnar (rectangular), and transitional (variable).
4. Simple squamous epithelium is adapted for diffusion and filtration and is found in lungs and kidneys. Endothelium lines the heart and blood vessels. Mesothelium lines body cavities and covers internal organs.
5. Simple cuboidal epithelium is adapted for secretion and absorption in kidneys and glands.

6. Simple columnar epithelium lines the digestive tract. Specialized cells containing microvilli perform absorption. Goblet cells perform secretion. In the respiratory tract, the cells are ciliated to move dust particles out of the body.
7. Stratified squamous epithelium is protective. It lines the upper digestive tract and forms the outer layer of skin.
8. Stratified cuboidal epithelium is found in adult sweat glands.
9. Stratified columnar epithelium protects and secretes. It is found in the male urethra and excretory ducts.
10. Transitional epithelium is found in the urinary bladder and is capable of stretching.
11. Pseudostratified epithelium is one layer but gives the appearance of having many layers. It lines excretory and respiratory structures where it protects and secretes.

Glandular epithelium

1. A gland is a single cell or a mass of epithelial cells adapted for secretion.
2. Exocrine glands (sweat, oil, and digestive glands) secrete into ducts. They are classified on the basis of structural complexity and physiology.
3. Endocrine glands secrete hormones directly into the blood.

CONNECTIVE TISSUE

1. Connective tissue is the most abundant body tissue. It has few cells, an extensive matrix, and a rich blood supply.
2. It functions in protection, support, and binding organs together.

Embryonal connective tissue

1. Mesenchyme, found almost exclusively in the embryo, forms all other kinds of connective tissue.
2. Mucous connective tissue is found only in the umbilical cord of the fetus, where it gives support.

Connective tissue proper

1. This tissue has a more or less fluid matrix. The fibroblast is the typical cell.
2. Loose connective tissue, or areolar tissue, is widely distributed. It contains three kinds of fibers (collagenous, elastic, and reticular). It also has several kinds of cells (fibroblasts, macrophages, plasma cells, mast cells, and white blood cells). Loose connective tissue forms the subcutaneous layer under the skin. It is present in mucous membranes and around blood vessels, nerves, and body organs. When the fibroblasts in loose connective tissue become infiltrated with fat, the tissue is then known as adipose tissue.
3. Dense connective tissue forms tendons, ligaments, and deep fasciae. All three are usually primarily composed of closely packed collagenous fibers. In tendons and ligaments, collagenous fibers are arranged in parallel bundles. In fasciae, these fibers are interwoven at various angles with each other. A few ligaments are composed of closely packed elastic fibers.

Cartilage

1. Cartilage has a gel-like matrix of collagenous and elastic fibers that contain chondrocytes.
2. Hyaline cartilage is found at the ends of bones, in the nose, and in respiratory structures. It is flexible, allows movement, and provides support.
3. Fibrocartilage connects the pelvic bones and the vertebrae. It provides strength.
4. Elastic cartilage maintains the shape of organs such as the external ear.

EPITHELIAL MEMBRANES

1. An epithelial membrane is an epithelial layer overlying a connective tissue layer. Examples are mucous, serous, and cutaneous.
2. Mucous membranes line cavities that open to the exterior, such as the digestive tract.
3. Serous membranes (pleura, pericardium, peritoneum) line closed cavities and the organs lying within the cavities. These membranes consist of the parietal and visceral portions.
4. The cutaneous membrane is the skin.
5. Synovial membranes line joint cavities.

Review questions and problems

1. Define a tissue. What are the four basic kinds of human tissues?
2. What characteristics are common to all epithelium? Distinguish covering and lining epithelium from glandular epithelium.
3. Discuss the classification of epithelium using layering and cell type as criteria.
4. For each of the following kinds of epithelium, briefly describe the microscopic appearance, location in the body, and functions: simple squamous, simple cuboidal, simple columnar, stratified squamous, stratified cuboidal, stratified columnar, transitional, and pseudostratified.
5. What is a gland? Distinguish between endocrine and exocrine glands. Describe the classification of exocrine glands according to structural complexity and physiology and give at least one example of each.
6. Enumerate the ways in which connective tissue differs from epithelium. How are connective tissues classified?
7. Compare embryonal connective tissue with adult connective tissue.
8. Describe the following connective tissues with regard to microscopic appearance, location in the body, and function: loose (areolar), adipose, collagenous, elastic, reticular, hyaline cartilage, fibrocartilage, and elastic cartilage.
9. Define the following kinds of membranes: mucous, serous, synovial, and cutaneous. Where is each located in the body? What are the functions?
10. Below are some descriptive statements for various tissues. Write the name of the tissue described.

    A stratified epithelium that permits distention.
    A single layer of flat cells concerned with filtration and absorption.
    Forms all other kinds of connective tissue.
    Specialized for fat storage.
    An epithelium with waterproofing qualities.
    Forms the framework of many organs.
    Produces perspiration, wax, oil, and digestive enzymes.
    Cartilage that shapes the external ear.
    Contains goblet cells and lines the intestine.
    Most widely distributed connective tissue.
    Forms tendons, ligaments, and aponeuroses.
    Specialized for the secretion of hormones.
    Provides support in the umbilical cord.
    Lines kidney tubules and is specialized for absorption and secretion.
    Permits extensibility of lung tissue.

Selected readings

Arey, L. B. *Human Histology.* 4th ed. Philadelphia: W. B. Saunders, 1974.

Basmajian, John V. *Primary Anatomy.* 6th ed. Baltimore: Williams and Wilkins, 1970. Pp. 4–8.

Bloom, W., and D. W. Fawcett. *A Textbook of Histology.* 10th ed. Philadelphia: W. B. Saunders, 1974.

Ham, Arthur W. *Histology.* 7th ed. Philadelphia: J. B. Lippincott, 1974.

Leeson, Thomas S., and C. Roland Leeson. *Histology.* 2d ed. Philadelphia: W. B. Saunders, 1970.

Reith, E. J., and M. H. Ross. *Atlas of Descriptive Histology.* 2d ed. New York: Harper and Row, 1970.

CHAPTER 4

THE SKIN AS A REPRESENTATIVE ORGAN

STUDENT OBJECTIVES

After you have read this chapter, you should be able to:

1. Define the skin as an organ and a component of the integumentary system
2. Explain how the skin is structurally divided into an epidermis and a dermis
3. List the structural layers of the epidermis and describe their functions
4. Explain the composition and functions of the dermis
5. Contrast the structure and functions of epidermal derivatives of the skin such as hair, glands, and nails
6. Define and describe the effects of a burn
7. Classify burns into first-, second-, and third-degree types
8. Define the "rule of nines" for estimating the extent of a burn
9. Define key terms associated with the integumentary system

An aggregation of tissues that performs a specific function is an organ. Recall that organs represent the next level of organization in the body. In considering the organ level of organization, we shall use the skin as an example. From organs, the next higher level of organization is a system—a group of organs that operate together to perform specialized functions. The skin and the organs derived from it, such as hair, nails, and glands, and several specialized receptors, constitute the **integumentary system** of the body. First, let us consider the skin as an organ.

THE SKIN AS AN ORGAN

The **skin** is an organ because it consists of tissues that are structurally joined together to perform specific activities. Most people view the skin as a simple thin covering that keeps the body together and gives it protection. A detailed analysis of the skin, however, reveals that it is quite complex in structure and performs several vital functions. In fact, this organ is essential for survival.

Considered by itself, the skin is the largest organ of the body. For the average adult, the skin occupies a surface area of approximately 7,620 sq cm (3,000 sq in.). It covers the body and protects the underlying tissues, not only from bacterial invasion but also from drying out and from harmful light rays. In addition to its protective function, the skin helps to control body temperature, prevents excessive loss of inorganic and organic materials, receives stimuli from the environment, stores chemical compounds, excretes water and salts, and synthesizes several important compounds.

Structurally, the skin consists of two principal parts (Figure 4–1). The outer, thinner portion, which is composed of epithelium, is called the **epidermis.** The epidermis is cemented to the inner, thicker, connective tissue part, which is called the **dermis.** Beneath the dermis is a *subcutaneous layer* of tissues. The combining form *sub* means under. The subcutaneous layer is also called the *superficial fascia,* and it consists of areolar and adipose tissues. Fibers from the dermis extend down into the superficial fascia and anchor the skin to the subcutaneous layer. The superficial fascia, in turn, is firmly attached to underlying tissues and organs.

The epidermis

The **epidermis** is composed of stratified epithelium in four or five cell layers, depending on its location in the body. (See Figure 4–1.) In areas where exposure to friction is greatest, such as the palms and soles, the epidermis has five layers. In all other parts of the body, the epidermis has four layers. The names of these layers from the inside outward are as follows:

1. *Stratum basale.* This is a single layer of columnar cells capable of continued cell division. The epidermis grows by the division of cells in the stratum basale and deep layers of the stratum spinosum, the next higher layer. As these cells multiply, they push up toward the surface. Their nuclei degenerate, and the cells die. Eventually, the cells are shed in the top layer of the epidermis.

2. *Stratum spinosum.* This layer of the epidermis, just above the stratum basale, contains 8 to 10 rows of polygonal (many-sided) cells that fit closely together. Since the surfaces of these cells may assume a prickly appearance when prepared for microscope examination, the layer is so named. The word *spinosum* means prickly. The stratum spinosum also helps in the continual production of new epidermis. The stratum basale and stratum spinosum are sometimes collectively referred to as the *stratum germinativum* to indicate the layers where new cells are germinated.

3. *Stratum granulosum.* This third layer of the epidermis consists of two or three rows of flattened cells that contain dark staining granules of a substance called keratohyaline. This compound is involved in the first step of keratin formation. Keratin is a waterproofing protein found in the top layer of the epidermis. The stratum granulosum contains cells whose nuclei are in various stages of degeneration. As these nuclei break down, the cells are no longer capable of carrying out vital metabolic reactions and die.

4. *Stratum lucidum.* This layer is quite pronounced in the thick skin of the palms and soles. It consists of three to four rows of clear, flat, dead cells that contain droplets of a translucent substance called eleidin. Eleidin is formed from keratohyaline and is eventually transformed to keratin. This layer is so named because of the translucent property of eleidin. The word root *lucidus* means clear.

5. *Stratum corneum.* This layer consists of 25 to 30 rows of flat, dead cells containing the protein *keratin.* The keratin serves as a waterproof covering. These cells are continuously shed and replaced. The stratum corneum serves as an effective barrier against light and heat waves, bacteria, and many chemicals. Be sure to examine the photomicrograph of the skin in Figure 4–2.

The color of the skin is due to a brown to black pigment called **melanin.** This pigment is found throughout the basale and spinosum layers and in the granulosum of all Caucasian people. In blacks,

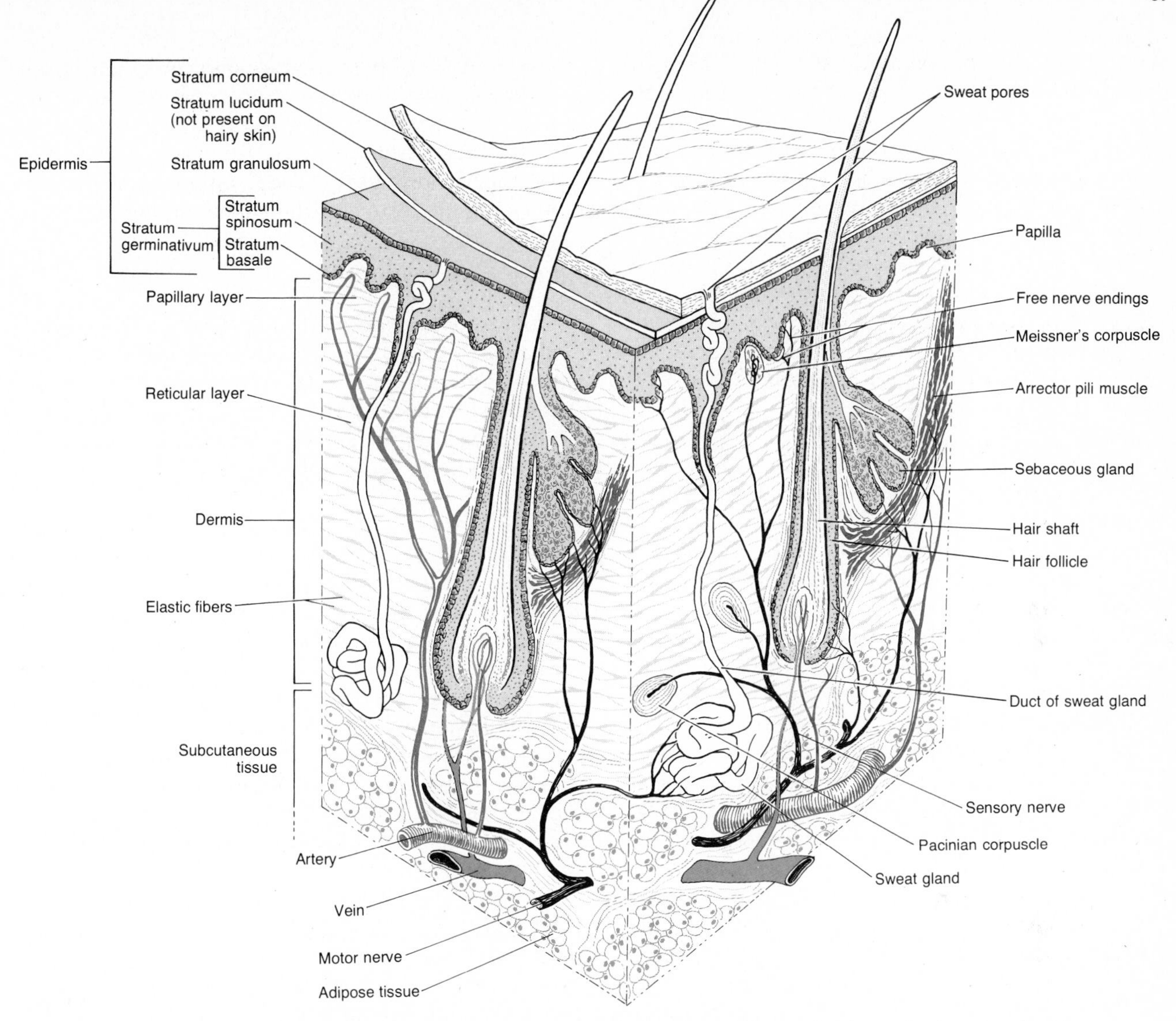

Figure 4–1. The skin. Note the structures in the epidermis and dermis and the underlying subcutaneous layer.

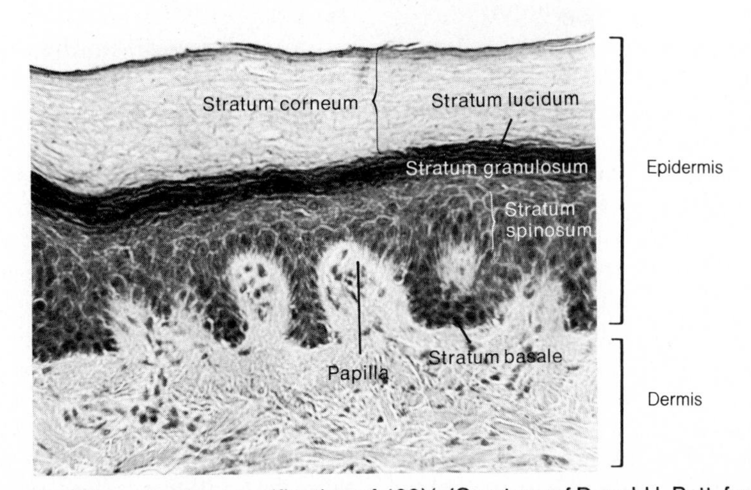

Figure 4–2. Photomicrograph of the skin at a magnification of 400X. (Courtesy of Donald I. Patt, from *Comparative Vertebrate Histology,* by Donald I. Patt and Gail R. Patt, Harper & Row, Publishers, Inc., New York, 1969.)

melanin is found in all epidermal layers. When the skin is exposed to ultraviolet radiation, both the amount and darkness of melanin increase, which causes tanning and further protects the body against radiation. Thus, melanin serves a very important protective function. Another pigment called **carotene** is found in the corneum and fatty areas of the dermis in Oriental people. Together, carotene and melanin account for the yellowish hue of their skin. The pink color of Caucasian skin is due to blood vessels in the dermis. The redness of the vessels is not heavily masked by pigment. The epidermis has no blood vessels.

The dermis

The second principal part of the skin, the **dermis,** is composed of connective tissue containing collagenous and elastic fibers. (See Figure 4–1.) Numerous blood vessels, nerves, glands, and hair follicles are also embedded in the dermis. The upper region of the dermis, which is about one-fifth of the total layer, is referred to as the *papillary region.* This part of the dermis is so named because its surface area is greatly increased by small, fingerlike projections called **papillae.** These structures project into the epidermis and contain loops of capillaries. In some cases, the papillae contain Meissner's corpuscles, which are nerve endings that are sensitive to touch. The dermis also contains nerve endings called Pacinian corpuscles, which are sensitive to deep pressure.

The ridges marking the external surface of the epidermis are caused by the size and arrangement of the papillae in the dermis. Some of the ridges cross at various angles and can be seen on the back surface of your hand. Other ridges on your palms and fingertips prevent slipping. The ridge patterns on the tips of the fingers and thumbs are different in each individual. Because of this, fingerprints can be taken and accurately used for purposes of identification.

The remaining portion of the dermis is called the *reticular region.* It is a dense, irregular, collagenous connective tissue. This area of the dermis contains many blood vessels and also contains collagenous and elastic fibers. The spaces between the interlacing fibers are occupied by adipose tissue and sweat glands. The reticular zone is attached to the organs beneath it, such as bone and muscle, by the subcutaneous layer.

It should now be obvious to you that the skin, despite its relatively simple physical appearance, is a very complex organ capable of carrying on numerous activities essential to life. The tissues of the skin are joined to form an organ that performs specific activities. These tissues consist of (1) the various epithelial layers of the epidermis that, together, protect, waterproof, and add new cells, and (2) the connective tissues of the dermis that protect, contain nerve endings for touch and pressure, and connect the epidermis to the subcutaneous layer.

EPIDERMAL DERIVATIVES

Organs that are derived from the skin, such as hair, glands, and nails, perform functions that are necessary and sometimes vital. Hair and nails offer further protection to the body, whereas the sweat glands perform the vital function of helping to regulate body temperature.

Hair

Certain growths of the epidermis variously distributed over the body are **hairs** or **pili.** Some of the regions of the body not covered by hair are the surfaces of the palms and undersides of the fingers, the back surfaces of the fingers from the first joint to the fingertips, the soles of the feet, the nipples, the labia minora, the glans penis, and the lips. The primary function of hair is protection. Though the protection is limited, hair guards the scalp from mechanical injury and from the sun's rays. The eyebrows and eyelashes protect the eyes from foreign particles. The hair in the nostrils and external ear canal also protects these structures from insects and dust particles.

Hairs are composed of a number of parts (Figure 4–3). Each hair consists of a free shaft and a root. The *shaft* is the visible portion of the hair projecting above the surface of the skin. The *root* is the portion of the hair below the surface of the skin that penetrates deep into the dermis. Surrounding the root is the *hair follicle,* which is made up of an internal zone of epithelium, the *internal root sheath,* and an external zone of epithelium, the *external root sheath.* The lower ends of each follicle are enlarged into an onion-shaped structure, the *bulb.* This structure contains an indentation, the *connective tissue* or *dermal papilla,* that is filled with loose connective tissue. The papilla contains many blood vessels and provides nourishment for the growing hair. The base of the bulb also contains a region of cells called the *matrix,* which is a germinal layer. The cells of the matrix produce new hairs by cell division when older hairs are shed. This replacement occurs within the same follicle. Hair grows about 1 mm (0.0394 in.) every 3 days. Hair loss in an adult is about 10 to 100 hairs each day. But the rate of growth may be altered by general health and illness.

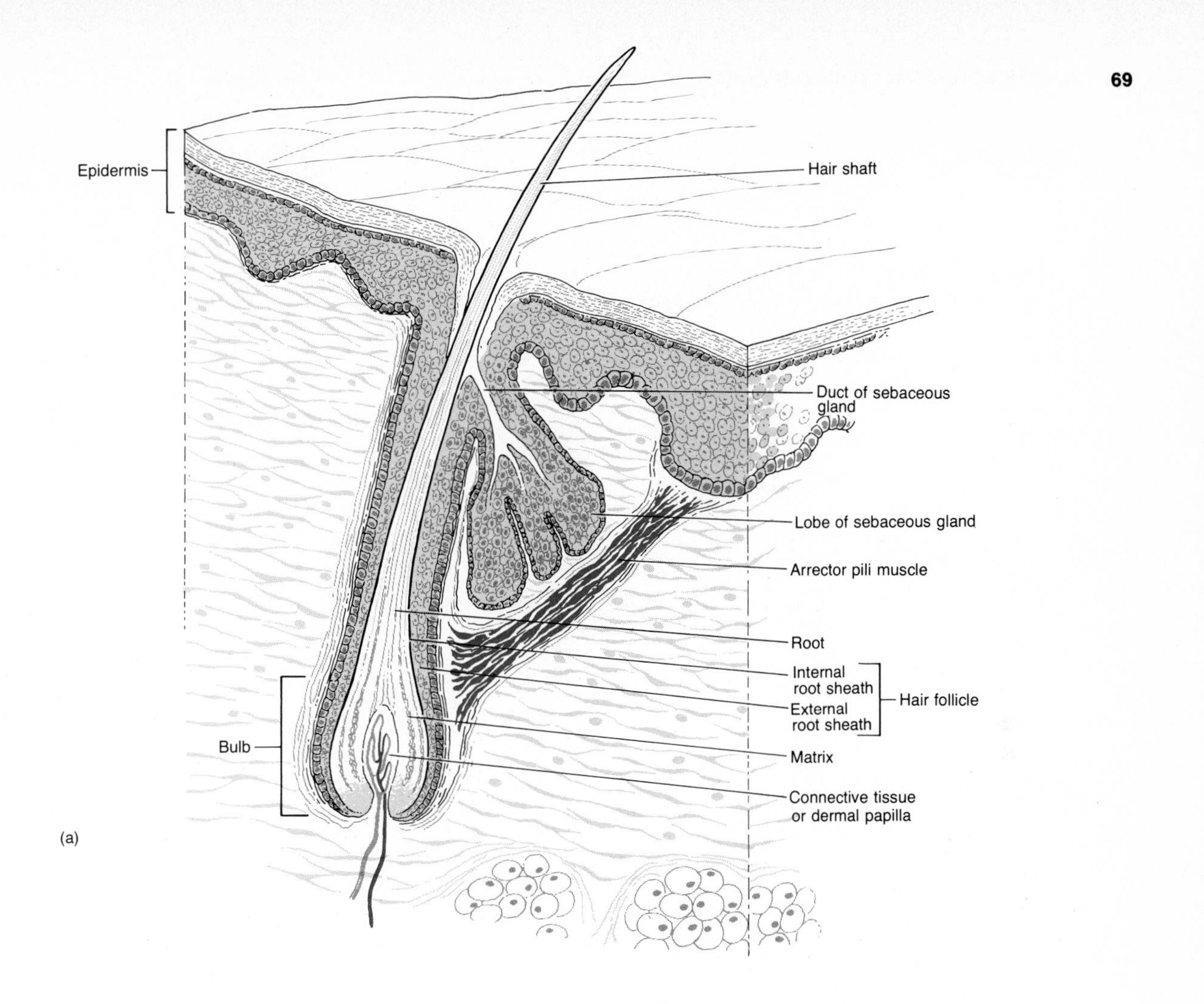

(a)

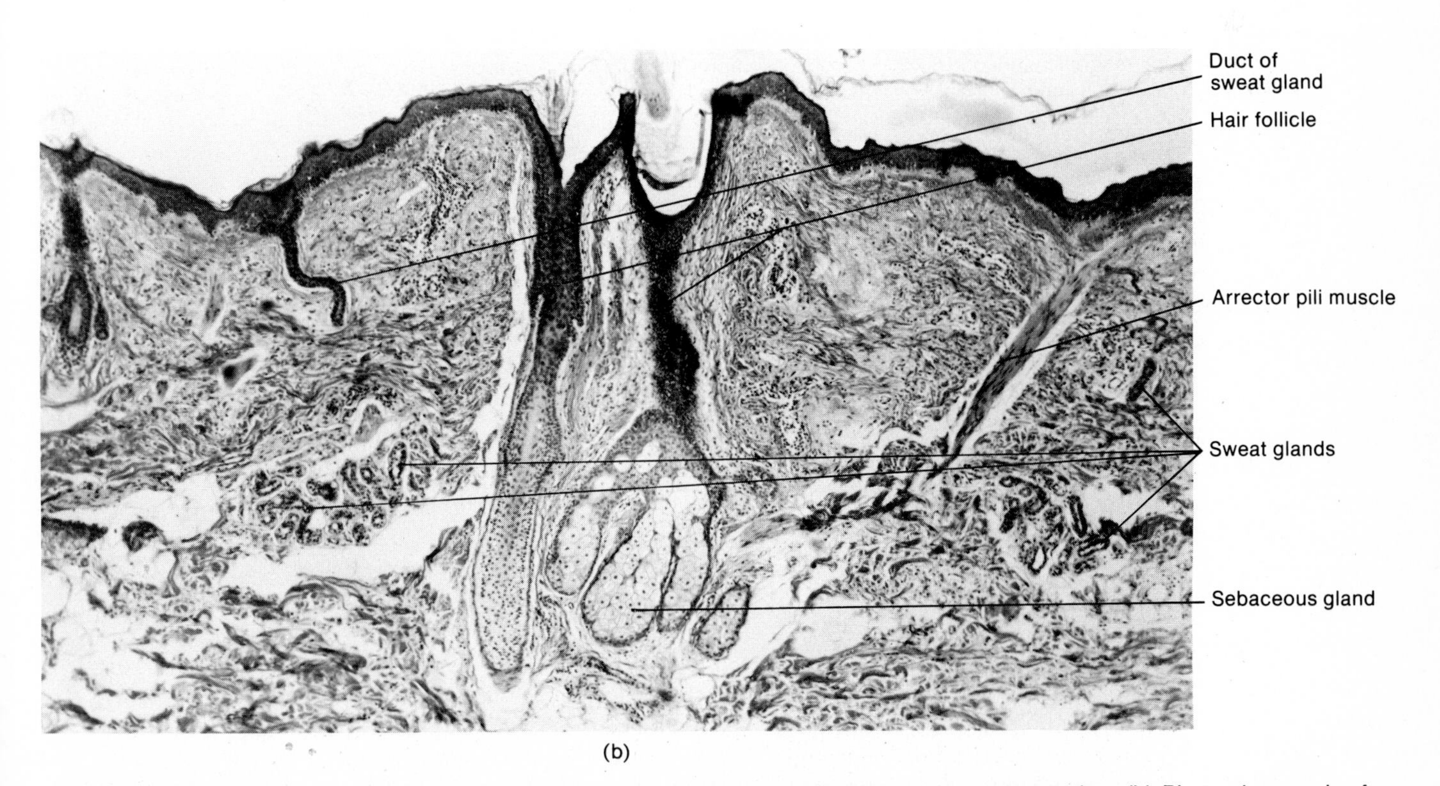

(b)

Figure 4–3. Principal parts of a hair and associated structures. (a) Diagrammatic representation. (b) Photomicrograph of a section from the skin of the face at a magnification of 65X.

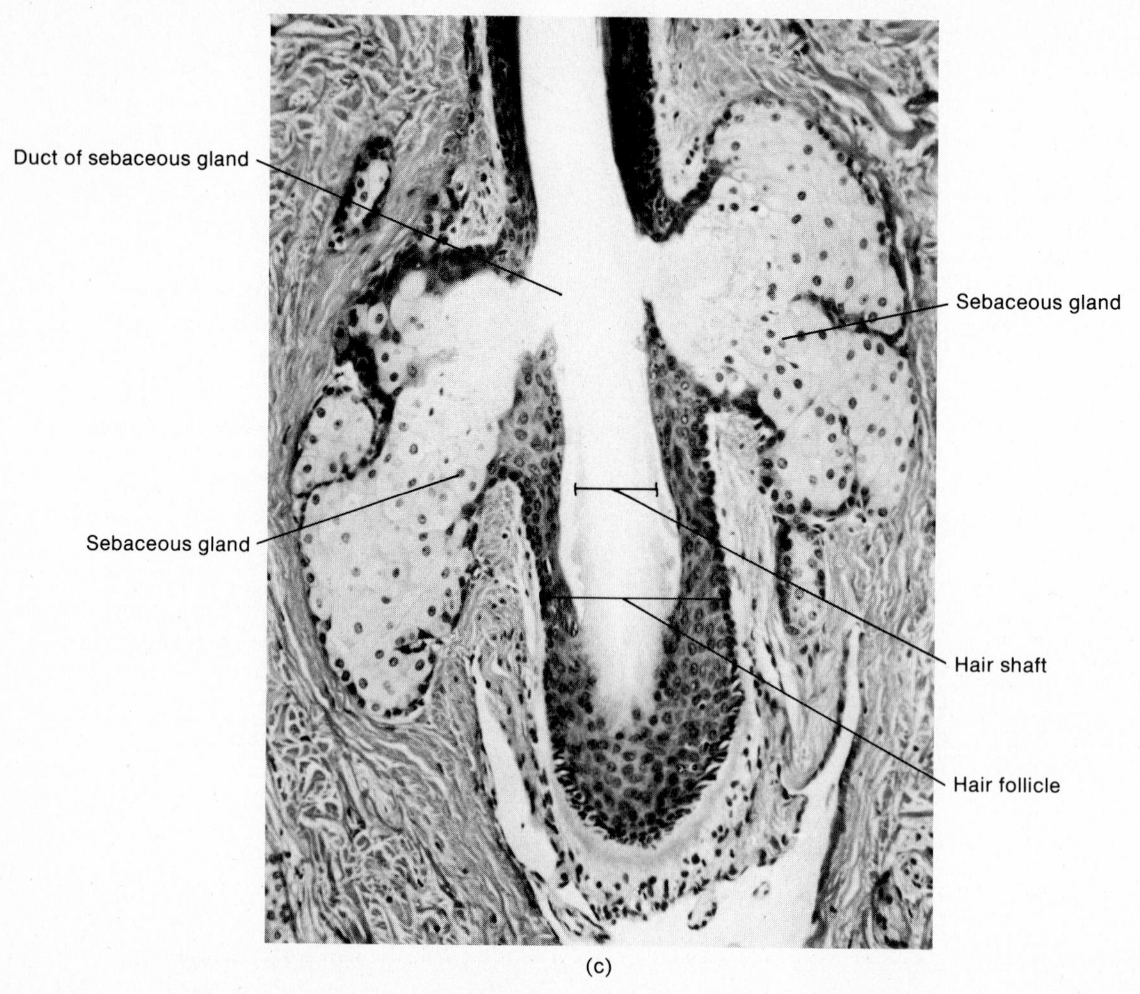

Figure 4–3 (cont.). Principal parts of a hair and associated structures. (c) Photomicrograph of two sebaceous glands opening into a hair follicle at a magnification of 160X. (Photomicrographs courtesy of Edward J. Reith, from *Atlas of Descriptive Histology,* by Edward J. Reith and Michael H. Ross, Harper & Row, Publishers, Inc., New York, 1970.)

Sebaceous glands, which will be discussed shortly, and bundles of smooth muscle are also associated with hair. These smooth muscles, called *arrector pili,* extend from the dermis of the skin to the side of the hair follicle. In its normal position hair is arranged at an angle to the surface of the skin. The arrector pili muscles, which are not voluntarily controlled, contract under stresses of fright and cold and pull the hairs into a more vertical position. This contraction results in "goosebumps" or "gooseflesh" because the skin around the shaft forms slight elevations. "Gooseflesh" traps heat between the skin and hair.

Glands

Two principal kinds of glands associated with the skin are sebaceous glands and sweat glands. **Sebaceous glands,** with only few exceptions, are connected to hair follicles. (See Figure 4–3.) They are simple branched acinar glands connected directly to the follicle by a short duct. These glands are absent in the palms and soles and differ in size and shape in other regions of the body. For example, they are fairly small in most areas of the trunk and extremities but are relatively large in the skin of the breasts, face, neck, and upper chest. The sebaceous glands secrete an oily substance called *sebum,* which is a mixture of fats, cholesterol, proteins, and inorganic salts. The functions of the sebaceous glands are to keep the hair from drying and becoming brittle and to form a protective film that prevents excessive evaporation of water from the skin. The sebum also keeps the skin soft and pliable. When sebaceous glands of the face become enlarged because of accumulated sebum, blackheads are produced. Since sebum is fine food for certain bacteria, this frequently results in pimples or boils. You might be interested to know that the color of blackheads is due to oxidized oil and not dirt.

Sweat (sudoriferous) glands are distributed throughout the skin except on the nail beds of the fingers and toes, margins of the lips, eardrum, and tips of the penis and clitoris. In contrast to sebaceous glands, sweat glands are most numerous in the skin of the palms and the soles. They are also found in abundance in the armpits and forehead. Each gland consists of a coiled end embedded in the subcutaneous tissue and a single tube that projects upward through the dermis and epidermis. This tube, actually the excretory duct, terminates in a pore at the surface of the epidermis. (See Figure 4–1.) The base of each sweat gland is surrounded by a network of small blood vessels. In the axillary region, sweat glands are one of the simple branched tubular type. In the remainder of the body, they are simple coiled tubular.

Perspiration, or *sweat,* is the substance produced by sweat glands. It is a mixture of water, salts (mostly NaCl), urea, uric acid, amino acids, ammonia, sugar, lactic acid, and ascorbic acid. Although perspiration helps eliminate waste materials from the body, its principal value is to help regulate body temperature.

Nails

Modified horny cells of the epidermis are referred to as **nails.** The cells form a clear, solid covering over the dorsal surfaces of the terminal portions of the fingers and toes. Each nail consists of a *nail body,* the attached visible portion of the nail; a *free edge,* the distal portion that overhangs the tip of the digit; and the *nail root,* the proximal portion of the nail that lies beneath a fold of skin (Figure 4–4). Most of the nail body is pink because of the underlying vascular tissue. The whitish portion of the body at its proximal end is a semilunar area called the *lunula.* It appears whitish because the vascular tissue underneath does not show through.

The fold of skin that extends around the proximal and lateral borders of the nail is known as the *nail fold,* and the skin beneath the nail constitutes the *nail bed.* The furrow between the two is the *nail groove.*

The *eponychium* ("cuticle") is a narrow band of epidermis that extends from the margin of the nail wall (lateral border), adhering to it. It occupies the proximal border of the nail and consists of stratum corneum. The thickened area of stratum corneum below the free edge of the nail is referred to as the *hyponychium.*

The epithelium of the posterior part of the nail bed is known as the *nail matrix.* Its function is to bring about the growth of nails. Essentially, what happens is that growth occurs by the transformation of the more superficial cells of the matrix into nail cells. In the process, the outer, harder layer is pushed forward over the stratum germinativum. The average growth in the length of fingernails is

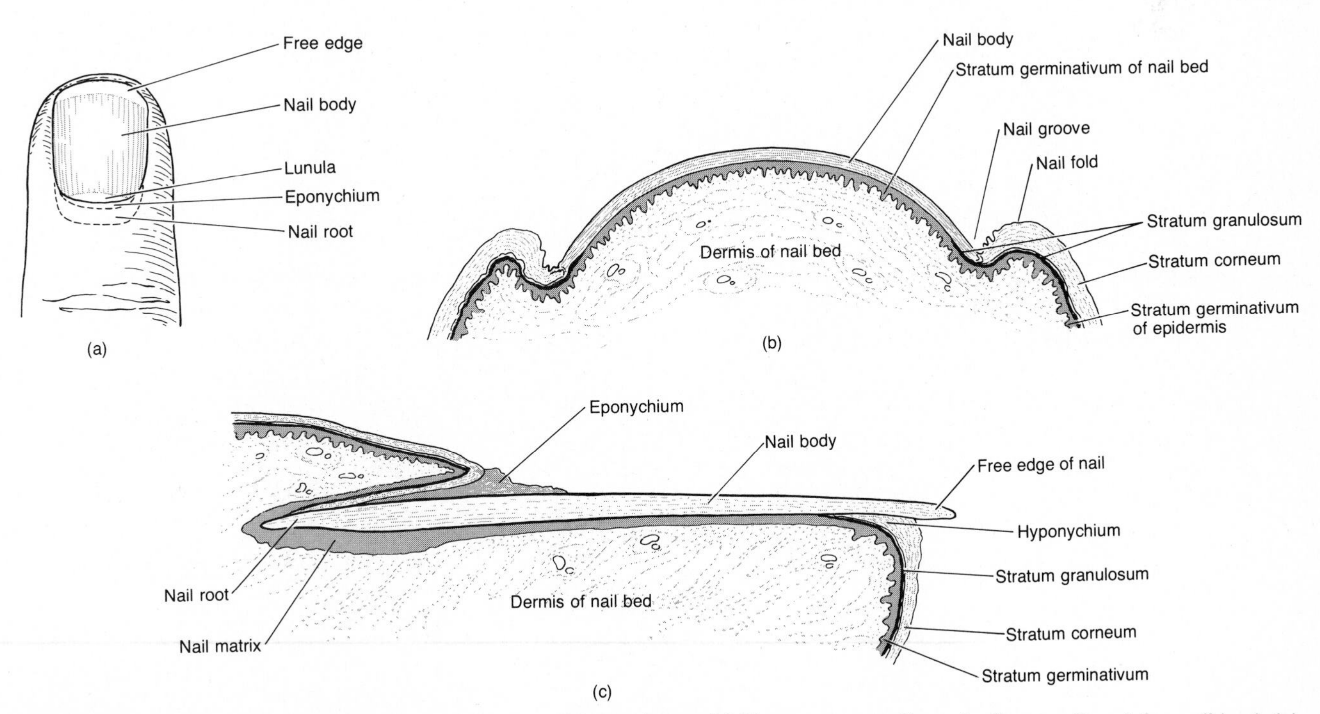

Figure 4–4. Structure of nails. (a) Fingernail viewed from above. (b) Transverse section of a fingernail and the nail bed. (c) Longitudinal section of a fingernail.

about 1 mm (0.0394 in.)/week. The growth rate is somewhat slower in toenails.

Just as the epidermis and dermis form the organ called the skin, the skin together with its accessory organs of hair, glands, and nails make up a system called the integumentary system. As a system, the skin and its derivatives protect underlying tissues from physical injury, harmful light rays, bacterial invasion, and drying out. The system receives stimuli from the environment, eliminates water and salts in the form of perspiration, and helps regulate body temperature.

APPLICATIONS TO HEALTH

Burns

Tissues may be damaged by thermal (heat), electrical, radioactive, or chemical agents. These agents can destroy the proteins in the exposed cells and thereby cause cell injury or death. Such damage results in a **burn.** The tissues that are directly or indirectly in contact with the environment, such as the skin or the linings of the respiratory and digestive tracts, are affected. Generally, however, the systemic effects of a burn are a greater threat to life than the local effects. Systemic effects are those that occur throughout the body—the term **systemic** referring to the whole body. **Local** effects are changes that occur in one area of the body. The systemic effects of burn include: (1) a large loss of water, plasma, and plasma proteins, which causes shock; (2) bacterial infection; (3) a slower circulation of blood; and (4) a decrease in urine production.

Burns are classified into three types: first-degree, second-degree, and third-degree. In *first-degree* burns, the damage is restricted to the epidermal layers of the skin, and symptoms are limited to local effects such as redness, tenderness, pain, and edema—the cardinal signs of inflammation. In *second-degree* burns, both the epidermal and portions of the dermal layer of the skin may be affected, but rapid regeneration of epithelium is still possible. Blisters containing elements of blood and lymph form on the skin surface or beneath the epidermis. Blisters beneath or within the epidermis are called *bullae.* In *third-degree* burns, both the epidermis and dermis are destroyed. The skin surface may be charred or white, or have patches of

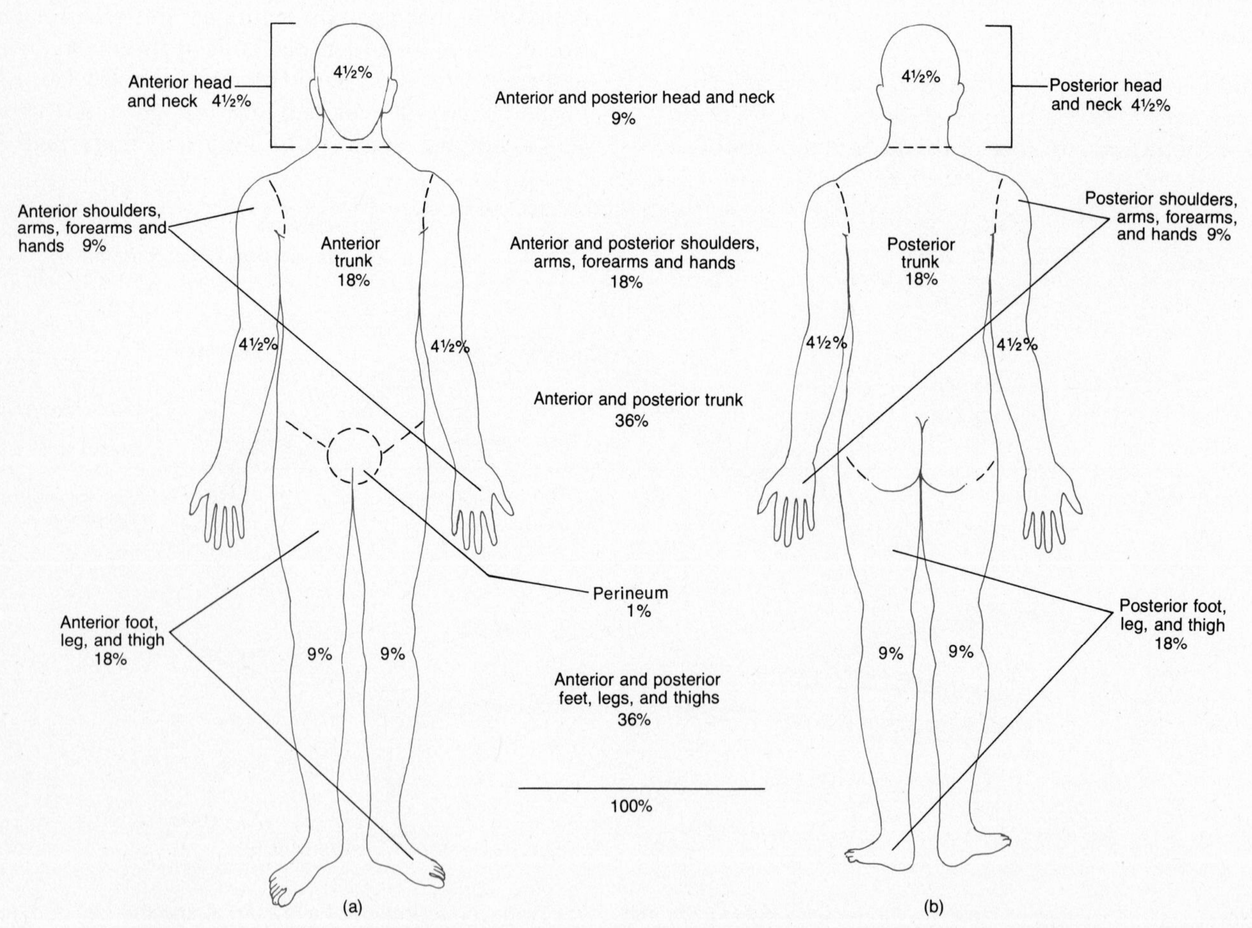

Figure 4–5. The "rule of nines" for estimating the extent of burns. (a) Anterior. (b) Posterior.

both. It is lifeless and insensitive to touch. Here, the regeneration of epithelium originates from the wound edges but it is a slow process. Even if skin grafting is started quickly, these wounds frequently contract and produce disfiguring or disabling scars.

A fairly accurate method for estimating the extent of a burn is to apply the *"rule of nines"*: (1) If the anterior and posterior surfaces of the head and neck are affected, the burn covers 9 percent of the body surface; (2) the anterior and posterior surfaces of each shoulder, arm, forearm, and hand also make up 9 percent of the body surface; (3) the anterior and posterior surfaces of the trunk including the buttocks make up 36 percent; (4) the anterior and posterior surfaces of each foot, leg, and thigh as far up as the buttocks make up 18 percent; and (5) the perineum makes up 1 percent. The perineum includes the anal region and the urogenital region (Figure 4–5).

Each chapter in this text which discusses a major system of the body will be followed by a glossary of key terms associated with that system. Both normal and pathological conditions of the system are included in these glossaries. You should familiarize yourself with the terms since they will play an essential role in your medical vocabulary.

Key terms associated with the integumentary system

Albinism (*alb* = white; *ism* = condition) Congenital (existing at birth) absence of pigment from the skin, hair, and parts of the eye.

Anhidrosis (*an* = without; *hidr* = sweating; *osis* = condition) Inability to sweat.

Callus An area of hardened and thickened skin that is usually seen in palms and soles and is due to pressure and friction.

Carbuncle A hard, round, deep, painful inflammation of the subcutaneous tissue that causes necrosis (deadness) and pus formation (abscess).

Comedo A collection of sebaceous material and dead cells in the hair follicle and excretory duct of the sebaceous gland. Usually found over the face, chest, and back, and more commonly during adolescence. Also called *blackhead* or *whitehead.*

Cyst (*cyst* = sac containing fluid) A sac with a distinct connective tissue wall, containing a fluid or other material.

Decubitus ulcer A bedsore. An ulcer formed due to continual pressure over the skin.

Epidermophytosis (*epi* = upon; *derm* = skin; *phyto* = pertaining to plants; *osis* = condition) Any fungus infection of the skin producing scaliness with itching. Called *athlete's foot* when it affects the feet.

Furuncle (boil) An abscess resulting from infection of a hair follicle.

Hypodermic (*hypo* = under; *derm* = skin) The area beneath the skin.

Intradermal (intra = within) Within the skin. Also called *intracutaneous.*

Melanoma (*melano* = dark colored; *oma* = tumor) A cancerous tumor consisting of melanocytes that produce skin pigment.

Nevi Round, pigmented, flat, or raised skin areas that may be present at birth or develop later. Varying in color from yellow-brown to black. Also called *moles* or *birthmarks.*

Nodule A large cluster of cells raised above the skin but extending deep into the tissues.

Papule A small round skin elevation varying in size from a pinpoint to that of a split pea. One example is a pimple.

Polyp A tumor on a stem found especially on mucous membranes.

Pustule A small round elevation of the skin containing pus.

Subcutaneous (*sub* = under; *cutis* = skin) Beneath the skin.

Wart A common, contagious, noncancerous epithelial tumor caused by a virus.

Chapter summary in outline

THE SKIN AS AN ORGAN

1. The skin and its accessory organs of hair, glands, and nails constitute the integumentary system.
2. The skin is the largest body organ. It performs the functions of protection, maintaining body temperature, picking up stimuli, and excretion.
3. The principal parts of the skin are the outer epidermis and inner dermis. The dermis overlies the subcutaneous layer.
4. The epidermal layers, from the inside outward, are the stratum basale, spinosum, granulosum, lucidum, and corneum. The basale and spinosum undergo continuous cell division and produce all other layers.
5. The dermis consists of a papillary region and a reticular region. The papillary region is connective tissue containing blood vessels, nerves, oil glands, hair follicles, and papillae. The reticular region is connective tissue containing fat and sweat glands.

EPIDERMAL DERIVATIVES

1. The epidermal derivatives of the skin are hair, sebaceous glands, sweat glands, and nails.
2. Hairs consist of a shaft above the surface, a root anchored in the dermis, and a hair follicle.
3. Sebaceous glands are connected to hair follicles. They secrete sebum, which moistens hair and waterproofs the skin.
4. Sweat glands produce perspiration, which carries wastes to the surface and helps regulate body temperature.
5. Nails are modified epidermal cells.

APPLICATIONS TO HEALTH

1. Skin burns are classified as first-degree (damage to the epidermis), second-degree (damage to both epidermis and dermis but rapid regeneration of epidermis is still possible), and third-degree (extreme damage to both epidermis and dermis).
2. The extent of burns can be found by using the "rule of nines."

Review questions and problems

1. Define an organ. In what respect is the skin an organ? What is the integumentary system?
2. List the principal functions of the skin.

3. Compare the structures of the epidermis and dermis. What is the subcutaneous layer?
4. List and describe the epidermal layers from the inside outward. What is the importance of each layer?
5. How is the dermis adapted to receive stimuli for touch, pressure, or pain?
6. Describe the structure of a hair. How are hairs moistened? What produces "goosebumps" or "gooseflesh"?
7. Contrast the locations and functions of sebaceous glands and sweat glands. What are the name and chemical components of the secretions of each?
8. From what layer of the skin do nails form? Describe the principal parts of a nail.
9. Define a burn. Classify burns according to degree. How is the "rule of nines" used? What steps may be employed in treating burns?
10. Refer to the glossary of key terms associated with the integumentary system. Be sure you can define each term.

Selected readings

Cunningham's Textbook of Anatomy. 11th ed., edited by G. J. Romanes. London: Oxford University Press, 1972. Pp. 795–803.

Gray, Henry. *Anatomy of the Human Body.* 29th ed., edited by Charles Mayo Goss. Philadelphia: Lea and Febiger, 1973. Chap. 14.

Montagna, William. *The Structure and Function of the Skin.* 2d ed. New York: Academic Press, 1962.

CHAPTER 5

OSSEOUS TISSUE

STUDENT OBJECTIVES

After you have read this chapter, you should be able to:

1. Describe the components of the skeletal system
2. Describe the functions of the skeletal system
3. Describe the gross features of a long bone
4. Describe the histological features of compact (dense) bone tissue
5. Compare the histological characteristics of spongy (cancellous) and compact (dense) bone
6. Define ossification
7. Contrast the steps involved in intramembranous and endochondral ossification
8. Interpret roentgenograms of normal ossification
9. Describe the processes of bone construction and destruction involved in bone replacement
10. Describe the conditions necessary for normal bone growth
11. Define rickets, osteomalacia, and scurvy as vitamin deficiency disorders
12. Contrast the causes and clinical symptoms associated with osteoporosis, Paget's disease, and osteomyelitis
13. Compare the origin of tumors of the skeletal system
14. Define key terms related to the skeletal system

Without the skeletal system, man would be little more than a slug. We would be unable to make coordinated movements, such as walking or grasping. The slightest knock on the head or chest could damage the brain or heart. It would even be impossible for us to chew our food. The framework of bones that protects our organs and allows us to move is called the **skeletal system.** In addition to bone, the skeletal system consists of cartilage, which is found in the nose, larynx, outer ear, and where bones attach to one another. The points at which the surfaces of bones attach to each other are called *joints*, or *articulations.*

The skeletal system performs several important, basic functions. First, it *supports* the soft tissues of the body so that the form of the body and an erect posture can be maintained. Second, the system *protects* delicate structures, such as the brain, the spinal cord, the lungs, the heart, and the major blood vessels in the thoracic cavity. Third, the bones serve as *levers* to which the muscles of the body are attached. When the muscles contract, the bones acting as levers produce *movement.* Fourth, the bones serve as *storage areas* for mineral salts—especially calcium and phosphorus. One last important feature of the skeletal system is *blood-cell production,* which occurs in the red marrow of the bones. This process is referred to as hematopoiesis.

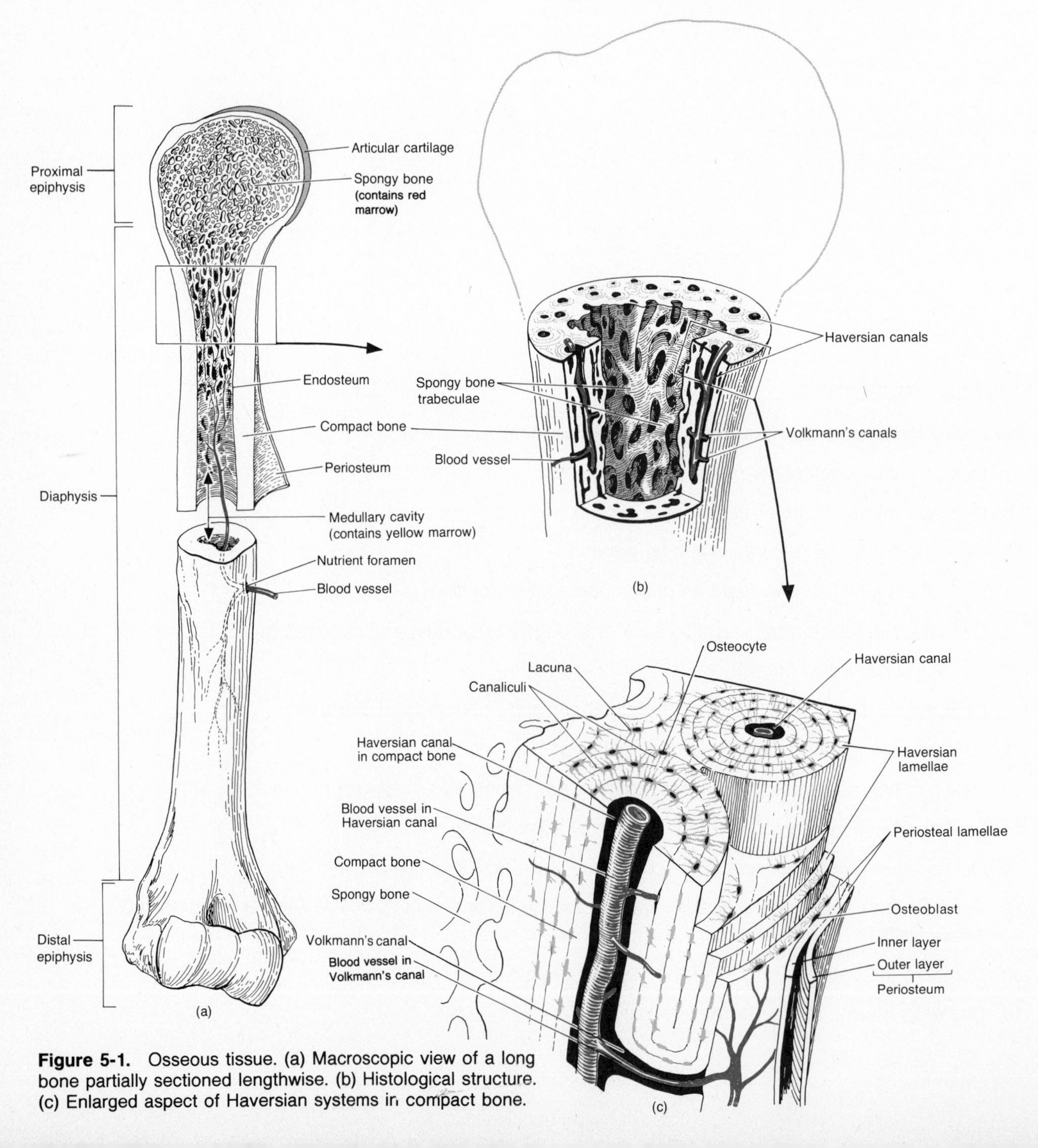

Figure 5-1. Osseous tissue. (a) Macroscopic view of a long bone partially sectioned lengthwise. (b) Histological structure. (c) Enlarged aspect of Haversian systems in compact bone.

A MICROSCOPIC VIEW

Structurally, the skeletal system consists of two types of connective tissue: cartilage and bone. In Chapter 3, we discussed the microscopic structure of cartilage. You may wish to review that discussion as you study the skeletal system. Here, our attention will be directed to discussing the microscopic structure of bone tissue.

Like other connective tissues, **bone,** or **osseous tissue,** contains a great deal of intercellular substance surrounding widely separated cells. Unlike other connective tissues, however, the intercellular substance of bone contains mineral salts, primarily calcium phosphate and calcium carbonate. These salts are responsible for the hardness of bone, which is thus said to be ossified. Embedded in the intercellular substance are collagenous fibers, which further reinforce the tissue.

The microscopic structure of bone may be analyzed by considering the anatomy of a long bone such as the humerus, or arm bone. As shown in Figure 5–1a, b, a typical long bone consists of the following parts: (1) the **diaphysis,** which is the shaft or main, long portion; (2) the **epiphyses,** which are the extremities or ends of the bone; (3) the **articular cartilage,** a thin layer of hyaline cartilage covering the epiphysis where the bone forms a joint with another bone; and (4) the **periosteum,** a dense, white, fibrous membrane covering the remaining surface of the bone. The periosteum, coming from the terms *peri,* meaning around, and *osteo,* meaning bone, consists of two layers. The outer, *fibrous layer* is composed of connective tissue containing blood vessels, lymphatic vessels, and nerves that run into the bone. The inner *osteogenic layer* contains elastic fibers, blood vessels, and **osteoblasts,** which are cells responsible for forming new bone during growth and repair (Figure 5–1c). The word *blast,* which is a part of many terms, means a germ or bud. It denotes an immature cell or tissue that develops into a more specialized form later on. The periosteum is essential for bone growth, repair, and nutrition. It also serves as a point of attachment for ligaments and tendons. A photomicrograph of the periosteum is shown in Figure 5–2. In addition to the four parts listed, a typical long bone also has: (5) a **medullary cavity,** or **marrow cavity,** which is the space within the diaphysis that contains the fatty *yellow marrow* in adults; and (6) an **endosteum,** which is a membrane that lines the medullary cavity and contains osteoclasts. These are cells that may assume a role in the resorption of bone.

Despite its macroscopic appearance, bone is not a solid, homogeneous substance. In fact, all bone is porous. In other words, it is full of pores or holes. As you will see, the pores contain living cells and blood vessels that supply the cells with nutrients. The pores also make bones lighter. Think of how much more energy you would expend if you had to

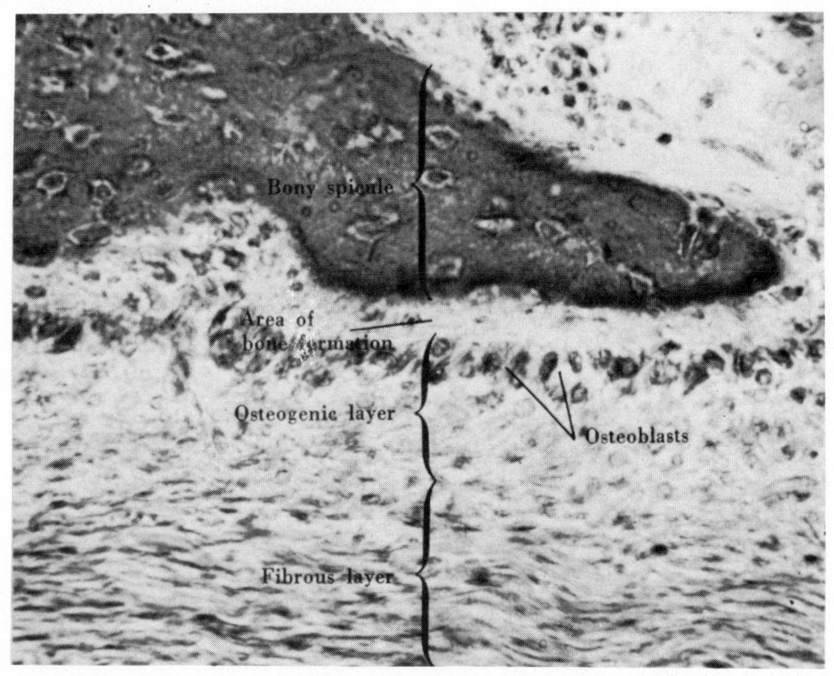

Figure 5-2. Photomicrograph of compact bone showing the osteogenic and fibrous layers of the periosteum at a magnification of 450X. (Courtesy of Donald I. Patt, from *Comparative Vertebrate Histology,* by Donald I. Patt and Gail R. Patt, Harper & Row, Publishers, Inc., New York, 1969.)

drag around solid bones. Depending upon the degree of porosity, the regions of a bone may be categorized as spongy or compact (Figure 5–1b). **Spongy,** or **cancellous,** bone tissue contains many large spaces filled with marrow. It makes up most of the bone tissue of short, flat, and irregularly shaped bones, and of the epiphyses of long bones. Spongy bone tissue not only makes bones lighter but also provides a storage area for marrow. **Compact,** or **dense,** bone tissue, by contrast, contains fewer spaces. It is deposited in a thin layer over the spongy bone tissue. It also composes most of the bone tissue of the diaphysis of long bones. Compact bone tissue provides protection and support and helps the long bones of the body resist the stress of weight that is placed on them.

We can compare the differences between spongy and compact bone tissue by looking at the highly magnified, transverse sections represented in Figure 5–1. One main difference between the bone tissues is that compact bone has a concentric-ring structure, whereas spongy bone does not. Blood vessels and nerves from the periosteum penetrate into the compact bone through *Volkmann's canals* (Figure 5–1c). The blood vessels of these canals connect with blood vessels and nerves of the medullary cavity and of the *Haversian canals.* The Haversian canals are circular cavities that run longitudinally through the bone. Around the Haversian canals are concentric rings of hard, calcified, intercellular substance. The concentric rings of this substance are called *lamellae.* Between the lamellae are small spaces called *lacunae,* where osteocytes are found. **Osteocytes** are mature osteoblasts that have lost their ability to produce new bone tissue. Radiating out in all directions from the lacunae are minute canals called *canaliculi,* which connect with other lacunae and, eventually, with the Haversian canal. Thus, an intricate network is formed throughout the bone. This branching network provides numerous routes for blood vessels so that nutrients can reach the osteocytes and wastes

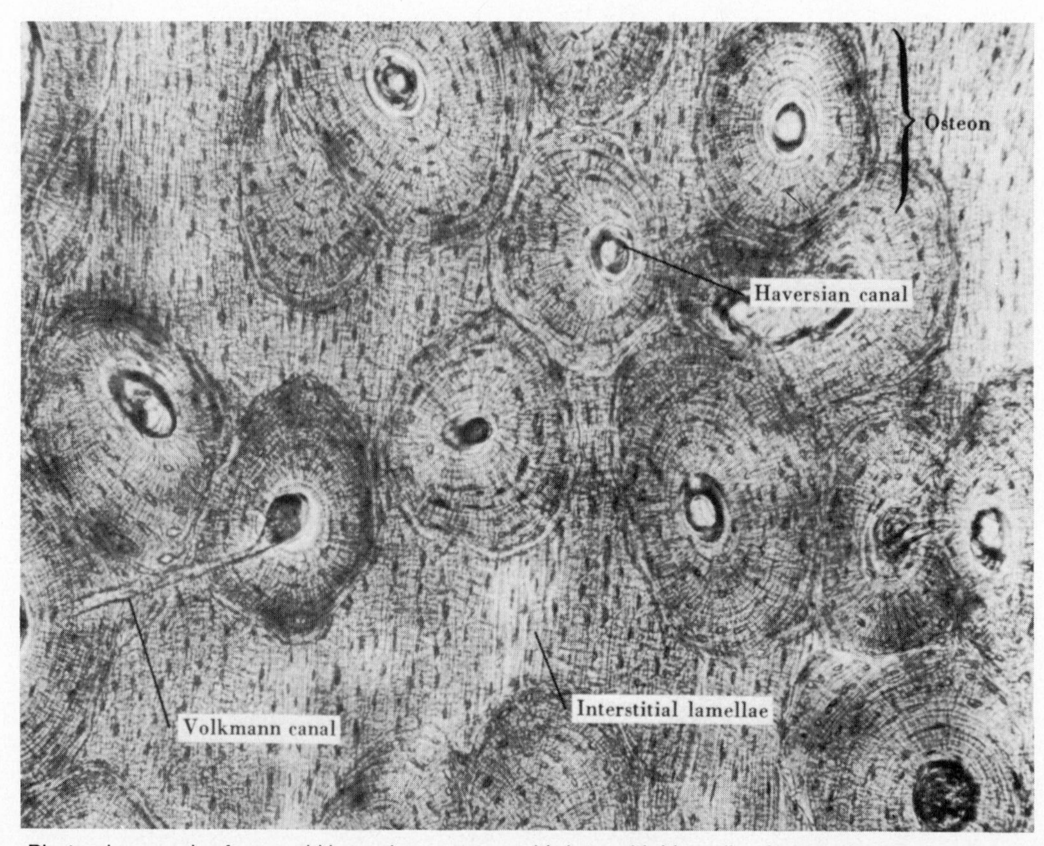

Figure 5-3. Photomicrograph of several Haversian systems with interstitial lamellae between systems at a magnification of 150X. If you look carefully, you can see the canaliculi as fine lines running radially, like the spokes of a wheel, and in circles within the Haversian systems. (Courtesy of Donald I. Patt, from *Comparative Vertebrate Histology,* by Donald I. Patt and Gail R. Patt, Harper & Row, Publishers, Inc., New York, 1969.)

can be removed. Each Haversian canal, with its surrounding lamellae, lacunae, osteocytes, and canaliculi, is called a **Haversian system** or **osteon.** The areas between Haversian systems contain *interstitial lamellae.* These also possess lacunae with osteocytes and canaliculi, but their lamellae are usually not connected to the Haversian systems. A photomicrograph of several Haversian systems is shown in Figure 5–3.

In contrast to compact bone, spongy bone does not contain Haversian systems. It consists of an irregular latticework of thin plates of bone called *trabeculae.* The spaces between the trabeculae of some bones are filled with *red marrow.* Within the trabeculae lie the small spaces called lacunae, which contain the osteocytes. Blood vessels from the periosteum penetrate into the spongy bone, and the osteocytes in the trabeculae receive their nourishment directly from the blood circulating through the red marrow cavities. The cells of red marrow are responsible for producing new blood cells. They belong to the circulatory system and will be described in a later chapter.

When most people think of bones, they visualize some sort of very hard, white material that has a variety of shapes, yet is pretty much the same in all people. However, if you consider the skeleton of an infant or a young child, certain differences become apparent. For example, it is common knowledge that it is very dangerous to drop an infant, especially on its head. This is because its bones are "soft," and the fall may change the shape of its head or damage its brain. Moreover, most of us know that the bones of a child are softer than those of an adult. The final shape and hardness of adult bone take many years to develop and are dependent upon a rather complex series of chemical changes. Let us now briefly examine how bones are formed and how they grow.

BONE FORMATION AND GROWTH

The process by which bone forms in the body is called **ossification.** The "skeleton" of a human embryo is composed of fibrous membranes and hyaline cartilage, both of which are shaped like bones and provide the medium in which ossification occurs (Figure 5–4). The actual process of ossification begins around the sixth week of embryonic life and continues until an individual reaches adulthood. Two kinds of bone formation are recognized. The first of these is called *intramembranous ossification.* This refers to the formation of bone directly on or within the fibrous membranes—thus, the terms *intra* meaning within, and *membranous* meaning membrane. The second kind, *endochondral (intracartilaginous) ossification,* refers to the formation of bone in cartilage. The term *endo* means within, and *chondro* means cartilage. These two kinds of ossification do *not* lead to any differences in the structure of mature bones. They simply indicate different methods of bone formation.

The first stage in the development of any bone is the coming together of embryonic connective tissue cells called mesenchymal cells, in the area where bone formation is about to begin. Next, the area is infiltrated by many small blood vessels. Soon thereafter, these cells increase in number and size. In some skeletal structures they become transformed into chondroblasts, and in other skeletal structures some become osteoblasts. The chondroblasts will be responsible for cartilage formation, whereas the osteoblasts will form bone tissue by either intramembranous or endochondral ossification.

Intramembranous ossification

Of the two types of bone formation, the simpler and more direct is **intramembranous ossification.** The flat bones of the roof of the skull, mandible (lower jawbone), and probably part of the clavicles are formed by intramembranous ossification. The essentials of this process are as follows.

Osteoblasts formed from mesenchymal cells cluster in the fibrous membrane. The site of such a cluster is called a *center of ossification.* The osteoblasts then secrete intercellular substances. These substances are partly composed of collagenous fibers that form a framework, or matrix, in which calcium salts are quickly deposited. The laying down of calcium salts is called *calcification.* When a cluster of osteoblasts is completely surrounded by the calcified matrix, it is called a *trabecula.* As trabeculae form in nearby ossification centers, they fuse into the open latticework characteristic of spongy bone. With the formation of successive layers of bone, some of the osteoblasts become entrapped in the minute spaces referred to as lacunae. The entrapped osteoblasts lose their ability to form bone and are called osteocytes. The spaces between the trabeculae fill with red marrow. The original connective tissue that surrounds the growing mass of bone then becomes the periosteum. The ossified area has now become true spongy bone. Eventually, the surface layers of the spongy bone will be reconstructed into compact bone. Much of this newly formed bone will be destroyed and reformed so that the bone may reach its final adult size and shape.

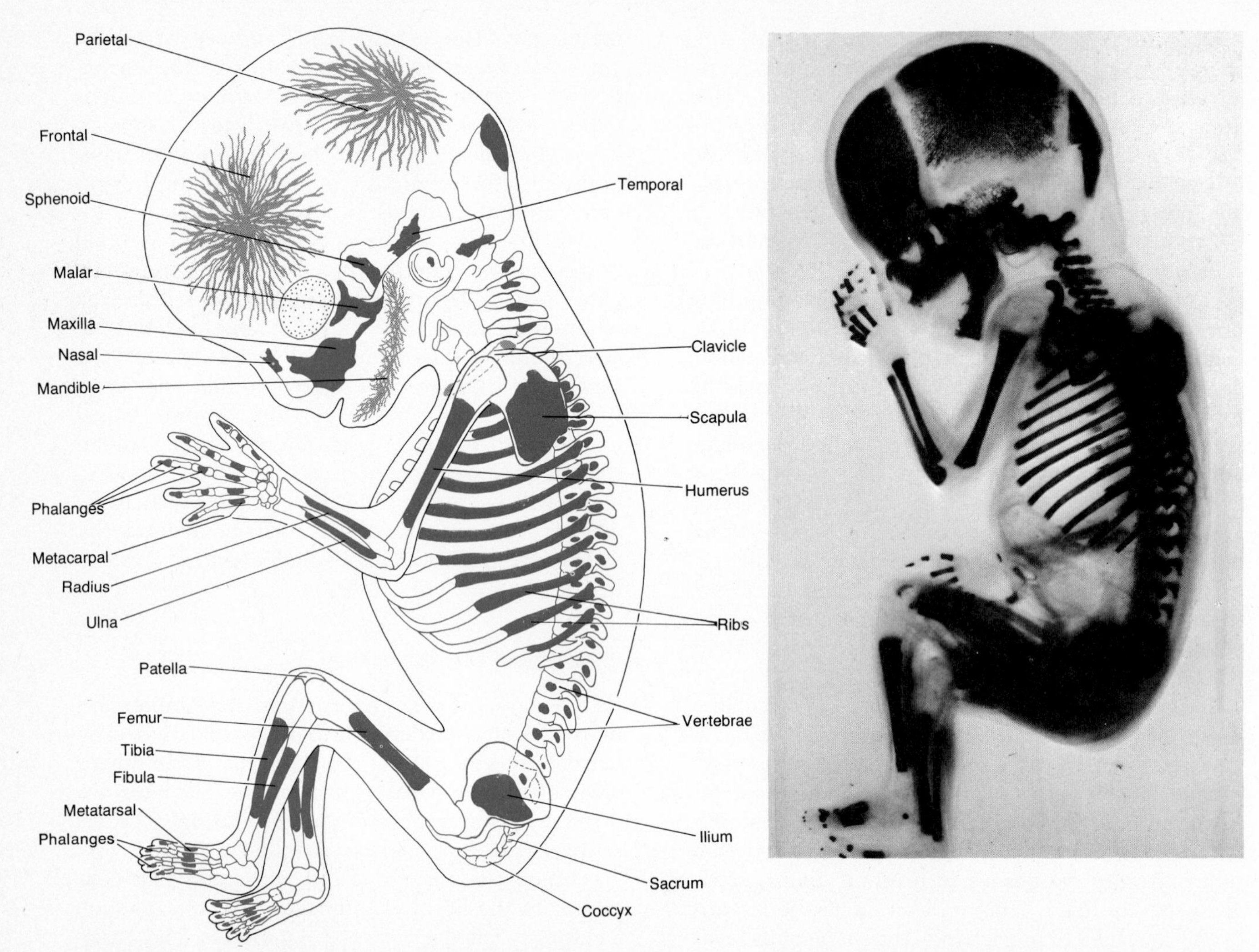

Figure 5-4. Diagrammatic representation of the 10-week-old human embryo shown in the photograph. Colored bones are those that develop by endochondral ossification. Gray bones are those that develop by intramembranous ossification. (Photograph courtesy of Carolina Biological Supply Company.)

Endochondral ossification

The replacement of cartilage by bone is called **endochondral ossification.** Most bones of the body, including the majority of the bones of the skull, are formed by endochondral ossification. Since this type of ossification is best observed in a long bone, we shall illustrate how it occurs in the tibia or shinbone (Figure 5–5).

Early in embryonic life, a cartilage model that more or less conforms to the shape of the future bone is laid down (Figure 5–5a). The cartilage model is covered by a membrane called the *perichondrium.* Midway along the shaft of this model, cells in the internal layer of the perichondrium enlarge and become osteoblasts. The cells begin to form a collar of spongy bone around the middle of the diaphysis of the cartilage model (Figure 5–5b). Once the perichondrium starts to form bone, it is called the periosteum. Simultaneously with the appearance of the bone collar, changes occur in the cartilage in the center of the diaphysis. In this area, the *primary ossification center* (Figure 5–5c), cartilage cells hypertrophy (increase in size). The cartilage cells hypertrophy because they accumulate glycogen for energy and produce enzymes to catalyze future chemical reactions. When the cartilage cells begin to secrete the enzyme, the intercellular substance becomes *calcified*–that is, minerals are deposited within it. Once the cartilage becomes calcified, nutritive materials required by the cartilage cells are no longer able to diffuse through the intercellular substance, and the cartilage cells die. As a result of the death of the cartilage cells, the intercellular substance begins to degenerate, leaving large cavities in the cartilage model. Blood vessels then grow along the spaces where cartilage cells were (Figure 5–5d) and open up the cavities further. Gradually, these spaces in the middle of the shaft connect with each other, and the marrow cavity is formed (Figure 5–5e).

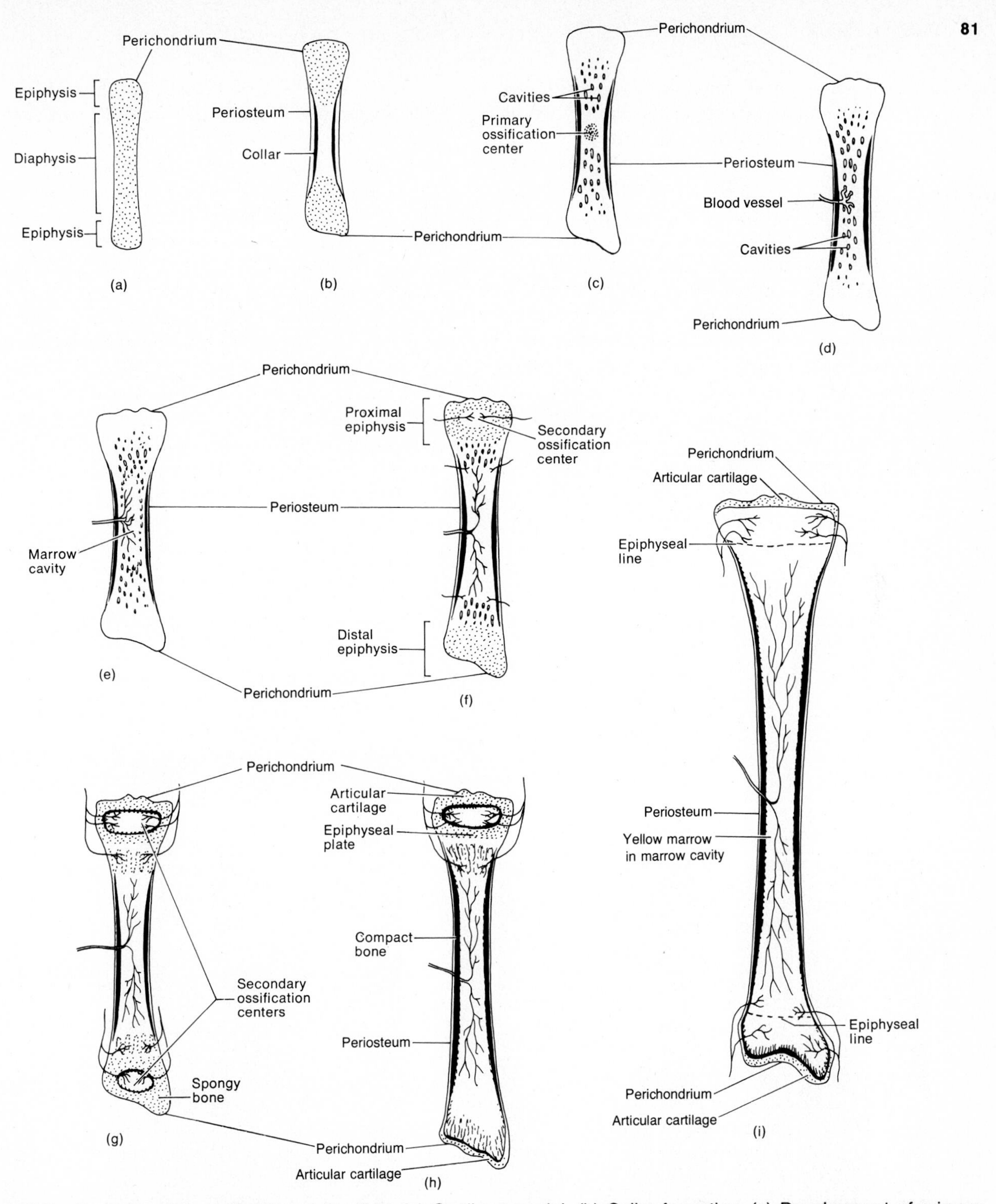

Figure 5-5. Endochondral ossification of the tibia. (a) Cartilage model. (b) Collar formation. (c) Development of primary ossification center. (d) Entrance of blood vessels. (e) Marrow-cavity formation. (f) Thickening and lengthening of the collar. (g) Formation of secondary ossification centers. (h) Remains of cartilage as the articular cartilage and epiphyseal plate. (i) Formation of the epiphyseal line.

As all these developmental changes are occurring, the osteoblasts of the periosteum deposit successive layers of bone on the outside so that the collar thickens, becoming thickest at the diaphysis (Figure 5–5f). The cartilage model continues to grow at its ends so that it steadily increases in length. Eventually, secondary ossification centers appear in the epiphyses and also lay down spongy bone. In the tibia, one secondary ossification center develops in the proximal epiphysis soon after birth. The other center develops in the distal epiphysis during the child's second year (Figure 5–5f, g).

After the two secondary ossification centers have formed, bone tissue has completely replaced cartilage, except in two regions. Cartilage continues to cover the articular surfaces of the epiphyses, in which case it is called articular cartilage. It also remains as a plate between the epiphysis and diaphysis, in which case it is called the **epiphyseal plate** (Figure 5–5h). The function of the epiphyseal plate is to allow the bone to increase in length until early adulthood. As the child grows, cartilage cells are produced by mitosis on the epiphyseal side of the plate. Cells are then destroyed and replaced by bone on the diaphyseal side of the plate. In this way, the thickness of the epiphyseal plate remains fairly constant, but the bone on the diaphyseal side increases in length. Growth in diameter occurs along with growth in length. In this process, the bone lining the marrow cavity is destroyed so that the cavity increases in diameter. At the same time, osteoblasts from the periosteum add new osseous tissue around the outside of the bone. Initially, diaphyseal and epiphyseal ossification produce all spongy bone. Later, by reconstruction, the outer region of spongy bone is reorganized into compact bone.

Around the age of 17, the epiphyseal cartilage cells stop multiplying, and the entire cartilage begins to be replaced by bone. Bone growth now stops. The remnant of the epiphyseal plate is called the **epiphyseal line** (Figure 5–5i). Ossification of all bones is usually completed by age 25. Figure 5–6 consists of a series of **roentgenograms,** or photographs taken with x-rays, that show ossification in the epiphyses of two long bones at the knee. Bones undergoing either intramembranous or endochondral ossification are continually remodeling themselves from the time that initial calcification occurs until the final structure appears. Compact bone is made by the reworking of spongy bone. The diameter of a long bone is increased by the destruction of the bone closest to the marrow cavity and the construction of new bone around the outside of the diaphysis. However, even after bones have reached their adult shapes and sizes, old bone is perpetually destroyed, and new osseous tissue is laid down in its place. We can now take a look at how an equilibrium is reached between the growth and destruction of adult bone.

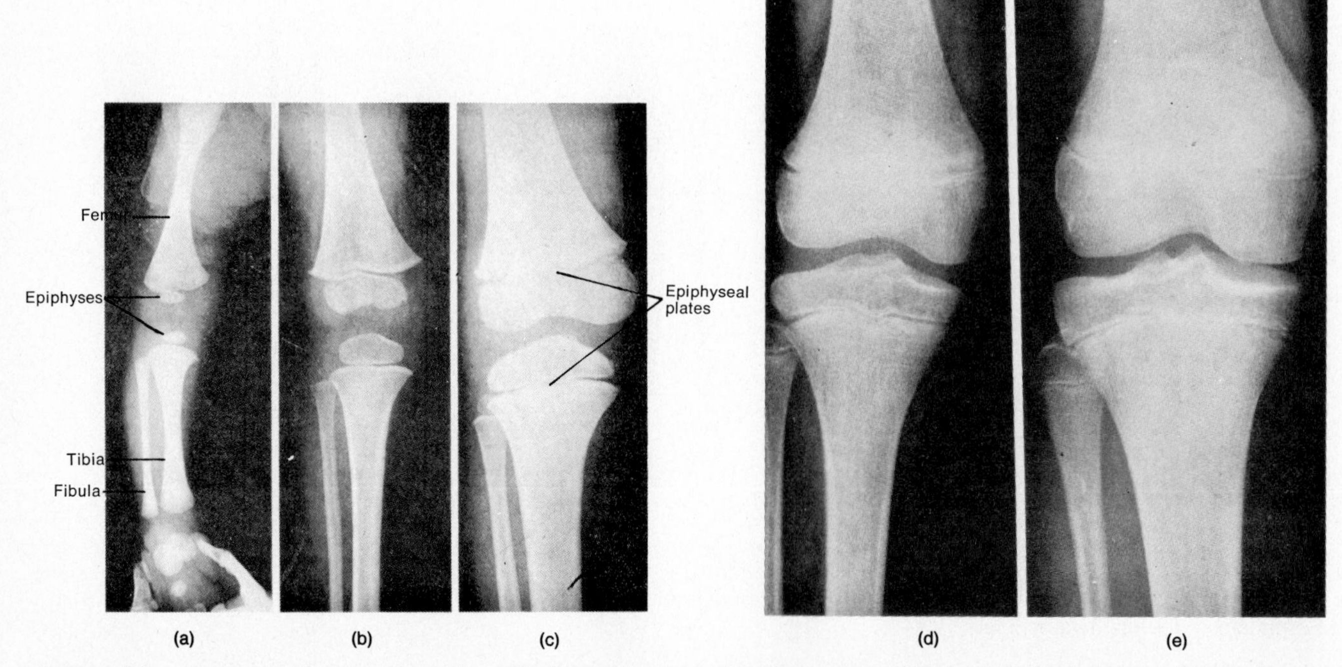

Figure 5-6. Roentgenograms of normal ossification at the knee. (a) One-month-old infant. The epiphyses at the knee are mostly cartilage. Ossification centers have formed for the femur and tibia. The space between the ends of the bones is occupied by epiphyseal cartilage. (b) Two-year-old child. Centers of ossification have grown. The white transverse zones marking the ends of the shafts are areas where mineral salts are deposited temporarily around the degenerating cartilage cells. (c) Five-year-old child. The epiphyses have assumed the shape of adult bones. The epiphyseal plates are clearly visible between the epiphyses and diaphyses of all three bones. (d) Eight-year-old child. The epiphyseal plates are still distinct as ossification continues. (e) Twelve-year-old child. The epiphyses have ossified almost completely. The epiphyseal plates are assuming the character of epiphyseal lines. (Courtesy of Lester W. Paul and John H. Juhl, *The Essentials of Roentgen Interpretation,* 3d ed., Harper & Row, Publishers, Inc., New York, 1972.)

BONE REPLACEMENT

Bone shares with the skin the unique feature of replacing itself throughout adult life. This remodeling of adult bone takes place at different rates in different parts of the body. For instance, the distal area of the femur (thighbone) replaces itself about every 4 months. By contrast, bone in certain areas of the shaft will not go through a complete turnover in the individual's life. Remodeling allows worn or injured bone to be removed and replaced with new tissue. It also allows bone to serve as the body's storage area for calcium. Many other tissues in the body need small amounts of calcium in order to perform their functions. For example, muscle needs calcium in order to contract. The muscle cells take their calcium from the blood. However, the blood itself needs calcium in order to clot. The blood continually trades off calcium with the bones, removing calcium when other tissues are not receiving enough of this element and resupplying the bones with calcium to keep them from getting soft.

The cells believed to be responsible for the destruction of bone tissue are called **osteoclasts,** the term *clast* coming from *klan,* meaning to break. In the healthy adult, a delicate homeostasis is maintained between the action of the osteoclasts and the removal of calcium, on the one hand, and the action of the bone-making osteoblasts and the deposition of calcium on the other. Should too much new tissue be laid down, the bones will become abnormally thick and heavy. If too much calcium is deposited in the bone, the surplus may form thick bumps, or spurs, on the bone that interfere with movement at joints. A loss of too much tissue or calcium weakens the bones and makes them break easily.

Normal bone growth in the young and bone replacement in the adult depend upon several factors. First, sufficient quantities of calcium and phosphorus, components of the chief salt that makes bone hard, must be included in the diet. Second, the individual must obtain sufficient amounts of vitamins A, C, and D. These substances are particularly responsible for the proper utilization of calcium and phosphorus by the body. Third, the body must manufacture the right amounts of the hormones responsible for bone-tissue activity.

Hormones regulate the growth and remodeling of bones. Certain hormones are responsible for the general growth of bones. Too much or too little of these hormones during childhood causes the person to be abnormally tall or short. Other hormones specialize in regulating the osteoclasts. And still others, especially the sex hormones, aid osteoblastic activity and thus promote the growth of new bone. The sex hormones act as a double-edged sword. They aid in the growth of new bone, but they also bring about the degeneration of all the cartilage cells in the epiphyseal plates. Due to the sex hormones, the typical adolescent goes through a spurt of growth around puberty, when sex-hormone production starts to increase. The individual then quickly finishes his or her growth as the epiphyseal cartilage disappears. Premature puberty can actually prevent the individual from reaching an average adult height because it brings with it premature degeneration of the plates. Still another kind of hormone, which is produced by the parathyroid glands in the neck, determines whether the blood will deposit calcium and phosphorus in osseous tissue or whether it will remove these elements from the bones. The names and activities of these hormones probably do not mean much to you now. For the time being, simply remember that hormones control bone growth and maintenance. In a later chapter, how the hormones accomplish these tasks will be explained.

APPLICATIONS TO HEALTH

Vitamin deficiencies

Vitamin D is important to normal bone growth and maintenance. Vitamin D is essential for the synthesis of a protein that transports the calcium obtained from foods across the lining of the intestine and into the extracellular fluid. When the body lacks this vitamin, it is unable to absorb calcium and phosphorus from foods. A deficiency of vitamin D produces rickets in children and a condition called osteomalacia in adults.

In the condition called **rickets,** epiphyseal cartilage cells cease to degenerate, and new cartilage continues to be produced. This causes the width of the epiphyseal cartilage to become wider than normal. At the same time, the soft matrix laid down by the osteoblasts in the diaphysis fails to calcify. As a result, the bones stay soft. When the child walks, the weight of the body causes the bones in the legs to bow. Malformations of the head, chest, and pelvis also occur.

The cure and prevention of rickets consists of adding generous amounts of calcium, phosphorus, and vitamin D to the diet. Exposing the skin to the ultraviolet rays of sunlight also helps the body manufacture additional vitamin D.

A deficiency of vitamin D in an adult means that the bones give up excessive amounts of calcium and phosphorus. This loss, called *demineralization,* is especially heavy in the bones of the pelvis, legs, and spine. Demineralization caused by vitamin D deficiency is called **osteomalacia,** the term *malacia* meaning softness. After the bones soften, the weight of the body produces a bowing of the leg bones, a shortening of the backbone, and a flattening of the pelvic bones. Osteomalacia mainly affects women who live on poor cereal diets devoid of milk, are seldom exposed to the sun, and have repeated pregnancies that deplete the body of calcium. The condition responds to the same treatment as rickets: if the disease is severe enough and threatens life, large doses of vitamin D should be given.

Osteoblast and osteoclast regulation

Osteoporosis is a bone disorder affecting middle-aged and elderly people. The majority of those affected are women over 60 years old. Between puberty and the middle years, the sex hormones maintain strong, healthy, osseous tissue by stimulating the osteoblasts to make new bone. However, after the menopause or "change of life," women produce less of the sex hormones. During old age, both men and women produce smaller amounts of these hormones. As a result, the osteoblasts become less active, and the bones become less compact and more porous—hence, the name of the disorder. Most elderly people suffer from some degree of osteoporosis. But it is apt to be most severe in women who experience a drastic cutback in female hormones after the menopause and in men and women who do not have enough calcium in their diets. Osteoporosis most often affects the spine. The femur is also a frequently affected area. As the spine fails to maintain itself, the individual shrinks in height and may develop a "dowager's hump." The porous bones also become too weak to stand much strain and fracture easily. Most physicians suggest that middle-aged and elderly people include more calcium in their diets, as well as increased amounts of vitamin A, C, and D. Hormone therapy is also sometimes a part of treatment or prevention.

Another disorder, called **Paget's disease,** is characterized by an irregular thickening and softening of the bones. It is rarely seen in people under the age of 50. The cause, or *etiology,* of the disease is unknown. But the bone-producing osteoblasts and the bone-destroying osteoclasts apparently become uncoordinated. In some areas, too much bone is produced, whereas in other areas too much old bone is removed. The balance between bone formation and bone destruction is altered. Paget's disease affects the skull, the pelvis, and the bones of the extremities. Very little can be done to alter the course of the disease.

Stress of infection

Osteomyelitis is the term that includes all the infectious diseases of bone. These diseases may be localized, or they may affect many bones. The infections may also involve the periosteum, marrow, and cartilage. Quite a few microorganisms may give rise to bone infection, but the most frequent are bacteria known as *Staphylococcus aureus,* which are commonly called "staph." These bacteria may reach the bone by various means: the bloodstream; an injury such as a fracture; or an infection, such as a soft-tissue abscess, a sinus infection, or an abscess of the tooth.

Before antibiotic treatment became available, osteomyelitis often became a long-lasting condition. It can destroy extensive areas of bone, spread to nearby joints, and, in rare cases, lead to death by producing abscesses in many parts of the body. Penicillin and other antibiotics have been effective in treating the disease and in preventing it from spreading through extensive areas of the bone.

Cancer of bone

When cells in some area of the body reproduce unusually quickly so that an excess of tissue develops, the excess is called a growth, or **tumor.** Tumors may be cancerous and fatal, or they can be quite harmless. A cancerous growth is called a **malignant tumor,** or **malignancy.** A noncancerous growth is called a **benign growth.**

The cells of malignant growths all have one thing in common: they reproduce continuously and often very quickly. This growth continues until the victim dies or until every malignant cell is removed through surgery or destroyed by other means. As the cancer grows, it runs out of room and begins to compete with normal tissues for space and nutrients. Eventually, the normal tissues will lose the fight and die. The organ will function less and less efficiently, until it finally ceases to function altogether. The cancer cells may also spread to regions of the body that are far away from the original, or *primary,* growth. Cancer of the breast, for instance, has a tendency to spread to the lungs. The spread of the cancer to other regions of the body is called **metastasis.** Metastasis occurs when a malignant cell breaks away from the growth, enters the bloodstream, and is carried through the body. Wherever the cell finally comes to rest, it will set up another tumor, called a *secondary* growth. Usually death is caused when a vital organ loses its fight against the cancer cells for room and nutrients. Pain develops when the growth impinges on nerves or blocks a passageway so that secretions build up pressure behind the blockage.

The name of the cancer is derived from the type of tissue in which it originally develops. *Sarcoma* is a general term for any cancer arising from connective tissue. For example, *osteogenic sarcomas* are malignant growths of osteoblasts. *Osteo,* of course, means bone; *genic* comes from the same words as *genesis* and *generate,* meaning origin. This is the most frequent type of childhood cancer, but it is rare in adults. Osteogenic sarcomas destroy normal bone tissue and eventually spread to other areas of the body. If the affected bone is not removed before metastasis occurs, death is inevitable. *Myelomas* are malignant tumors occurring in the bone marrow of middle-aged and older people. The term *myelos* means marrow, and *oma* is the term for tumor. These tumors interfere with the blood-cell-producing function of bone marrow and cause anemia. They also destroy normal bone tissue and eventually spread throughout the body. *Chondrosarcoma* is a cancerous growth of the cartilage.

Benign tumors are composed of cells that do not metastasize. That is, the growth remains in one small area of the body and does not spread to other organs. All or part of a benign tumor may be removed so that cells can be examined under a microscope to make sure they are not malignant. This procedure is called a biopsy. A benign tumor is also removed if it gets in the way of a normal body function or if it causes disfiguration. Like malignancies, benign tumors are named after the type of cell from which they originate. The names of these tumors usually contain the combining form *oma.* For instance, an osteoblastoma is a benign growth of osteoblasts.

Key terms associated with the skeletal system

Achondroplasia (*a* = without; *chondro* = cartilage; *plasia* = growth) Imperfect ossification within cartilage of long bones during fetal life; also called fetal rickets.
Brodie's abscess Infection in the spongy tissue of a long bone, with a small inflammatory area.
Craniotomy (*cranium* = skull; *tome* = a cutting) Any surgery that requires cutting through the bones surrounding the brain.
Necrosis (*necros* = death; *osis* = condition) Death of tissues or organs; in the case of bone, necrosis results from deprivation of blood supply; could result from fracture, extensive removal of periosteum in surgery, exposure to radioactive substances, and other causes.
Osteitis Inflammation or infection of bone.
Osteoarthritis (*arthro* = joint) A degenerative condition of bone and also the joint.
Osteoblastoma (*oma* = tumor) A benign tumor of the osteoblasts.
Osteochondroma A benign tumor of the bone and cartilage.
Osteoma A benign bone tumor.
Osteomyelitis Infection that involves bone marrow.
Osteosarcoma (*sarcoma* = connective-tissue tumor) A malignant tumor composed of osseous tissue.
Pott's disease Inflammation of the backbone, caused by the microorganism that produces tuberculosis.

Chapter summary in outline

1. The skeletal system consists of all bones attached at joints, cartilage between joints, and cartilage found elsewhere (nose, larynx, and outer ear).
2. The functions of the skeletal system include support, protection, movement, mineral storage, and blood-cell formation.

A MICROSCOPIC VIEW

1. Parts of a typical long bone are the diaphysis (shaft), epiphyses (ends), articular cartilage, periosteum, medullary (marrow) cavity, and endosteum.
2. Spongy, or cancellous, bone has many marrow-filled pores and does not contain Haversian systems. It consists of trabeculae containing osteocytes and lacunae. It lies deep to compact bone.
3. Compact, or dense, bone has fewer pores and is made up of structural units called Haversian systems. Compact bone lies external to spongy bone and composes most of the bone tissue of the diaphysis.

BONE FORMATION AND GROWTH

1. Bone forms by a process called ossification, which begins when mesenchymal cells become transformed into osteoblasts.
2. Intramembranous ossification occurs within fibrous membranes of the embryo.
3. Endochondral ossification occurs within a cartilage model.
4. The primary ossification center of a long bone is in the diaphysis. Osteoblasts lay down bone, and cartilage degenerates, leaving cavities that merge to form the marrow cavity. Next ossification occurs in the epiphyses, where bone replaces cartilage, except for the epiphyseal plate.
5. In both types of ossification, spongy bone is laid down first. Compact bone is later reconstructed from spongy bone.

BONE REPLACEMENT

1. Bone growth is dependent upon a balance between bone formation and destruction.
2. Old bone is constantly destroyed by osteoclasts, whereas new bone is constructed by osteoblasts. This process is called remodeling.
3. Normal growth depends on calcium, phosphorus, and vitamins (A, C, and D) and is controlled by hormones.

APPLICATIONS TO HEALTH

1. Rickets is a vitamin D deficiency in children in which the body does not absorb calcium and phosphorus. The bones soften and bend under the body's weight.
2. Osteomalacia is a vitamin D deficiency in adults that leads to demineralization.

3. With osteoporosis, the amount and strength of bone tissue decreases due to decreases in hormone output.
4. Paget's disease is the irregular thickening and softening of bones in which osteoclast and osteoblast activities are imbalanced.
5. Osteomyelitis is a term for the infectious diseases of bones, marrow, and periosteum. It is frequently caused by "staph" bacteria.
6. Tumors may be classified as malignant (cancerous) or benign (noncancerous). Tumors of the skeletal system include osteogenic sarcomas (arise in bone), chondrosarcomas (arise in cartilage), and myelomas (arise in marrow).

Review questions and problems

1. Define the skeletal system. What are the five principal functions of the system?
2. Diagram the parts of a long bone, and list the functions of each part. What is the difference between compact and spongy bone tissue? Diagram the microscopic structure of bone.
3. What is meant by ossification? When does the process begin and end?
4. Distinguish between the two principal kinds of ossification.
5. Outline the major events involved in intramembranous and endochondral ossification.
6. What is a roentgenogram?
7. List the primary factors involved in bone growth.
8. How does osteoblast activity, in balance with osteoclast activity, control the replacement of bone?
9. Define rickets and osteomalacia in terms of symptoms, cause, and treatment. What do these two diseases have in common?
10. What are the principal symptoms of osteoporosis, osteomyelitis, and Paget's disease? What is the etiology of each?
11. Define the tumors associated with the skeletal system according to origin. Contrast malignant and benign tumors. What is metastasis?
12. Refer to the glossary of key terms associated with the skeletal system. Be sure that you can define each term.

Selected readings

Basmajian, John V. *Grant's Method of Anatomy.* 9th ed. Baltimore: Williams and Wilkins, 1975. Pp. xii–xvii.

______. *Primary Anatomy.* 6th ed. Baltimore: Williams and Wilkins, 1970. Pp. 19–22.

CHAPTER 6

THE SKELETAL SYSTEM: AXIAL SKELETON

STUDENT OBJECTIVES

After you have read this chapter, you should be able to:

1. Define the four principal types of bones in the skeleton
2. Note the relationship between bone structure and function
3. Describe the various kinds of markings on the surfaces of bones
4. Relate the structure of the marking to the function it performs
5. Describe the components of the axial and appendicular skeleton
6. Identify the bones of the skull and the major markings associated with each
7. Identify the sutures and fontanels of the skull
8. Identify the paranasal sinuses of the skull in projection diagrams and roentgenograms
9. Identify the principal foramina of the skull
10. Identify the bones of the vertebral column
11. List the defining characteristics and curves of each region of the vertebral column
12. Identify the bones of the thorax and their principal markings

The **skeletal system** forms the framework of the body. For this reason, a familiarity with the names, shapes, and positions of individual bones will help you to understand some of the anatomy and physiology of the other organ systems. For example, movements such as throwing a ball, typing, and walking require the coordinated use of bones and muscles. In order to understand how muscles produce different movements, you need to learn the parts of the bones to which the muscles are attached. Efficient functioning of the respiratory system is also highly dependent on normal bone structure. The bones in the nasal cavity form a series of passageways that help clean, moisten, and warm inhaled air. Furthermore, the bones of the thorax are specially shaped and positioned so that the chest can expand during inhalation. Many bones also serve as reference landmarks to students of anatomy as well as to surgeons. Blood vessels and nerves often run along bones. These organs can be located more easily if the bone is identified first. The superior portions of the lungs are found just inferior to the clavicle. The bottom of the rib cage can be used as a landmark in finding the diaphragm and liver.

We shall study bones by examining the various regions of the body. For instance, we shall look at the skull first and see how the bones of the skull relate to each other. Then we shall move on to the chest. This regional approach will allow you to see how all the many bones of the body relate to each other.

TYPES OF BONES

The bones of the body may be classified into one of the four principal types on the basis of their shape: (1) long, (2) short, (3) flat, and (4) irregular. **Long bones** have greater length than width and consist of a diaphysis (shaft) and two epiphyses (extremities). They are more or less curved for greater strength. Curvature of these bones is rather important for body support. A curved bone is structurally designed to absorb the stress of body weight at several different points so that the stress is evenly distributed. If such bones were straight, the weight of the body would be unevenly distributed, and the bone would fracture very easily. Long bones have considerably more compact bone than spongy bone and thus have a further structural adaptation to their weight-bearing or leverage function. Examples of long bones include bones of the thighs, legs, toes, arms, forearms, and fingers. Refer to Figure 5–1 for a description of the parts of a long bone. **Short bones** are somewhat cube-shaped, and differences in length and width are not important. Their texture is spongy throughout except at the surface, where there is a thin layer of compact bone. Examples of short bones are the wrist and ankle bones. **Flat bones** are generally thin and flat and are composed of two more or less parallel plates of compact bone enclosing a layer of spongy bone. The term *diploe* is applied to the spongy bone of the cranial bones. Flat bones afford considerable protection and provide extensive areas for muscle attachment. Examples of flat bones include the cranial bones, which protect the brain, the sternum and ribs, which protect organs in the thorax, and the scapulas. **Irregular bones** have very complex shapes and cannot be grouped into any of the three categories just described. They also vary in the amount of spongy and compact bone present. Such bones are the vertebrae and some facial bones.

In addition to these four principal types of bones, two other types are recognized. **Wormian,** or **sutural, bones** are small clusters of bones between the joints of certain cranial bones. Their number is quite variable among different individuals. **Sesamoid bones** are small bones found in various tendons where a lot of pressure develops—for instance, in the wrist. These bones, like the Wormian bones, are also variable in number. Two sesamoid bones, the patellas, or kneecaps, are present in all individuals.

SURFACE MARKINGS

If you look at the surfaces of bones, you will see various kinds of **markings.** The structure of many of these markings indicates their functions. For instance, long bones that bear a great deal of weight have very large, rounded ends that can form sturdy joints. Other bones have depressions that receive the rounded ends. Roughened areas serve for the attachment of muscles, tendons, and ligaments. Grooves in the surfaces of bones provide a roadbed for the passage of blood vessels, and openings occur where blood vessels and nerves pass in and out of the bone. Exhibit 6–1 describes the different kinds of markings and their functions. As you learn the names and parts of bones in this chapter and the next, see if you can define each bone marking and indicate its function.

DIVISIONS OF THE SKELETAL SYSTEM

The adult human skeleton consists of approximately 206 bones that are grouped in two principal divisions: the **axial** and the **appendicular.** The *axis,* or center, of the human body is a straight line that runs along the center of gravity of the body. This imaginary line runs through the head and down to the space between the feet. The midsagittal section is drawn through this line. The axial division of the skeleton consists of the bones that lie around the axis. These include the ribs and the breastbone and the bones of the skull and backbone. The appendicular division contains the bones of the free *appendages,* which are the upper and lower extremities, plus the bones called *girdles,* which connect the free appendages to the axial skeleton. The 80 bones of the axial and the 126 bones of the appendicular divisions are typically grouped as shown in the following outline:

I. Axial skeleton	
A. Skull	
1. Cranium	8
2. Face	14
B. Hyoid (above the larynx)	1
C. Ossicles* (ear bones), 3 in each ear	6
D. Vertebral column	26
E. Thorax	
1. Sternum	1
2. Ribs	24
	80
II. Appendicular skeleton	
A. Shoulder girdles	
1. Clavicle	2
2. Scapula	2
B. Free upper extremities	
1. Humerus	2
2. Ulna	2
3. Radius	2
4. Carpals	16
5. Metacarpals	10
6. Phalanges	28
C. Pelvic girdle	
1. Coxal, hip, or pelvic bone	2
D. Free lower extremities	
1. Femur	2
2. Fibula	2
3. Tibia	2
4. Patella	2
5. Tarsals	14
6. Metatarsals	10
7. Phalanges	28
	126

* Although the ossicles are not considered part of the axial or appendicular skeleton, but rather as a separate group of bones, they are placed with the axial skeleton for convenience.

Now that you have a general idea as to the organization of the skeleton into axial and appendicular divisions, refer to Figure 6–1 to see how the two divisions are put together to form your skeleton. In order to distinguish the two divisions from each other, the bones of the axial skeleton are shown in gray. As you examine the anterior and posterior views of the skeleton, be sure that you can locate the following regions of the skeleton: skull, cranium, face, hyoid bone, vertebral column, thorax, shoulder girdle, upper extremity, pelvic girdle, and lower extremity.

Instead of learning all the bones of the skeleton at one time, we are going to divide your learning efforts into two phases. In this chapter, you will learn the bones of the axial skeleton. In the next chapter, you will learn the bones of the appendicular skeleton (along with some basic differences between a male and female skeleton), fractures, and the events associated with fracture repair. To assist you to learn the bones of the skeleton, a number of Exhibits have been designed. Each contains a description of the bones plus an accompanying diagram. The bones of the skull are described in Exhibits 6–2 through 6–19. The foramina of the skull are summarized in Exhibit 6–20. The rest of the bones of the axial skeleton are described in Exhibits 6–21 through 6–30.

Two terms that will be useful in studying these Exhibits have been mentioned briefly but will be reviewed again. As you remember, the term **articulation** means joint. Therefore, to say that a bone **articulates** with another bone is to say that the two bones form a joint. The other term, **roentgenogram,** is a film exposed to x-rays. **Roentgenography** is the use of x-rays. Roentgenograms of some normal bones are included with these Exhibits. After studying these representative roentgenograms, you will have a more comprehensive understanding of how they are used by the physician. With roentgenograms, a doctor can determine the location and extent of a fracture, see what progress has been made in fracture repair, and diagnose certain bone diseases.

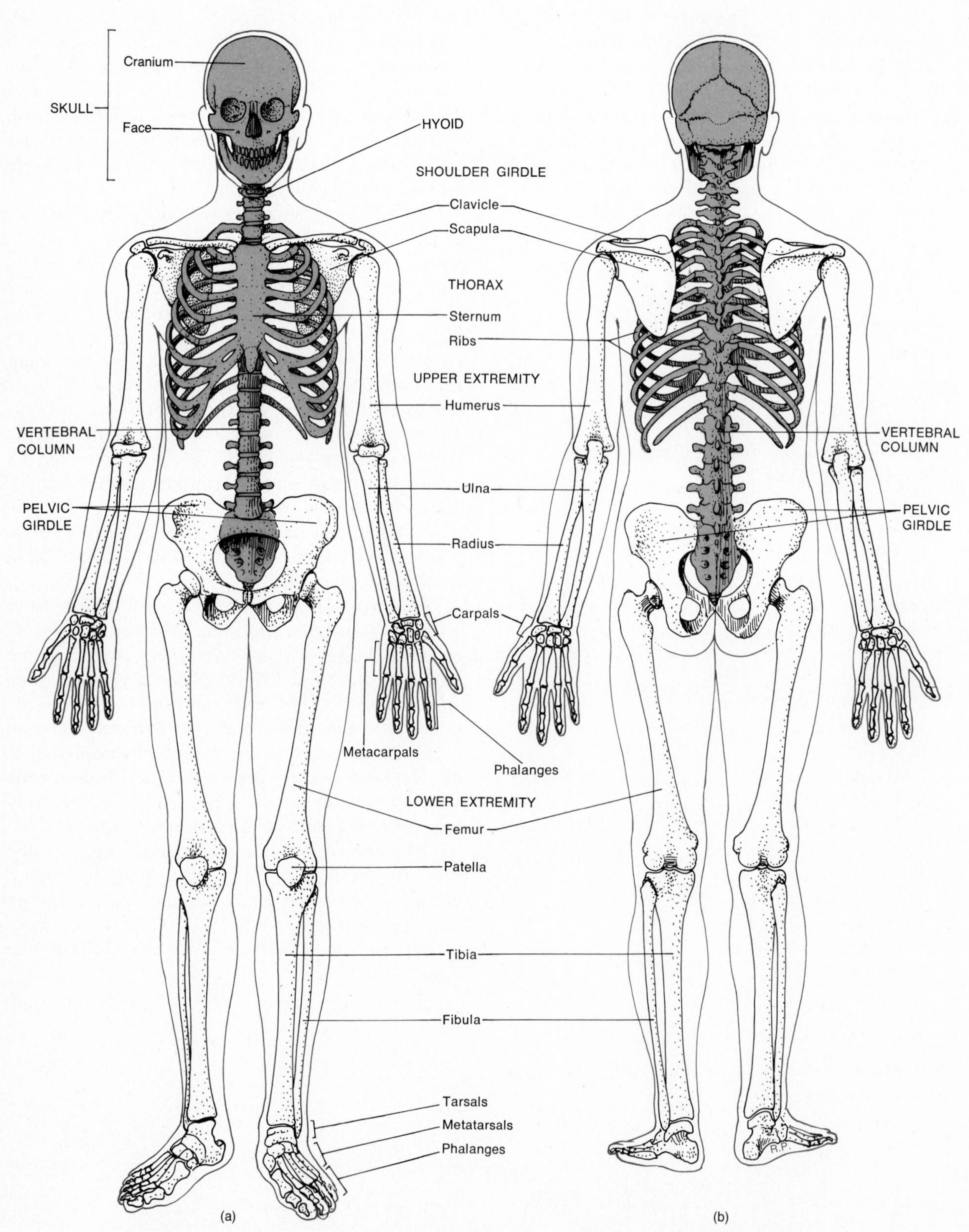

Figure 6–1. Divisions of the skeletal system. (a) Anterior view. (b) Posterior view.

Exhibit 6–1. BONE MARKINGS

MARKING	DESCRIPTION	EXAMPLE
A. Depressions and openings		
1. **Fissure**	A narrow, cleftlike opening between adjacent parts of bones through which blood vessels and nerves pass	Superior orbital fissure of the sphenoid bone (Figure 6–2)
2. **Foramen** (*foramen* = hole)	A rounded opening through which blood vessels, nerves, and ligaments pass	Infraorbital foramen of the maxilla (Figure 6–2)
3. **Meatus** (canal)	A tubelike passageway running within a bone	External auditory meatus of the temporal bone (Figure 6–2)
4. **Paranasal sinus** (*sin* = cavity)	An air-filled cavity within a bone connected to the nasal cavity	Frontal sinus of the frontal bone (Figure 6–8)
5. **Groove** or **sulcus** (*sulcus* = ditchlike groove)	A furrow or groove that accommodates a soft structure such as a blood vessel, nerve, or tendon	Intertubecular groove of the humerus (Figure 7–4)
6. **Fossa** (*fossa* = basinlike depression)	A depression in or on a bone	Mandibular fossa of the temporal bone (Figure 6–4)
B. Process	Any prominent, roughened projection	The mastoid process of the temporal bone (Figure 6–2)
Processes that form joints		
1. **Condyle**	A relatively large, convex articular prominence (knucklelike)	Medial condyle of the femur (Figure 7–10)
2. **Head**	A rounded articular projection supported on the constricted portion (neck) of a bone	Head of the femur (Figure 7–10)
3. **Facet**	A flat or shallow articular surface	Articular facet for tubercle of rib on a vertebra (Figure 6–14)
Processes to which tendons, ligaments, and other connective tissues attach		
1. **Tubercle** (*tuber* = knob)	A small, rounded process	Greater tubercle of the humerus (Figure 7–4)
2. **Tuberosity**	A large, rounded, usually roughened process	Ischial tuberosity of the hipbone (Figure 7–8)
3. **Trochanter**	A very large, blunt projection found only on the femur	Greater trochanter of the femur (Figure 7–10)
4. **Crest**	A prominent border or ridge on a bone	Iliac crest of the hipbone (Figure 7–7)
5. **Line**	A less prominent ridge	Linea aspera of the femur (Figure 7–10)
6. **Spinous process** (**spine**)	A sharp, slender process	Spinous process of a vertebra (Figure 6–12)
7. **Epicondyle** (*epi* = above)	A prominence above or on a condyle	Medial epicondyle of the femur (Figure 7–10)

Exhibit 6–2. GENERAL DESCRIPTION OF THE SKULL (see Figure 6–2)

The **skull**, which contains 22 bones, rests upon the superior end of the vertebral column and is composed of two sets of bones: cranial bones and facial bones. The **cranial bones** enclose and protect the brain and the organs of sight, hearing, and balance. The eight cranial bones are the frontal bone, parietal bones (2), temporal bones (2), the occipital bone, sphenoid, and ethmoid.

There are 14 **facial bones**, or bones of the face. These include: the nasal bones (2), maxillae (2), zygomatic bones (2), the mandible, lacrimal bones (2), palatine bones (2), inferior conchae (2), and the vomer. Be sure you can locate all the cranial and facial bones in the anterior, lateral, and medial views of the skull. The names of the bones are indicated in capital letters.

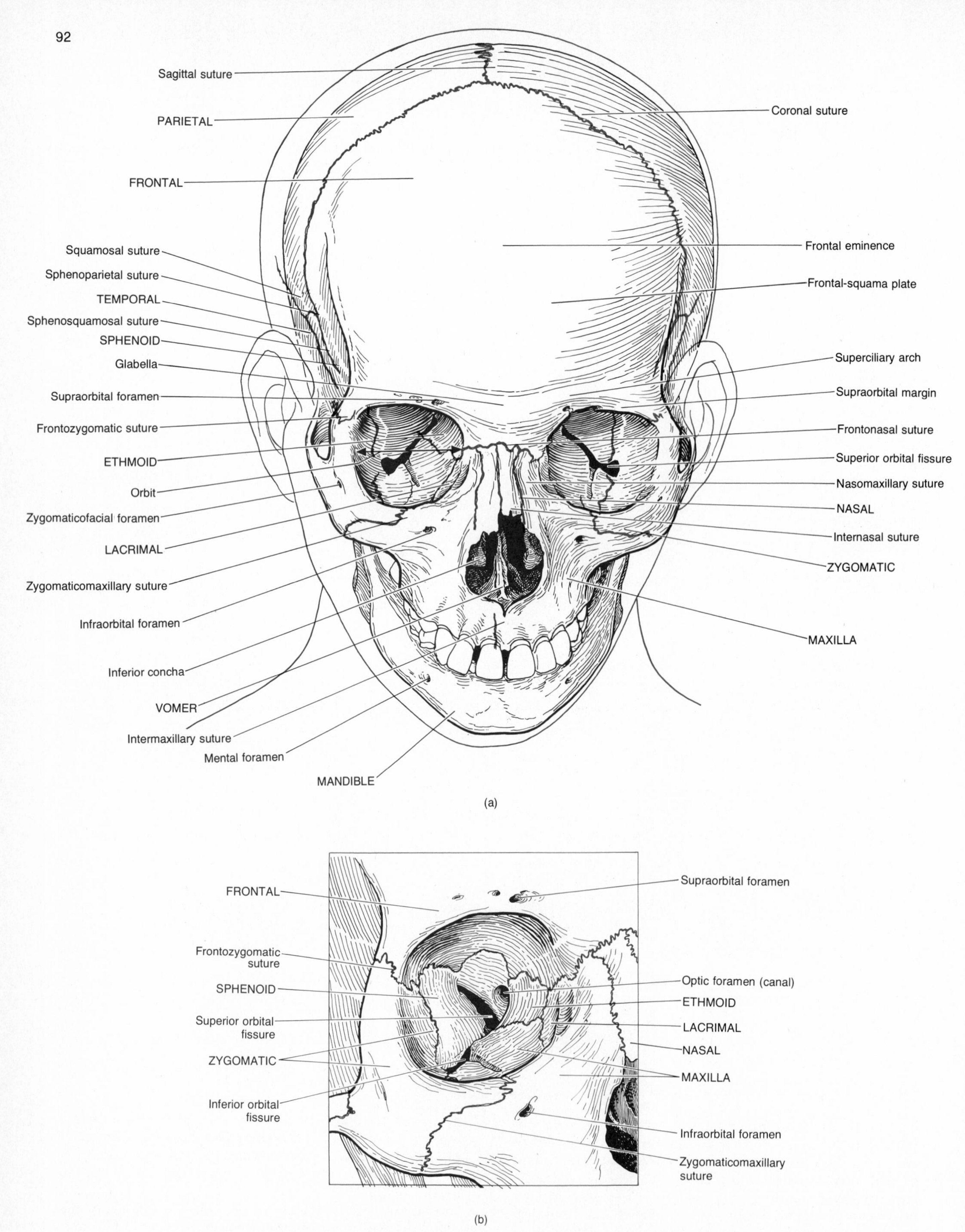

Figure 6–2. The skull. (a) Anterior view. (b) Detail of the right orbit in anterior view. (c) Photograph of skull in anterior view. (d) Photograph of detail of the right orbit in anterior view. (Photographs courtesy of Lenny Patti.)

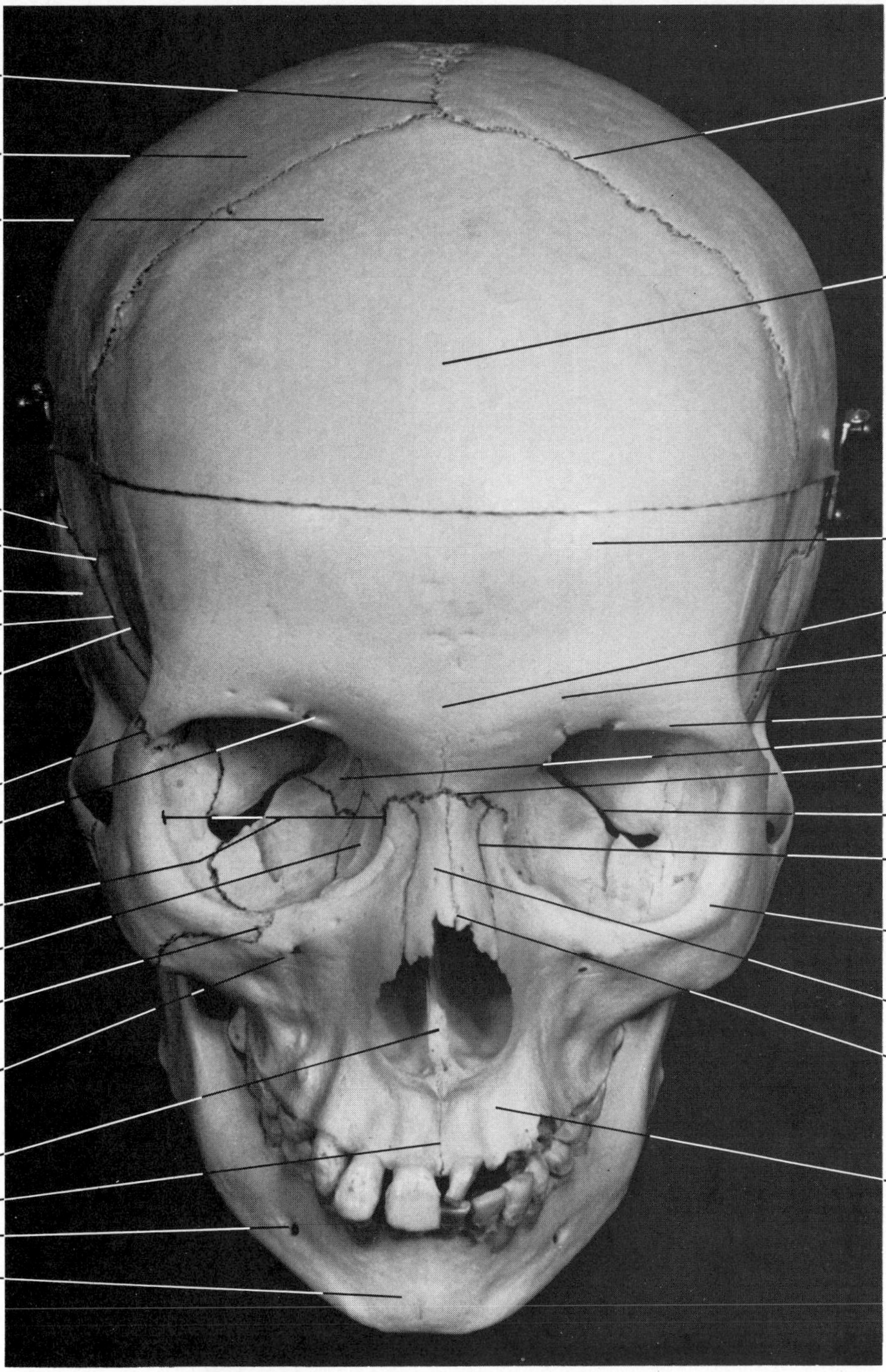

Sagittal suture
Coronal suture
PARIETAL
FRONTAL
Frontal eminence
Squamosal suture
Sphenoparietal suture
Frontal-squama plate
TEMPORAL
Sphenosquamosal suture
Glabella
SPHENOID
Superciliary arch
Supraorbital margin
ETHMOID
Frontozygomatic suture
Frontonasal suture
Supraorbital foramen
Superior orbital fissure
Nasomaxillary suture
Orbit
ZYGOMATIC
LACRIMAL
Zygomaticomaxillary suture
NASAL
Internasal suture
Infraorbital foramen
VOMER
MAXILLA
Intermaxillary suture
Mental foramen
MANDIBLE

(c)

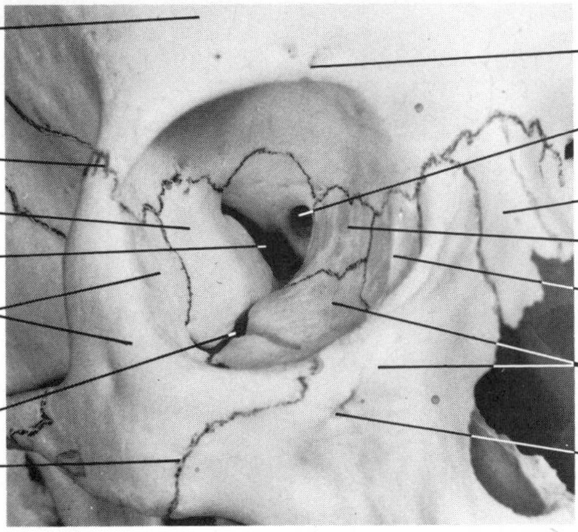

FRONTAL
Supraorbital foramen
Optic foramen (canal)
Frontozygomatic suture
NASAL
SPHENOID
ETHMOID
Superior orbital fissure
LACRIMAL
ZYGOMATIC
MAXILLA
Inferior orbital fissure
Infraorbital foramen
Zygomaticomaxillary suture

(d)

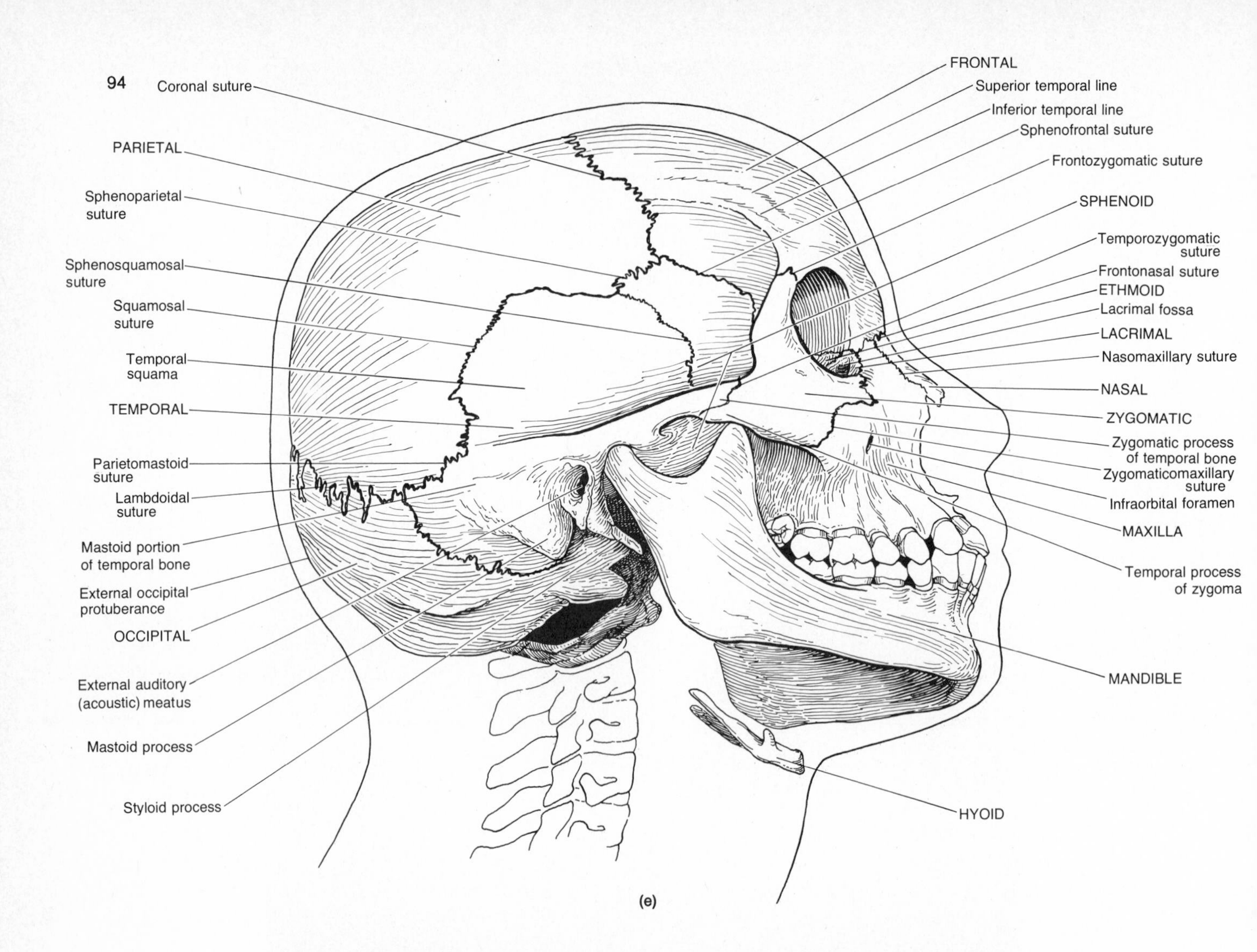

(e)

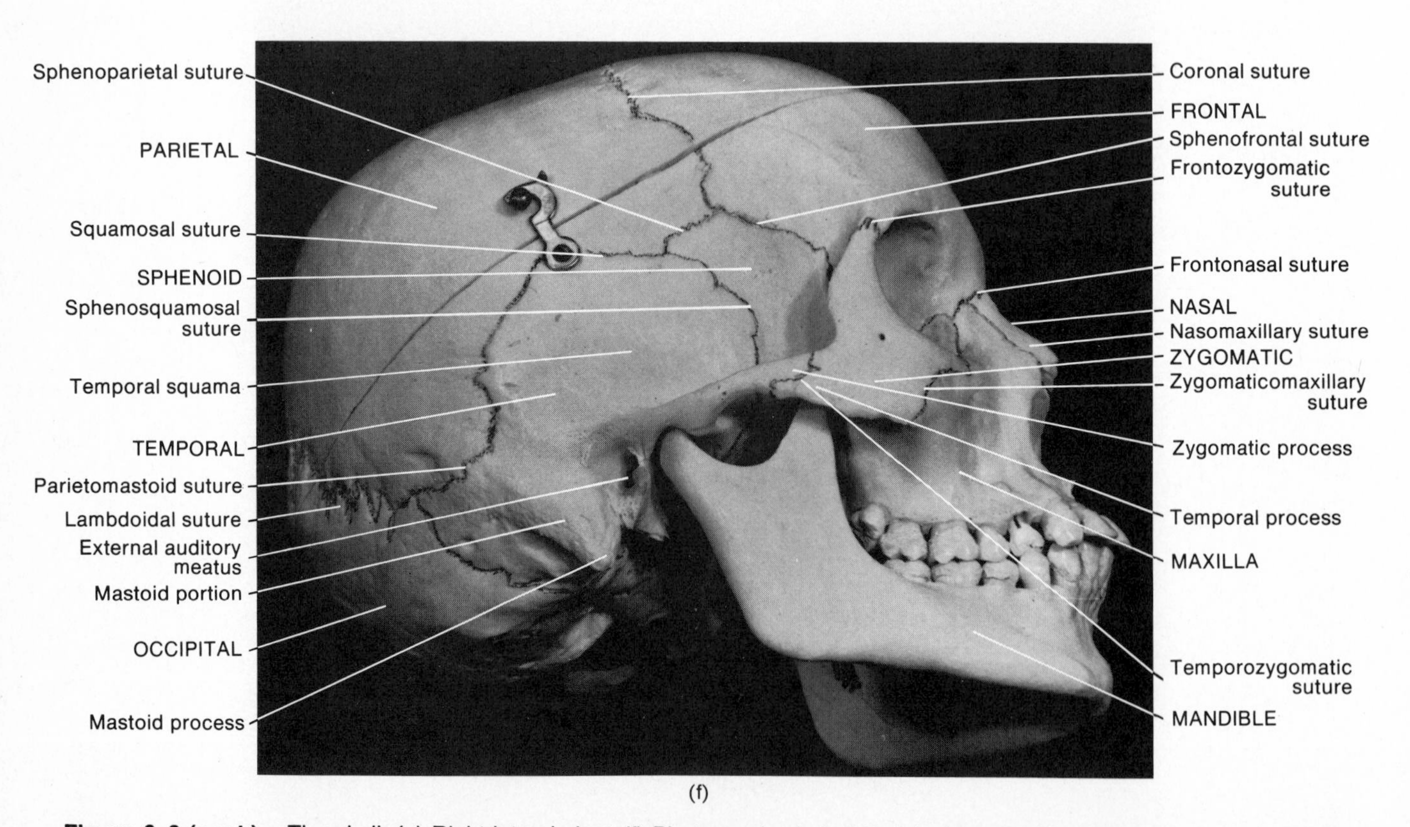

(f)

Figure 6–2 (cont.). The skull. (e) Right lateral view. (f) Photograph of skull in right lateral view. (Courtesy of Lenny Patti.)

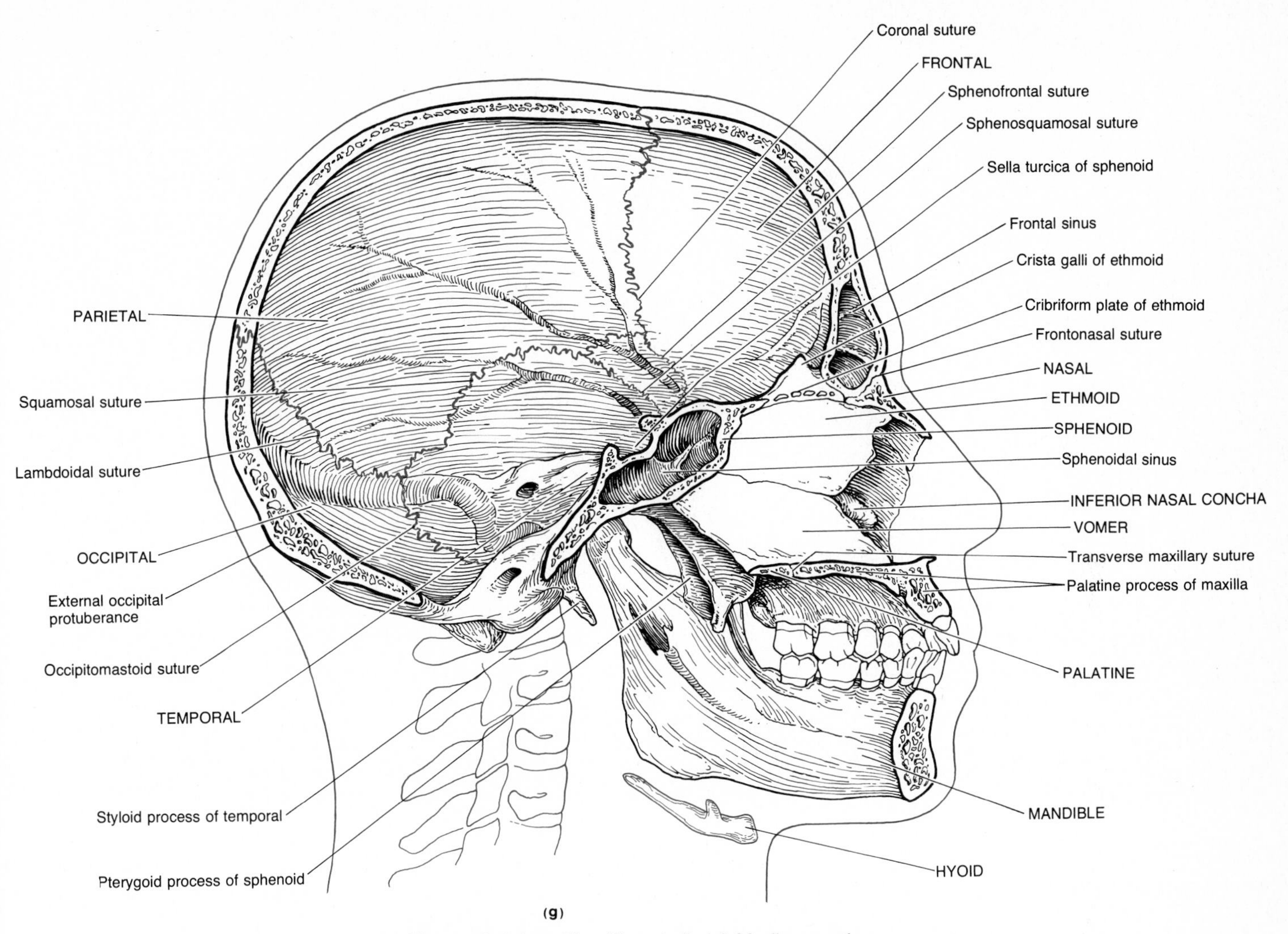

Figure 6–2 (cont.). The skull. (g) Median section.

Exhibit 6–3. SUTURES OF THE SKULL

The term **suture** means seam. It is an immovable joint found only between skull bones. Very little connective tissue is found between the bones of the suture. Four prominent skull sutures include: (1) the **coronal suture** between the frontal bone and the two parietal bones, (2) the **sagittal suture** between the two parietal bones, (3) the **lambdoidal suture** between the parietal bones and the occipital bone, and (4) the **squamosal suture** between the parietal bones and the temporal bones. Refer to Figures 6–2 and 6–3 for the locations of these sutures. Several other sutures are also shown. Their names are typically descriptive of the bones they connect. For example, the frontonasal suture is between the frontal bone and the nasal bones. These sutures are indicated in Figures 6–2 to 6–5.

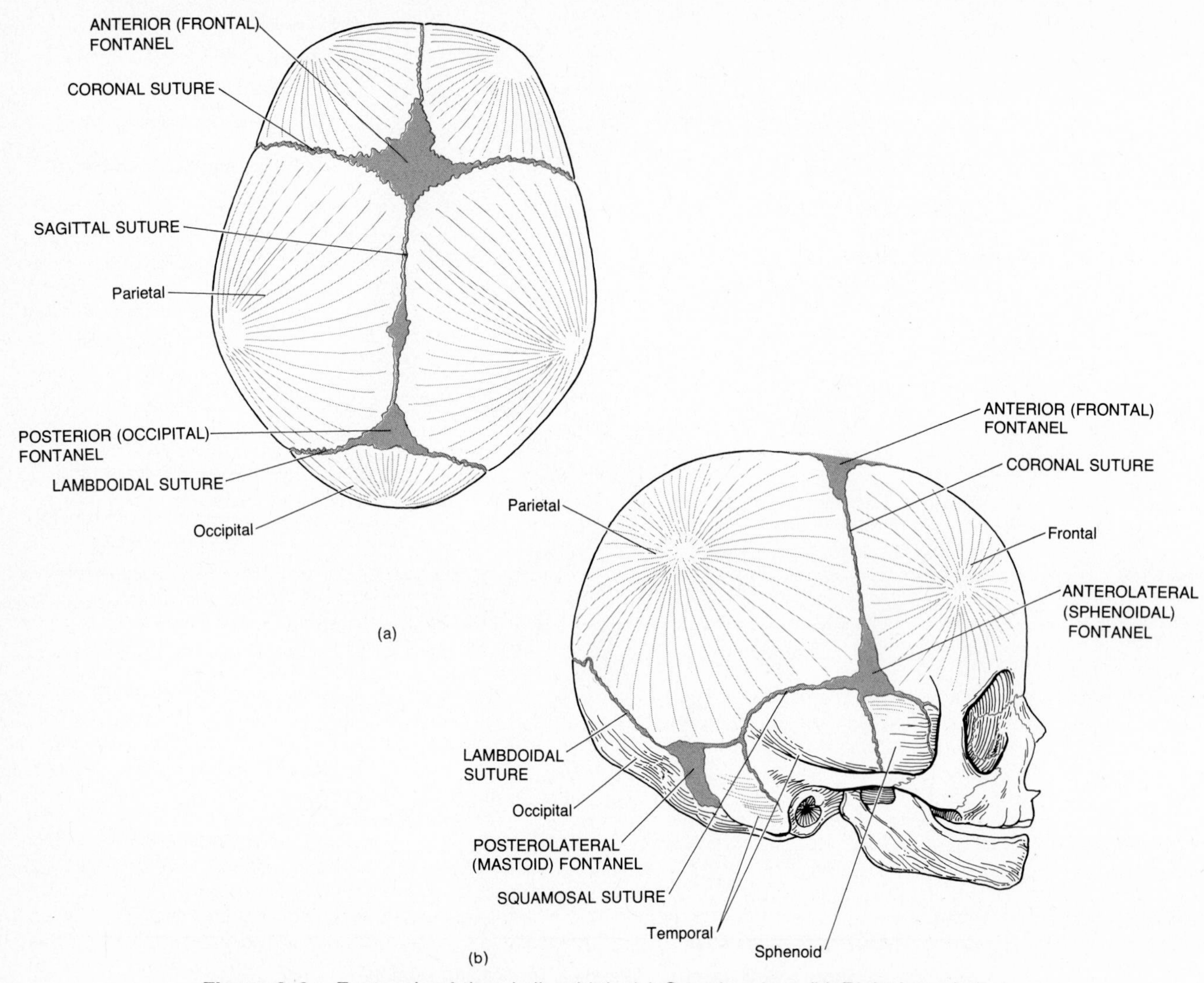

Figure 6–3. Fontanels of the skull at birth. (a) Superior view. (b) Right lateral view.

Exhibit 6–4. FONTANELS OF THE SKULL (see Figure 6–3)

The "skeleton" of a newly formed embryo consists of cartilage or fibrous membrane structures shaped like bones. Gradually the cartilage or fibrous membrane is replaced by bone. At birth, membrane-filled spaces called **fontanels,** meaning fountains, are found between cranial bones. These so-called "soft spots" are areas where the bone-making process is not yet complete. They allow the skull to be compressed during birth. Physicians find the fontanels helpful in determining the position of the infant's head prior to delivery. Although an infant may have many fontanels at birth, the form and location of six of them are fairly constant.

1. The **anterior (frontal) fontanel** is located between the angles of the two parietal bones and the two segments of the frontal bone. This fontanel is roughly diamond-shaped, and it is the largest of the six fontanels. It usually closes at 18 to 24 months. In the disorder called *microcephalus,* meaning small head, the fontanel closes earlier than normal. The brain consequently does not have enough room to grow, and mental retardation results. In *hydrocephalus,* meaning water on the brain, pressure from excess fluids inside the skull may cause the fontanel to remain open.

2. The **posterior (occipital) fontanel** is situated between the two parietal bones and the occipital bone. This fontanel is considerably smaller than the anterior fontanel. It is diamond-shaped and generally closes about 2 months after birth.

3. The **anterolateral (sphenoidal) fontanels** are two in number. One is located on each side of the skull at the junction of the frontal, parietal, temporal, and sphenoid bones. These fontanels are quite small and irregular in shape. They normally close by the third month after birth.

4. The **posterolateral (mastoid) fontanels** are also two in number. One is situated on each side of the skull at the junction of the parietal, occipital, and temporal bones. These fontanels are somewhat irregularly shaped. They start to close 1 or 2 months after birth, but closure is not generally complete until the age of 1 year.

Exhibit 6–5. BONES OF THE SKULL: THE FRONTAL BONE

General Description: The **frontal bone** forms the anterior part of the cranium, or the forehead, the superior portion of the *orbits,* or eye sockets, and most of the anterior part of the cranial floor. At birth, the frontal bone consists of left and right parts that unite soon thereafter.

Markings: If you examine the anterior and lateral views of the skull in Figure 6–2, you will note the **squama,** or vertical plate. The term *squam* means scale. This scale-like plate gradually slopes down from the coronal suture, then turns abruptly downward. On either side of the midline of the body, it projects slightly to form the **frontal eminences.** Inferior to each eminence is a horizontal ridge caused by the projection of the frontal sinuses posterior to the eyebrow. This is called the **superciliary arch.** Between the eminences and the arches just superior to the nose is a flattened area, the **glabella.** A thickening of the frontal bone inferior to the superciliary arches is called the **supraorbital margin.** From this margin the frontal bone extends posteriorly to form the roof of the orbit and part of the floor of the cranial cavity. Within the supraorbital margin, slightly medial to its midpoint, is a hole called the **supraorbital foramen,** the term *foramen* meaning passageway. The supraorbital nerve and artery pass through this foramen. The **frontal sinuses** (see Figure 6–8) lie deep to the superciliary arches. These mucosa-lined cavities act as sound chambers to provide the voice with resonance.

Exhibit 6–6. BONES OF THE SKULL: THE PARIETAL BONES

General Description: The two **parietal bones,** *paries* meaning wall, form the greater portion of the sides and roof of the cranial cavity.

Markings: The external surface contains two slight ridges that may be observed by looking at the lateral view of the skull in Figure 6–2. These are the **superior temporal line** and a less conspicuous **inferior temporal line.** The internal surface has many eminences and depressions that accommodate the blood vessels supplying the brain.

Exhibit 6–7. BONES OF THE SKULL: THE TEMPORAL BONES

General Description: The two **temporal bones** form the inferior sides of the cranium and part of the cranial floor. The term *tempora* pertains to the temples.

Markings: Looking at the lateral view of the skull in Figure 6–2, you will notice the **squama** or **squamous portion** — a thin, large, expanded area that forms the anterior and superior part of the temple. Projecting from the inferior portion of the squama is the **zygomatic process,** which articulates with the temporal process of the zygomatic bone. The zygomatic process of the temporal bone together with the temporal process of the zygomatic bone constitutes the **zygomatic arch.** If you look at the floor of the cranial cavity, shown in Figure 6–5, you will see the **petrous portion** of the temporal bone. This portion is triangular and located at the base of the skull between the sphenoid and occipital bones. The petrous portion contains the internal ear, the essential part of the organ of hearing. It also contains the **carotid foramen (canal)** through which the internal carotid artery passes. This is shown in Figure 6–4. If you look posterior to the foramen, between it and the occipital bone, you will find the **jugular foramen (fossa)** through which the internal jugular vein and the hypoglossal nerve (IX), vagus nerve (X), and accessory nerve (XI) pass. Between the squamous and petrous portions is a socket called the **mandibular fossa.** This part of the temporal bone articulates with the condyle of the lower jaw. It is seen best in Figure 6–4. If you now examine the lateral view of the skull in Figure 6–2, you will see the **mastoid portion** of the temporal bone, located posterior and inferior to the external auditory meatus, or ear canal. In the adult, this portion of the bone contains a number of air spaces called **mastoid air "cells."** These spaces are separated from the brain only by thin bony partitions. If *mastoiditis,* or the inflammation of these bony cells, occurs, the infection may spread to the brain or its outer covering. The mastoid air cells do not drain as do the paranasal sinuses. The **mastoid process** is a rounded projection of the temporal bone posterior to the external auditory meatus. It serves as a point of attachment for several neck muscles. Near the posterior border of the mastoid process is the **mastoid foramen** through which a vein to the transverse sinus and a small branch of the occipital artery to the dura mater pass (see Figure 6–4). The **external auditory meatus** is a canal in the temporal bone that leads to the middle ear. The **styloid process** projects downward from the undersurface of the temporal bone and serves as a point of attachment for muscles and ligaments of the tongue and neck. Between the styloid process and the mastoid process is the **stylomastoid foramen** which transmits the facial nerve (VII) and the stylomastoid artery (see Figure 6–4).

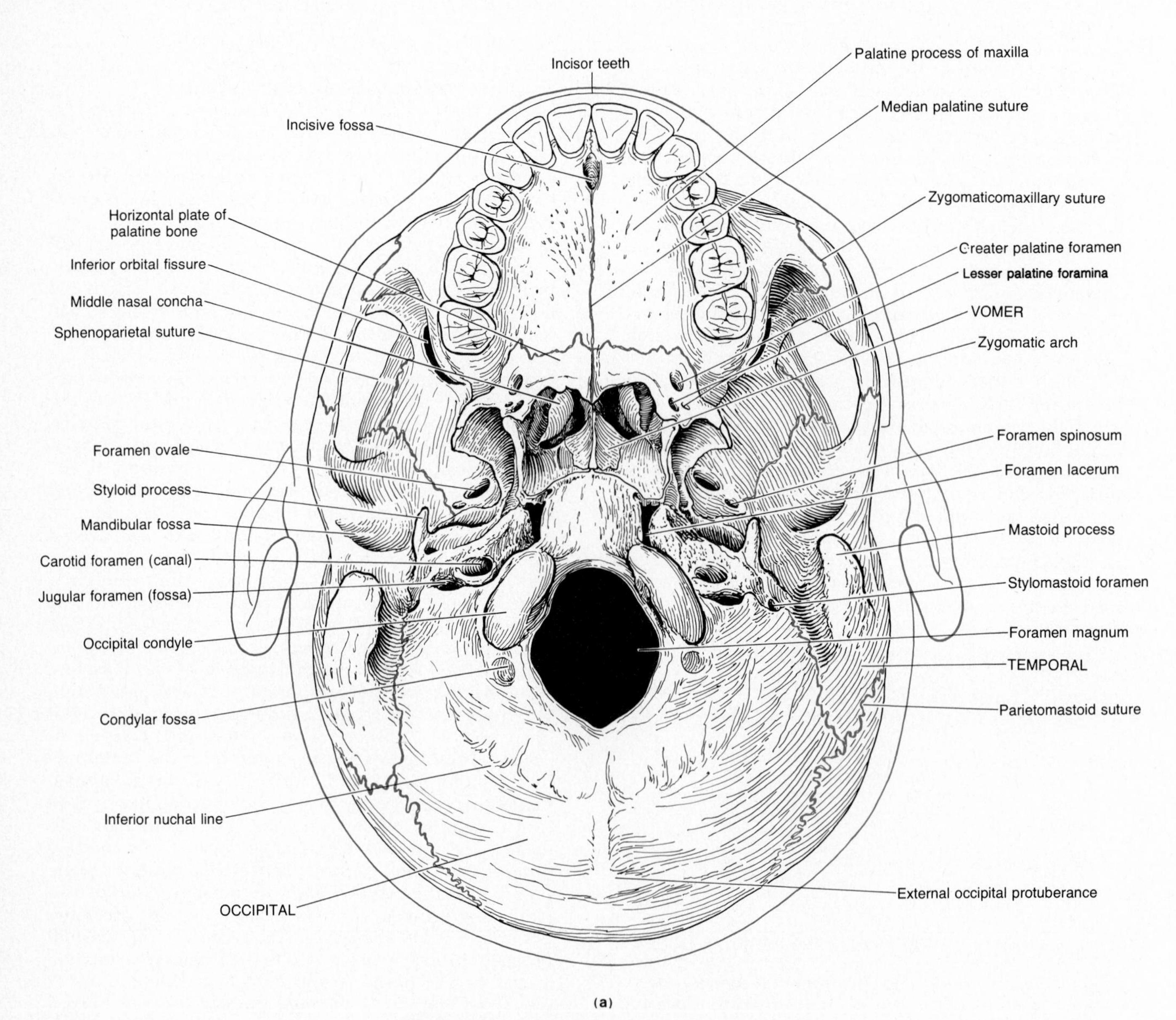

(a)

Figure 6–4. The skull. (a) Inferior view.

Exhibit 6–8. BONES OF THE SKULL: THE OCCIPITAL BONE (see Figure 6–4)

General Description: The **occipital bone** forms the posterior part and a good portion of the base of the cranium.

Markings: The **foramen magnum** is a large hole in the inferior part of the bone through which the medulla oblongata and its membranes, the accessory nerve (XI), and the vertebral and spinal arteries pass. The **occipital condyles** are oval-shaped processes with convex surfaces, one on either side of the foramen magnum, which articulate with depressions on the first cervical vertebra. At the base of the condyles is the **hypoglossal fossa** through which the hypoglossal nerve (XII) passes (see Figure 6–5). The **external occipital protuberance** is a prominent projection on the posterior surface of the bone just superior to the foramen magnum. You can feel this structure as a definite bump on the back of your head, just above your neck. The protuberance is also visible in Figure 6–2e, f.

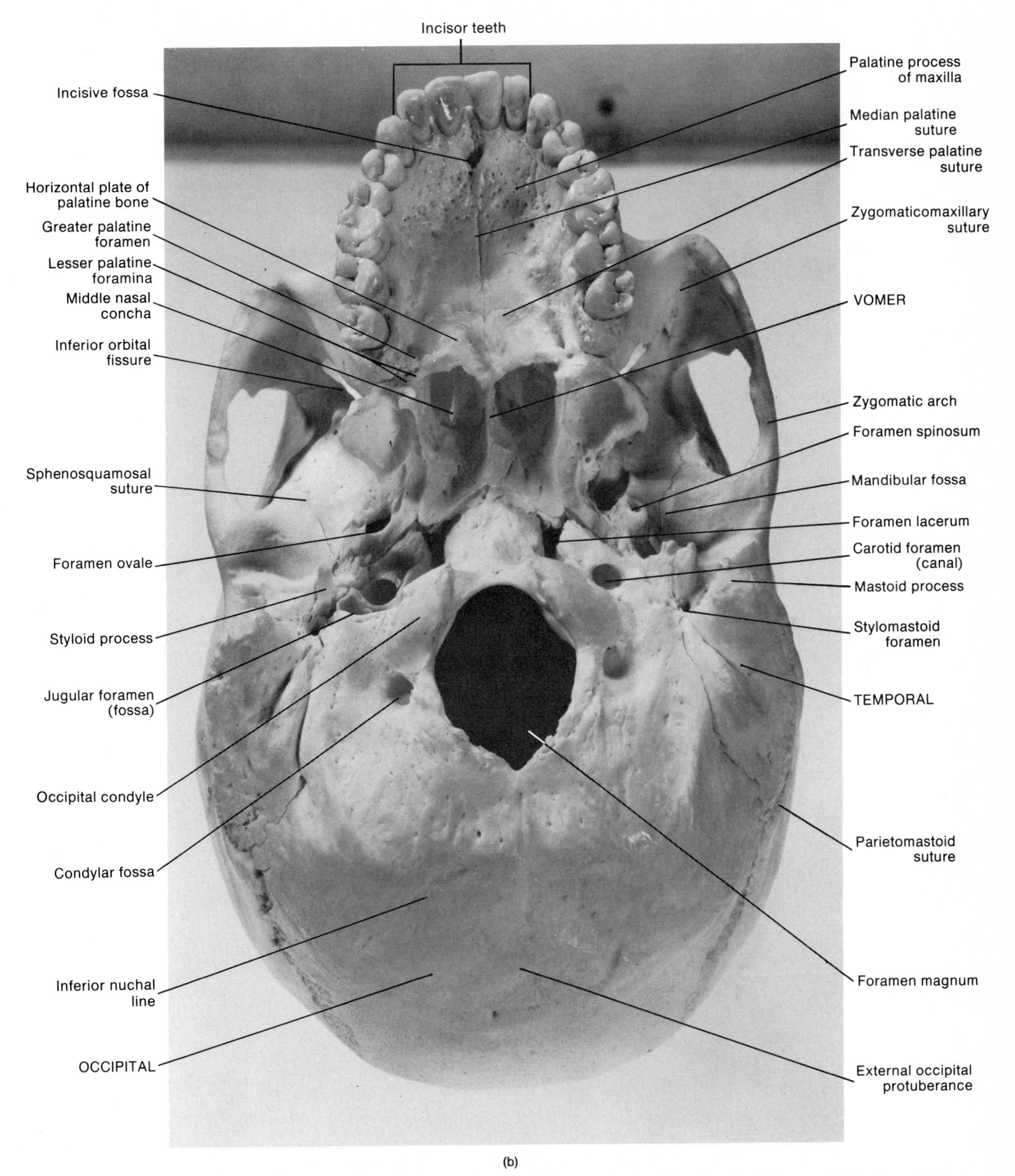

(b)

Figure 6–4 (cont.). The skull. (b) Photograph of the skull in inferior view. (Courtesy of Vincent P. Destro, Mayo Foundation.)

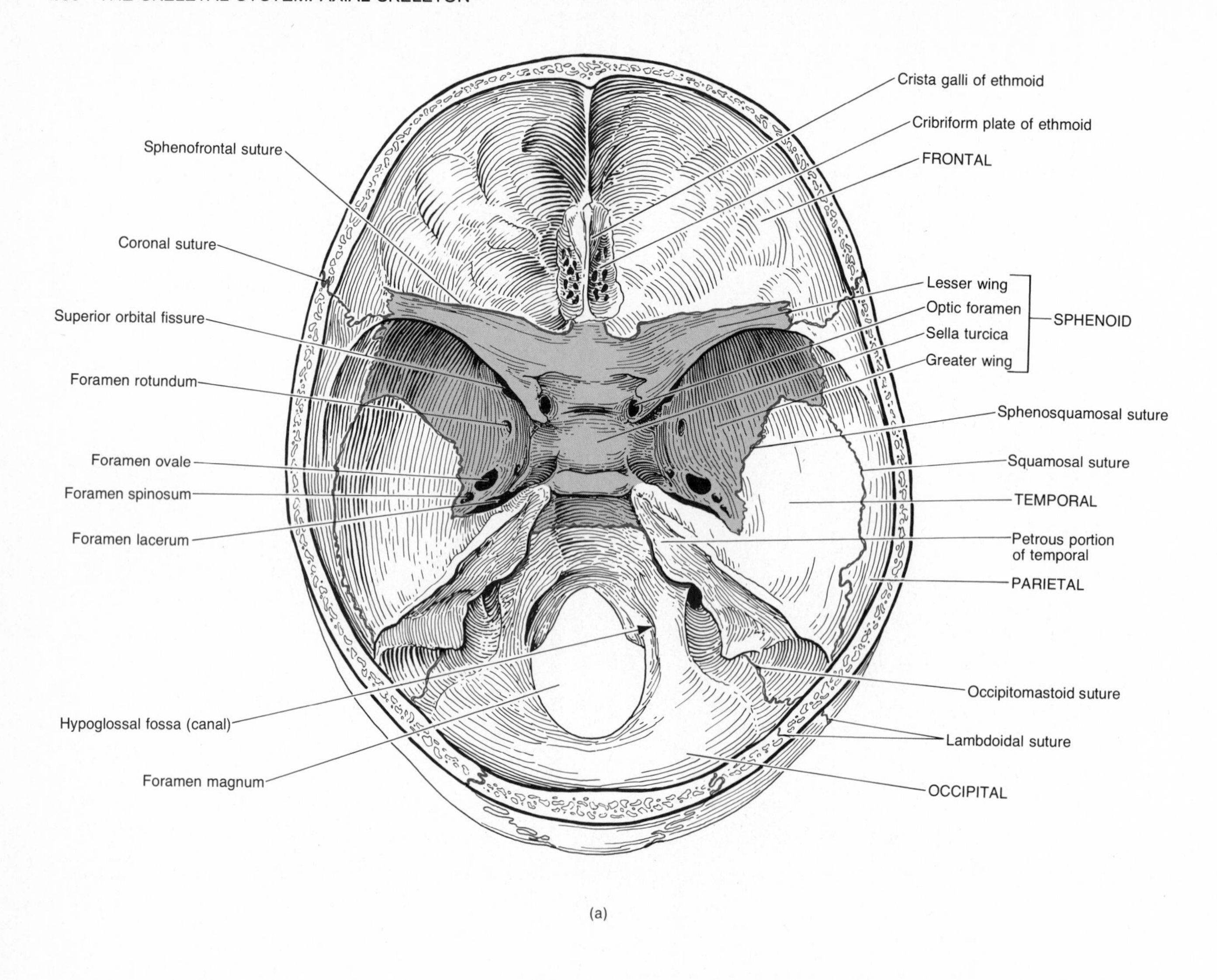

(a)

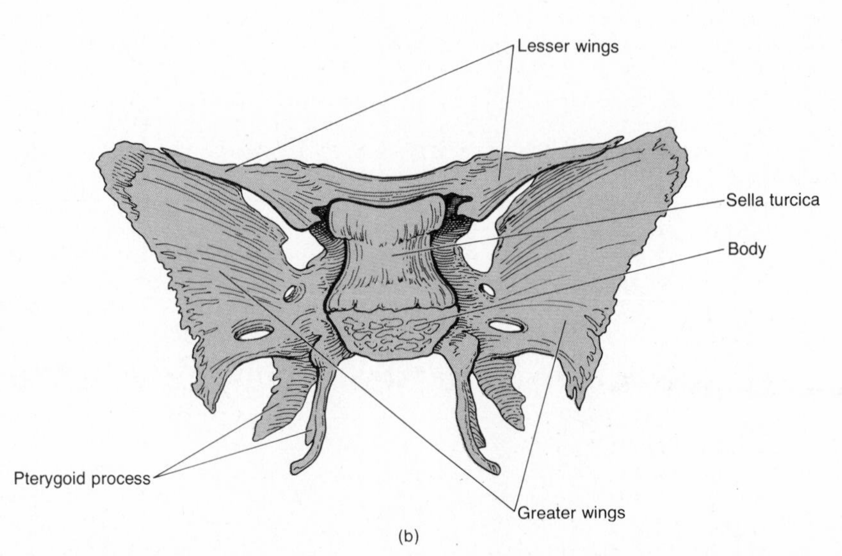

(b)

Figure 6–5. The sphenoid bone. (a) Viewed in the floor of the cranium from above. (b) Posterior view.

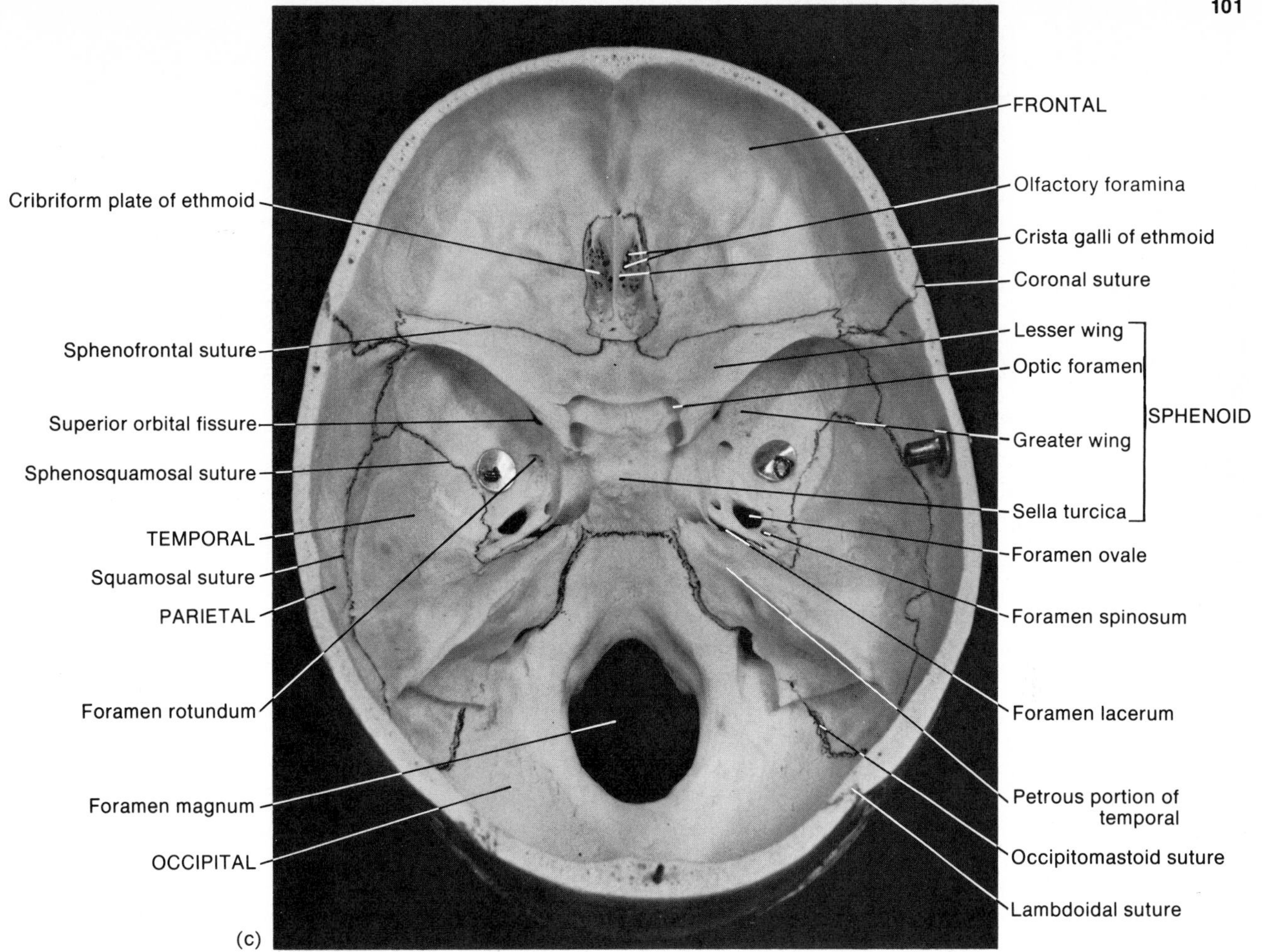

Figure 6–5 (cont.). The sphenoid bone. (c) Photograph of the sphenoid bone viewed in the floor of the cranium from above. (Courtesy of Lenny Patti.)

Exhibit 6–9. BONES OF THE SKULL: THE SPHENOID BONE (see Figure 6–5)

General Description: The **sphenoid bone** is situated at the anterior part of the base of the skull. The combining form *spheno* means wedge. This bone is referred to as the "keystone" of the cranial floor because it binds the other cranial bones together. If you view the floor of the cranium from above, you will note that the sphenoid articulates with the temporal bones laterally, the ethmoid and frontal bones anteriorly, and the occipital bone posteriorly. It lies posterior and slightly superior to the nose and forms part of the floor and sidewalls of the eye socket. The shape of the sphenoid is frequently described as a bat with outstretched wings.

Markings: The **body** of the sphenoid is the cubelike central portion between the ethmoid and occipital bones. It contains two large air spaces, the **sphenoidal sinuses,** which drain into the nasal cavity. (See Figure 6–8.) On the superior surface of the sphenoid body is a depression called the **sella turcica,** meaning Turk's Saddle. This depression houses the pituitary gland. The **greater wings** of the sphenoid are lateral projections from the body and form the anterolateral floor of the cranium. The greater wings also form part of the sidewall of the skull just anterior to the temporal bone. The **lesser wings** are anterior and superior to the greater wings. They form part of the floor of the cranium and the posterior part of the roof of the orbit, or eye socket. Between the body and lesser wing, you can locate the **optic foramen** through which pass the optic nerve (II) and ophthalmic artery. Just lateral to the body between the greater and lesser wings is a somewhat triangular gap called the **superior orbital fissure.** It is an opening for the oculomotor nerve (III), trochlear nerve (IV), ophthalamic branch of the trigeminal nerve (V), and abducens nerve (VI). This fissure may also be seen in the anterior view of the skull in Figure 6–2. On the inferior part of the sphenoid bone you can see the **pterygoid processes.** These structures project inferiorly from the points where the body and greater wings unite. The pterygoid processes form part of the lateral walls of the nasal cavity. At the base of the lateral pterygoid process in the greater wing is the **foramen ovale** through which the mandibular branch of the trigeminal nerve (V) passes. Another foramen, the **foramen spinosum,** at the posterior angle of the sphenoid transmits the middle meningeal vessels. The **foramen lacerum** is bounded anteriorly by the sphenoid bone, posteriorly by the petrous portion of the temporal bone, and medially by the sphenoid and occipital bones. It transmits the internal carotid artery and the meningeal branch of the ascending pharyngeal artery. A final foramen associated with the sphenoid bone is the **foramen rotundum** through which the maxillary branch of the trigeminal nerve (V) passes.

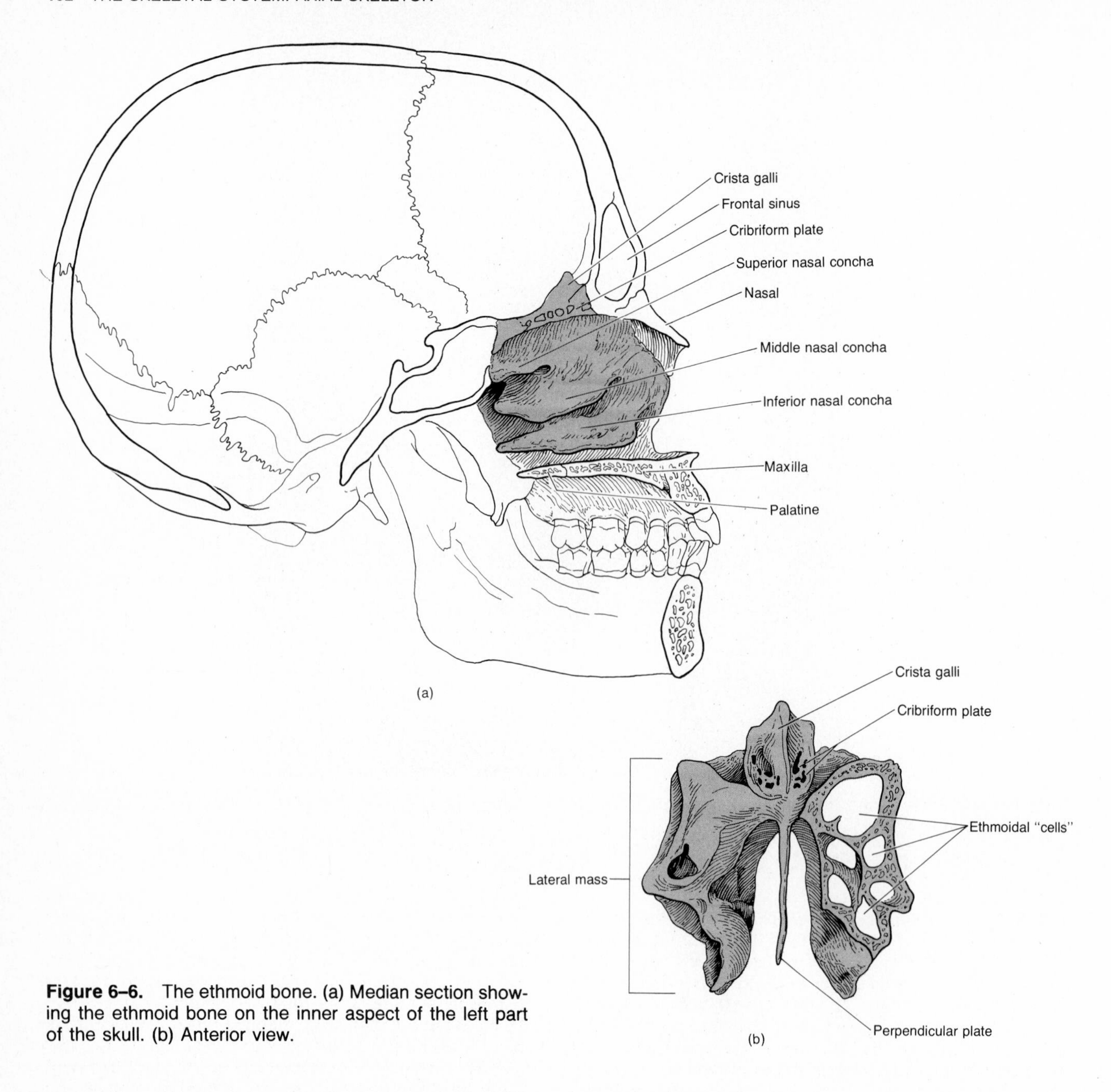

Figure 6–6. The ethmoid bone. (a) Median section showing the ethmoid bone on the inner aspect of the left part of the skull. (b) Anterior view.

Exhibit 6–10. BONES OF THE SKULL: THE ETHMOID BONE (see Figure 6–6)

General Description: The **ethmoid bone** is a light, spongy bone located anterior to the sphenoid and posterior to the nasal bones. The combining form *ethmos* means sieve. This bone helps form the anterior portion of the cranial floor, the medial wall of the orbits, the superior portions of the nasal septum, or partition, and most of the sidewalls of the nasal roof. The ethmoid is the principal supporting structure of the nasal cavity.

Markings: Its lateral masses or **labyrinths** compose most of the wall between the nasal cavities and the orbits. They contain several air spaces, or "cells," which together form the **ethmoidal sinuses.** The sinuses are shown in Figure 6–8. The **perpendicular plate** (Figure 6–7) forms the superior portion of the nasal septum. The **cribriform plate,** or horizontal plate, lies in the anterior floor of the cranium and forms the roof of the nasal cavity. The cribriform plate contains the **olfactory foramina** through which the olfactory nerves (I) pass. These nerves function in smell (see Figure 6–5). Projecting upward from the horizontal plate is a triangular process called the **crista galli,** which means Cock's Comb. This structure serves as a point of attachment for the membranes that cover the brain. On either side of the nasal septum, the labyrinths contain two thin scroll-shaped bones. These are called the **superior nasal concha** and the **middle nasal concha.** The conchae allow for the efficient circulation and filtration of inhaled air before it passes into the lungs. See Figure 6–2 also.

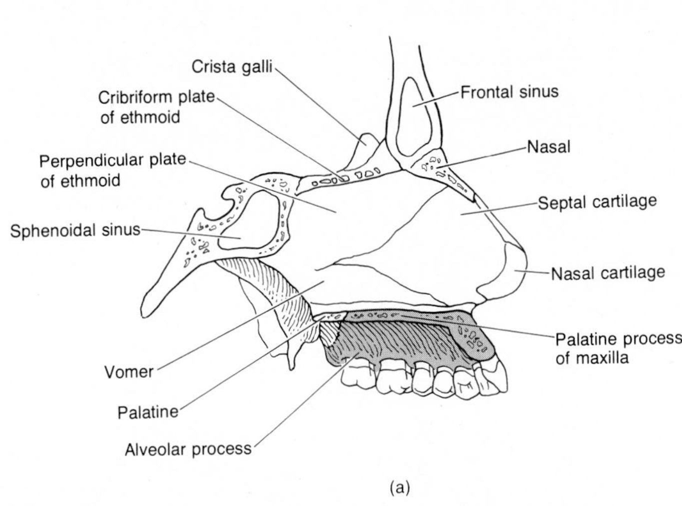

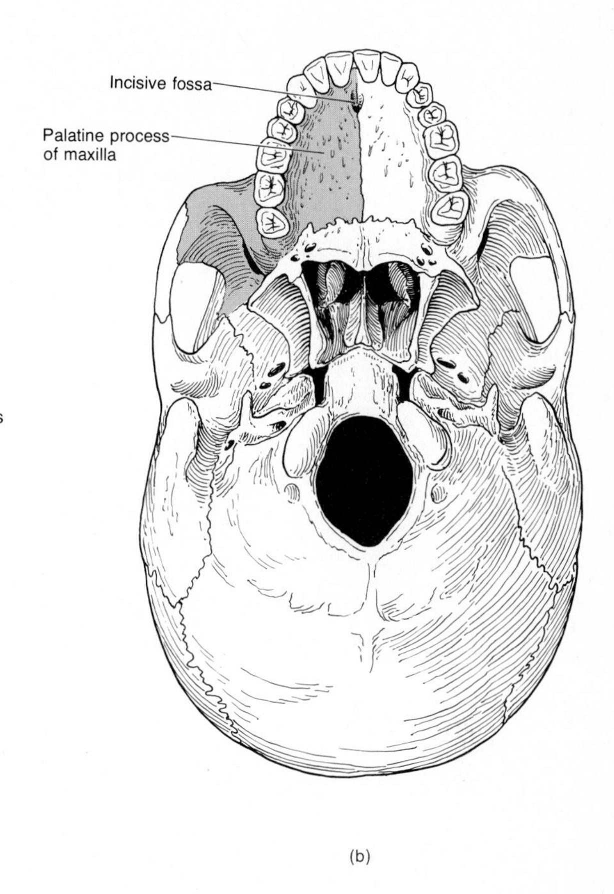

Figure 6–7. The maxillae. (a) Median view of the left maxilla. (b) Inferior view of the skull, showing the maxillae.

Exhibit 6–11. BONES OF THE SKULL: THE NASAL BONES

General Description: The paired **nasal bones** are small, oblong bones located side by side at the middle and superior part of the face. Their fusion forms the superior part of the bridge of the nose. The inferior portion of the nose, indeed the major portion, consists of cartilage. See Figures 6–7 and 6–2.

Exhibit 6–12. BONES OF THE SKULL: THE MAXILLAE (see Figure 6–7)

General Description: The paired maxillary bones unite to form the upper jawbone. The **maxillae** articulate with every bone of the face except the mandible, or lower jawbone. They form part of the floor of the orbits, part of the roof of the mouth, and part of the sidewalls and floor of the nose. The two portions of the maxillary bones unite, and the fusion is normally completed before birth. If the palatine processes of the maxillary bones do not unite before birth, a condition called **cleft palate** results. Another form of the condition, called **harelip,** involves a split in the upper lip. Harelip is often associated with cleft palate. Depending on the extent and position of the cleft, activities such as speech and swallowing may be affected.

Markings: A maxillary bone contains a **maxillary sinus** that empties into the nose. See Figure 6–8. The **alveolar process,** *alveolus* meaning hollow, contains the bony sockets into which the teeth are set. The **palatine process** is a horizontal projection of the maxilla that forms the anterior and larger part of the hard palate, or anterior portion of the roof of the mouth. The **infraorbital foramen,** which can be seen in the anterior view of the skull in Figure 6–2, is a hole in the maxilla inferior to the orbit. The infraorbital nerve and artery are transmitted through this opening. Another prominent fossa in the maxilla is the **incisive fossa** just posterior to the incisor teeth. Through it pass branches of the descending palatine vessels and the nasopalatine nerve. A final fossa associated with the maxilla and sphenoid bone is the **inferior orbital fissure.** It is located between the greater wing of the sphenoid and maxilla (see Figure 6–4). It transmits the maxillary branch of the trigeminal nerve (V) and the infraorbital vessels.

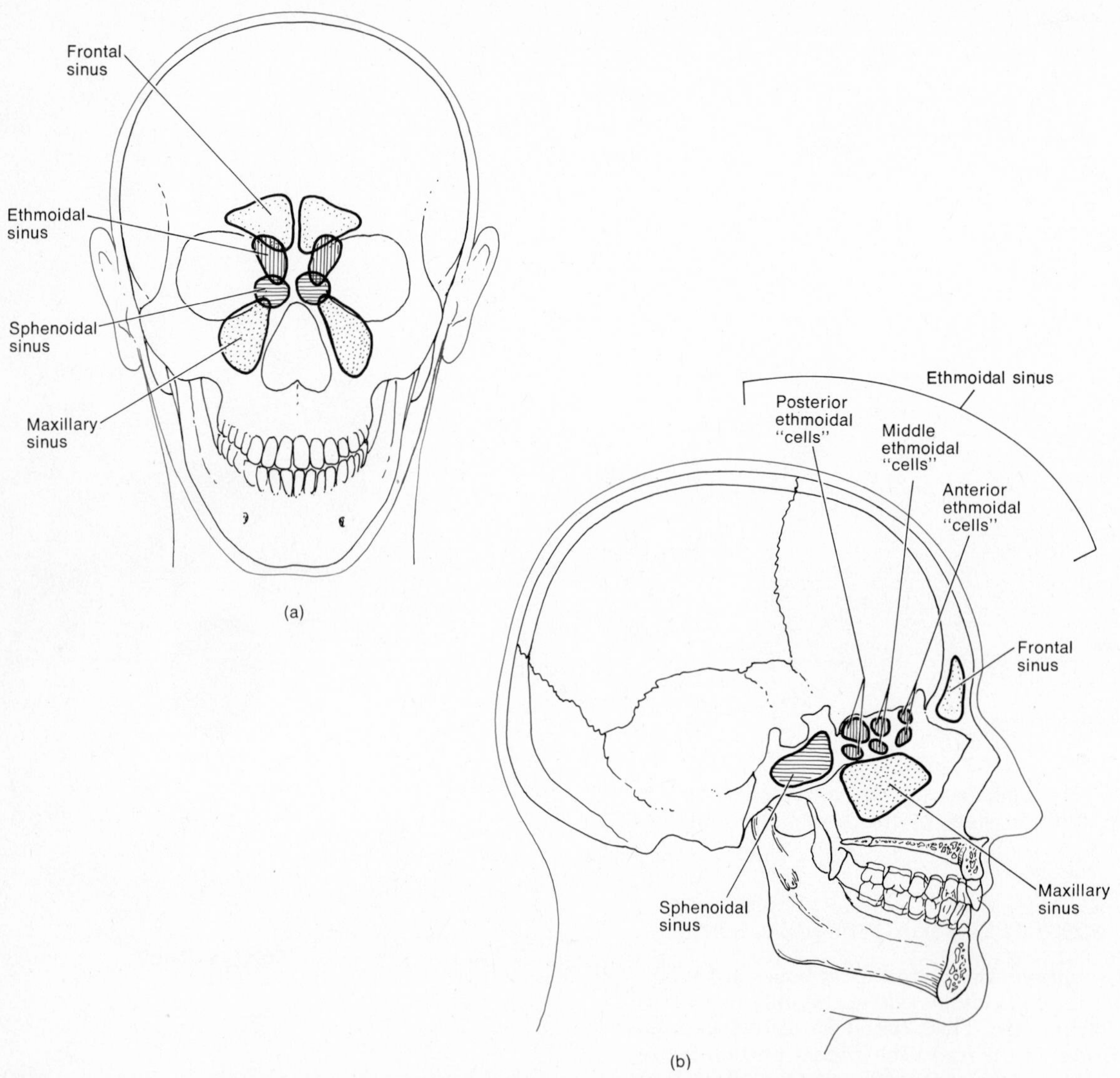

Figure 6–8. The paranasal sinuses. (a) Anterior view. (b) Right lateral view.

Exhibit 6–13. PARANASAL SINUSES OF THE SKULL (see Figure 6–8)

While discussing several cranial bones, we have referred to sinuses. These cavities, called **paranasal sinuses,** are located within certain bones near the nasal cavity. The paranasal sinuses are lined with mucous membranes that are continuous with the lining of the nasal cavity. Cranial bones containing paranasal sinuses are the frontal bone, the sphenoid, the ethmoid, and the maxillae. The ethmoid sinus consists of a series of small cavities called ethmoid "cells," which range in number from 3 to 18.

Secretions produced by the mucous membranes of the paranasal sinuses drain into the nasal cavity. An inflammation of the membranes due to an allergic reaction or infection is called *sinusitis.* If the membranes swell enough to block drainage into the nasal cavity, fluid pressure builds up in the paranasal sinuses, and a common sinus headache results.

Exhibit 6–14. BONES OF THE SKULL: THE ZYGOMATIC BONES

General Description: The two **zygomatic,** or **malar, bones** are commonly referred to as the cheekbones. They form the prominences of the cheeks and part of the outer wall and floor of the orbits. These bones can be seen in the lateral view of the skull in Figure 6–2.

Markings: The **temporal process** of the zygomatic bone projects posteriorly and articulates with the zygomatic process of the temporal bone. These two processes form the **zygomatic arch.** A foramen associated with the zygomatic bone is the **zygomaticofacial foramen** near the center of the bone (see Figure 6–2). It transmits the zygomaticofacial nerve and vessels.

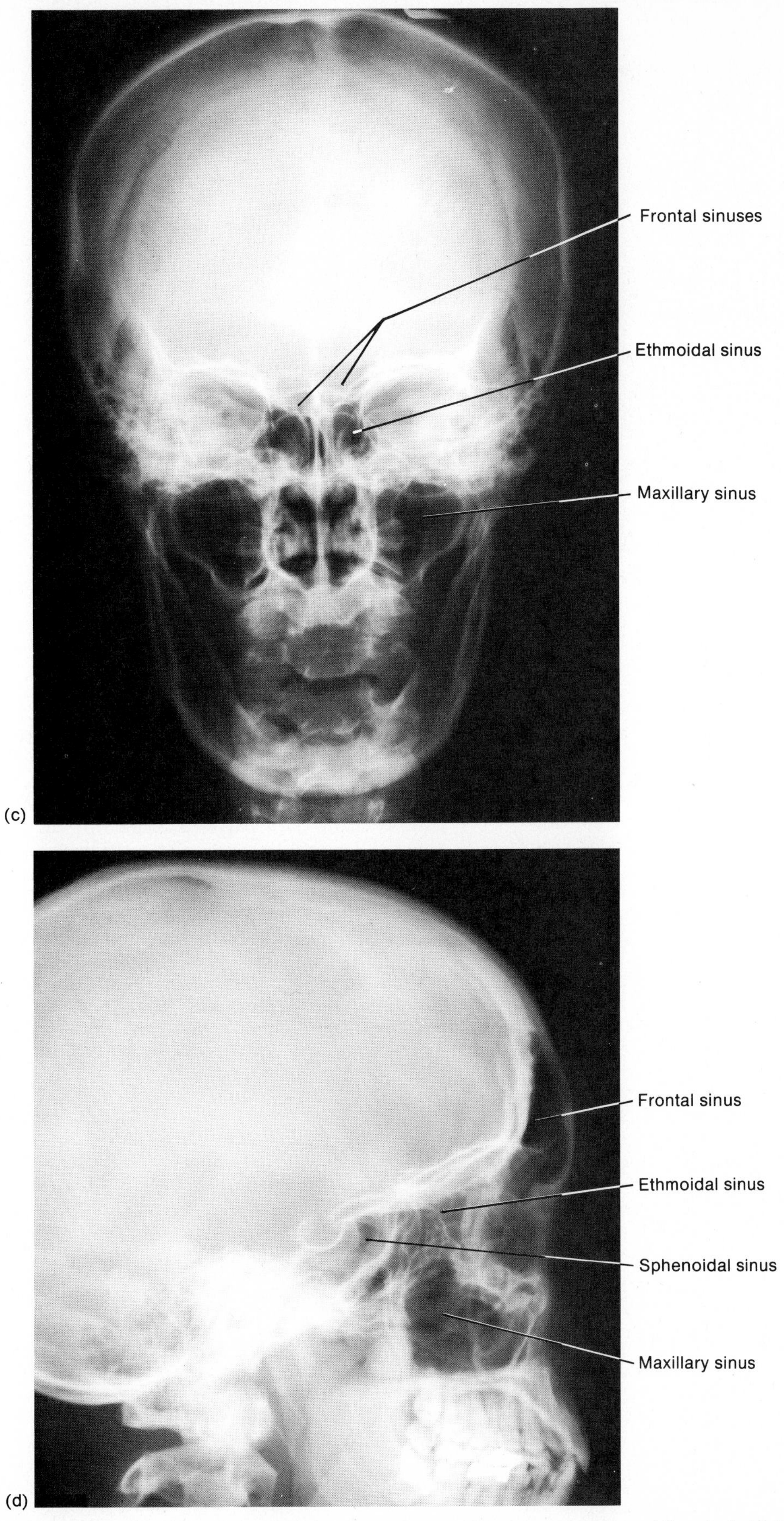

Figure 6–8 (cont.). The paranasal sinuses. (c) Anteroposterior roentgenogram of the skull. (d) Right lateral roentgenogram of the skull. (Courtesy of Eastman Kodak Company.)

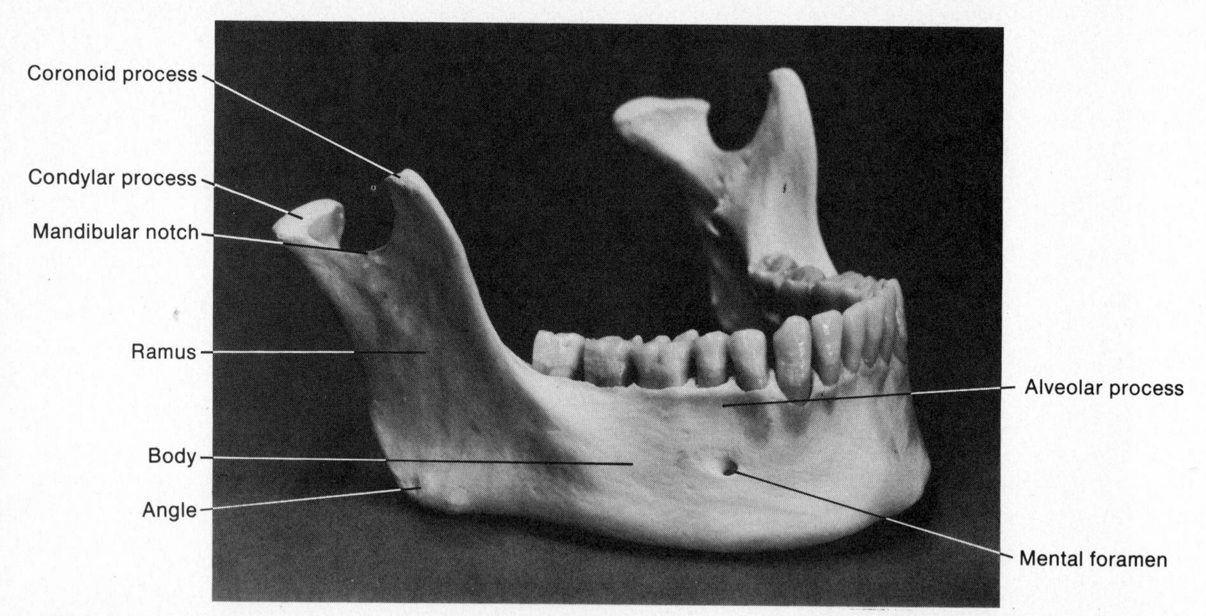

Figure 6–9. Right lateral photograph of the mandible. (Courtesy of Vincent P. Destro, Mayo Foundation.)

Exhibit 6–15. BONES OF THE SKULL: THE MANDIBLE (see Figure 6–9)

General Description: The **mandible** or lower jawbone is the largest, strongest facial bone. It is the only movable bone in the skull.

Markings: In the lateral view you can see that the mandible consists of a curved, horizontal portion called the **body** and two perpendicular portions called the **rami.** The **angle** of the mandible is the area where each ramus meets the body. Each ramus has a **condylar process** that articulates with the mandibular fossa of the temporal bone. It also has a **coronoid process** to which the temporal muscle attaches. The depression between the coronoid and condylar processes is called the **mandibular notch.** The **mental foramen,** the term *mentum* meaning chin, is approximately below the first molar tooth. The mental nerve and vessels pass through this opening. Dentists inject anesthetics through this foramen. The **alveolar process,** like that of the maxillae, is an arch containing the sockets for the teeth. Another foramen associated with the mandible is the **mandibular foramen** on the medial surface of the ramus. It transmits the inferior alveolar nerve and vessels.

Exhibit 6–16. BONES OF THE SKULL: THE LACRIMAL BONES

General Description: The paired **lacrimal bones** are thin bones roughly resembling a fingernail in size and shape. The term *lacrima* means tear. These bones are posterior and lateral to the nasal bones in the medial wall of the orbit. They can be seen in the anterior and lateral views of the skull in Figure 6–2. The lacrimal bones help to form the medial wall of the nasal cavity. They also contain the **lacrimal foramina** through which the tear ducts pass into the nasal cavity (see Figure 6–2). The lacrimal bones are the smallest bones of the face.

Exhibit 6–17. BONES OF THE SKULL: THE PALATINE BONES

General Description: The two **palatine bones** are L-shaped and form the posterior portion of the hard palate, part of the floor and lateral walls of the nasal cavities, and a small portion of the floor of the orbit. The posterior portion of the hard palate is formed by the **horizontal plates** of the palatine bones. These can be seen in Figures 6–4 and 6–2g. Two foramina associated with the palatine bones are the greater and lesser palatine foramina. The **greater palatine foramen,** at the posterior angle of the hard palate, transmits the greater palatine nerve and descending palatine vessels (see Figure 6–4). The **lesser palatine foramina,** usually two or more on each side, are posterior to the greater palatine foramina. They transmit the lesser palatine nerve. See Figure 6–4.

Exhibit 6–18. BONES OF THE SKULL: THE INFERIOR CONCHAE

General Description: Refer to the views of the skull in Figures 6–2a and 6–6a. The two **inferior conchae** are scroll-like bones that project into the nasal cavity inferior to the superior and middle conchae of the ethmoid bone. The name of these bones is derived from *concha,* meaning shell. They serve the same function as the superior and middle conchae. This is, they allow for the circulation and filtration of air before it passes into the lungs. The inferior conchae are separate bones and are not part of the ethmoid.

General Description: The **vomer,** which means plowshare, is a roughly triangular bone that forms the inferior part of the nasal septum. It is clearly seen in the anterior view of the skull in Figure 6–2. Its inferior border articulates with the cartilage septum that divides the nose into a right and left nostril. Its superior border articulates with the perpendicular plate of the ethmoid bone. The structures, then, that form the nasal septum or partition are the perpendicular plate of the ethmoid, the septal cartilage, and the vomer. If the vomer is pushed to one side — that is, deviated — the nasal chambers are of unequal size. See the skull viewed from below (Figure 6–4) for another view of the vomer.

Exhibit 6–20. SUMMARY OF FORAMINA OF THE SKULL

FORAMEN	LOCATION	STRUCTURES PASSING THROUGH
Carotid (Figure 6–4)	Petrous portion of temporal	Internal carotid artery
Greater palatine (Figure 6–4)	Posterior angle of hard palate	Greater palatine nerve and descending palatine vessels
Hypoglossal (Figure 6–5)	Superior to base of occipital condyles	Hypoglossal nerve and branch of ascending pharyngeal artery
Incisive (Figure 6–7)	Posterior to incisor teeth	Branches of descending palatine vessels and nasopalatine nerve
Inferior orbital (Figure 6–4)	Between greater wing of sphenoid and maxilla	Maxillary branch of trigeminal nerve (V), zygomatic nerve, and infraorbital vessels
Infraorbital (Figure 6–2 a)	In maxilla inferior to orbit	Infraorbital nerve and artery
Jugular (Figure 6–4)	Posterior to carotid canal between petrous portion of temporal and occipital	Internal jugular vein, glossopharyngeal nerve (IX), vagus nerve (X), and accessory nerve (XI)
Lacerum (Figure 6–5)	Bounded anteriorly by sphenoid, posteriorly by petrous portion of temporal, and medially by the sphenoid and occipital	Internal carotid artery and branch of ascending pharyngeal artery
Lacrimal (Figure 6–2e)	Lacrimal bone	Lacrimal (tear) duct
Lesser palatine (Figure 6–4)	Posterior to greater palatine foramen	Lesser palatine nerves
Magnum (Figure 6–4)	Occipital bone	Medulla oblongata and its membranes, the accessory nerve (XI), and the vertebral and spinal arteries
Mandibular	Medial surface of ramus of mandible	Inferior alveolar nerve and vessels
Mastoid (Figure 6–4)	Posterior border of mastoid process of temporal bone	Vein to transverse sinus and branch of occipital artery to dura mater
Mental (Figure 6–9)	Inferior to second premolar tooth in mandible	Mental nerve and vessels
Olfactory (Figure 6–5)	Cribriform plate of ethmoid	Olfactory nerve (I)
Optic (Figure 6–5)	Between upper and lower portions of small wing of sphenoid	Optic nerve (II) and ophthalmic artery
Ovale (Figure 6–5)	Greater wing of sphenoid	Mandibular branch of trigeminal nerve (V)
Rotundum (Figure 6–5)	Junction of anterior and medial parts of sphenoid	Maxillary branch of trigeminal nerve (V)
Spinosum (Figure 6–5)	Posterior angle of sphenoid	Middle meningeal vessels
Stylomastoid (Figure 6–4)	Between styloid and mastoid processes of temporal	Facial nerve (VII) and stylomastoid artery
Superior orbital (Figure 6–5)	Between greater and lesser wings of sphenoid	Oculomotor nerve (III), trochlear nerve (IV), ophthalmic branch of trigeminal nerve (V), and abducens nerve (VI)
Supraorbital (Figure 6–2a)	Supraorbital margin of orbit	Supraorbital nerve and artery
Zygomaticofacial (Figure 6–2a)	Zygomatic bone	Zygomaticofacial nerve and vessels

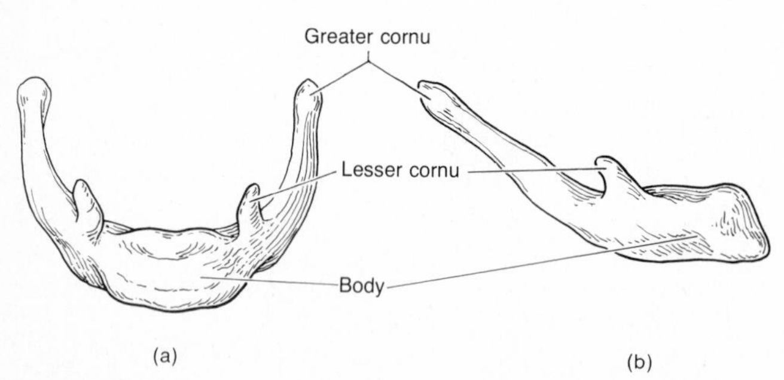

Figure 6–10. The hyoid bone. (a) Anterior view. (b) Right lateral view.

Exhibit 6–21. THE HYOID BONE (see Figure 6–10)

General Description. The single **hyoid bone,** *hyoid* meaning U-shaped, is a unique component of the axial skeleton because it does not articulate with any other bone. Rather, it is suspended from the styloid process of the temporal bone by ligaments. The hyoid is located in the neck between the mandible and larynx. It supports the tongue and affords attachment for some of its muscles. Refer to the anterior and lateral views of the skull in Figure 6–2 to see the position of the hyoid bone in the body.

Markings: The hyoid consists of a horizontal **body** and paired projections called the **lesser cornu** and the **greater cornu.** Muscles attach to these paired projections.

Exhibit 6–22. THE VERTEBRAL COLUMN: GENERAL CONSIDERATIONS (see Figure 6–11)

Together with the sternum and ribs, the **vertebral column,** or **spine,** constitutes the skeleton of the **trunk** of the body. The vertebral column is composed of a series of bones called **vertebrae.** And in the average adult, the column measures about 71 cm (28 in.). In effect, the vertebral column is a strong, flexible rod that moves anteriorly, posteriorly, and laterally. It encloses and protects the spinal cord, supports the head, and serves as a point of attachment for the ribs and for the muscles of the back. Between the vertebrae are openings called **intervertebral foramina.** The nerves that connect the spinal cord to various parts of the body run out of these openings.

The adult vertebral column typically contains 26 vertebrae. These are distributed as follows: 7 **cervical vertebrae** in the neck region; 12 **thoracic vertebrae** posterior to the thoracic cavity; 5 **lumbar vertebrae** supporting the small of the back; 5 **sacral vertebrae** fused into one bone called the **sacrum;** and **coccygeal vertebrae** fused into one or two bones called the **coccyx.** Prior to the fusion of the sacral and coccygeal vertebrae, the total number of vertebrae is 33. Between the vertebrae are fibrocartilaginous **intervertebral discs.** These discs form strong joints and permit various movements of the column. A number of conditions, including injury, disease, and old age, may cause an intervertebral disc to rupture or protrude. Such a "slipped" disc is rather painful because nerves passing through the intervertebral foramina are irritated.

When viewed from the side, the vertebral column shows four **curves.** When seen from the anterior view, these are alternately convex, meaning they curve out toward the viewer, and concave, meaning they curve away from the viewer. The curves of the column, like the curves in a long bone, are very important because they increase its strength, help to maintain balance in the upright position, absorb shocks from walking, and help protect the column from fracture.

In the fetus, the four curves of the vertebrae are not present. There is only a single curve that is anteriorly concave. At about the third postnatal month when an infant begins to hold its head erect, the **cervical curve** develops. Later, when the child stands erect and walks, the **lumbar curve** develops. The cervical and lumbar curves are convex anteriorly. Because they are modifications of the fetal positions, they are called **secondary curves.** The other two curves, the **thoracic curve** and the **sacral curve,** are anteriorly concave. Since they retain the anterior concavity of the fetus, they are referred to as **primary curves.**

As a result of various conditions, the normal curves of the column may become exaggerated. In such cases, they are called **curvatures.** For example, muscular paralysis on either side of the spine, poor posture, or disease may result in **scoliosis,** the term *scolio* meaning bent. With this condition, there is a curvature toward the left or right side of the body. Scoliosis may occur in any region of the backbone. An exaggerated anterior concavity of the thoracic region is called **kyphosis,** or hunchback. **Lordosis,** or swayback, refers to an exaggerated anterior convexity in the lumbar region. Kyphosis and lordosis may be caused by tuberculosis of the vertebrae, rickets, or poor posture.

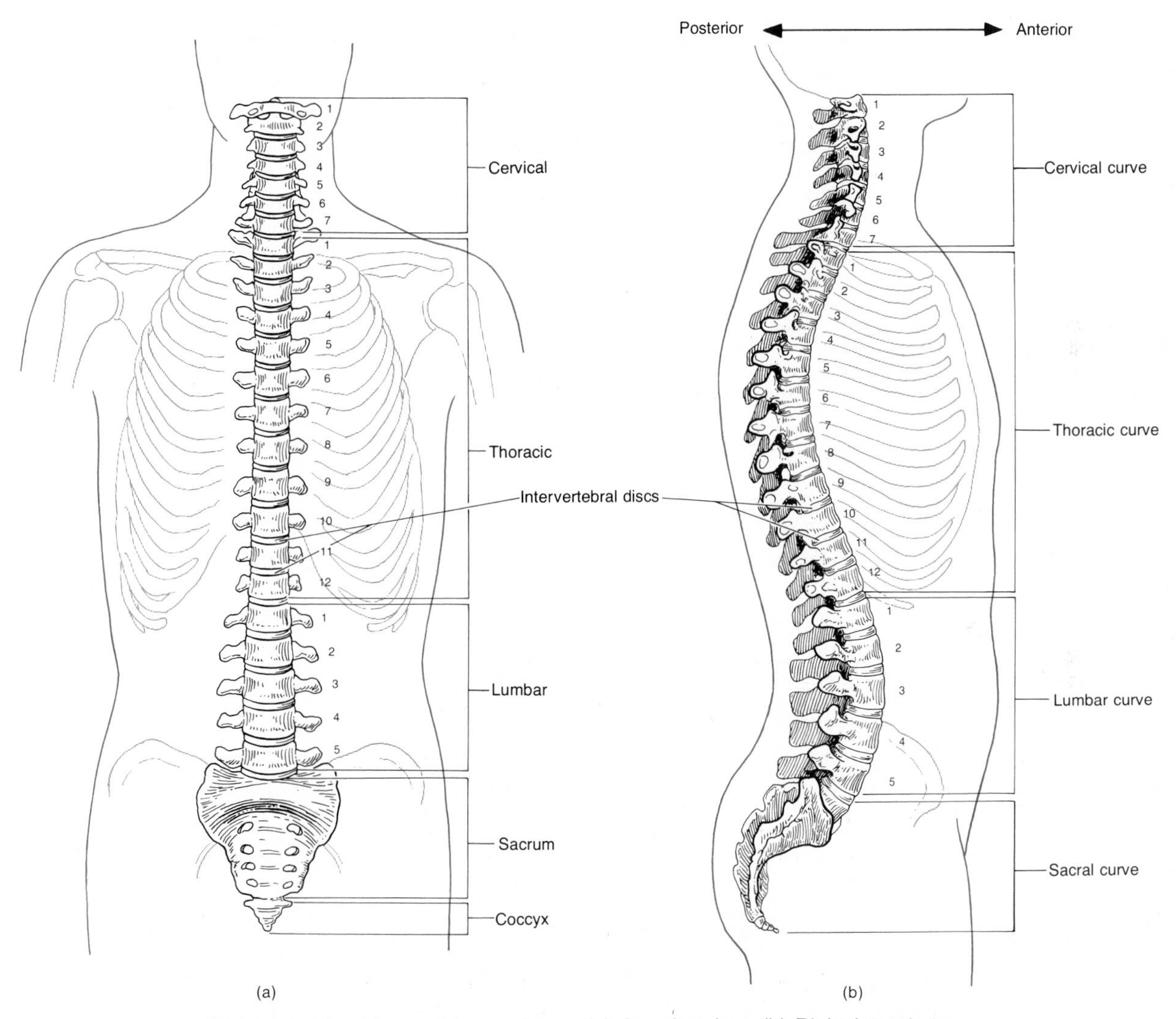

Figure 6–11. The vertebral column. (a) Anterior view. (b) Right lateral view.

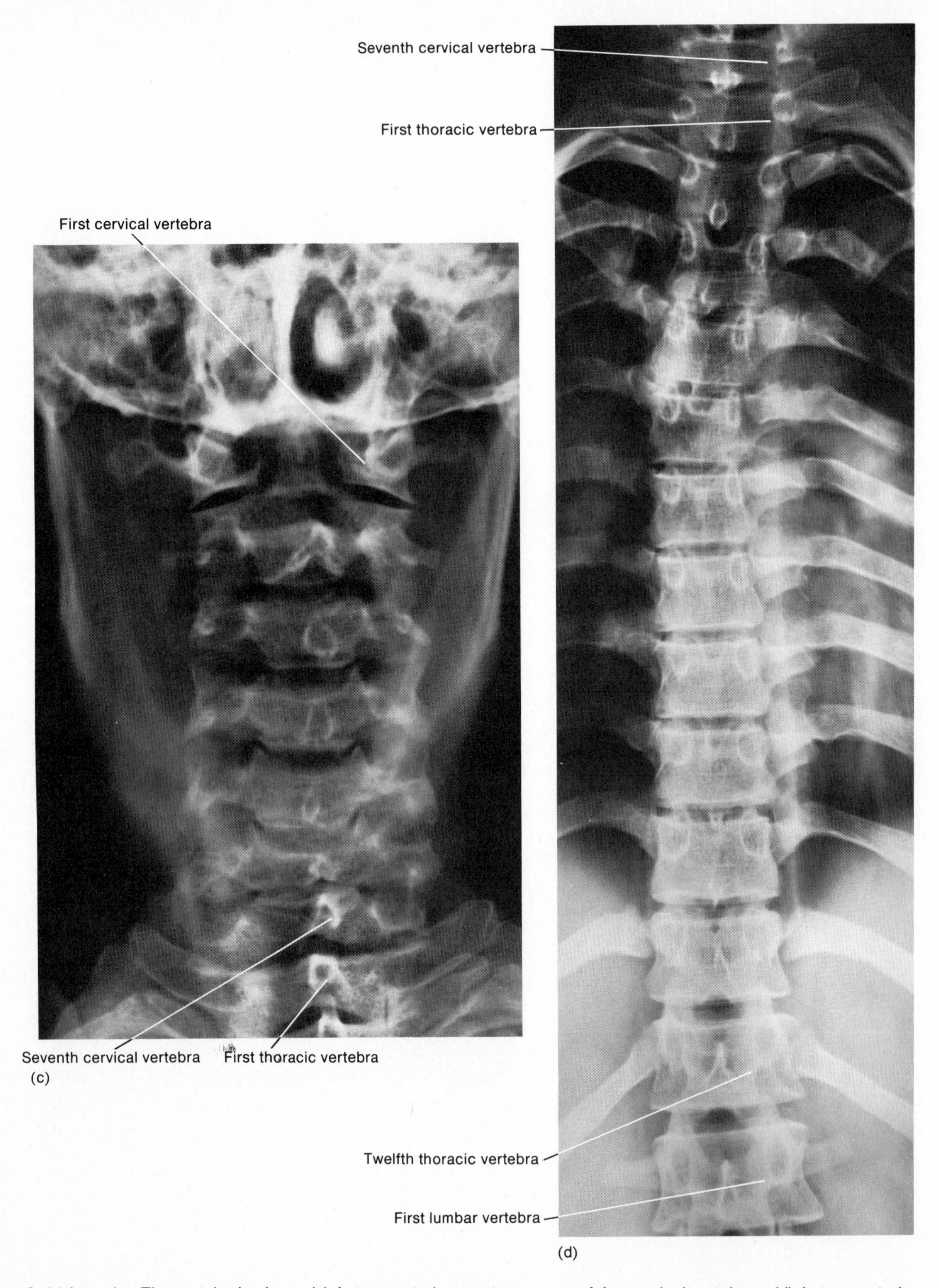

Figure 6–11 (cont.). The vertebral column. (c) Anteroposterior roentgenogram of the cervical vertebrae. (d) Anteroposterior roentgenogram of the thoracic vertebrae. (e) Anteroposterior roentgenogram of the lumbar vertebrae. (f) Lateral roentgenogram of the sacrum. (Courtesy of Eastman Kodak Company.)

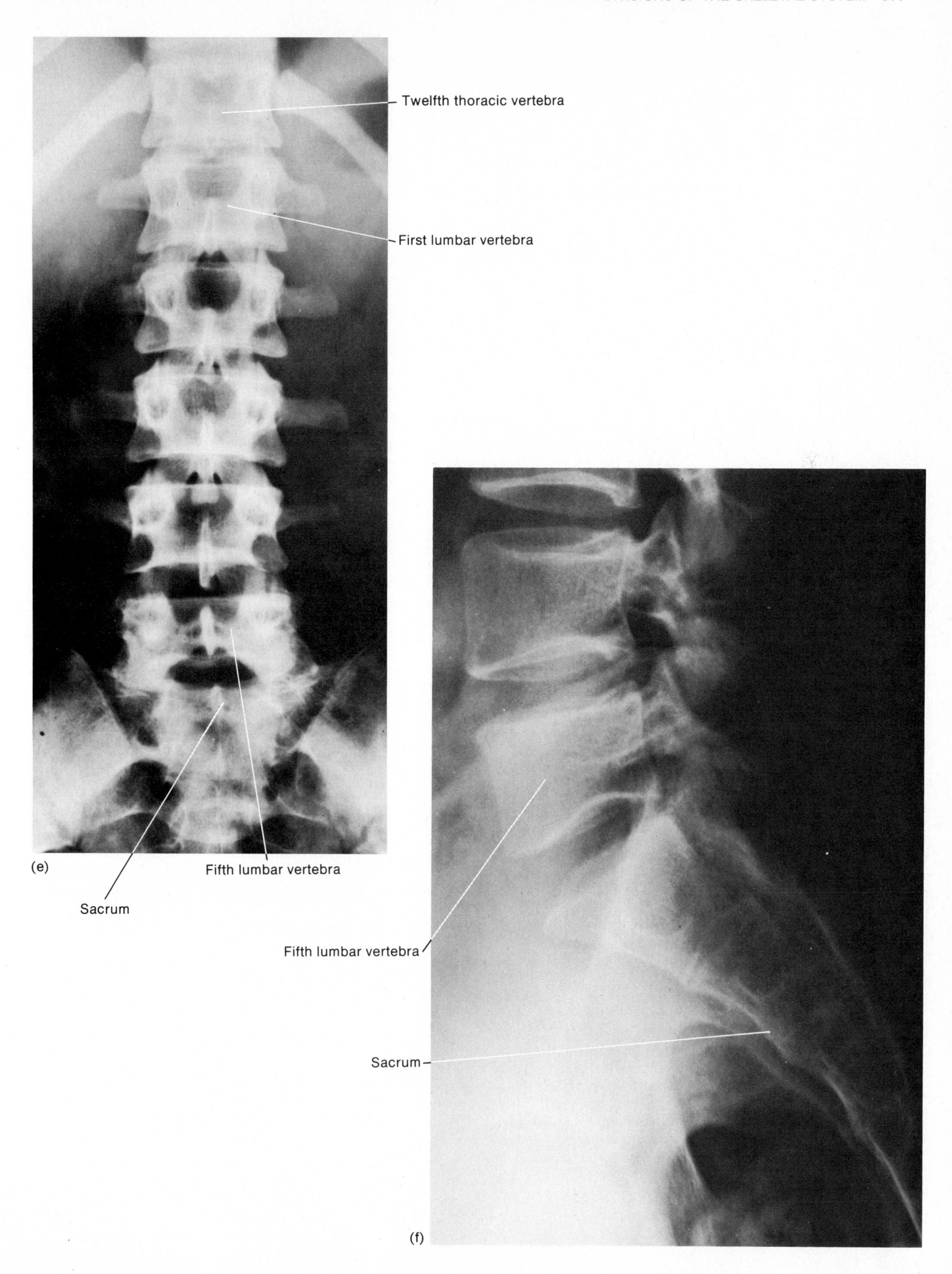

Twelfth thoracic vertebra
First lumbar vertebra
(e)
Fifth lumbar vertebra
Sacrum
Fifth lumbar vertebra
Sacrum
(f)

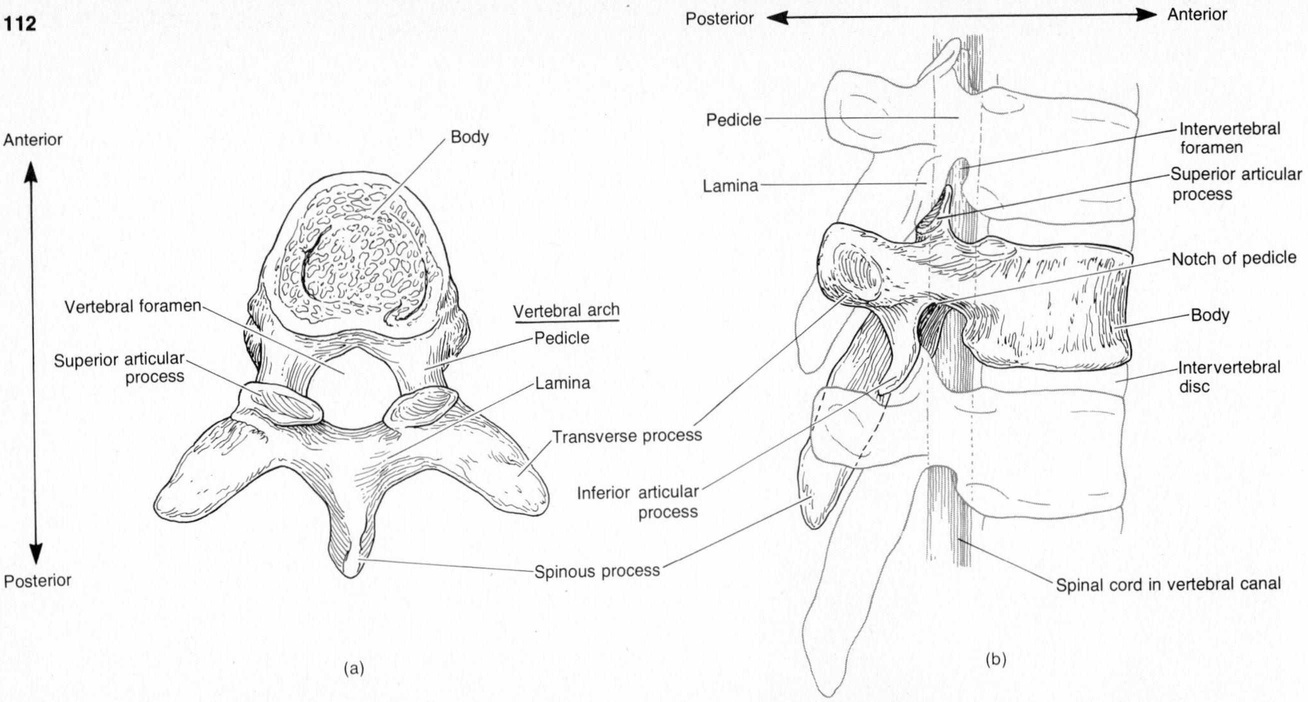

Figure 6–12. A typical vertebra. (a) Superior view. (b) Right lateral view.

Exhibit 6–23. THE VERTEBRAL COLUMN: STRUCTURE OF A TYPICAL VERTEBRA (see Figure 6–12)

All the vertebrae of the column are basically similar in structure. But there are differences in size, shape, and specialized details. A typical vertebra consists of the following components that may be viewed from above and laterally:

1. The **body** is the thick, disc-shaped anterior portion that is the weight-bearing part of a vertebra. Its superior and inferior surfaces are roughened for the attachment of intervertebral discs. The anterior and lateral surfaces contain foramina for blood vessels.

2. The **vertebral,** or **neural, arch** extends posteriorly from the body of the vertebra. Along with the body of the vertebra, it surrounds the spinal cord. It is formed by two short, thick processes, the **pedicles,** which project posteriorly to unite with the **laminae.** The space that lies between the vertebral arch and body contains the spinal cord. This space is known as the **vertebral foramen.** The vertebral foramina of all vertebrae together form the **vertebral** or **spinal canal.** The pedicles are notched superiorly and inferiorly in such a way that, when they are arranged in the column, there is an opening between vertebrae on each side of the column. This opening, the **intervertebral foramen,** permits the passage of nerves to and from the spinal cord.

3. Seven **processes** arise from the vertebral arch. At the point where a lamina and pedicle join, there is a **transverse process** extending laterally on each side. A single **spinous process** projects posteriorly and inferiorly from the junction of the laminae forming the **spine.** These three processes serve as points of muscular attachment. The function of the remaining four processes is to form joints with other vertebrae. The two **superior articular processes** of a vertebra articulate with the vertebra immediately superior to it. The two **inferior articular processes** of a vertebra articulate with the vertebra inferior to it.

Exhibit 6–24. THE VERTEBRAL COLUMN: THE CERVICAL REGION (see Figure 6–13)

When viewed from above, it can be seen that the bodies of **cervical vertebrae** are smaller than those of the thoracic vertebrae. The arches, however, are larger. The spinous processes of the second through sixth cervical vertebrae are often *bifid*—that is, cleft in two. Each transverse process contains an opening, the **transverse foramen.** The vertebral artery and its accompanying vein and nerve filaments pass through this opening.

The first two cervical vertebrae differ considerably from the others. The first cervical vertebra, the **atlas,** is so named because it supports the head. Essentially, the atlas is a ring of bone with **anterior** and **posterior arches** and large **lateral masses.** It lacks a body and a spinous process. The superior surfaces of the lateral masses, called **superior articular surfaces,** are concave and articulate with the occipital condyles of the occipital bone. This articulation permits the movement seen when nodding the head "yes." The inferior surfaces of the lateral masses, the **inferior articular surfaces,** articulate with the second cervical vertebra. The transverse processes and **transverse foramina** of the atlas are quite large.

The second cervical vertebra, the **axis,** does have a **body.** A peglike process called the **dens,** or odontoid process, projects up through the ring of the atlas. This makes a pivot on which the atlas and head rotate. This arrangement permits movement from side to side as in shaking the head to mean "no."

The third through sixth cervical vertebrae correspond to the structural pattern of the typical cervical vertebra shown. The seventh cervical vertebra, however, is somewhat different. It is called the **vertebra prominens** and is marked by a large, nonbifid spinous process that may be seen and felt at the base of the neck. (See Exhibit 1–5 and Figure 6–11.)

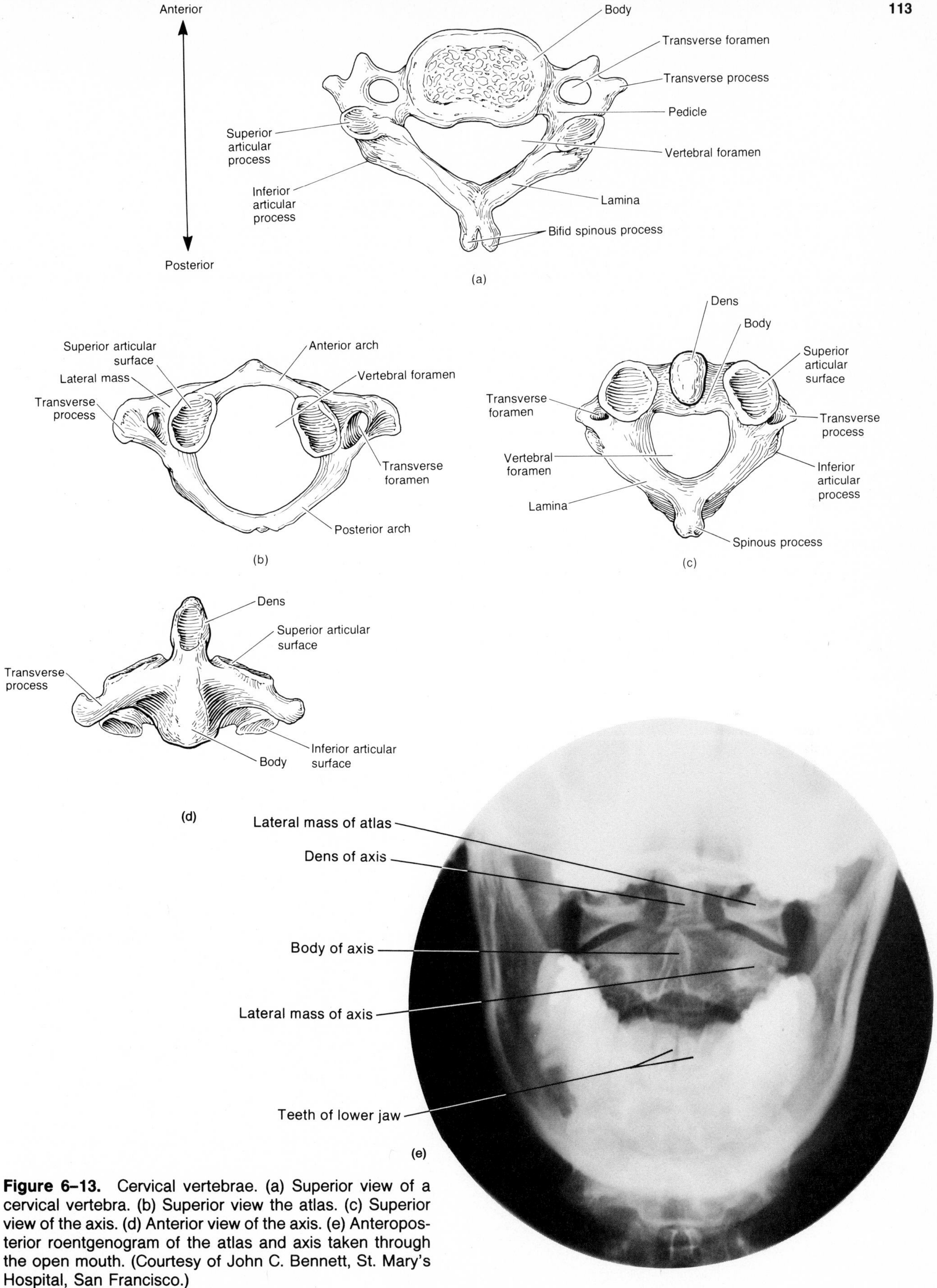

Figure 6–13. Cervical vertebrae. (a) Superior view of a cervical vertebra. (b) Superior view the atlas. (c) Superior view of the axis. (d) Anterior view of the axis. (e) Anteroposterior roentgenogram of the atlas and axis taken through the open mouth. (Courtesy of John C. Bennett, St. Mary's Hospital, San Francisco.)

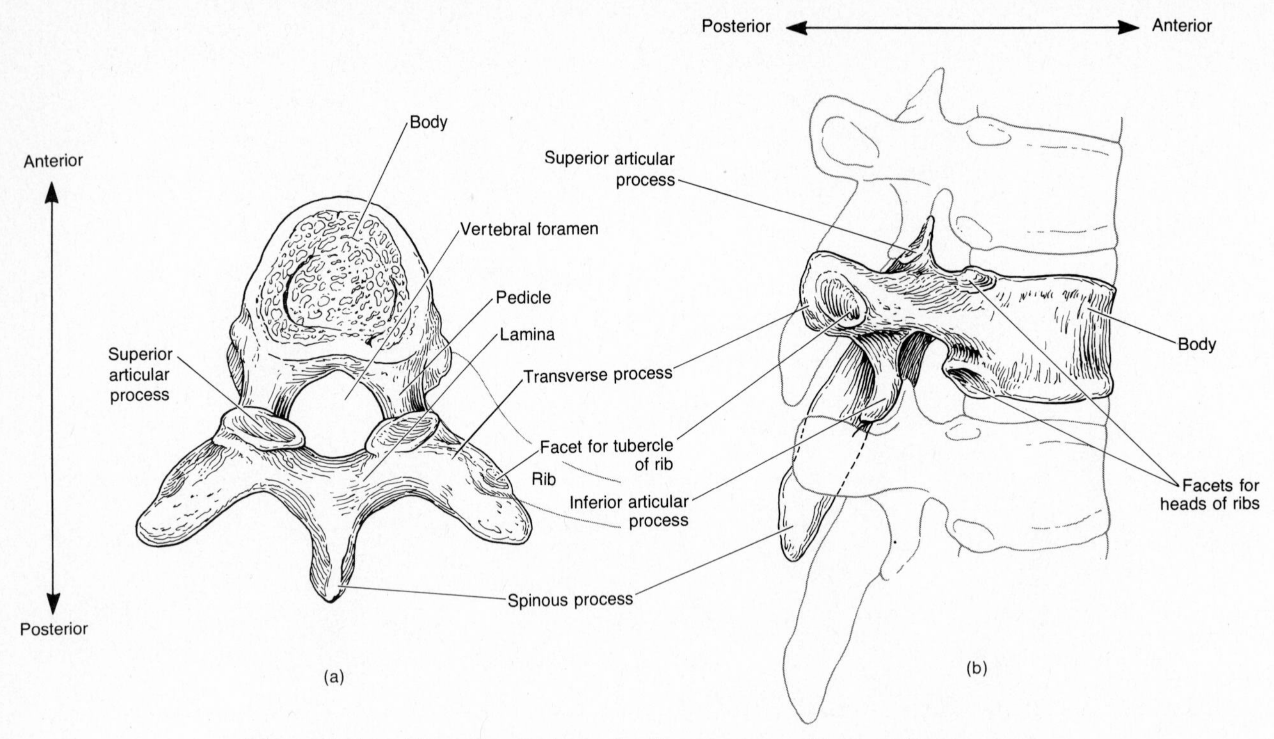

Figure 6–14. Thoracic vertebrae. (a) Superior view. (b) Right lateral view.

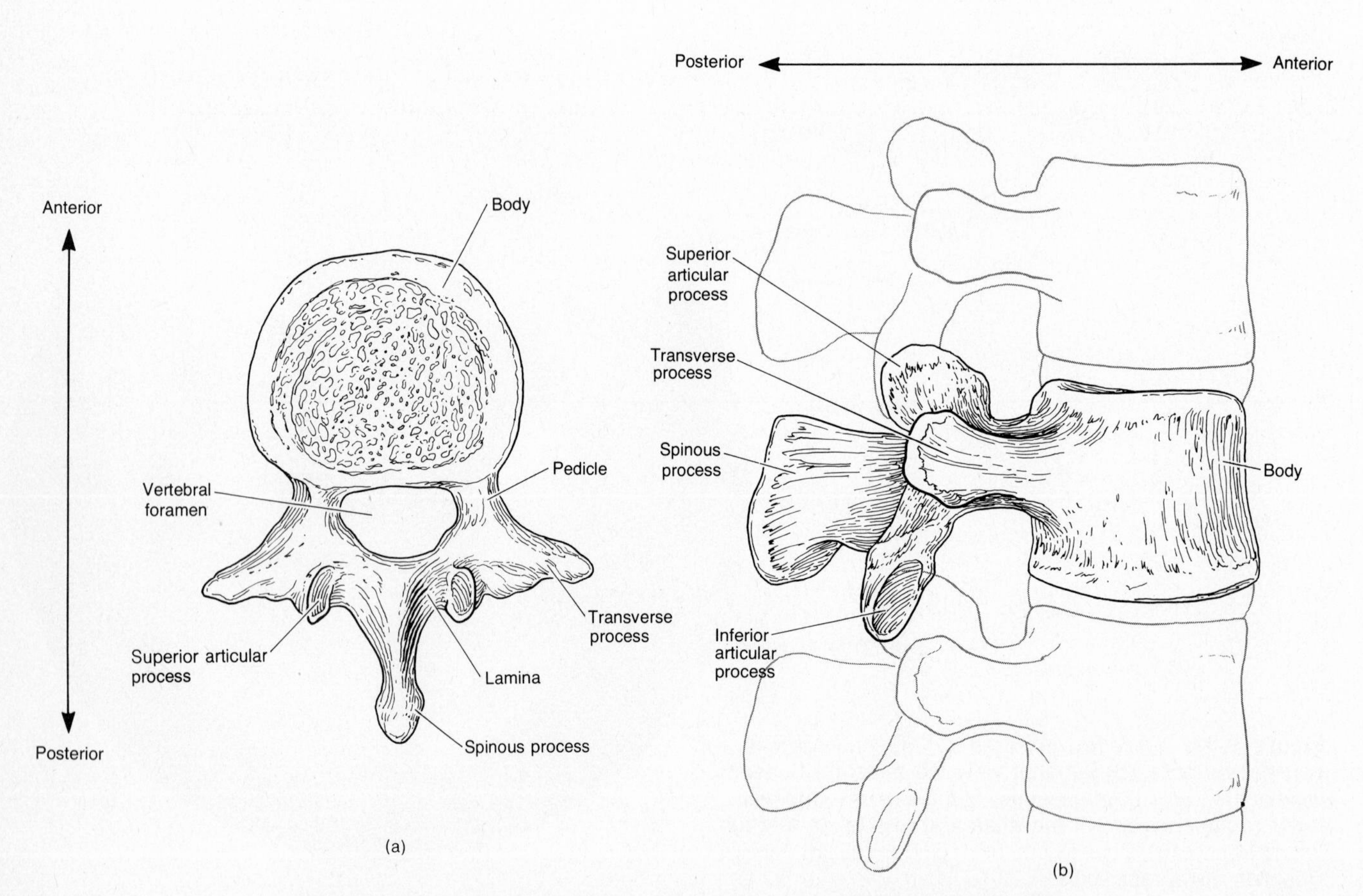

Figure 6–15. Lumbar vertebrae. (a) Superior view. (b) Right lateral view.

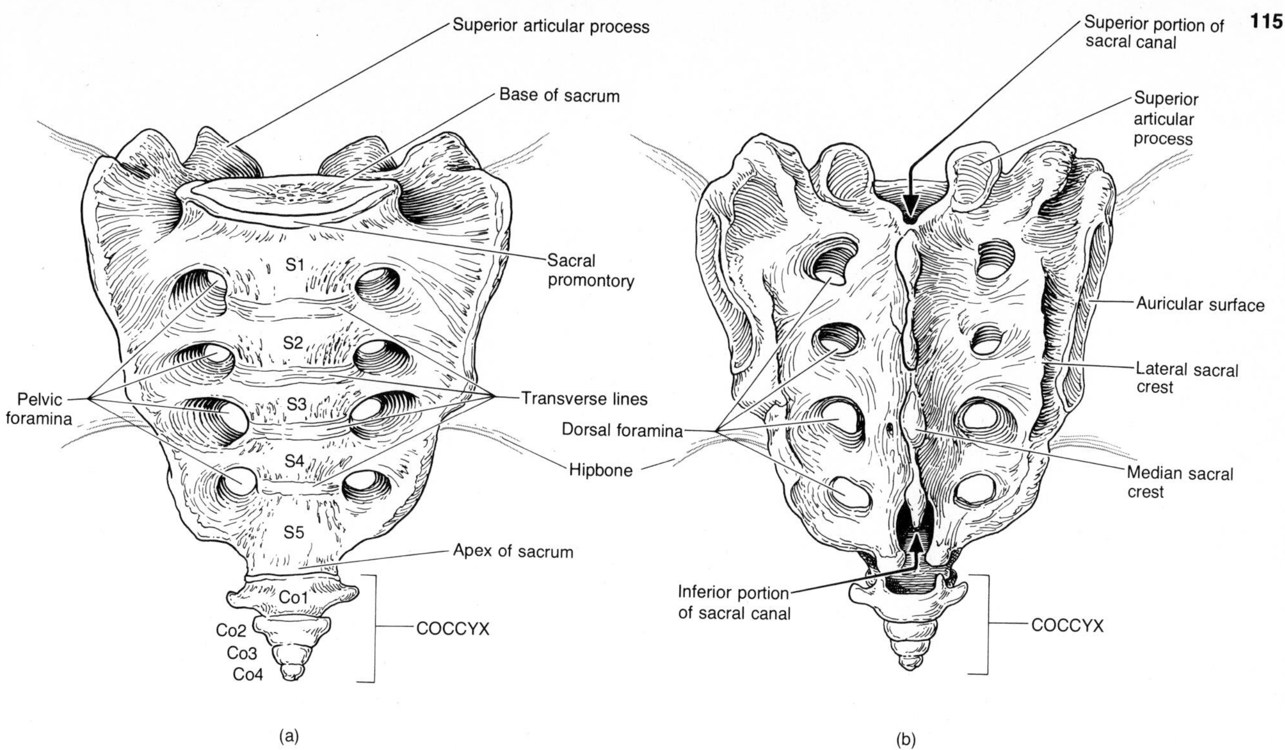

Figure 6–16. The sacrum and coccyx. (a) Anterior view. (b) Posterior view.

Exhibit 6–25. THE VERTEBRAL COLUMN: THE THORACIC REGION (see Figure 6–14)

Viewing a typical **thoracic vertebra** from above, you can see that it is considerably larger and stronger than a vertebra of the cervical region. In addition, the spinous process on each vertebra is long and pointed and projects inferiorly. Thoracic vertebrae also have longer and heavier transverse processes than cervical vertebrae. Except for the eleventh and twelfth thoracic vertebrae, the transverse processes have **facets** for articulating with the tubercles of the ribs.

Exhibit 6–26. THE VERTEBRAL COLUMN: THE LUMBAR REGION (see Figure 6–15)

The **lumbar vertebrae** are the largest and strongest in the entire column. Their superior articular processes are directed medially instead of superiorly. And their inferior articular processes are directed laterally instead of inferiorly. Their various projections are short and thick, and the spinous process is heavy for attachment of the large back muscles.

Exhibit 6–27. THE VERTEBRAL COLUMN: THE SACRUM AND COCCYX (see Figure 6–16)

The **sacrum** is a triangular bone formed by the union of five sacral vertebrae. These are indicated in the figure as S1 through S5. It serves as a strong foundation for the pelvic girdle. It is positioned at the posterior portion of the pelvic cavity between the two hipbones. Anterior and posterior views of the bone are shown here. The concave anterior side of the sacrum faces the pelvic cavity. It is smooth and contains four **transverse lines** that mark the joining of the bodies of the vertebrae. At the ends of the lines are four pairs of **pelvic foramina.** The convex, posterior surface of the sacrum is irregular. It contains a **median sacral crest,** a **lateral sacral crest,** and four pairs of **dorsal foramina.** These foramina communicate with the pelvic foramina through which nerves and blood vessels pass. The **sacral canal** is a continuation of the vertebral canal. The superior border of the sacrum exhibits an anteriorly projecting border, the **sacral promontory.** It is an important obstetrical landmark for measurements of the pelvis. An imaginary line running from the superior surface of the symphysis pubis to the sacral promontory separates the abdominal and pelvic cavities. Laterally, the sacrum has a large **auricular surface** for articulating with the ilium of the hipbone. Its superior articular process articulates with the fifth lumbar vertebra.

A clinical condition associated with the lamina of the vertebrae, especially those in the lumbosacral region, is referred to as *spina bifida.* In this condition, there is an imperfect union of the two sides of the lamina, leaving a cleft in the arch. The membranes and spinal cord may protrude through the cleft, forming a "tumor" on the back.

The **coccyx** is also triangular in shape and is formed by the fusion of the coccygeal vertebrae, usually the last four. These are indicated in Figure 6–16 as Co1 through Co4. It articulates superiorly with the sacrum. The coccyx is the most rudimentary part of the column, representing the vestige of a tail.

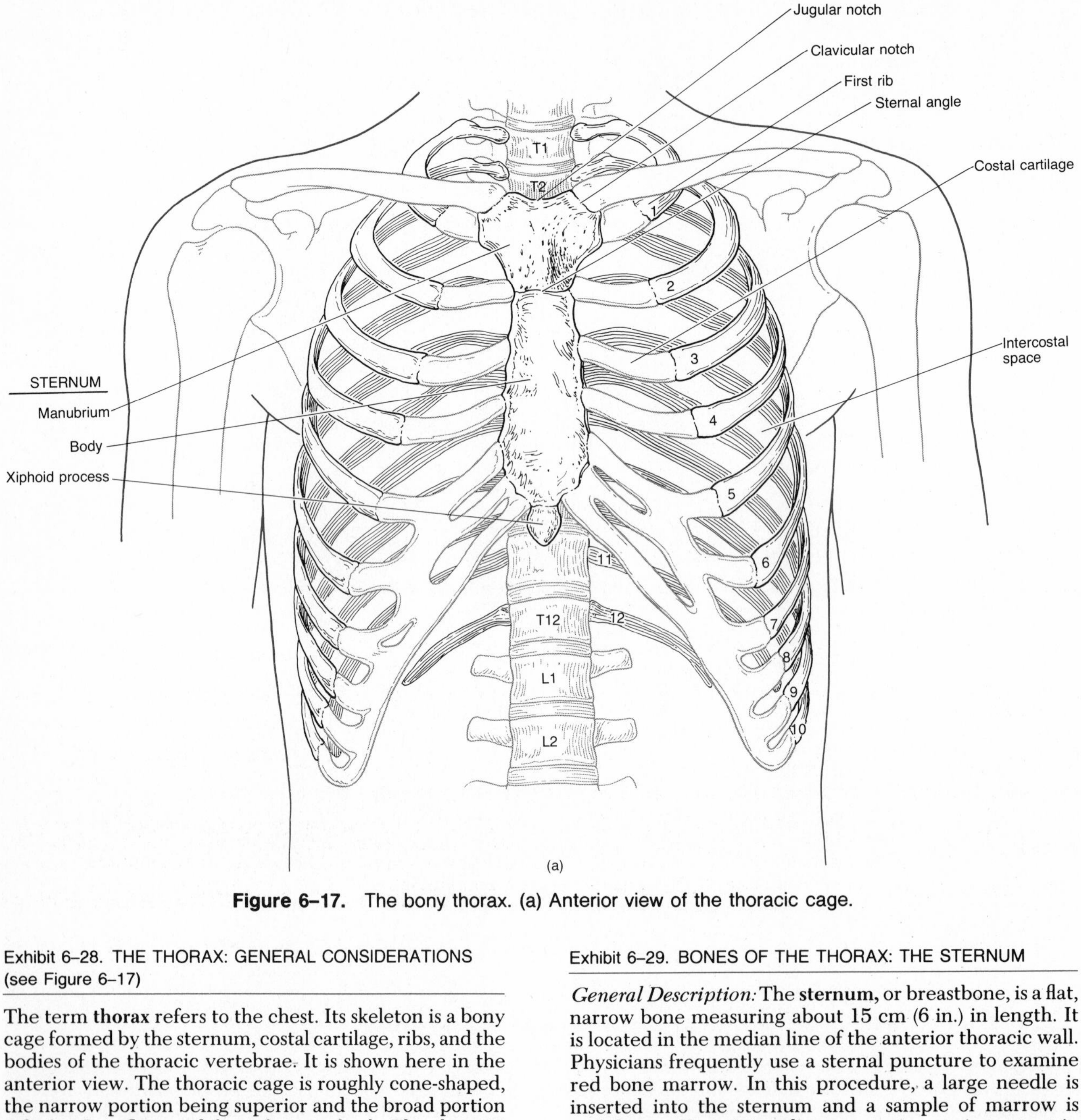

(a)

Figure 6–17. The bony thorax. (a) Anterior view of the thoracic cage.

Exhibit 6–28. THE THORAX: GENERAL CONSIDERATIONS (see Figure 6–17)

The term **thorax** refers to the chest. Its skeleton is a bony cage formed by the sternum, costal cartilage, ribs, and the bodies of the thoracic vertebrae. It is shown here in the anterior view. The thoracic cage is roughly cone-shaped, the narrow portion being superior and the broad portion inferior. It is flattened from front to back. The thoracic cage encloses and protects the organs in the thoracic cavity. It also provides support for the bones of the shoulder girdle and upper extremities.

Exhibit 6–29. BONES OF THE THORAX: THE STERNUM

General Description: The **sternum,** or breastbone, is a flat, narrow bone measuring about 15 cm (6 in.) in length. It is located in the median line of the anterior thoracic wall. Physicians frequently use a sternal puncture to examine red bone marrow. In this procedure, a large needle is inserted into the sternum and a sample of marrow is withdrawn. Examining the marrow is very important in the diagnosis of blood disorders.

Markings: The sternum (see Figure 6–17) consists of three basic portions: (1) the **manubrium,** which is a triangular, superior portion; (2) the **body,** which is the middle, largest portion; and (3) the **xiphoid process,** which is the inferior, smallest portion. The manubrium has a depression on its superior surface called the **jugular notch.** On each side of the jugular notch are **clavicular notches** that articulate with the medial ends of the clavicles. The manubrium also articulates with the first and second ribs. The body of the sternum articulates with the second through tenth ribs. These bones articulate either directly or indirectly. The xiphoid process has no ribs attached to it but affords attachment for some abdominal muscles.

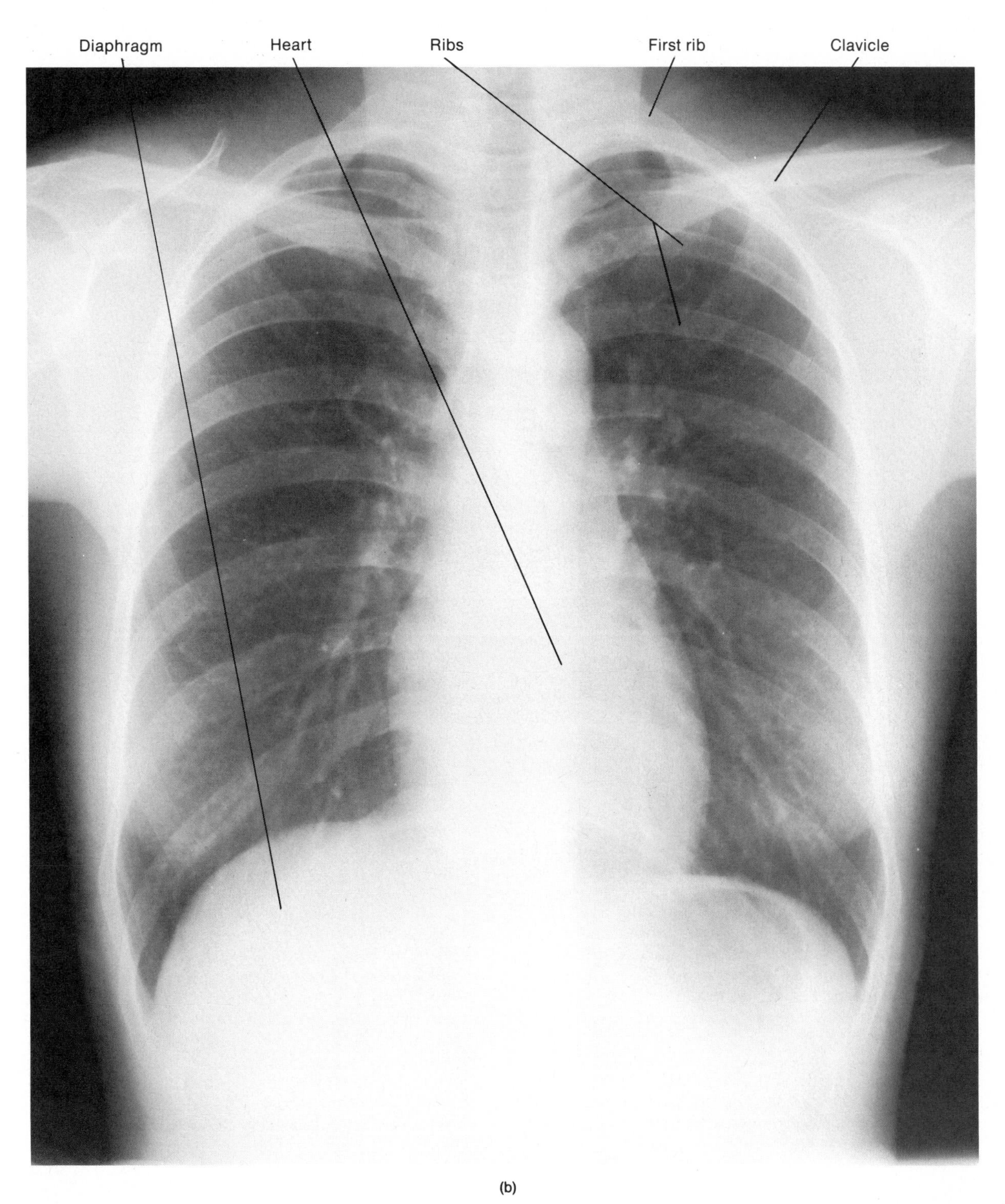

(b)

Figure 6–17 (cont.). The bony thorax. (b) Anteroposterior roentgenogram of the thorax. (Courtesy of John C. Bennett, St. Mary's Hospital, San Francisco.)

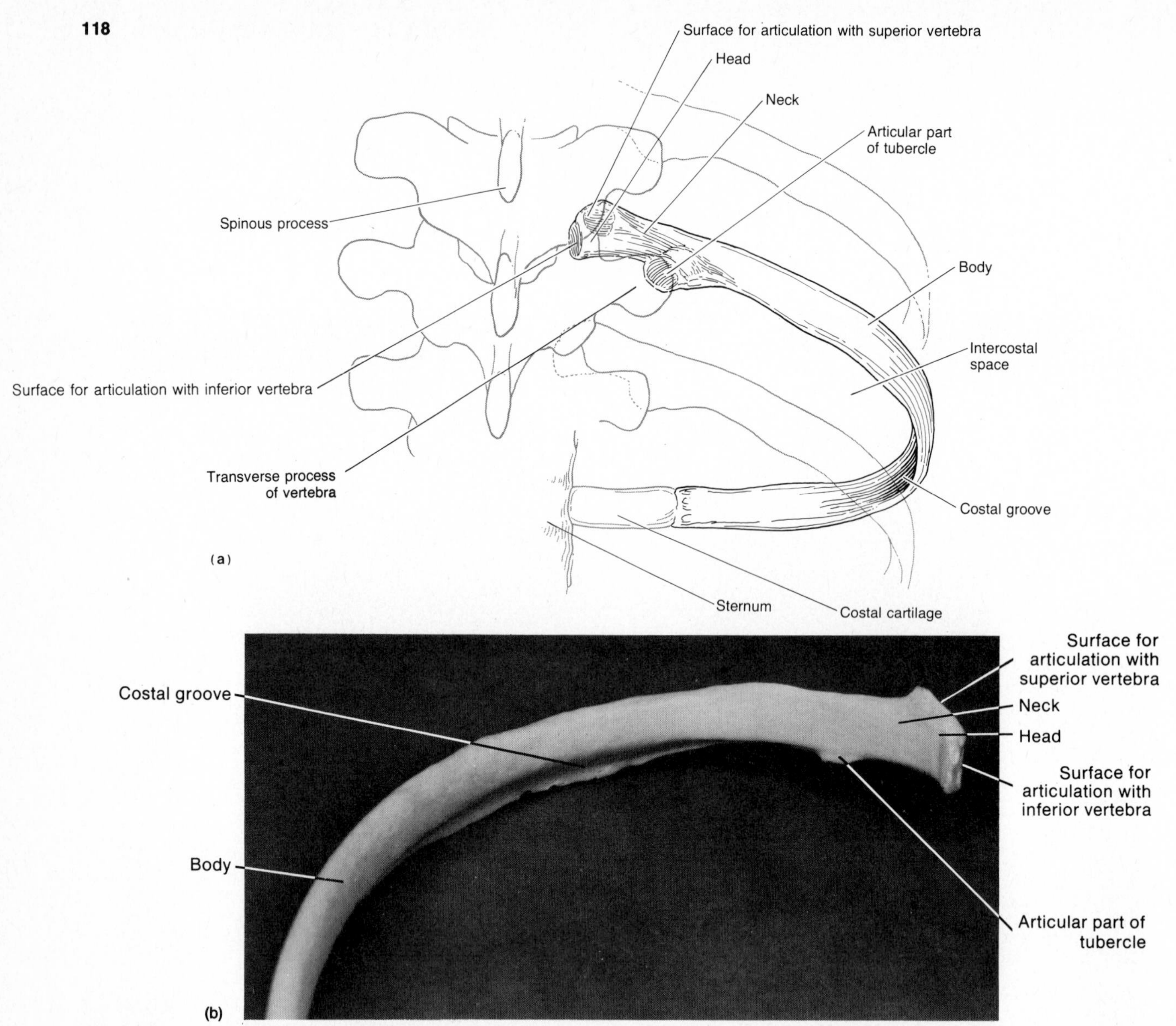

Figure 6–18. A typical rib. (a) Viewed from below and behind. (b) Photograph of the inner aspect of a portion of the fifth right rib. (Courtesy of Vincent P. Destro, Mayo Foundation.)

Exhibit 6–30. BONES OF THE THORAX: THE RIBS (see Figure 6–18)

General Description: Twelve pairs of **ribs** are located on either side of the thoracic cavity. (See Figure 6–17.) The ribs increase in length from the first through seventh. Then, they decrease in length to the twelfth rib. Each rib articulates posteriorly with its corresponding thoracic vertebra. The first through seventh ribs are also attached directly to the sternum by a strip of hyaline cartilage, called **costal cartilage.** The term *costa* means rib. These ribs are called **true ribs.** The remaining five pairs of ribs are referred to as **false ribs** because their costal cartilages do not attach directly to the sternum. Instead, the cartilages of the eighth, ninth, and tenth ribs attach to each other and then to the cartilage of the seventh rib. The eleventh and twelfth ribs are also designated as **floating ribs** because their anterior ends do not attach even indirectly to the sternum. They attach to the muscles of the body wall instead.

Markings: Although there is some variation in rib structure, we will examine the parts of a typical rib when viewed from the right side from behind. The **head** of a typical rib is a projection at the posterior end of the rib. The **neck** is a constricted portion just lateral to the head. A knoblike structure just on the posterior surface where the neck joins the body is called a **tubercle.** The **body,** or **shaft,** is the main part of the rib. The inner surface of the rib has a **costal groove** that shelters blood vessels and a nerve. The posterior portion of the rib is connected to a vertebra by its head and tubercle. The head fits into a facet on the body of a vertebra, and the tubercle articulates with the transverse process of the vertebra. Each of the second through ninth ribs articulates with the bodies of two adjacent vertebrae. The first, tenth, eleventh, and twelfth ribs articulate with only one vertebra each. On the eleventh and twelfth ribs, there is no articulation between the tubercles and the transverse processes of their corresponding vertebrae. Spaces between ribs are called **intercostal spaces.**

Chapter summary in outline

TYPES OF BONES

1. On the basis of shape, bones are classified as long, short, flat, or irregular.
2. Wormian or sutural bones are found between the sutures of certain cranial bones. Sesamoid bones develop in tendons or ligaments.

SURFACE MARKINGS

1. Markings are definitive areas on the surfaces of bones.
2. Each marking is structured for a specific function such as joint formation, muscle attachment, or allowing nerves and blood vessels to pass.
3. Examples of markings are fissure, foramen, meatus, fossa, process, condyle, head, facet, tuberosity, crest, and spine.

AXIAL SKELETON

1. The axial skeleton consists of bones arranged along the longitudinal axis. Parts of the axial skeleton are the skull, hyoid, ossicles, vertebral column, sternum, and ribs.
2. The appendicular skeleton consists of bones of the girdles and the upper and lower extremities. Parts of the appendicular skeleton are the shoulder girdle, bones of the free upper extremities, pelvic girdles, and bones of the free lower extremities.

Sutures and fontanels of the skull

1. Sutures are immovable joints between bones of the skull. Examples are coronal, sagittal, lambdoidal, and squamosal sutures.
2. Fontanels are membrane-filled spaces between the cranial bones of fetuses and infants. Major fontanels are the anterior, posterior, anterolateral, and posterolateral.

Bones of the skull

1. The skull consists of the cranium and the face. It is composed of 22 bones.
2. The eight cranial bones include the frontal, parietal, temporal, occipital, sphenoid, and ethmoid bones.
3. The 14 facial bones are the nasal, maxillae, zygomatic, mandible, lacrimal, palatine, inferior conchae, and vomer.

Paranasal sinuses of the skull

1. Paranasal sinuses are cavities in bones of the skull. They are lined by mucous membranes and communicate with the nasal cavity.
2. The cranial bones containing paranasal sinuses are the frontal, sphenoid, ethmoid, and maxilla.

The vertebral column

1. Much of the vertebral column, the sternum, and the ribs constitute the skeleton of the trunk.
2. The bones of the column are the cervical vertebrae (7), thoracic vertebrae (12), lumbar vertebrae (5), and the sacrum and coccyx.
3. The column contains primary curves (thoracic and sacral) and secondary curves (cervical and lumbar). These curves give strength, support, and balance.

The thorax

1. The thoracic skeleton consists of the sternum, the ribs and costal cartilages, and the thoracic vertebrae.
2. The thorax protects vital organs in the chest area.

Review questions and problems

1. What are the four principal kinds of bones? Give an example of each.
2. What are surface markings? Describe and give an example of each.
3. Distinguish between the axial and appendicular skeletons. What subdivisions and bones are contained in each?
4. What bones comprise the skull? The cranium? The face?
5. Define a suture. What are the four principal sutures of the skull? Where are they located?
6. What is a fontanel? Describe the location of the six major fontanels.
7. What is a paranasal sinus? Give examples of cranial bones that contain such sinuses.
8. Identify each of the following: sinusitis, cleft palate, harelip, and mastoiditis.
9. What is the hyoid bone? In what respect is it unique? What is its function?
10. What bones form the skeleton of the trunk? Distinguish between the number of nonfused vertebrae found in the adult vertebral column and in that of a child.
11. What is a curve in the vertebral column? How are primary and secondary curves differentiated? What is a curvature? Give three examples of curvatures.
12. What bones form the skeleton of the thorax? What are the functions of the thoracic skeleton? What is a sternal puncture?

Selected readings

Basmajian, John V. *Grant's Method of Anatomy.* 9th ed. Baltimore: Williams and Wilkins, 1975. Pp. 3–11.

______. *Primary Anatomy.* 6th ed. Baltimore: Williams and Wilkins, 1970. Pp. 22–27.

Cunningham's Textbook of Anatomy. 11th ed., edited by G. J. Romanes. London: Oxford University Press, 1972. Pp. 85–146.

Dawson, Helen L. *Basic Human Anatomy.* 2d ed. New York: Appleton-Century-Crofts, 1974. Pp. 20–28, 32–49.

Gray, Henry. *Anatomy of the Human Body.* 29th ed., edited by Charles Mayo Goss. Philadelphia: Lea and Febiger, 1973, Pp. 100–196.

Leeson, Thomas S., and C. Roland Leeson. *Histology.* 2d ed. Philadelphia: W. B. Saunders, 1970. Chap. 7.

CHAPTER 7

THE SKELETAL SYSTEM: APPENDICULAR SKELETON

STUDENT OBJECTIVES

After you have read this chapter, you should be able to:

1. Identify the bones of the shoulder girdle and their major markings
2. Identify the upper extremity and its component bones and their markings
3. Identify the components of the pelvic girdle and their principal markings
4. Identify the lower extremity and its component bones and their markings
5. Define the structural features and importance of the arches of the foot
6. Compare the principal structural differences between male and female skeletons
7. Define a fracture and list 13 kinds of fractures
8. Describe the sequence of events involved in fracture repair
9. Interpret sequential roentgenograms of fracture repair

With the essential features of the bones of the axial skeleton under your belt, let us now examine the bones of the appendicular skeleton. The primary concerns of this chapter are to study the bones of the girdles and extremities, compare the differences between male and female skeletons, describe several kinds of fractures, and discuss the steps involved in fracture repair.

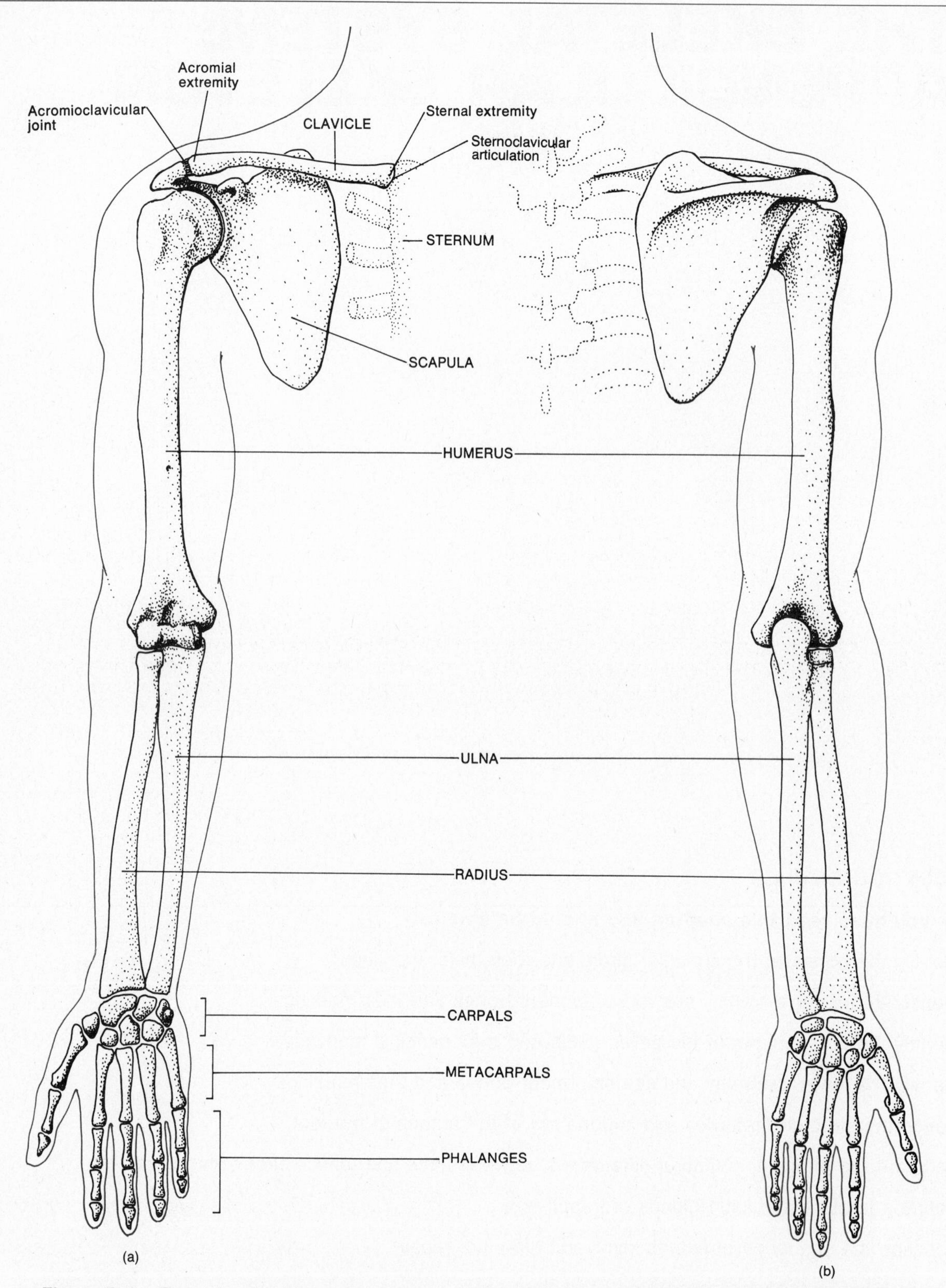

Figure 7–1. The right shoulder girdle and free upper extremity. (a) Anterior view. (b) Posterior view.

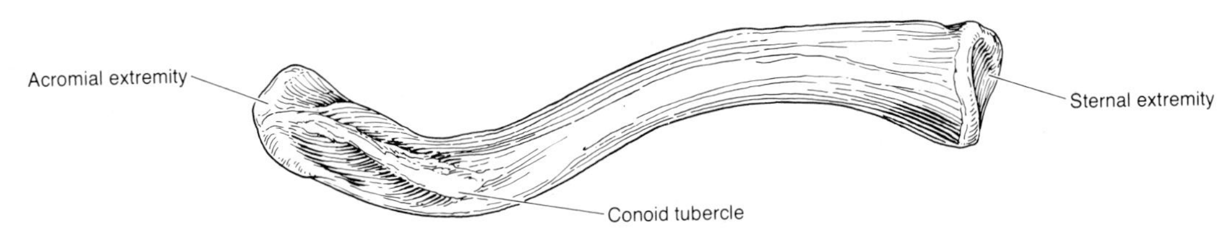

Figure 7–2. The right clavicle viewed from below.

Exhibit 7–1. THE SHOULDER GIRDLE: GENERAL CONSIDERATIONS (see Figure 7–1)

The **shoulder,** or **pectoral, girdles** serve to attach the bones of the free upper extremities to the axial skeleton. Structurally, each of the two shoulder girdles consists of only two bones: a clavicle and a scapula. The shoulder girdles have no articulation with the vertebral column. The clavicle is the anterior component of the shoulder girdle and articulates with the sternum at the sternoclavicular joint. The posterior component, the scapula, which is positioned freely within a complex musculature, articulates with a clavicle and with the humerus. Although the shoulder joints are weak, they are freely movable and allow movements in many directions.

Exhibit 7–2. THE SHOULDER GIRDLE: THE CLAVICLES (see Figure 7–2)

General Description: The **clavicles,** or collarbones, are long slender bones with a double curvature. The two bones lie horizontally in the superior and anterior part of the thorax superior to the first rib.

Markings: The medial end of a clavicle, the **sternal extremity,** is rounded and articulates with the sternum. The broad, flat, lateral end, the **acromial extremity,** articulates with the acromion of a scapula. This joint is called the **acromioclavicular joint.** Refer to Figure 7–1 for a view of these articulations. The **conoid tubercle** on the inferior surface of the lateral end of the bone serves as a point of attachment for a ligament.

Exhibit 7–3. THE SHOULDER GIRDLE: THE SCAPULAE (see Figure 7–3)

General Description: The **scapulae,** or shoulder blades, are large, triangular, flat bones situated in the dorsal part of the thorax between the levels of the second and seventh ribs. Their medial borders are located about 5 cm (2 in.) from the vertebral column.

Markings: A sharp ridge, the **spine,** runs diagonally across the posterior surface of the flattened, triangular **body.** The end of the spine projects as a flattened, expanded process called the **acromion.** This process articulates with the clavicle. Inferior to the acromion is a depression called the **glenoid cavity.** This cavity articulates with the head of the humerus to form the shoulder socket. The thin edge of the body near the vertebral column is the **medial** or **vertebral border.** The thick edge closer to the arm is the **lateral** or **axillary border.** The medial and lateral borders join at the **inferior angle.** The superior edge of the scapular body is called the **superior border.** At the lateral end of the superior border is a projection of the anterior surface called the **coracoid process** to which muscles attach. Above and below the spine are two fossae: the **supraspinatous fossa** and the **infraspinatous fossa,** respectively. Both serve as surfaces of attachment for shoulder muscles. On the anterior surface is a slightly hollowed-out area called the **subscapular fossa,** also a surface of attachment for shoulder muscles.

Exhibit 7–4. THE FREE UPPER EXTREMITIES: GENERAL CONSIDERATIONS

The **free upper extremities** consist of 60 bones. The skeleton of the right free upper extremity is shown in Figure 7–1. It includes a humerus in each arm, an ulna and radius in each forearm, carpals, or wrist bones, metacarpals, which are the palm bones, and finger bones or phalanges of each hand.

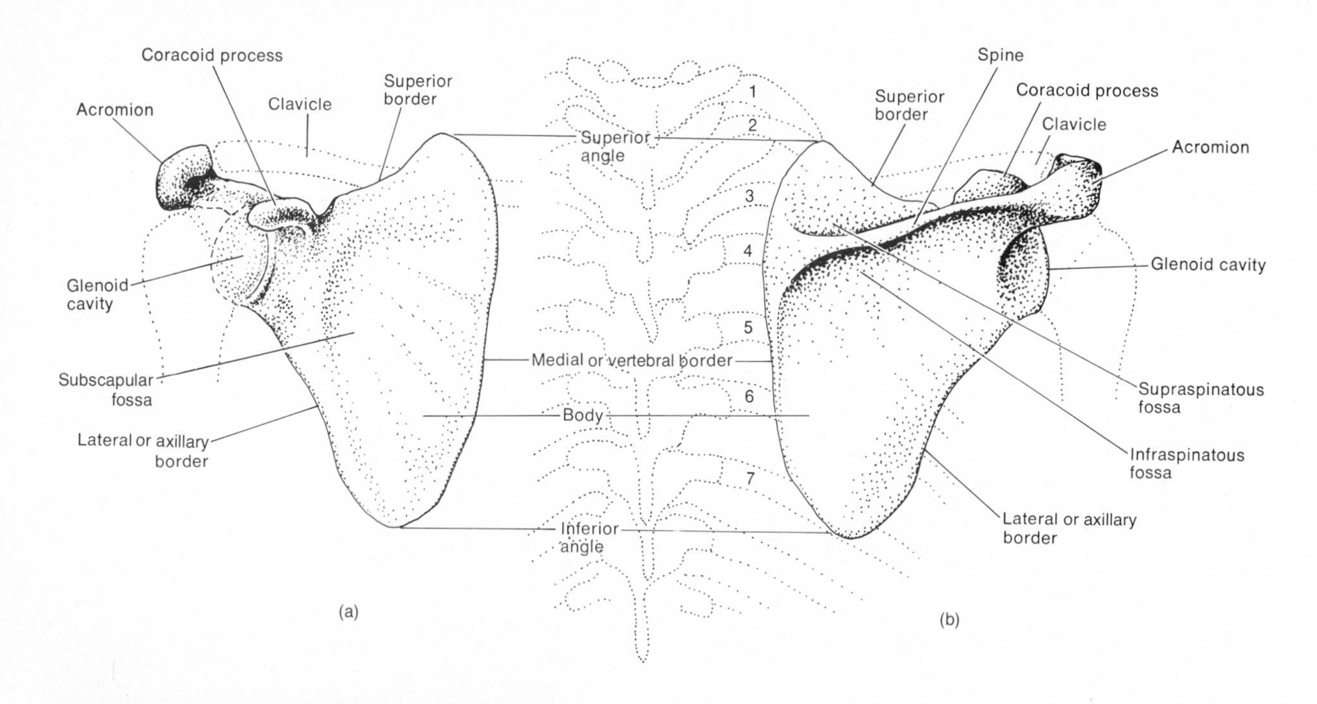

Coracoid process
Acromion
Clavicle
Superior border
Glenoid cavity
Subscapular fossa
Lateral or axillary border
Superior angle
Medial or vertebral border
Body
Inferior angle
1
2
3
4
5
6
7
Spine
Superior border
Coracoid process
Clavicle
Acromion
Glenoid cavity
Supraspinatous fossa
Infraspinatous fossa
Lateral or axillary border
(a)
(b)

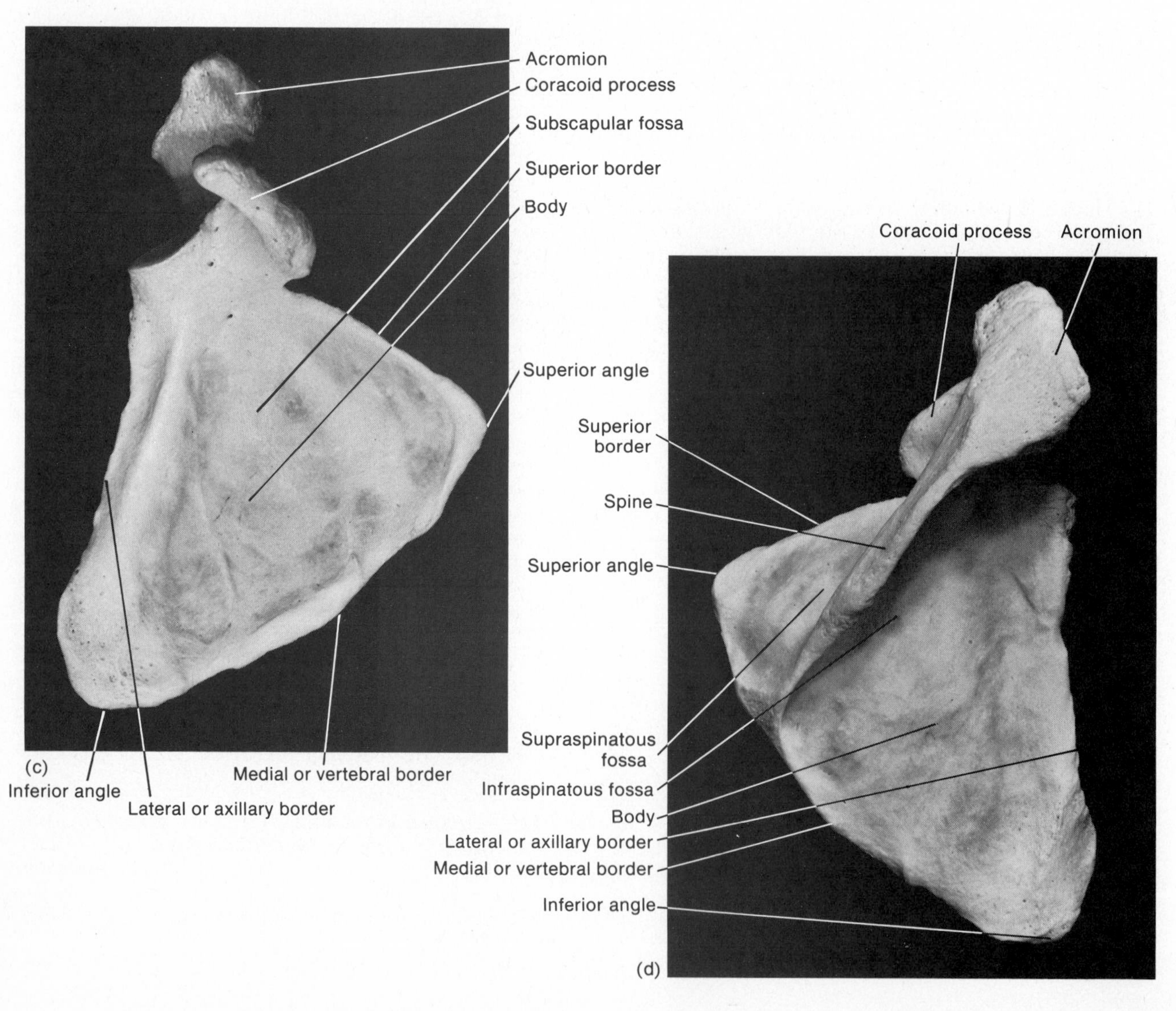

Acromion
Coracoid process
Subscapular fossa
Superior border
Body
Superior angle
(c)
Inferior angle
Lateral or axillary border
Medial or vertebral border
Coracoid process
Acromion
Superior border
Spine
Superior angle
Supraspinatous fossa
Infraspinatous fossa
Body
Lateral or axillary border
Medial or vertebral border
Inferior angle
(d)

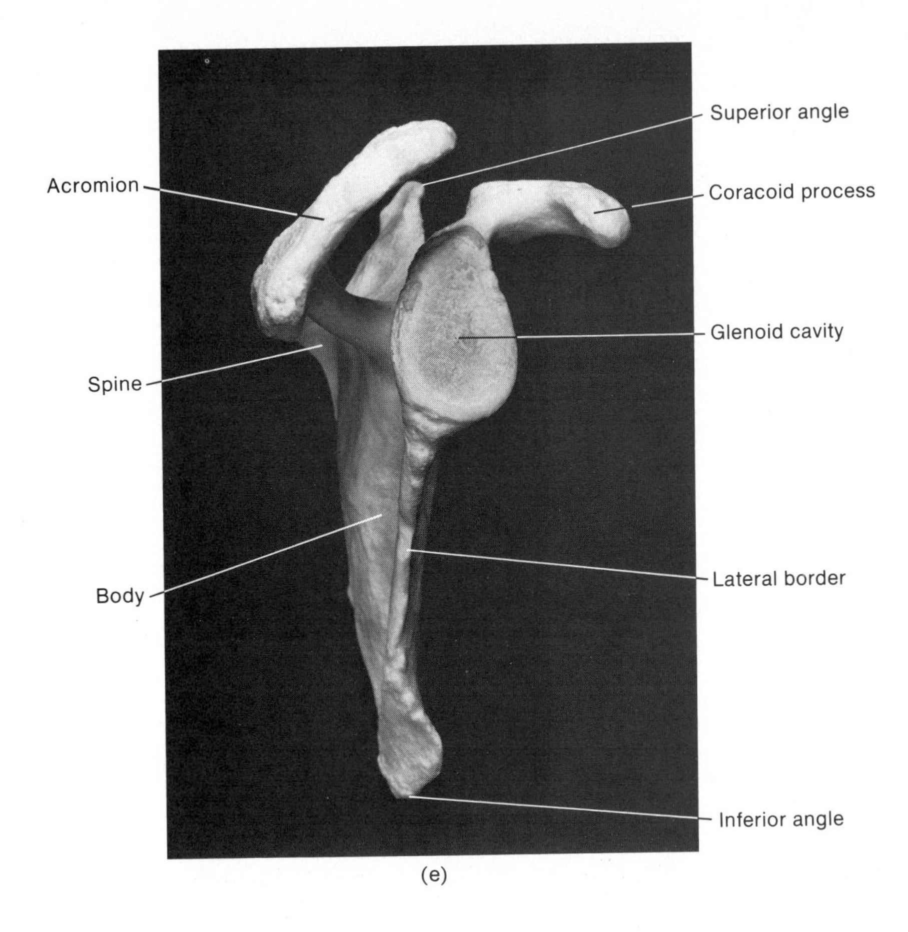

(e)

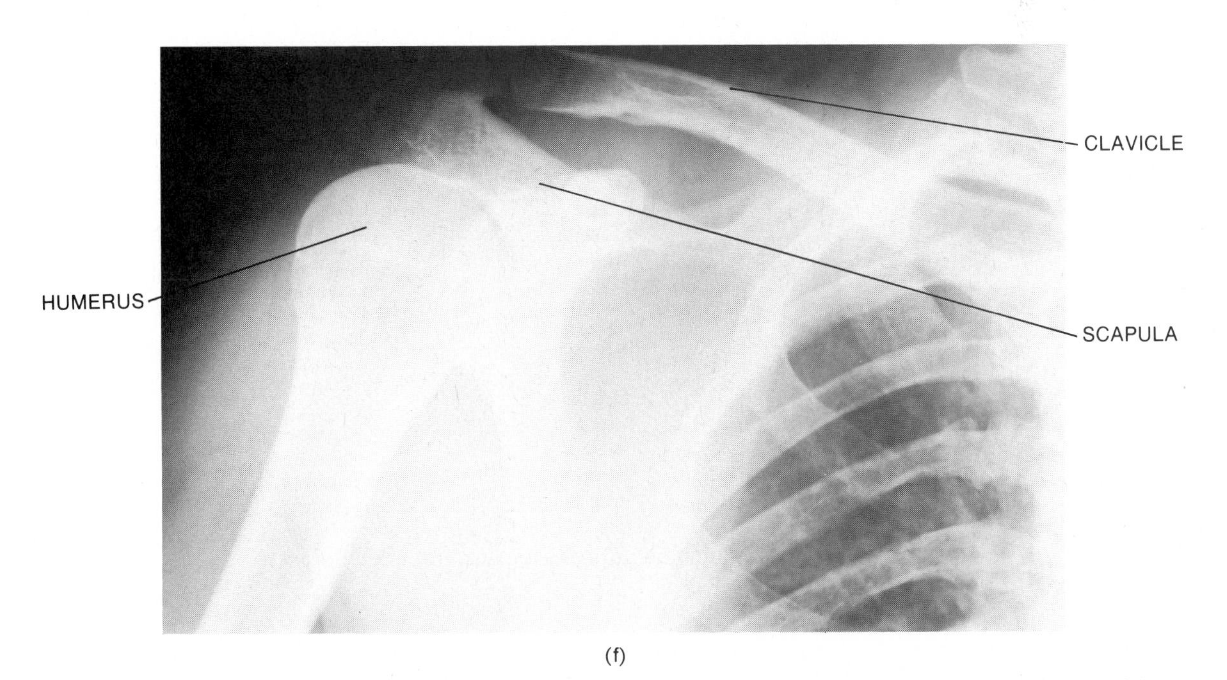

(f)

Figure 7–3. The right scapula. (a) Anterior view. (b) Posterior view. (c) Photograph in anterior view. (d) Photograph in posterior view. (e) Photograph in lateral border view. (Photographs courtesy of Lenny Patti.) (f) Anteroposterior roentgenogram of the right shoulder girdle. (Courtesy of John C. Bennett, St. Mary's Hospital, San Francisco.)

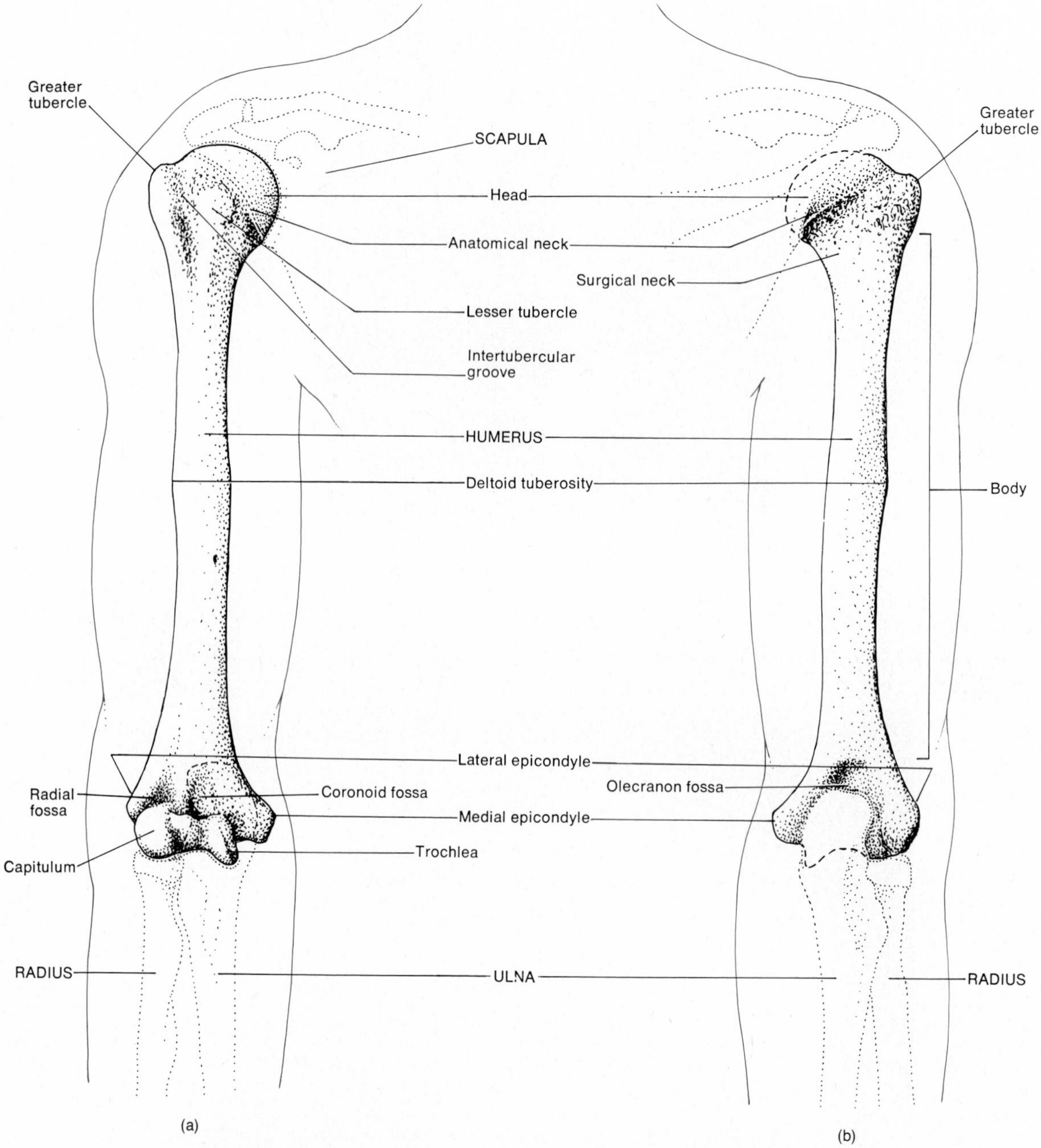

Figure 7–4. The right humerus. (a) Anterior view. (b) Posterior view.

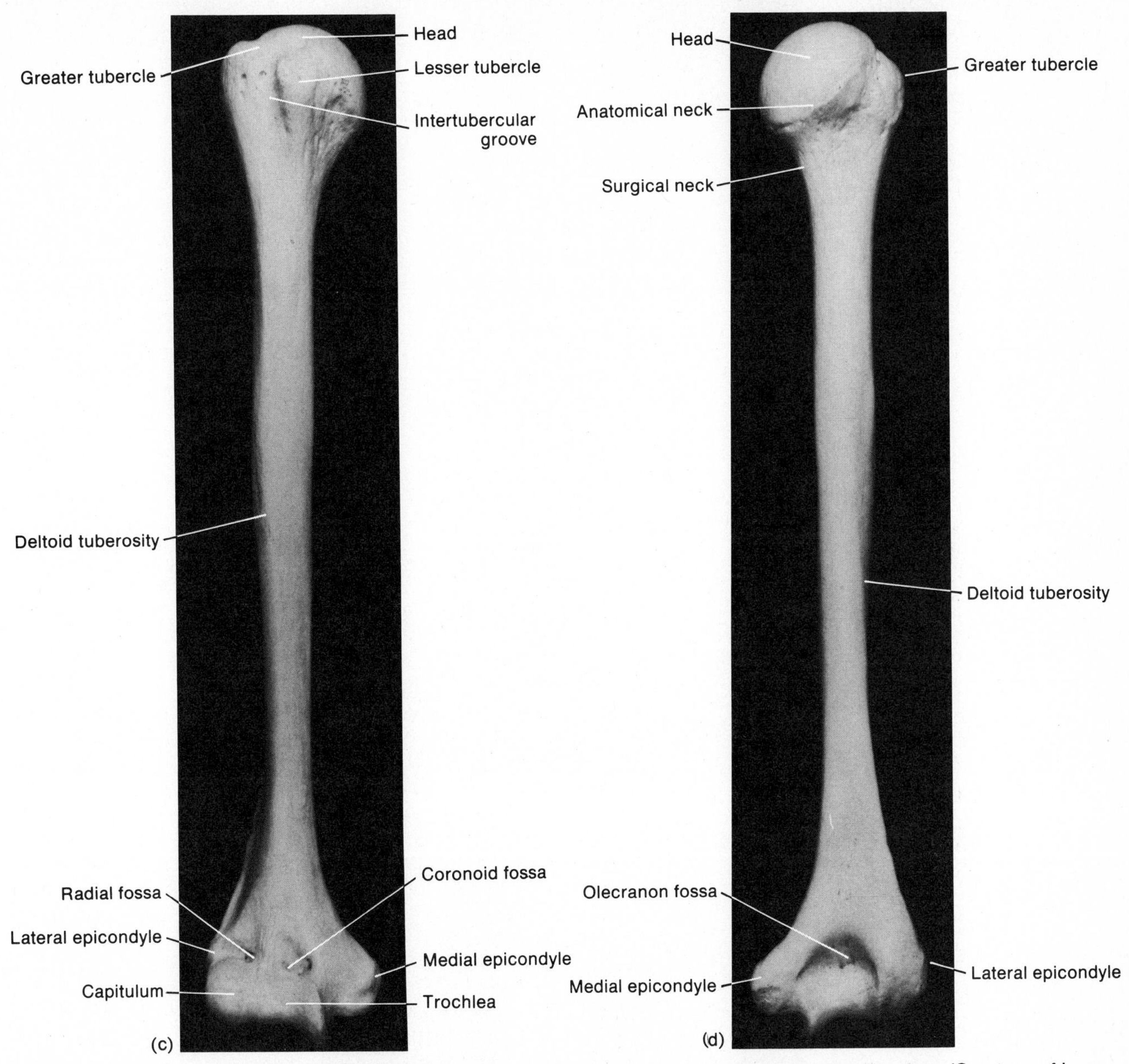

Figure 7–4 (cont.). The right humerus. (c) Photograph in anterior view. (d) Photograph in posterior view. (Courtesy of Lenny Patti.)

Exhibit 7–5. THE FREE UPPER EXTREMITY: THE HUMERUS (see Figure 7–4)

General Description: The **humerus,** or arm bone, is the longest and largest bone of the free upper extremity. It articulates proximally with the scapula and distally at the elbow with both the ulna and radius.

Markings: The proximal end of the humerus consists of a **head** that articulates with the glenoid cavity of the scapula. It also has an **anatomical neck,** which is an oblique groove just distal to the head. The **greater tubercle** is a lateral projection distal to the neck. The **lesser tubercle** is an anterior projection. Between these tubercles runs an **intertubercular groove.** The **surgical neck** is a constricted portion just distal to the tubercles and is so named because of its liability to fracture. The **body** or shaft of the humerus is cylindrical at its proximal end. It gradually becomes triangular and is flattened and broad at its distal end. About midway down the shaft, there is a roughened, V-shaped area called the **deltoid tuberosity.** This serves as a point of attachment for the deltoid muscle. The following parts are found in the distal end of the humerus: The **capitulum** is a rounded knob that receives the head of the radius when the forearm is flexed. The **radial fossa** is a depression that articulates with the head of the radius when the arm is flexed. The **trochlea** is a pulleylike surface that articulates with the ulna. The **coronoid fossa** is an anterior depression that receives part of the ulna when the forearm is flexed. The **olecranon fossa** is a posterior depression that receives the olecranon of the ulna when the forearm is extended. The **medial epicondyle** and **lateral epicondyle** are rough projections on either side of the distal end.

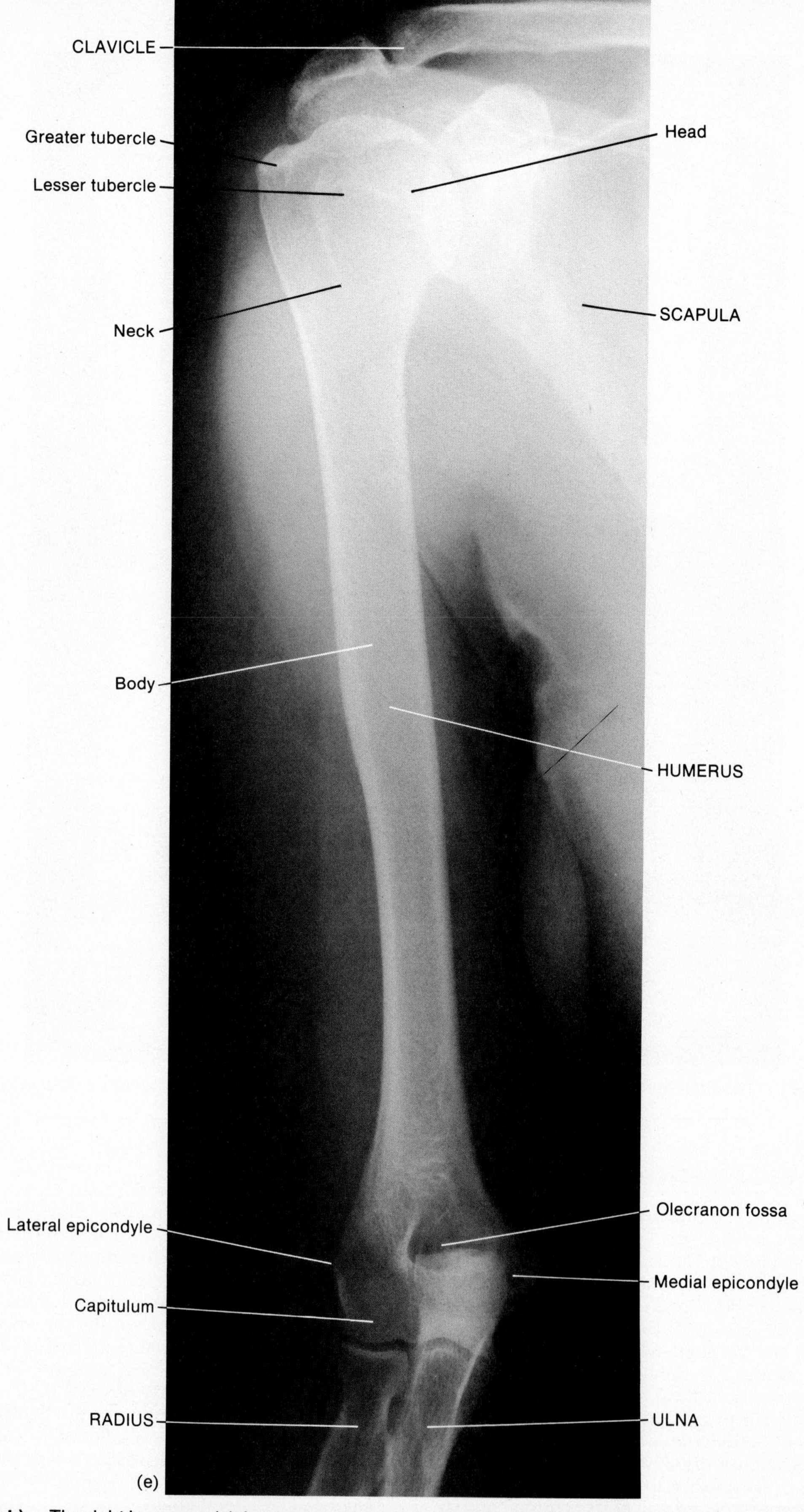

Figure 7–4 (cont.). The right humerus. (e) Anteroposterior roentgenogram. (Courtesy of John C. Bennett, St. Mary's Hospital, San Francisco.)

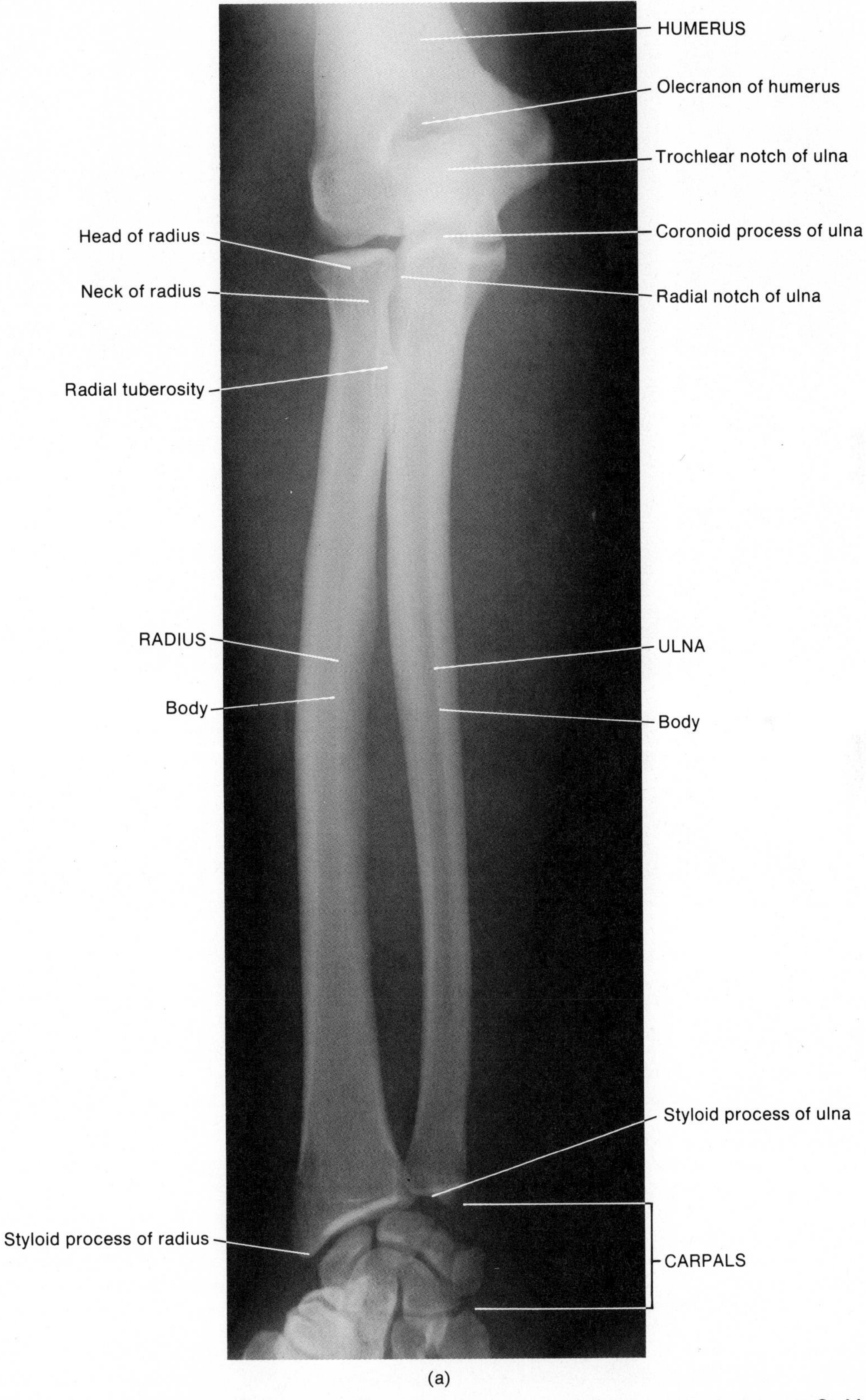

Figure 7–5. The right ulna and radius. (a) Anteroposterior roentgenogram. (Courtesy of John C. Bennett, St. Mary's Hospital, San Francisco.)

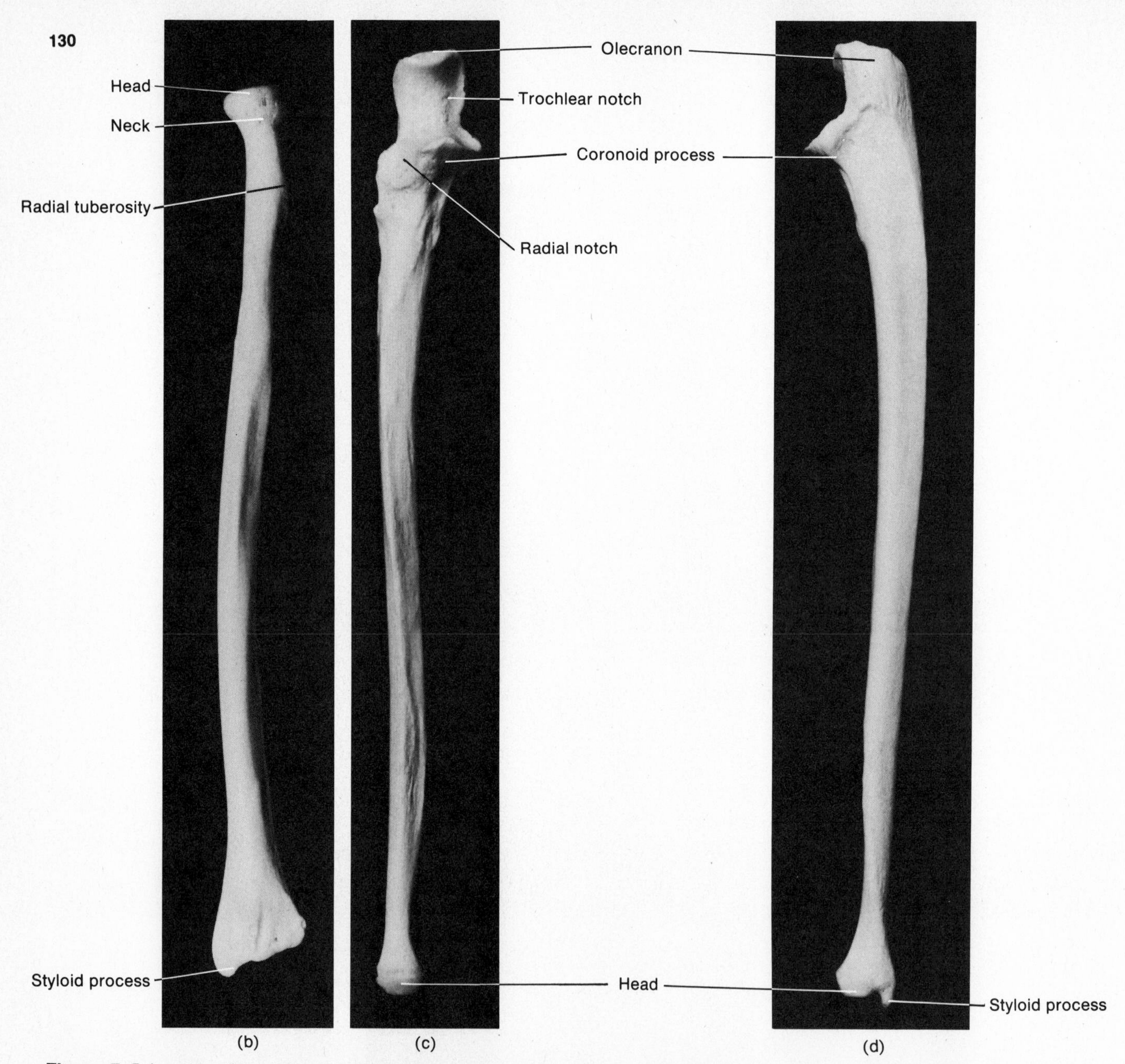

Figure 7–5 (cont.). The right ulna and radius. (b) Photograph of the radius in anterior view. (c) Photograph of the ulna in anterior view. (d) Photograph of the ulna in lateral view. (Courtesy of Lenny Patti.)

Exhibit 7–6. THE FREE UPPER EXTREMITY: THE ULNA AND RADIUS (see Figure 7–5)

General Description of the Ulna: The **ulna** is the medial bone of the forearm. In other words, it is located on the side of the little finger.

Markings: The proximal end of the ulna presents an **olecranon,** which forms the prominence of the elbow. The **coronoid process** is an anterior projection that, together with the olecranon, receives the trochlea of the humerus. The **trochlear notch** is a curved area between the olecranon and the coronoid processes. The trochlea of the humerus fits into this notch. The **radial notch** is a depression located laterally and inferiorly to the trochlear notch. It receives the head of the radius. The distal end of the ulna consists of a **head** that is separated from the wrist by a fibrocartilage disc. A **styloid process** is on the posterior side of the distal end.

General Description of the Radius: The **radius** is the lateral bone of the forearm. That is, it is situated on the thumb side.

Markings: The proximal end of the radius has a disc-shaped **head** that articulates with the capitulum of the humerus and radial notch of the ulna. It also has a raised, roughened area on the medial side called the **radial tuberosity.** This is a point of attachment for the biceps muscle. The shaft of the radius widens distally to form a concave inferior surface that articulates with two bones of the wrist called the lunate and navicular bones. Also at the distal end is a **styloid process** on the lateral side and a medial, concave **ulnar notch** for articulation with the distal end of the ulna. A very common fracture of the radius, called a *Colles' fracture* occurs about 2½ cm (1 in.) up from the distal end of the bone.

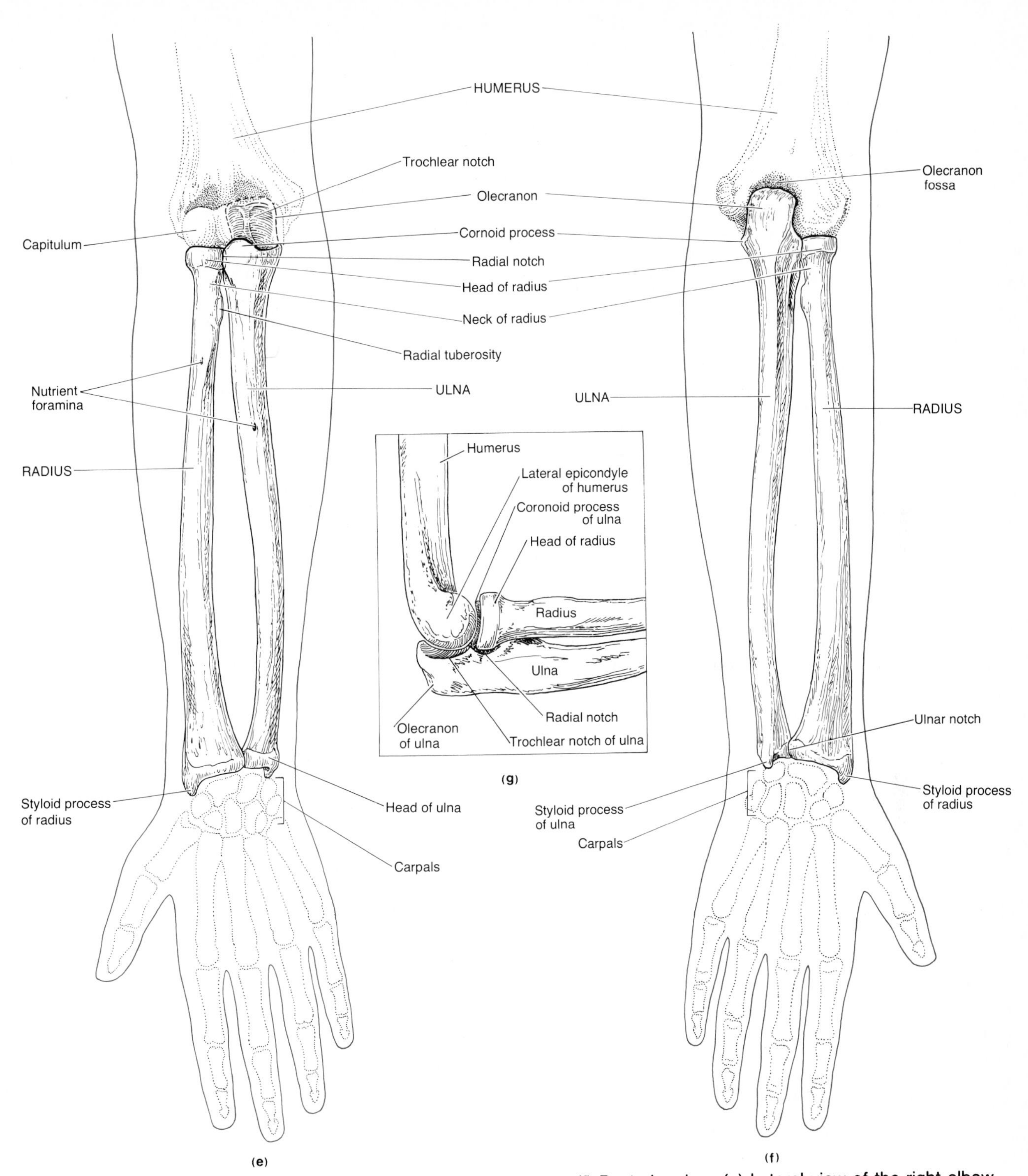

Figure 7–5 (cont.). The right ulna and radius. (e) Anterior view. (f) Posterior view. (g) Lateral view of the right elbow.

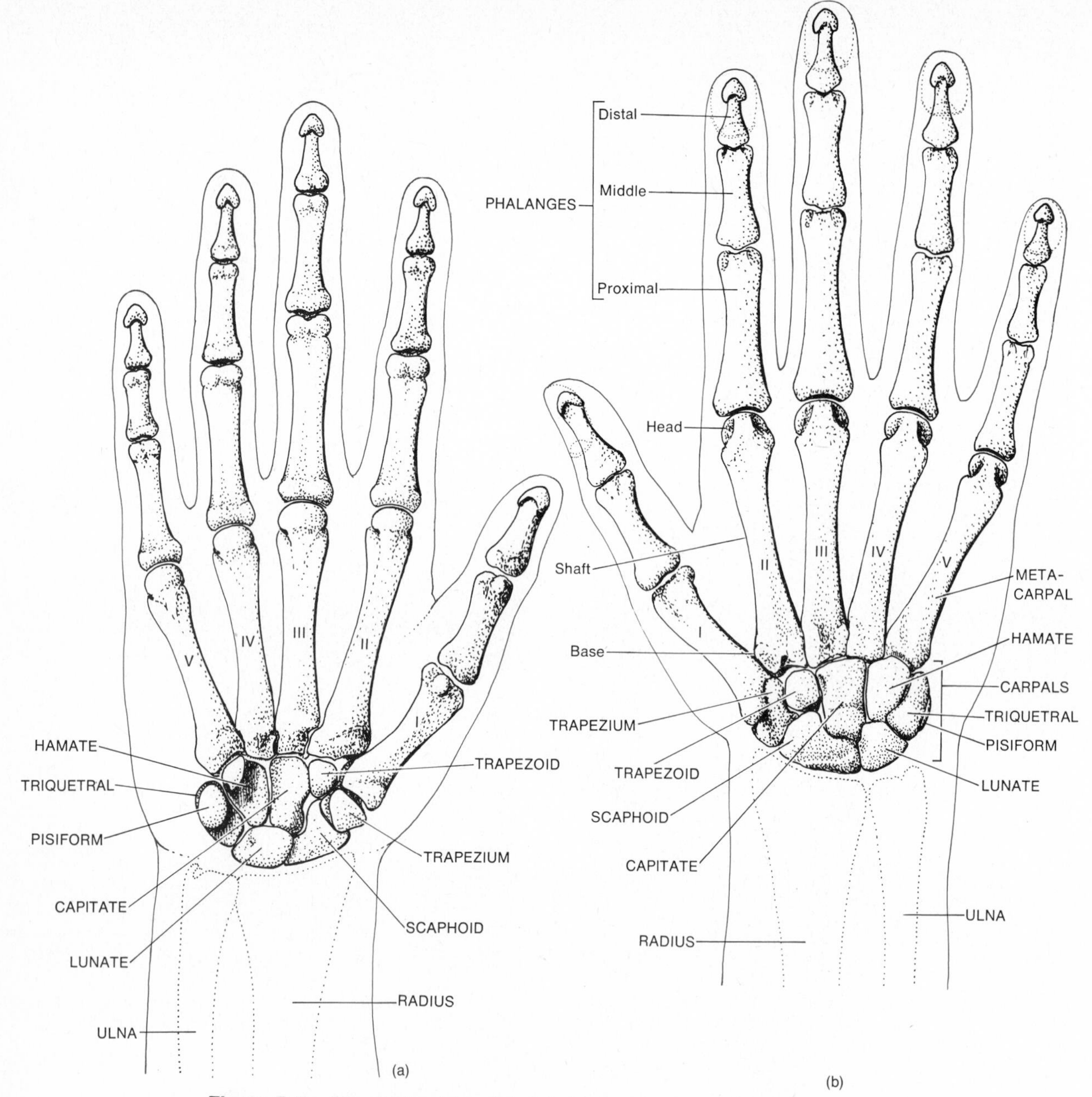

Figure 7–6. The right wrist and hand. (a) Anterior view. (b) Posterior view.

Exhibit 7–7. THE FREE UPPER EXTREMITY: THE WRIST AND HAND (see Figure 7–6)

Carpus: The **carpus,** or wrist, consists of eight small bones united to each other by ligaments. The bones are arranged in two transverse rows, with four bones in each row. The proximal row, from the medial to lateral position, consists of the following bones: **pisiform; triquetral,** or **triangular; lunate;** and **scaphoid,** or **navicular.** In about 70 percent of the cases involving carpal fractures, only the scaphoid is involved. The distal row of bones, from the medial to lateral position, consists of the following: **hamate; capitate; trapezoid,** or **lesser multangular;** and **trapezium,** or **greater multangular.**

Metacarpus: The five bones of the **metacarpus** constitute the palm of the hand. Each metacarpal bone consists of a proximal **base,** a **shaft,** and a distal **head.** The metacarpal bones are numbered I to V, starting with the lateral bone. The bases articulate with the distal row of carpal bones and with one another. The heads articulate with the proximal phalanges of the fingers. The heads of the metacarpals are the "knuckles," which are readily visible when the fist is clenched.

Phalanges: The **phalanges,** or bones of the fingers, number 14 in each hand. There are two phalanges in the first digit, the thumb, and three phalanges in each of the remaining four digits. The first row of phalanges, the **proximal row,** articulates with the metacarpal bones and second row of phalanges. The second row of phalanges, the **middle row,** articulates with the proximal row and the third row. Finally, the third row of phalanges, the **distal row,** articulates with the middle row. A single finger bone is referred to as a **phalanx.**

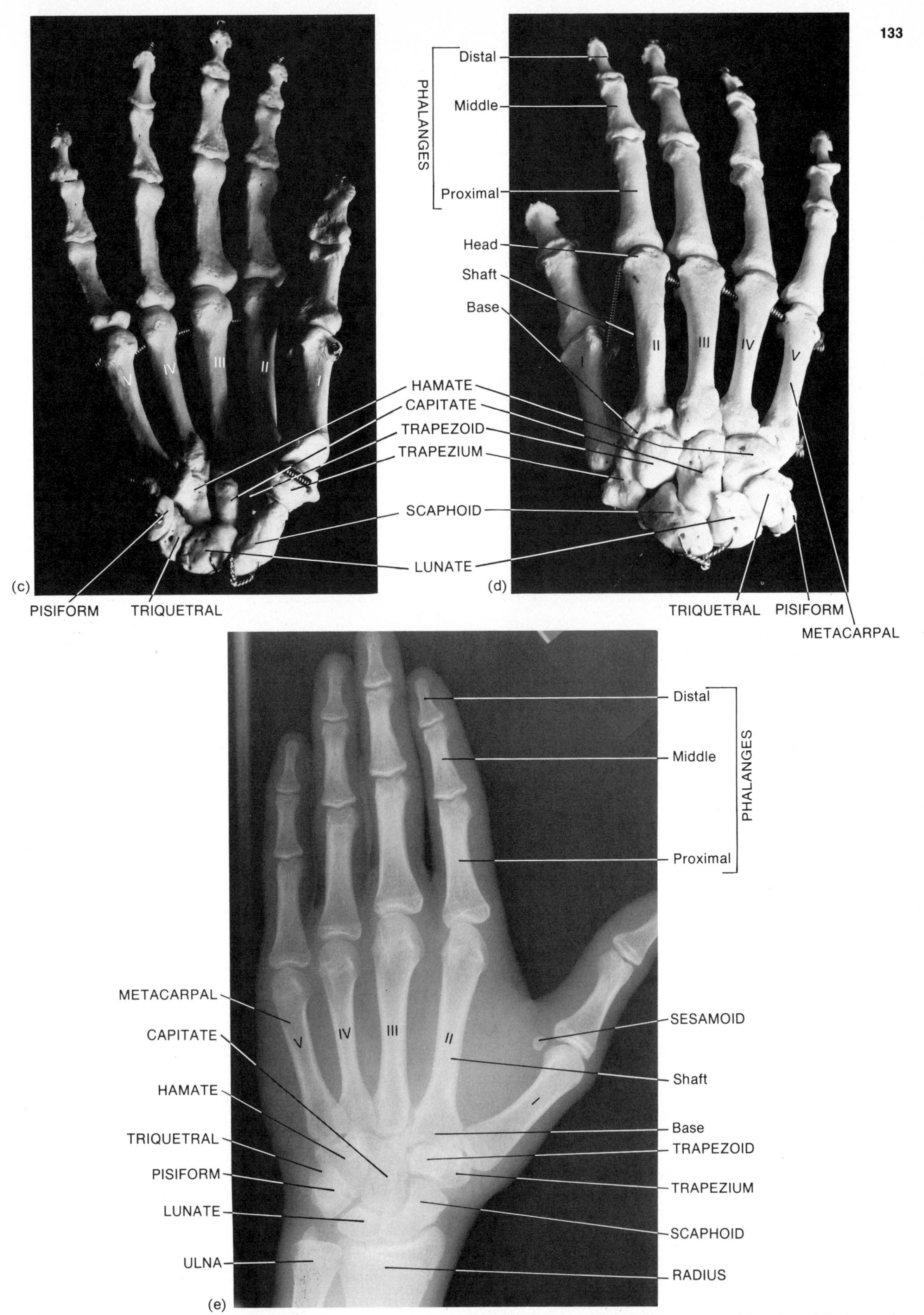

Figure 7–6 (cont.). The right wrist and hand. (c) Photograph in anterior view. (d) Photograph in posterior view. (Photographs courtesy of Lenny Patti.) (e) Anteroposterior roentgenogram. Note the sesamoid bone. (Courtesy of John C. Bennett, St. Mary's Hospital, San Francisco.)

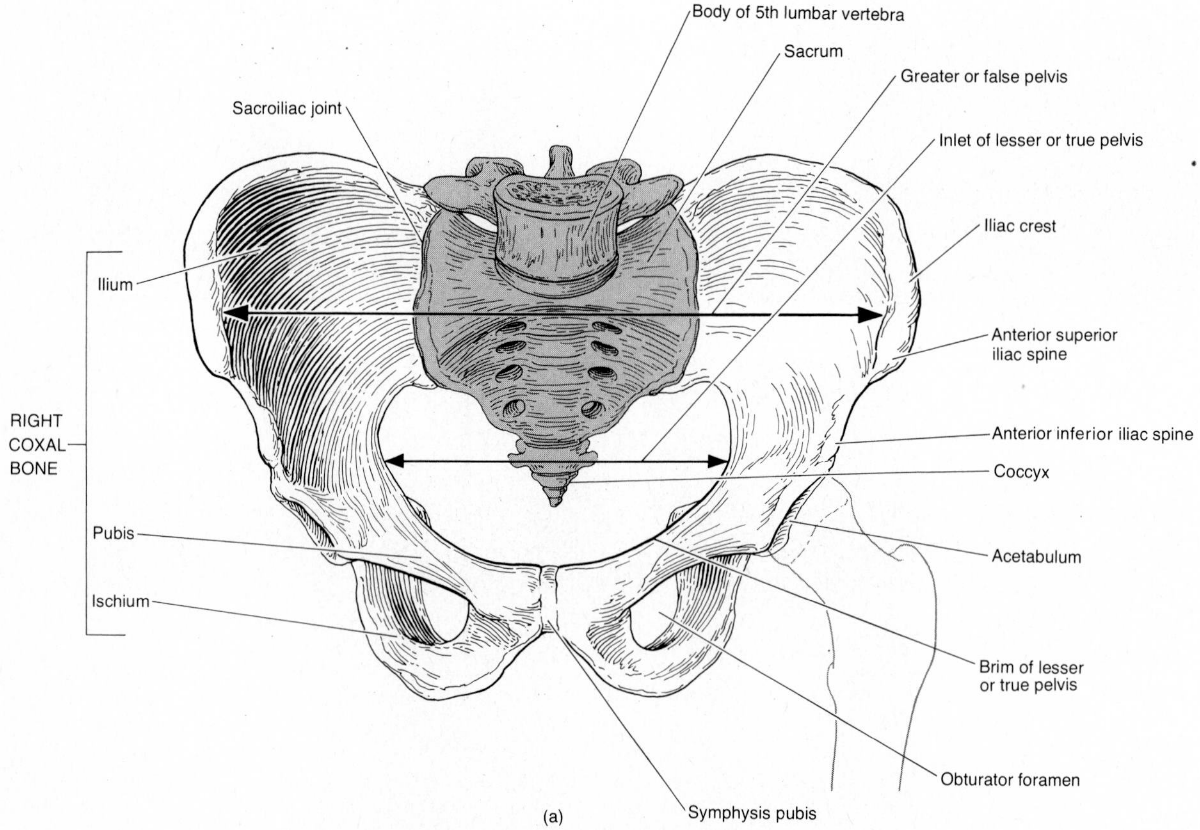

(a)

Figure 7–7. The pelvic girdle. (a) Anterior view.

Exhibit 7–8. THE PELVIC GIRDLE: GENERAL CONSIDERATIONS (see Figure 7–7)

The **pelvic girdle** consists of the two **coxal bones,** commonly called the pelvic or hipbones. The pelvic girdle provides a strong and stable support for the free lower extremities on which the weight of the body is carried. The coxal bones are united to each other anteriorly at the symphysis pubis. They unite posteriorly to the sacrum.

Together with the sacrum and coccyx, the pelvic girdle forms the basinlike structure called the **pelvis.** The pelvis is divided into a greater pelvis and a lesser pelvis. The **greater** or **false pelvis** is the expanded portion situated superior to the narrow bony ring called the **brim of the pelvis.** The greater pelvis consists laterally of the two ilia and posteriorly of the superior portion of the sacrum. There is no bony component in the anterior aspect of the greater pelvis. Rather, the front is formed by the walls of the abdomen. The **lesser** or **true pelvis** is inferior and posterior to the pelvic brim. It is constructed of parts of the ilium, pubis, sacrum, and coccyx. The lesser pelvis contains a superior opening called the **pelvic inlet** and an inferior opening, called the **pelvic outlet.** *Pelvimetry* is the measurement of the size of the inlet and outlet of the birth canal. Measurement of the pelvic cavity is important to the physician since the baby must pass through the lesser pelvis at birth.

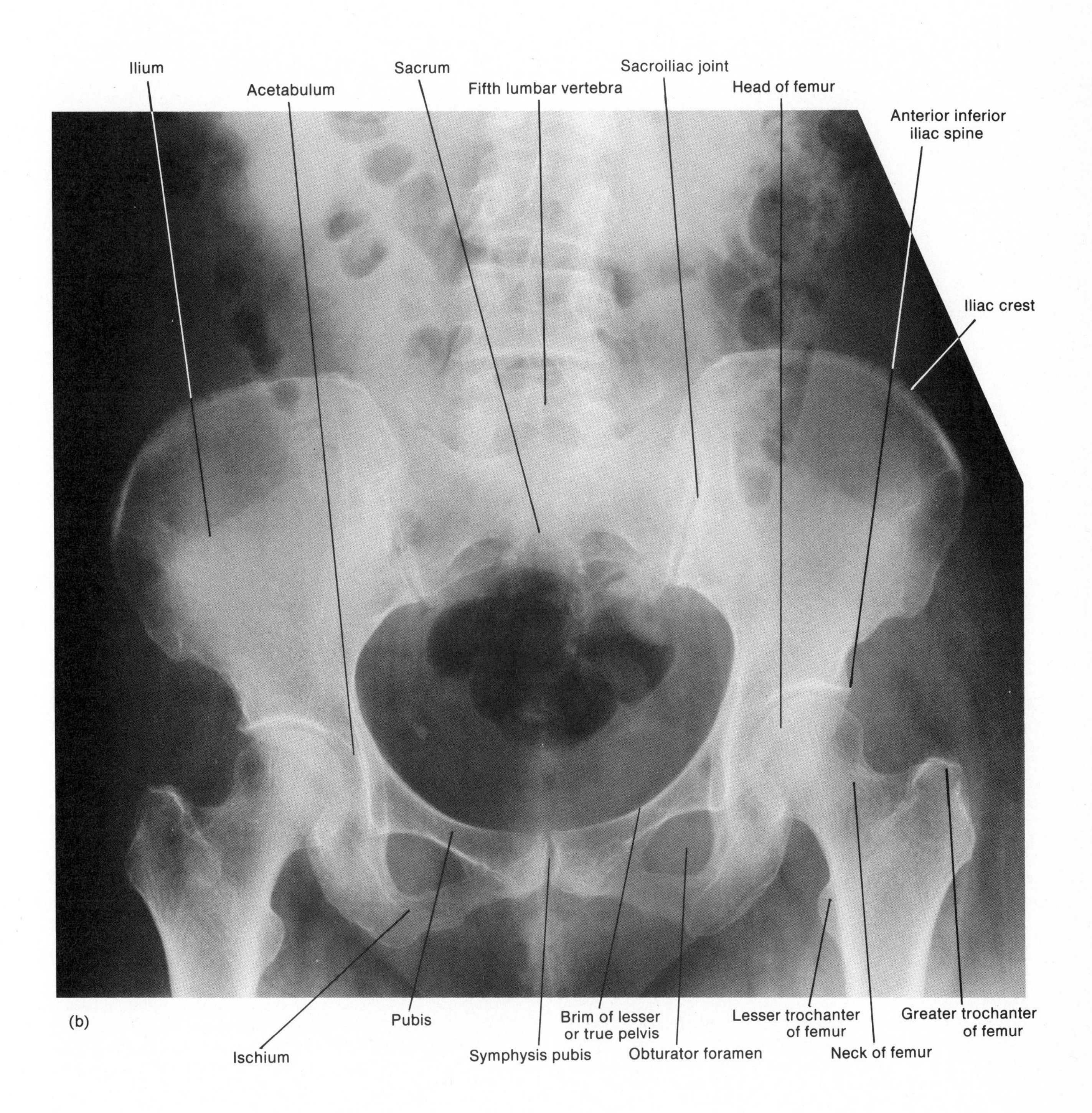

Figure 7–7 (cont.). The pelvic girdle. (b) Anteroposterior roentgenogram. (Courtesy of John C. Bennett, St. Mary's Hospital, San Francisco.)

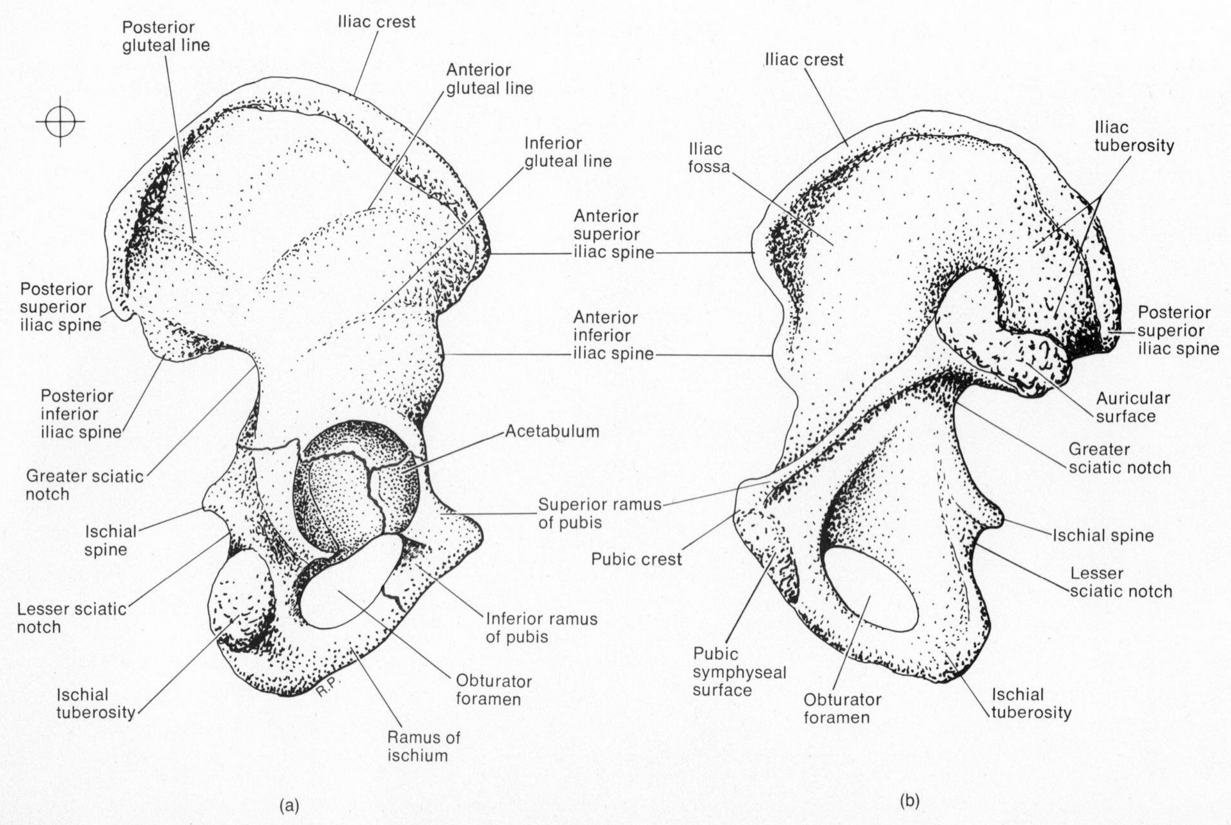

Figure 7–8. The right coxal bone. (a) Lateral view. (b) Medial view. The lines of fusion are not actually visible in an adult bone.

Exhibit 7–9. THE PELVIC GIRDLE: THE COXAL BONES (see Figure 7–8)

General Description: The **coxal bones** of a newborn consist of three components: a superior **ilium,** an inferior and anterior **pubis,** and an inferior and posterior **ischium.** Eventually, the three separate bones fuse into one. The area of fusion is a deep, lateral fossa called the **acetabulum.** This structure is the socket for the head of the femur. Although the adult coxae are both single bones, it is common to discuss the bones as if they consisted of three portions.

Markings: The ilium is the largest of the three subdivisions of the coxal bone. Its superior border, the **iliac crest,** ends anteriorly in the **anterior superior iliac spine.** The **anterior inferior iliac spine** is located inferior to the anterior superior spine. Posteriorly, the iliac crest ends in the **posterior superior iliac spine.** The **posterior inferior iliac spine** is just inferior. The spines serve as points of attachment for muscles of the abdominal wall. Just inferior to the posterior inferior iliac spine is the **greater sciatic notch.** The internal surface of the ilium seen from the medial side is the **iliac fossa.** It is a concavity where the iliacus muscle attaches. Posterior to this fossa are the **iliac tuberosity,** a point of attachment for the sacroiliac ligament, and the **auricular surface,** which articulates with the sacrum. The other conspicuous markings of the ilium are three arched lines on its gluteal (buttock) surface called the **posterior gluteal line,** the **anterior gluteal line,** and the **inferior gluteal line.** Between these lines the gluteal muscles are attached.

The ischium is the inferior posterior portion of the coxal bone. It contains a prominent **ischial spine,** a **lesser sciatic notch** under the spine, and an **ischial tuberosity.** The rest of the ischium, the **ramus,** joins with the pubis and together they surround the **obturator foramen.**

The pubis is the anterior and inferior part of the coxal bone. It consists of a **superior ramus,** an **inferior ramus,** and a body that enters into the symphysis pubis. The **symphysis pubis** is the joint between the two coxal bones. It consists of fibrocartilage. It can be seen in Figure 7–7. The **acetabulum** is the socket formed by the ilium, ischium, and pubis. Two-fifths of the acetabulum is formed by the ilium, two-fifths by the ischium, and one-fifth by the pubis.

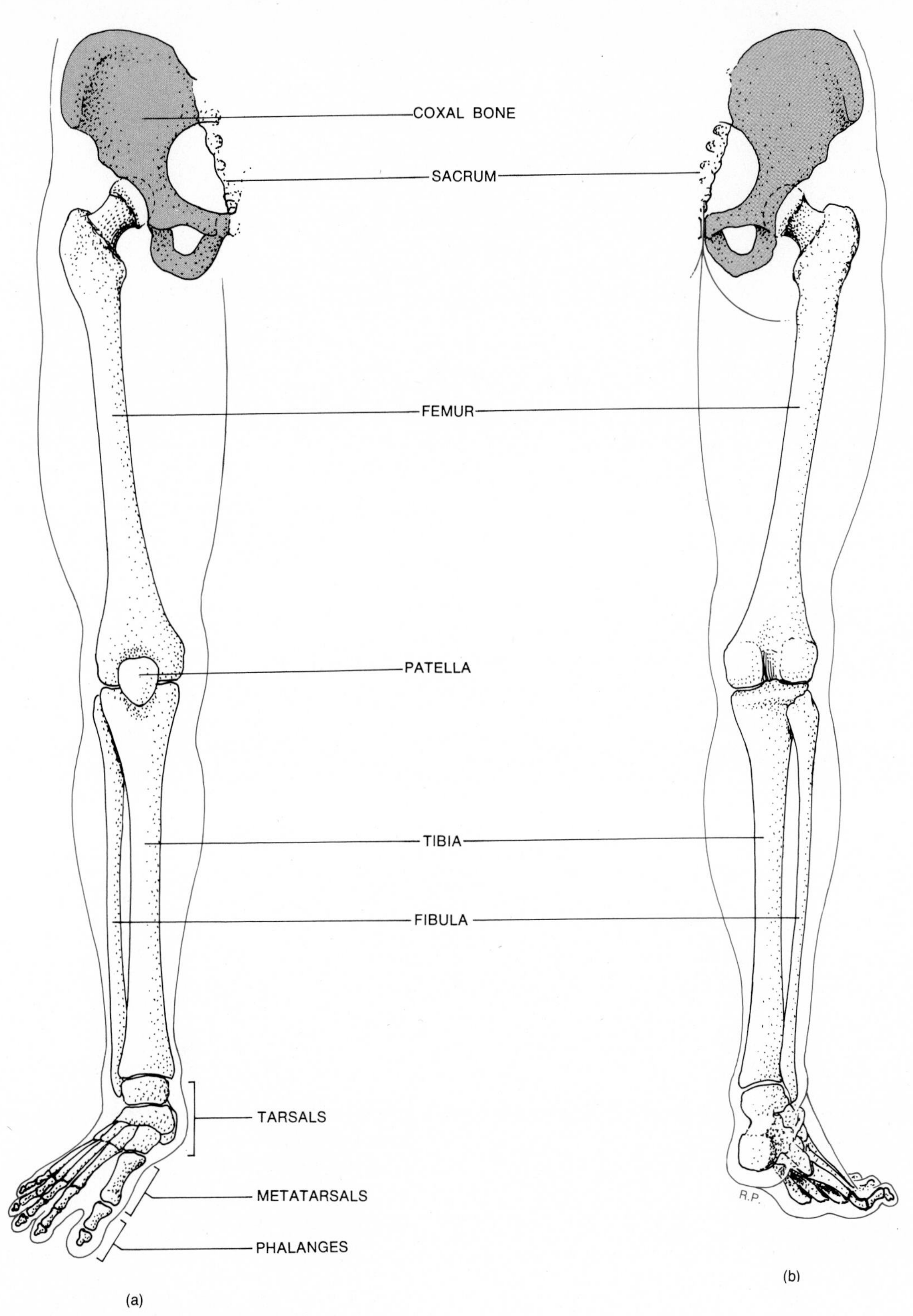

Figure 7–9. The right pelvic girdle and free lower extremity. (a) Anterior view. (b) Posterior view.

Exhibit 7–10. THE FREE LOWER EXTREMITIES: GENERAL CONSIDERATIONS (see Figure 7–9)

The **free lower extremities** are made up of 60 bones. These include the femur of each thigh, each kneecap, the fibula and tibia in each leg, the ankle bones in each ankle, and the metatarsals and phalanges of each foot.

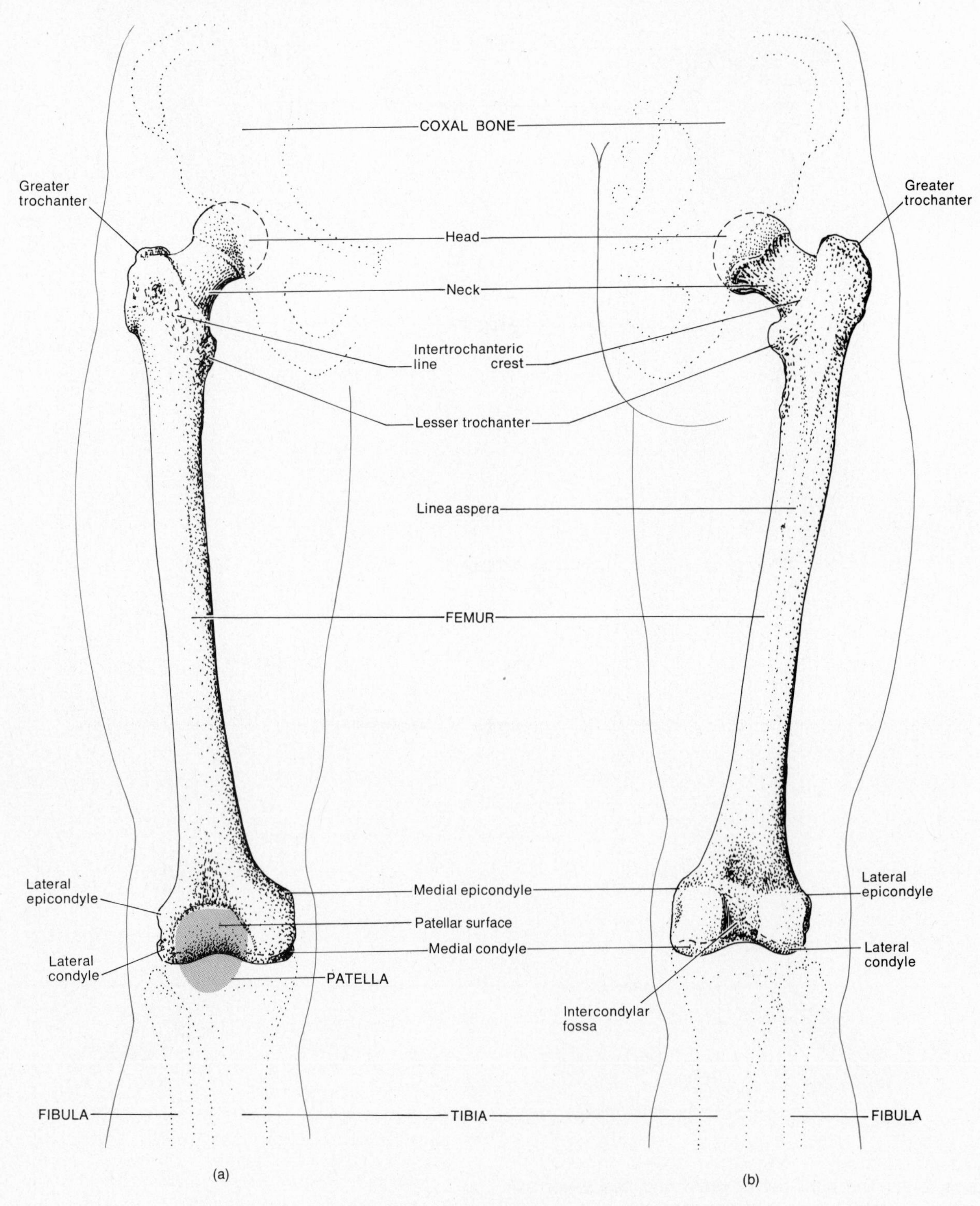

Figure 7–10. The right femur. (a) Anterior view. (b) Posterior view.

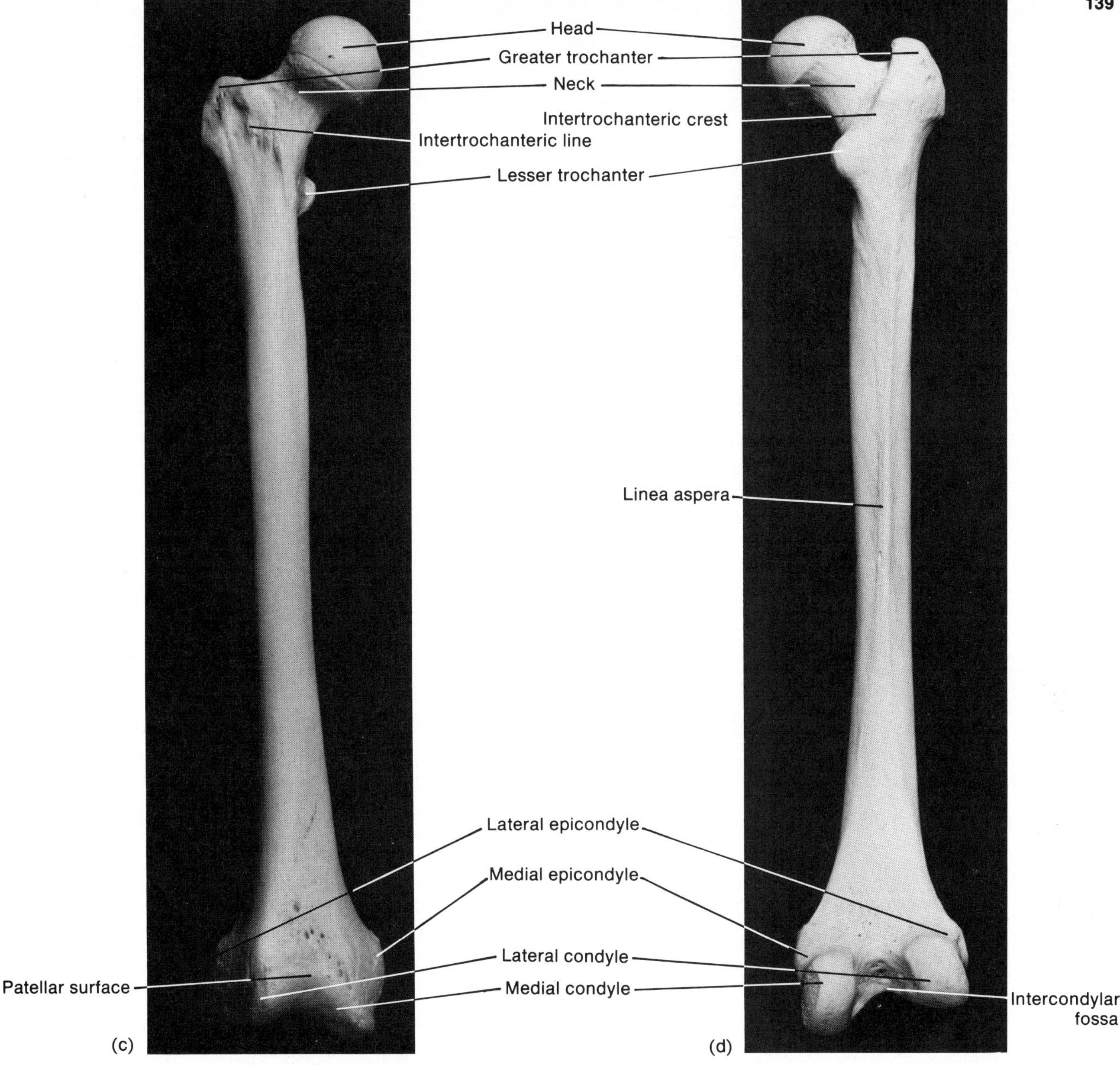

Figure 7–10 (cont.). The right femur. (c) Photograph in anterior view. (d) Photograph in posterior view. (Courtesy of Lenny Patti.)

Exhibit 7–11. THE FREE LOWER EXTREMITY: THE FEMUR (see Figure 7–10)

General Description: The **femur,** or thigh bone, is the longest and heaviest bone in the body. Its proximal end articulates with the coxal bone. And its distal end articulates with the tibia and patella. The shaft of the femur bows medially so that it approaches the femur of the opposite thigh. As a result of this convergence, the knee joints are brought nearer to the line of gravity of the body. The degree of convergence is greater in the female because the female pelvis is broader.

Markings: The proximal end of the femur consists of a rounded **head** that articulates with the acetabulum of the coxal bone. The **neck** of the femur is a constricted region distal to the head. A fairly common fracture in elderly people occurs at the neck of the femur. Apparently the neck becomes so weak that it fails to support the body weight. The **greater trochanter** and **lesser trochanter** are projections that serve as points of attachment for some of the muscles of the thigh and buttock. Between the trochanters on the anterior surface is a narrow **intertrochanteric line.** Between the trochanters on the posterior surface is an **intertrochanteric crest.**

The shaft of the femur contains a roughened vertical ridge on its posterior surface called the **linea aspera.** This serves for the attachment of several thigh muscles.

The distal end of the femur is expanded and includes the **medial condyle** and **lateral condyle.** These articulate with the tibia. A depressed area between the condyles on the posterior surface is called the **intercondylar fossa.** The **patellar surface** is located between the condyles on the anterior surface. Lying superior to the condyles are the **medial epicondyle** and **lateral epicondyle.**

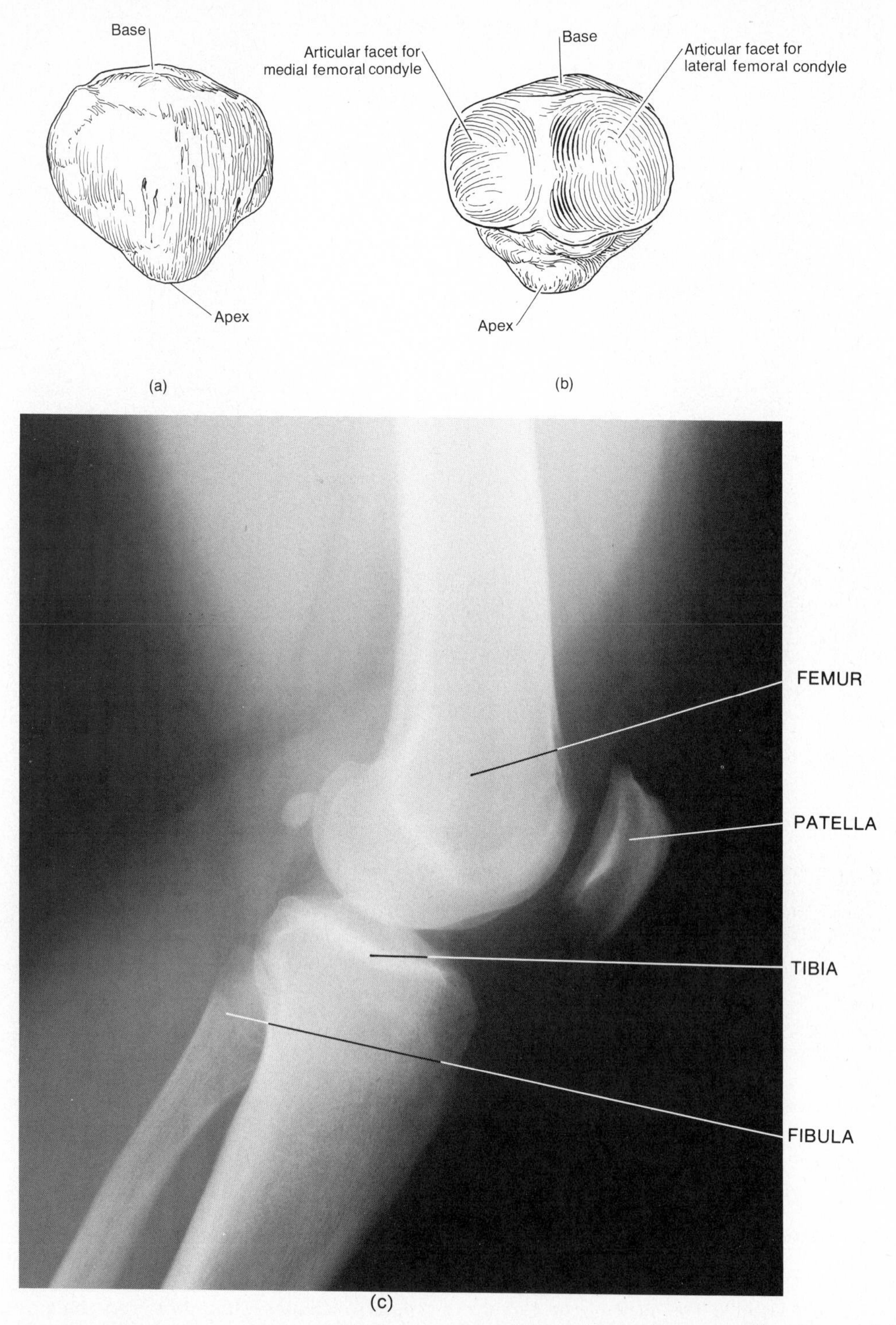

Figure 7–11. The right patella. (a) Anterior view. (b) Posterior view. (c) Lateral roentgenogram. (Courtesy of John C. Bennett, St. Mary's Hospital, San Francisco.)

Exhibit 7–12. THE FREE LOWER EXTREMITY: THE PATELLA (see Figure 7–11)

The **patella,** or kneecap, is a small, triangular bone anterior to the knee joint. It develops in the tendon of the quadriceps femoris muscle. A bone that forms in a tendon, such as the patella, is called a **sesamoid** bone. The broad superior end of the patella is called the **base.** And the pointed inferior end is referred to as the **apex.** The posterior surface contains two articular surfaces. These are the **articular facets** for the medial and lateral condyles of the femur.

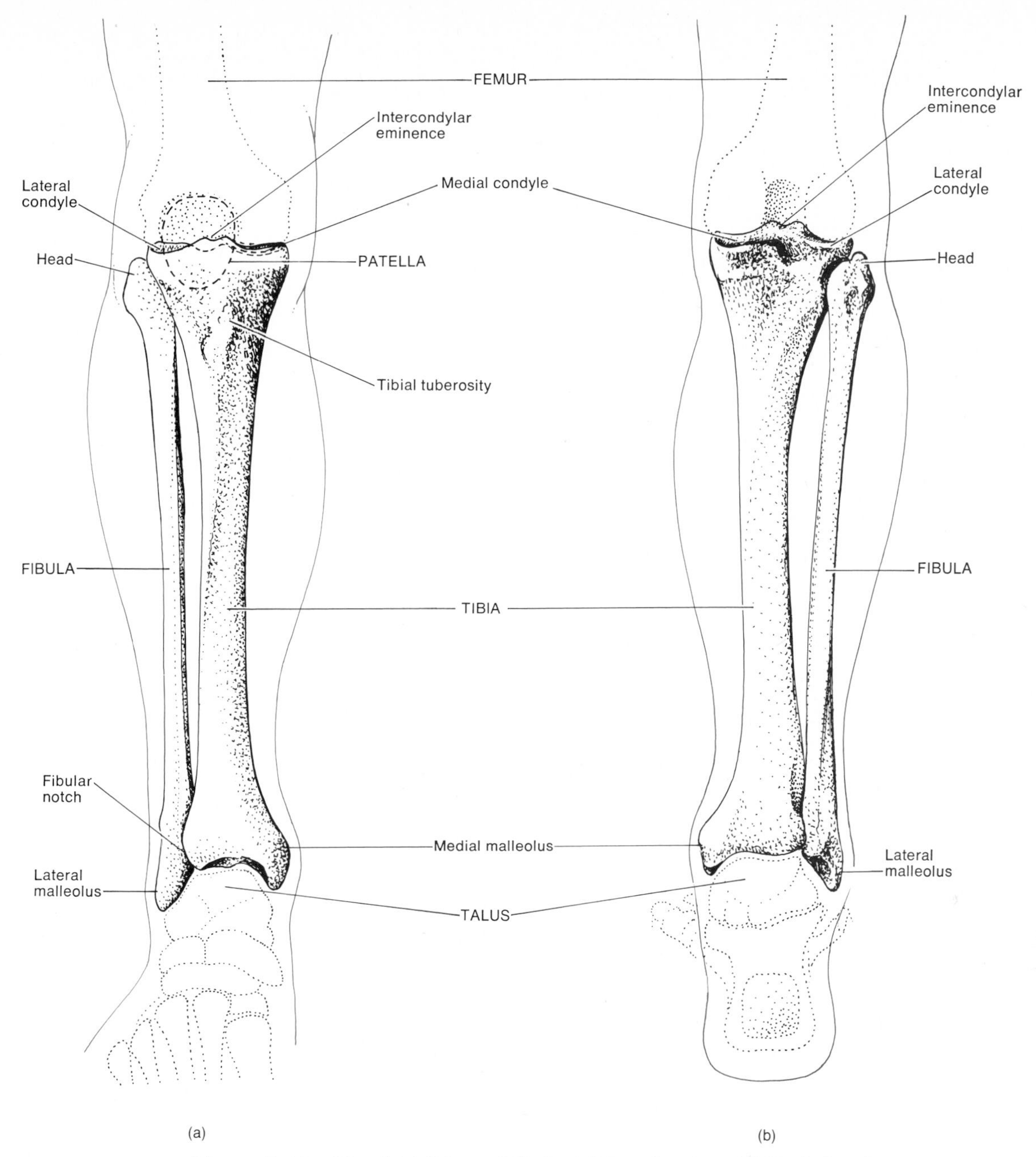

Figure 7–12. The right tibia and fibula. (a) Anterior view. (b) Posterior view.

Exhibit 7–13. THE FREE LOWER EXTREMITY: THE TIBIA AND FIBULA (see Figure 7–12)

General Description of the Tibia: The **tibia,** or shinbone, is the larger, medially placed bone of the lower leg. It bears the brunt of the weight on the leg. The tibia articulates at its proximal end with the femur and at its distal end with the fibula of the leg and talus of the ankle.

Markings: The proximal end of the tibia is expanded into a **lateral condyle** and a **medial condyle.** These articulate with the condyles of the femur. The slightly concave condyles are separated by an upward projection called the **intercondylar eminence.** The **tibial tuberosity** on the anterior surface is a point of attachment for the patellar ligament.

The distal end of the tibia contains a **medial malleolus.** This structure articulates with the talus bone of the ankle and forms the prominence that you can feel on the medial surface of your ankle. The **fibular notch** articulates with the fibula.

General Description of the Fibula: The **fibula** is lateral and runs parallel to the tibia. It is considerably smaller than the tibia.

Markings: The **head** of the fibula, the proximal end, articulates with the tibia. The distal end has a projection called the **lateral malleolus,** which articulates with the talus bone of the ankle. This forms the prominence on the lateral surface of the ankle. The inferior portion of the fibula also articulates with the tibia at the **fibular notch.** A fracture of the lower end of the fibula with injury to the tibial articulation is called a *Pott's fracture.*

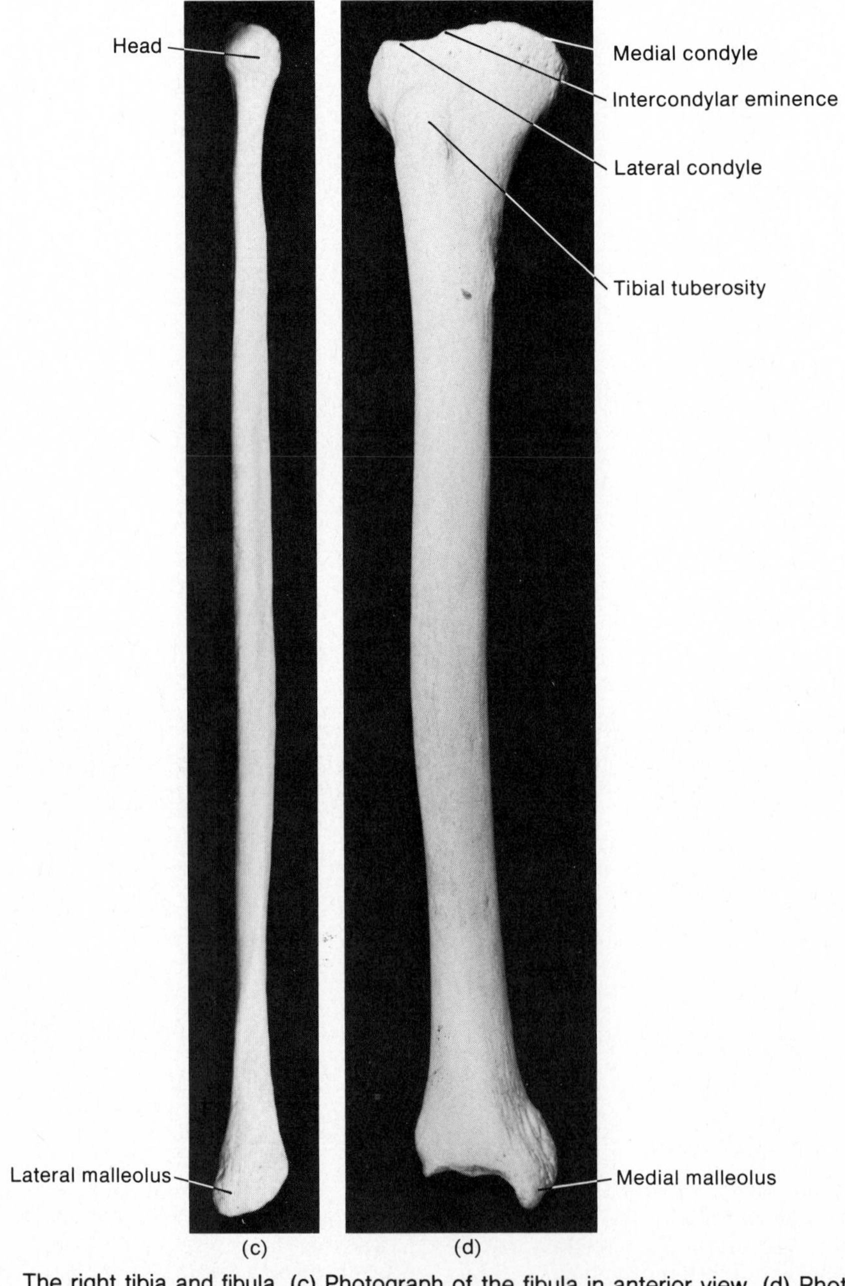

Figure 7–12 (cont.). The right tibia and fibula. (c) Photograph of the fibula in anterior view. (d) Photograph of the tibia in anterior view. (Photographs courtesy of Lenny Patti.) (e) Anteroposterior roentgenogram of the proximal ends. (f) Anteroposterior roentgenogram of the distal ends. (Roentgenograms courtesy of John C. Bennett, St. Mary's Hospital, San Francisco.)

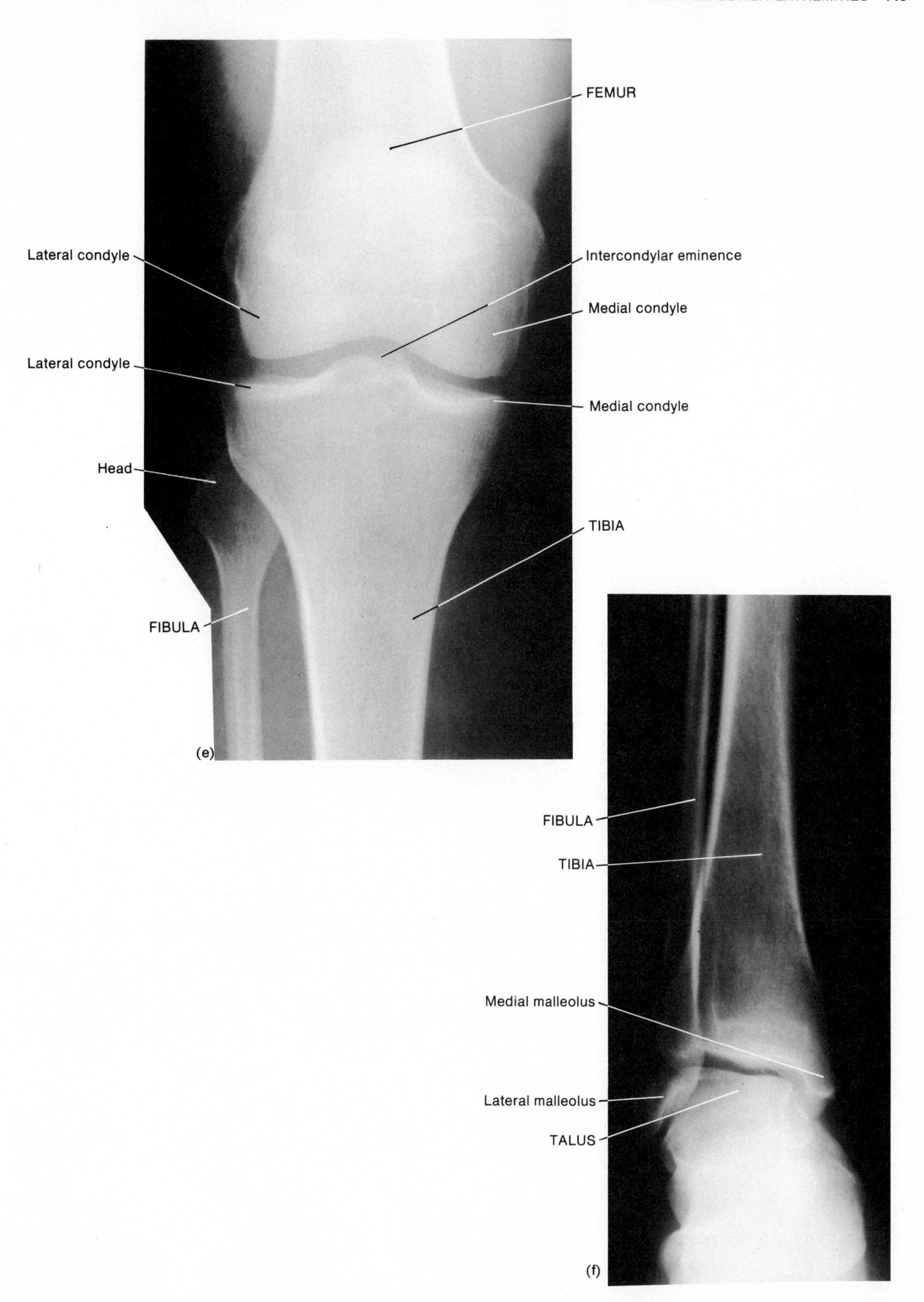
FEMUR
Lateral condyle
Intercondylar eminence
Medial condyle
Lateral condyle
Medial condyle
Head
TIBIA
FIBULA
(e)
FIBULA
TIBIA
Medial malleolus
Lateral malleolus
TALUS
(f)

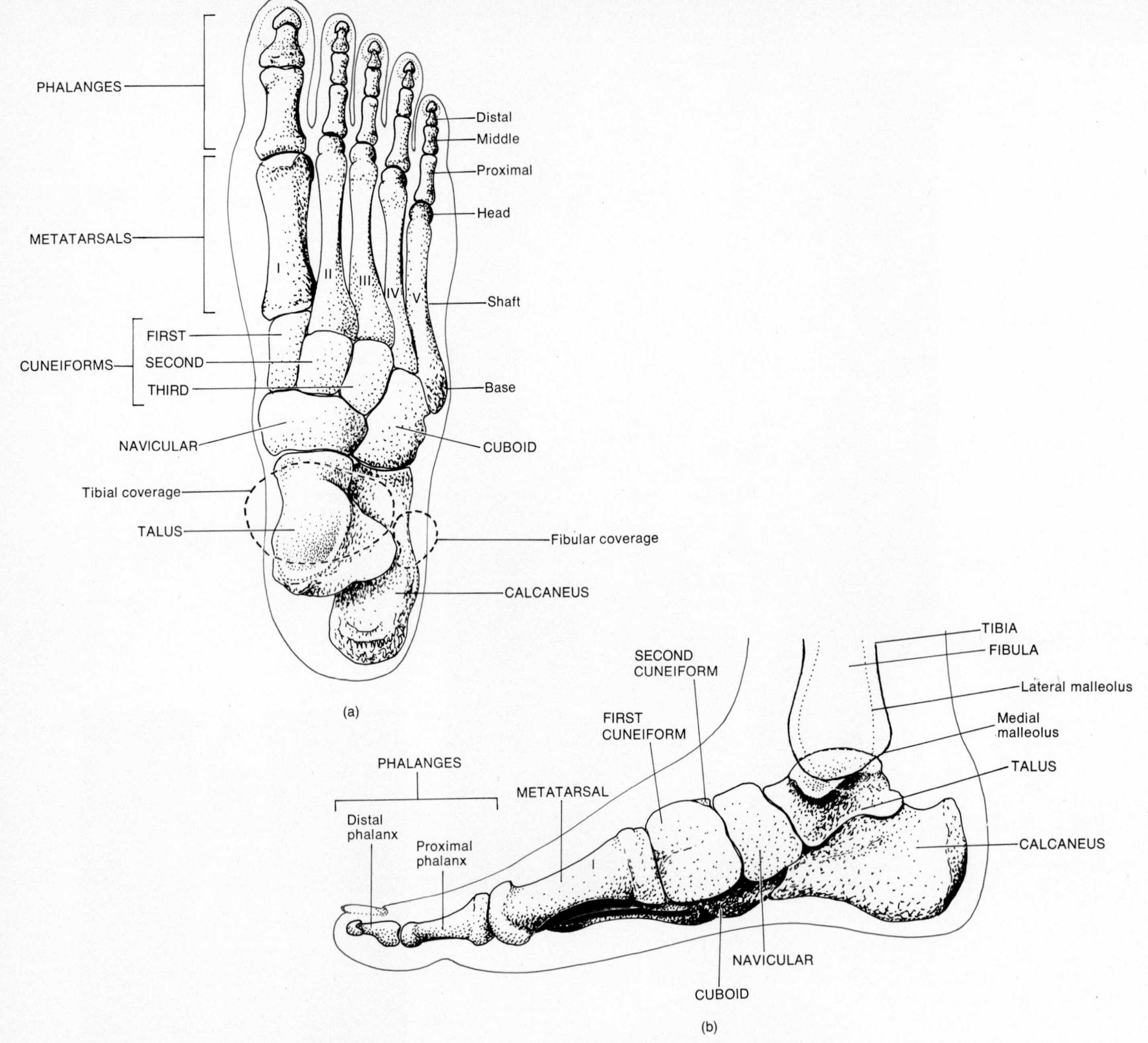

Figure 7–13. The right foot. (a) Superior view. (b) Medial view.

Exhibit 7–14. THE FREE LOWER EXTREMITY: THE TARSUS, METATARSUS, AND PHALANGES (see Figure 7–13)

The **tarsus** is a collective designation for the seven bones of the ankle. The term *tarsos* pertains to a broad, flat surface. The **talus** and **calcaneus** are located on the posterior part of the foot. The anterior part contains the **cuboid, navicular,** and three **cuneiform** bones called the first (medial), second (intermediate), and third (lateral) cuneiform. The talus, the uppermost tarsal bone, is the only bone of the foot that articulates with the fibula and tibia. It is surrounded on one side by the medial malleolus of the tibia and on the other side by the lateral malleolus of the fibula. During walking, the talus initially bears the entire weight of its extremity. About half the weight is then transmitted to the calcaneus. The remainder is transmitted to the other tarsal bones. The calcaneus, or heel bone, is the largest and strongest tarsal bone.

The **metatarsus** consists of five metatarsal bones numbered I to V from the medial to lateral position. The metatarsals articulate proximally with the first, second, and third cuneiform bones and with the cuboid. Distally, they articulate with the proximal row of phalanges. The first metatarsal is thicker than the others because it bears more weight.

The **phalanges** of the foot resemble those of the hand both in number and arrangement. The first toe, or big toe, has two large, heavy phalanges. The other four toes each have three phalanges. These are the proximal middle, and distal phalanges.

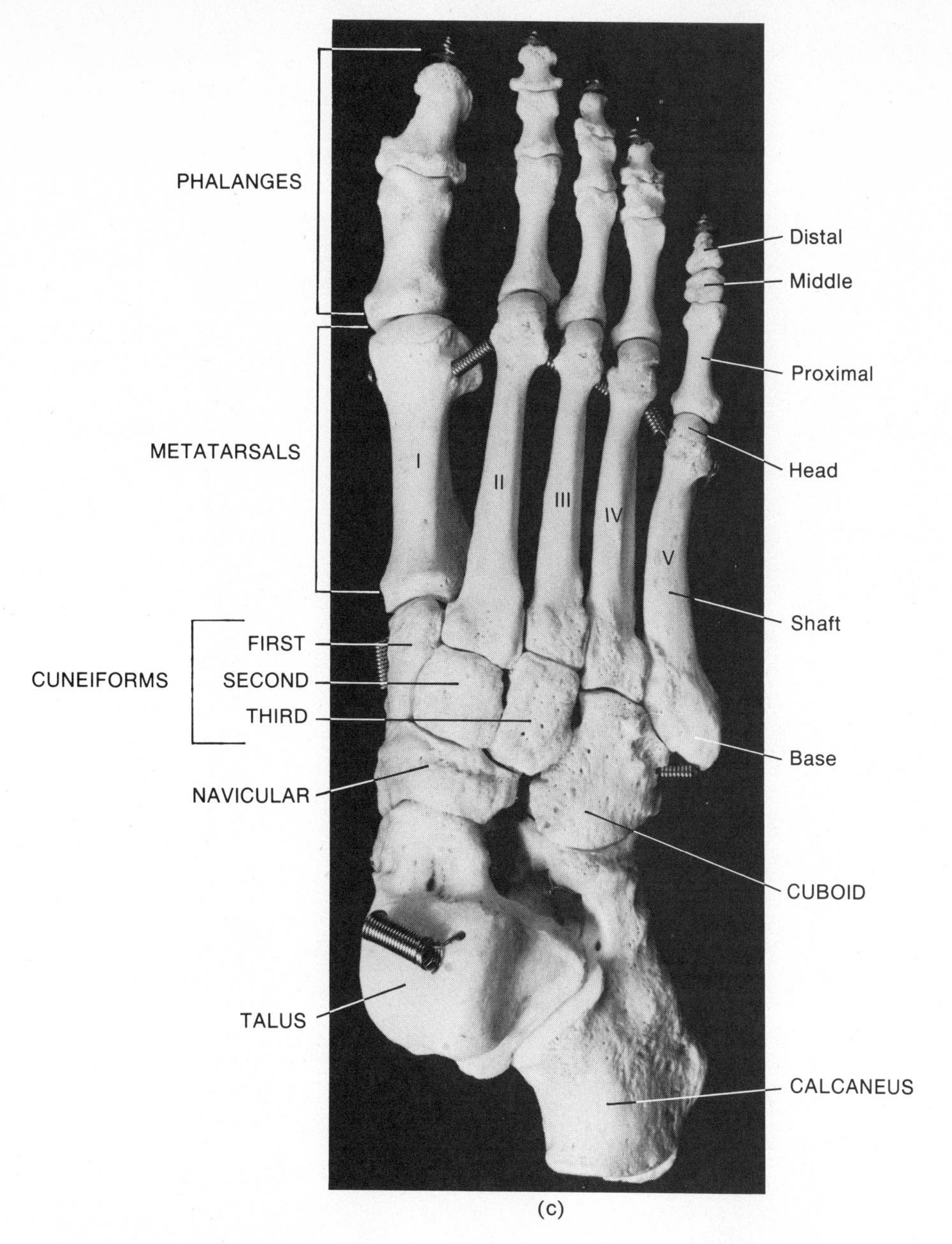

(c)

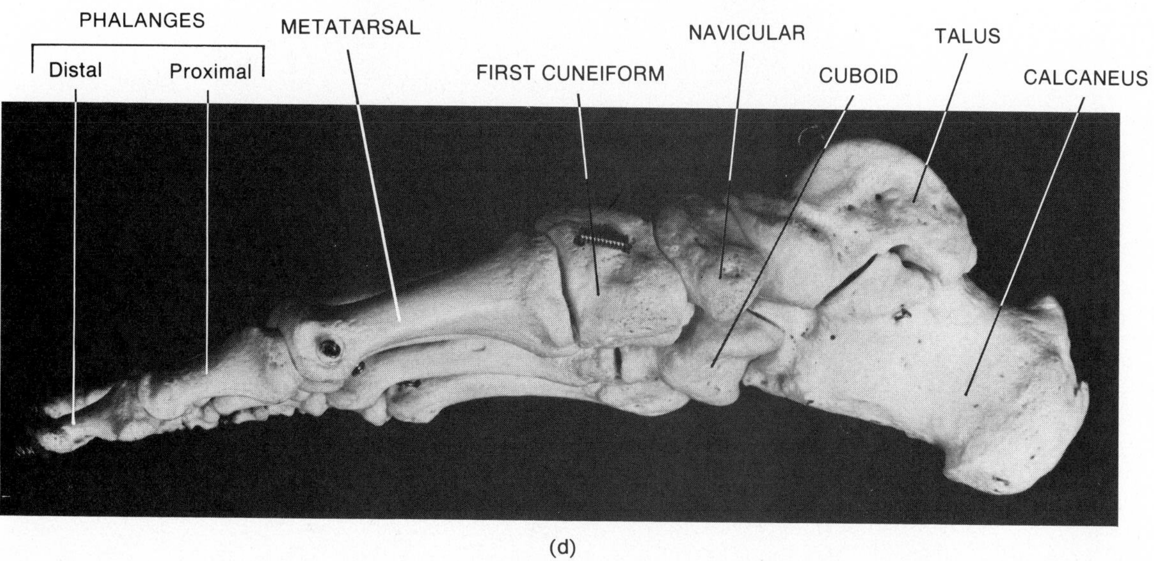

(d)

Figure 7–13 (cont.). The right foot. (c) Photograph in superior view. (d) Photograph in medial view. (Courtesy of Lenny Patti.)

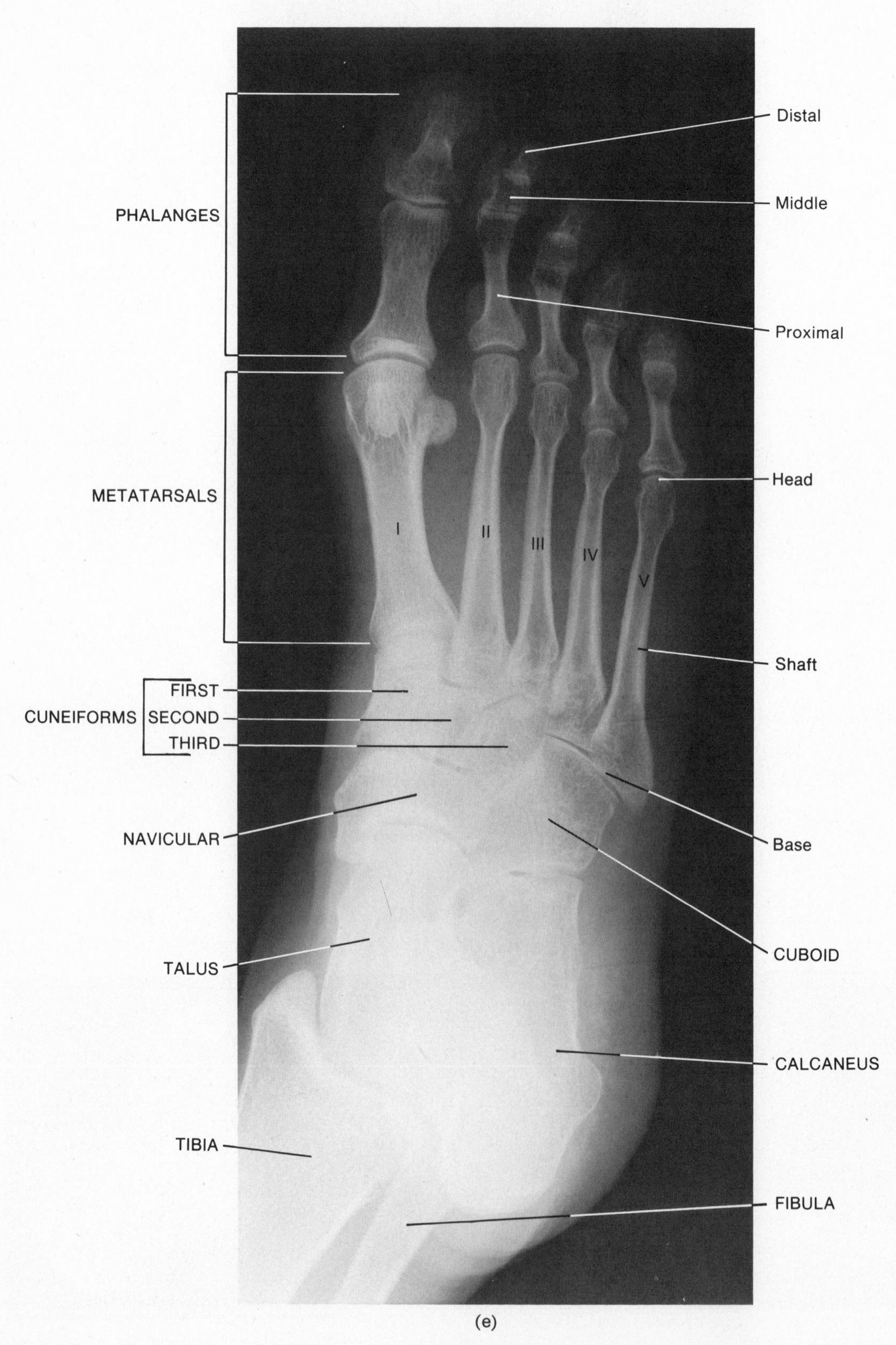

(e)

Figure 7–13 (cont.). The right foot. (e) Oblique roentgenogram. (Courtesy of John C. Bennett, St. Mary's Hospital, San Francisco.)

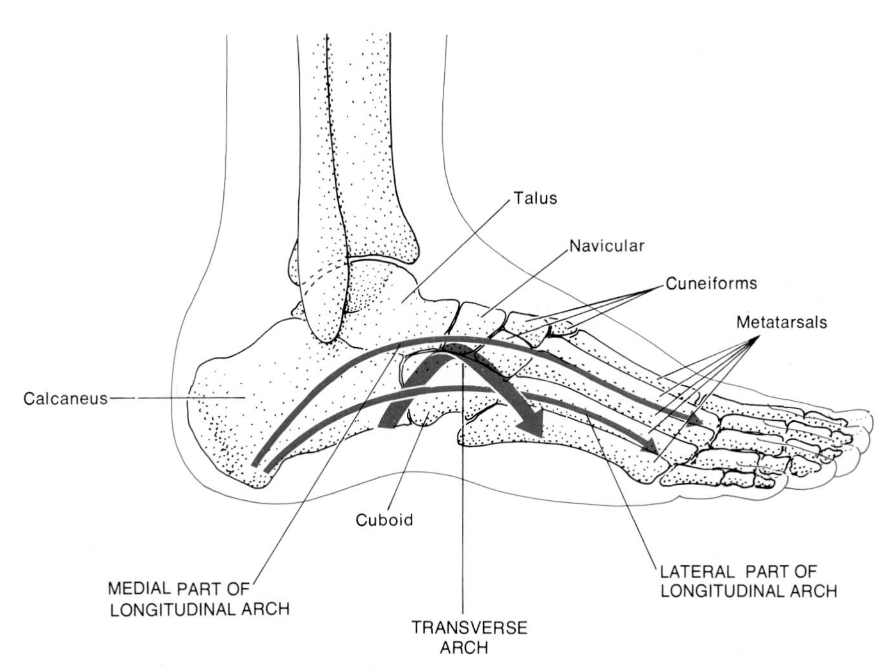

Figure 7–14. Lateral view of the arches of the right foot.

Exhibit 7–15. THE FREE LOWER EXTREMITY: ARCHES OF THE FOOT (see Figure 7–14)

The bones of the foot are arranged in two **arches.** These arches enable the foot to support the weight of the body and provide leverage while walking. The arches are not rigid. They yield as weight is applied and spring back when the weight is lifted.

The **longitudinal arch** has two parts. Both consist of tarsal and metatarsal bones arranged to form an arch from the anterior to the posterior part of the foot. The **medial,** or inner, part of the longitudinal arch originates at the calcaneus. It rises to the talus, and descends anteriorly through the navicular, the three cuneiforms, and the three medial metatarsals. The talus is the keystone of this arch. The **lateral,** or outer, part of the longitudinal arch also begins at the calcaneus. It rises in the cuboid and descends to the two lateral metatarsals. The cuboid is the keystone of the arch.

The **transverse arch** is formed by the calcaneus, navicular, cuboid, and the posterior parts of the five metatarsals.

The bones composing the arches are held in position by ligaments and tendons. If these ligaments and tendons are weakened, the height in the longitudinal arch may decrease or "fall." The result is *flatfoot.* A *bunion* is an abnormal lateral displacement of the big toe from its natural position. One cause is wearing shoes that are too small.

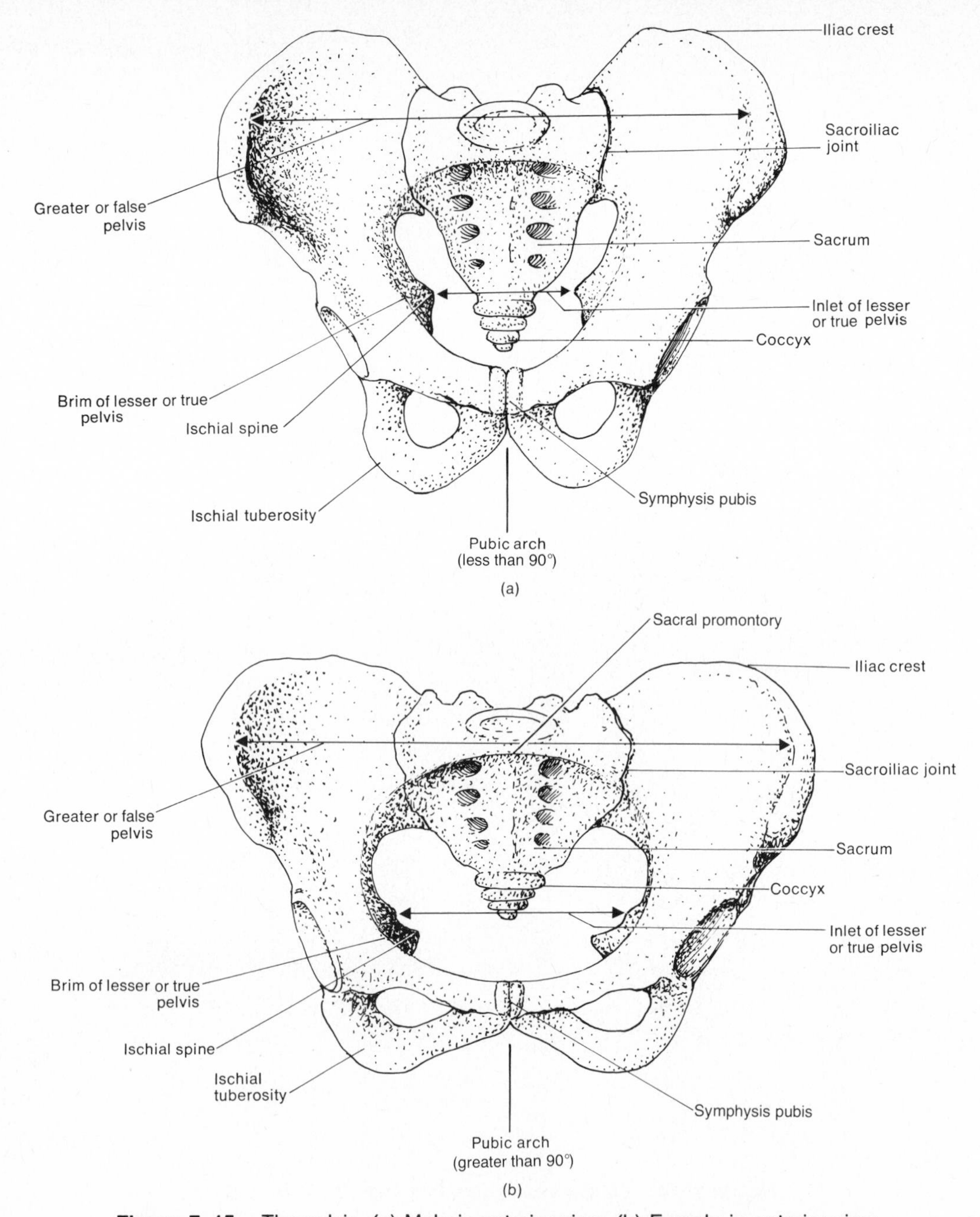

Figure 7–15. The pelvis. (a) Male in anterior view. (b) Female in anterior view.

COMPARISON BETWEEN MALE AND FEMALE SKELETONS

The bones of the male are generally larger and heavier than those of the female. And the articular ends are thicker when compared with the shafts. In addition, some of the muscles of the male are larger than those of the female. Consequently, the male skeleton has larger tuberosities, lines, and ridges for the attachment of the larger muscles.

One marked difference between male and female skeletons can be seen in the pelvis. The main differences between the male pelvis and the female pelvis concern adaptations directly related to the childbearing function (Figure 7–15). The female pelvis is wider, shallower, and lighter in structure than that of the male. The ilia of the female flare out to the sides and give her broader hips. The inlet of the true pelvis of the female is nearly oval, whereas that of the male is triangular. The sacrum of the female is shorter, wider, and less curved than that of the male. And the female coccyx is more movable. The ischial spines and tuberosities of the female turn outward and are further apart than in the male. The pubic arch thus forms an obtuse angle rather than an acute angle as in the male.

All characteristics contribute to the wider outlet of the true pelvis in the female. This, of course, accommodates the birth of the child. In addition, the ligaments of the sacroiliac joint stretch during pregnancy and childbirth. More room is provided for the developing fetus, and delivery is made easier.

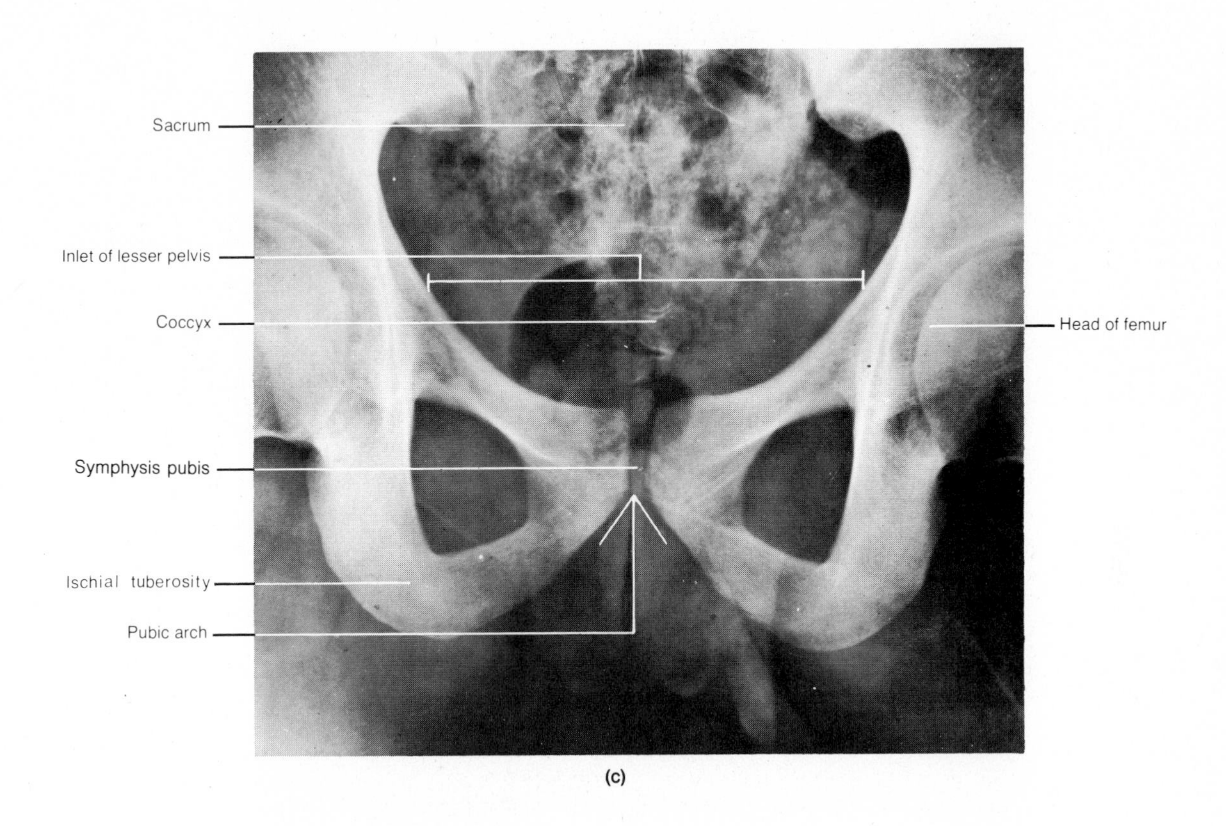

(c)

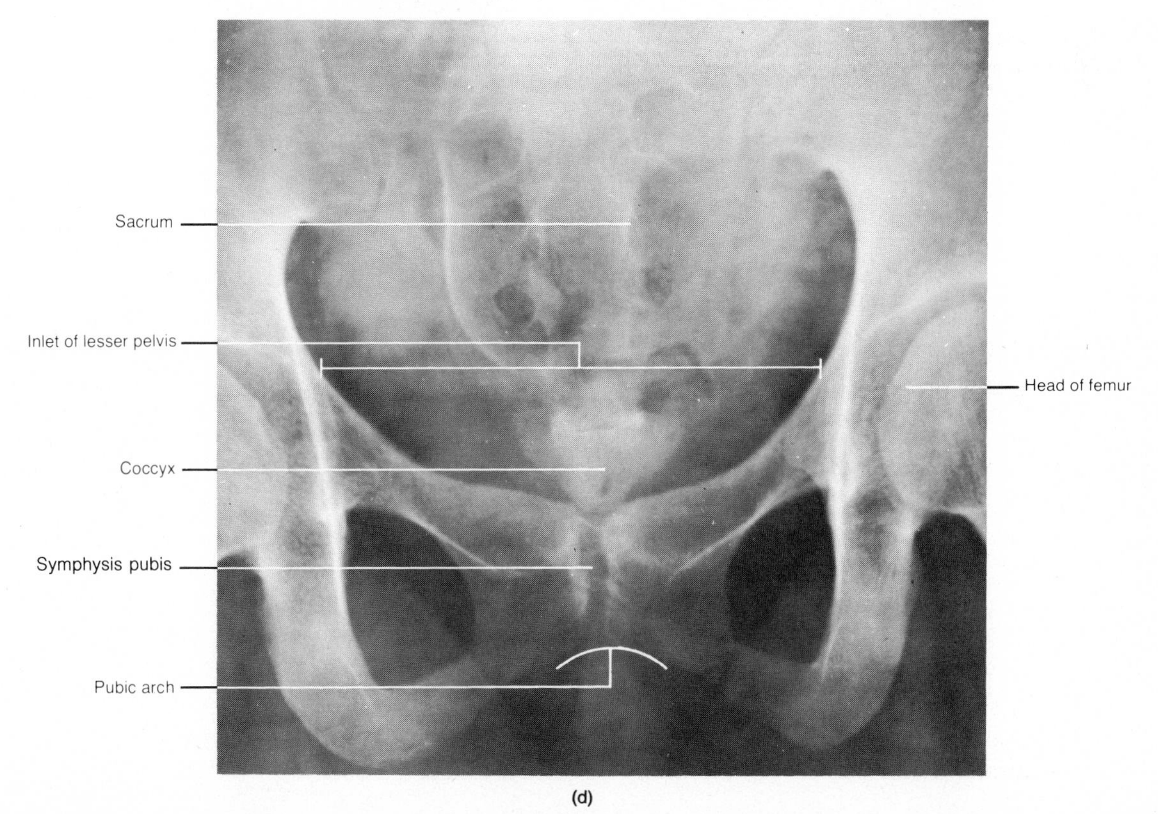

(d)

Figure 7–15 (cont.). The pelvis. (c) Anteroposterior roentgenogram of the male pelvis. (d) Anteroposterior roentgenogram of the female pelvis. (Courtesy of Eastman Kodak Company.)

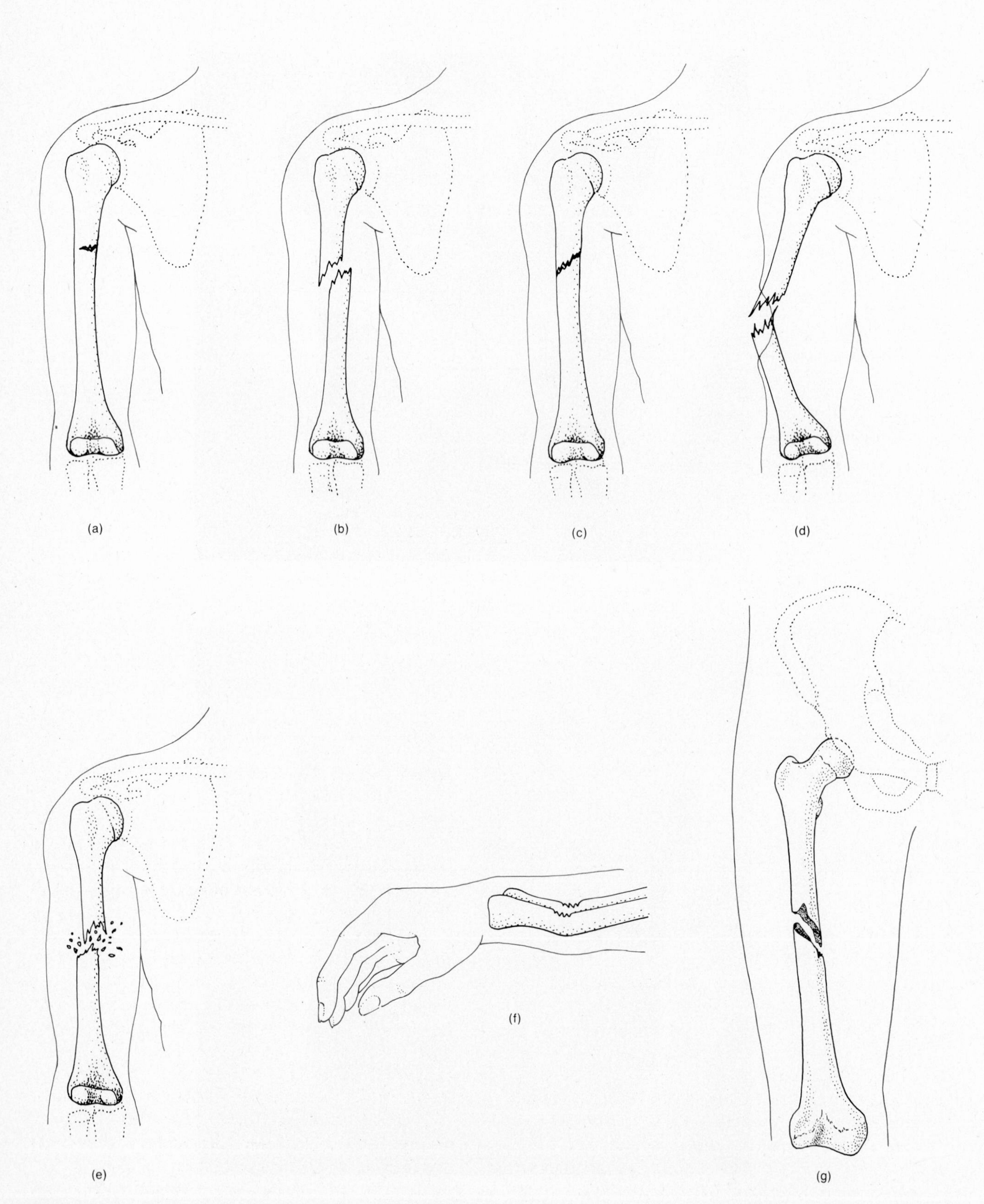

Figure 7–16. Types of fractures. (a) Partial. (b) Complete. (c) Simple. (d) Compound. (e) Comminuted. (f) Greenstick. (g) Spiral. (h) Transverse. (i) Impacted. (j) Pott's. (k) Colles'.

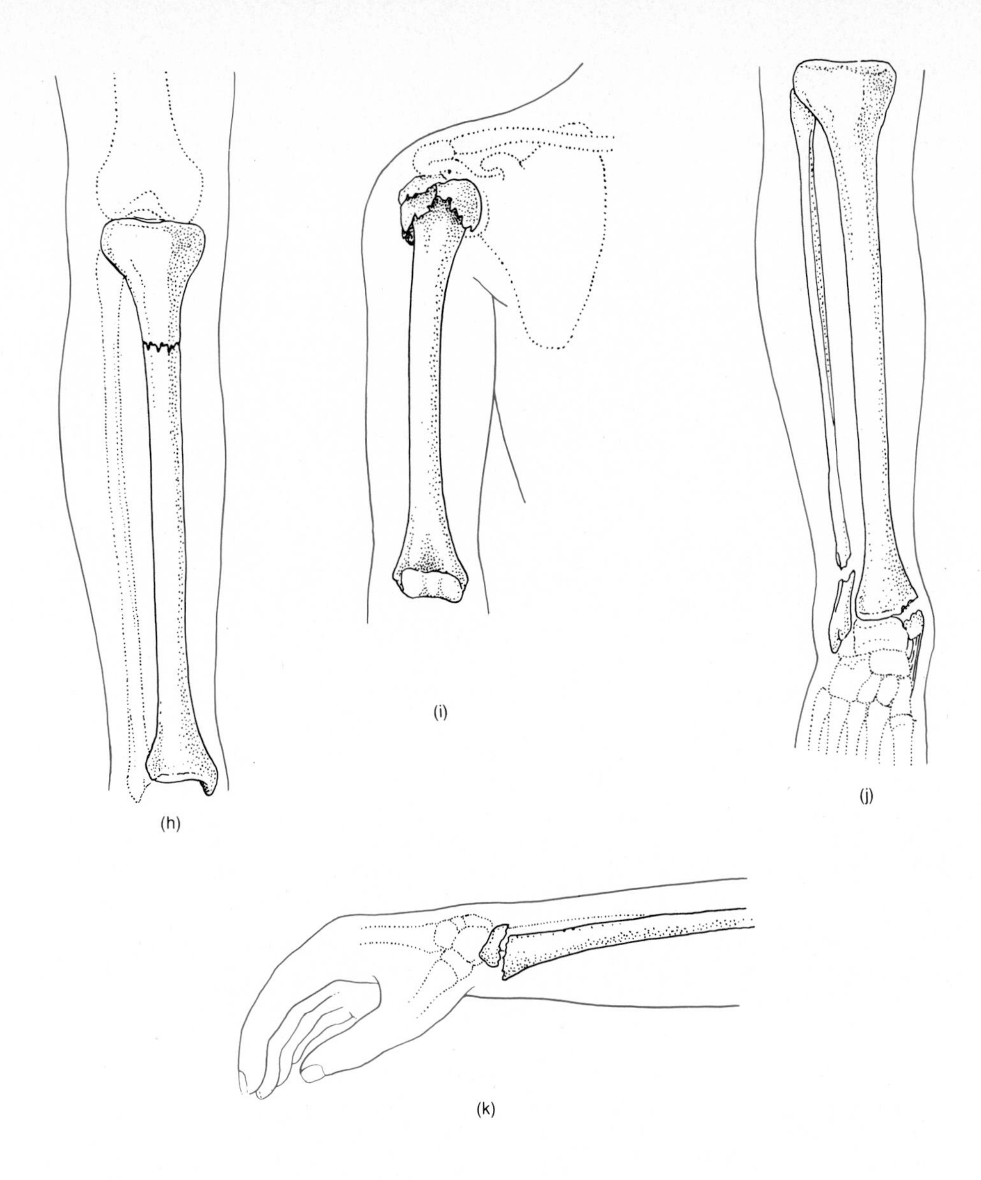

APPLICATIONS TO HEALTH

Fractures

In simplest terms, a **fracture** may be defined as any break in a bone. Although fractures of the bones of the extremities may be classified in several different ways, we shall use the following scheme for purposes of our discussion:

1. With a *partial* fracture, the break across the bone is incomplete (Figure 7–16a).
2. In a *complete* fracture, the break occurs across the entire bone. The bone is completely broken in two (Figure 7–16b).
3. A *simple* fracture is also called a *closed* fracture. The fractured bone does not break through the skin (Figure 7–16c).
4. In a *compound,* or *open,* fracture, the broken ends of the fractured bone protrude through the skin (Figure 7–16d).
5. With a *comminuted* fracture, the bone is splintered at the site of impact, and smaller fragments of bone are found between the two main fragments (Figure 7–16e).
6. A *greenstick* fracture is a partial fracture in which one side of the bone is broken and the other side of the bone bends (Figure 7–16f).
7. A *spiral* fracture is one in which the bone has been twisted apart (Figure 7–16g).
8. A *transverse* fracture is a fracture at right angles to the long axis of the bone (Figure 7–16h).
9. An *impacted* fracture is one in which one fragment is firmly driven into the other (Figure 7–16i).
10. A *Pott's* fracture is a fracture of the distal end of the fibula, with serious injury of the distal tibial articulation. Usually, there is also a chipping off of a portion of the lateral malleolus or rupture of the lateral ligament (Figure 7–16j).
11. A *Colles'* fracture is a fracture of the distal end of the radius in which the distal fragment is displaced posteriorly (Figure 7–16k).

12. A *displaced* fracture is a fracture in which the anatomical alignment of the bone fragments is not preserved.
13. A *nondisplaced* fracture is a fracture in which the anatomical alignment of the bone fragments has not been disrupted.

Unlike the skin, which may repair itself in days, or muscle, which may mend in weeks, a bone sometimes requires months to heal. A fractured femur, for example, may take 6 months to heal because sufficient calcium to strengthen and harden new bone is deposited very gradually. In addition, bone cells grow and reproduce slowly. Also the blood supply to bone is poor, which partially explains the difficulty in healing an infected bone.

The following steps occur in the repair of a fracture (Figure 7–17):

1. As a result of the fracture, blood vessels crossing the fracture line are broken. These vessels are found in the periosteum, Haversian systems, and marrow cavity. As blood pours from the torn ends of the vessels, it coagulates and forms a clot in and about the site of the fracture. This clot is called a **fracture hematoma.** It usually occurs 6 to 8 hours after the injury (Figure 7–17a). Since the circulation of blood ceases when the fracture hematoma forms, bone cells and periosteal cells at the fracture line die.

2. A growth of new bone tissue develops in and around the fractured area (Figure 7–17b). This new tissue is called a **callus.** It forms a bridge between separated areas of bone so that they are united. The callus that forms around the outside of the fracture is called an *external callus.* The callus formed between the two ends of bone fragments and between the two marrow cavities is called the *internal callus.*

About 48 hours after a fracture occurs, the cells that ultimately repair the fracture become actively mitotic. These cells come from the osteogenic layer of the periosteum, the endosteum of the marrow cavity, and the bone marrow. As a result of their accelerated mitotic activity, the cells of the three regions grow toward the fracture. During the first week following the fracture, the cells of the endosteum and bone marrow form new trabeculae in the marrow cavity close to the line of fracture. This is the internal callus. Over the next few days, osteogenic cells of the periosteum form a collar around each bone fragment. The collar, or external callus, is replaced by trabeculae. The trabeculae of the calli are joined to living and dead portions of the original bone fragments.

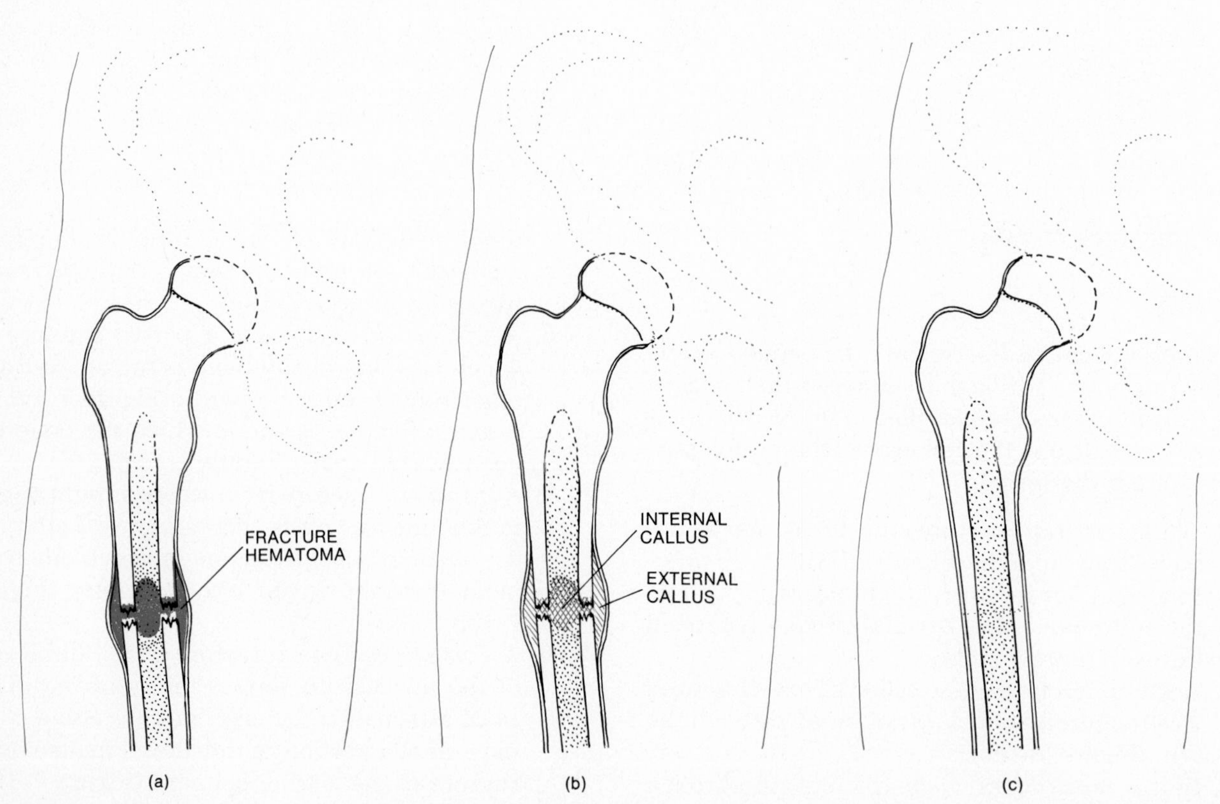

Figure 7–17. Fracture repair. (a) Formation of fracture hematoma. (b) Formation of external and internal calli. (c) Completely healed fracture.

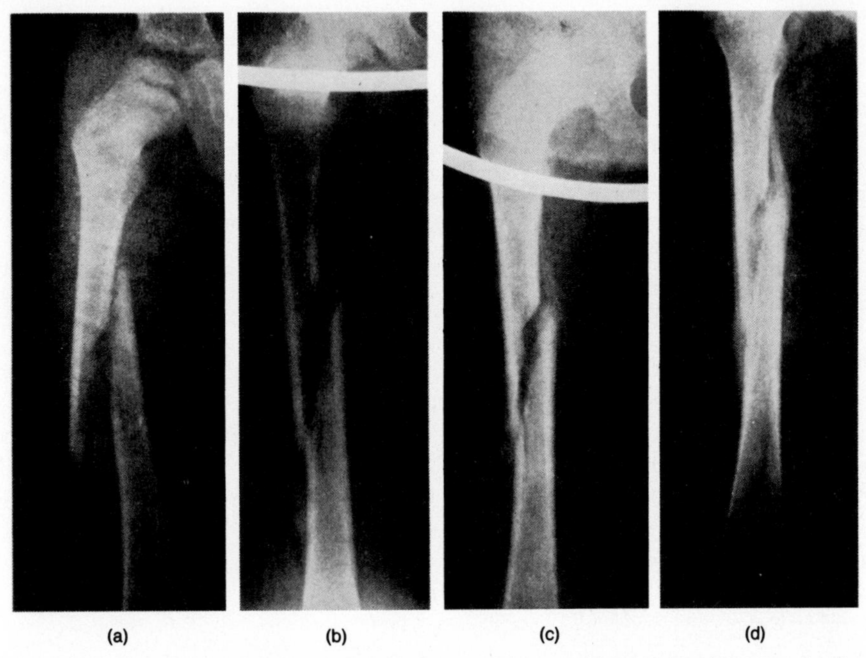

Figure 7–18. Roentgenograms of the repair of a fractured femur. (a) Immediately after fracture. (b) Two weeks later, a hazy external callus is visible around the margins of the fracture. (c) About 3½ weeks after the fracture, the internal and external calli begin to bridge the separated fragments. (d) Almost 9 weeks after the fracture, the bridge between fragments is fairly well developed. In time, remodeling will occur and the fracture will be repaired. (Courtesy of Ralph C. Frank, Eau Claire, Wisconsin, from Lester W. Paul and John H. Juhl, *The Essentials of Roentgen Interpretation,* 3d ed., Harper & Row, Publishers, Inc., New York, 1972.)

3. The final phase of fracture repair is the **remodeling** of the calli. In the remodeling process (Chapter 6), the dead portions of the original fragments are gradually resorbed. Compact bone replaces spongy bone around the periphery of the fracture (Figure 7–17c). In some cases, the healing is so complete that the fracture line is undetectable, even by x-ray. However, a thickened area on the surface of the bone usually remains as evidence of the fracture site. The healing of a fracture of the femur is illustrated by sequential roentgenograms in Figure 7–18.

Now that you have a fairly good understanding of the bones of the axial and appendicular skeletons, you can relate your knowledge to some of the surface anatomy that you learned in Chapter 1. Refer to Exhibits 1–3 through 1–7 and determine which bony surface landmarks are palpable. As you do this, be sure that you can palpate each on your own body. You might want to work with a partner to palpate the dorsally located bony landmarks.

You are now prepared to study the manner in which bones are attached to each other. This attachment, which is called a joint or articulation, determines how bones move.

Chapter summary in outline

APPENDICULAR SKELETON

The shoulder girdle

1. A shoulder girdle or pectoral girdle consists of a clavicle and scapula.
2. It attaches the free upper extremity to the trunk.

The free upper extremity

1. The bones of a free upper extremity include the humerus, ulna, radius, carpals, metacarpals, and phalanges.

The pelvic girdle

1. The pelvic girdle consists of two coxal bones or hipbones.
2. It attaches the free lower extremities to the trunk.

The free lower extremity

1. The bones of a free lower extremity include the femur, tibia, fibula, tarsus, metatarsals, and phalanges.

Arches of the foot

1. The arches of the foot are bones arranged for support and leverage.
2. The two parts of the longitudinal arch are the higher medial and lower lateral arches. The other arch is the transverse arch.

COMPARISON BETWEEN MALE AND FEMALE SKELETONS

1. The bones in a male are generally larger than the female's bones.
2. The female pelvis is adapted for pregnancy and childbirth.

APPLICATIONS TO HEALTH

1. A fracture is any break in a bone.
2. The types of fractures include: partial, complete, simple, compound, comminuted, greenstick, spiral, transverse, impacted, Pott's, Colles', displaced, and nondisplaced.
3. Fracture repair consists of forming a fracture hematoma, forming a callus, and remodeling.

Review questions and problems

1. What is a shoulder girdle? What are the bones of the free upper extremity? What is a pelvic girdle? What are the bones of the free lower extremity?
2. Define an arch of the foot. What is its function? Distinguish between a longitudinal arch and a transverse arch.
3. How do bunions and flatfeet arise?
4. What are the principal structural differences between male and female skeletons?
5. What is a fracture? Distinguish 13 principal kinds.
6. Outline the steps involved in fracture repair.

Selected readings

Cunningham's Textbook of Anatomy. 11th ed., edited by G. J. Romanes. London: Oxford University Press, 1972. Pp. 146–202.

Dawson, Helen L. *Basic Human Anatomy.* 2d ed. New York: Appleton-Century-Crofts, 1974. Pp. 49–65.

Gray, Henry. *Anatomy of the Human Body.* 29th ed., edited by Charles Mayo Goss. Philadelphia: Lea and Febiger, 1973. Pp. 196–268.

CHAPTER 8

ARTICULATIONS

STUDENT OBJECTIVES

After you have read this chapter, you should be able to:

1. Define an articulation and identify the factors that determine the degree of movement at a joint
2. Contrast the structure, kind of movement, and location of fibrous, cartilaginous, and synovial joints
3. Describe the detailed structure of a synovial joint
4. Discuss and compare the movements possible at various kinds of synovial joints
5. Describe the causes and symptoms of common joint disorders, including arthritis, bursitis, tendinitis, and intervertebral disc abnormality
6. Define key terms associated with joints

Bones are much too rigid to bend. In fact, if the skeleton of your body were composed of a continuous mass of ossified tissue, you would have less flexibility than a two-by-four. Fortunately, the skeletal system consists of many separate bones, which are held together at joints by flexible types of connective tissue. All movements that change the positions of the bony parts of the body, such as the extremities, take place at joints. Joints are necessary for numerous activities that we take for granted, for example, holding a pencil, turning the head, swinging the arms, driving a car, and dancing. You can understand the importance of joints if you imagine how a cast over the knee joint prevents flexing of the leg or how a splint on a finger limits the ability to manipulate small objects.

The term **articulation** or **joint** refers to a point of contact between bones or between cartilage and bones. In every case, the structure of a joint determines its function. Some joints permit no movement. Others permit a slight degree of movement. And still others afford relatively unrestricted movement. In general, the more closely the bones fit together, the stronger the joint. However, at tightly fitted joints, movement is restricted. The greater the degree of movement, the looser the fit. Unfortunately, loosely fitted joints are more prone to dislocation. Movement at joints is also determined by the flexibility of the connective tissue that binds the bones together and by the position of ligaments, muscles, and tendons. First, we shall look at the various kinds of joints in the body. Later in the chapter, we can discuss joint disorders.

CLASSIFICATION OF JOINTS

The various joints of the body may be classified on the basis of function and structure. The functional classification of joints takes into account the degree of movement they permit. Functionally, joints are classified as **synarthroses,** which are immovable joints; **amphiarthroses,** which are slightly movable joints; and **diarthroses,** which are freely movable joints.

The structural classification of joints is based upon the presence or absence of a space between the bones called the *joint cavity* and the kind of connective tissue that binds the bones together. Structurally, joints are classified as **fibrous,** in which there is no joint cavity and the bones are held together by fibrous connective tissue; **cartilaginous,** in which there is no joint cavity and the bones are held together by cartilage; and **synovial,** in which the joint contains a synovial cavity.

As the various kinds of joints are discussed, the structural classification outline will be used. However, continual reference will also be made to the functional classification.

Fibrous joints

Fibrous joints lack a joint cavity, and the articulating bones are held close together by fibrous connective tissue. They allow little or no movement. The two types of fibrous joints are sutures and syndesmoses (Figure 8–1). **Sutures** are found between bones of the skull. Some sutures consist of interlocking, jagged margins of the bone that fit together like

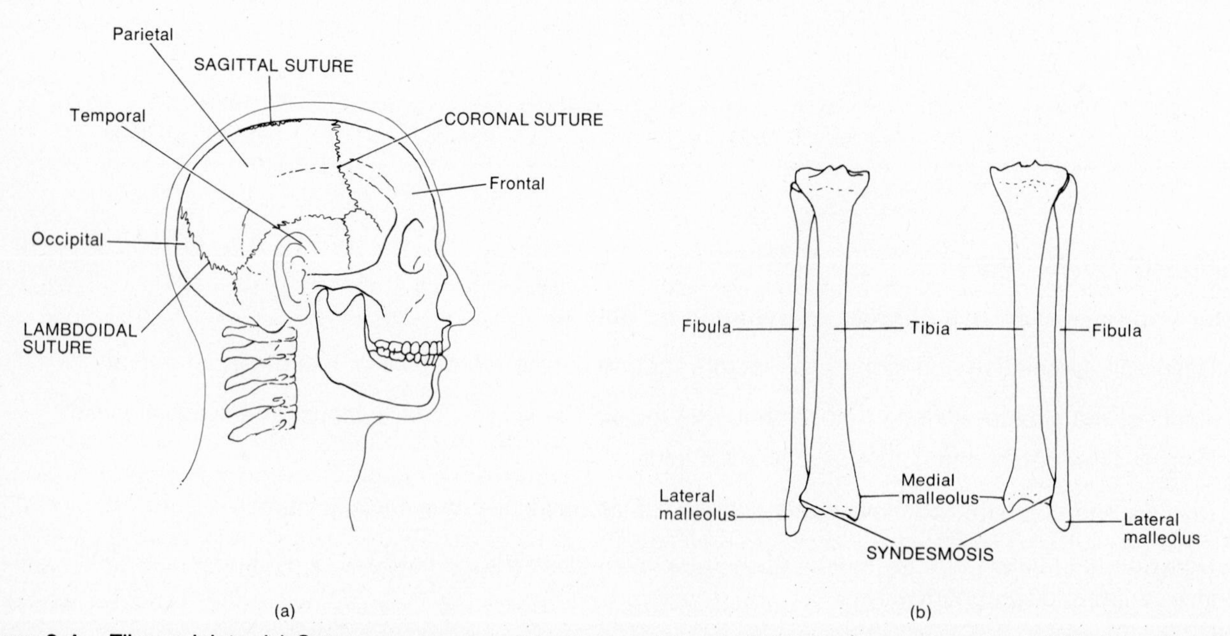

Figure 8-1. Fibrous joints. (a) Sutures as seen in a lateral view of the skull. (b) Syndesmosis at the tibiofibular articulation seen in anterior (left) and posterior (right) views.

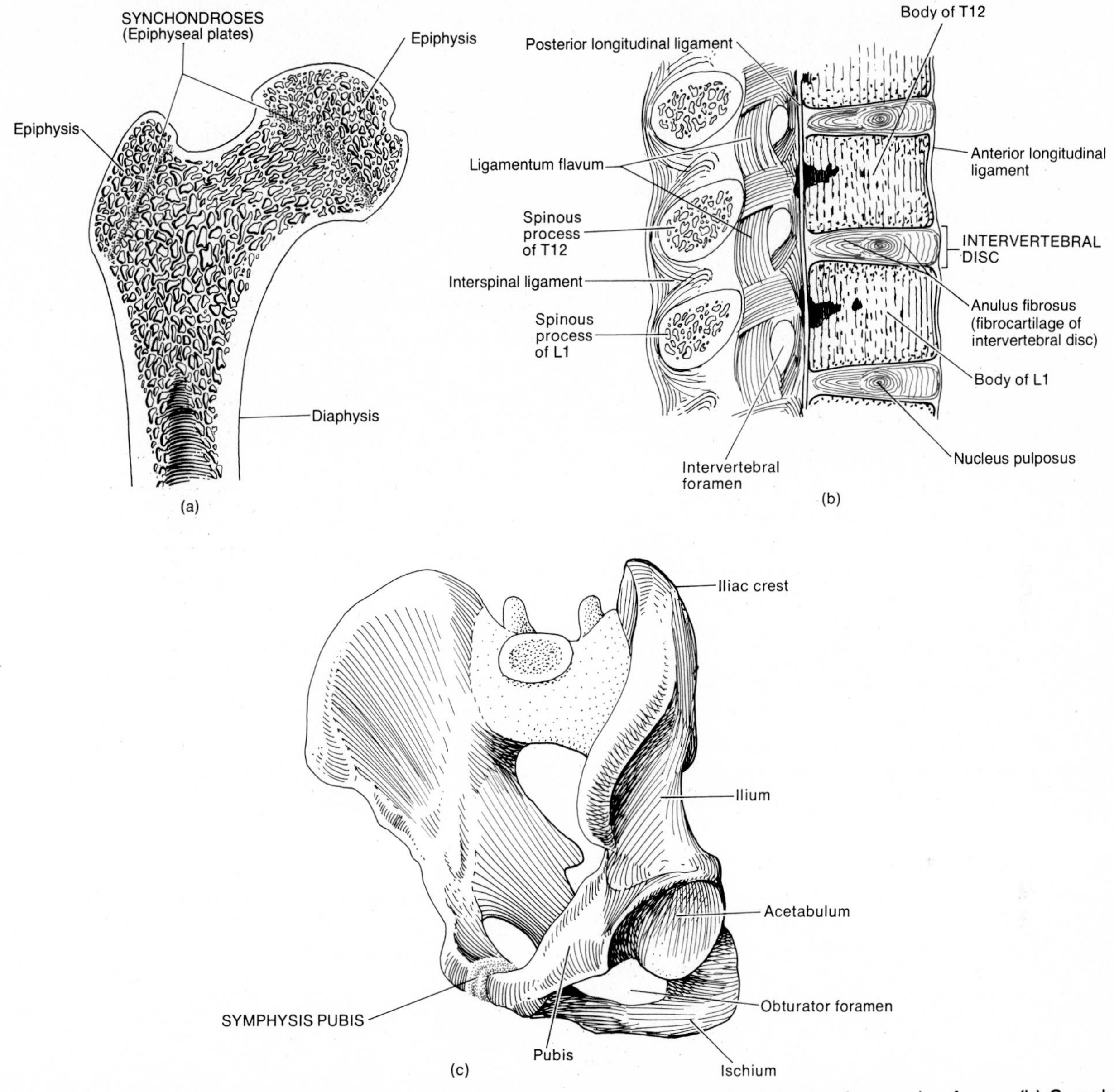

Figure 8-2. Cartilaginous joints. (a) Synchondrosis between the diaphysis and epiphysis of a growing femur. (b) Symphysis joint between the bodies of vertebrae seen in sagittal section. (c) Symphysis joint (symphysis pubis) between the coxal bones seen in an oblique view.

a jigsaw puzzle. In other sutures, the margins of the bones overlap. In either case, the bones are very slightly separated by a thin layer of fibrous tissue. As a result of this tight fit, such joints are immovable and are classified as synarthroses. Some sutures, present during growth, are replaced by bone in the adult. In this case, they are referred to as **synostoses,** in which there is complete fusion of bone across the suture line. One example of a synostosis is the joint between the left and right sides of the frontal bone. Synostoses are synarthrotic. The bone surfaces of a **syndesmosis** are united by dense fibrous tissue. The joint is very slightly movable because the bones are more separated from each other than they are in a suture. Thus, it is an amphiarthrotic joint. An example of a syndesmosis-type joint is the distal articulation of the tibia with the fibula.

Cartilaginous joints

Another type of joint that has no joint cavity is the **cartilaginous joint.** Here, the articulating bones are tightly connected by cartilage. Like fibrous joints, they allow little or no movement (Figure 8–2). A **synchondrosis** is a cartilaginous joint in which the connecting material is hyaline cartilage, as in the epiphyseal plate. Such a joint is found between the epiphysis and diaphysis of a growing bone and is immovable. Thus, it is synarthrotic. Since the hyaline cartilage is eventually replaced by bone when growth ceases, the joint is temporary. It is replaced by a synostosis. A **symphysis** is a cartilaginous joint in which the connecting material is a broad, flat disc of fibrocartilage. This kind of joint is found between bodies of vertebrae. A portion of the intervertebral disc is cartilaginous material. The symphysis pubis between the anterior surfaces of the coxal bones is another example of a symphysis-type joint. These joints are very slightly movable, or amphiarthrotic.

Synovial joints

When a joint cavity is present, the articulation is called a **synovial joint** (Figure 8–3). The cavity, called a **synovial cavity,** is a space between the articulating bones. Because of this cavity and because no tissue exists between the articulating surfaces of the bones, synovial joints are freely movable. By function, synovial joints are diarthrotic. Synovial joints are surrounded by a capsule of dense fibrous connective tissue which is continuous with the periosteum of the articulating bones. This capsule is called the *joint* or *fibrous capsule.* It protects and, in some cases, strengthens the joint. These joints are also characterized by a layer of hyaline cartilage, called *articular cartilage.* Articular cartilage covers the surfaces of the articulating bones but does not bind the bones together. These joints also have a *synovial membrane,* which lines the walls of the cavity and secretes synovial fluid to lubricate the joint. Synovial joints are held together by bands of collagenous fibers called *ligaments.* A ligament attaches to processes on one of the articulating bones, runs alongside the joint, and attaches to processes on the other bone.

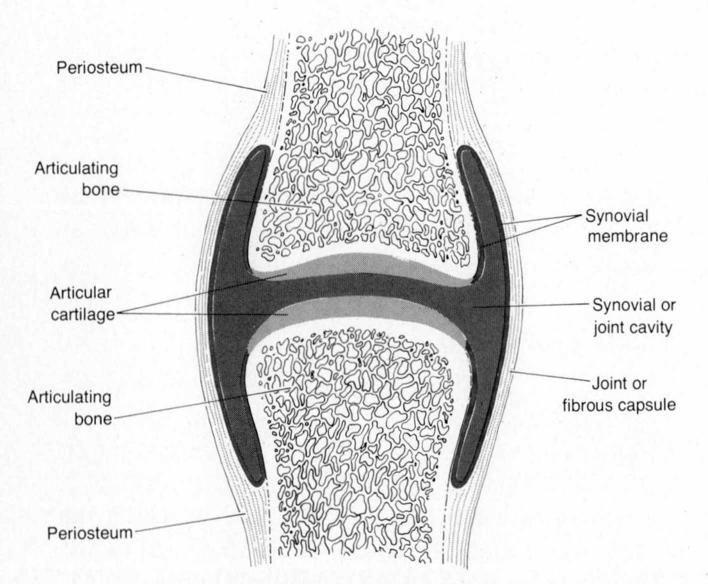

Figure 8-3. A synovial joint.

Synovial joints are free of the limitations of fibrous and cartilaginous joints. In synovial joints, movement is determined by the location of the ligaments, by the muscles and their tendons, and by the presence of other bones that might get in the way of particular movements. The knee joint, which is one of the largest articulations in the body, illustrates the basic structure of a synovial joint and the restrictions on its movement.

You can examine Figure 8–4 and note the following relationships among ligaments, tendons, and fibrocartilage in the knee joint:

1. Externally the joint is strengthened by muscles and tendons. These include the tendon of the quadriceps femoris muscle in front (Figure 8–4a) and the gastrocnemius muscle in back (Figure 8–4c). The patella lies within the tendon of the quadriceps femoris.
2. The thickened portion of the quadriceps femoris tendon between the top of the patella and the tibia is called the patellar ligament (Figure 8–4a). This ligament strengthens the anterior portion of the joint and prevents the lower leg from being flexed backward too far.
3. On either side of the joint are the fibular and tibial collateral ligaments. The fibular collateral ligament, between the femur and fibula, strengthens the lateral side of the joint. The tibial collateral ligament, between the femur and tibia, strengthens the medial side of the joint. Both ligaments prohibit any side-to-side movement at the joint (Figure 8–4a, d).
4. The oblique popliteal ligament is located on the posterior surface of the joint (Figure 8–4c). It starts in a tendon that lies over the tibia and runs upward and laterally to the lateral side of the femur. It supports the back of the knee and prevents hyperextension – bending the knee in the direction opposite to which it normally bends.
5. Internally, the joint is strengthened by the cruciate ligaments (Figure 8–4b, d). The anterior cruciate ligament passes posteriorly and laterally from the tibia and attaches to the femur. The posterior cruciate ligament passes anteriorly and medially from the tibia and attaches to the femur. Both ligaments are believed to stabilize the knee joint during its movements.
6. Between the articular cartilage of the femur and tibia are **menisci,** concentric wedge-shaped pieces of fibrocartilage (Figure 8–4b, d). These are called the lateral meniscus and the medial meniscus. The menisci afford support for the continuous weight placed on the knee joint.

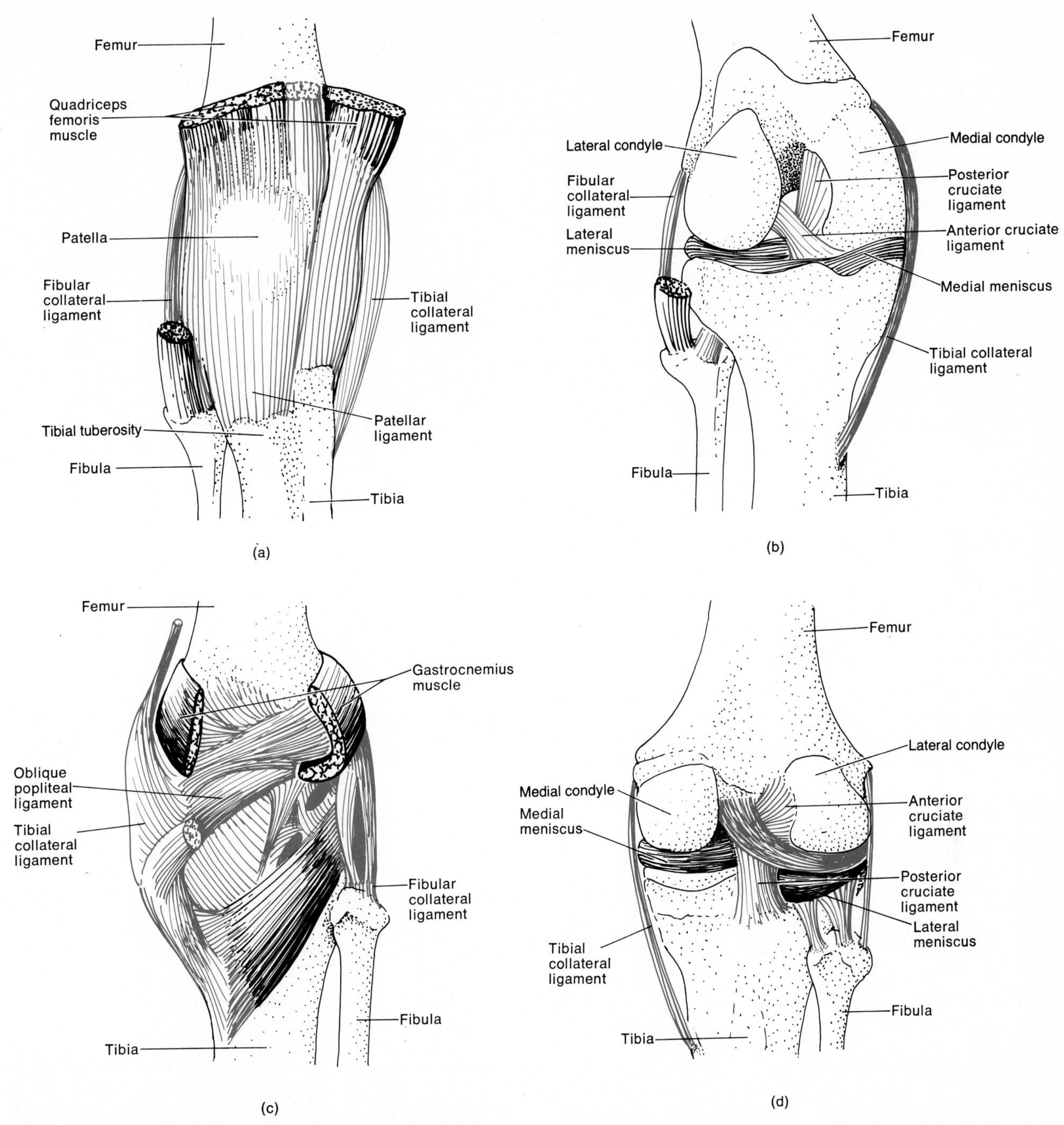

Figure 8-4. Relationship of ligaments, tendons, and menisci to the knee joint. (a) Anterior, superficial view of the knee. (b) Anterior view of the knee with many superficial structures removed. (c) Posterior, superficial view of the knee. (d) Posterior view of the knee with many superficial structures removed.

Their surfaces also assist in rotation of the knee. A tearing of the menisci, commonly called torn cartilage, occurs frequently among athletes.

The various movements of the body could cause friction to develop between moving parts. Saclike structures called **bursae** are situated in the body tissues to reduce this friction. These sacs resemble joints in that their walls consist of connective tissue lined by a synovial membrane. They are also filled with synovial fluid. Bursae are located between the skin and bone in places where the skin rubs over the bone. They are also found between tendons and bones, muscles and bones, and between ligaments and bones. As fluid-filled sacs, they act as cushions and ease the movement of one part of the body over another. An inflammation of a bursa is called **bursitis.**

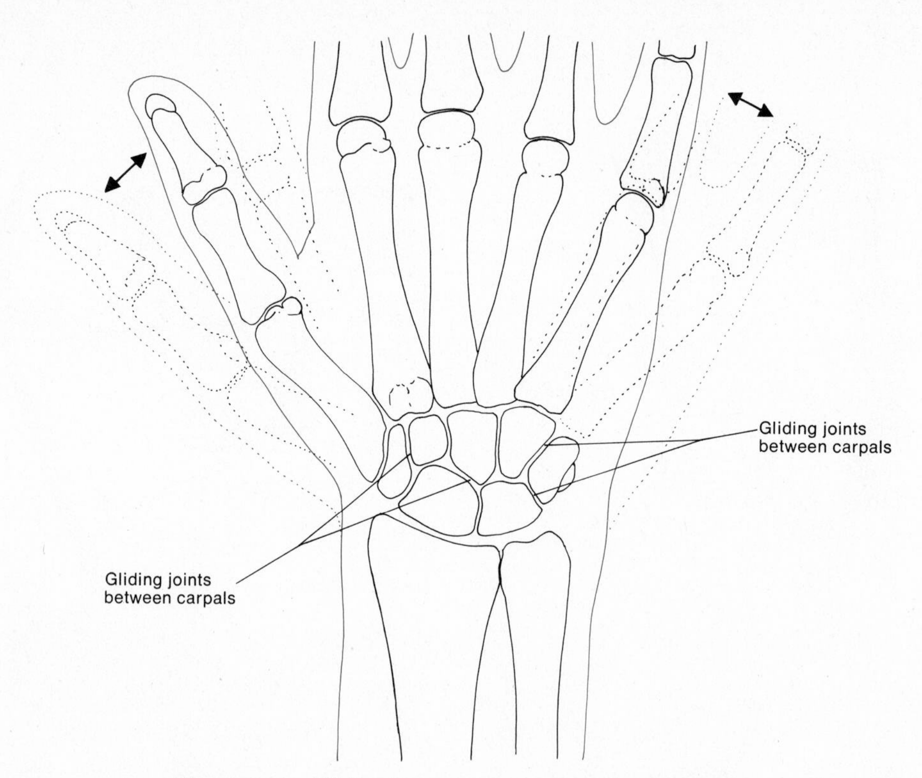

Figure 8-5. Gliding movement between carpal bones.

Movements at synovial joints

The movements possible at synovial joints may be categorized into five principal kinds: gliding movements, angular movements, rotation, circumduction, and special movements.

GLIDING. A **gliding movement** is the simplest kind that can occur in a joint. One surface moves back and forth and from side to side over another surface without angular or rotary movements. Some joints that glide are those between the carpals and between the tarsals. The heads and tubercles of ribs glide on the bodies and transverse processes of vertebrae (Figure 8–5).

ANGULAR. **Angular movements** increase or decrease the angle between bones. Among the angular movements are flexion, extension, abduction, and adduction (Figure 8–6). **Flexion** usually involves a decrease in the angle between the anterior surfaces of the articulating bones. An exception to this definition is flexion of the knee and the toe joints in which there is a decrease in the angle between the posterior surfaces of the articulating bones. Examples include bending the head forward, where the joint is between the occipital bone and the atlas, bending the elbow, and bending the knee. Flexion of the foot at the ankle joint is referred to as **dorsiflexion. Extension** involves an increase in the angle between the anterior surface of the articulating bones, with the two exceptions noted above. Extension restores a body part to its anatomical position after it has been flexed. Examples of extension are returning the head to the anatomic position after flexion, straightening out the arm after flexion, and straightening out the leg after flexion. Continuation of extension beyond the anatomic position, as in bending the head backward, is called **hyperextension.** Extension of the foot at the ankle joint is called **plantar flexion.**

Abduction usually means the movement of a bone *away from* the midline of the body. An example of abduction is moving the arms upward and away from the body until they are held straight out at right angles to the chest. In the case of the fingers and toes, however, the midline of the body is not used as the line of reference. Abduction of the fingers is a movement away from an imaginary line drawn through the middle finger—in other words, spreading the fingers. Abduction of the toes is relative to an imaginary line drawn through the second toe. **Adduction** is usually the movement of a part *toward* the midline of the body. Returning the arms to the sides after abduction is adduction. As in abduction, adduction of the fingers is relative to the middle finger. Adduction of the toes is relative to the second toe.

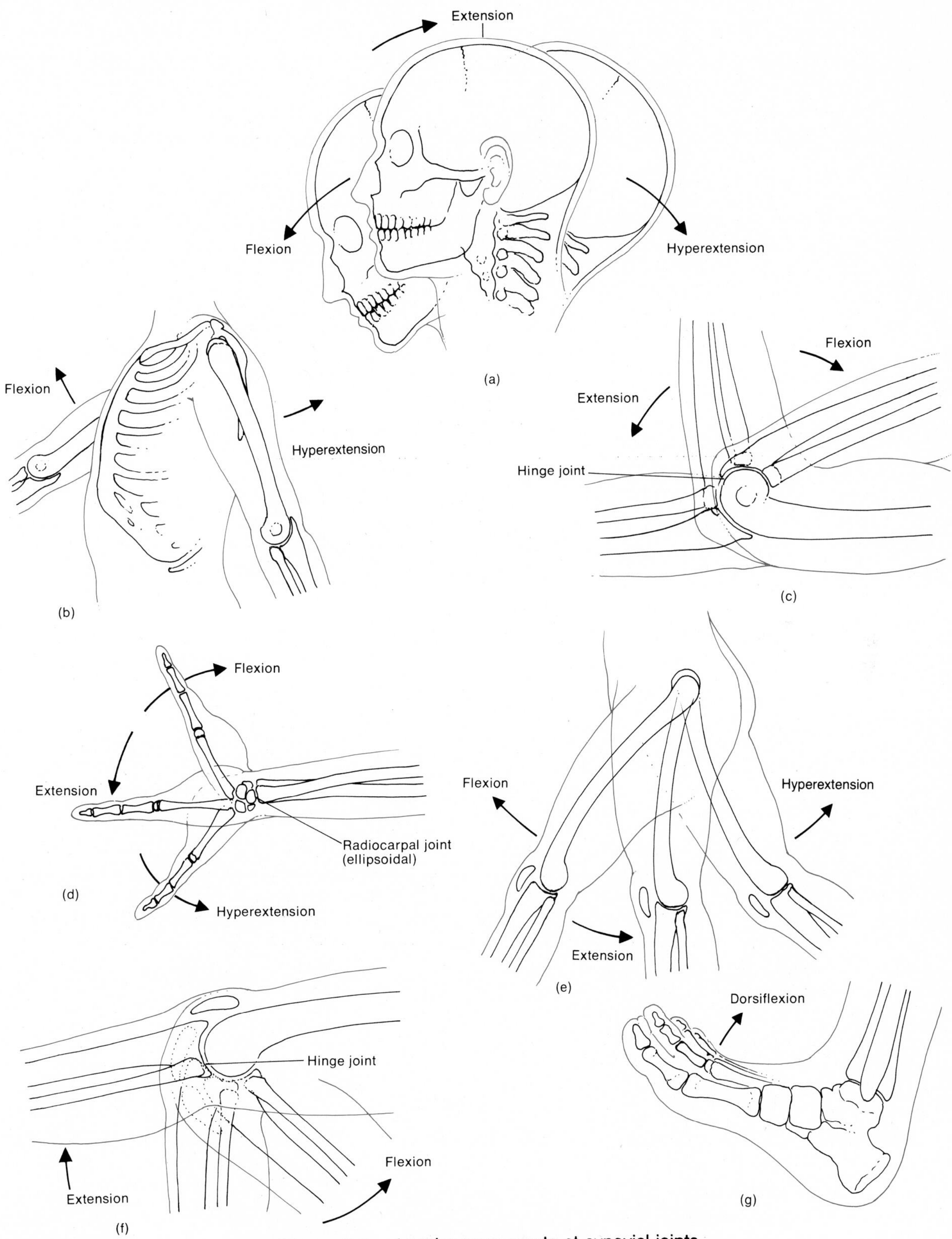

Figure 8-6. Angular movements at synovial joints.

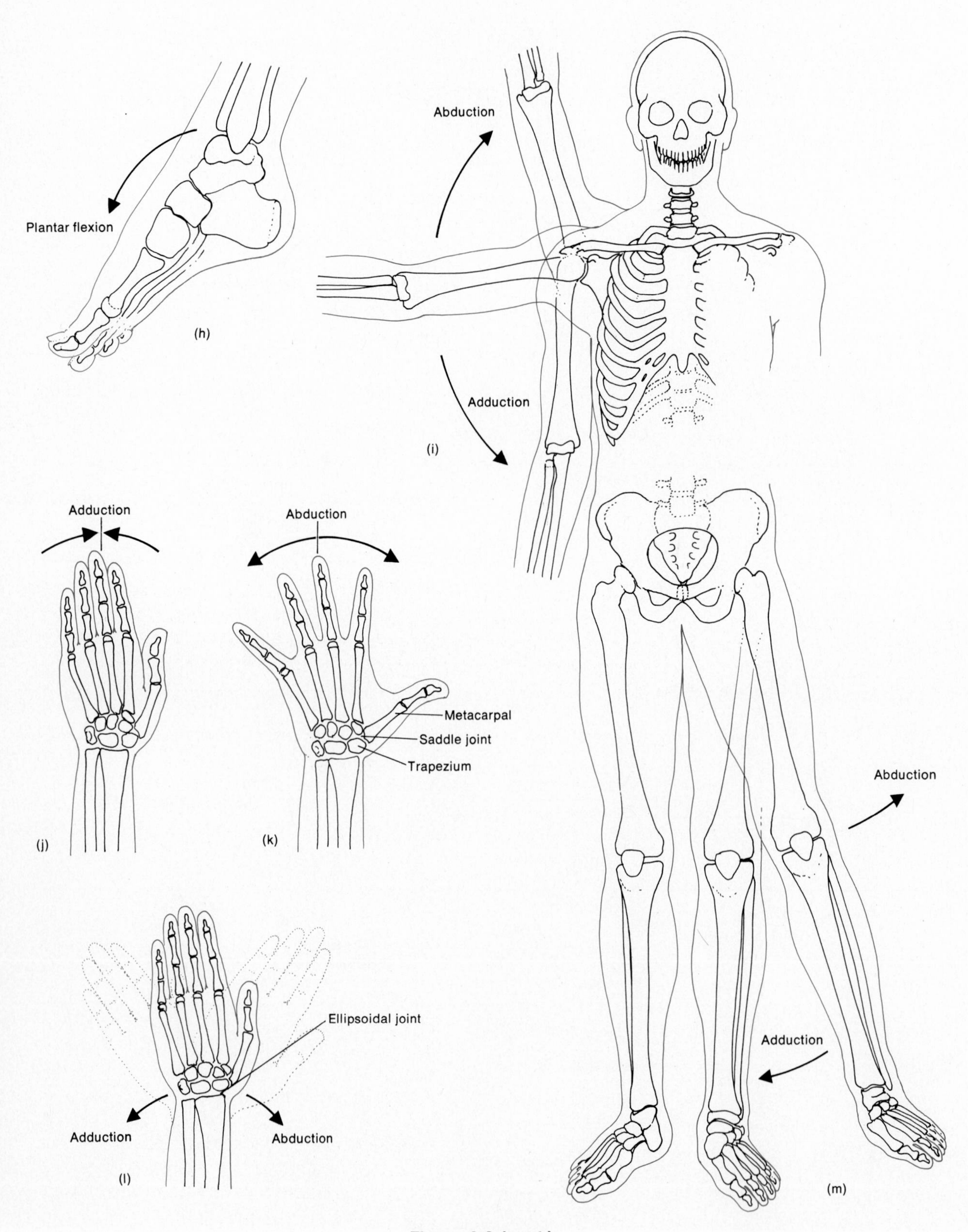

Figure 8-6 (cont.).

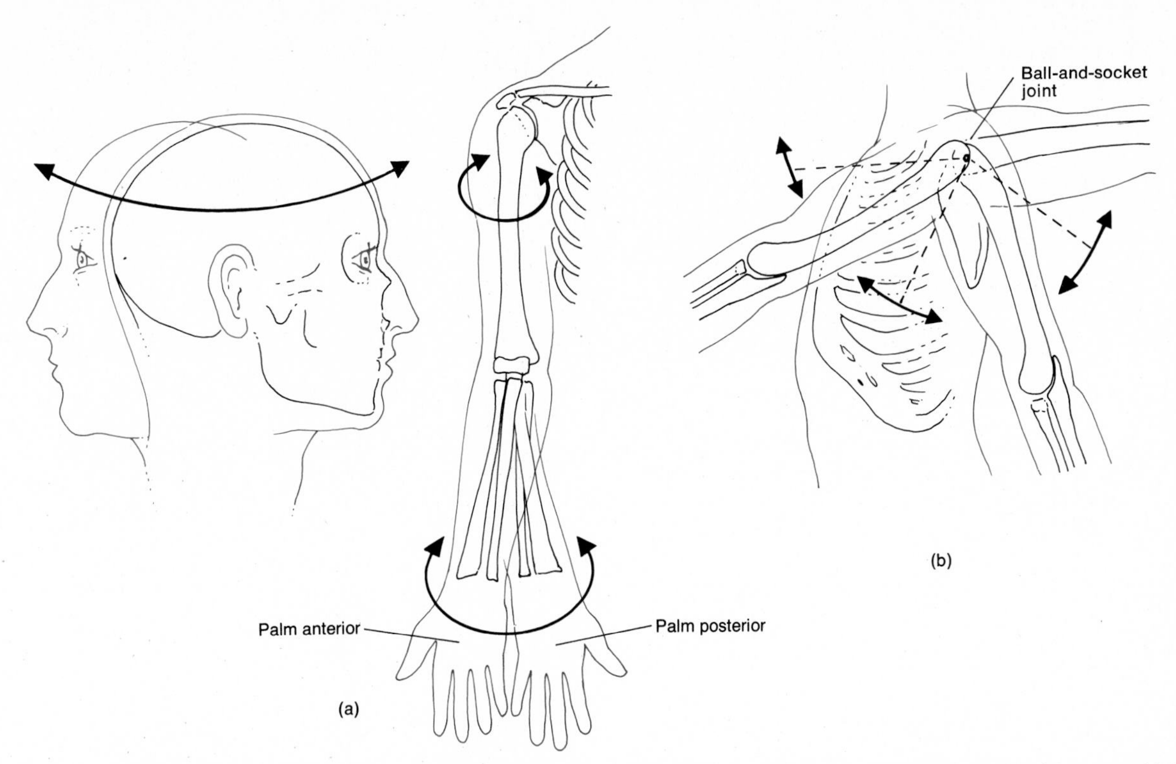

Figure 8-7. Rotation and circumduction. (a) Rotation at the atlas-axis joint (left) and rotation of the humerus (right). (b) Circumduction of the humerus at the shoulder joint.

ROTATION. **Rotation** means moving a bone around its own axis. During rotation, no other motion is permitted. We rotate the atlas around the odontoid process of the axis when we turn the head from side to side to signify "no." Moving from the shoulder and turning the forearm up, then palm down, and then palm up again is an example of slight rotation of the forearm (Figure 8–7a).

CIRCUMDUCTION. **Circumduction** is a movement in which the distal end of a bone moves in a circle while the proximal end remains relatively stable. The bone draws a cone in the air. Circumduction typically involves flexion, abduction, adduction, extension, and rotation. An example is moving the outstretched arm in a circle to wind up for a pitch (Figure 8–7b).

SPECIAL. **Special movements** are those that are found only at the joints indicated in Figure 8–8. They include the following:

Inversion is the movement of the sole of the foot inward at the ankle joint. **Eversion** is the movement of the sole of the foot outward at the ankle joint. **Protraction** is the movement of the mandible or clavicle forward on a plane parallel to the ground. Thrusting out the jaw is protraction of the mandible. Bringing your arms forward until the elbows touch each other requires protraction of the clavicle. **Retraction** is the movement of a protracted part of the body backward on a plane parallel to the ground. Pulling the lower jaw back in line with the upper jaw is retraction. **Supination** is a movement of the forearm in which the palm of the hand is turned forward (anterior). If you wish to practice supination, first flex your arm at the elbow so that rotation of the humerus in the shoulder joint is prevented. **Pronation** is a movement of the flexed forearm in which the palm of the hand is turned backward (posterior).

Types of synovial joints

Even though all synovial joints are basically similar in structure, variations exist in the shape of the articulating surfaces. Accordingly, synovial joints are divided into six subtypes. These include gliding, hinge, pivot, ellipsoidal, saddle, and ball-and-socket joints.

The articulating surfaces of **gliding joints** are usually flat. Only side-to-side and back-and-forth movements are permitted. Since this kind of joint allows movements in two planes, the joint is referred to as *biaxial.* Twisting and rotation are inhibited at gliding joints generally because either the arrangement of the ligaments or the presence

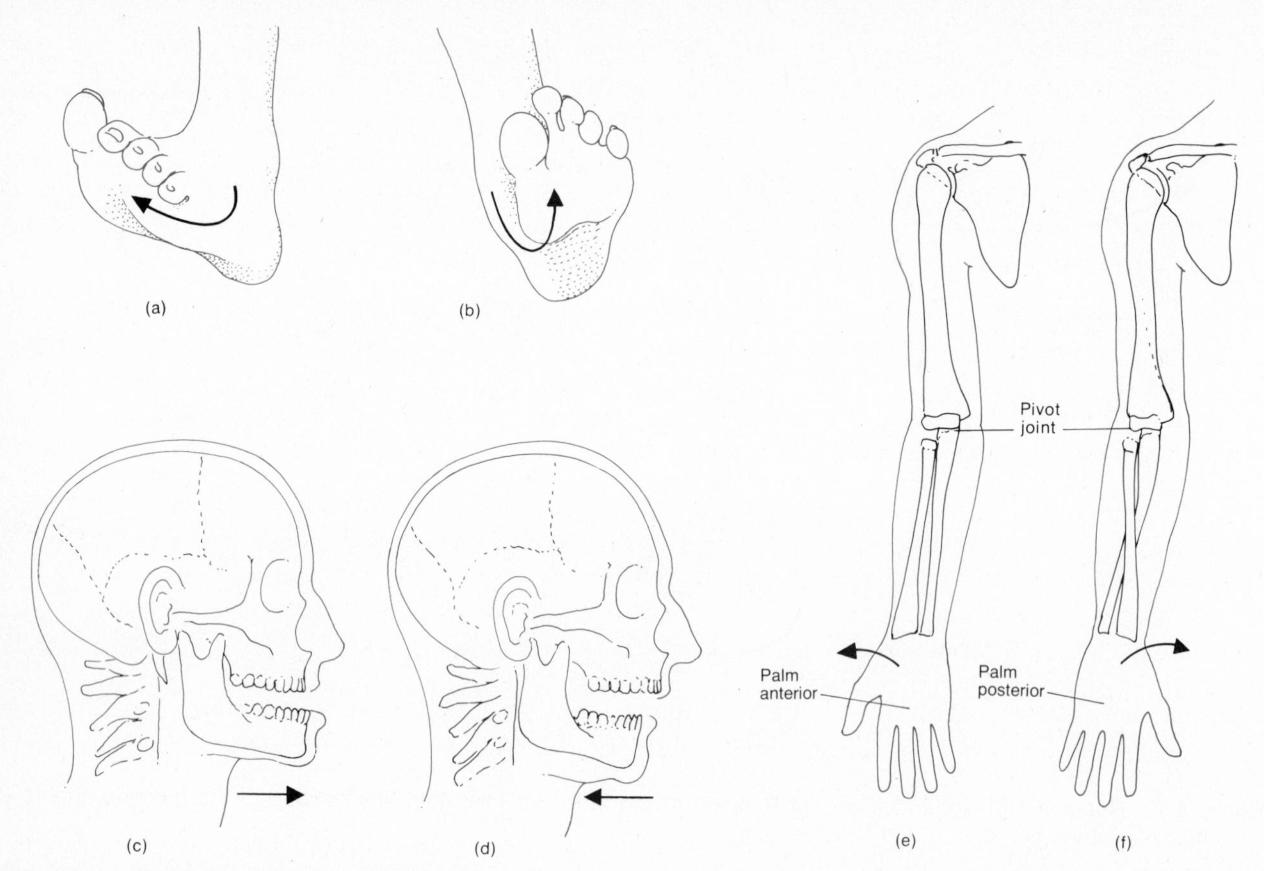

Figure 8-8. Special movements. (a) Inversion. (b) Eversion. (c) Protraction. (d) Retraction. (e) Supination. (f) Pronation.

of other bones close by restricts the range of movement. Examples are the joints found between carpal bones, tarsal bones, the sternum and clavicle, and the scapula and clavicle. (See Figure 8–5.)

Hinge joints are characterized by a convex surface of one bone that fits into a concave surface of another bone. Movement is primarily in a single plane and is usually flexion and extension. The joint is therefore referred to as *monaxial.* The motion is similar to that of a hinged door. Examples of hinge joints include the elbow, knee, ankle, and interphalangeal joints. The movement afforded by a hinge joint is illustrated by flexion and extension at the elbow and knee. (See Figure 8–6c, f.)

In a **pivot joint,** a rounded, pointed, or conical surface of one bone articulates with a shallow depression of another bone. The primary movement permitted is rotation, and the joint is therefore monaxial. Examples include the joints between the atlas and axis and between the proximal ends of the radius and ulna. Movement at a pivot joint is illustrated by supination and pronation of the palms. (See Figure 8–8c, f.)

Ellipsoidal joints are structured so that an oval-shaped condyle of one bone fits into an elliptical cavity of another bone. Since the joint permits side-to-side and back-and-forth movements, it is biaxial. The joint at the wrist between the radius and carpals is ellipsoidal. The movement permitted by such a joint is illustrated when you flex and extend and abduct and adduct the wrist. (See Figure 8–6d, l.)

In a **saddle joint,** the articular surfaces of both of the bones is saddle-shaped – in other words, concave in one direction and convex in the other direction. Essentially, the saddle joint is a modified ellipsoidal joint in which the movement is somewhat freer. Movements at a saddle joint are side to side and back and forth. Thus, the joint is biaxial. The joint between the trapezium and metacarpal of the thumb is an example of a saddle joint. (See Figure 8–6k.)

Ball-and-socket joints consist of a ball-like surface of one bone fitting into a cuplike depression of another bone. Such a joint permits *triaxial* movement. That is, there is movement in three planes of motion: flexion-extension, abduction-adduction, and rotation. Examples of ball-and-socket joints are the shoulder joint and the hip joint. The range of movements afforded at a ball-and-socket joint is illustrated by circumduction of the arm. (See Figure 8–7b.)

A summary of joints is presented in Exhibit 8–1. Note that the classification shown is based primarily on the anatomy of the joints. It was indicated in the beginning of the chapter that joints could also be classified according to the degree of movement that they permit. If we simply rearrange the anatomical classification in Exhibit 8–1 into a classification based on movement, we arrive at the following:

A. Synarthroses: Immovable joints
   1. Suture
   2. Synchondrosis
B. Amphiarthroses: Slightly movable joints
   1. Symphysis
   2. Syndesmosis
C. Diarthroses: Freely movable joints
   1. Gliding
   2. Hinge
   3. Pivot
   4. Ellipsoidal
   5. Saddle
   6. Ball-and-socket

APPLICATIONS TO HEALTH

Various disorders may occur in and around the joints. Structures associated with the joints, such as muscles, tendons, articular cartilage, synovial membranes, and bursae, may be affected.

Arthritis

A disease that has plagued mankind through the ages is **arthritis.** Signs of the disorder have been found in the bones of the Java man and in the mummies of Egypt. The term arthritis actually refers to at least 25 different diseases, the most common being rheumatoid arthritis, osteoarthritis, and gout. However, all these ailments are characterized by inflammation in one or more joints. Inflammation, pain, and stiffness may also be present in neighboring parts of the body, such as the muscles near the joint.

Exhibit 8–1. SUMMARY OF JOINTS

TYPE	DESCRIPTION	MOVEMENT	EXAMPLES
Fibrous	No joint cavity; bones held together by a thin layer of fibrous tissue or dense fibrous tissue		
Suture	Found only between bones of the skull; articulating bones are separated by a thin layer of fibrous tissue	None—synarthrotic	Lambdoidal suture between occipital and parietal bones
Syndesmosis	Articulating bones united by dense fibrous tissue	Slight—amphiarthrotic	Distal ends of tibia and fibula
Cartilaginous	No joint cavity; articulating bones united by cartilage		
Synchondrosis	Connecting material is hyaline cartilage	None—synarthrotic	Temporary joint between the diaphysis and epiphyses of a long bone
Symphysis	Connecting material is a broad, flat disc of fibrocartilage	Slight—amphiarthrotic	Intervertebral joints and symphysis pubis
Synovial	Joint cavity and articular cartilage present; synovial membrane lines cavity	Freely movable—diarthrotic	
Gliding	Articulating surfaces usually flat	Biaxial (flexion-extension, abduction-adduction)	Intercarpal and intertarsal joints
Hinge	Spool-like surface fits into a concave surface	Monaxial (flexion-extension)	Elbow, knee, ankle, and interphalangeal joints
Pivot	Rounded, pointed, or concave surface fits into a shallow depression	Monaxial (rotation)	Atlas-axis and radioulnar joints
Ellipsoidal	Oval-shaped condyle fits into an elliptical cavity	Biaxial (flexion-extension, abduction-adduction)	Radiocarpal joint
Saddle	Articular surfaces concave in one direction and convex in opposite direction	Biaxial (flexion-extension, abduction-adduction)	Carpometacarpal joint of thumb
Ball-and-socket	Ball-like surface fits into a cuplike depression	Triaxial (flexion-extension, abduction-adduction, rotation)	Shoulder and hip joints

The causes of arthritis are unknown. In some cases, it has followed the stress of sprains, infections, and joint injury. Some researchers believe that the cause is a bacterium or virus, whereas others suspect an allergy. There are those who believe the nervous system or hormones are involved, whereas some suspect a disorder of the metabolic system. Still others believe that certain types of prolonged psychologic stress, such as inhibited hostility, can upset homeostatic balance and bring on arthritic attacks.

Rheumatoid arthritis

Rheumatoid arthritis is a disease that is usually limited to the destruction of joints and their surrounding structures, such as muscles and tendons. In about 80 percent of the cases, the onset of the disease occurs between the ages of 20 and 50. Women are affected three times as often as men. About one-third of all arthritic people have the rheumatoid type of arthritis.

The primary symptom of rheumatoid arthritis is inflammation of the synovial membrane (Figure 8–9). The membrane thickens, and synovial fluid accumulates. The resulting pressure causes pain and tenderness. The membrane then produces an abnormal tissue called **pannus,** which grows onto the surface of the articular cartilage. The pannus formation sometimes erodes the cartilage completely. When the cartilage is destroyed, fibrous tissue grows out of the exposed bone ends. The tissue ossifies and fuses the joint so that it is immovable. This is the ultimate crippling effect of rheumatoid arthritis. Most cases do not progress this far. But the range of motion of the joint is greatly inhibited by the severe inflammation and swelling.

Osteoarthritis

A degenerative joint disease that is far more common than rheumatoid arthritis, and usually less damaging, is **osteoarthritis.** This type of arthritis is one of the oldest diseases known. It apparently results from a combination of aging, irritation of the joints, and normal wear and tear. Symptoms rarely occur before the age of 40.

Osteoarthritis is a chronic inflammation of the articular cartilage of some of the joints. The inflammation causes swelling, stiffness, and pain. The cartilage slowly degenerates. And as the bone ends become exposed, they lay down little bumps, or spurs, of new osseous tissue. These spurs decrease the space of the joint cavity and put restrictions on movement. Unlike rheumatoid arthritis, osteoarthritis usually affects only the articular cartilage. The synovial membrane is very rarely destroyed, and other tissues are unaffected. The disorder does not cause the articulating bones to fuse.

Gouty arthritis

Uric acid is a waste product given off during the metabolism of the nucleic acid purine. Normally, all the acid is quickly excreted in the urine. And, in fact, it gives urine its name. The person who suffers from *gout* either produces excessive amounts of uric acid or is not able to excrete normal amounts. The result is a buildup of uric acid in the blood. This

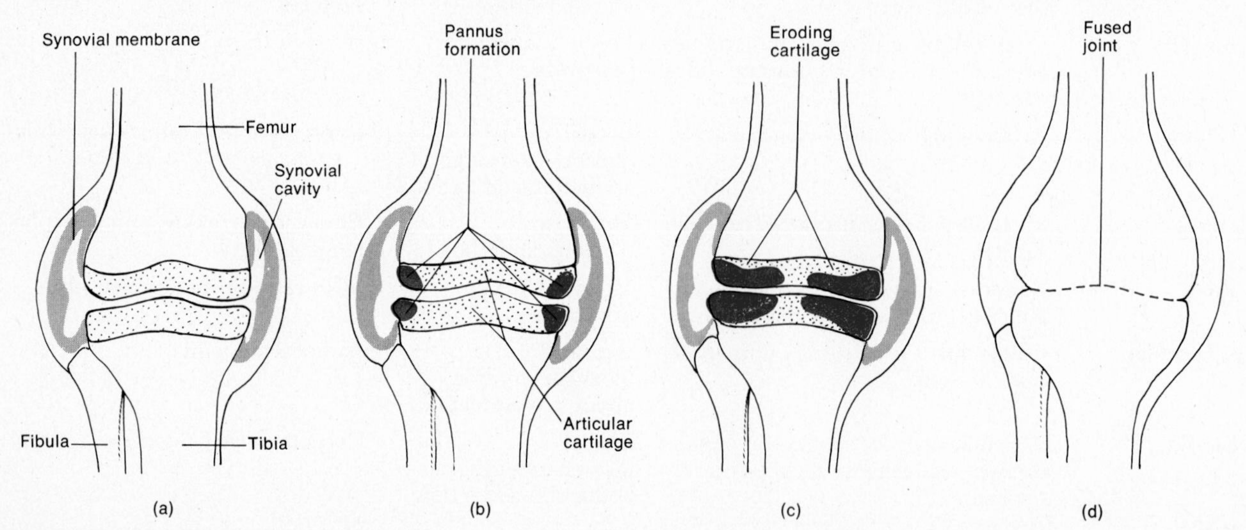

Figure 8-9. Progressive stages of rheumatoid arthritis at the knee joint. (a) Inflammation of the synovial membrane. (b) Early stage of pannus formation and erosion of articular cartilage. (c) Advanced stage of pannus formation and further erosion of cartilage. (d) Obliteration of joint cavity and fusion of bones.

excess acid then reacts with sodium to form a salt called sodium urate. Crystals of this salt are deposited in soft tissues, some favorite spots being the kidneys and the cartilage of the ears and joints.

In **gouty arthritis,** sodium urate crystals are deposited in the soft tissues of the joints. The crystals irritate the cartilage, causing inflammation, swelling, and acute pain. Eventually, the crystals destroy all the joint tissues. If the disorder is not treated, the ends of the articulating bones fuse, and the joint becomes immovable.

Gout occurs primarily in males of any age. It is believed to be the cause of 2 to 5 percent of all chronic joint diseases. Numerous studies indicate that gout is sometimes caused by an abnormal gene. The gene instructs the body to manufacture purine by a mechanism that gives off unusually large amounts of uric acid. Diet and environmental factors such as stress and climate are also suspected causes of gout.

Although other forms of arthritis cannot be treated with complete success, gout has almost been conquered. The advent of various drugs has been the reason. One chemical called colchicine has been used on and off since the sixth century to relieve the pain, swelling, and tissue destruction that occurs during attacks of gouty arthritis. This chemical is derived from the variety of crocus plant from which we also get the spice saffron. Other drugs, which either inhibit uric-acid production or help the kidneys eliminate excess uric acid, are used to prevent further attacks. The drug allopurinal is used to treat gout because it prevents the formation of uric acid without interfering with purine synthesis.

Bursitis

An acute or chronic inflammation of a bursa is called **bursitis.** The condition may be caused by *trauma,* which is a wound or injury, by an acute or chronic infection including syphilis and tuberculosis, or by rheumatoid arthritis. It is characterized by pain, swelling, tenderness, and limitation of any motion that involves the inflamed bursa. The patient usually recovers in 1 to 2 weeks, but the condition may become chronic.

Tendinitis

A disorder involving the inflammation of a tendon and synovial membrane at a joint is called **tendinitis** or **tenosynovitis.** The shoulder, the elbow (tennis elbow), the joints in the fingers (trigger finger), and their associated tendons are most often the sites of inflammation.

Intervertebral disc abnormality

Intervertebral discs are located between the bodies of adjacent vertebrae from the axis to the sacrum. Each disc is composed of an outer fibrous ring consisting of fibrocartilage called the *anulus fibrosus* and an inner soft, pulpy, highly elastic structure called the *nucleus pulposus.* (See Figure 8–2b.) The intervertebral discs serve as shock absorbers. Under compression, they become flatter and broader and bulge from their intervertebral spaces. The discs between the fourth and fifth lumbar vertebrae and between the fifth lumbar vertebra and sacrum are subject to great compressional forces. Should the anterior and posterior ligaments of the discs become injured or weakened, it is possible for the disc to become herniated—that is, the pressure developed in the nucleus pulposus is so great that it ruptures the surrounding fibrocartilage. If this occurs, the nucleus pulposus may protrude posteriorly or into one of the adjacent vertebral bodies. This is referred to as a "slipped" disc, and the result is **intervertebral disc abnormality.** Most often the nucleus pulposus slips posteriorly toward the spinal cord and spinal nerves. This puts pressure on the roots of the nerves that run out of the spinal cord. The pressure on the nerves causes considerable, and sometimes very acute, pain. If the root of the sciatic nerve, which runs from the spinal cord to the foot, is pressured, the pain radiates down the back of the thigh, through the calf, and occasionally into the foot. If pressure is put on the spinal cord, destruction of nervous tissue can be quite serious.

Surgical treatment may involve removing the disc and fusing together the two vertebrae. The fusion, of course, limits back movement somewhat. Surgery is considered only if the disc is destroying nervous tissue, or if the chronic pain and inflammation cannot be alleviated by less drastic methods.

Key terms associated with joints

Ankylosis (*ankyl* = bent; *osis* = condition) Severe or complete loss of movement at a joint.
Arthralgia (*algia* = pain) Pain in a joint.
Arthrosis A disease of a joint; also means articulation or joint.
Bursectomy (*ectomy* = removal of) Removal of a bursa.
Chondritis Inflammation of a cartilage.
Dislocation Displacement of a bone from its natural position in a joint.
Sprain Tearing of tendons and ligaments.
Synovitis Inflammation of a synovial membrane in a joint.

Chapter summary in outline

CLASSIFICATION OF JOINTS

1. A joint or articulation is a point of contact between two bones.
2. Closely fitting bones are strong but not freely movable. Loosely fitting joints are weaker but freely movable.
3. Joints may be classified on the basis of their anatomy or the degree of movement they permit.

Fibrous joints

1. Bones held by fibrous connective tissue, with no joint cavity, are fibrous joints.
2. These joints include immovable or synarthrotic sutures (found in the skull) and slightly movable or amphiarthrotic syndesmoses (such as the tibiofibular articulation).

Cartilaginous joints

1. Bones held together by cartilage, with no joint cavity, are cartilaginous joints.
2. These joints include immovable or synarthrotic synchondroses united by hyaline cartilage (temporary cartilage between diaphysis and epiphyses) and partially movable or amphiarthrotic symphyses united by fibrocartilage (the symphysis pubis).

Synovial joints

1. These joints contain a joint cavity, articular cartilage, and synovial membranes. They are held together by ligaments and tendons.
2. All synovial joints are freely movable or diarthrotic.
3. Types of synovial joints include gliding joints (wrist bones), hinge joints (elbow), pivot joints (radioulnar), ellipsoidal joints (radiocarpal), saddle joints (carpometacarpal), and ball-and-socket joints (shoulder and hip).
4. Planes of movement at synovial joints include the monaxial, biaxial, and triaxial planes.
5. Types of movements at synovial joints include gliding movements, angular movements, rotation, circumduction, and special movements, such as inversion, eversion, protraction, retraction, supination, and pronation.

APPLICATIONS TO HEALTH

1. Arthritis is the inflammation of a joint. Types of arthritis include rheumatoid arthritis, osteoarthritis, and gouty arthritis.
2. Bursitis is the inflammation of a bursa.
3. Tendinitis is the inflammation of the tendon and synovial membrane at a joint.
4. Intervertebral disc abnormality, called "slipped" disc, involves misplacement of the nucleus pulposus of a disc by injury or chronic irritation.

Review questions and problems

1. Define an articulation. What factors determine the degree of movement at joints?
2. Distinguish among the three kinds of joints. List the subtypes of joints. Be sure to include structure, degree of movement, and specific examples.
3. Using the knee as a typical joint, explain the components of a synovial joint. Indicate the relationship of ligaments and tendons to the strength of the joint and restrictions on movement.
4. What are bursae? What is their function? Define bursitis.
5. Define the following principal movements: gliding, angular, rotation, circumduction, and special. Name a joint where each occurs.
6. Have your partner assume the anatomical position and execute each of the movements at joints discussed in the text. Reverse roles, and see if you can execute the same moves.
7. Contrast monaxial, biaxial, and triaxial planes of movement. Give examples of each, and name a joint at which each occurs.
8. What is arthritis? Distinguish among the following kinds: rheumatoid arthritis, osteoarthritis, and gouty arthritis. What is a pannus?
9. Define tendinitis and intervertebral disc abnormality. Why is a "slipped" disc usually so painful?
10. Refer to the glossary of key terms associated with joints. Be sure that you can define each term.

Selected readings

Basmajian, John V. *Primary Anatomy*. 6th ed. Baltimore: Williams and Wilkins, 1970. Chap. 3.
Cunningham's Textbook of Anatomy. 11th ed., edited by G. J. Romanes. London: Oxford University Press, 1972. Pp. 206–258.
Dawson, Helen L. *Basic Human Anatomy*. 2d ed. New York: Appleton-Century-Crofts, 1974. Chap. 4.
Gray, Henry. *Anatomy of the Human Body*. 29th ed., edited by Charles Mayo Goss. Philadelphia: Lea and Febiger, 1973. Chap. 5.

CHAPTER 9

MUSCLE TISSUE

STUDENT OBJECTIVES

After you have read this chapter, you should be able to:

1. List the distinguishing properties of muscle tissue
2. Compare the location, microscopic appearance, nervous control, and functions of the three kinds of muscle tissue
3. Define epimysium, perimysium, and endomysium
4. Define tendons and aponeuroses and their modes of attachment to muscles
5. Describe the relationship of blood vessels and nerves to skeletal muscles
6. Identify the histological characteristics of skeletal muscle
7. Describe the physiological importance of the motor unit
8. Describe the physiology of contraction by listing the events associated with the sliding-filament hypothesis
9. Describe the source of energy for muscular contraction
10. Define the all-or-none principle of muscular contraction
11. Define such common muscular disorders as fibrosis, muscular dystrophy, myasthenia gravis, and tumors
12. Compare spasms, cramps, convulsions, tetany, and fibrillation as abnormal muscular contractions
13. Define key terms associated with the muscular system

In the preceding chapters on the skeletal system, we discussed how bones are connected in various ways to form joints. Although bones and joints provide leverage and form the framework of the body, they are not capable of moving the body by themselves. Motion is an essential body function that is made possible by the contraction of muscles. Muscle tissue constitutes about 40 to 50 percent of the total body weight and is composed of highly specialized cells having four striking characteristics. One of these features, **irritability,** is the ability of muscle tissue to receive and respond to stimuli. A stimulus is a change in the internal or external environment that discharges an impulse by a nerve cell. A second characteristic of muscle is **contractility,** the ability to shorten and thicken, or contract, when a sufficient stimulus is received. Muscle tissue also exhibits **extensibility,** which means that it can be stretched when pulled. You will see later that many skeletal muscles are arranged in opposing pairs. While one is contracting, the other is undergoing extension. The final characteristic of muscle tissue is **elasticity,** the ability of muscle to return to its original shape after contraction or extension. Other tissues of the body also exhibit irritability, extensibility, and elasticity, but only muscle can contract. Through contraction, muscle performs the three important functions: (1) motion, (2) maintaining posture, and (3) producing heat.

The most obvious kinds of motions of the body are walking, running, and moving from one place to another. Other obvious types of movements, such as grasping a pencil or nodding the head, are limited to one or more parts of the body. These kinds of movement rely on the integrated functioning of the bones, joints, and muscles that are attached to the bones. Less noticeable kinds of motion produced by muscles are the beating of the heart, the churning of food in the stomach, the pushing of food through the intestines, the contraction of the gallbladder to release bile, and the contraction of the urinary bladder to expel urine.

In addition to performing the function of movement, muscle tissue also enables the body to maintain posture. The contraction of skeletal muscles holds the body in stationary positions, such as standing and sitting. Completely relax the muscles of your legs and torso, and see how quickly you flop to the floor!

The third function of muscle tissue is heat production. Skeletal muscle contractions produce heat and are thereby important in maintaining normal body temperature. The mechanisms involved in muscle contraction will be discussed later in greater detail.

KINDS OF MUSCLE TISSUE

Three kinds of muscle tissue are recognized: (1) **skeletal,** (2) **visceral,** and (3) **cardiac.** These three types of muscle tissue are further described and categorized on the basis of location, presence of microscopic striations, and nervous control. Skeletal muscle tissue, which is named after its location, is found attached to bones. It is *striated* muscle tissue because striations, or bandlike structures, are visible when the tissue is examined under a microscope. Finally, it is a *voluntary* muscle tissue because it can be made to contract by conscious, or voluntary, control. Smooth muscle is located in the walls of hollow internal structures, such as blood vessels, the stomach, and the intestines. It is, therefore, described as *visceral* muscle tissue. Smooth muscle tissue is *nonstriated,* which is the reason why it is called smooth. It is *involuntary* muscle tissue because its contraction is usually not under the conscious control of the individual. Cardiac muscle tissue forms the walls of the heart and is named after its location. Cardiac muscle tissue is also striated and is usually involuntary muscle tissue. In sum, all muscle tissues are classified in the following ways: (1) skeletal, striated, or voluntary muscle, (2) cardiac, striated, or involuntary muscle, and (3) smooth, visceral, or involuntary muscle. We shall now examine each of these groups of muscle tissue in some detail.

SKELETAL MUSCLE TISSUE

In order to understand the structure of skeletal muscle tissue, you will need to have some knowledge of its connective tissue components, its nerve and blood supply, and its histology, or microscopic structure.

Let us first examine what is meant by the term fascia and how it is related to a skeletal muscle.

Fascia

The term **fascia** is applied to a sheet or broad band of fibrous connective tissue underneath the skin or around muscles and other organs of the body. Fasciae may be divided into three types: (1) superficial, (2) deep, and (3) subserous. The *superficial fascia,* or *subcutaneous layer,* is immediately deep to the skin. It covers the entire body and varies in thickness in different regions. For example, on the back or dorsum of the hand it is quite sparse, whereas over the inferior abdominal wall it is relatively thick. The superficial fascia is actually composed of two layers. The outer layer, the panniculus adiposus, usually contains fat and varies considerably in thickness. The inner, membranous layer is thin and elastic. Between the two layers of superficial fascia are found arteries, veins, lymphatics, nerves, the mammary glands, and the facial muscles. Hair follicles, sweat glands, and sebaceous glands are embedded in the superficial fascia.

The *deep fascia* is by far the most extensive of the three types of fascia. Unlike the superficial fascia, the deep fascia does not contain fat. The deep fascia lines the body wall and extremities, and it holds muscles together and separates them into functioning groups. The outer layer of deep fascia, called the external investing layer, is located under the superficial fascia and covers the trunk, neck, a portion of the head, and the extremities. The internal investing layer is the deepest layer of deep fascia and is located on the inside of the body wall of the trunk. Here it forms a portion of the walls of the body cavities. It is covered internally by the subserous fascia and serous membrane. The intermediate membranes, located between the external investing layer and internal investing layer, form compartments around muscles and structures of the body wall.

The *subserous fascia* is found between the internal investing layer of deep fascia and the serous membrane. It covers the external surfaces of viscera in the thoracic, abdominal, and pelvic cavities.

Connective tissue component

Skeletal muscles are further protected, strengthened, and attached to other structures by several connective tissue components. The entire muscle is usually wrapped with a substantial quantity of fibrous connective tissue called the **epimysium** (Figure 9–1a). When the muscle is cut in cross section, one can see that invaginations of the epimysium

Blood vessels
Perimysium
Epimysium
Muscle fibers or cells
Fasciculus or fascicle
Perimysium
Blood capillaries
Endomysium
Portion of a nerve fiber
Blood capillary
Motor end plate
(a)

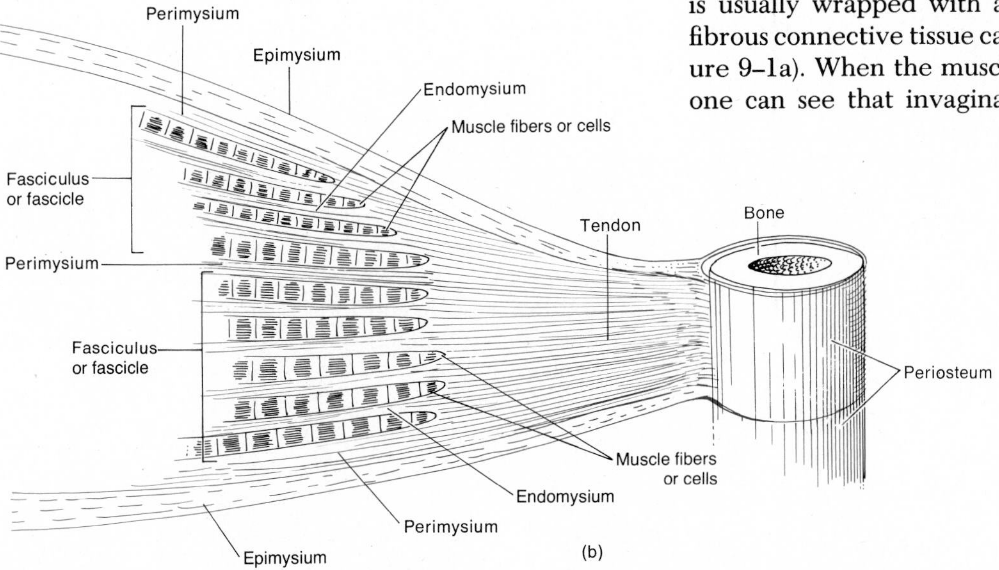

Figure 9–1. Connective tissue relationships to skeletal muscle tissue. (a) Cross section of a skeletal muscle showing connective tissue components. (b) Longitudinal section of a muscle illustrating the attachment of a tendon to the periosteum of bone.

divide the muscle into bundles called *fasciculi* or *fascicles.* These invaginations of the epimysium are called the **perimysium.** In turn, invaginations of the perimysium, called **endomysium,** penetrate into the interior of each fascicle and separate each muscle cell. The epimysium, perimysium, and endomysium are all continuous with the connective tissue that attaches the muscle to some other structure, such as bone or another muscle (Figure 9–1b). For example, all three elements may be extended beyond the muscle cells as a *tendon,* which is a cord of connective tissue that attaches a muscle to the periosteum of a bone. In other cases, the connective tissue elements may extend as a broad, flat tendon called an *aponeurosis.* This structure also attaches to the coverings of a bone or another muscle. When a muscle contracts, the tendon or aponeurosis and its corresponding bone or muscle are pulled toward the contracting muscle. It is in this way that skeletal muscles produce movement.

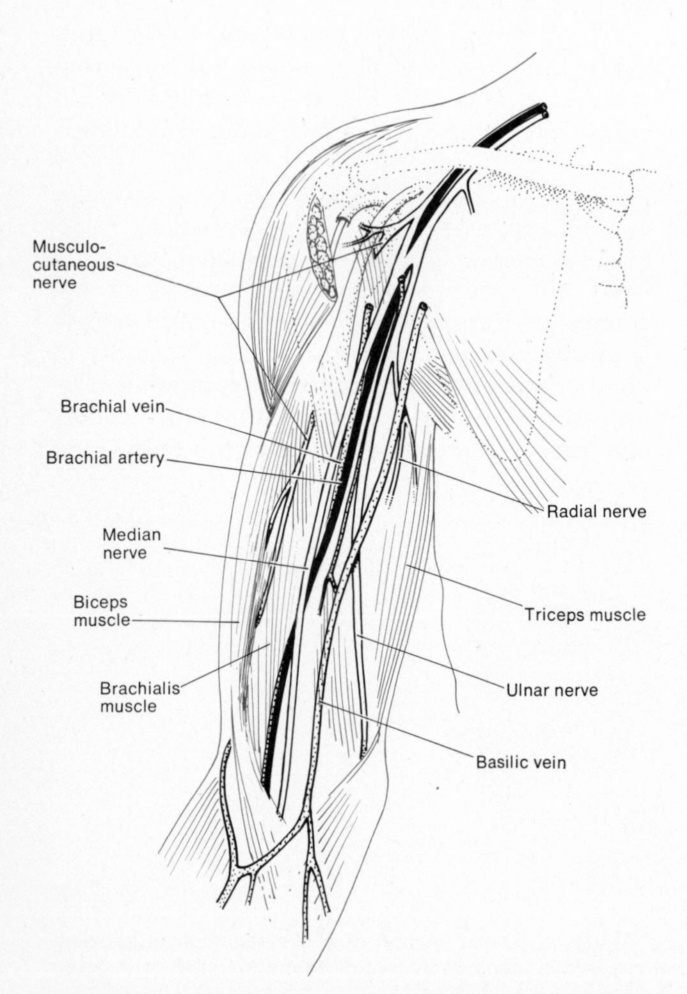

Figure 9–2. Relationship of blood vessels and nerves to skeletal muscles of the right arm seen in anterior view.

Nerve and blood supply

Skeletal muscles as organs are well supplied with nerves and blood vessels. This heavy infiltration of nervous and circulatory tissues is directly related to contractions, the chief characteristic of muscle. In order for a skeletal muscle to contract, it must first be stimulated by an impulse from a nerve cell. A skeletal muscle cell without a functional nerve connection cannot operate. Muscle contraction also requires a great deal of energy. To carry out the reactions that produce this energy, muscle cells must be supplied with large amounts of nutrients and oxygen. The waste products of these energy-producing reactions must also be eliminated. In fact, muscle action is greatly hindered by any drastic lowering of the blood supply.

Generally, an artery and one or two veins accompany a nerve into a skeletal muscle. The larger branches of the blood vessel accompany the nerve branches through the connective tissue of the muscle (Figure 9–2). Microscopic blood vessels called capillaries are arranged in the endomysium. Each muscle cell is thus in close contact with one or more capillaries. (See Figure 9–1a.) Each muscle cell also makes contact with a portion of a nerve cell.

Histology

Muscle tissue that is attached to and responsible for the movement of bones is generally referred to as **skeletal muscle tissue.** When typical skeletal muscle tissue (Figure 9–3a) is teased apart and viewed microscopically, it can be seen that it consists of many elongated, cylindrical cells called **muscle fibers** (Figure 9–3b). These fibers lie parallel to each other and range from 10 to 100 μm in diameter. Each muscle fiber (Figure 9–3c) is surrounded by a plasma membrane called the **sarcolemma.** The combining form *sarco* means flesh, and *lemma* means sheath or husk. The sarcolemma contains a quantity of cytoplasm called **sarcoplasm.** Within the sarcoplasm of a fiber and lying close to the sarcolemma are many nuclei and a number of mitochondria. Also found in a fiber is the **sarcoplasmic reticulum,** a tubelike network of roughly parallel vesicles (Figure 9–3d). Running transversely through the fiber and perpendicularly to the sarcoplasmic reticulum are **T tubules.** The tubules connect with the sarcoplasmic reticulum and open to the outside of the fiber. A **triad** consists of a T tubule and the segments of sarcoplasmic reticulum on both sides of it.

A highly magnified view of a skeletal muscle fiber reveals the presence of threadlike structures, about 1 μm in diameter, called **myofibrils** (Figure 9–3c to e). The prefix *myo* means muscle. These

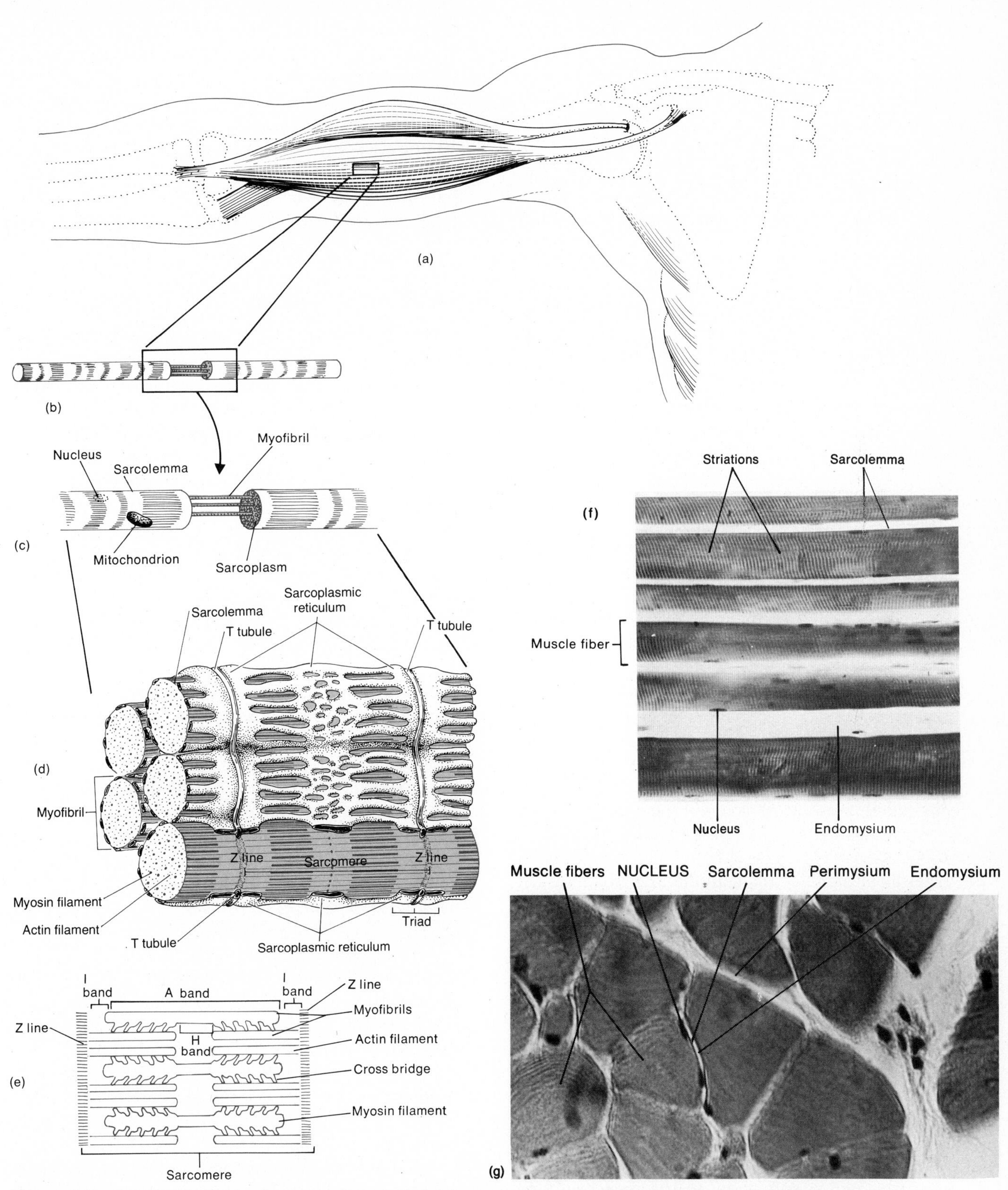

Figure 9–3. Levels of structural organization of skeletal muscle tissue. (a) Entire muscle. (b) Enlarged aspect of a single muscle fiber with a portion removed to show internal structure. (c) Further enlargement of a single muscle fiber showing more detail. (d) Several myofibrils showing the relationship of internal structures to a sarcomere and Z lines. (e) Details of a sarcomere showing myosin and actin filaments and various internal zones. (f) Photomicrograph of several skeletal muscle fibers in longitudinal section at a magnification of 640X. (Courtesy of Edward J. Reith, from *Atlas of Descriptive Histology,* by Edward J. Reith and Michael H. Ross, Harper & Row, Publishers, Inc., New York, 1970.) (g) Photomicrograph of several skeletal muscle fibers in cross section at a magnification of 200X. (Courtesy of Victor B. Eichler, Wichita State University.)

myofibrils (Figure 9–3c, d) run longitudinally through the fiber and consist of two kinds of still smaller filaments. The thin actin filaments are composed of the protein **actin.** The thick myosin filaments consist of the protein **myosin** (Figure 9–3e).

Note that the filaments of a myofibril do not extend the entire length of a muscle fiber; they are stacked in definite compartments called **sarcomeres.** Sarcomeres are partitioned by separations called **Z lines,** which are narrow zones of dense material. In a relaxed muscle cell, thick and thin filaments overlap for part of their respective lengths and form a dark, dense band called the *anisotropic* or *A band.* A light-colored, less dense area called the *isotropic* or *I band* is composed of thin filaments only. This combination of alternating dark and light bands gives the muscle fiber its striped or striated appearance. A narrow *H band* contains thick filaments only and is located in the middle of the A band.

CONTRACTION OF SKELETAL MUSCLE

In order for a skeletal muscle to contract, it must have a supply of energy. It must be stimulated by a nerve. And it must be provided with calcium ions. We shall now discuss in detail the contraction of skeletal muscle.

The motor unit

In order for contraction of skeletal muscle to occur, a stimulus must be applied to the muscle tissue. Such a stimulus is normally transmitted by nerve cells called neurons. As you will see in Chapter 12, a neuron has a threadlike process called a fiber, or axon, that may run as far as 91 cm (3 ft) to a muscle. A bundle of such fibers belonging to many different neurons composes a nerve. A neuron that transmits a stimulus to muscle tissue is called a motor neuron. Upon entering a skeletal muscle, the axon of the motor neuron branches. These branches then make contact with the individual muscle cells. The area of contact between a neuron and a muscle fiber is called a **motor end plate, neuromuscular junction,** or **myoneural junction** (Figure 9–4). When a nerve impulse reaches a motor end plate, small vesicles in the terminal branches of the nerve fiber release a chemical called acetylcholine. This chemical transmits the nerve impulse from the neuron, across the motor end plate, to the muscle fibers, thus stimulating the fibers to contract.

A motor neuron, together with the muscle cells it stimulates, is referred to as a **motor unit.** On an average, a single motor neuron innervates about 150 muscle fibers. This means that stimulation of one neuron will cause the simultaneous contraction of about 150 muscle fibers. In addition, all the muscle fibers of a motor unit contract and relax together. Muscles that control very fine, precise movements, such as the eye muscles, have less than 10 muscle fibers to each motor unit. Muscles of the body that are responsible for gross movements, such as the muscles of the buttocks, may have as many as 200 or more muscle fibers in each motor unit.

Physiology of contraction

When a nerve impulse reaches the neuromuscular junction, the neuron releases acetylcholine, which causes an electrical change in the sarcolemma. The electrical change travels over the surface of the sarcolemma and into the T tubules. When the impulse is conveyed from the T tubules to the sarcoplasmic reticulum, the reticulum, in some yet unexplained manner, releases calcium ions into the sarcoplasm surrounding the myofibrils. These calcium ions trigger the contractile process. Muscle contraction lasts only as long as calcium ions are present in the sarcoplasm. When the nerve impulse is over, the calcium ions recombine with the reticulum, and muscle contraction ceases.

How calcium ions cause a muscle to contract is still somewhat of a puzzle. The most commonly accepted explanation is the **sliding-filament hypothesis.** We know that the movement of a muscle fiber is caused by the contraction of the myofibrils in the fiber. We also know that the myofibrils contain a protein called troponin and that contraction requires actin, myosin, calcium ions, and energy. According to the sliding-filament hypothesis, also called the myosin-filament hypothesis, the myosin filaments are intersected by a series of cross bridges. When calcium is released by the sarcoplasmic reticulum, the cross bridges connect with actin filaments (see Figure 9–3e). The bridges pull the actin filaments of each sarcomere inward toward each other until their approaching ends overlap (Figure 9–5). As the actin filaments slide inward, the Z lines are drawn toward the A band, and the sarcomere is shortened. The shortening of the sarcomeres and myofibrils causes the shortening of the muscle fibers.

Researchers believe that the protein troponin, which is found in the filaments, prevents any interaction between actin and myosin in a noncontracting muscle. When a nerve impulse reaches the sarcoplasmic reticulum and triggers the release of calcium ions, the ions combine with the troponin,

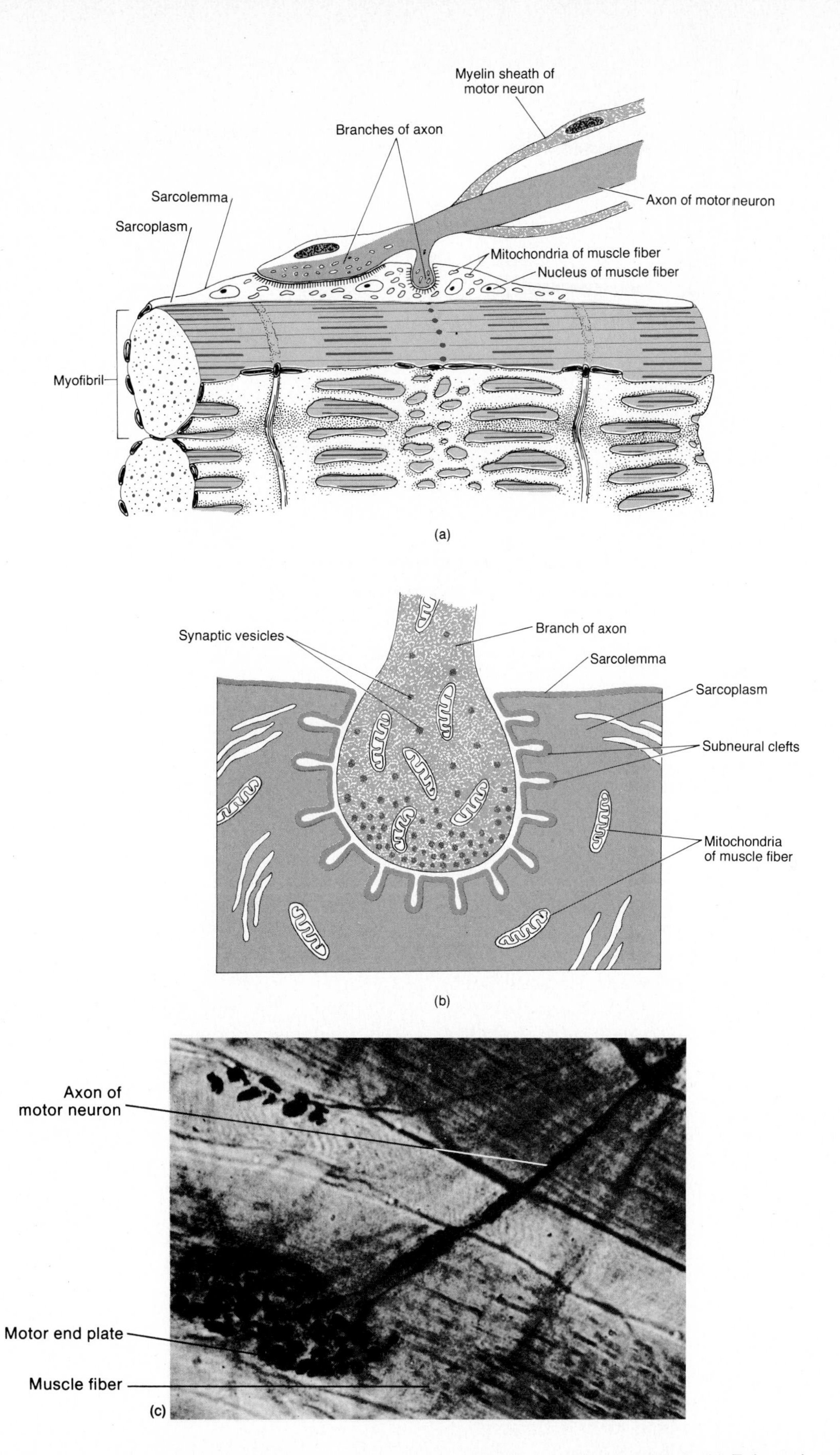

Figure 9–4. A motor end plate. (a) Diagrammatic representation seen by a light microscope. (b) Enlarged aspect seen by an electron microscope. (c) Photomicrograph at a magnification of 220X. (Courtesy of Donald I. Patt, from *Comparative Vertebrate Histology,* by Donald I. Patt and Gail R. Patt, Harper & Row, Publishers, Inc., New York, 1969.)

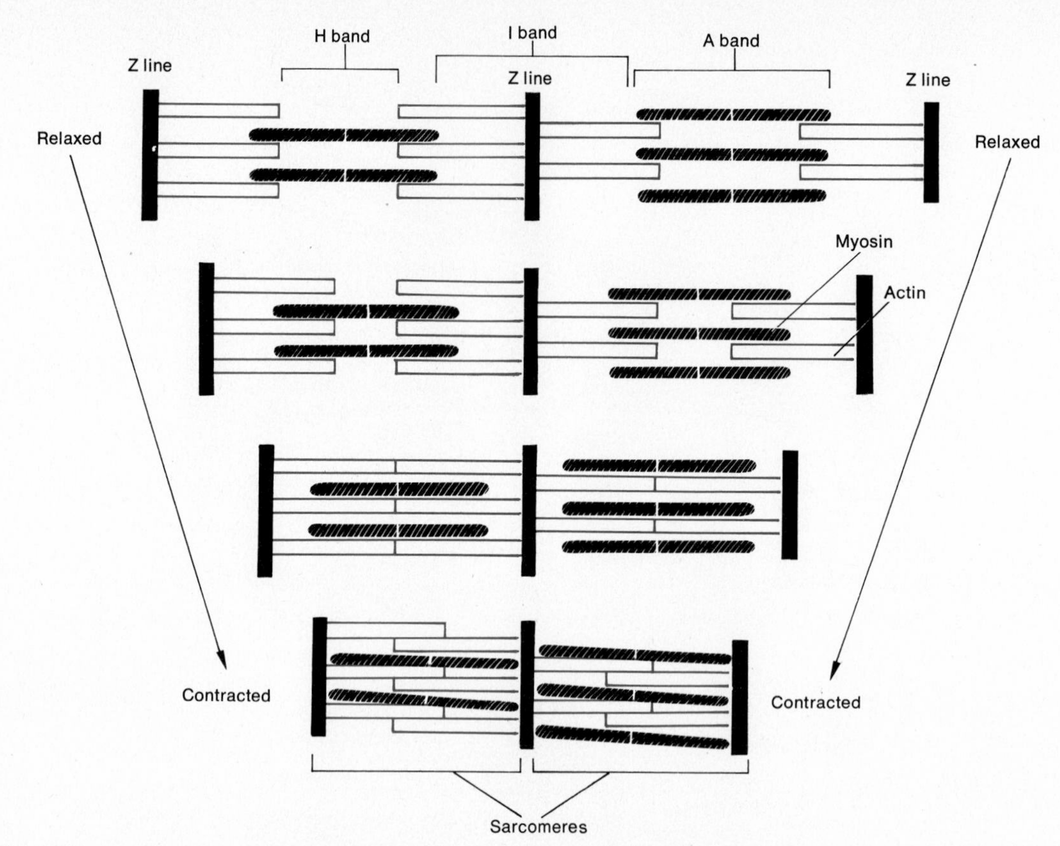

Figure 9–5. Sliding-filament hypothesis of muscular contraction. Position of various zones in relaxed and contracted sarcomeres and illustration of sliding actin filaments.

rendering it inactive. The myosin and actin are then free to interact. At the same time, the nerve impulse somehow stimulates the breakdown of adenosine triphosphate (ATP). ATP is a molecule found universally in living systems. Its essential function is to store energy for cellular activities. When certain chemical bonds of ATP are broken, the energy released is used by cells to carry on their basic functions. In muscle cells, the energy that is released by ATP breakdown is used for the attachment of the cross bridges and the sliding of the actin filaments. After the impulse ends, the calcium ions return to the sarcoplasmic reticulum. The troponin breaks the cross-bridge attachments, and the actin filaments slip back outward to their original place. The sarcomeres are thereby returned to their resting lengths, and the muscle resumes its resting shape.

Energy for contraction

Contraction of a muscle requires work, and in order for work to be done, energy is needed. Attached to the actin filaments of a muscle fiber is the high-energy compound ATP. When a nerve impulse stimulates a muscle fiber, ATP breaks down into ADP + P, and energy is released. As far as we know, ATP is always the immediate source of energy for muscle contraction.

Like the other cells of the body, muscle cells synthesize ATP as follows:

$$\text{ADP} + \text{P} + \text{energy} \rightarrow \text{ATP}$$

The energy for replenishing ATP is derived from the breakdown of digested foods. However, unlike most other cells of the body, muscle fibers alternate between a high degree of activity and virtual inactivity. When a muscle is contracting, its energy requirements are high, and the synthesis of ATP is accelerated. If the exercise is strenuous, ATP is used up even faster than it can be manufactured. Thus, muscles must be able to build up a reserve supply of energy. They do this in two ways. A resting muscle needs relatively little energy and produces much more ATP than it can use. At first, the muscle fiber stores the excess ATP on the myosin filaments. When the fiber runs out of storage space for the ATP molecules, it then combines the remainder of the ATP with a substance called *creatine.* Creatine can accept a high-energy phosphate from ATP to become the high-energy compound *creatine phosphate.*

$$\text{ATP} + \text{creatine} \rightarrow \text{creatine phosphate} + \text{ADP}$$

Creatine phosphate is produced only when the muscle fibers are resting. During strenuous contraction, the reaction reverses itself, as shown in the following equation:

$$\text{ADP} + \text{creatine phosphate} \rightarrow \text{ATP} + \text{creatine}$$

The all-or-none principle

According to the **all-or-none principle,** the muscle fibers of a motor unit will contract to their fullest extent or will not contract at all, providing conditions remain constant. In other words, *muscle fibers* do not partially contract. The principle does not imply that the strength of the contraction is the same every time the fiber is stimulated. For example, the strength of contraction may be decreased by fatigue, lack of nutrients, or lack of oxygen. The weakest stimulus from a neuron that can initiate a contraction is called a **threshold** or **liminal stimulus.** A stimulus of lesser intensity, or one that cannot initiate contraction, is referred to as a **subthreshold** or **subliminal stimulus.**

CARDIAC MUSCLE TISSUE

The principal constituent of the wall of the heart is **cardiac muscle tissue.** It has the same striated appearance as skeletal muscle tissue, but, unlike skeletal muscle tissue, it is involuntary. The cells of cardiac muscle tissue are roughly quadrangular in shape and have only a single nucleus (Figure 9–6). The individual fibers are covered by a thin, poorly defined sarcolemma, and the internal myofibrils produce the characteristic striations. Cardiac muscle cells have the same basic arrangement of actin, myosin, and sarcoplasmic reticulum that is found in skeletal muscle cells. In addition, they contain a system of transverse tubules similar to the T tubules of skeletal muscle. However, the nuclei in cardiac cells are centrally located as compared to the peripheral location of nuclei in skeletal muscle. And whereas the fibers of skeletal muscle are arranged in a parallel fashion, those of cardiac muscle branch freely with other fibers to form two separate networks. The muscular walls and septum of the upper chambers of the heart (atria) compose one network. The muscular walls and septum of the lower chambers of the heart (ventricles) compose the other network. When a single fiber of either network is stimulated, all the fibers in the network become stimulated as well. Thus, each network contracts as a functional unit. The fibers of each network were once thought to be fused together into a multinucleated mass called a syncytium. But it is now known that each fiber in a network is separated from the next fiber by an irregular transverse thickening of the sarcolemma called an **intercalated disc.** These discs strengthen the cardiac muscle tissue.

Under normal conditions, cardiac muscle tissue contracts rapidly, continuously, and rhythmically about 72 times a minute, never stopping. This is a major physiological difference between cardiac and skeletal muscle tissue. Another difference between skeletal and cardiac muscle tissue is the source of stimulation. Skeletal muscle tissue ordinarily contracts only when it is stimulated by a nerve impulse. In contrast, cardiac muscle tissue can contract without any nerve stimulation. Its source of stimulation is a conducting tissue of specialized muscle that lies only within the heart. About 72 times a minute, this tissue sends out electrical impulses that stimulate cardiac contraction. Nerve stimulation merely causes the conducting tissue to increase or decrease its rate of discharge.

SMOOTH MUSCLE TISSUE

Like cardiac muscle tissue, **smooth muscle tissue** is usually involuntary. Unlike either cardiac or skeletal muscle tissue, however, it is nonstriated. A single fiber of smooth muscle tissue is spindle-shaped, and within the fiber is a single, oval, centrally located nucleus (Figure 9–7). Smooth muscle cells also contain actin and myosin filaments, but the filaments are not as neatly arranged as in skeletal and cardiac muscle tissue. This is the reason why well-differentiated striations do not occur.

Two kinds of smooth muscle tissue, visceral and multiunit, are recognized. The more common type is called *visceral muscle tissue.* It is found in wraparound sheets that form part of the walls of the hollow viscera such as the stomach, intestines, uterus, and urinary bladder. The terms *smooth muscle tissue* and *visceral muscle tissue* are sometimes used interchangeably. The fibers in visceral muscle tissue are arranged in a branching pattern and are tightly bound together to form a continuous network (Figure 9–7a). When a neuron stimulates one fiber, the impulse travels over the other fibers so that contraction occurs in a wave over many adjacent fibers.

The second kind of smooth muscle tissue, called *multiunit smooth muscle tissue,* consists of individual fibers each with its own motor-nerve endings (Figure 9–7a). Whereas stimulation of a single visceral muscle fiber causes contraction of many adjacent fibers, stimulation of a single multiunit fiber causes contraction of only that fiber. In this

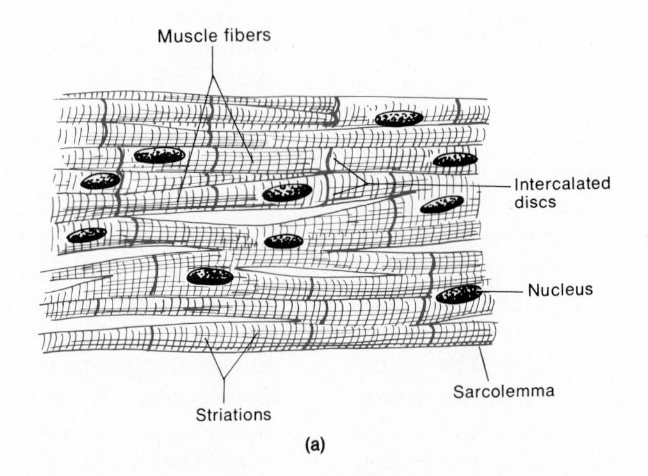

(a)

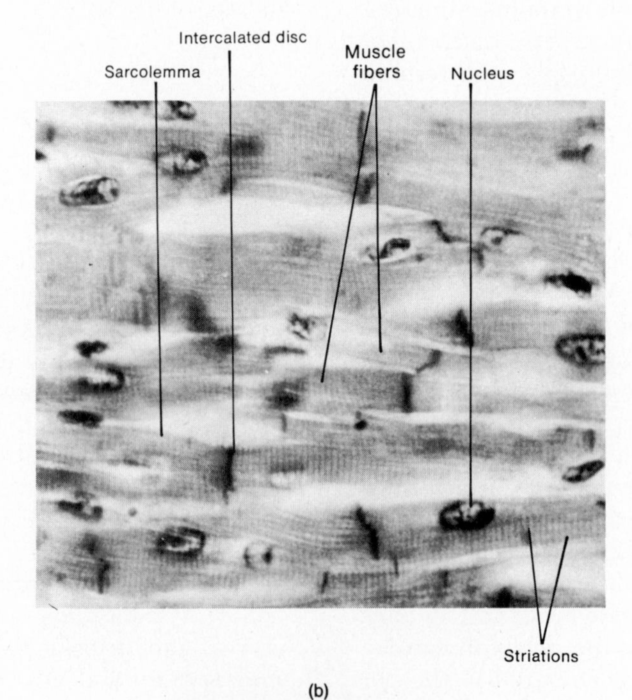

(b)

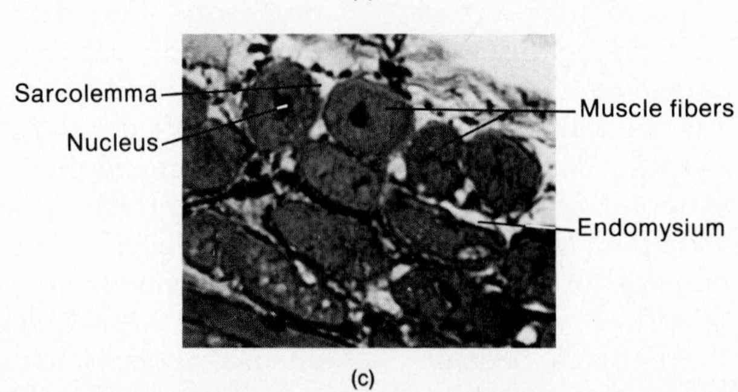

(c)

Figure 9–6. Histology of cardiac muscle tissue. (a) Diagrammatic representation. (b) Photomicrograph of cardiac muscle tissue in longitudinal section at a magnification of 640X. (Courtesy of Edward J. Reith, from *Atlas of Descriptive Histology,* by Edward J. Reith and Michael H. Ross, Harper & Row, Publishers, Inc., New York, 1970.) (c) Photomicrograph of cardiac muscle tissue in cross section at a magnification of 100X. (Courtesy of Victor B. Eichler, Wichita State university.)

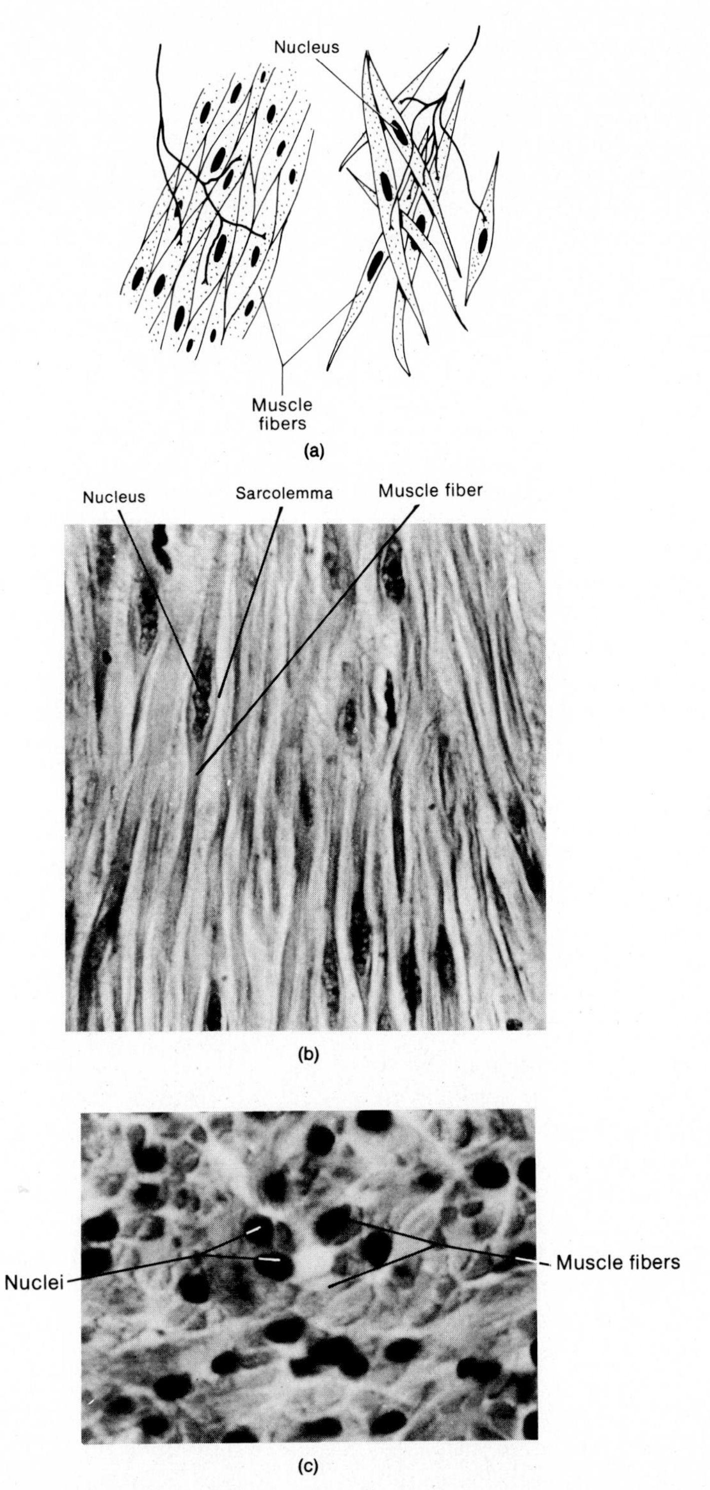

Figure 9–7. Histology of smooth muscle tissue. (a) Diagrammatic representation of visceral smooth muscle tissue (left) and multiunit smooth muscle tissue (right). (b) Photomicrograph of several smooth muscle fibers in longitudinal section at a magnification of 640X. (Courtesy of Edward J. Reith, from *Atlas of Descriptive Histology,* by Edward J. Reith and Michael H. Ross, Harper & Row, Publishers, Inc., New York, 1970.) (c) Photomicrograph of several smooth muscle fibers in cross section at a magnification of 200X. (Courtesy of Victor B. Eichler, Wichita State University.)

respect, multiunit muscle tissue is like skeletal muscle tissue. Multiunit smooth muscle tissue is found in the walls of blood vessels, in the arrector pili muscles attached to hair follicles, and in the intrinsic muscles of the eye–the iris, for example.

Both kinds of smooth muscle tissue contract more slowly and relax more slowly than skeletal muscle tissue. The reason is probably that the actin and myosin filaments of smooth muscle are so poorly arranged. Moreover, whereas skeletal muscle cells contract as individual units, visceral muscle cells contract one after another as the impulse spreads from one to another.

APPLICATIONS TO HEALTH

Fibrosis and fibrositis

The formation of fibrous tissue in places where it normally does not exist is called **fibrosis.** Muscle cells cannot undergo mitosis, and dead muscle cells are normally replaced with fibrous connective tissue. Fibrosis, then, is often a consequence of muscle injury or degeneration.

Fibrositis is an inflammation of fibrous tissue. If it occurs in the lumbar region and low back, it is referred to as *lumbago.* Fibrositis is a common condition characterized by pain, stiffness, or soreness of fibrous tissue, especially in the muscle coverings. It is not destructive or progressive. But it may persist for years, or it may disappear spontaneously. Attacks of fibrositis may follow an injury, repeated muscular strain, or prolonged muscular tension. *Myositis* is an inflammation of muscle cells.

Muscular dystrophy and myasthenia gravis

The term *atrophy* refers to a reduction in the size of an organ. Muscular atrophy may result from a decrease in the size of individual muscle cells or destruction of muscle cells. Muscular atrophy may be caused by disuse, either because the individual fails to exercise the muscle or because nerve impulses do not reach the muscle fibers. Muscular atrophy may also be caused by a *myopathy,* which is a general term for the diseases that attack and destroy muscle tissue. Muscular atrophy is always accompanied by some degree of muscular weakness. If the connections with nervous tissue are destroyed, the fibers cease to contract at all.

The term **muscular dystrophy** applies to a number of inherited myopathies. The word *dystrophy* means degeneration. The disease is characterized by degeneration of the individual muscle cells, which leads to a progressive atrophy of the muscle. Usually the voluntary skeletal muscles are weakened equally on both sides of the body, whereas the internal muscles, such as the diaphragm, are not affected. Histologically, the changes that occur include variation in muscle fiber size, degeneration of fibers, and deposition of fat.

The cause of muscular dystrophy has been variously attributed to a genetic defect, faulty metabolism of potassium, protein deficiency, and inability of the body to utilize creatine.

Myasthenia gravis is an atrophy of the skeletal muscles caused by an abnormality at the motor end plate that prevents the muscle fibers from contracting. Recall that motor neurons stimulate the skeletal muscle fibers to contract by releasing a chemical called acetylcholine. Myasthenia gravis is caused by failure of the neurons to release acetylcholine or by excess amounts of cholinesterase, a chemical that interferes with the action of acetylcholine. As the disease progresses, more motor end plates become affected. The muscle becomes increasingly weaker and may eventually cease to function altogether.

The cause of myasthenia gravis is unknown. It is more common in females, occurring most frequently between the ages of 20 and 50. The muscles of the face and neck are most apt to be involved. Initial symptoms include a weakness of the eye muscles and difficulty in swallowing. Later, the individual has trouble chewing and talking. Eventually, the muscles of the limbs may become involved. Death may result from paralysis of the respiratory muscles, but usually the disorder does not progress to this stage.

Abnormal contractions

One kind of abnormal contraction of a muscle is **spasm,** a sudden, involuntary contraction of short duration. A **cramp** is a painful, spasmodic contraction of a muscle. **Convulsions** are violent, involuntary contractions of a whole group of muscles. Convulsions occur when motor neurons are stimulated by factors such as fever, poisons, hysteria, and changes in body chemistry due to the withdrawal of certain drugs. The stimulated neurons send many seemingly "senseless" impulses to the muscle fibers. **Fibrillation** is the uncoordinated contraction of individual muscle fibers so that the muscle fails to contract smoothly. Cardiac muscle is most prone to this abnormality.

Key terms associated with the muscular system

Gangrene Death of a muscle that results from almost complete interruption of its blood supply.
Myology Study of muscles.
Myomalacia (*malaco* = soft) Softening of a muscle.
Myopathy Any disease of muscle tissue.
Myosclerosis (*scler* = hard) Hardening of a muscle.
Myospasm Spasm of a muscle.
Myotonia Increased muscular irritability and contractility with decreased power of relaxation; tonic spasm of the muscle.
Sprain The forcible twisting of a joint causing rupture of a ligament or causing the ligament to loosen at one of its attachments.
Trichinosis A myositis caused by the parasitic worm *Trichinella spiralis,* which may be found in the muscles of people, rats, and pigs. Humans contract the disease by eating infected pork that is insufficiently cooked.
Volkmann's contracture (*contra* = against) Permanent contraction of a muscle due to replacement of destroyed muscle cells with fibrous tissue that lacks ability to stretch. Destruction may occur from interference with circulation caused by a tight bandage, a piece of elastic, or a cast.
Wryneck or torticollis Contracted state of one of the neck muscles; produces twisting of the neck and an unnatural position of the head.

Chapter summary in outline

CHARACTERISTICS

1. Irritability is the property of receiving and responding to stimuli.
2. Contractility is the ability to shorten and thicken, or contract.
3. Extensibility is the ability to be stretched or extended.
4. Elasticity is the ability to return to original shape after contraction or extension.

KINDS OF MUSCLE TISSUE

1. Skeletal muscle is attached to bones. It is striated and voluntary.
2. Smooth muscle is located in viscera. It is nonstriated and usually involuntary.
3. Cardiac muscle forms the walls of the heart. It is striated and usually involuntary.

SKELETAL MUSCLE TISSUE

Fascia

1. Fascia refers to a sheet or broad band of fibrous connective tissue beneath the skin or around muscles and organs of the body.
2. Types of fascia are superficial, deep, and subserous.

Connective tissue components

1. The entire muscle is covered by the epimysium. Fasciculi or fascicles are covered by perimysium. Fibers are covered by endomysium.
2. Tendons and aponeuroses attach muscle to bone or to other muscles.

Nerve and blood supply

1. Nerves convey impulses, and blood provides nutrients and oxygen and removes wastes for contraction.

Histology

1. Skeletal muscle tissue consists of fibers (cells) covered by a sarcolemma. The fibers contain sarcoplasm, nuclei, sarcoplasmic reticulum, and many mitochondria.
2. Each fiber contains myofibrils (actin and myosin filaments). The myofibrils are compartmentalized into sarcomeres.

CONTRACTION OF SKELETAL MUSCLE

The motor unit

1. A motor neuron transmits the stimulus to a muscle for contraction.
2. The area of contact between a motor neuron and muscle fiber is a motor end plate.
3. A motor neuron and the muscle fibers it stimulates form a motor unit.

Physiology of contraction

1. A nerve impulse travels over the sarcolemma and enters the T tubules and sarcoplasmic reticulum.
2. Calcium ions released by the sarcoplasmic reticulum trigger the contractile process.
3. Actual contraction is brought about when the actin filaments slide toward each other.

Energy for contraction

1. The source of energy is ATP.
2. When muscles are resting, ATP combines with creatine to form creatine phosphate, which breaks down to produce ATP when muscles contract strenuously.

The all-or-none principle

1. Muscle fibers of a motor unit contract to their fullest extent for the prevailing conditions or not at all.
2. The weakest stimulus capable of causing contraction is a threshold or liminal stimulus.
3. A stimulus not capable of inducing contraction is a subthreshold or subliminal stimulus.

CARDIAC MUSCLE TISSUE

1. This muscle tissue is found only in the heart. It is striated and involuntary.
2. Cells are quadrangular and usually contain one centrally placed nucleus.
3. The fibers form a continuous, branching network that contracts as a functional unit.
4. Intercalated discs are found at the junction between cells and provide strength and aid impulse conduction.

SMOOTH MUSCLE TISSUE

1. Smooth muscle tissue is found in viscera, the skin, eyeballs, and blood vessels. It is nonstriated and involuntary.
2. Visceral smooth muscle tissue is found in the walls of viscera, and the fibers are arranged in a network.
3. Multiunit smooth muscle tissue is found in blood vessels, the eye, and the skin. The fibers operate singly rather than as a unit.

APPLICATIONS TO HEALTH

Fibrosis and fibrositis

1. Fibrosis is the formation of fibrous tissue where it normally does not exist; it frequently occurs in damaged muscle tissue.
2. Fibrositis is an inflammation of fibrous tissue. If it occurs in the lumbar region, it is called lumbago. Myositis is muscle tissue inflammation.

Muscular dystrophy and myasthenia gravis

1. Muscular dystrophy is a hereditary disease of muscles characterized by degeneration of individual muscle cells.
2. Myasthenia gravis is a disease exhibiting great muscular weakness and fatigability resulting from improper neuromuscular transmission.

Abnormal contractions

1. Abnormal contractions include spasms, cramps, convulsions, and fibrillation.

Review questions and problems

1. How is the skeletal system related to the muscular system? What are the three kinds of motion that are accomplished by the muscular system?
2. What are the four characteristics of muscle tissue?
3. What criteria are employed for distinguishing the three kinds of muscle tissue?
4. Define epimysium, perimysium, endomysium, tendon, and aponeurosis. Describe the nerve and blood supply to a muscle.
5. Define fascia. Distinguish between superficial fascia, deep fascia, and subserous fascia on the basis of location.
6. Discuss the microscopic structure of skeletal muscle tissue.
7. In considering the contraction of skeletal muscle tissue, describe the following:
   (a) motor unit
   (b) sliding-filament hypothesis
   (c) importance of calcium and troponin
   (d) sources of energy
8. What is the all-or-none principle? Relate it to a liminal and subliminal stimulus.
9. Compare cardiac and smooth muscle tissue with regard to microscopic structure, functions, and locations.
10. Define fibrosis. What is one of its causes? Define myositis.
11. What is muscular dystrophy?
12. What is myasthenia gravis? In this disease, why do the muscles not contract normally?
13. Define each of the following abnormal muscular contractions: spasm, cramp, convulsion, and fibrillation.
14. Refer to the glossary of key terms associated with the muscular system. Be sure that you can define each term.

Selected readings

Basmajian, John V. *Muscles Alive.* 3d ed. Baltimore: Williams and Wilkins, 1974.

Ham, Arthur W. *Histology.* 7th ed. Philadelphia: J. B. Lippincott, 1974.

Leeson, Thomas S., and C. Roland Leeson. *Histology.* 2d ed. Philadelphia: W. B. Saunders, 1970. Chap. 9.

CHAPTER 10

THE MUSCULAR SYSTEM

STUDENT OBJECTIVES

After you have read this chapter, you should be able to:

1. Describe the relationship between bones and skeletal muscles in producing body movements
2. Define a lever and fulcrum and compare the three classes of levers on the basis of placement of the fulcrum, effort, and resistance
3. Describe most body movements as activities of groups of muscles by explaining the roles of the prime mover, antagonist, and synergist
4. Define the criteria employed in naming skeletal muscles
5. Identify the principal skeletal muscles in different regions of the body by name, origin, insertion, and action
6. Compare the common sites for intramuscular injections

When speaking of muscle tissue, we are referring to all the contractile tissues of the body, that is, skeletal, cardiac, and smooth muscle. But when we speak of the **muscular system,** we refer to the *skeletal* muscle system, which is composed of skeletal muscle tissue and connective tissues that make up individual muscle organs, such as the biceps. Cardiac and smooth muscle tissues are classified with other organ systems. For instance, cardiac muscle tissue is found in the heart, an organ of the circulatory system. Smooth muscle tissue of the intestine is part of the digestive system. And smooth muscle tissue of the urinary bladder is part of the urinary system. In this chapter, we shall discuss the muscular system only. We shall do this by taking a look at how skeletal muscles produce movement and by describing and locating some of the principal skeletal muscles.

HOW MUSCLES PRODUCE MOVEMENT

Skeletal muscles produce movements by pulling on tendons, which, in turn, pull on bones. The majority of muscles cover at least one joint and are attached to the articulating bones that form the joint (Figure 10–1a). When such a muscle contracts, it pulls on the articulating bones, and this pull draws one of the articulating bones toward the other. The two articulating bones usually do not move equally in response to the contraction. One of them is kept pretty much in its original position because other muscles contract to pull it in the opposite direction or because the structure of the bone makes it less movable. Ordinarily, the attachment of a muscle tendon to the more stationary bone is called the **origin.** The attachment of the other muscle tendon to the more movable bone is referred to as the **insertion.** A good analogy for remembering this is a spring on a door. The part of the spring attached to the door represents the insertion, whereas the part of the spring attached to the door frame represents the origin. The fleshy portion of the muscle between the tendons of the origin and insertion is referred to as the **belly,** or **gaster.** In the appendages, especially, the origin is proximal, and the insertion is distal. In addition, muscles that move a part of the body generally do not cover the moving part. Note in Figure 10–1a, for example, that al-

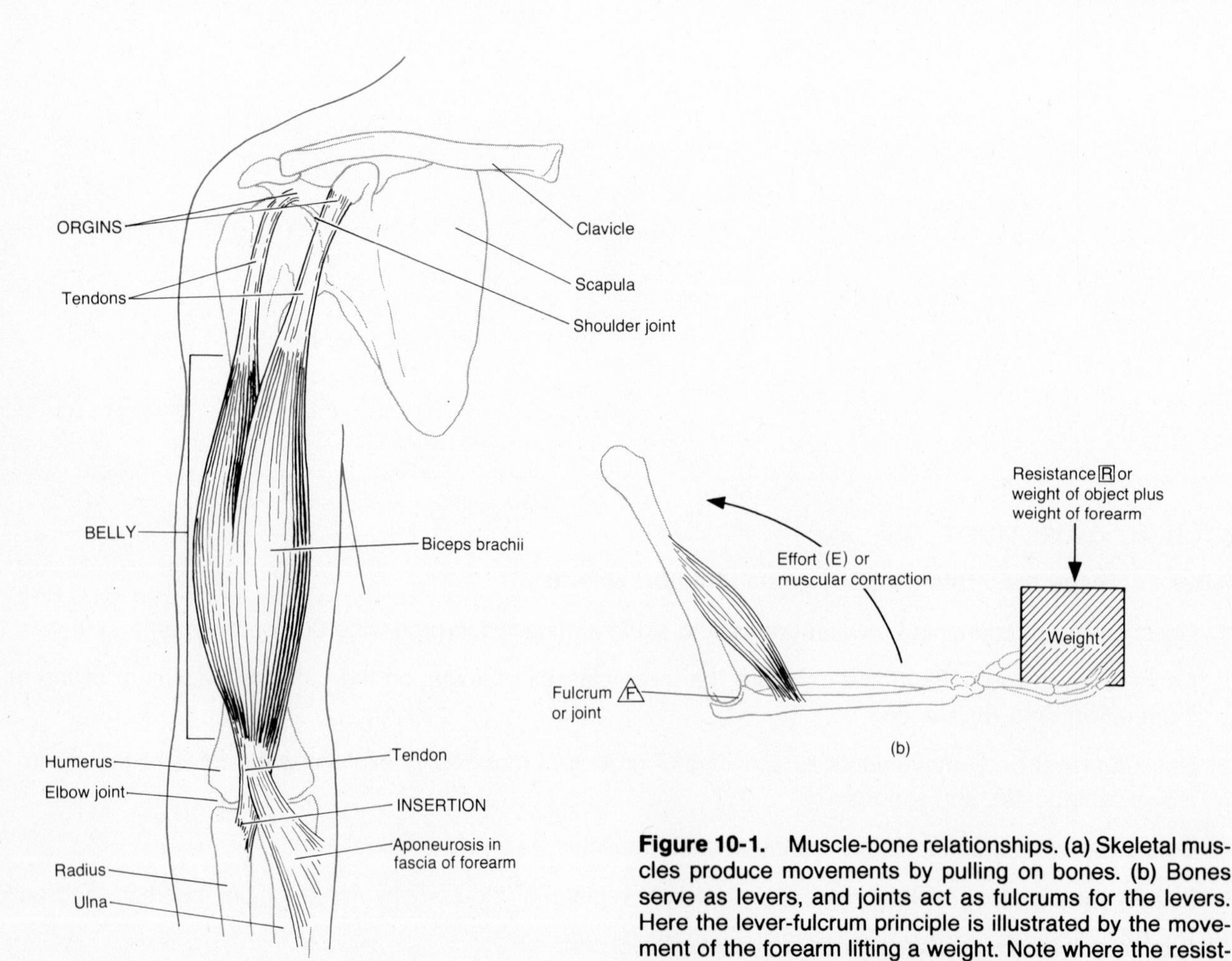

Figure 10-1. Muscle-bone relationships. (a) Skeletal muscles produce movements by pulling on bones. (b) Bones serve as levers, and joints act as fulcrums for the levers. Here the lever-fulcrum principle is illustrated by the movement of the forearm lifting a weight. Note where the resistance and effort are applied in this example.

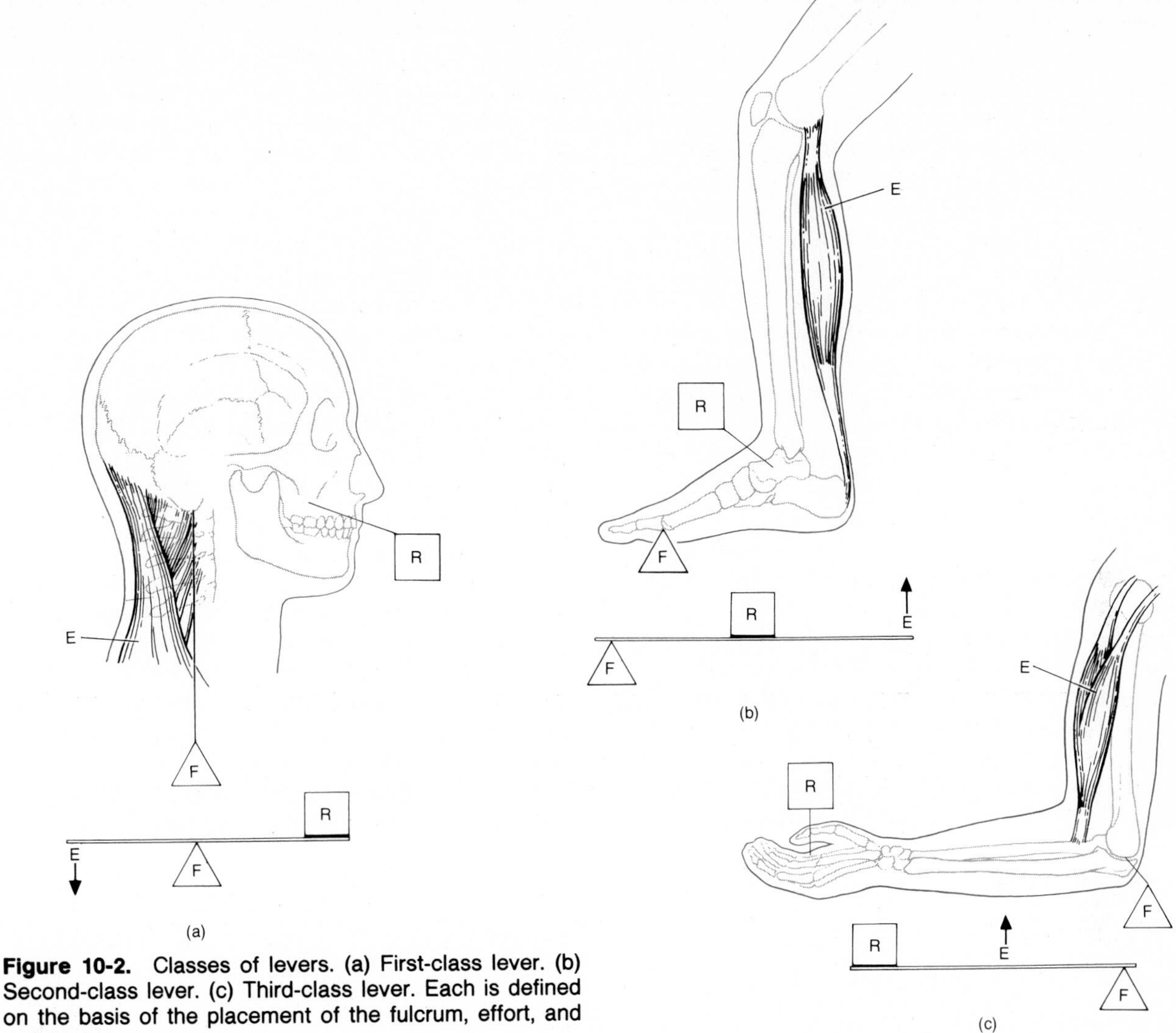

Figure 10-2. Classes of levers. (a) First-class lever. (b) Second-class lever. (c) Third-class lever. Each is defined on the basis of the placement of the fulcrum, effort, and resistance.

though contraction of the biceps muscle moves the forearm, the major portion of the muscle lies over the humerus of the arm.

In bringing about a body movement, bones act as levers, and joints function as fulcrums of these levers. A **lever** may be defined as a rigid rod that moves about on some fixed point or support called a **fulcrum.** A fulcrum may be symbolized as F. A lever is acted upon at two different points by two different forces: (1) the *resistance,* which can be symbolized as R, and (2) the *effort,* which can be symbolized as E. The resistance may be regarded as something to be overcome or balanced, whereas the effort is exerted to overcome the resistance. The resistance may be the weight of the body that is to be moved, such as an arm or leg, or some object to be lifted, or both. The muscular effort (contraction) is applied to the bone at the insertion of the muscle and brings about the work of motion. As an example, consider the biceps flexing the forearm at the elbow as a weight is lifted (Figure 10–1b). In this case, when the forearm is raised, the elbow is the fulcrum. The weight of the forearm plus the weight in the hand is the resistance. And the shortening of the biceps pulling the forearm upward is the effort.

Levers are categorized into three types on the basis of the relative placement of the fulcrum, effort, and resistance. In a **first-class lever,** the fulcrum is placed between the effort and resistance (Figure 10–2a). An example of a first-class lever is a seesaw. Examples of first-class levers in the body are not abundant, however. One example is the head resting on the vertebral column. When the head is raised, the facial portion of the skull is the resistance. The joint between the atlas and occipital bone (atlanto-occipital joint) is the fulcrum. And the muscles of the back in contraction represent the effort.

Second-class levers are similar to the mechanics of a wheelbarrow. In this case, the fulcrum is at one end. The effort is at the opposite end. And the resistance is in between (Figure 10–2b). Most authorities agree that there are no examples of second-class levers in the body. Some, however, consider that raising the body on the toes (resistance) and utilizing the ball of the foot as the fulcrum is an example. Here, the calf muscles pull the heel upward as it shortens (effort).

Third-class levers are the most common kinds of levers in the body and consist of the fulcrum at one end, the resistance at the opposite end, and the effort in between (Figure 10–2c). A common example is flexing the forearm at the elbow. Here, the weight of the forearm is the resistance, the contraction of the biceps is the effort, and the elbow joint is the fulcrum.

In general, most movements are coordinated by the activity of several skeletal muscles. In fact, skeletal muscles usually act in groups rather than individually. As an example, let us examine flexing the forearm at the elbow. A muscle that causes a desired action is referred to as the **agonist** or **prime mover.** In this instance, the biceps brachii is the agonist. Simultaneously with the contraction of the biceps brachii, another muscle, called the **antagonist,** is relaxing. In this movement, the triceps brachii serves as the antagonist. The antagonist has an effect opposite to that of the agonist. That is, the antagonist relaxes and gives way to the movement of the agonist. If we consider the extension of the forearm at the elbow, the triceps brachii would assume the role of the agonist, and the biceps brachii would act as antagonist. It will become obvious to you that most joints are operated by antagonistic groups of muscles. Still other muscles called **synergists,** or **fixators,** assist the agonist by reducing undesired action or unnecessary movements in the less mobile articulating bone. While flexing the forearm, the synergists, in this case the deltoid and pectoralis major muscles, would hold the arm and shoulder in a suitable position for the flexing action. Whereas the deltoid abducts the humerus, the pectoralis major adducts and medially rotates the humerus. Essentially, synergists contract at the same time as the prime mover and help the prime mover produce a more effective movement.

NAMING MUSCLES

Most of the almost 700 skeletal muscles are named on the basis of one or more distinctive criteria. If you understand these criteria, you will find it much easier to learn and remember the names of individual muscles.

Some muscles are named on the basis of *action.* Listed here are the principal actions of muscles, their definitions, and examples of muscles that perform the actions. For convenience, the actions are grouped as antagonistic pairs where possible.

MUSCLES NAMED ACCORDING TO ACTION

ACTION	DEFINITION	EXAMPLE
Flexor	Usually decreases the anterior angle at a joint	Flexor carpi radialis
Extensor	Usually increases the anterior angle at a joint	Extensor carpi ulnaris
Abductor	Moves a bone away from the midline	Abductor hallucis
Adductor	Moves a bone closer to the midline	Adductor longus
Levator	Produces an upward movement	Levator scapulae
Depressor	Produces a downward movement	Depressor labii inferioris
Supinator	Turns the palm upward or to the anterior	Supinator
Pronator	Turns the palm downward or to the posterior	Pronator teres
Dorsiflexor	Flexes the foot at the ankle joint	Tibialis anterior
Plantar flexor	Extends the foot at the ankle joint	Plantaris
Invertor	Turns the sole of the foot inward	Tibialis anterior
Evertor	Turns the sole of the foot outward	Peroneus tertius
Sphincter	Decreases the size of an opening	Pyloric sphincter between stomach and duodenum
Tensor	Makes a body part more rigid	Tensor fasciae latae
Rotator	Moves a bone around its longitudinal axis	Obturator

Still another criterion for naming muscles is the *direction of the muscle fibers.* There are, for example, rectus (straight), transverse, and oblique muscles. Rectus fibers usually run parallel to the midline of the body. Transverse fibers run perpendicular to the midline. And oblique fibers are diagonal to the midline. Muscles named according to these three directions include the rectus abdominis, transversus abdominis, and external oblique, respectively.

Another criterion employed is *location.* For example, the temporalis is so named because of its proximity to the temporal bone. The tibialis anterior is located near the tibia. *Size* is also commonly employed. For instance, the term *maximus* means largest, and *minimus* means smallest. *Longus* means long, and *brevis* means short. Examples include the gluteus maximus, gluteus minimus, adductor longus, and peroneus brevis.

Some muscles such as the biceps, triceps, and quadriceps are named on the basis of the *number of origins* they have. For instance, the biceps has two origins. The triceps has three, and the quadriceps has four. Other muscles are named on the basis of *shape.* Common examples include the deltoid (meaning triangular) and trapezius (meaning trapezoid). Muscles may also be named after their *origin* and *insertion.* One example is the sternocleidomastoid, which originates on the sternum and clavicle and inserts at the mastoid process of the temporal bone. The stylohyoideus originates on the styloid process of the temporal bone and inserts at the hyoid bone.

PRINCIPAL MUSCLES OF THE BODY

In the following pages, Exhibits 10–1 through 10–16 are provided for you to learn the principal skeletal muscles of the body. The Exhibits contain a listing of the muscles in terms of their origins, insertions, actions, and innervations. Diagrams of the muscles under consideration are also included, and the names of the muscles are indicated by capital letters. The prefixes, roots, suffixes, and definitions that explain the derivation of the muscles' names are included with the muscles' names. By reading them, your task of learning the names of muscles will be easier and more fun. It is strongly suggested that you make frequent reference to Chapters 6 and 7 in order to review your knowledge of bone markings since they serve as points of origin and insertion for muscles. By no means have all the muscles of the body been included. Only those considered important for introductory courses will be discussed. Innervation refers to the distribution or supply of nerves to a part of the body. Although the innervations may not mean much to you at this point, you will see their importance after you have read Chapters 13 and 14.

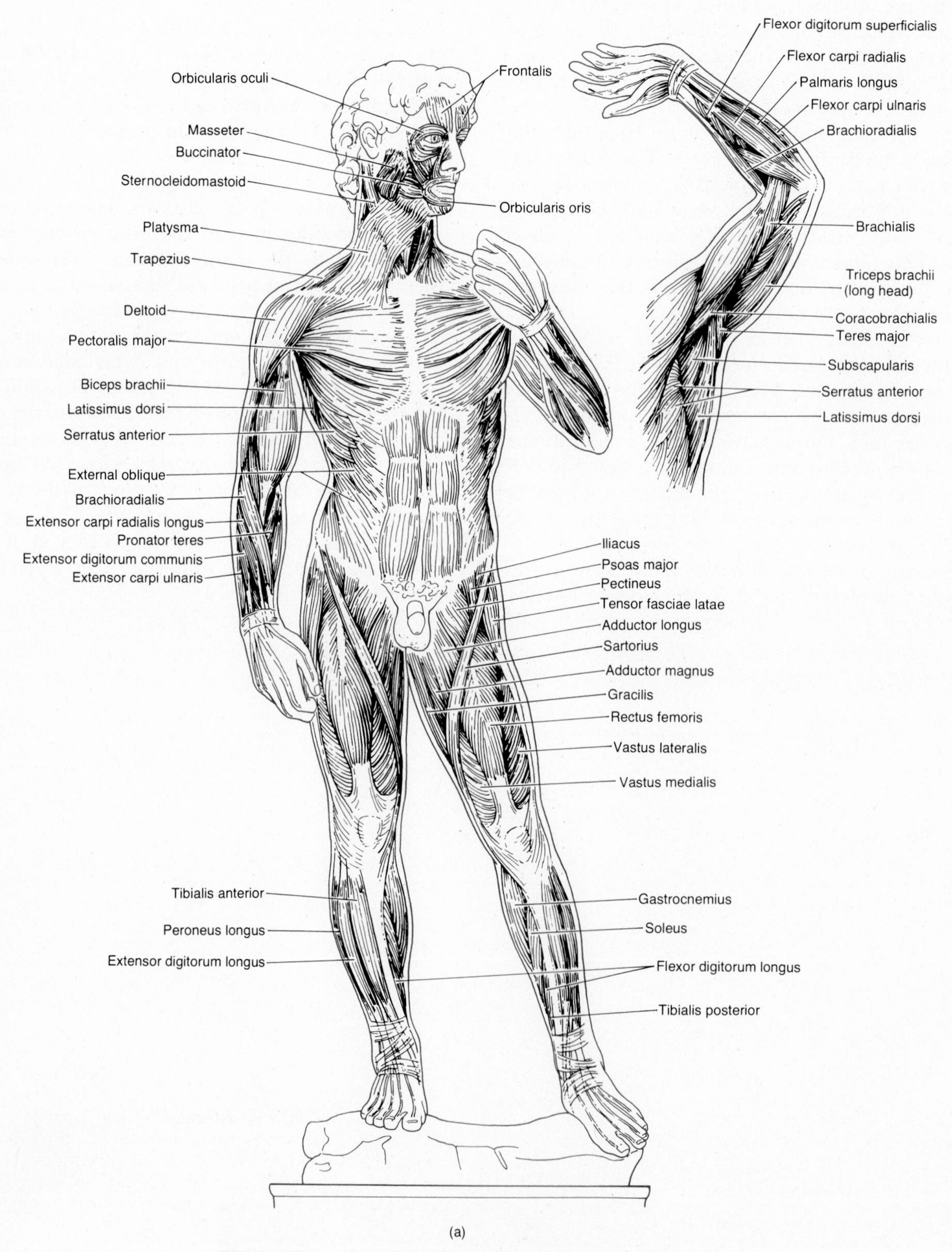

Figure 10-3. Principal superficial muscles. (a) Anterior view.

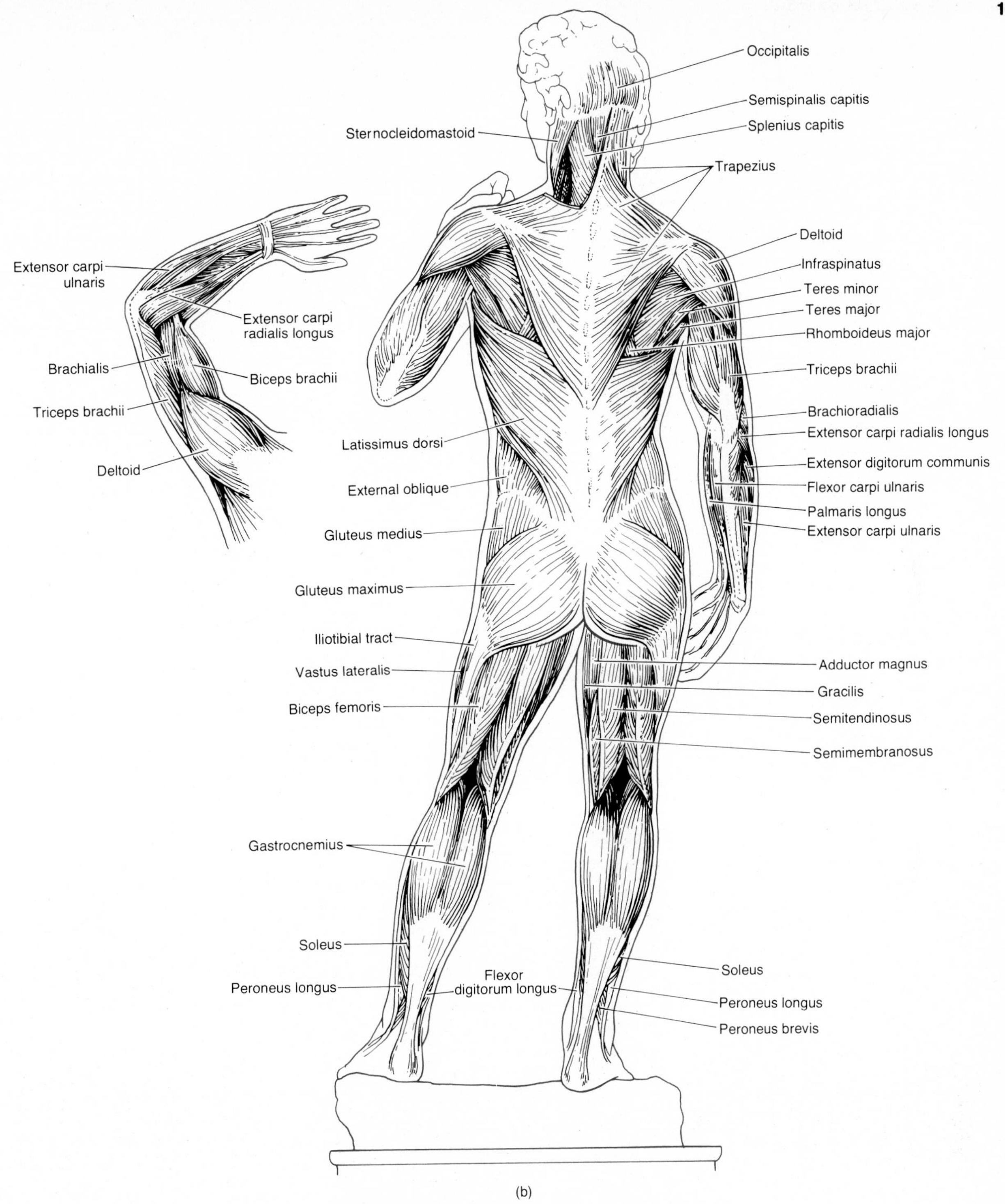

Figure 10-3 (cont.). (b) Posterior view.

Exhibit 10–1. THE MUSCULAR SYSTEM: AN OVERVIEW (see Figure 10–3)

Figure 10–3 shows general anterior and posterior views of the muscular system. Do not try to memorize all these muscles yet. As you study groups of muscles in subsequent Exhibits, refer to Figure 10–3 to see how each grouping is related to all others.

I have attempted, as much as possible, to indicate whether the muscles are superficial or deep, anterior or posterior, and medial or lateral. I have also tried to show the relationship of the muscles under consideration to other muscles in the area you are studying. If you have mastered the various criteria used in naming muscles, the names and actions of many muscles will have more meaning.

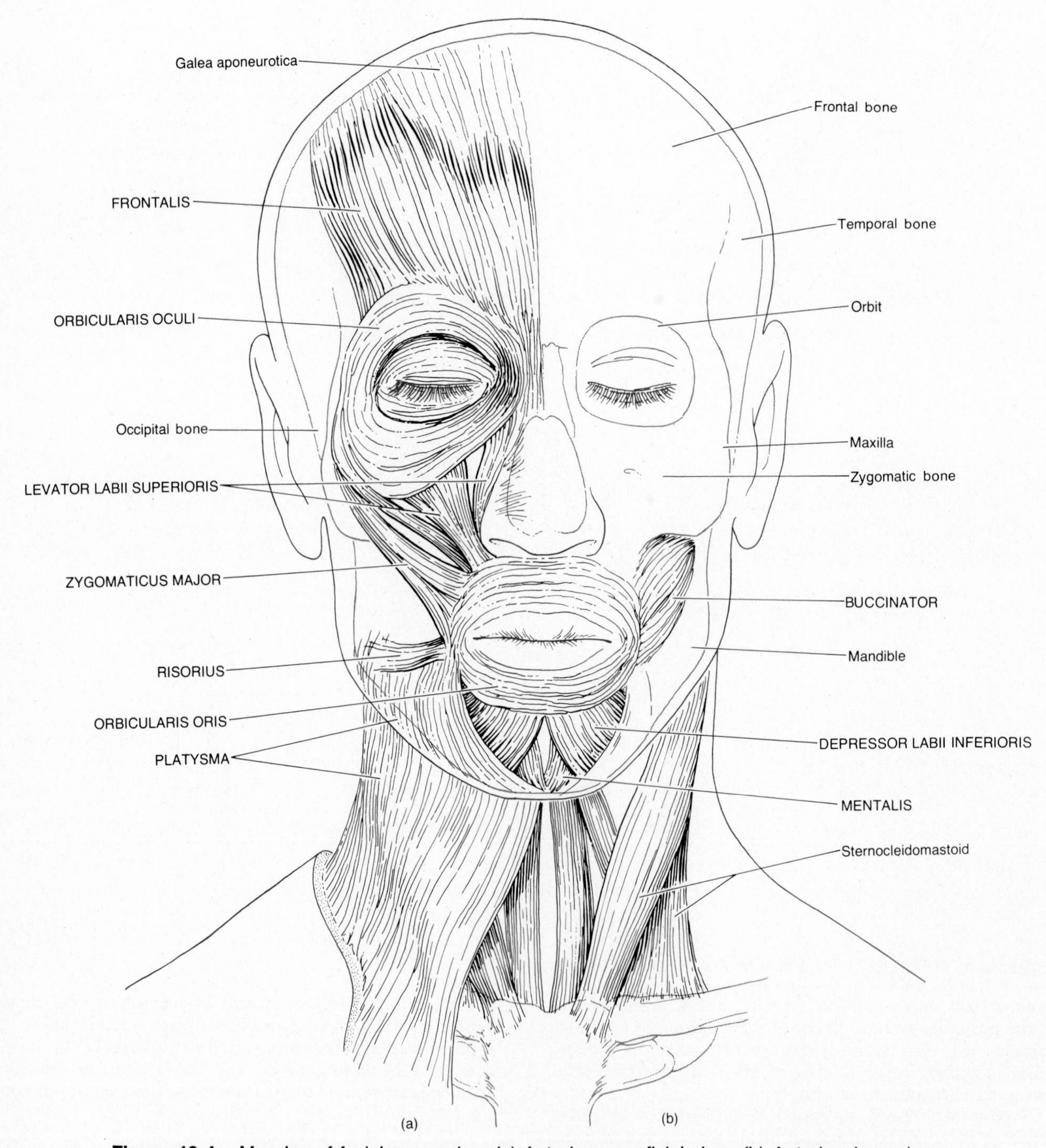

Figure 10-4. Muscles of facial expression. (a) Anterior superficial view. (b) Anterior deep view.

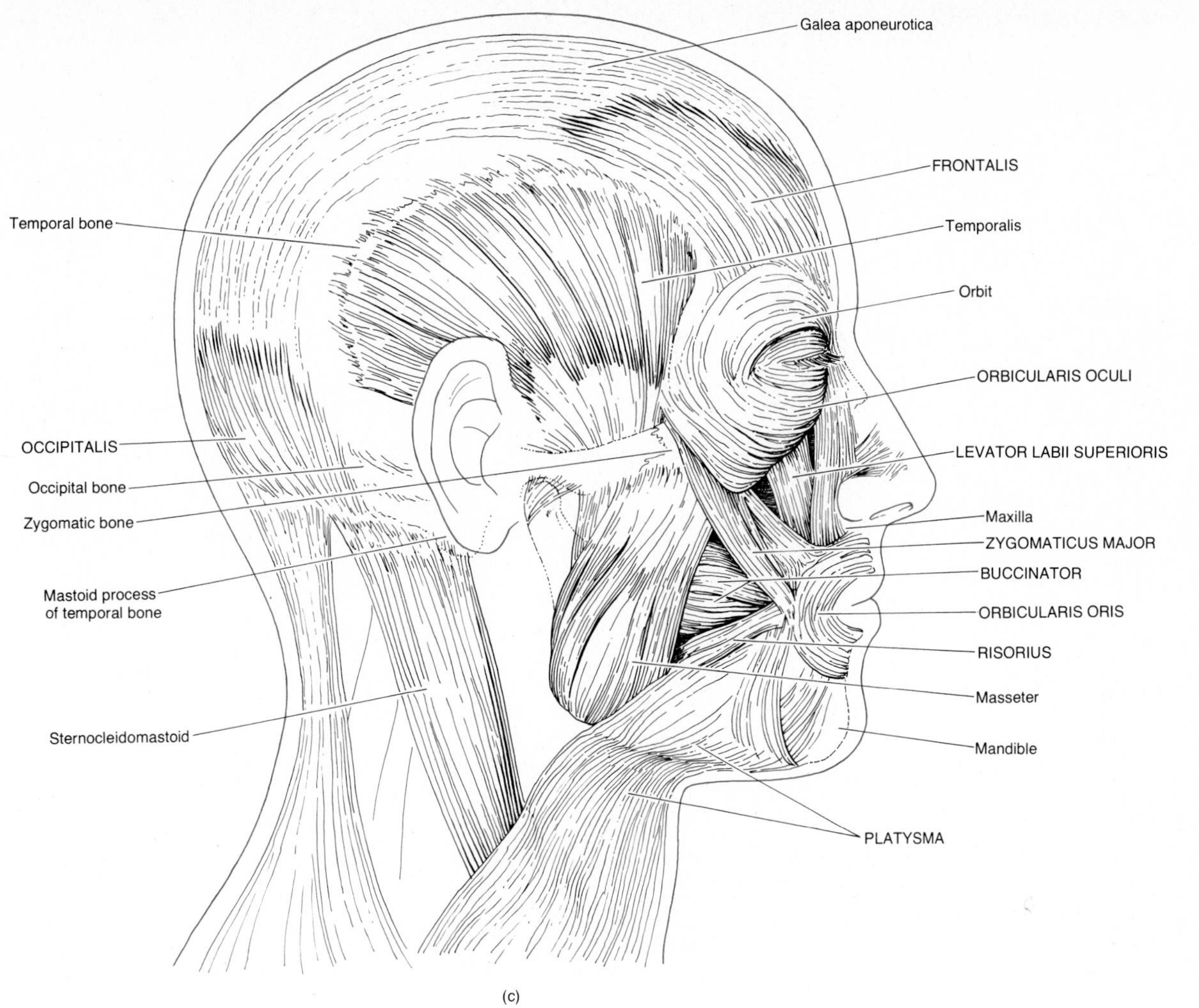

Figure 10-4 (cont.). Muscles of facial expression. (c) Right lateral superficial view.

Exhibit 10–2. MUSCLES OF FACIAL EXPRESSION (see Figure 10–4)

MUSCLE	ORIGIN	INSERTION	ACTION	INNERVATION
Epicranius (*epi* = over; *crani* = skull)	This muscle is divisible into two portions: the frontalis over the frontal bone and the occipitalis over the occipital bone. The two muscles are united by a strong aponeurosis, the galea aponeurotica, which covers the superior and lateral surfaces of the skull.			
Frontalis (*front* = forehead)	Galea aponeurotica	Skin superior to supraorbital line	Draws scalp forward, raises eyebrows, and wrinkles forehead horizontally	Facial nerve (VII)
Occipitalis (*occipito* = base of skull)	Occipital bone and mastoid process of temporal bone	Galea aponeurotica	Draws scalp backward	Facial nerve (VII)
Orbicularis oris (*orb* = circular; *or* = mouth)	Muscle fibers surrounding opening of mouth	Skin at corner of mouth	Closes lips, compresses lips against teeth, protrudes lips, and shapes lips during speech	Facial nerve (VII)
Zygomaticus major (*zygomatic* = cheek bone; *major* = greater)	Zygomatic bone	Skin at angle of mouth and orbicularis oris	Draws angle of mouth upward and outward as in smiling or laughing	Facial nerve (VII)

Exhibit 10–2 (cont.)

MUSCLE	ORIGIN	INSERTION	ACTION	INNERVATION
Levator labii superioris (*levator* = raises or elevates; *labii* = lip; *superioris* = upper)	Inferior to infraorbital foramen of maxilla	Skin at angle of mouth and orbicularis oris	Elevates (raises) upper lip	Facial nerve (VII)
Depressor labii inferioris (*depressor* = depresses or lowers; *inferioris* = lower)	Mandible	Skin of lower lip	Depresses (lowers) lower lip	Facial nerve (VII)
Buccinator (*bucc* = cheek)	Alveolar processes of maxilla and mandible	Orbicularis oris	Major cheek muscle; compresses cheek as in blowing air out of mouth and causes the cheeks to cave in, producing the action of sucking	Facial nerve (VII)
Mentalis (*mentum* = chin)	Mandible	Skin of chin	Elevates and protrudes lower lip and pulls skin of chin up as in pouting	Facial nerve (VII)
Platysma (*platy* = flat, broad)	Fascia over deltoid and pectoralis major muscles	Mandible, muscles around angle of mouth, and skin of lower face	Draws outer part of lower lip downward and backward as in pouting	Facial nerve (VII)
Risorius (*risor* = laughter)	Fascia over parotid (salivary) gland	Skin at angle of mouth	Draws angle of mouth laterally as in tenseness	Facial nerve (VII)
Orbicularis oculi (*ocul* = eye)	Medial wall of orbit	Circular path around orbit	Closes eye	Facial nerve (VII)

Exhibit 10–3. MUSCLES THAT MOVE THE MANDIBLE (see Figure 10–5)

MUSCLE	ORIGIN	INSERTION	ACTION	INNERVATION
Masseter (*maseter* = chewer)	Maxilla and zygomatic arch	Angle and ramus of mandible	Elevates mandible as in closing the mouth and protracts (protrudes) mandible	Mandibular branch of trigeminal nerve (V)
Temporalis (*tempora* = temples)	Temporal bone	Coronoid process of mandible	Elevates and retracts mandible	Temporal nerve from mandibular division of trigeminal nerve (V)
Medial pterygoid (*medial* = closer to midline; *pterygoid* = like a wing; pterygoid plate of sphenoid)	Medial surface of lateral pterygoid plate of sphenoid; maxilla	Angle and ramus of mandible	Elevates and protracts mandible and moves mandible from side to side	Mandibular branch of trigeminal nerve (V)
Lateral pterygoid (*lateral* = farther from midline)	Great wing and lateral surface of lateral pterygoid plate of sphenoid	Condyle of mandible; temporomandibular articulation	Protracts mandible, opens mouth, and moves mandible from side to side	Mandibular branch of trigeminal nerve (V)

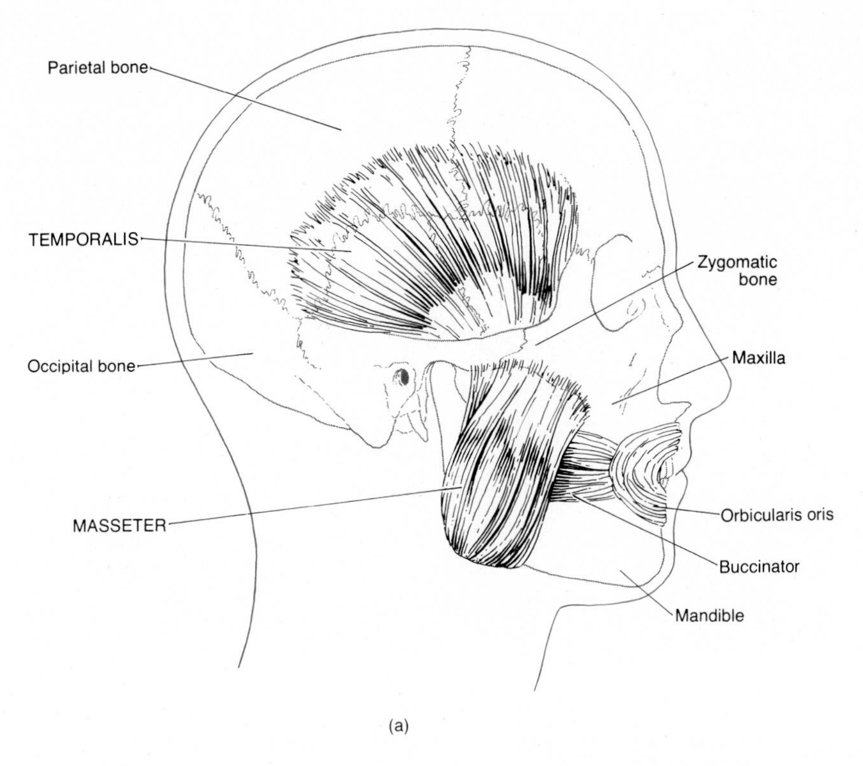

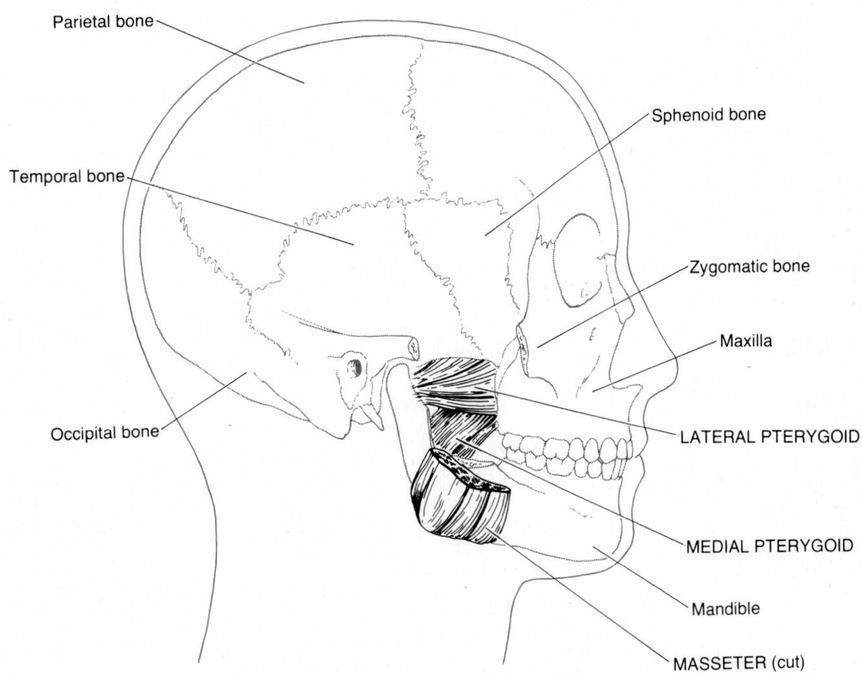

Figure 10-5. Muscles that move the mandible. (a) Superficial right lateral view. (b) Deep right lateral view.

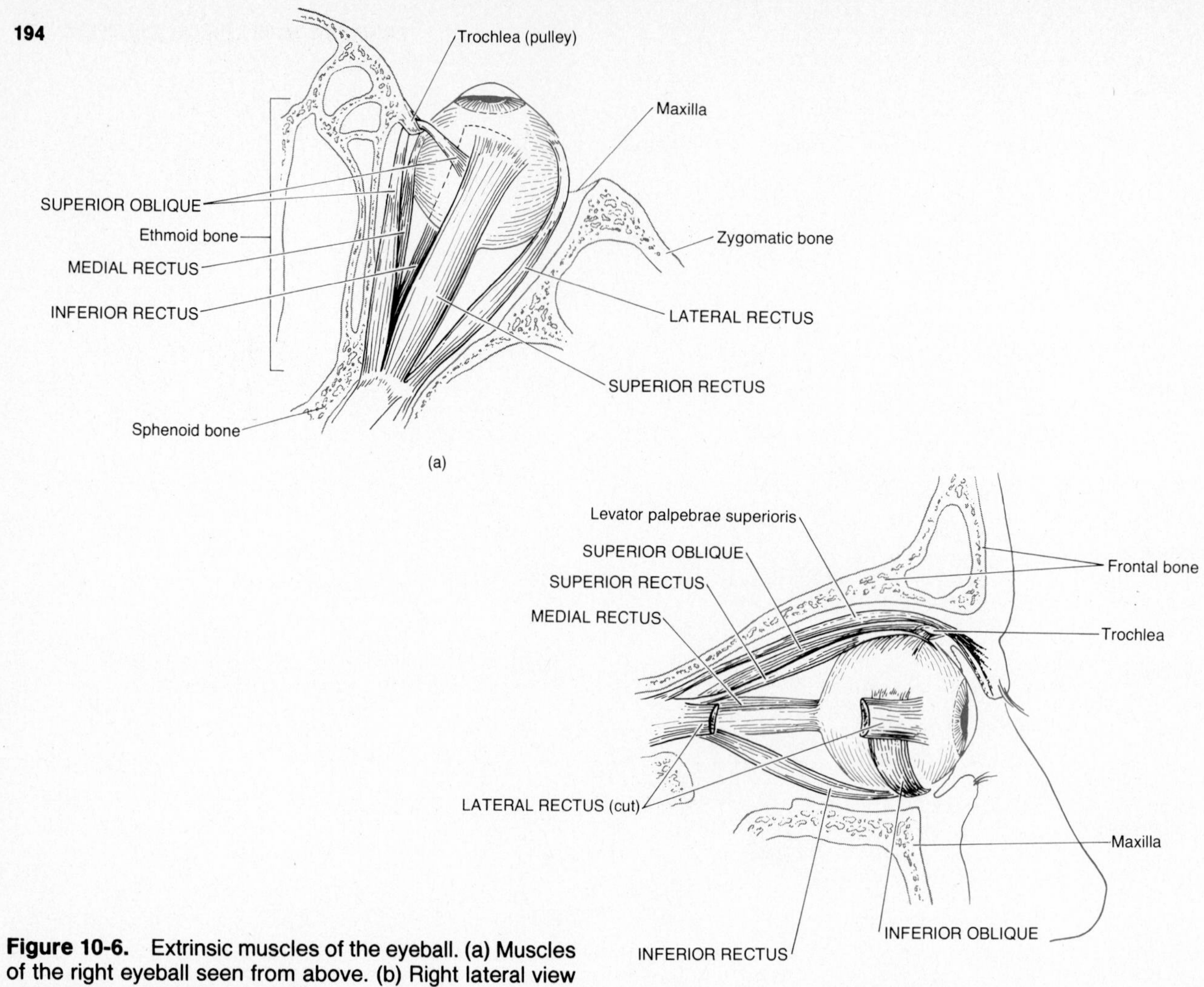

Figure 10-6. Extrinsic muscles of the eyeball. (a) Muscles of the right eyeball seen from above. (b) Right lateral view of muscles of the right eyeball.

Exhibit 10–4. MUSCLES THAT MOVE THE EYEBALL—THE EXTRINSIC MUSCLES* (see Figure 10–6)

MUSCLE	ORIGIN	INSERTION	ACTION	INNERVATION
Superior rectus (*superior* = above; *rectus* = in this case, muscle fibers running parallel to long axis of eyeball)	Orbital cavity	Superior and central part of eyeball	Rolls eyeball upward	Oculomotor nerve (III)
Inferior rectus (*inferior* = below)	Orbital cavity	Inferior and central part of eyeball	Rolls eyeball downward	Oculomotor nerve (III)
Lateral rectus	Orbital cavity	Lateral side of eyeball	Rolls eyeball laterally	Abducens nerve (VI)
Medial rectus	Orbital cavity	Medial side of eyeball	Rolls eyeball medially	Oculomotor nerve (III)
Superior oblique (*oblique* = in this case, muscle fibers running diagonally to long axis of eyeball)	Orbital cavity	Eyeball between superior and lateral recti	Rotates eyeball on its axis; directs cornea downward and laterally; note that it moves through a ring of fibrocartilagenous tissue called the trochlea (*trochlea* = pulley)	Trochlear nerve (IV)
Inferior oblique	Maxilla (front of orbital cavity)	Eyeball between superior and lateral recti	Rotates eyeball on its axis; directs cornea upward and laterally	Oculomotor nerve (III)

* Muscles situated on the outside of the eyeball.

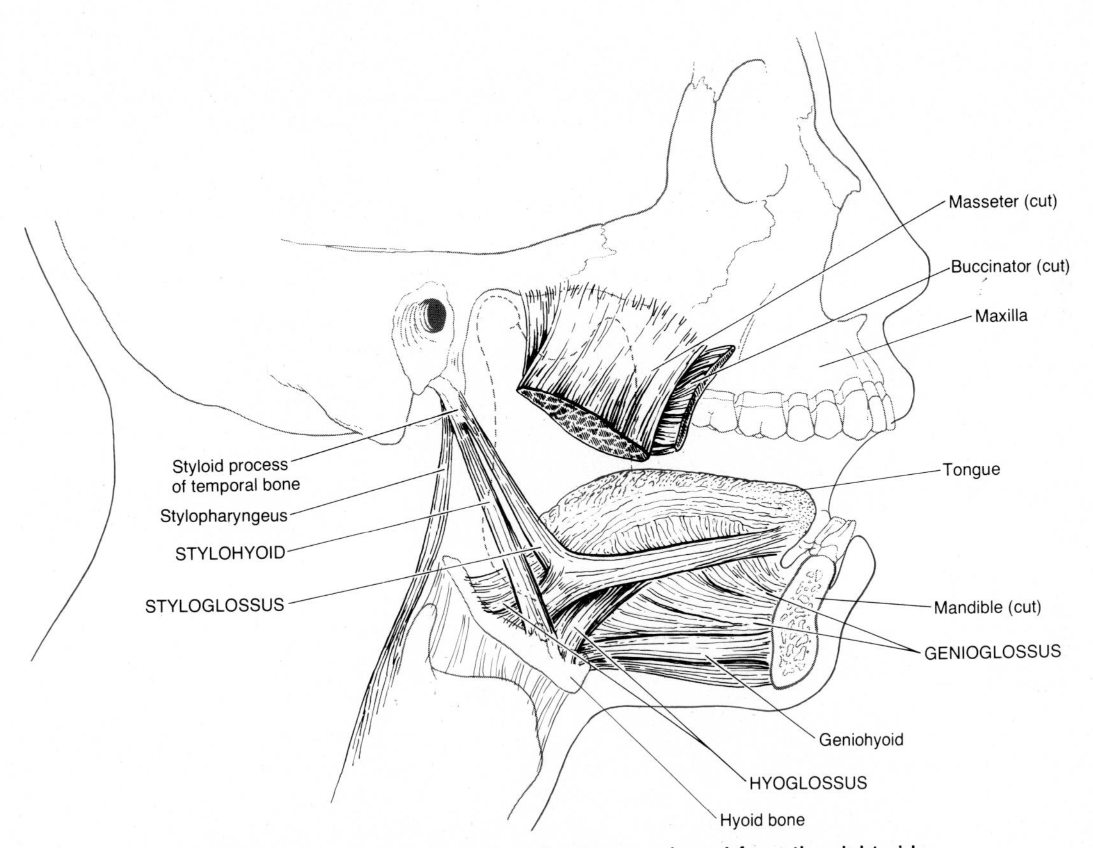

Figure 10-7. Muscles that move the tongue viewed from the right side.

Exhibit 10–5. MUSCLES THAT MOVE THE TONGUE (see Figure 10–7)

MUSCLE	ORIGIN	INSERTION	ACTION	INNERVATION
Genioglossus (*geneion* = chin; *glossus* = tongue)	Mandible	Undersurface of tongue and hyoid bone	Depresses and thrusts tongue forward (protraction)	Hypoglossal nerve (XII)
Styloglossus (*stylo* = stake or pole; styloid process of temporal)	Styloid process of temporal bone	Side and under surface of tongue	Elevates tongue and draws it backward (retraction)	Hypoglossal nerve (XII)
Stylohyoid (*hyoeides* = U-shaped; pertaining to hyoid bone)	Styloid process of temporal bone	Body of hyoid bone	Elevates and retracts tongue	Facial nerve (VII)
Hyoglossus	Body of hyoid bone	Side of tongue	Depresses tongue and draws down its sides	Hypoglossal nerve (XII)

Exhibit 10–6. MUSCLES THAT MOVE THE HEAD

MUSCLE	ORIGIN	INSERTION	ACTION	INNERVATION
Sternocleidomastoid (*sternum* = breast-bone; *cleido* = clavicle; *mastoid* = mastoid process of temporal) (see Exhibit 10–9)	Sternum and clavicle	Mastoid process of temporal bone	Contraction of both muscles flexes the head on the chest; contraction of one muscle rotates face toward side opposite contracting muscle	Accessory nerve (XI); C2–C3
Semispinalis capitis (*semi* = half; *spine* = spinous process; *caput* = head) (see Exhibit 10–13)	Articular process of seventh cervical vertebra and transverse processes of first six thoracic vertebrae	Occipital bone	Both muscles extend head; contraction of one muscle rotates face toward same side as contracting muscle	Dorsal rami of spinal nerves
Splenius capitis (*splenion* = bandage) (see Exhibit 10–13)	Ligamentum nuchae and spines of seventh cervical vertebra and first four thoracic vertebrae	Occipital bone and mastoid process of temporal bone	Both muscles extend head; contraction of one rotates it to same side as contracting muscle	Dorsal rami of middle and lower cervical nerves
Longissimus capitis (*longissimus* = longest) (see Exhibit 10–13)	Transverse processes of last four cervical vertebrae	Mastoid process of temporal bone	Extends head and rotates face toward side opposite contracting muscle	Dorsal rami of middle and lower cervical nerves

Exhibit 10–7. MUSCLES THAT ACT ON THE ANTERIOR ABDOMINAL WALL (see Figure 10–8)

MUSCLE	ORIGIN	INSERTION	ACTION	INNERVATION
Rectus abdominis (*abdomino* = belly)	Pubic crest and symphysis pubis	Cartilage of fifth to seventh ribs and xiphoid process	Flexes vertebral column	Branches of 7–12 intercostal nerves
External oblique (*external* = closer to the surface)	Lower eight ribs	Iliac crest; linea alba (midline aponeurosis)	Both muscles compress abdomen; one side alone bends vertebral column laterally	Branches of 8–12 intercostal nerves, iliohypogastric and ilioinguinal nerves
Internal oblique (*internal* = farther from the surface)	Iliac crest, inguinal ligament, and thoracolumbar fascia	Cartilage of last three or four ribs	Compresses abdomen; one side alone bends vertebral column laterally	Branches of 8–12 intercostal nerves, iliohypogastric, and ilioinguinal nerves
Transversus abdominis (*transverse* = muscle fibers run transversely to midline)	Iliac crest, inguinal ligament, lumbar fascia, and cartilages of last six ribs	Xiphoid process, linea alba, and pubis	Compresses abdomen	Branches of 7–12 intercostal nerves, iliohypogastric, and ilioinguinal nerves

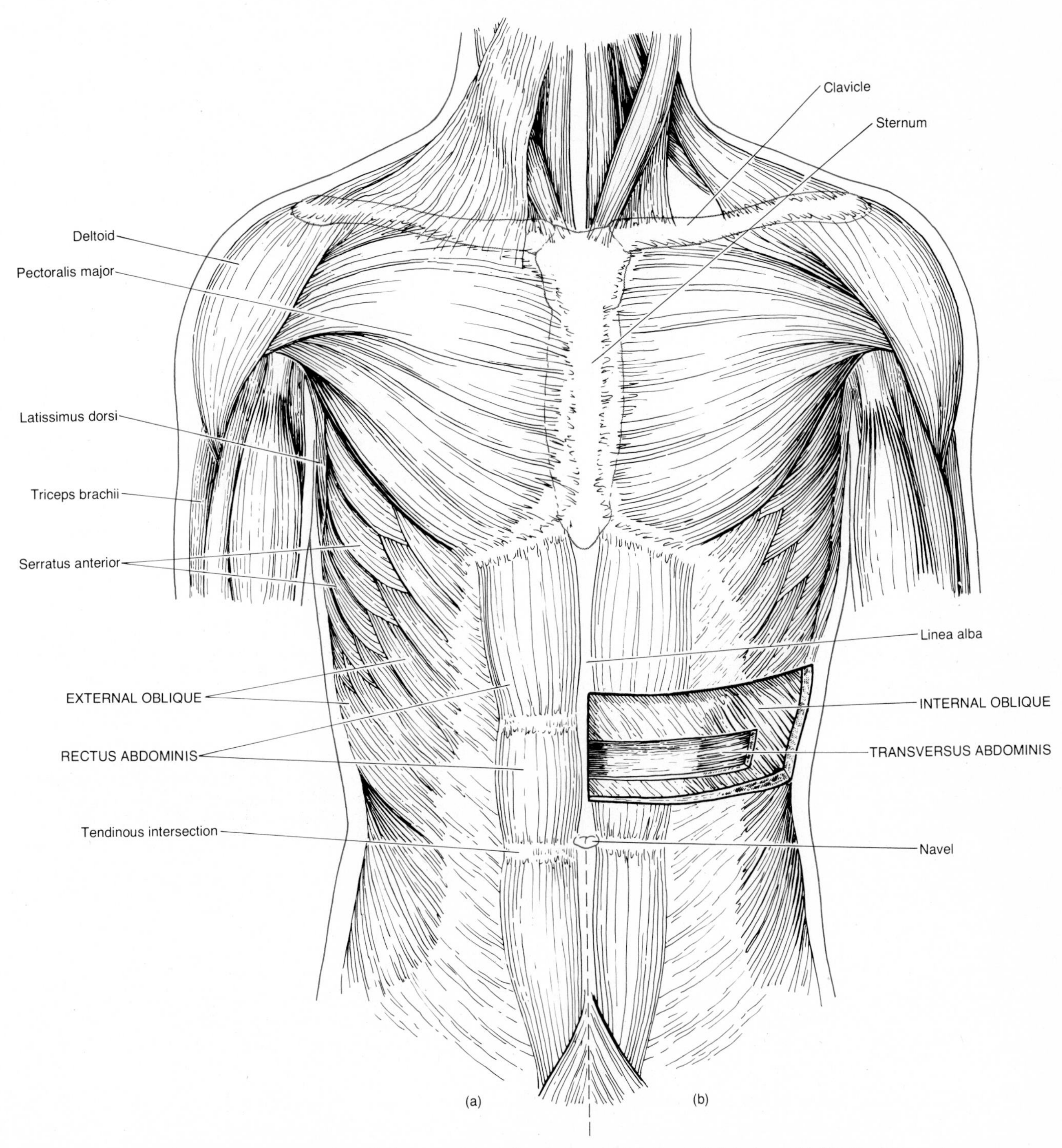

Figure 10-8. Muscles of the anterior abdominal wall. (a) Superficial view. (b) Deep view.

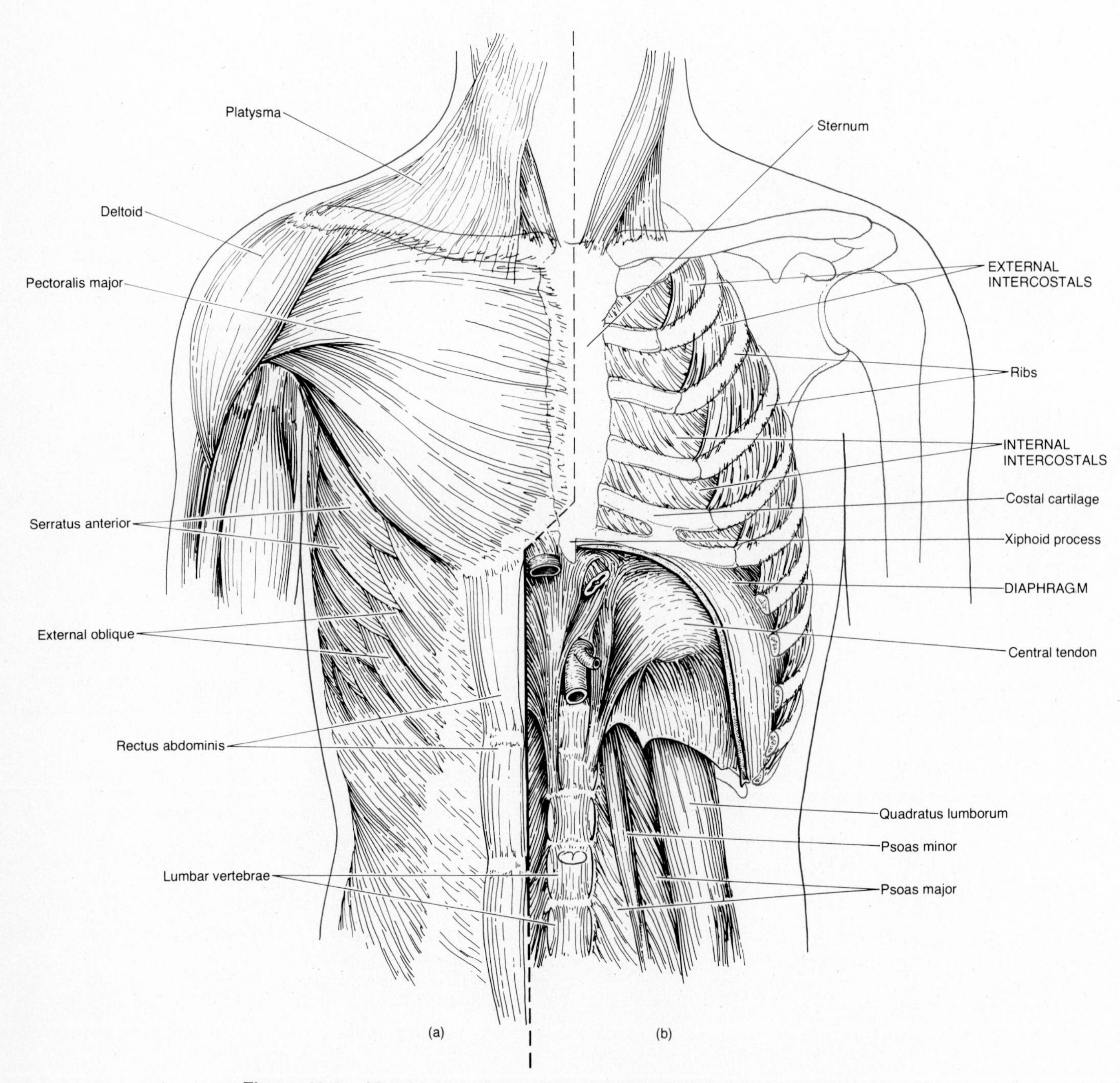

Figure 10-9. Muscles used in breathing. (a) Superficial view. (b) Deep view.

Exhibit 10–8. MUSCLES USED IN BREATHING (see Figure 10–9)

MUSCLE	ORIGIN	INSERTION	ACTION	INNERVATION
Diaphragm (*dia* = across, between; *phragm* = fence)	Xiphoid process, costal cartilages of last six ribs, and lumbar vertebrae	Central tendon	Forms floor of thoracic cavity; contraction pulls central tendon downward and increases vertical length of thorax during inspiration	Phrenic nerve
External intercostals (*inter* = between; *costa* = rib)	Inferior border of rib above	Superior border of rib below	Elevate ribs during inspiration and thus increase lateral and anteroposterior dimensions of the thorax	Intercostal nerves
Internal intercostals	Superior border of rib below	Inferior border of rib above	Draw adjacent ribs together during expiration and thus decrease the lateral and anteroposterior dimensions of the thorax	Intercostal nerves

Exhibit 10–9. MUSCLES THAT MOVE THE SHOULDER GIRDLE (see Figure 10–10)

MUSCLE	ORIGIN	INSERTION	ACTION	INNERVATION
Subclavius (*sub* = under; *clavius* = clavicle)	First rib	Clavicle	Depresses clavicle	Brachial plexus (C5–C6)
Pectoralis minor (*pectus* = breast, chest, thorax; *minor* = lesser)	Third through fifth ribs	Coracoid process of scapula	Depresses scapula, rotates shoulder joint upward, and raises third through fifth ribs during forced inspiration when scapula is fixed	Medial pectoral nerve
Serratus anterior (*serratus* = serrated; *anterior* = front)	Upper eight or nine ribs	Vertebral border and inferior angle of scapula	Moves scapula forward and rotates scapula upward	Long thoracic nerve
Trapezius (*trapezoides* = trapezoid-shaped)	Occipital bone and spines of seventh cervical and all thoracic vertebrae	Acromion process of clavicle and spine of scapula	Elevates clavicle, adducts scapula, and elevates or depresses scapula	Acessory nerve (XI); C3–C4
Levator scapulae (*levator* = raises; *scapulae* = scapula)	Upper four or five cervical vertebrae	Vertebral border of scapula	Elevates scapula	Dorsal scapular nerve (C3–C5)
Rhomboideus major (*rhomboides* = rhomboid-shaped or diamond-shaped)	Spines of second to fifth thoracic vertebrae	Vertebral border of scapula	Moves scapula backward and upward and slightly rotates it downward	Dorsal scapular nerve
Rhomboideus minor	Spines of seventh cervical and first thoracic vertebrae	Superior angle of scapula	Adducts scapula	Dorsal scapular nerve

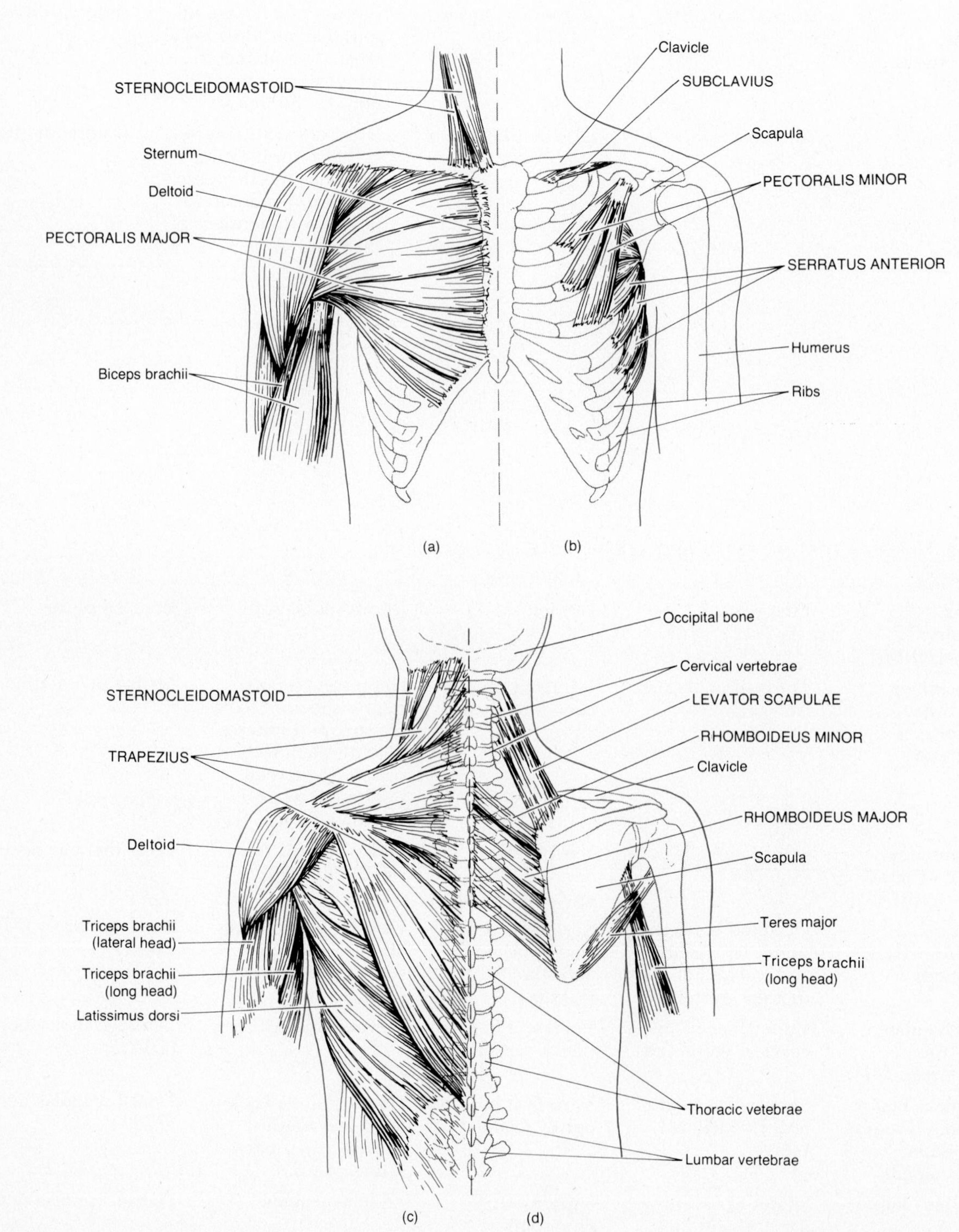

Figure 10-10. Muscles that move the shoulder girdle. (a) Anterior superficial view. (b) Anterior deep view. (c) Posterior superficial view. (d) Posterior deep view.

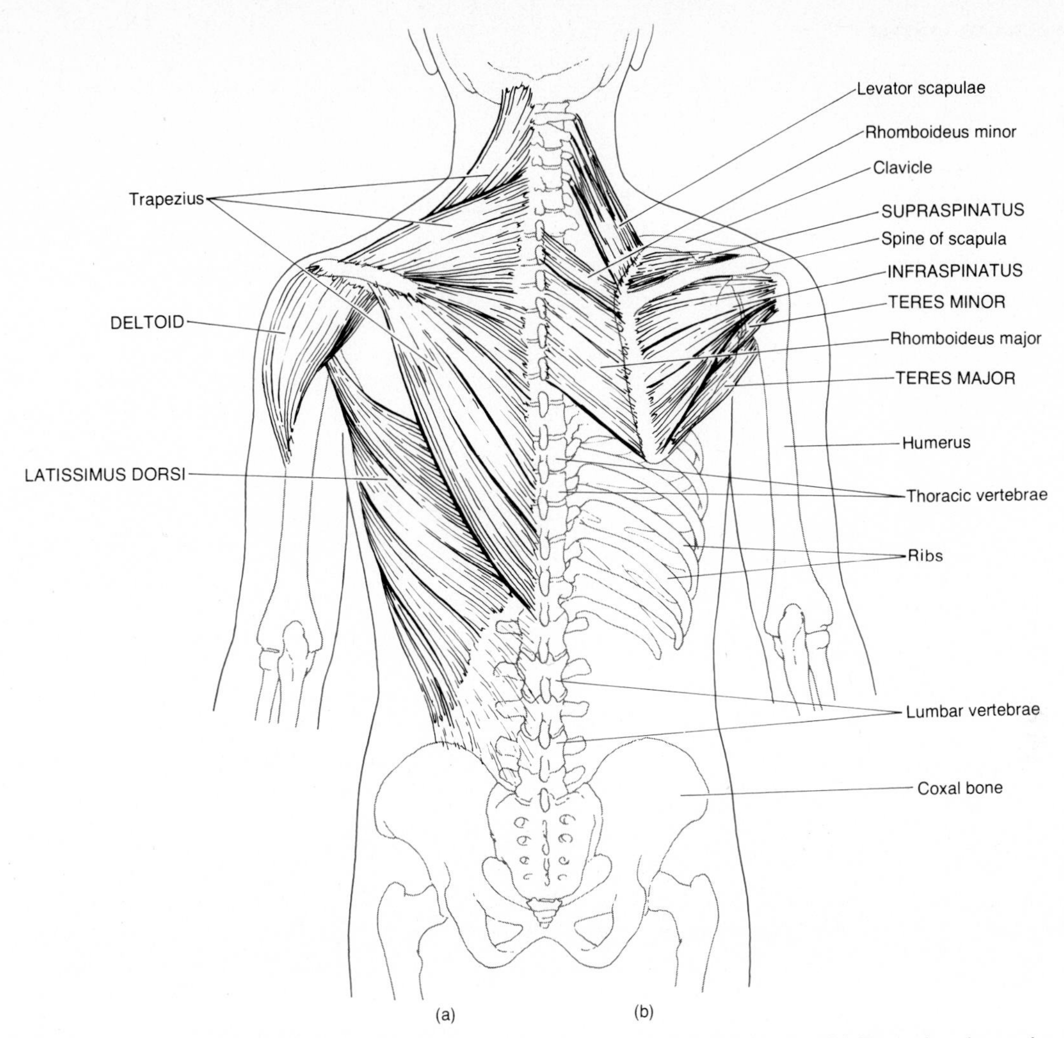

Figure 10-11. Muscles that move the humerus. (a) Posterior superficial view. (b) Posterior deep view.

Exhibit 10–10. MUSCLES THAT MOVE THE HUMERUS (see Figure 10–11)

MUSCLE	ORIGIN	INSERTION	ACTION	INNERVATION
Pectoralis major (see Figure 10–10)	Clavicle, sternum, cartilage of second to sixth ribs	Greater tubercle of humerus	Flexes, adducts, and rotates arm medially	Medial and lateral pectoral nerve
Latissimus dorsi (*dorsum* = back)	Spines of lower six thoracic vertebrae, lumbar vertebrae, crests of sacrum and ilium, lower four ribs	Intertubercular groove of humerus	Extends, adducts, and rotates arm medially; draws shoulder downward and backward	Thoracodorsal nerve
Deltoid (*delta* = triangular-shaped)	Clavicle and acromion process and spine of scapula	Lateral surface of body of humerus	Abducts arm	Axillary nerve
Supraspinatus (*supra* = above; *spinatus* = spine of scapula)	Fossa superior to spine of scapula	Greater tubercle of humerus	Assists deltoid muscle in adducting arm	Suprascapular nerve
Infraspinatus (*infra* = below)	Fossa inferior to spine of scapula	Greater tubercle of humerus	Rotates humerus laterally	Suprascapular nerve
Teres major (*teres* = long and round)	Inferior angle of scapula	Distal to lesser tubercle of humerus	Extends humerus and draws it down; assists in adduction and medial rotation of arm	Lower subscapular nerve
Teres minor	Axillary border of scapula	Greater tubercle of humerus	Rotates humerus laterally	Axillary nerve

Before you move on to Exhibit 10–11 (Muscles That Move the Forearm), refer to Figure 1–3b. That figure is a cross section of the trunk through the heart and lungs. Several of the muscles that you have studied up to this point are shown in the cross-sectional view. Figure 1–3b will add to your understanding of how muscles are arranged with respect to each other from external to internal. It will also show you how muscles are oriented with regard to bones and viscera.

Exhibit 10–11. MUSCLES THAT MOVE THE FOREARM (see Figure 10–12)

MUSCLE	ORIGIN	INSERTION	ACTION	INNERVATION
Biceps brachii (*biceps* = two heads of origin; *brachion* = arm)	Long head originates from tuberosity above glenoid cavity, and short head originates from coracoid process of scapula	Radial tuberosity	Flexes and supinates forearm	Musculocutaneous nerve
Brachialis	Anterior surface of humerus	Tuberosity and coronoid process of ulna	Flexes forearm	Musculocutaneous, radial, and median nerves
Brachioradialis (*radialis* = radius) (see Figure 10–13 also)	Supracondyloid ridge of humerus	Superior to styloid process of radius	Flexes forearm	Radial nerve
Triceps brachii (*triceps* = three heads of origin)	Long head originates from infraglenoid tuberosity of scapula, lateral head originates from lateral and posterior surface of humerus superior to radial groove, medial head originates from posterior surface of humerus inferior to radial groove	Olecranon of ulna	Extends forearm	Radial nerve
Supinator (*supination* = turning palm upward or anteriorly)	Lateral epicondyle of humerus, ridge on ulna	Oblique line of radius	Supinates forearm	Deep radial nerve
Pronator teres (*pronation* = turning palm downward or posteriorly)	Medial epicondyle of humerus, coronoid process of ulna	Midlateral surface of radius	Pronates forearm	Median nerve

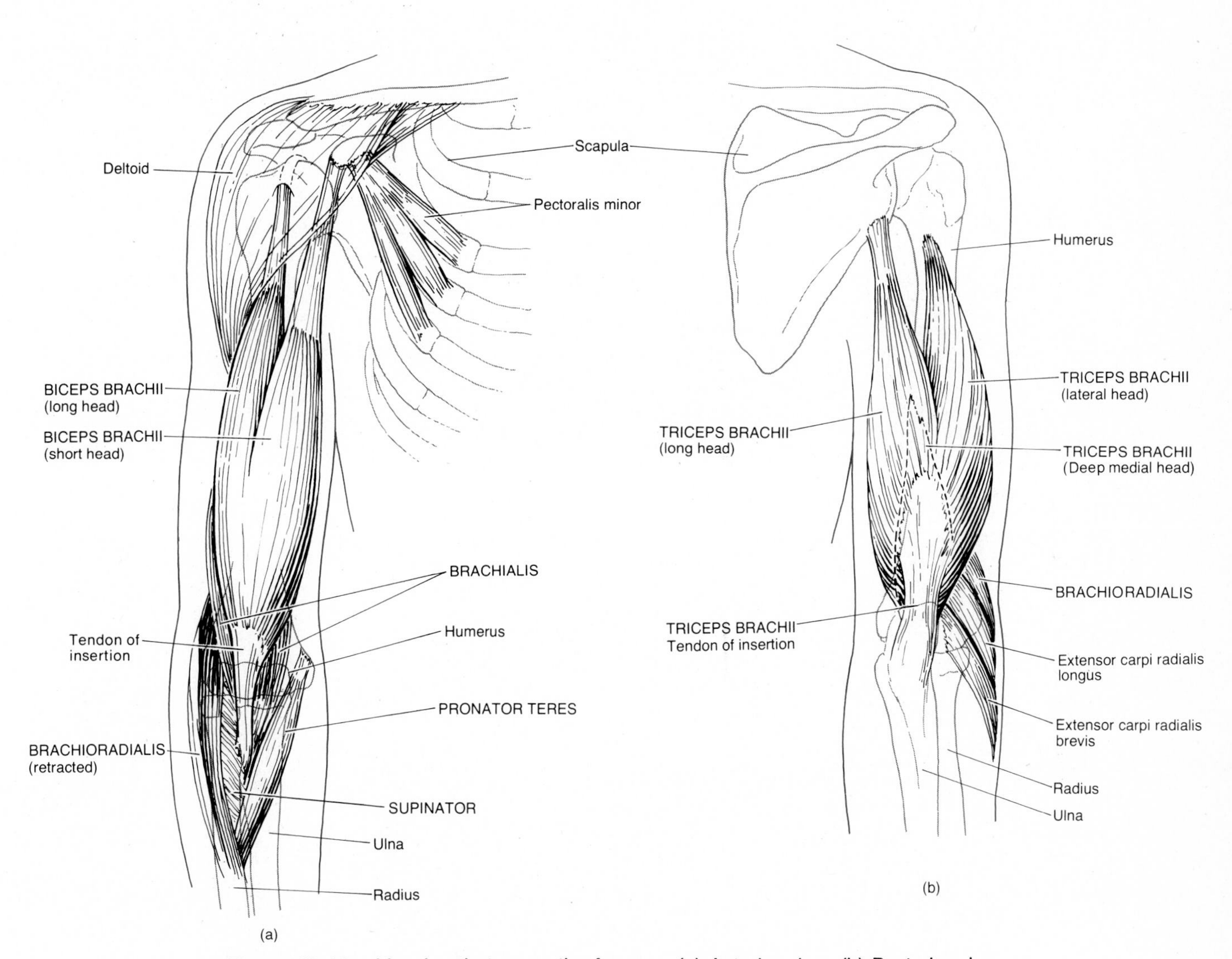

Figure 10-12. Muscles that move the forearm. (a) Anterior view. (b) Posterior view.

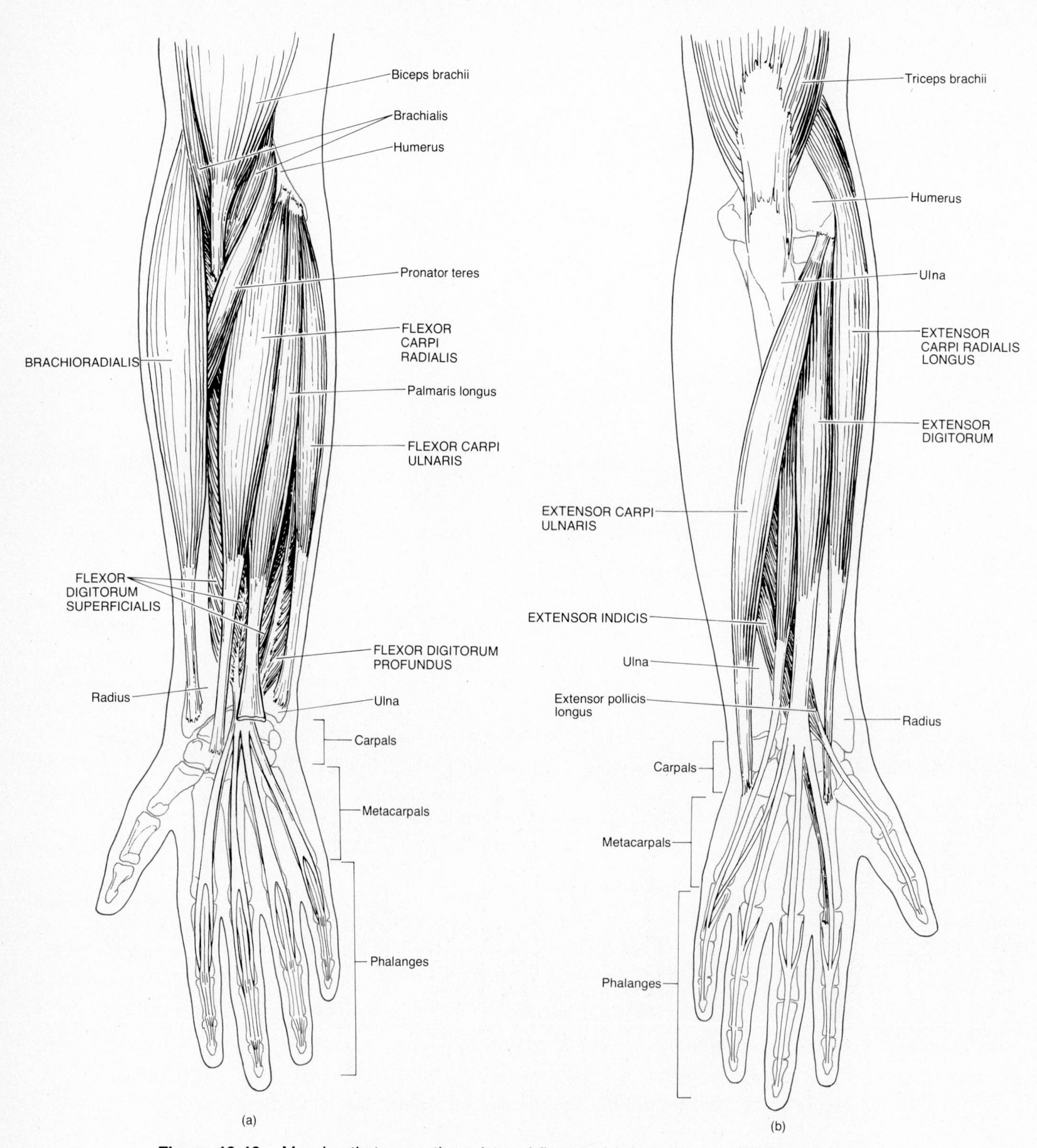

Figure 10-13. Muscles that move the wrist and fingers. (a) Anterior view. (b) Posterior view.

Exhibit 10–12. MUSCLES THAT MOVE THE WRIST AND FINGERS (see Figure 10–13)

MUSCLE	ORIGIN	INSERTION	ACTION	INNERVATION
Flexor carpi radialis (*flexor* = decreases angle; *carpus* = wrist)	Medial epicondyle of humerus	Second and third metacarpals	Flexes and abducts wrist	Median nerve
Flexor carpi ulnaris (*ulnaris* = ulna)	Medial epicondyle of humerus and upper dorsal border of ulna	Pisiform, hamate, and fifth metacarpal	Flexes and adducts wrist	Ulnar nerve
Extensor carpi radialis longus (*extensor* = increases angle at a joint; *longus* = long)	Lateral epicondyle of humerus	Second metacarpal	Extends and abducts wrist	Radial nerve
Extensor carpi ulnaris	Lateral epicondyle of humerus and dorsal border of ulna	Fifth metacarpal	Extends and adducts wrist	Deep radial nerve
Flexor digitorum profundus (*digit* = finger or toe; *profundus* = deep)	Anterior medial surface of body of ulna	Bases of distal phalanges	Flexes distal phalanges of each finger	Median and ulnar nerves
Flexor digitorum superficialis (*superficialis* = superficial)	Medial epicondyle of humerus, coronoid process of ulna, oblique line of radius	Middle phalanges	Flexes middle phalanges of each finger	Median nerve
Extensor digitorum	Lateral epicondyle of humerus	Middle and distal phalanges of each finger	Extends phalanges	Deep radial nerve
Extensor indicis (*indicis* = index)	Dorsal surface of ulna	Tendon of extensor digitorum into index finger	Extends index finger	Deep radial nerve

Exhibit 10–13. MUSCLES THAT MOVE THE VERTEBRAL COLUMN (see Figure 10–14)

MUSCLE	ORIGIN	INSERTION	ACTION	INNERVATION
Rectus abdominis (see Figure 10–8)	Body of pubis of coxal bone	Cartilage of fifth through seventh ribs	Flexes vertebral column at the lumbar spine as in doing a sit-up	7–12 intercostal nerves
Quadratus lumborum (*quad* = four; *lumb* = lumbar region)	Iliac crest	Twelfth rib and upper four lumbar vertebrae	Flexes vertebral column laterally	T12–L1
Sacrospinalis (erector spinae)	This muscle consists of three posterior groupings: iliocostalis, longissimus, and spinalis. These groups, in turn, consist of a series of overlapping muscles. The iliocostalis group is laterally placed. The longissimus group is intermediate in placement. And the spinalis is medially placed.			
LATERAL				
Iliocostalis lumborum (*ilium* = flank; *lumbus* = loin)	Iliac crest	Lower six ribs	Extends lumbar region of vertebral column	Dorsal rami of lumbar nerves
Iliocostalis thoracis (*thorax* = chest)	Lower six ribs	Upper six ribs	Maintains erect position of spine	Dorsal rami of thoracic nerves
Iliocostalis cervicis (*cervix* = neck)	First six ribs	Transverse processes of fourth to sixth cervical vertebrae	Extends cervical region of vertebral column	Dorsal rami of cervical nerves
INTERMEDIATE				
Longissimus thoracis	Transverse processes of lumbar	Transverse processes of all thoracic and upper lumbar vertebrae, and the ninth and tenth ribs	Extends thoracic region of vertebral column	Dorsal rami of spinal nerves
Longissimus cervicis	Transverse processes of fourth and fifth thoracic vertebrae	Transverse processes of second to sixth cervical vertebrae	Extends cervical region of vertebral column	Dorsal rami of spinal nerves
Longissimus capitis	Transverse processes of upper four thoracic vertebrae	Mastoid process of temporal bone	Extends head and rotates it to opposite side	Dorsal rami of middle and lower cervical nerves
MEDIAL				
Spinalis thoracis	Spines of upper lumbar and lower thoracic vertebrae	Spines of upper thoracic vertebrae	Extends vertebral column	Dorsal rami of spinal nerves

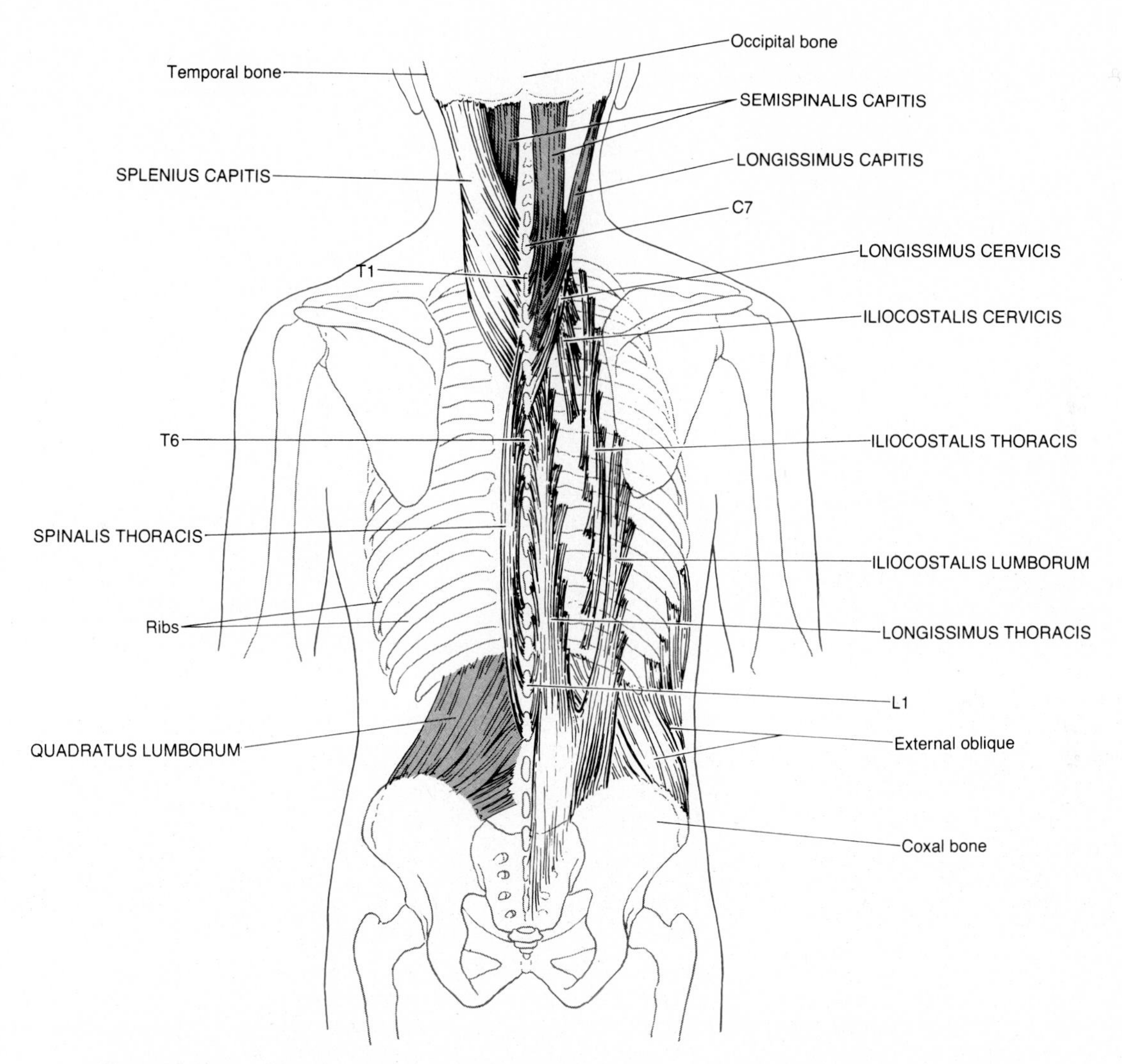

Figure 10-14. Muscles that move the vertebral column. The deep muscles are shaded.

Exhibit 10–14. MUSCLES THAT MOVE THE FEMUR (see Figure 10–15)

MUSCLE	ORIGIN	INSERTION	ACTION	INNERVATION
Psoas major (*psoa* = muscle of loin)	Transverse processes and bodies of lumbar vertebrae	Lesser trochanter of femur	Flexes and rotates femur laterally; flexes vertebral column	L2–L3
Iliacus (*iliac* = ilium)	Iliac fossa	Tendon of psoas major	Flexes and rotates femur laterally; slight flexion of vertebral column	Femoral nerve
Gluteus maximus (*gloutos* = buttock; *maximus* = largest)	Iliac crest, sacrum, coccyx, and aponeurosis of sacrospinalis	Iliotibial tract of fascia lata and gluteal tuberosity of femur	Extends and rotates femur laterally	Inferior gluteal nerve
Gluteus medius (*media* = middle)	Ilium	Greater trochanter of femur	Abducts and rotates femur laterally	Superior gluteal nerve
Gluteus minimus (*minimus* = small)	Ilium	Greater trochanter of femur	Abducts and rotates femur medially	Superior gluteal nerve
Tensor fasciae latae (*tensor* = makes tense; *fascia* = band; *latus* = broad, wide	Iliac crest	Iliotibial tract of fascia lata	Flexes and abducts femur	Superior gluteal nerve
Adductor longus (*adductor* = moves a part closer to the midline)	Crest and symphysis	Linea aspera of femur	Adducts, rotates, and flexes femur	Obturator nerve
Adductor brevis (*brevis* = short)	Inferior ramus of pubis	Linea aspera of femur	Adducts, rotates, and flexes femur	Obturator nerve
Adductor magnus (*magnus* = large)	Inferior ramus of pubis; ischium to ischial tuberosity	Linea aspera of femur	Adducts, flexes, and extends femur (anterior part flexes, posterior part extends)	Obturator nerve
Piriformis (*pirum* = pear; *forma* = shape)	Sacrum	Greater trochanter of femur	Rotates femur laterally and abducts it	S2 or S1–S2
Obturator internus (*obturator* = closed because it arises over the obturator foramen, which is closed by a heavy membrane; *internal* = inside)	Margin of obturator foramen, pubis, and ischium	Greater trochanter of femur	Rotates thigh laterally and abducts thigh	Obturator nerve
Pectineus (*pecten* = comb-shaped)	Fascia of pubis	Pectineal line of femur	Flexes, adducts, and rotates thigh laterally	Femoral nerve

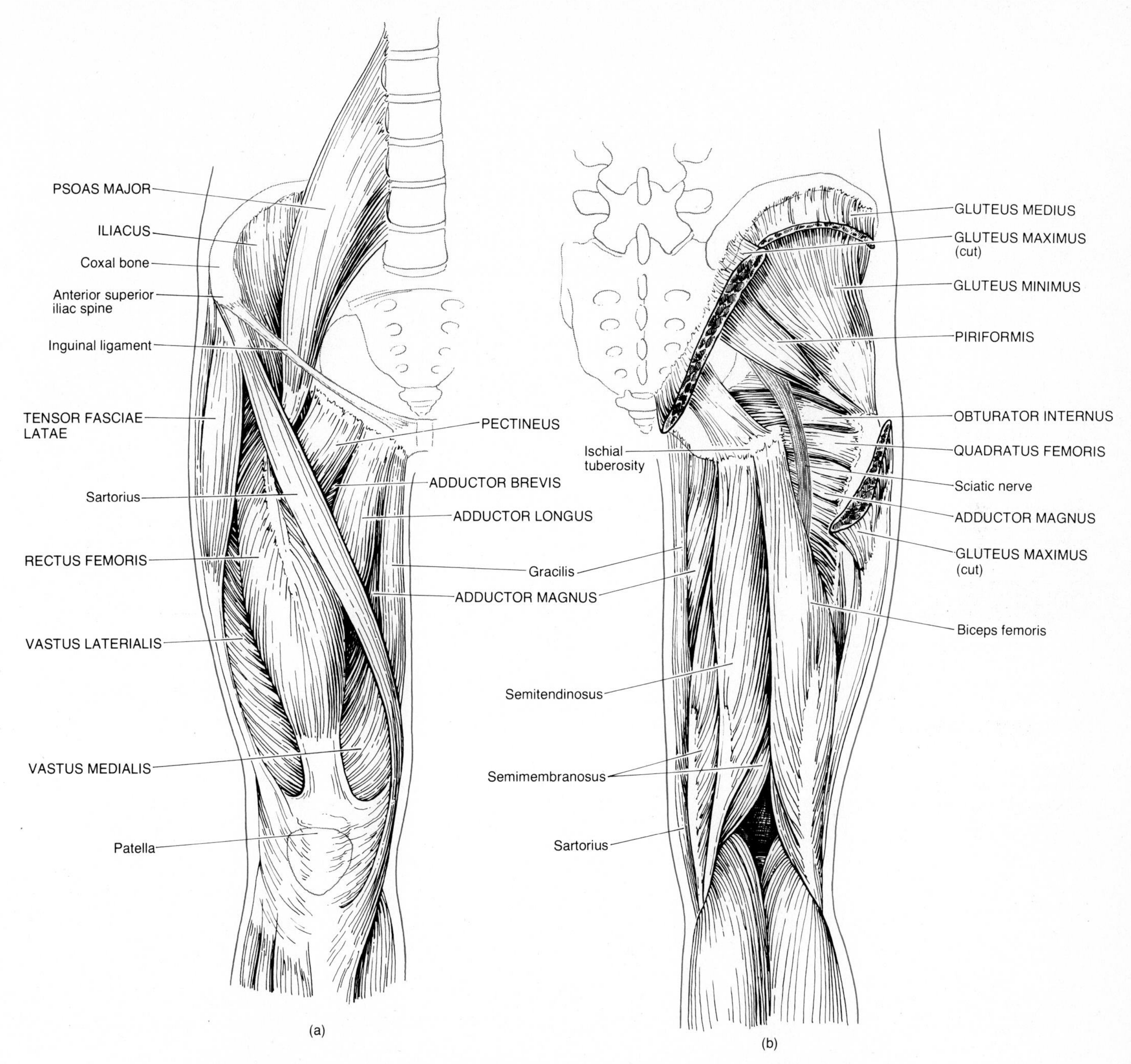

Figure 10-15. Muscles that move the femur. (a) Anterior view. (b) Posterior view.

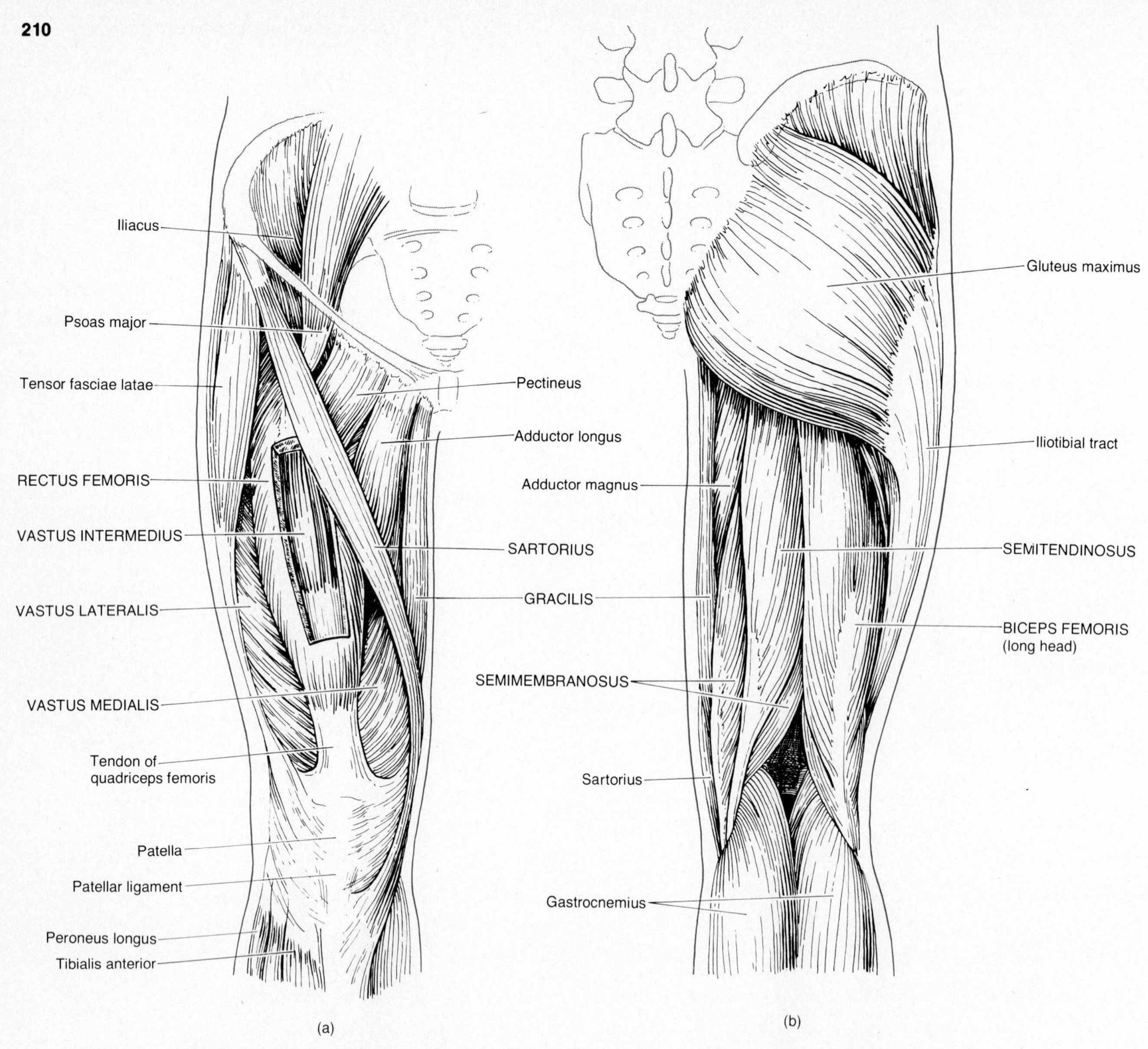

Figure 10-16. Muscles that act on the knee joint. (a) Anterior view. (b) Posterior view.

Exhibit 10–15. MUSCLES THAT ACT ON THE KNEE JOINT (see Figure 10–16)

MUSCLE	ORIGIN	INSERTION	ACTION	INNERVATION
Quadriceps femoris	A composite muscle that includes four distinct parts, usually described as four separate muscles. The common tendon from the patella to the tibial tuberosity is known as the patellar ligament.			
Rectus femoris (*rectus* = fibers parallel to midline; *femoris* = femur)	Anterior inferior spine of ilium	Upper border of patella		Femoral nerve
Vastus lateralis (*vastus* = vast, large; *lateralis* = lateral)	Greater trochanter and linea aspera of femur	Upper border and sides of patella; tibial tuberosity through the patellar ligament (tendon of quadriceps)	All four heads extend knee; rectus portion alone also flexes thigh	Femoral nerve
Vastus medialis (*medialis* = medial)	Linea aspera of femur			Femoral nerve
Vastus intermedius (*intermedius* = middle)	Anterior and lateral surfaces of body of femur			Femoral nerve

Exhibit 10–15 (cont.)

MUSCLE	ORIGIN	INSERTION	ACTION	INNERVATION
Hamstrings	A collective designation for three separate muscles.			
Biceps femoris	Long head arises from ischial tuberosity, and short head arises from linea aspera of femur	Head of fibula and lateral condyle of tibia	Flexes knee and extends thigh	Tibial nerve from sciatic nerve
Semitendinosus (*semi* = half; *tendo* = tendon)	Ischial tuberosity	Proximal part of body of tibia	Flexes knee and extends thigh	Tibial nerve from sciatic nerve
Semimembranosus (*membran* = membrane)	Ischial tuberosity	Medial condyle of tibia	Flexes knee and extends thigh	Tibial nerve from sciatic nerve
Gracilis (*gracilis* = slender)	Symphysis pubis and pubic arch	Medial surface of body of tibia	Flexes knee and adducts thigh	Obturator nerve
Sartorius (*sartor* = tailor; refers to cross-legged position in which tailors sit)	Anterior superior spine of ilium	Medial surface of body of tibia	Flexes knee; flexes thigh and rotates it laterally, thus crossing leg	Femoral nerve

Exhibit 10–16. MUSCLES THAT MOVE FOOT AND TOES (see Figure 10–17)

MUSCLE	ORIGIN	INSERTION	ACTION	INNERVATION
Gastrocnemius (*gaster* = belly; *kneme* = leg)	Lateral and medial condyles of femur and capsule of knee	Calcaneus	Plantar flexes foot	Tibial nerve
Soleus (*soleus* = sole of foot)	Head of fibula and medial border of tibia	Calcaneus	Plantar flexes foot	Tibial nerve
Peroneus longus (*perone* = fibula)	Head and body of fibula and lateral condyle of tibia	First metatarsal and first cuneiform bone	Plantar flexes and everts foot	Superficial peroneal nerve
Peroneus brevis	Body of fibula	Fifth metatarsal	Plantar flexes and everts foot	Superficial peroneal nerve
Peroneus tertius (*tertius* = third)	Distal third of fibula	Fifth metatarsal	Dorsally flexes and everts foot	Deep peroneal nerve
Tibialis anterior (*tibialis* = tibia)	Lateral condyle and body of tibia	First metatarsal and first cuneiform	Dorsally flexes and inverts foot	Deep peroneal nerve
Tibialis posterior (*posterior* = back)	Interosseus membrane between tibia and fibula	Second, third, and fourth metatarsals; navicular; third cuneiform, cuboid	Plantar flexes and inverts foot	Tibial nerve
Flexor digitorum longus (*digitorum* = digit, finger, or toe)	Tibia	Distal phalanges of four outer toes	Flexes toes and plantar flexes and inverts foot	Tibial nerve
Extensor digitorum longus	Lateral condyle of tibia and anterior surface of fibula	Middle and distal phalanges of four outer toes	Extends toes and dorsally flexes and everts foot	Deep peroneal nerve

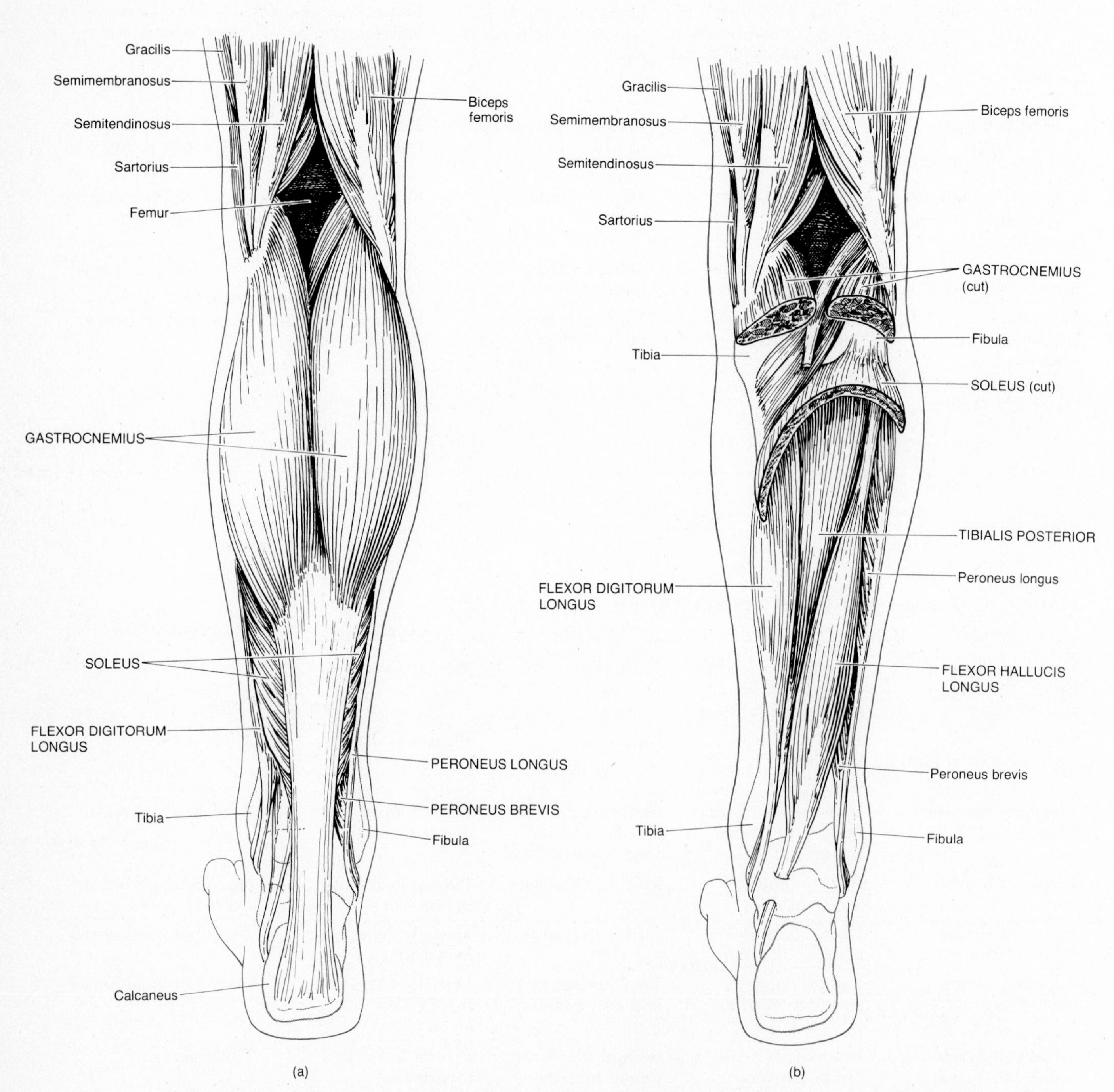

Figure 10-17. Muscles that move the foot and toes. (a) Superficial posterior view. (b) Deep posterior view.

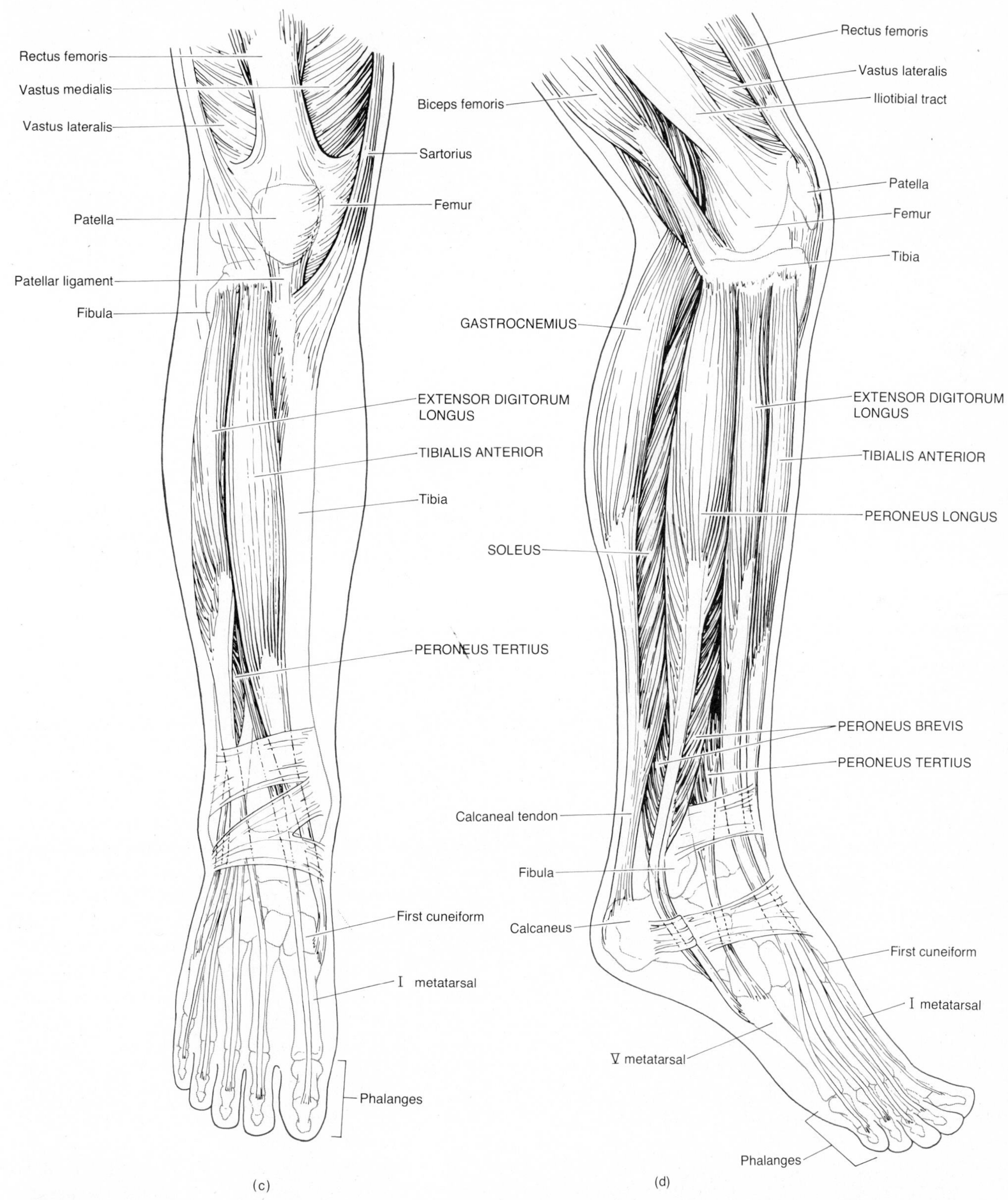

Figure 10-17 (cont.). Muscles that move the foot and toes. (c) Superficial anterior view. (d) Superficial lateral view.

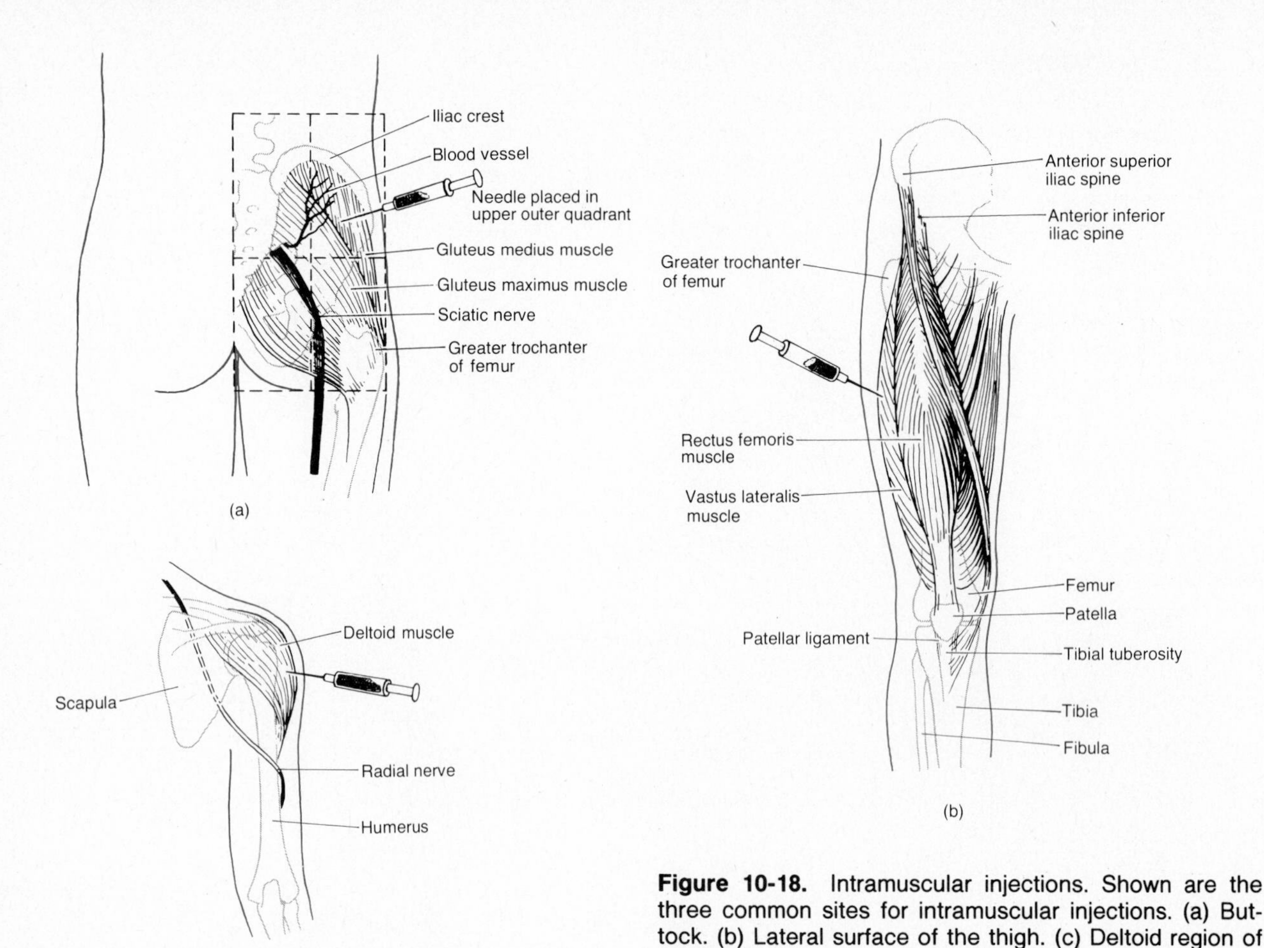

Figure 10-18. Intramuscular injections. Shown are the three common sites for intramuscular injections. (a) Buttock. (b) Lateral surface of the thigh. (c) Deltoid region of the arm.

INTRAMUSCULAR INJECTIONS

The way a particular drug is administered is determined primarily by the physical and chemical properties of the drug, the area of the body in which the drug is supposed to act, and the speed of response required. Generally, drugs are administered for one of two effects: local or systemic. With a *local* effect, the action of the drug is confined to the site of application. With a *systemic* effect, the drug is first absorbed into the blood and then diffuses into body tissues far away from the site of administration. Local effects are achieved by applying a medication to the surface of the tissue being treated. For instance, lotions, ointments, wet dressings, and certain anti-inflammatories may be put on the surface of the skin. Mucous membranes of the eyes, nose, mouth, throat, respiratory tract, reproductive tract, and urinary passageways may be treated locally with sprays, swabs, irrigations, douches, and suppositories.

Drugs designed to produce systemic effects may be administered several ways, depending upon the nature and amount of the drug, the speed of response desired, and the general condition of the patient. Methods of administration include swallowing, placing the drug under the tongue and allowing it to dissolve, inserting the drug rectally, inhaling, and injecting the drug with a needle. All the ways that a drug may be administered with a needle are described by the term **parenteral.** Among the parenteral routes of administration are (1) *intradermal* or *intracutaneous,* in which the needle penetrates the epidermis and is inserted into the dermis; (2) *subcutaneous* or *hypodermic,* in which the needle is inserted into the subcutaneous layer; (3) *intramuscular,* in which the needle is inserted into a muscle; (4) *intravenous,* in which the needle is inserted directly into a vein; and (5) *intraspinal,* in which the needle is inserted into the vertebral canal. At this point, we shall discuss intramuscular injections only. Other parenteral routes will be considered later.

When an intramuscular injection is given, it is made through the skin and subcutaneous tissue into the muscle itself. Intramuscular injections are preferred when prompt absorption is desired, when larger doses than can be given cutaneously are indicated, or if the drug is too irritating to give subcutaneously. The common sites for intramuscular injections include the buttock, lateral side of the thigh, and the deltoid region of the arm. Muscles in these areas, especially the gluteal muscles in the buttock, are fairly thick. Because of the large number of muscle fibers and extensive fascia, the drug has a larger surface area for absorption. Absorption is further promoted by the extensive blood supply to muscles. Ideally, intramuscular injections should be given deep within the muscle and away from major nerves and blood vessels.

For many intramuscular injections, the preferred site is the buttock (Figure 10–18a). In giving such an injection, the buttock should be divided into four quadrants and the upper outer quadrant used as the injection site. The iliac crest serves as a good landmark for this quadrant. The spot for injection should be about 5 to 7½ cm (2 to 3 in.) below the iliac crest. The upper outer quadrant is chosen because, in this area, the muscle is quite thick with few nerves. For this reason, there is not much chance of injuring the sciatic nerve. Injury to the nerve can cause paralysis of the lower extremity. The probability of injecting the drug into a blood vessel is also remote in this area. After the needle is inserted, the plunger should be pulled up for a few seconds. If the syringe fills with blood, this indicates that the needle is in a blood vessel, and a different injection site on the opposite buttock should be chosen.

Injections given in the lateral side of the thigh are inserted into the midportion of the vastus lateralis muscle (Figure 10–18b). This site is determined by using the knee and greater trochanter of the femur as landmarks. The midportion of the muscle is located by measuring a handbreadth above the knee and a handbreadth below the greater trochanter.

The deltoid injection is given in the midportion of the muscle about two to three fingerbreadths below the acromion of the scapula and lateral to the axilla (Figure 10–18c).

When you finished your study of bones, it was suggested that you return to Exhibits 1–3 through 1–7 to relate your knowledge of bones to surface landmarks. Now that you have finished your study of muscles, return again to those same Exhibits in Chapter 1 to relate your knowledge of muscles to surface landmarks. From the Exhibits, determine which muscles are palpable and be sure that you can palpate each on your own body. Once again, you might want to work with a partner to palpate the dorsally located muscles.

Chapter summary in outline

HOW MUSCLES PRODUCE MOVEMENT

1. Skeletal muscles produce movements by pulling on bones.
2. The stationary attachment is the origin; the movable attachment is the insertion.
3. Bones serve as levers and joints as fulcrums.
4. Levers are acted upon by two different forces: resistance and effort. There are first-class, second-class, and third-class levers.
5. The agonist produces the desired action. The antagonist produces an opposite action. The synergist assists the agonist by reducing unnecessary movements.

NAMING MUSCLES

1. Skeletal muscles are named on the basis of one or more distinctive criteria, including direction of fibers, location, size, number of heads, shape, origin, insertion, and action.

INTRAMUSCULAR INJECTIONS

1. Drugs are generally administered for either local or systemic effects.
2. The term *parenteral* is applied to all ways that drugs are administered with a needle.
3. Parenteral routes may be intradermal, subcutaneous, intramuscular, intravenous, or intraspinal.
4. Advantages of intramuscular injections are prompt absorption, use of larger doses than can be given cutaneously, and minimal irritation.
5. Common sites for intramuscular injections are the buttock, lateral side of the thigh, and deltoid region of the arm.

Review questions and problems

1. What is meant by the muscular system? Explain fully.
2. Using the terms origin, insertion, and belly in your discussion, describe how skeletal muscles produce body movements by pulling on bones.
3. What is a lever? Fulcrum? Apply these terms to the body, and indicate the nature of the forces that act on levers.
4. Describe the three classes of levers, and provide one example for each in the body.
5. Define the role of the agonist, antagonist, and synergist in producing body movements.
6. At the beginning of this chapter, several criteria for naming muscles were discussed. These were direction of fibers, location, size, number of heads, shape, origin, insertion, and action. Select at random the names of some muscles presented in Exhibits 10–2 through 10–16 and see if you can determine the criterion or criteria employed for each. In addition, refer to the prefixes, suffixes, roots, and definitions in each Exhibit as a guide. Select as many muscles as you wish, as long as you feel you understand the concept involved.

7. Discuss the muscles and their actions involved in facial expression.
8. What muscles would you use to do the following: *(a)* frown; *(b)* pout; *(c)* show surprise; *(d)* show your upper teeth; *(e)* pucker your lips; *(f)* squint; *(g)* blow up a balloon; *(h)* smile?
9. What are the principal muscles that move the mandible? Give the function of each.
10. What would happen if you lost tone in the masseter and temporalis muscles?
11. What muscles move the eyeball? In which direction does each muscle move the eyeball?
12. Describe the action of each of the muscles acting on the tongue.
13. What tongue, facial, and mandibular muscles would you use when chewing a piece of gum?
14. What muscles are responsible for moving the head, and how do they move the head?
15. Which of the muscles listed above would you use to signify "yes" and "no" by moving your head?
16. What muscles accomplish compression of the anterior abdominal wall?
17. What are the principal muscles involved in breathing? What are their actions?
18. In what directions is the shoulder girdle drawn? What muscles accomplish these movements?
19. What muscles are used to *(a)* raise your shoulders, *(b)* lower your shoulders, *(c)* join your hands behind your back, *(d)* join your hands in front of your chest?
20. What movements are possible at the shoulder joint? What muscles accomplish these movements?
21. What muscles move the forearm? In which directions do these movements occur?
22. What muscles move the forearm and what actions are used when striking a match?
23. Discuss the various movements possible at the wrist and fingers. What muscles accomplish these movements?
24. How many muscles and actions of the wrist and fingers used when writing can you list?
25. Discuss the various muscles and movements of the vertebral column.
26. Can you perform an exercise that would involve the use of each of the muscles listed in Exhibit 10–13?
27. What muscles accomplish movements of the femur? What actions are produced by these muscles?
28. Review in your mind the various movements involved in your favorite kind of dancing. What muscles listed in Exhibit 10–14 would you be using and what actions would you be performing?
29. What muscles act at the knee joint? What kinds of movements do these muscles perform?
30. Determine the muscles and their actions listed in Exhibit 10–15 that you would use in climbing a ladder to a diving board, diving into the water, swimming the length of a pool, and then sitting at pool side.
31. Discuss the muscles that plantar flex, evert, pronate, dorsiflex, and supinate the foot.
32. In which directions are the toes moved? What muscles bring about these movements?
33. Distinguish between local and systemic drug actions. How are drugs applied locally?
34. What routes of administration are used for drugs that produce systemic effects?
35. Define parenteral. What are the various routes of parenteral administration?
36. What are the advantages of intramuscular injections?
37. Describe how you would locate the sites for an intramuscular injection in the buttock, lateral side of the thigh, and deltoid region of the arm.

Selected readings

Cunningham's Textbook of Anatomy. 11th ed., edited by G. J. Romanes. London: Oxford University Press, 1972. Pp. 269–398.

Dawson, Helen L. *Basic Human Anatomy.* 2d ed. New York: Appleton-Century-Crofts, 1974. Pp. 88–131.

Gray, Henry. *Anatomy of the Human Body.* 29th ed., edited by Charles Mayo Goss. Philadelphia: Lea and Febiger, 1973. Pp. 378–527.

CHAPTER 11

THE BLOOD VASCULAR AND LYMPHATIC SYSTEMS

STUDENT OBJECTIVES

After you have read this chapter, you should be able to:

1. Define the principal physical characteristics of blood and its functions in the body
2. Identify the components of blood
3. List the functions of the formed elements in blood
4. Describe the composition and functions of blood plasma
5. Contrast the chemical differences between interstitial fluid and lymph
6. Describe the location of the heart
7. Describe the composition of the pericardial sac and heart wall
8. Identify the blood vessels, chambers, and valves of the heart
9. Describe the initiation and conduction of a nerve impulse through the conduction system of the heart
10. Label and explain the normal deflection waves of an electrocardiogram
11. Describe the sounds of the heart and their clinical significance
12. Describe the surface-projection landmarks of the heart
13. Contrast between the structure and function of arteries, arterioles, capillaries, venules, and veins
14. Identify the principal arteries and veins of systemic circulation
15. Describe the route of blood in coronary circulation
16. Describe the importance and route of blood involved in hepatic portal circulation
17. Identify the major blood vessels of pulmonary circulation
18. Contrast fetal and adult circulation
19. Identify the components and functions of the lymph vascular system
20. Compare the structure of veins and lymphatics
21. Describe the structure and function of lymph nodes
22. Contrast the functions of the tonsils, spleen, and thymus gland as lymphatic organs
23. Describe the path taken by circulating lymph
24. Identify several common blood disorders, blood vessel disorders, and heart disorders
25. Define key terms related to blood, the blood vascular system, and the lymphatic system

The cells of the human body have developed an ability to perform highly specialized functions. This specialization is usually accompanied by the loss of other, frequently vital functions. Thus, the more specialized a cell becomes, the less capable it is of carrying on an independent existence. For instance, a specialized cell is less capable of protecting itself from extreme temperatures, toxic chemicals, and changes in pH, that is, the degree of acidity or alkalinity. It often cannot go looking for food or devour whole bits of food. And, if it is firmly implanted in a tissue, it cannot move away from its own wastes. These vital functions must be performed for the cell. The substance that bathes the cell and carries out these functions is called interstitial fluid. It is also known as intercellular or tissue fluid.

The interstitial fluid, in turn, must be serviced by the blood and lymph. The blood picks up oxygen from the lungs, nutrients from the digestive tract, hormones from the endocrine glands, and enzymes from still other parts of the body. The blood then transports these substances to all the tissues where they diffuse into the interstitial fluid. In the interstitial fluid, the substances are passed on to the cells and exchanged for wastes.

The blood must service all the tissues of the body. This also means that it can be an ideal medium for the transport of disease-causing organisms throughout the body. To protect itself from such disease spread, however, the body has a lymphatic system. This system is a collection of vessels containing a fluid called lymph. The lymph picks up materials, including wastes, from the interstitial fluid, cleanses it of bacteria, and returns the materials and wastes to the blood. The blood then carries the wastes to the lungs, kidneys, and sweat glands, where they are eliminated from the body. The blood also takes wastes to the liver, where they are detoxified and recycled.

Blood inside blood vessels, interstitial fluid around body cells, and lymph inside lymph vessels constitute the *internal environment* of the human organism. Because the cells are too specialized to adjust to more than very limited changes in their environment, the internal environment must be kept as constant as possible. This condition we have called homeostasis. In preceding chapters, we have discussed how the internal environment was kept in homeostasis. Now we shall take a look at that environment itself.

The blood, heart, and blood vessels constitute the **blood vascular** or **cardiovascular system.** The lymph, lymph vessels, and lymph glands make up the **lymph vascular system.** Together, the two systems are called the **circulatory system.** Let us first take a look at those substances called blood, interstitial fluid, and lymph.

BLOOD

The red body fluid that flows through all the vessels except the lymph vessels is called **blood.** Arterial blood is blood that is leaving the heart and generally tends to be bright red. Venous blood is blood that is returning to the heart and is also red, but generally somewhat darker. Blood is a viscous fluid, which means that it is thicker and more adhesive than water. As you know from practical experience, thick fluids flow more slowly than thin, watery ones. Water is considered to have a viscosity of 1. The viscosity of blood, by comparison, ranges from 4.5 to 5.5. This means that it flows 4½ to 5½ times more slowly than water. The adhesive quality of blood, or its stickiness, may be felt by touching it. Blood is also slightly heavier than water. Other physical characteristics of blood include a necessary temperature of about 38°C (100.4°F), a pH range of 7.35 to 7.45, which is slightly alkaline, and a 0.85 to 0.90 percent concentration of salt (NaCl). Blood constitutes about 8 percent of the total body weight. The blood volume of an average-sized man is between 5 to 6 liters (5 to 6 qt). The blood volume in an average-sized woman is about 4 to 5 liters.

Despite its rather simple physical appearance, blood is an exceedingly complex liquid that performs a number of important functions. Among the important functions are:

1. The transportation of oxygen from the lungs to all cells of the body
2. The transportation of carbon dioxide from the cells of the body to the lungs
3. The transportation of nutrients from the digestive organs to the cells of the body
4. The transportation of waste products from the blood vascular system to the kidneys
5. The transportation of hormones from endocrine glands to cells of the body
6. The transportation of enzymes to various cells of the body
7. The regulation of body pH through buffers dissolved in the blood and amino acids of proteins
8. The regulation of normal body temperature because it contains such a large volume of water (an excellent heat absorber and coolant)
9. The regulation of the water content of cells, principally through dissolved sodium ions
10. The prevention of body fluid loss through the clotting mechanism

11. Protection against toxins and foreign microbes through special combat-unit cells in the blood and lymph

Microscopically, blood is composed of two portions: plasma, which is a liquid containing dissolved substances, and formed elements, which are cells and cell-like bodies suspended in the plasma. We shall look first at the formed elements and then discuss plasma.

Formed elements

In clinical practice, the most common classification of the **formed elements** of the blood is the following:

I. Erythrocytes, or red blood cells
II. Leucocytes, or white blood cells
 A. Granular leucocytes (granulocytes)
  1. Neutrophils
  2. Eosinophils
  3. Basophils
 B. Agranular leucocytes (agranulocytes)
  1. Lymphocytes
  2. Monocytes
III. Thrombocytes, or platelets

Some of the more important facts about each of the formed elements will now be considered.

Where do these formed elements come from? The process by which blood cells are formed is called *hemopoiesis* or *hematopoiesis*. During embryonic and fetal life, there are no clear-cut centers for blood cell production. For example, the yolk sac, liver, spleen, thymus gland, lymph nodes, and bone marrow all participate at various times in producing the formed elements. In the adult, however, we can pinpoint the production process precisely. Bone marrow (myeloid tissue) is responsible for the production of red blood cells, granular leucocytes, and platelets. On the other hand, lymphatic tissue – which includes spleen, tonsils, and lymph nodes – and myeloid tissue are responsible for the production of agranular leucocytes.

Erythrocytes

Microscopically, *red blood cells,* or **erythrocytes,** appear as biconcave discs averaging about 7.7 μm in diameter (Figure 11–1). Mature red blood cells are quite simple in structure. They lack a nucleus and can neither reproduce nor carry on extensive metabolic activities. The interior of the cell contains some cytoplasm; protein; lipid substances, including cholesterol; and a red pigment called hemoglobin. *Hemoglobin,* which constitutes about 33 percent of the cell volume, is responsible for the red color of blood.

The function of the erythrocytes is to combine with oxygen and carbon dioxide and transport them through the blood vessels. Red blood cells are highly specialized for this purpose. The hemoglobin molecule consists of a protein called globin and a pigment called heme, which contains iron. As the erythrocyte passes through the lungs, each of the four iron atoms in the hemoglobin molecule combines with a molecule of oxygen. The oxygen is

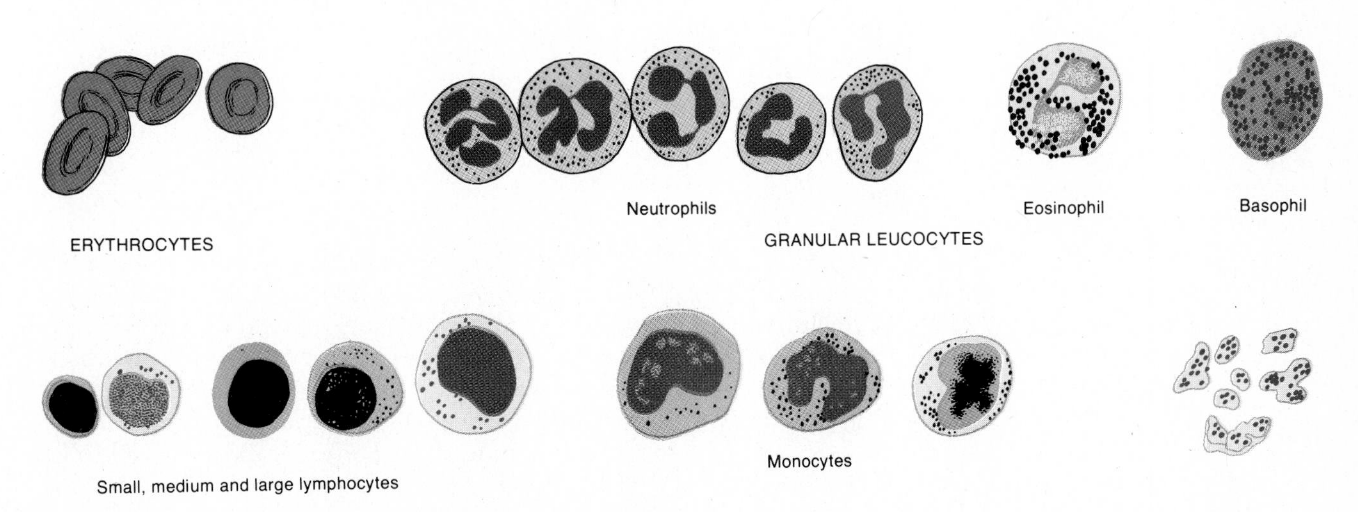

Figure 11–1. Cells from normal human blood.

transported in this state to other tissues of the body. In the tissues, the iron-oxygen reaction reverses, and the oxygen is released to diffuse into the interstitial fluid. Can you guess why iron is sometimes prescribed for "tired blood"? On the return trip, the globin portion combines with a molecule of carbon dioxide from the interstitial fluid. This complex is transported and released in the lungs. Red blood cells contain a great many hemoglobin molecules in order to increase their carrying capacity. One estimate is 280 million molecules of hemoglobin per erythrocyte. The shape of a red blood cell also increases its carrying capacity. A biconcave structure has a much greater surface area than, say, a sphere or cube. The erythrocyte thus provides the maximum surface area for the diffusion of gas molecules that pass through the membrane to combine with hemoglobin.

As we mentioned, a red blood cell does not live long. Its cell membrane becomes fragile, and the cell is nonfunctional in about 120 days. The main reason for its short life is its inability to replace enzymes. The blood, however, contains inordinate numbers of these cells. A healthy male has about 5.4 million red blood cells per cubic millimeter of blood. A healthy female has about 4.8 million red blood cells per cubic millimeter. To maintain normal quantities of erythrocytes, the body must produce new mature cells at the astonishing rate of 2 million per second. In the adult, production takes place in the red bone marrow in the spongy bone of the cranium, ribs, sternum, bodies of vertebrae, and proximal epiphyses of the humerus and femur. The process by which erythrocytes are formed is called **erythropoiesis.**

Leucocytes

Unlike red blood cells, **leucocytes,** or *white blood cells,* have nuclei and do not contain hemoglobin. In addition, they are far less numerous, averaging from 5,000 to 9,000 cells per cubic millimeter of blood. Red blood cells, therefore, outnumber white blood cells about 700 to 1. Leucocytes fall into two major groups (Figure 11–1). The first group contains the *granular leucocytes.* These develop from bone marrow. They have granules in the cytoplasm and possess lobed nuclei. Three kinds of granular leucocytes exist. These are the neutrophils, the eosinophils, and the basophils. The second principal group of leucocytes is called the agranular leucocytes. They develop from lymphatic tissue. When they are placed under a light microscope, no cytoplasmic granules can be seen. Their nuclei are more or less spherical. The two kinds of agranular leucocytes are lymphocytes and monocytes.

The general function of the leucocytes is to combat inflammation and infection. Some leucocytes are actively **phagocytotic.** This means that they can ingest bacteria and dispose of dead matter. Most leucocytes also possess, to some degree, the ability to crawl through the minute spaces between the cells that form the walls of capillaries, the smallest blood vessels, and through connective and epithelial tissue. This movement, the same kind that is exhibited by amoebas, is called **diapedesis.** First, a part of the cell membrane stretches out in an armlike projection. Then the cytoplasm and nucleus flow into the projection. Finally, the rest of the membrane snaps up into place. Another projection is made, and so on, until the cell has "crawled" to its destination.

If a tissue of the body becomes injured or infected, blood vessels in the immediate vicinity dilate and bring more blood to the affected area. This results in inflammation, which is characterized by redness, heat, swelling, pain, and often loss of function. Leucocytes become active and migrate by diapedesis in large numbers through capillary walls into the affected tissue. As leucocytes engulf the bacteria, some **suppuration,** or pus formation, may occur. Essentially, pus consists of dead and living bacteria, white blood cells, cellular debris, and blood fluid. If the leucocytes destroy the invading bacteria effectively, the affected area returns to normal, and the process of repair takes over. If the leucocytes are ineffective, suppuration increases, and the infection may also spread.

The diagnosis of an injury or infection within the body may involve a **differential count.** In this procedure, the number of each kind of white cell in 100 white blood cells is counted. A normal differential count might appear as follows:

Neutrophils	60–70%
Eosinophils	2–4%
Basophils	0.5–1%
Lymphocytes	20–25%
Monocytes	3–8%
Total	100%

When interpreting the results of a differential count, particular attention is paid to the neutrophils. The neutrophils are the most active in response to tissue destruction. More often than not, a high neutrophil count indicates that the damage is caused by invading bacteria. An increase in the number of monocytes generally indicates a chronic (of long duration) infection such as tuberculosis. It is hypothesized that monocytes take longer to reach the site of infection than neutrophils, but once they arrive they do so in larger numbers and destroy more microbes than neutrophils. High eosinophil

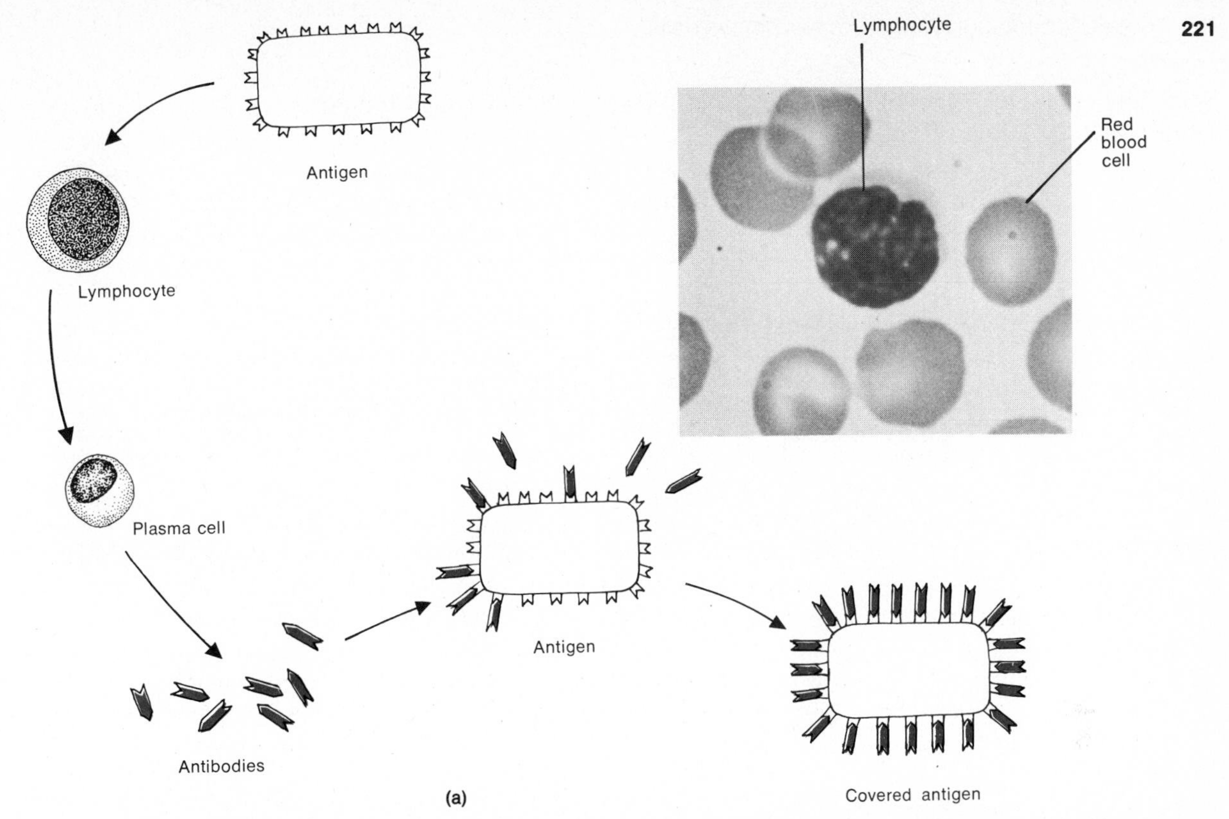

Figure 11–2. Antigen-antibody response. (a) An antigen entering the body stimulates a lymphocyte to develop into an antibody-producing plasma cell. The antibodies attach to the antigen, cover it, and render it harmless. (b) Photomicrograph of a lymphocyte at a magnification of 1800X. (Courtesy of Edward J. Reith, from *Atlas of Descriptive Histology,* by Edward J. Reith and Michael H. Ross, Harper & Row, Publishers, Inc., New York, 1970.)

counts typically indicate allergic conditions since eosinophils are believed to combat allergens, the causative agents of allergies. Basophils are also believed to be involved in allergic reactions.

The term **leucocytosis** refers to an increase in the number of white blood cells. If the increase exceeds 10,000, a pathological condition is usually indicated. A decrease below normal in the number of white blood cells is termed **leucopenia.**

Some leucocytes afford protection in another, extremely important way. These are the lymphocytes, which are involved in the production of antibodies. **Antibodies** are special proteins that inactivate antigens. An **antigen** is any type of protein that the body is not capable of synthesizing. Antigens, in other words, are "foreign" proteins that are introduced into the body by any number of ways. For example, many of the proteins that make up the cell structures and enzymes of bacteria are antigens. The toxins released by bacteria are also antigens. When antigens enter the body, they react chemically with substances in the lymphocytes and stimulate some of the lymphocytes to become *plasma cells* (Figure 11–2). The plasma cells then produce **antibodies,** which are globulin-type proteins that attach to antigens, much like enzymes attach to substrates. Like enzymes, a particular type of antibody will generally attach only to a particular type of antigen. However, unlike enzymes, which enhance the reactivity of the substrate, antibodies "cover" their antigens so the antigens cannot come in contact with other chemicals in the body. In this way, bacterial poisons can be "sealed up" and rendered harmless. The bacteria themselves are destroyed as the antibodies "walk off" with the proteins in their cell membranes. This process is called the **antigen-antibody response.** The antigen-antibody response allows us to combat infection. It gives us immunity to some diseases. And, as you will find out, it is responsible for blood types, for allergies, and for the body's rejection of organs transplanted from an individual with a different genetic makeup.

Foreign bacteria exist everywhere in the environment and have continuous access to the body through the mouth, nose, and pores of the skin. Furthermore, many cells, especially those of epithelial tissue, age and die, and their remains must be disposed of daily. Even when the body is healthy, the leucocytes actively ingest bacteria and debris. However, a leucocyte can phagocytose only a certain number of substances before the substances interfere with the normal metabolic activities of the leucocyte and bring on its death. Consequently, the

Exhibit 11–1. CHEMICAL COMPOSITION AND DESCRIPTION OF THE SUBSTANCES IN PLASMA

CONSTITUENT	DESCRIPTION
A. Water	Constitutes about 92 percent of plasma and is the liquid portion of blood. Ninety percent of the water is derived from absorption from the digestive tract, and 10 percent from the metabolism of nutrients. The functions of the water are to act as a solvent and suspending medium for the solid components of blood and to absorb heat.
B. Solutes	
1. Proteins	Plasma proteins constitute about 7 to 9 percent of the solutes in plasma.
Albumins	These constitute 55 to 64 percent of the plasma proteins and are the smallest of the plasma proteins. Albumins are produced by the liver and provide the blood with viscosity, a factor related to the maintenance and regulation of blood pressure. Albumins also exert a considerable osmotic pressure. This helps maintain the water balance between the blood and tissues and helps regulate the volume of blood.
Globulins	These proteins constitute about 15 percent of the plasma proteins. They are the protein group to which antibodies belong. One of the more important subgroups is the gamma globulins, which attack measles and hepatitis viruses, the tetanus bacterium, and possibly poliomyelitis virus.
Fibrinogen	This protein represents only a small fraction of plasma proteins (4 percent). It is also produced by the liver and plays an essential role in the blood-clotting mechanisms.
2. Nonprotein nitrogen (NPN) substances	These compounds are the substances in plasma that contain nitrogen but are not proteins. Among such compounds are urea, uric acid, creatine, creatinine, and ammonium salts. They represent breakdown products of protein metabolism and are carried by the blood to the organs of excretion.
3. Food substances	Once foods are broken down in the digestive tract, the products of digestion are passed into the blood for distribution to all cells of the body. These products include amino acids (from proteins), glucose (from carbohydrates), and fats (from lipids).
4. Regulatory substances	The two principal regulatory substances are enzymes and hormones. Enzymes are produced by the cells of the body and catalyze chemical reactions. Hormones, produced by endocrine glands, regulate growth and developmental processes in the body.
5. Respiratory gases	The respiratory gases, oxygen and carbon dioxide, are carried by the blood. It should be noted, however, that these gases are more closely associated with the hemoglobin of red blood cells than the plasma itself.
6. Electrolytes	A number of ions constitute the inorganic salts of plasma. The cations or positive ions include Na^+, K^+, Ca^{2+}, and Mg^{2+}. The anions or negative ions include Cl^-, PO_4^{3-}, SO_4^{2-}, and HCO_3^-. The salts in plasma function in the maintenance of proper osmotic pressure, the regulation of normal pH, and the maintenance of the proper physiological balance between the tissues and the blood.

life span of most leucocytes is very short. During times of health, some white blood cells will live a couple of days. During a period of infection they may live for only a few hours.

Thrombocytes

Thrombocytes, or *platelets,* are disc-shaped bodies enclosed by a membrane and without a nucleus (see Figure 11–1). They average from 2 to 4 μm in diameter. Between 250,000 and 500,000 platelets appear in each cubic millimeter of blood.

The function of the platelets is to prevent fluid loss by initiating a chain of reactions that results in blood clotting. Like most of the other formed elements of the blood, platelets have a short life span, probably only about 1 week. This short life span is due to the facts that platelets are "used up" in clotting and that they are just too simple to carry on much metabolic activity. When the formed elements–erythrocytes, leucocytes, and thrombocytes–are removed, the liquid portion of the blood is left.

Plasma

When the formed elements are removed from blood, a straw-colored liquid called **plasma** is left. Exhibit 11–1 outlines the chemical composition of plasma. However, a few of the solutes should be pointed out. About 7 to 9 percent of the solutes are proteins. Some of these proteins are also found elsewhere in the body, but when they occur in blood, they are called *plasma proteins.* Albumins, which constitute the majority of plasma proteins, are responsible for the viscosity or thickness of blood. Along with the electrolytes, albumins also regulate blood volume by preventing all the water in the blood from diffusing into the interstitial fluid. Recall that water moves by osmosis from an area of low solute (high water) concentration to an area of high solute (low water) concentration. Globulins, which are antibody proteins released by plasma cells, form a small component of the plasma proteins. Gamma globulin is especially well known because it is able to form an antigen-antibody complex with the proteins of the hepatitis and measles viruses and the tetanus bacterium. Fibrinogen, a third plasma protein, takes part in the blood-clotting mechanism, along with the platelets. *Serum* is the term applied to blood plasma minus its fibrinogen.

INTERSTITIAL FLUID AND LYMPH

For all practical purposes interstitial fluid and lymph are the same. The major differences between the two is location. When the fluid bathes the cells, it is called **interstitial fluid,** or **tissue fluid.** When it flows through the lymphatic vessels, it is called **lymph** (Figure 11-3). Both fluids are similar in composition to plasma. The principal chemical difference is that they contain less protein because the larger protein molecules are not easily filtered through the cells that form the walls of the blood vessels. Keep in mind that whole blood does not flow into the tissue spaces; it remains within closed vessels. Certain constituents of the plasma do move, however, and once they move out of the blood, they are collectively called interstitial fluid. The transfer of materials between blood and interstitial fluid occurs by osmosis, diffusion, and filtration across the cells that make up the capillary walls. Both interstitial fluid and lymph contain variable numbers of leucocytes. Leucocytes can enter the tissue fluid by diapedesis, and the lymphoid tissue itself produces nongranular leucocytes. However, interstitial fluid and lymph both lack erythrocytes and platelets.

Other substances, especially organic molecules, in interstitial fluid and lymph vary in kinds and amounts in relation to the location of the sample analyzed. The lymph vessels of the digestive tract, for example, contain a great deal of lipid that has been absorbed from food.

The circulatory system can be likened to an intricate transportation network within the body. Actually, the circulatory system consists of two principal divisions, the blood vascular or cardiovascular system and the lymph vascular system. The first consists of the blood, heart, and blood vessels, and the second consists of lymph, lymph vessels, lymph nodes, and lymphatic organs.

We will now examine the center of your body's blood vascular system, the heart. This is the pump that maintains circulation in the blood vascular system.

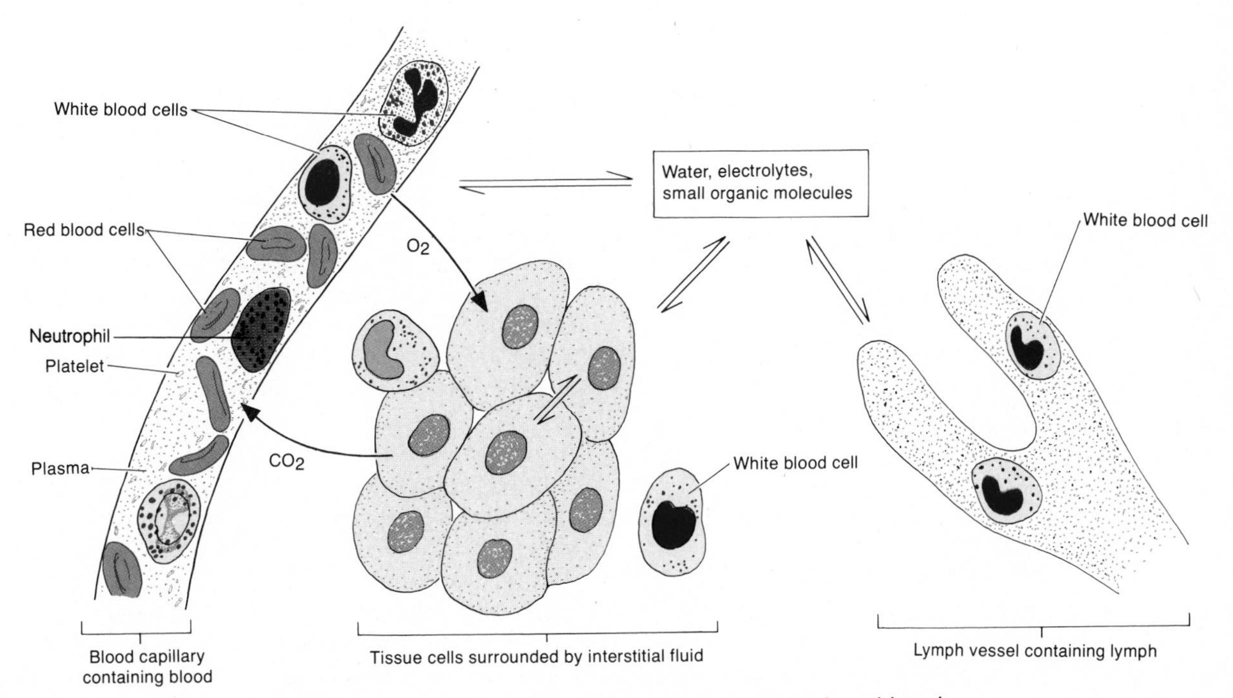

Figure 11-3. Composition of blood, interstitial fluid, and lymph.

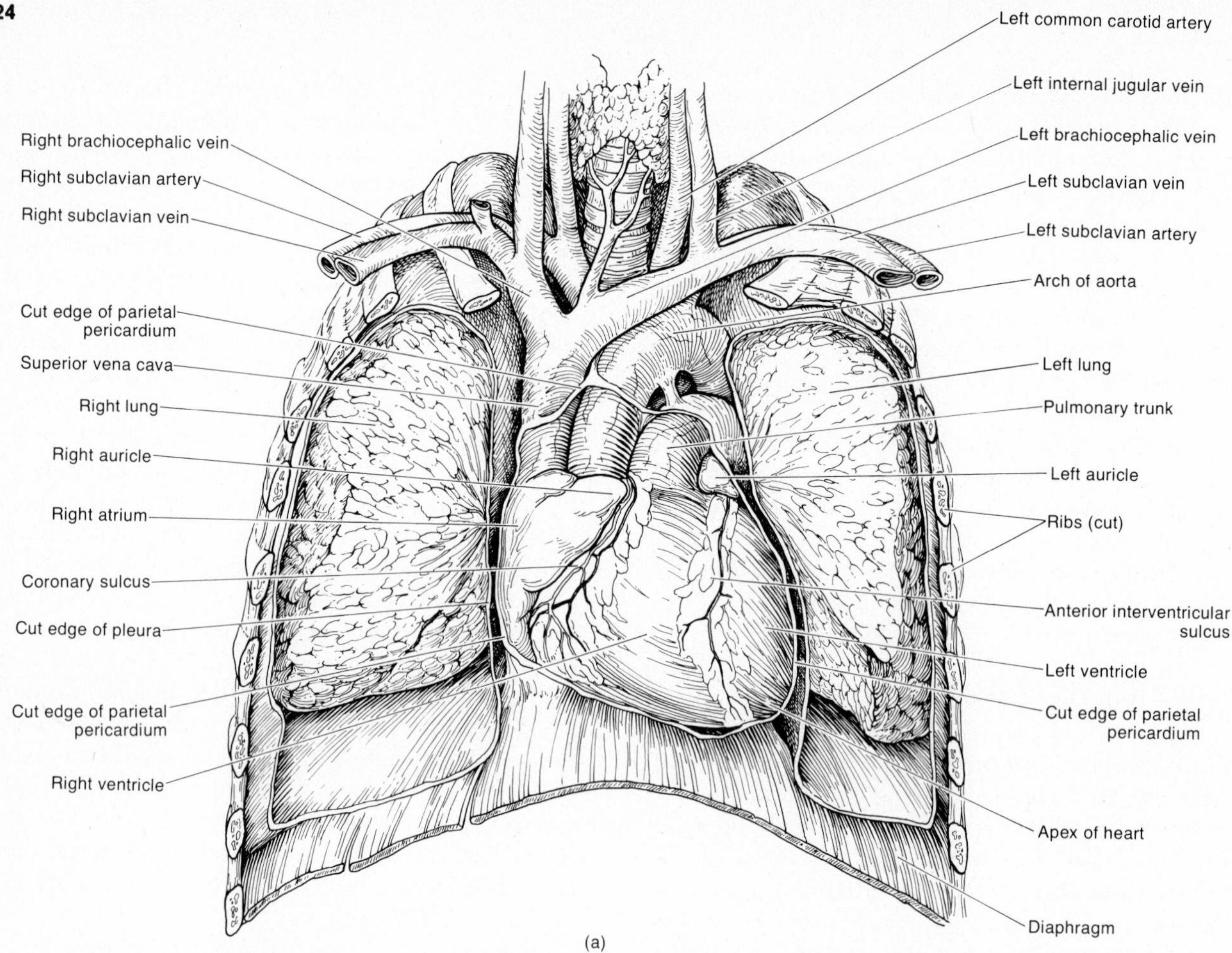

Figure 11–4. The heart. (a) Position of the heart and associated blood vessels in the thoracic cavity.

THE HEART

The **heart** is a hollow, muscular organ that pumps the blood through the vessels. It is situated obliquely between the lungs in the mediastinum, and about two-thirds of its mass lies to the left of the midline of the body (Figure 11–4a). The heart is shaped like a blunt cone about the size of a closed fist that is, 12 cm (5 in.) long, 9 cm (3½ in.) wide at its broadest point, and 6 cm (2½ in.), thick. A roentgenogram of a normal heart is provided in Figure 11–4b. Compare it with the adjacent diagram. Its pointed end, called the *apex,* projects downward, forward, and to the left, and lies superior to the central depression of the diaphragm. Its broad end, or *base,* projects upward, backward, and to the right, and lies just inferior to the second rib. The major parts of the heart to be considered here are the parietal pericardium, the walls and chambers, and the valves.

The parietal pericardium (pericardial sac)

The heart is enclosed in a loose-fitting serous membrane called the **parietal pericardium** or **pericardial sac** (Figure 11–4a, c). The parietal pericardium consists of two layers referred to as the fibrous layer and the serous layer. The *fibrous layer* or *fibrous pericardium* is the outer layer and consists of a very heavy fibrous connective tissue. The fibrous pericardium is attached to the large blood vessels entering and leaving the heart, to the diaphragm, and to the inside of the sternal wall of the thorax. It also adheres to the parietal pleurae. The fibrous pericardium prevents overdistension of the heart, provides a tough protective membrane around the heart, and anchors the heart in the mediastinum. The inner layer of the parietal pericardium is referred to as the *serous layer* or *serous pericardium.* This is a thinner, more delicate membrane that is continuous with the epicardium at the base of the heart and around the large blood vessels. The **epicardium** or **visceral pericardium** is the thin, transparent outer layer of the wall of the heart that is composed of fibrous tissue and mesothelium. Between the serous pericardium and the epicardium is a potential space called the *pericardial cavity.* The cavity contains a watery fluid, known as pericardial fluid, which prevents friction between the membranes as the heart moves. An inflammation of the parietal pericardium is called *pericarditis.*

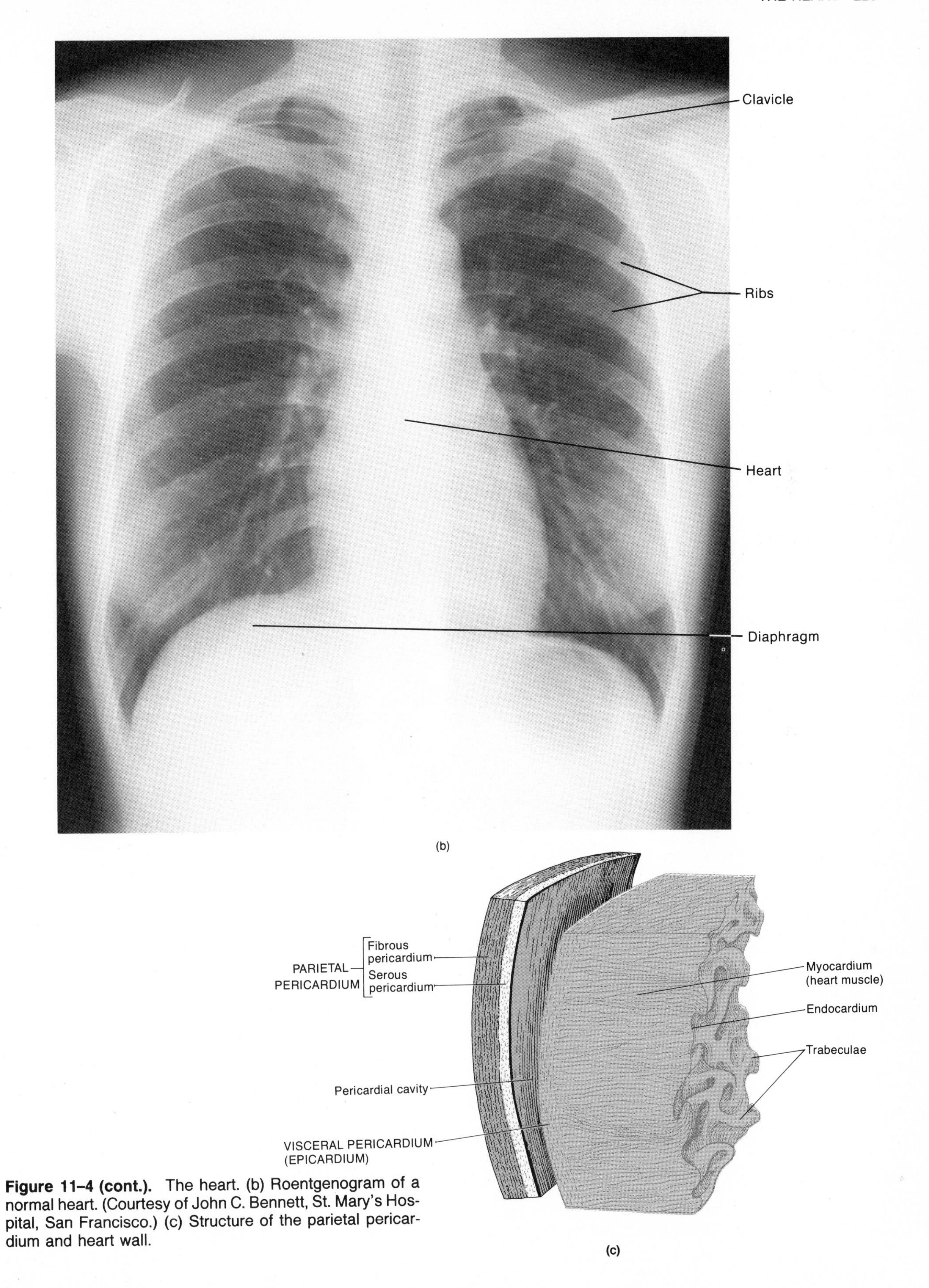

Figure 11–4 (cont.). The heart. (b) Roentgenogram of a normal heart. (Courtesy of John C. Bennett, St. Mary's Hospital, San Francisco.) (c) Structure of the parietal pericardium and heart wall.

Walls and chambers

The wall of the heart (Figure 11–4c) is divided into three portions: (1) the epicardium, or external layer; (2) the myocardium, or middle layer; and (3) the endocardium, or inner layer. The **epicardium** is the same as the visceral pericardium, which has just been described. The **myocardium,** which is cardiac muscle tissue, comprises the bulk of the heart. As you will recall from Chapter 9, cardiac muscle fibers are involuntary, striated, and branched, and the tissue is arranged in interlacing bundles of fibers. The myocardium is responsible for the actual contraction of the heart. Inflammation of the myocardium is referred to as *myocarditis.* The **endocardium** is a thin layer of endothelium overlying a thin layer of connective tissue which is pierced by tiny blood vessels and some bundles of smooth muscle. It lines the inside of the myocardium and covers the valves of the heart and the tendons that hold the valves open. It is continuous with the endothelial lining of the large blood vessels of the heart. Inflammation of the endocardium is called *endocarditis.*

The interior of the heart is divided into four spaces or chambers, which receive the circulating blood (Figure 11–5). The two upper chambers are called the right and left **atria.** The lining of the atria is smooth except for the anterior parts of the wall and the auricle that contain projecting muscle bundles, the *musculi pectinati.* These bundles give the lining of the auricles a ridged appearance. The atria are separated by a partition called the *interatrial septum.* On the posterior wall of the interatrial septum is an oval-shaped depression, the *fossa ovalis,* which corresponds to the foramen ovale of the fetal heart. This is described as part of the fetal circulation. Each atrium has an appendage called an *auricle,* so named because of its resemblance to a dog's

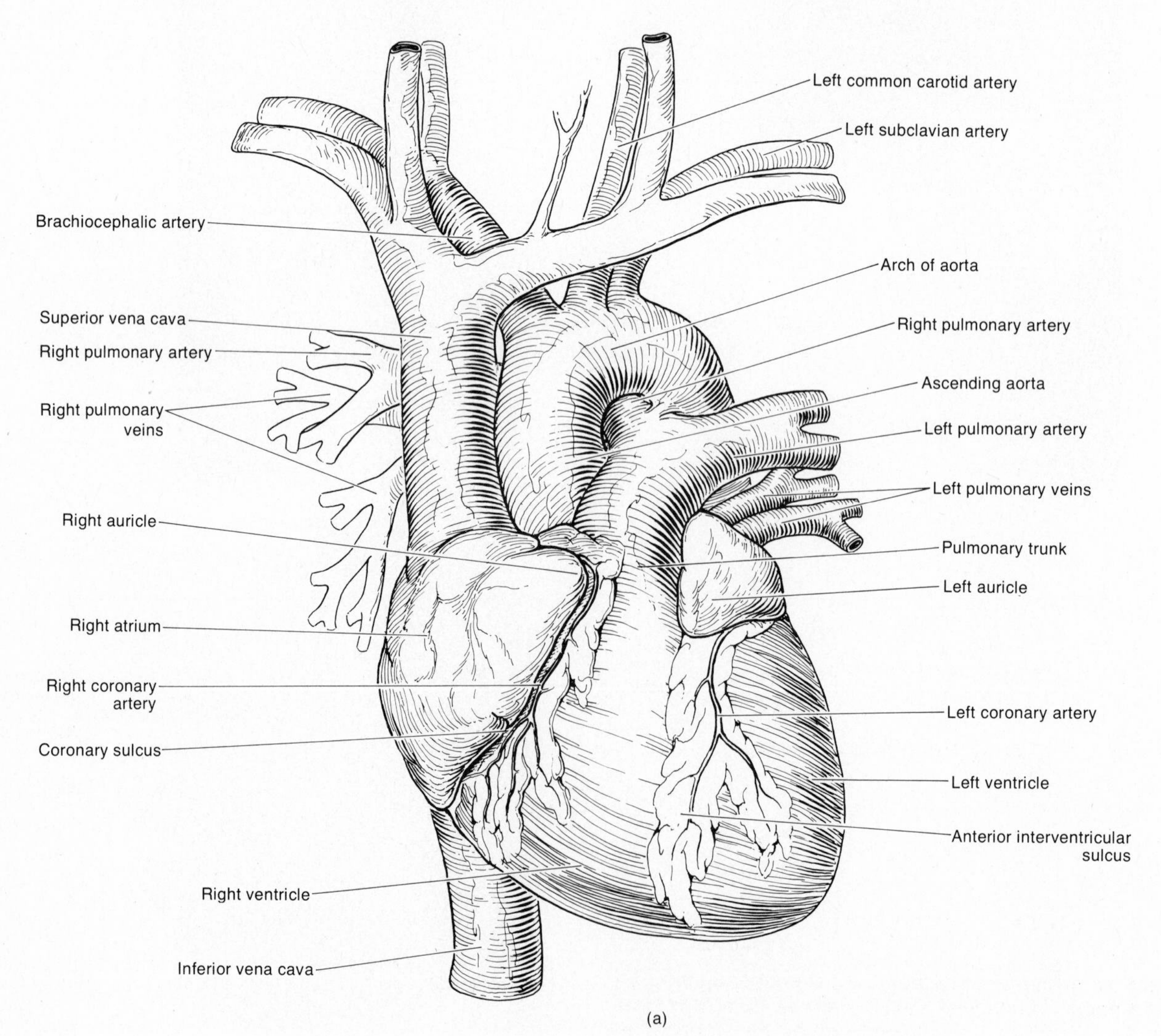

Figure 11–5. The structure of the heart. (a) Anterior external view.

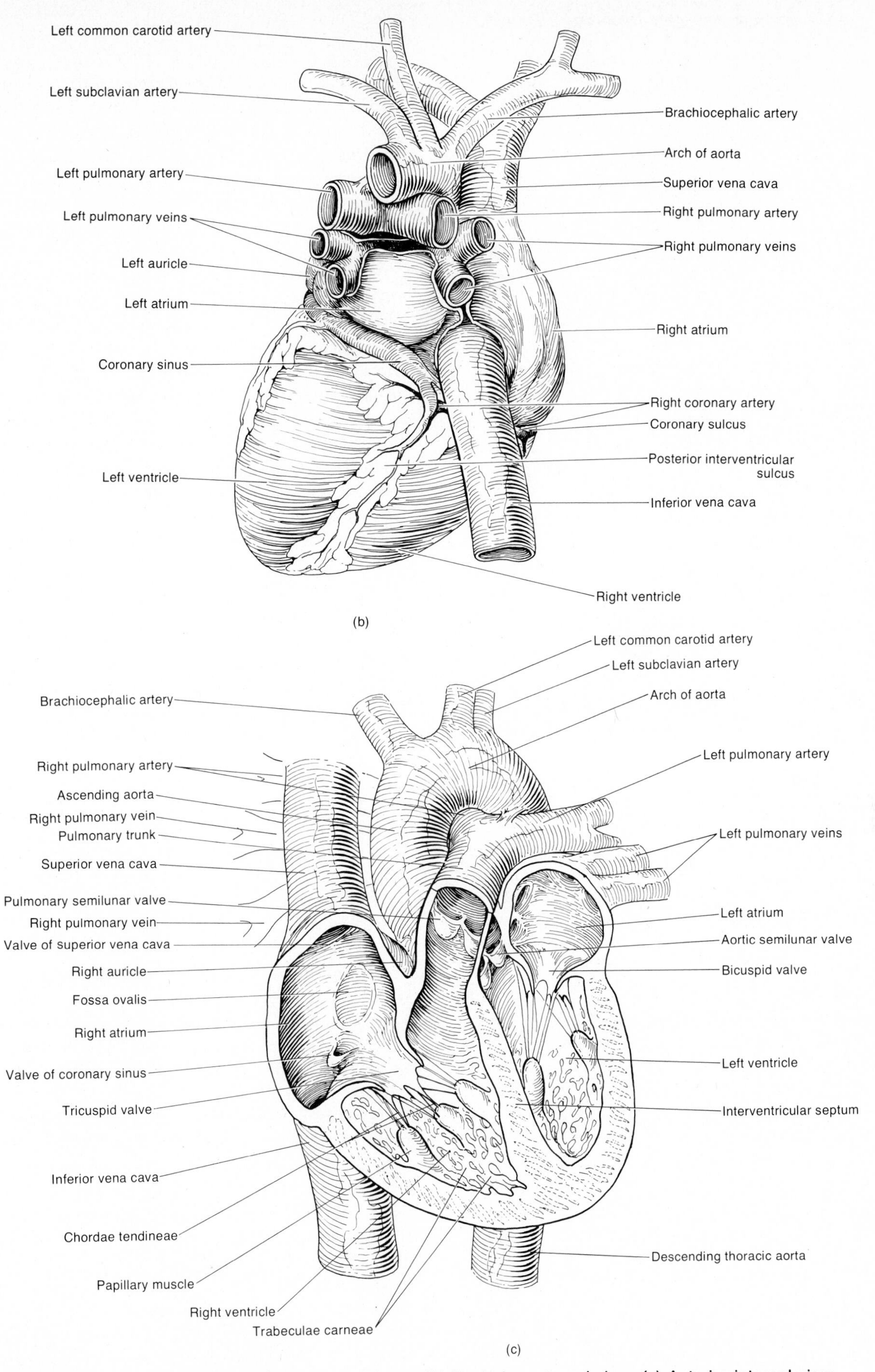

Figure 11–5 (cont.). The structure of the heart. (b) Posterior external view. (c) Anterior internal view.

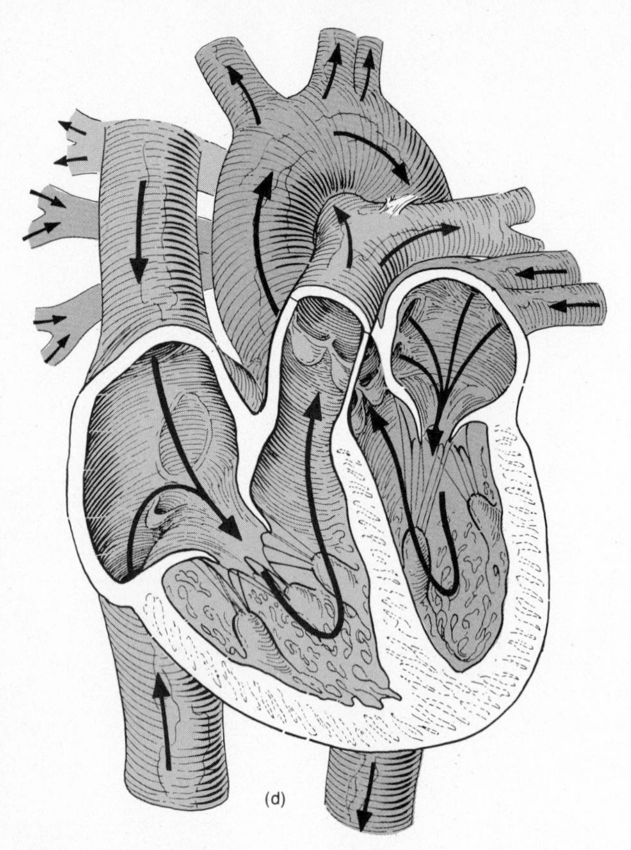

Figure 11–5 (cont.). The structure of the heart. (d) Path of blood through the heart.

ear. The auricle helps to increase the surface area of the atrium. The two lower chambers, the right and left **ventricles,** are separated by an *interventricular septum.* Externally, a groove known as the *coronary sulcus* separates the atria from the ventricles. It encircles the heart and houses the coronary sinus and circumflex branch of the left coronary artery. The *anterior interventricular sulcus* and *posterior interventricular sulcus* separate the right and left ventricles externally. These sulci contain coronary blood vessels and a variable amount of fat (see Figure 11–4a). The right atrium receives blood from all parts of the body except the lungs. It receives the blood through three veins. One of these veins is the *superior vena cava,* which brings blood from the upper portion of the body. Another of the veins is the *inferior vena cava,* which brings blood from the lower portions of the body. The third vein is the *coronary sinus,* which drains blood from the vessels supplying the walls of the heart. The right atrium then squeezes the blood into the right ventricle, which pumps it into the *pulmonary trunk.* The pulmonary trunk then divides into a *right* and *left pulmonary artery,* each of which carries blood to the lungs. In the lungs, the blood releases its carbon dioxide and takes on oxygen. It returns to the heart via four *pulmonary veins* that empty into the left atrium. The blood is then squeezed into the left ventricle and exits from the heart through the *ascending aorta.* From here aortic blood is passed into the *arch of the aorta, descending thoracic aorta,* and *abdominal aorta.* These blood vessels transport the blood to all body parts except the lungs.

If you look closely at the sectioned heart in Figure 11–5, you will see that the sizes of the four chambers vary according to their functions. The right atrium, which must collect blood coming from almost all parts of the body, is slightly larger than the left atrium, which receives only the blood coming from the lungs. The thickness of the walls of the chambers varies too. The atria are relatively thin-walled because they need only enough cardiac muscle tissue to squeeze the fluid into the ventricles. The right ventricle has a much thicker layer of myocardium since it must send blood to the lungs and around back to the left atrium. The left ventricle has the thickest walls since it must pump blood through literally miles of vessels in the head, trunk, and extremities. If you examine the cross-sectional view through the mediastinum in Figure 1–3, you can see the relationship of the heart, in cross section, to the surrounding structures.

Valves

As each chamber of the heart contracts, it pushes a portion of blood into a ventricle or out of the heart through an artery. But as the walls of the chambers relax, some structure must prevent the blood from flowing back into the chamber. That structure is a **valve.**

Atrioventricular valves lie between the atria and their ventricles (Figure 11–5). The atrioventricular valve between the right atrium and right ventricle is called the *tricuspid valve* because it consists of three flaps, or cusps. These flaps are fibrous tissues that grow out of the walls of the heart and are covered with endocardium. The pointed ends of the cusps project into the ventricle. Other names for this valve are the right atrioventricular or right AV valve. Cords called *chordae tendineae* connect the pointed ends to small conical projections, the *papillary muscles* (muscular columns), that are located on the inner surface of the ventricles. The irregular surface of ridges and folds of the myocardium in the ventricles is known as the *trabeculae carneae.* The chordae tendineae and their muscles keep the flaps pointing in the direction of the blood flow. As the atrium relaxes and the ventricle squeezes the blood out of the heart, any blood that is driven back toward the atrium is

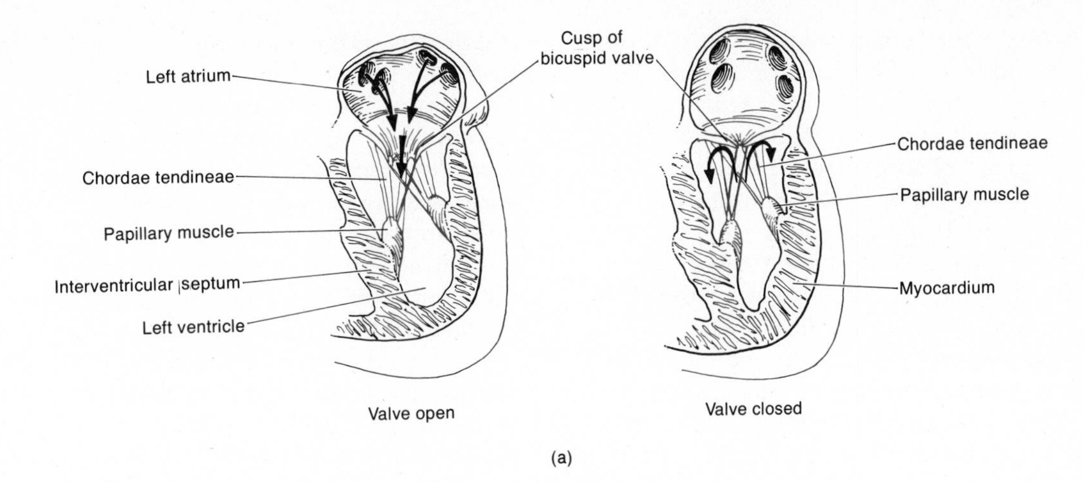

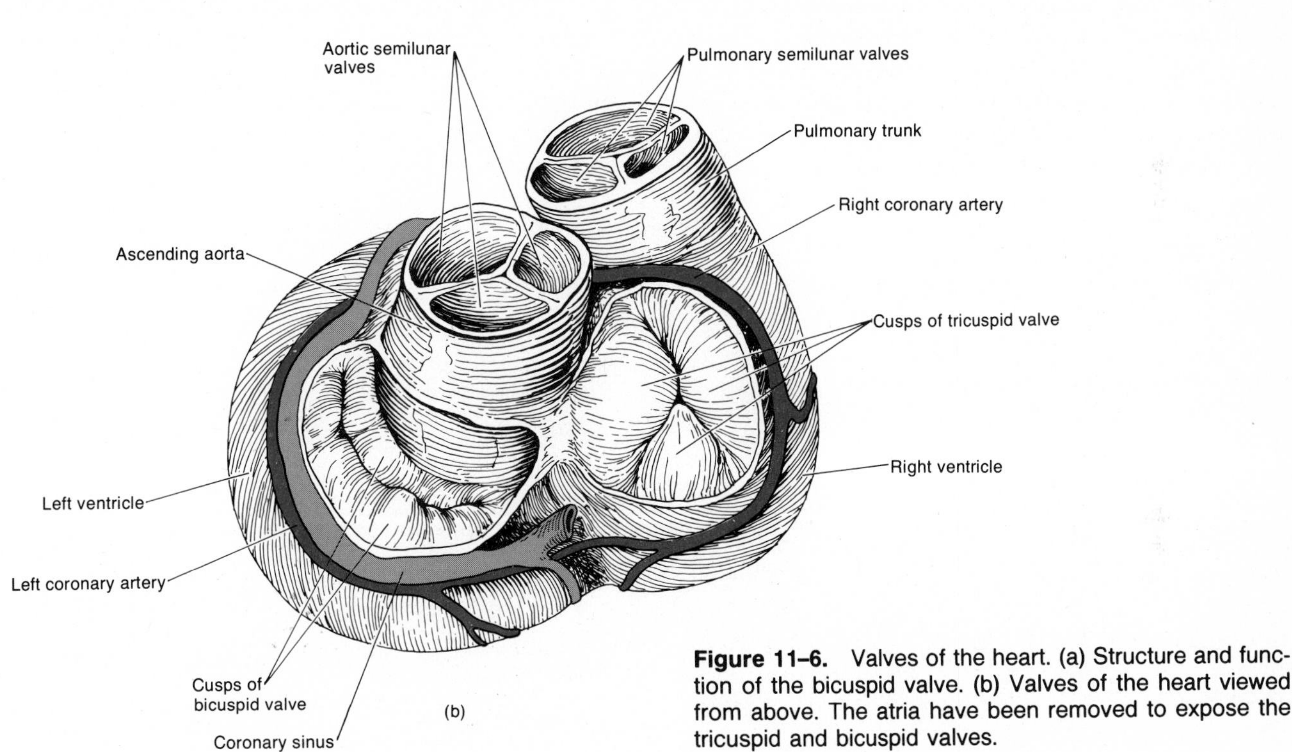

Figure 11–6. Valves of the heart. (a) Structure and function of the bicuspid valve. (b) Valves of the heart viewed from above. The atria have been removed to expose the tricuspid and bicuspid valves.

pushed between the flaps and the walls of the ventricle (Figure 11–6a). This drives the cusps upward until their edges meet and close the opening. The atrioventricular valve between the left atrium and left ventricle is called the *bicuspid* or *mitral valve.* It has two cusps that work in the same way as the cusps of the tricuspid valve. The bicuspid valve is also known as the left atrioventricular or left AV valve.

Each of the arteries that leaves the heart has a valve that prevents blood from flowing back into the heart. These are the **semilunar valves**—*semilunar* meaning half-moon or crescent-shaped. The *pulmonary semilunar valve* lies in the opening where the pulmonary artery leaves the right ventricle. The *aortic semilunar value* is situated at the opening between the left ventricle and the aorta (Figure 11–5). Both valves consist of three semilunar cusps. Each cusp is attached by its convex or inwardly curved margin to the wall of its artery. The free borders of the cusps curve outward and project into the opening inside the blood vessel (Figure 11–6b). Like the atrioventricular valves, the semilunar valves permit the blood to flow in only one direction. In this case, the flow is from the ventricles into the arteries.

The conduction system

The heart is innervated by the autonomic nervous system, but the autonomic neurons only increase or decrease the time it takes to complete a cardiac cycle. The walls of the chambers can go on contracting and relaxing, contracting and relaxing, without any direct stimulus from the nervous system. This is because the heart has an intrinsic regulating system called the **conduction system.** The conduction system is composed of specialized tissues that generate and distribute the electrical impulses which stimulate the cardiac muscle fibers to contract. These tissues are the sinu-atrial node, the atrioventricular node, the atrioventricular bundle, the bundle branches, and the Purkinje fibers. The cells of the conduction system develop during embryological life from certain cardiac muscle cells. These cells lose their ability to contract and become specialists in impulse transmission.

A **node** is a compact mass of conducting cells. The **sinu-atrial node,** known as the **SA node** or **pacemaker,** is located in the right atrium inferior to the opening of the superior vena cava (Figure 11–7a). The SA node initiates each cardiac cycle and thereby sets the basic pace for the heart rate. This is why it is commonly called the pacemaker. However, the rate set by the SA node may be altered by nervous impulses from the autonomic nervous system or by certain chemicals such as thyroid hormone, epinephrine, or acetylcholine. Once an electrical impulse is initiated by the SA node, the impulse spreads out over both atria and causes them to contract. From here, the impulse passes to the **atrioventricular (AV) node,** located near the inferior portion of the interatrial septum. From the AV node, a tract of conducting fibers called the **atrioventricular bundle** or **bundle of His,** runs to the top of the interventricular septum. It then continues down both sides of the septum as the **right** and **left bundle branches.** The bundle of His distributes the charge over the medial surfaces of the ventricles. Actual contraction of the ventricles is stimulated by the **Purkinje fibers.** The Purkinje fibers are branches that emerge from the bundle branches and pass into the cells of the myocardium.

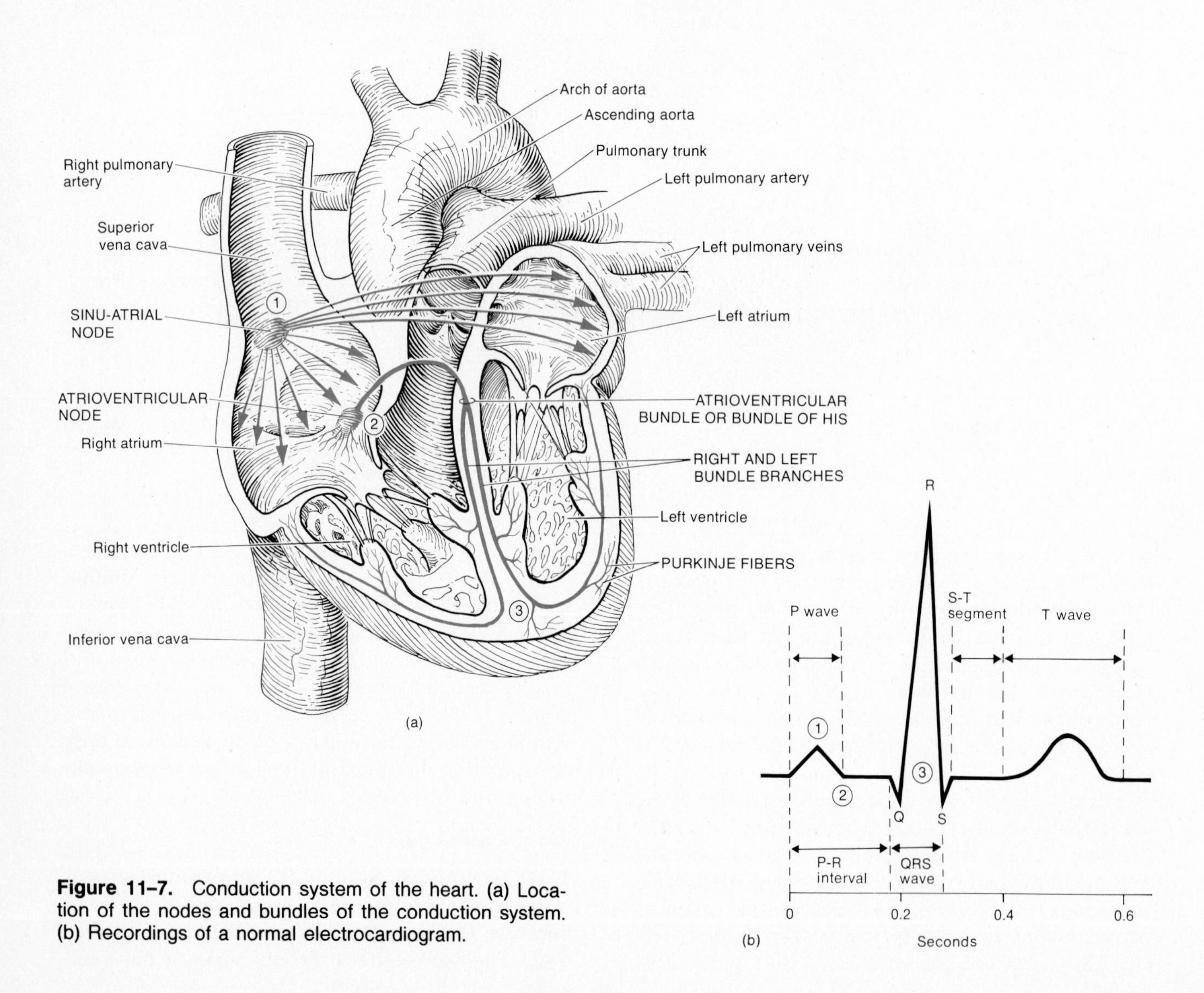

Figure 11–7. Conduction system of the heart. (a) Location of the nodes and bundles of the conduction system. (b) Recordings of a normal electrocardiogram.

The electrocardiogram

Impulse transmission through the conduction system generates electrical currents that may be detected on the surface of the body. A recording of the electrical changes that accompany the cardiac cycle is called an **electrocardiogram (ECG).** The instrument used to record the changes is an *electrocardiograph.*

Each portion of the cardiac cycle produces a different electrical impulse. These are transmitted from the electrodes to a recording needle that graphs the impulses as a series of up-and-down waves called *deflection waves.* In a typical record (Figure 11–7b), three clearly recognizable waves accompany each cardiac cycle. The first wave, called the **P wave,** is a small upward wave. It indicates the spread of an impulse from the SA node through the muscle of the two atria. A fraction of a second after the P wave begins, the atria contract. Following this, there is a complex called the **QRS wave.** It begins as a downward deflection, continues as a large, upright, triangular wave, and ends as a downward wave at its base. This deflection represents the spread of the electrical impulse through the ventricles. The third recognizable deflection is a dome-shaped wave called the **T wave.** This wave indicates ventricular repolarization.

In reading an electrocardiogram, it is exceedingly important to note time relationships between various waves. For example, refer to Figure 11–7b and note the P-R interval. This interval, measured from the beginning of the P wave to the beginning of the Q wave, represents the conduction time from the beginning of atrial excitation to the beginning of ventricular excitation. The P-R interval is the time required for an impulse to travel through the atria and atrioventricular node, to the atrioventricular bundle, and Purkinje fibers. The lengthening of this interval indicates partial blockage of conduction at the atrioventricular bundle. Other intervals and their significance are also indicated in the figure. The ECG is invaluable in diagnosing abnormal cardiac rhythms and conduction patterns, detecting the presence of fetal life, and determining multiple pregnancies.

AUTONOMIC CONTROL. The pacemaker receives nerves from both the parasympathetic and sympathetic divisions of the autonomic nervous system. The parasympathetic neurons originate in the **cardioinhibitory center** of the medulla of the brain and travel in the vagus nerve to the heart. Stimulation of the pacemaker by the parasympathetic fibers slows down the cardiac cycle. The sympathetic pathway originates in the **cardioacceleratory center** of the medulla. It travels in a tract down the spinal cord and then passes over sympathetic nerves to the heart. Sympathetic stimulation counteracts parasympathetic stimulation and quickens the heartbeat. When neither of the cardiac centers is stimulated by sensory neurons, the cardioacceleratory center tends to dominate. The result is that sympathetic fibers have a free reign to speed up heart rate until receptors intervene to stimulate the cardioinhibitory center.

Sounds of the heart

The first sound, which can be described as a **lubb** (o͞o) sound, is a comparatively long, booming sound. The lubb is the sound of the atrioventricular valves closing soon after ventricular systole begins. The second sound, which is heard as a short, sharp sound, can be described as a **dupp** (ŭ) sound. Dupp is the sound of the semilunar valves closing toward the end of ventricular systole. A pause about two times longer comes between the second sound and the first sound of the next cycle. Thus, the cardiac cycle can be heard as a lubb, dupp, pause; lubb, dupp, pause; lubb, dupp, pause. This is the sound of the heartbeat. But note that it comes from the closure of the valves and not from the contraction of the heart muscle.

Heart sounds provide valuable information about the valves of the heart. If the sounds are peculiar, they are referred to as **murmurs.** Murmurs are frequently the noise made by a little blood bubbling back up into an atrium because of the failure of one of the atrioventricular valves to close properly. However, murmurs do not always indicate a valve problem, and many have no clinical significance.

Although the heart sounds are produced in part by the closure of the valves, they are not necessarily best heard over these valves. Instead, each sound tends to be clearest in the location where the vascular chamber distal to the valve in terms of blood flow lies closest to the surface of the body. These locations are illustrated in Figure 11–8.

Surface projection of the heart

Although the shape and position of the heart are subject to individual variation, certain landmarks are useful for identification (Figure 11–8). As indicated earlier, the *apex* projects downward and to the left. This site is recognized as the position of the heartbeat. Anteriorly, the apex is in the fifth intercostal space, about 7.5 to 8 cm (about 3 in.) from the midline of the body. The apex is situated at the

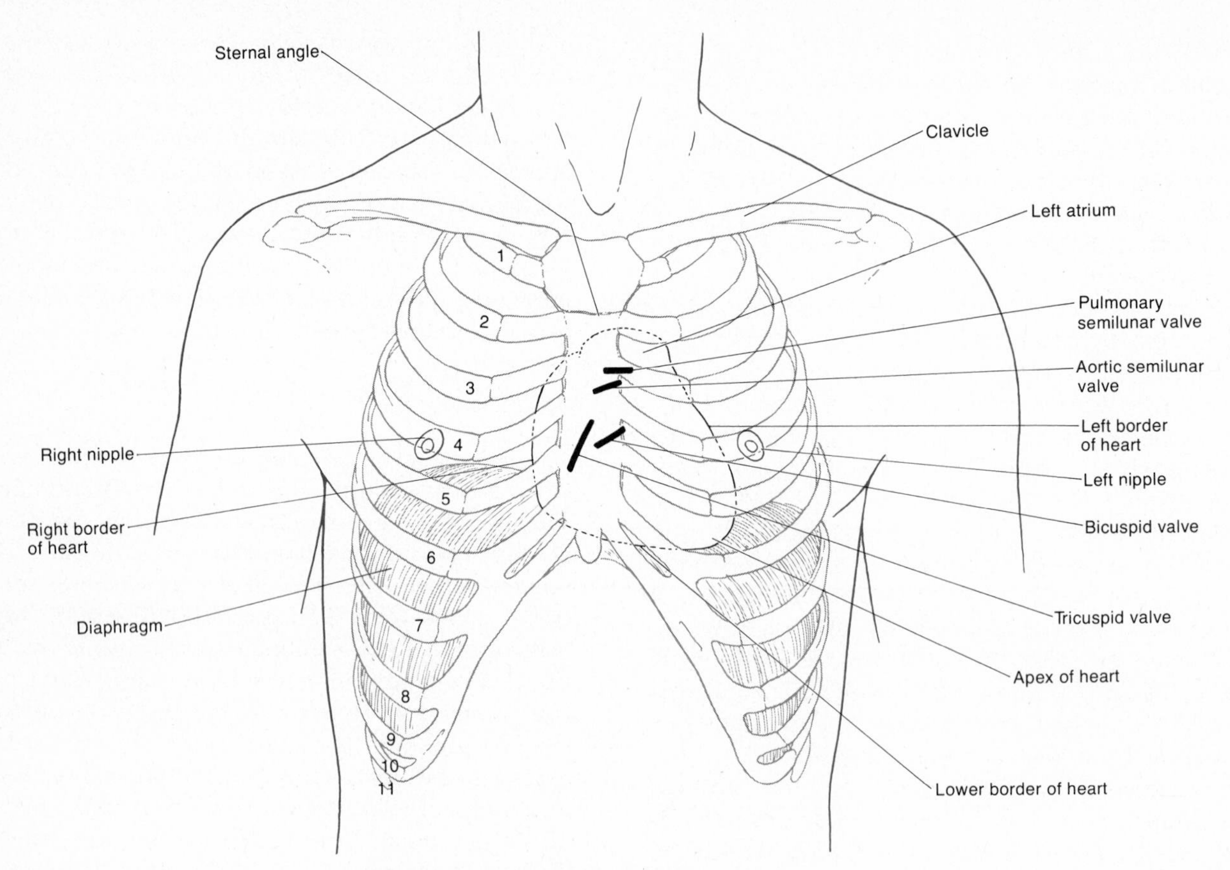

Figure 11–8. Surface projection of the heart.

junction of the left and lower borders of the heart. The *left border* is formed almost entirely by the left ventricle, although the left atrium forms part of the upper end of the border. The left border of the heart is indicated approximately as a curved line drawn from the left fifth costochondral junction to the left second costochondral junction. The *lower border* of the heart corresponds to a line drawn from the apex through the xiphisternal articulation.

The *base* of the heart projects upward and to the right. It is formed by the atria, mostly the left atrium. It lies opposite the fifth to ninth thoracic vertebrae. Anteriorly, the base corresponds to a surface line drawn from a point 1 cm (about ½ in.) below the left second chondrosternal articulation and 3 cm (about 1 in.) from the midline to another point 3 cm from the midline and 1 cm above the right third chondrosternal articulation. The *right border* corresponds to a curved line drawn from the xiphisternal articulation to about the middle of the right third costal cartilage.

The valves of the heart may also be identified by surface projection. For example, the pulmonary and aortic semilunar valves are represented on the surface by a line about 2.5 cm (1 in.) in length. The pulmonary semilunar valve lies horizontally behind the inner end of the left third costal cartilage and the adjoining part of the sternum. The aortic semilunar valve is placed obliquely behind the left side of the sternum, opposite the third intercostal space. The tricuspid valve lies behind the sternum, extending from the midline at the level of the fourth costal cartilage down toward the right sixth chondrosternal junction. The bicuspid lies behind the left side of the sternum obliquely at the level of the fourth costal cartilage. It is represented by a line about 3 cm in length.

BLOOD VESSELS

The blood vessels form a network of tubes that carry blood away from the heart, transport it to the tissues of the body, and then return it to the heart. If you examine the general plan of these vessels in Figure 11–9, you will note that blood vessels are called either arteries, arterioles, capillaries, venules, or veins. **Arteries** are the vessels that carry blood from the heart to the tissues. Two large arteries leave the heart and divide into medium-sized vessels that head toward the various regions of the body. The medium-sized arteries, in turn, divide into small arteries which, in turn, divide into vessels called **arterioles.** As the arterioles enter a tissue, they branch into countless numbers of microscopic vessels called **capillaries.** Through the walls of the capillaries, substances are exchanged between the blood and body tissues. Before leaving the tissue, groups of capillaries reunite to form small veins called **venules.** These, in turn, merge to form progressively larger tubes—the veins themselves. **Veins,** in other words, are blood vessels that convey blood from the tissues back to the heart. Since blood vessels require oxygen and nutrients just like other tissues of the body, they also have blood vessels in their own walls called **vasa vasorum.**

Arteries

Arteries and veins are fairly similar in construction (Figure 11–9a, d, e). They both have walls constructed of three coats and a hollow inner core, called a *lumen,* through which the blood flows. Arteries, however, are considerably thicker and stronger than veins. The pressure in an artery is always greater than in a vein. The inner coat of an arterial wall is called the *tunica interna* or *intima.* It is composed of a lining of endothelium (simple squamous epithelium) that is in contact with the blood. It also has an overlying layer of areolar connective tissue and an outer layer of elastic tissue called the internal elastic membrane. The middle coat, or *tunica media,* is usually the thickest layer. It consists of elastic fibers and smooth muscle. The outer coat, the *tunica externa* or *adventitia,* is composed principally of loose connective tissue. The tunica externa contains elastic and collagenous fibers and a few smooth muscle fibers. An external elastic membrane may separate the tunica media from the tunica externa.

As a result of the structure of the middle coat, especially, arteries have two very important properties: elasticity and contractility. When the ventricles of the heart contract and eject blood into the large arteries, the arteries expand to contain the extra blood volume. Then, as the ventricles relax, the elastic recoil of the arteries forces the blood onward. The contractility of an artery is a function of its smooth muscle. The smooth muscle is arranged in rings around the lumen, resembling somewhat the shape of a donut. As the muscle contracts, it squeezes the wall more tightly around the lumen and consequently narrows the area through which the blood flows. Such a decrease in the size of the lumen is called *vasoconstriction.* The nerves responsible for the action are termed *vasoconstrictor nerves.* Conversely, if all the muscle fibers relax, the size of the arterial lumen increases. This is called *vasodilation,* and the nerves mediating the response are called *vasodilator nerves.*

The contractility of arteries also serves a minor function in stopping bleeding. The blood flowing through an artery is under a great deal of pressure. If you accidentally cut an artery, you will notice that the blood flows out in rapid spurts. Great quantities of blood can be quickly lost from a broken artery. When an artery is cut, its walls constrict so that blood does not flow out of it quite so rapidly. However, there is a limit to how much vasoconstriction can help.

Most parts of the body receive branches from more than one artery. In such areas the distal ends of the vessels unite and the junction of two or more vessels supplying the same body region is called an **anastomosis.** Anastomoses may also occur between the origins of veins and between arterioles and venules. Anastomoses between arteries provide alternate routes by which blood can reach a given tissue or organ. Thus, if a vessel is occluded by disease, injury, or surgery, circulation to a part of the body is not necessarily stopped. The alternate route of blood to a body part through an anastomosis is known as **collateral circulation.** An alternate blood route may also be from nonanastomosing vessels that supply the same region of the body.

Capillaries

Capillaries are microscopic vessels measuring about 4 to 12 μm in diameter. They connect arterioles with venules (Figure 11–9b). The function of the capillaries is to permit the exchange of nutrients and gases between the blood and the interstitial fluid. The structure of the capillaries is admirably suited for this purpose. First, the capillary walls are composed of only a single layer of cells (endothelium). This means that a substance in the blood must diffuse through the plasma membranes of just one cell in order to reach the interstitial fluid. It should be noted again that this vital exchange of materials occurs only through capillary walls. The thick walls of arteries and veins present too great a barrier for diffusion to occur. Capillaries are also well suited to their function in that they branch to form an extensive *capillary network* throughout the tissue. The network increases the surface area through which diffusion can take place and thereby allow a rapid exchange of large quantities of materials.

Though capillaries lack the elastic connective fibers of arteries, their walls are still capable of distention. This permits them to adjust to the amount and force of blood flowing through them.

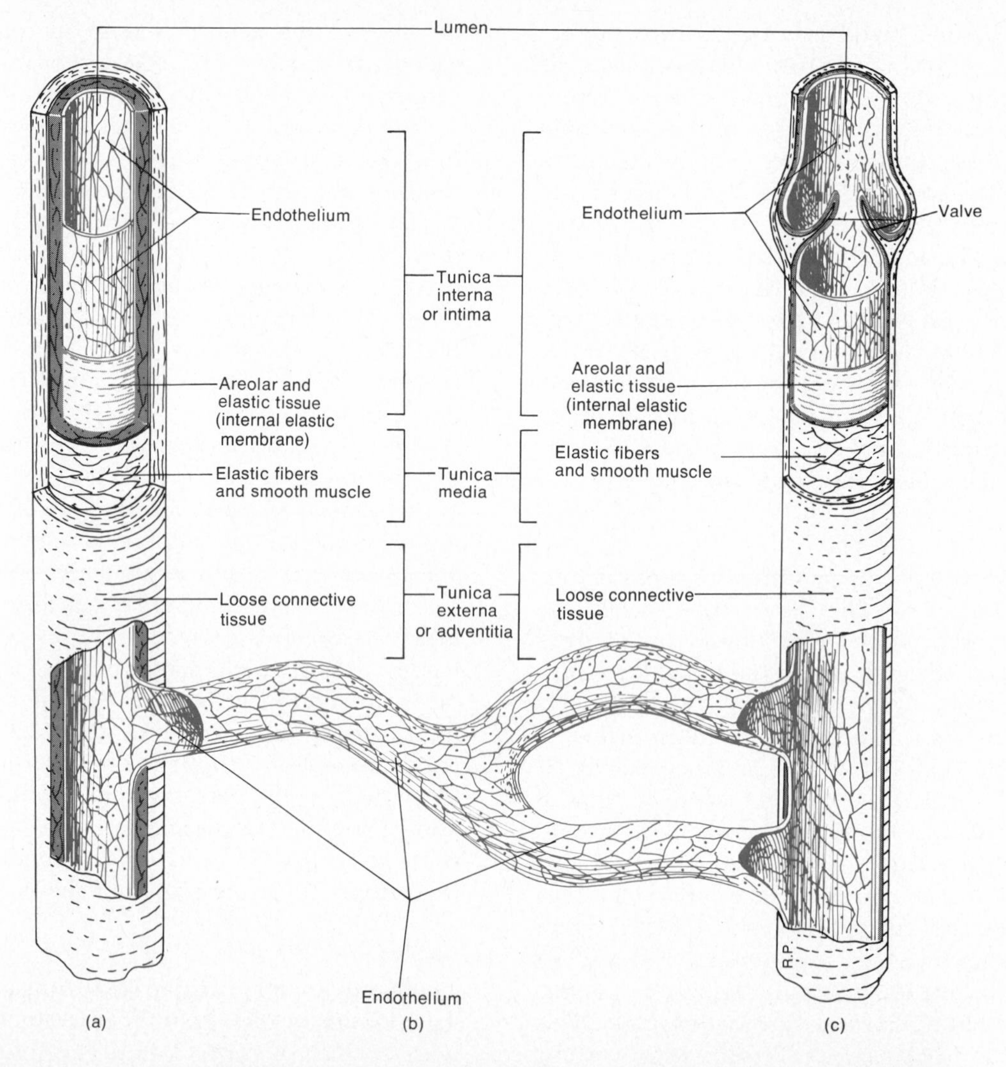
Lumen
Endothelium
Tunica interna or intima
Endothelium
Valve
Areolar and elastic tissue (internal elastic membrane)
Areolar and elastic tissue (internal elastic membrane)
Elastic fibers and smooth muscle
Tunica media
Elastic fibers and smooth muscle
Loose connective tissue
Tunica externa or adventitia
Loose connective tissue
Endothelium
R.P.
(a)
(b)
(c)

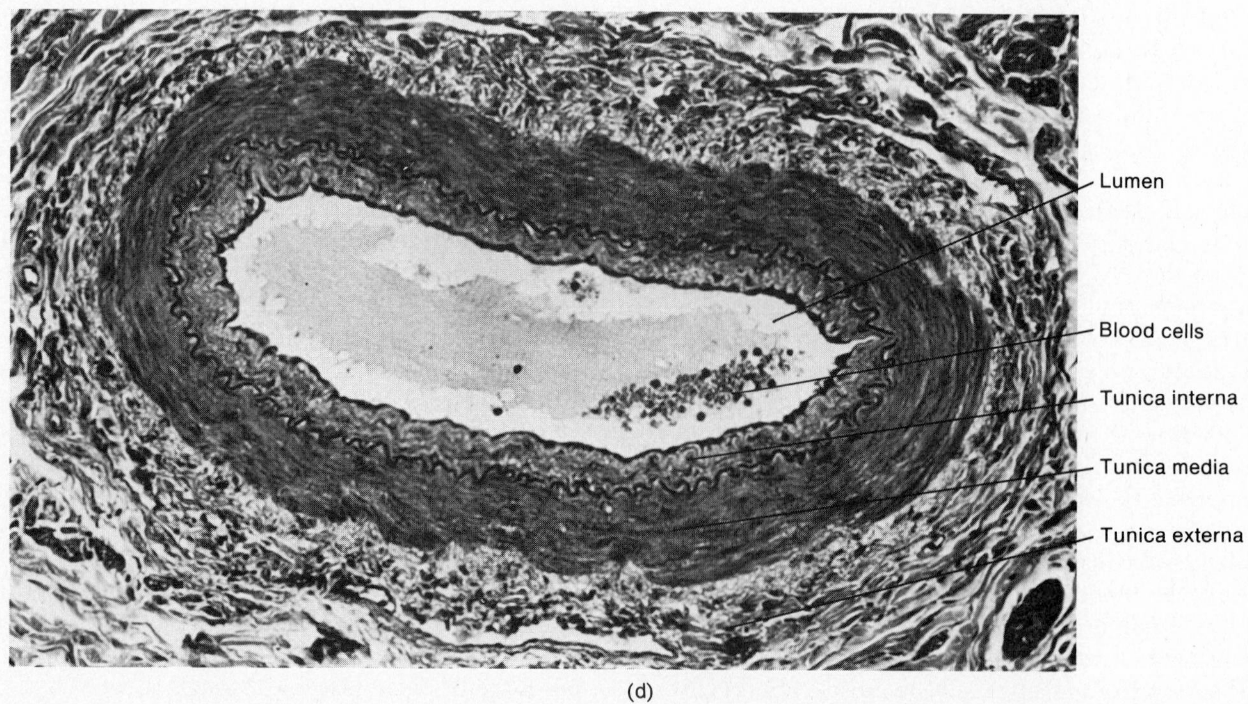
Lumen
Blood cells
Tunica interna
Tunica media
Tunica externa

(d)

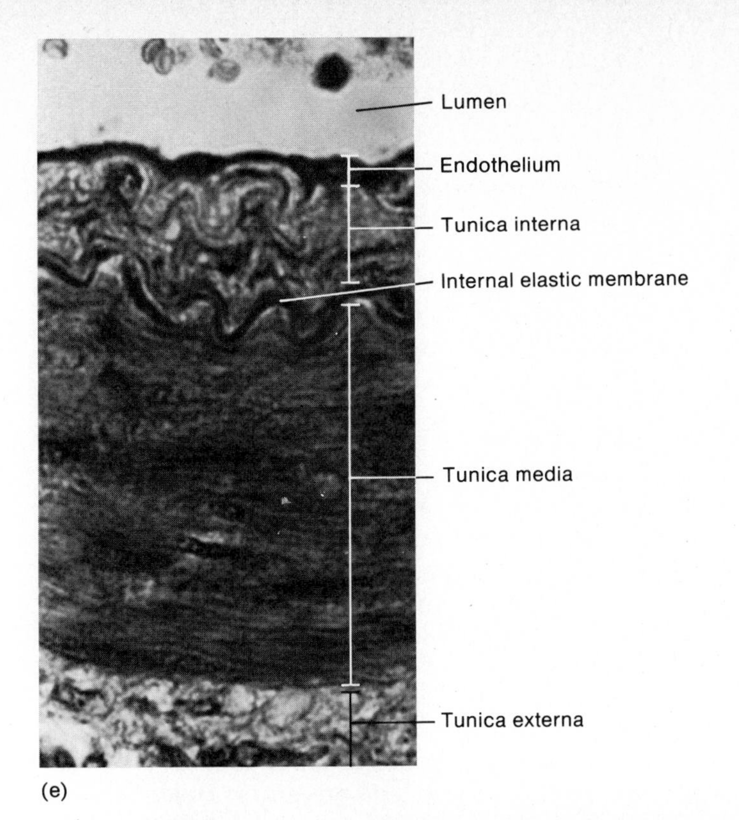

(e)

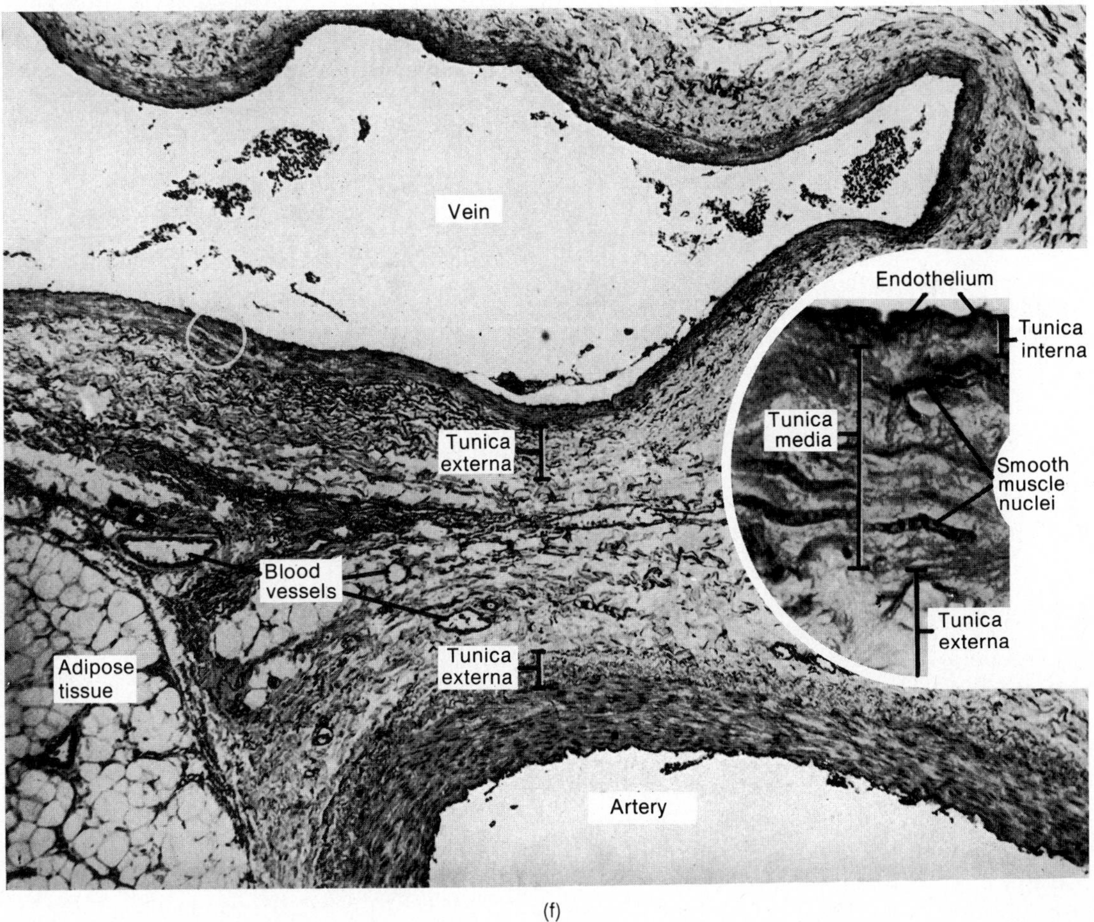

(f)

Figure 11–9. (a-c) Structure of blood vessels. (a) Artery. (b) Capillary. (c) Vein. The relative size of the capillary is enlarged for emphasis. (d-f) Histology of arteries and veins. (d) Cross section of an artery at a magnification of 50X. (e) Enlarged aspect of an artery at a magnification of 200X. (f) Comparison of the structure of an artery and its accompanying vein at a magnification of 65X. The inset is an enlarged aspect of the layers of a vein at a magnification of 640X. (Photomicrographs d and e courtesy of Victor B. Eichler, Wichita State University. Photomicrograph f courtesy of Edward J. Reith, from *Atlas of Descriptive Histology,* by Edward J. Reith and Michael H. Ross, Harper & Row, Publishers, Inc., New York, 1970.)

Veins

Veins are composed of essentially the same three coats as arteries, but they have considerably less elastic tissue and smooth muscle (Figure 11–9c). However, veins do contain more white fibrous tissue. They are also distensible enough to adapt to variations in the volume and pressure of blood passing through them. If you cut a vein, you will notice that the blood leaves the vessel in an even flow rather than in the rapid spurts characteristic of arteries. This is because by the time the blood leaves the capillaries and moves into the veins, it has lost a great deal of its pressure. Most of the structural differences between arteries and veins reflect this pressure difference. For example, veins do not need to have walls that are as strong as the walls of their corresponding arteries. The low pressure in veins, however, has its disadvantages. For instance, when you stand, the pressure pushing blood up the veins in your legs is barely enough to balance the force of gravity pushing it back down. For this reason, many veins, especially those in the limbs, contain valves that prevent any backflow. Normal valves ensure the flow of blood toward the heart.

Some people inherit weak valves. When the valves in a vein are weak, large quantities of blood are forced by gravity back down into distal parts of the vein. This overloads the vein and pushes the walls outward. After repeated overloading, the walls lose their elasticity and become permanently stretched and flabby like an overused rubber band. A vein that has been damaged in this way is called a *varicose vein.* A varicosed wall is not able to exert a firm resistance against the blood. Instead of moving upward, some of the blood tends to accumulate in the pouched-out area of the vein, causing it to swell and forcing fluid out into the surrounding tissue. The veins that lie close to the external surfaces of the legs are highly susceptible to varicosities. Veins that lie deeper are not as vulnerable because surrounding skeletal muscles prevent their walls from overstretching. Varicosities are also common in the veins that lie in the walls of the rectum. These are called *hemorrhoids.*

The arteries, arterioles, capillaries, venules, and veins are organized into definite routes in order to circulate the blood throughout the body. We can now look at the basic routes the blood takes as it is transported through its vessels.

CIRCULATORY ROUTES

Looking at Figure 11–10, you will see that there are a number of basic *circulatory routes* through which the blood travels. The largest route by far is the *systemic circulation.* This includes all the oxygenated blood that leaves the left ventricle through the aorta and returns to the right atrium after traveling to all the organs including the nutrient arteries to the lungs. Two of the many subdivisions of the systemic circulation are the *coronary circulation,* which supplies the myocardium of the heart, and the *hepatic portal circulation,* which runs from the digestive tract to the liver. Blood leaving the aorta and traveling through the systemic arteries is a bright red color. As it moves through the capillaries, it loses its oxygen and takes on carbon dioxide. The carbon dioxide gives the blood in the systemic veins its dark red color. When blood returns to the heart from the systemic route, it then goes out of the right ventricle through the *pulmonary circulation* to the lungs. In the lungs, it loses its carbon dioxide and takes on oxygen. It is now a bright red color again. It returns to the left atrium of the heart and reenters the systemic circulation. A third major route is one that exists only in the fetus and contains special structures that allow the developing human to exchange materials with its mother. This is the *fetal circulation,* which will be described later.

Systemic circulation

The flow of blood from the left ventricle to all parts of the body except the lungs and back to the right atrium is called the **systemic circulation.** The purpose of systemic circulation is to carry oxygen and nutrients to body tissues and to remove carbon dioxide and other wastes from the tissues. All systemic arteries branch from the *aorta,* which arises from the left ventricle of the heart. (See Figure 11–11.) As the aorta emerges from the left ventricle, it passes upward deep to the pulmonary artery. At this point, it is called the *ascending aorta.* The ascending aorta divides to send two coronary branches off to the heart muscle. Then the ascending aorta turns to the left and then downward. As it makes the turn, it is called the *arch of the aorta.* As it runs down to the fourth lumbar vertebra, it is called the *descending aorta.* The descending aorta lies close to the vertebral bodies, passes through the diaphragm to the fourth lumbar vertebra, and terminates at this level by dividing into two *common iliac arteries,* which carry blood to the lower extremities. The section of the descending aorta between the arch of aorta and the diaphragm is also referred to as the *thoracic aorta.* The section between the diaphragm and the common iliac arteries is termed the *abdominal aorta.* Each section of the aorta gives off arteries that continue to branch

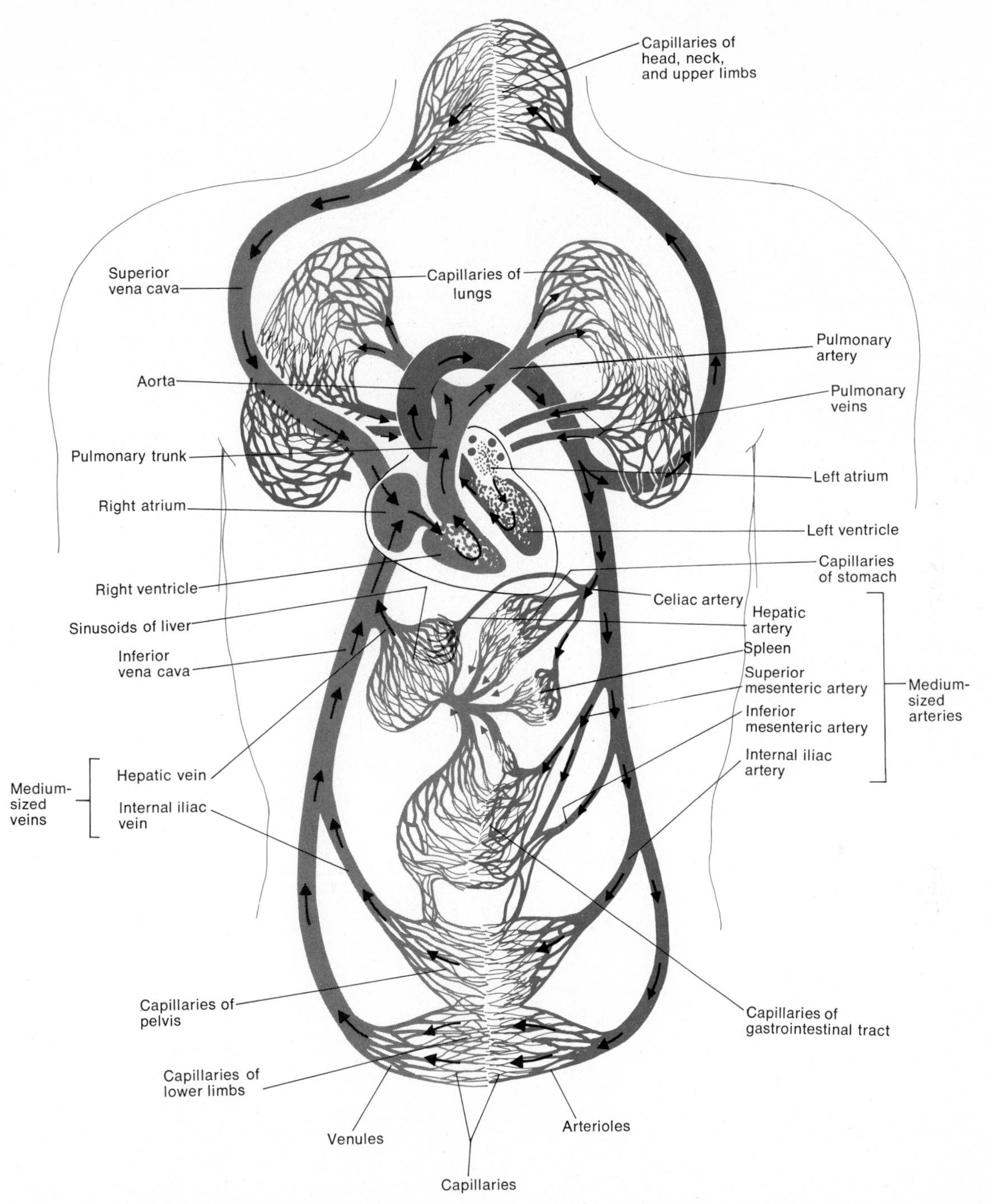

Figure 11–10. Circulatory routes. Systemic circulation is indicated by heavy black arrows; pulmonary circulation by thin black arrows; and hepatic portal circulation by thin colored arrows. Refer to Figure 11–21 for the details of coronary circulation and Figure 11–24 for the details of fetal circulation.

into distributing arteries leading to organs and finally into the capillaries that pierce the tissues.

Blood is returned to the heart through the systemic veins. All the veins of the systemic circulation flow into either the *superior* or *inferior venae cavae* or the *coronary sinus.* They in turn empty into the right atrium. The principal arteries and veins of the systemic circulation are described and illustrated in Exhibits 11–2 to 11–13 and Figures 11–11 to 11–20, which follow, and in the subsequent sections on coronary and portal circulation.

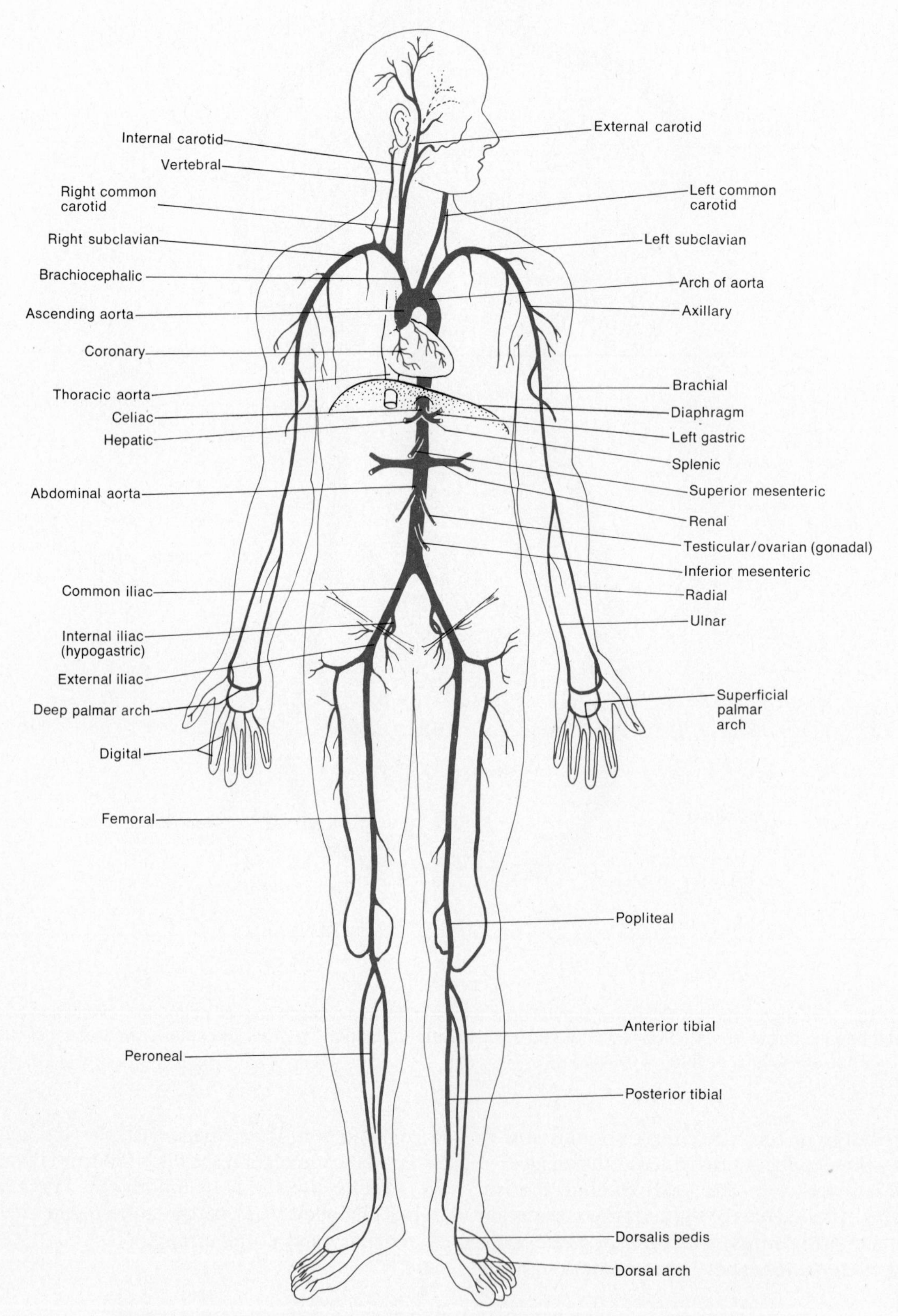

Figure 11–11. Anterior view of the aorta and its principal arterial branches.

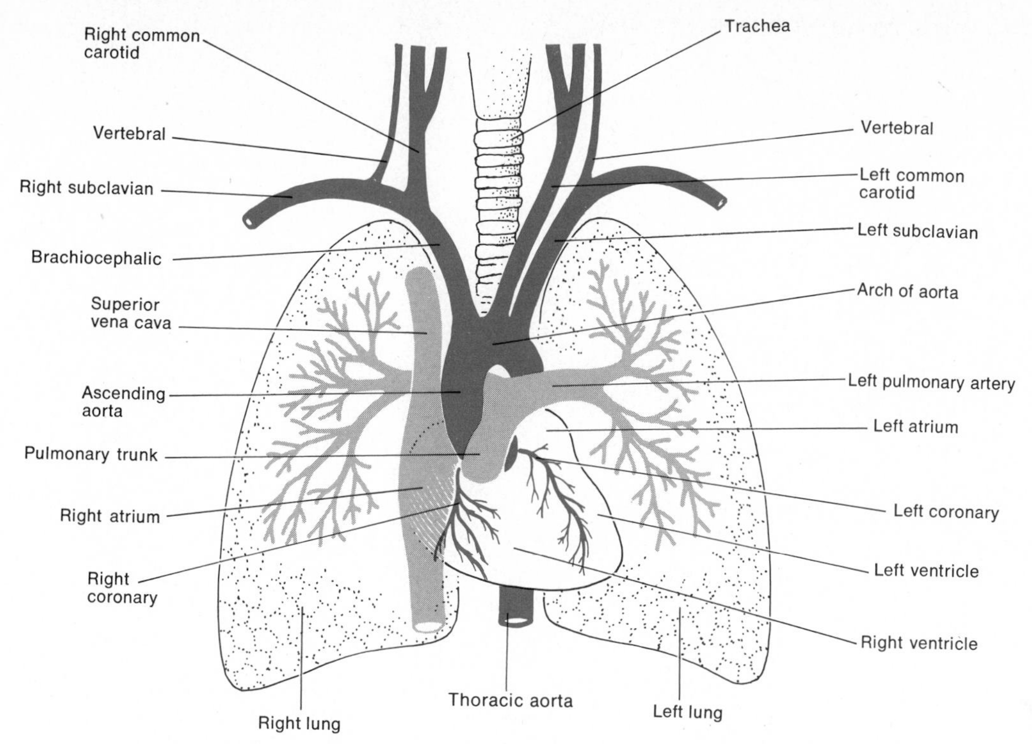

Figure 11–12. Anterior view of ascending aorta and its arterial branches.

Exhibit 11–2. THE AORTA AND ITS BRANCHES (see Figure 11–11)

DIVISION OF AORTA	ARTERIAL BRANCH	REGION SUPPLIED
Ascending aorta	Right and left coronary	Heart
Arch of aorta	Brachiocephalic → Right common carotid	Right side of head and neck
	Brachiocephalic → Right subclavian	Right upper extremity
	Left common carotid	Left side of head and neck
	Left subclavian	Left upper extremity
Thoracic aorta	Intercostals	Intercostal and chest muscles, pleurae
	Superior phrenics	Posterior and superior surfaces of diaphragm
	Bronchials	Bronchi of lungs
	Esophageals	Esophagus
	Inferior phrenics	Inferior surface of diaphragm
Abdominal aorta	Celiac → Common hepatic	Liver
	Celiac → Left gastric	Stomach and esophagus
	Celiac → Splenic	Spleen, pancreas, and stomach
	Superior mesenteric	Small intestine, cecum, ascending and transverse colons
	Suprarenals	Adrenal glands
	Renals	Kidneys
	Gonadals → Testiculars	Testes
	Gonadals → Ovarians	Ovaries
	Inferior mesenteric	Transverse, descending, and sigmoid colons; rectum
	Common iliacs → External iliacs	Lower extremities
	Common iliacs → Internal iliacs (hypogastrics)	Uterus, prostate, muscles of buttocks

Exhibit 11–3. THE ASCENDING AORTA (see Figure 11–12)

BRANCH	DESCRIPTION AND REGION SUPPLIED
Coronary arteries	The two **coronary arteries** are branches that arise from the ascending aorta just superior to the aortic semilunar valve. They form a crown around the heart giving off branches to the atrial and ventricular myocardium. (See Figure 11–21 for the details of coronary circulation.)

Exhibit 11–4. THE ARCH OF THE AORTA (see Figure 11–13)

BRANCH	DESCRIPTION AND REGION SUPPLIED
Brachiocephalic	The **brachiocephalic artery** is the first branch off the arch of the aorta. It subdivides to form the right subclavian artery and right common carotid artery. (See Figure 11–12.) The **right subclavian artery** extends from the brachiocephalic to the first rib, passes into the armpit, or **axilla,** and supplies the arm, forearm, and hand. See Figure 11–13a. This artery is a good example of the convention of giving the same vessel different names as it passes through different regions. The continuation of the right subclavian into the axilla is called the **axillary artery.** From here, it continues into the upper arm as the **brachial artery.** At the bend of the elbow the brachial artery divides into the medial **ulnar** and lateral **radial arteries.** These vessels pass down to the palm, one on each side of the forearm. In the palm, branches of the two arteries anastomose to form two palmar arches—the **superficial palmar arch** and the **deep palmar arch.** From these arches arise the **digital arteries,** which supply the fingers and the thumb. Before passing into the axilla, the right subclavian gives off an important branch to the brain called the **vertebral artery.** See Figure 11–13b. The right vertebral artery passes through the foramina of the transverse processes of the cervical vertebrae and enters the skull through the foramen magnum to reach the undersurface of the the brain. Here, it unites with the left vertebral artery to form the **basilar artery.** Anastomoses of the left and right internal carotids along with the basilar artery form an arterial circle at the base of the brain called the **circle of Willis.** See Figure 11–13c. From this anastomosis arises the arteries supplying the brain. Essentially, the circle of Willis is formed by the union of the **anterior cerebral arteries** (branches of the internal carotids) and the **posterior cerebral arteries** (branches of the basilar artery). The posterior cerebral arteries are connected with the internal carotids by the **posterior communicating arteries.** The anterior cerebral arteries are connected by the **anterior communicating arteries.** The circle of Willis is important because it equalizes blood pressure to the brain and provides alternate routes for blood to the brain should arteries become diseased or damaged.
Right common carotid	The **right common carotid artery** passes upward. At the upper level of the larynx, it divides into the **right external** and **right internal carotid arteries.** See Figure 11–13b. The external carotid supplies the right sides of the thyroid gland, tongue, throat, face, ear, scalp, and dura mater. The internal carotid supplies the brain, right eye, and right sides of the forehead and nose.
Left common carotid	The **left common carotid** branches directly from the arch of the aorta. (See Figure 11–12.) Corresponding to the right common carotid, it divides into basically the same branches with the same names—except that the arteries are now labeled "left" instead of "right."
Left subclavian	The **left subclavian artery** is the third branch off the arch of the aorta. (See Figure 11–12.) It distributes blood to the left vertebral artery and to the vessels of the left upper extremity. The arteries branching from the left subclavian are named like those of the right subclavian.

Exhibit 11–5. THE THORACIC AORTA

The **thoracic aorta** runs from the fourth to twelfth thoracic vertebrae. See Figure 11–14. Along its course, it sends off numerous small arteries to the viscera and skeletal muscles of the chest. The *visceral branches* supply the pericardium around the heart, the bronchial tubes that lead from the windpipe to the lungs, the cells of the lungs (but not the areas of the lungs that oxygenate blood), the gullet, and the tissue lining the mediastinum. The *parietal branches* supply the chest muscles, diaphragm, and mammary glands.

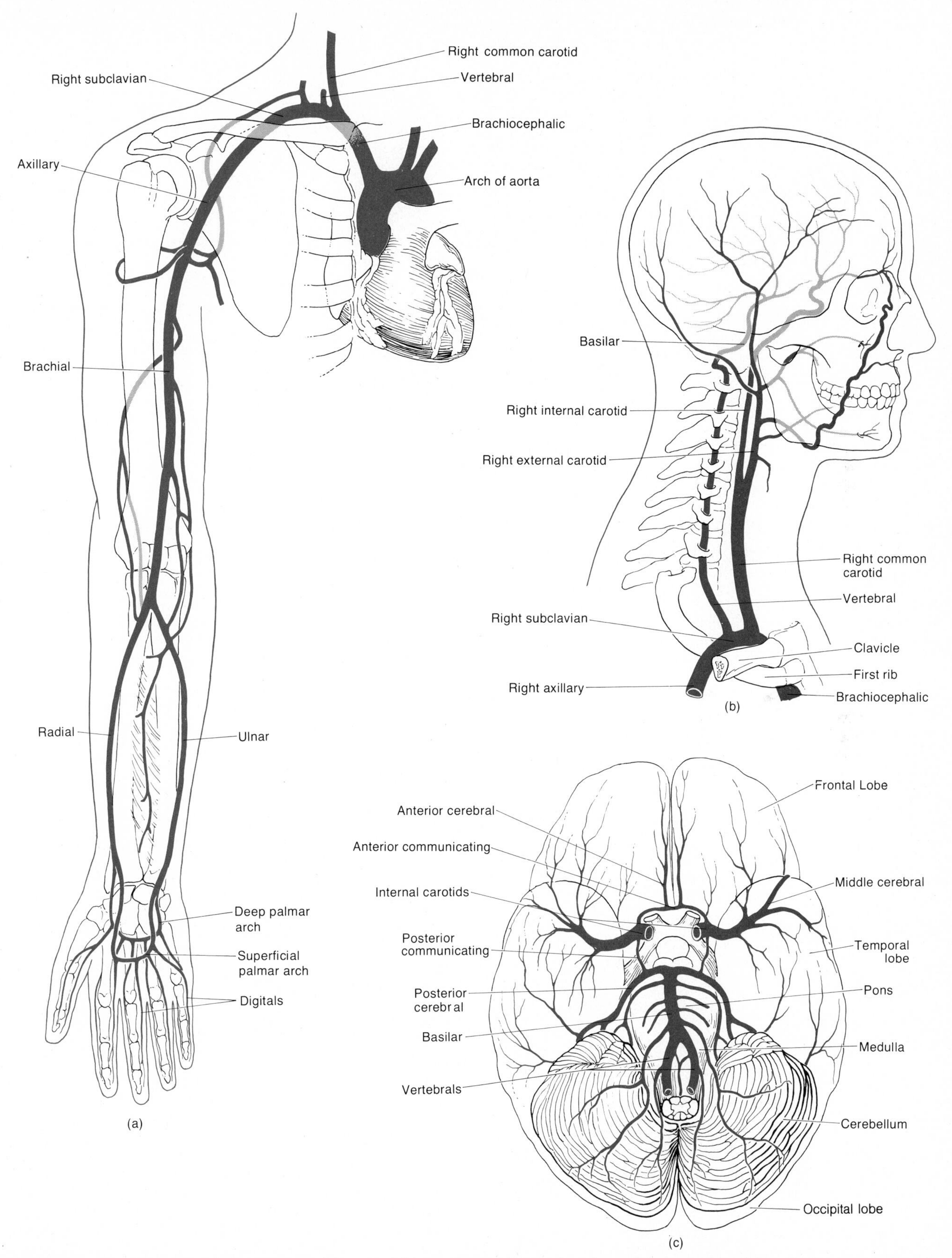

Figure 11–13. The arch of the aorta and its arterial branches. (a) Anterior view of the arteries of the right upper extremity. (a) Right lateral view of the arteries of the neck and head. (c) Arteries of the base of the brain.

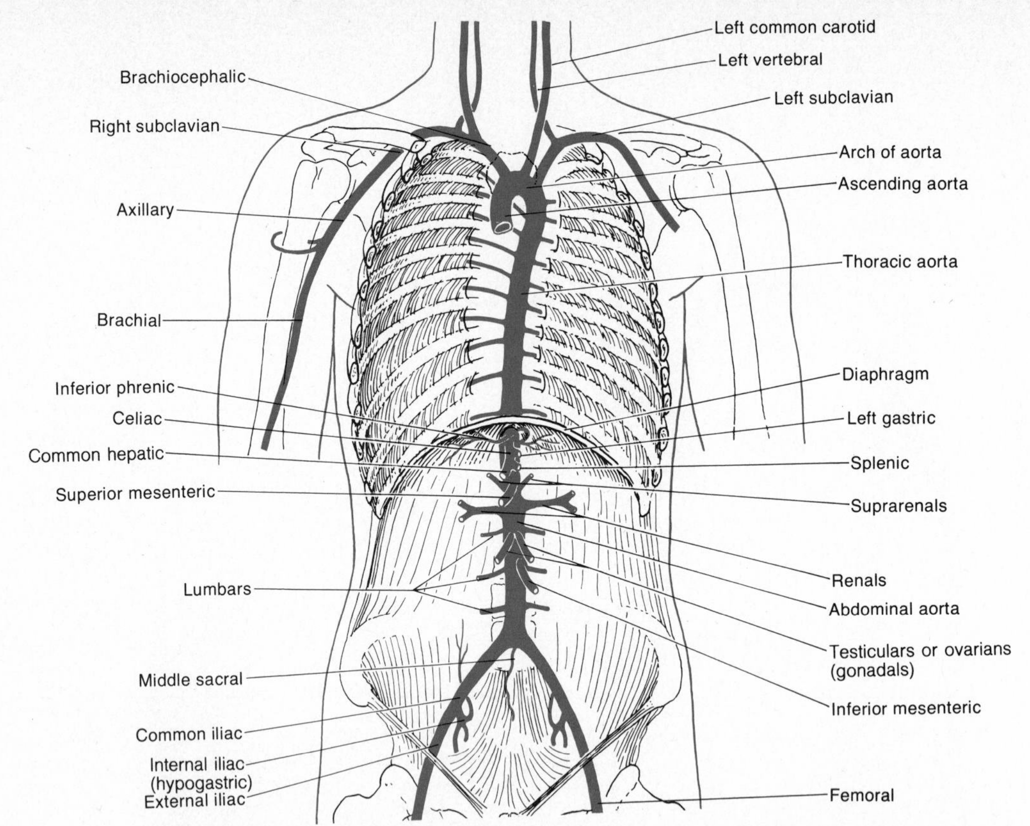

Figure 11–14. Anterior view of the abdominal aorta and its arterial branches.

Exhibit 11–6. THE ABDOMINAL AORTA (see Figure 11–14)

BRANCH	DESCRIPTION AND REGION SUPPLIED
Visceral	
Celiac	The **celiac artery** (trunk) is the first aortic branch below the diaphragm. It has three branches: (1) **common hepatic artery,** which supplies the tissues of the liver; (2) **left gastric artery,** which supplies the stomach; and (3) **splenic artery,** which supplies the spleen, pancreas, and stomach.
Superior mesenteric	The **superior mesenteric artery** distributes blood to the small intestine and to part of the large intestine.
Suprarenals	The right and left **suprarenal arteries** supply blood to the adrenal glands.
Renals	The right and left **renal arteries** carry blood to the kidneys.
Testiculars	The right and left **testicular arteries** extend into the scrotum and terminate in the testes.
Ovarians	The right and left **ovarian arteries** are distributed to the ovaries.
Inferior mesenteric	The **inferior mesenteric artery** supplies the major part of the large intestine and the rectum.
Parietal	
Inferior phrenics	The **inferior phrenic arteries** are distributed to the undersurface of the diaphragm.
Lumbars	The **lumbar arteries** supply the spinal cord and its meninges and the muscles and skin of the lumbar region of the back.
Middle sacral	The **middle sacral artery** supplies the sacrum, coccyx, gluteus maximus muscles, and rectum.

Exhibit 11–7. ARTERIES OF THE PELVIS AND LOWER EXTREMITIES (see Figure 11–15)

BRANCH	DESCRIPTION AND REGION SUPPLIED
	At about the level of the fourth lumbar vertebra, the abdominal aorta divides into the right and left **common iliac arteries.** Each of these, in turn, passes downward about 5 cm (2 in.) and gives rise to two branches: the internal iliac and external iliac.
Internal iliacs	The **internal iliac** or **hypogastric arteries** form branches that supply the gluteal muscles, medial side of each thigh, urinary bladder, rectum, prostate gland, uterus, and vagina.
External iliacs	The **external iliac arteries** diverge through the pelvis, enter the thighs, and here become the right and left **femoral arteries.** Both femorals send branches back up to the genitals and to the wall of the abdomen. Other branches run to the muscles of the thigh. The femoral continues down the medial and posterior side of the thigh at the back of the knee joint, where it becomes the **popliteal artery.** Between the knee and ankle, the popliteal runs down the back of the leg and is called the **posterior tibial artery.** Below the knee, the **peroneal artery** branches off the posterior tibial to supply the structures on the medial side of the fibula and calcaneus. In the calf, another artery, the **anterior tibial artery,** branches off the popliteal and runs along the front of the leg. At the ankle, it becomes the **dorsalis pedis artery.** At the ankle, the posterior tibial divides into the **medial** and **lateral plantar arteries.** These arteries anastomose with the dorsalis pedis and supply blood to the foot.

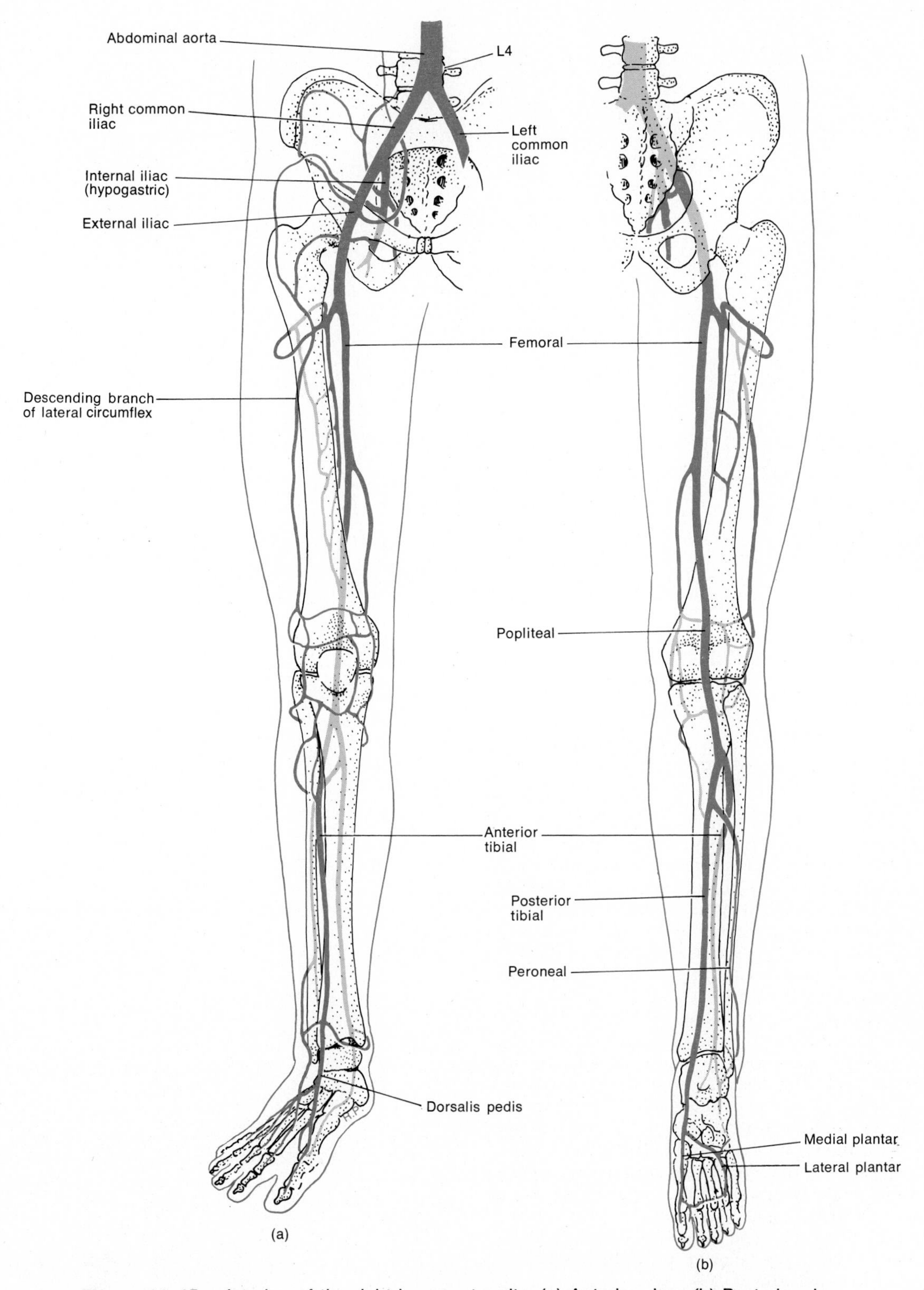

Figure 11–15. Arteries of the right lower extremity. (a) Anterior view. (b) Posterior view.

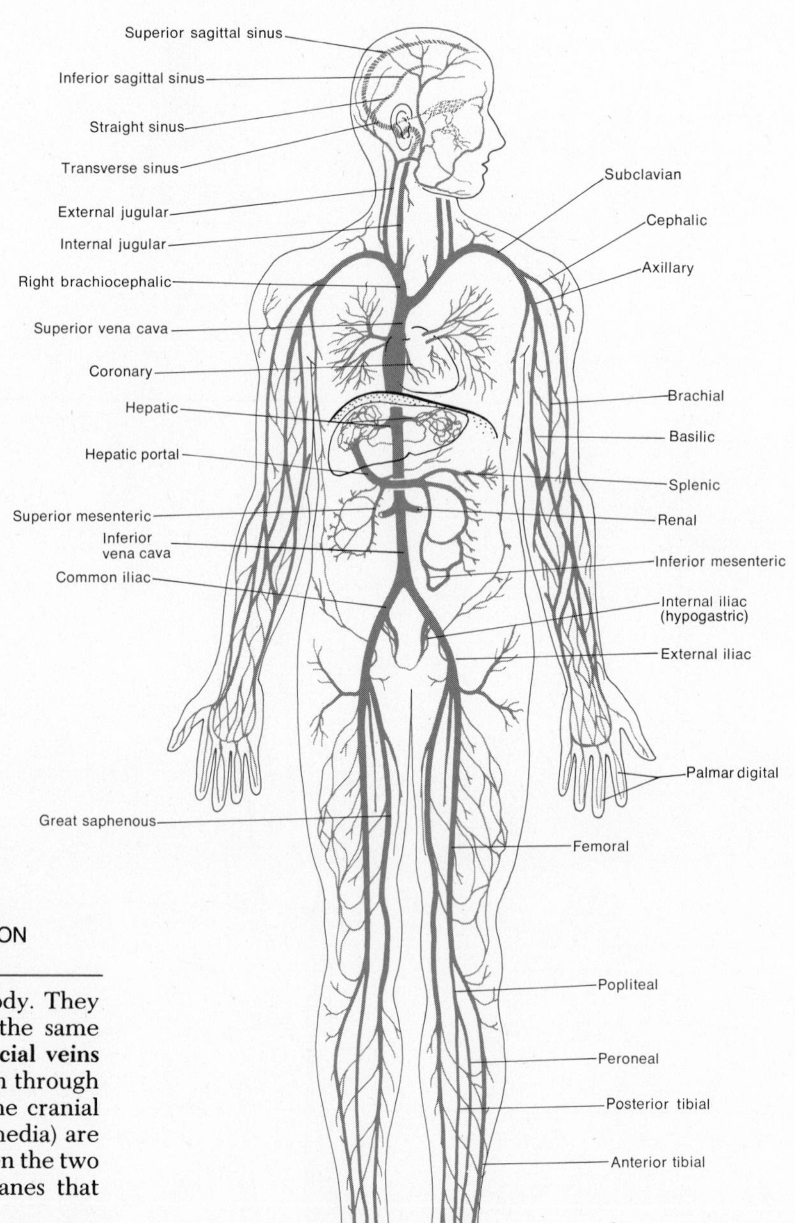

Figure 11–16. Anterior view of the principal veins.

Exhibit 11–8. VEINS OF THE SYSTEMIC CIRCULATION (see Figure 11–16)

Deep veins are located deep within the body. They usually accompany arteries, and many have the same names as their corresponding arteries. **Superficial veins** are located just below the skin and may be seen through the skin. Veins that have very thin walls in the cranial cavity (they lack a tunica externa and tunica media) are called **venous sinuses.** Venous sinuses lie between the two layers of the dura mater, one of the membranes that covers the brain.

All the systemic veins return blood to the right atrium of the heart through one of three large vessels: (1) the coronary sinus, (2) the superior vena cava, and (3) the inferior vena cava. The return flow from the coronary arteries is taken up by the **cardiac veins,** which empty into a large vein of the heart called the **coronary sinus.** From here, the blood empties into the right atrium of the heart. The veins that empty into the **superior vena cava** are the veins of the head and neck, upper extremities, thorax, and the azygos veins. (The azygos veins are discussed in Exhibit 11–11.) The veins that empty into the **inferior vena cava** are the veins of the abdomen, pelvis, lower extremities, and the azygos veins.

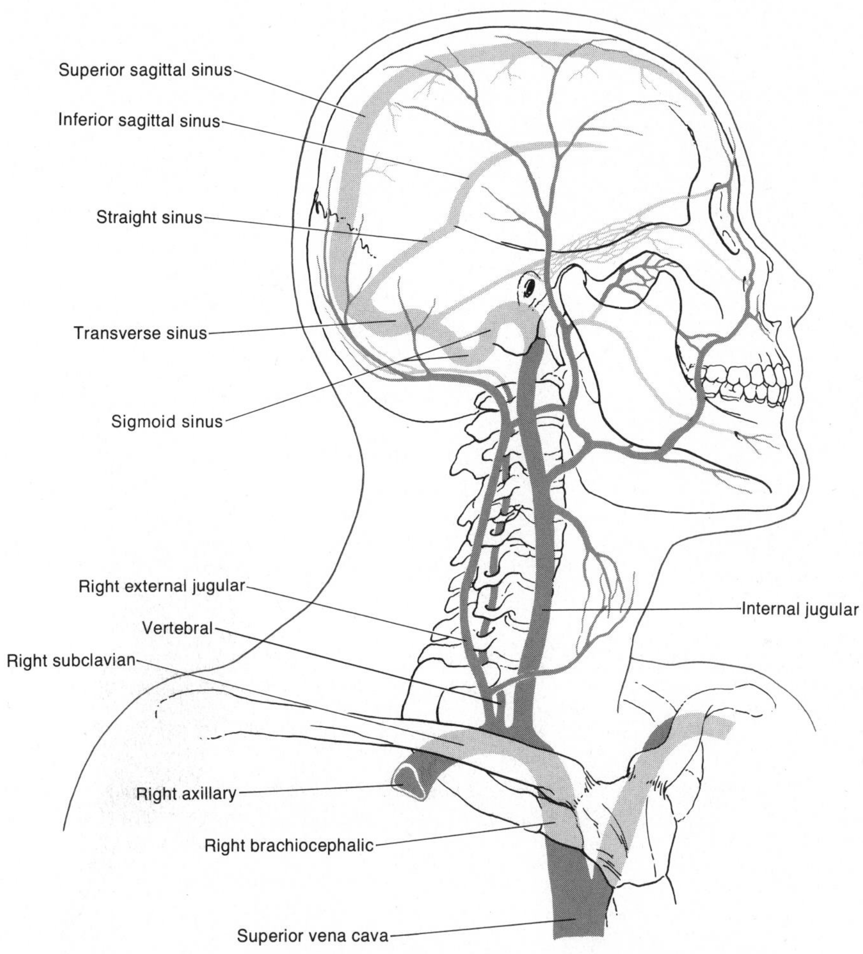

Figure 11–17. Right lateral view of the veins of the neck and head.

Exhibit 11–9. VEINS OF THE HEAD AND NECK (see Figure 11–17)

VEIN	DESCRIPTION AND REGION DRAINED
Internal jugulars	The right and left **internal jugular veins** arise as a continuation of the **transverse (lateral) sinuses** at the base of the skull. The sinuses are located between the layers of the dura mater and receive blood from the brain. Other sinuses that drain into the internal jugular include the **superior sagittal sinus,** the **inferior sagittal sinus,** the **straight sinus,** and **sigmoid sinuses.** The internal jugulars descend on either side of the neck. They receive blood from the superior part of the face and neck and pass behind the clavicles, where they join with the right and left **subclavian veins.** The unions of the internal jugulars and subclavians form the right and left **brachiocephalic veins.** From here the blood flows into the **superior vena cava.**
External jugulars	The left and right **external jugular veins** run down the neck, along the outside of the internal jugulars. They drain blood from the parotid (salivary) glands, facial muscles, scalp, and other superficial structures into the **subclavian veins.**

Exhibit 11–10. VEINS OF THE UPPER EXTREMITIES (see Figure 11–18)

VEIN	DESCRIPTION AND REGION DRAINED
	Blood from the upper extremity is returned to the heart by deep and superficial veins. Both sets of veins contain valves.
Deep veins	The deep veins run alongside arteries and are called the **radial, ulnar, brachial, axillary,** and **subclavian veins.**
Superficial veins	The superficial veins anastomose extensively with each other and with the deep veins. They include the following:
Cephalics	The **cephalic vein** of each upper extremity begins in the **dorsal arch** and winds upward around the radial border of the forearm. At a point just below the elbow, it unites with the accessory cephalic vein to form the cephalic vein of the upper extremity. It eventually empties into the axillary vein.
Basilics	The **basilic vein** of each upper extremity originates in the ulnar part of the dorsal arch. It extends along the posterior surface of the ulna to a point below the elbow where it joins the **median cubital vein.** If a vein must be punctured for an injection, transfusion, or removal of a blood sample, the median cubitals are the preferred veins.
Axillaries	The **axillary vein** is a continuation of the basilic. It ends at about the first rib, where it becomes the subclavian.
Subclavians	The right and left **subclavian veins** unite with the internal jugulars to form the **brachiocephalic veins.** The thoracic duct of the lymphatic system flows into the left subclavian vein at its junction with the internal jugular. On the right, the right lymphatic duct enters the right subclavian vein at the corresponding junction.

Exhibit 11–11. VEINS OF THE THORAX (see Figure 11–19)

VEIN	DESCRIPTION AND REGION DRAINED
	The principal thoracic vessels that empty into the superior vena cava are the brachiocephalic veins and the azygos veins.
Brachiocephalic	The right and left **brachiocephalic veins,** formed by the union of the subclavians and internal jugulars, drain blood from the head, neck, upper extremities, mammary glands, and upper thorax. The brachiocephalics unite to form the **superior vena cava.**
Azygos	The **azygos veins,** in addition to collecting blood from various parts of the thorax, serve as a bypass for the inferior vena cava that drains blood from the lower part of the body. Several small veins directly link the azygos veins with the vena cava. And the large veins that drain the lower extremities and abdomen dump some of their blood into the azygos. If the inferior vena cava or hepatic portal vein becomes obstructed, the azygos veins can return blood from the lower part of the body to the superior vena cava. The three azygos veins are as follows:
Azygos vein	The **azygos vein** lies in front of the vertebral column, slightly to the right of the midline. It begins as a continuation of the right ascending lumbar vein. And it has connections with the inferior vena cava, right common iliac, and lumbar veins. The azygos receives blood from the right intercostal veins that drain the chest muscles; from the hemiazygos and accessory hemiazygos veins (discussed below); from several esophageal, mediastinal, and pericardial veins; and from the right bronchial vein. The vein ascends to the fourth thoracic vertebra, arches over the right lung, and empties into the superior vena cava.
Hemiazygos vein	The **hemiazygos vein** is in front of the vertebral column and slightly to the left of the midline. It begins as a continuation of the left ascending lumbar vein. It receives blood from the lower four or five intercostal veins and from some esophageal and mediastinal veins. At about the level of the ninth thoracic vertebra, it connects with the azygos vein.
Accessory hemiazygos vein	The **accessory hemiazygos vein** is also in front and to the left of the vertebral column. It receives blood from three or four intercostal veins and from the left bronchial vein. It joins the azygos at the level of the eighth thoracic vertebra.

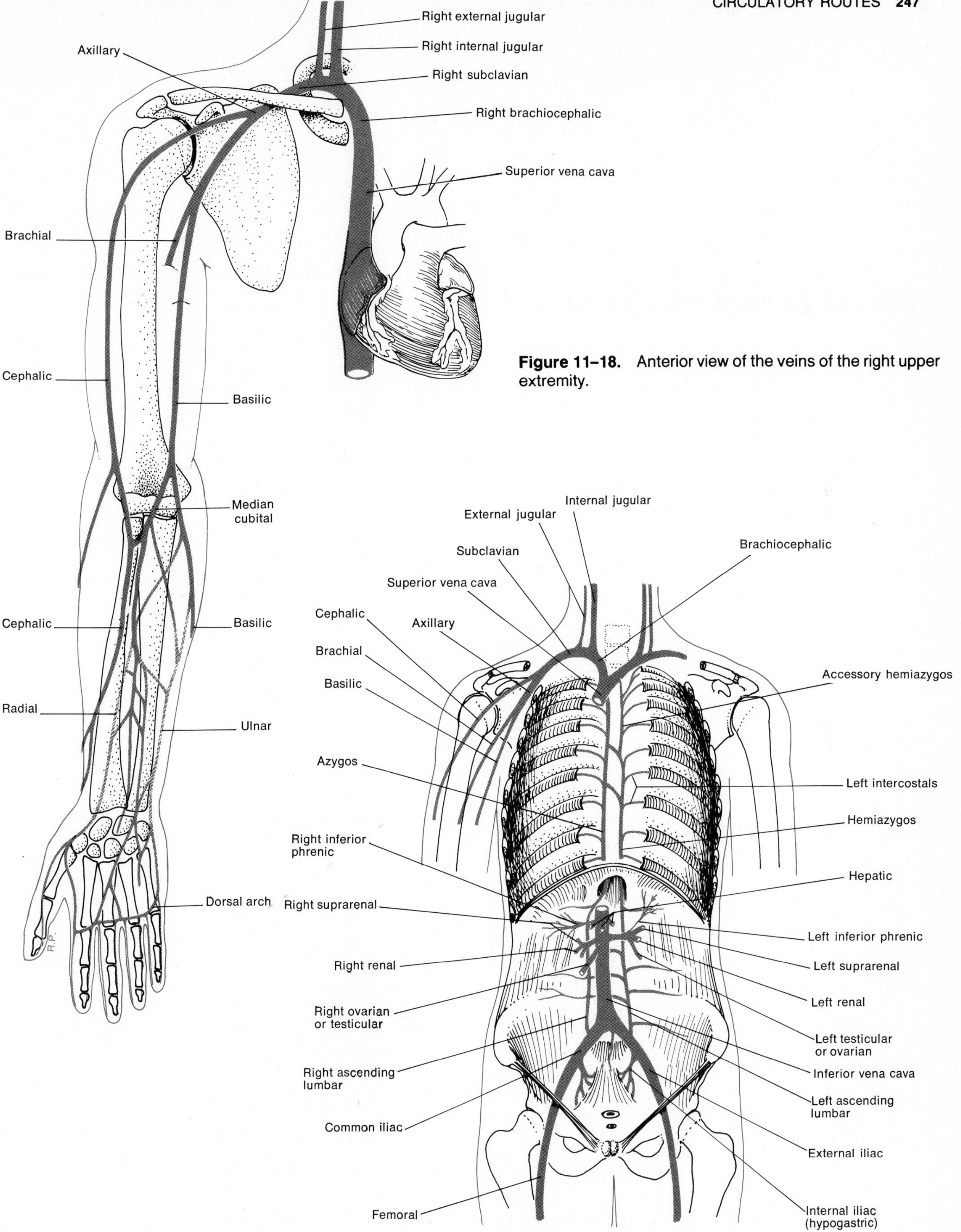

Figure 11–18. Anterior view of the veins of the right upper extremity.

Figure 11–19. Anterior view of the veins of the thorax, abdomen, and pelvis.

Exhibit 11–12. VEINS OF THE ABDOMEN AND PELVIS

VEIN	DESCRIPTION AND REGION DRAINED
Inferior vena cava	The **inferior vena cava** is the largest vein of the body. (See Figure 11–19.) It is formed by the union of the two **common iliac veins** that drain the lower extremities and abdomen. It extends upward through the abdomen and thorax to the right atrium. Numerous small veins enter the inferior vena cava. For the most part, they carry return flow from branches of the abdominal aorta, and their names correspond to the names given to arteries. Among these are: the left and right **renal veins** from the kidneys; the right **testicular vein** from the testes (the left testicular vein empties into the left renal vein); the right **ovarian vein** from the ovaries (the left ovarian vein also empties into the left renal vein); the right **suprarenal veins** from the adrenal glands (the left suprarenal vein empties into the left renal vein); the right **inferior phrenic vein** from the diaphragm (the left inferior phrenic vein sends a tributary to the left renal vein); and the **hepatic veins** from the liver. In addition, a series of roughly parallel **lumbar veins** drain blood from both sides of the posterior abdominal wall. The lumbars connect at right angles with the right and left ascending lumbar veins, which form the origin of the corresponding azygos or hemiazygos vein. The lumbars drain some of their blood into the ascending lumbars and then run to the inferior vena cava, where they release the remainder of their flow.

Exhibit 11–13. VEINS OF THE LOWER EXTREMITIES (see Figure 11–20)

VEIN	DESCRIPTION AND REGION DRAINED
	Blood from each lower extremity is returned by a superficial and deep set of veins. The superficials are formed from extensive anastomoses close to the surface, whereas the deep veins follow the large arterial trunks. Valves are present in both sets.
Superficial veins	The principal superficial veins are the great saphenous and the small saphenous. Both veins, especially the great saphenous, frequently become varicosed.
Great saphenous	The **great saphenous vein,** the longest vein in the body, begins at the medial end of the **dorsal venous arch** of the foot. It passes in front of the medial malleolus and then upward along the medial aspect of the leg and thigh. It receives tributaries from superficial tissues and connects with deep veins as well. It empties into the femoral vein in the region of the groin.
Small saphenous	The **small saphenous vein** begins at the lateral end of the dorsal venous arch of the foot. It passes behind the lateral malleolus and ascends under the skin of the back of the leg. It receives blood from the foot and posterior portion of the leg. And it empties into the popliteal vein behind the knee.
Deep veins	Among the deep veins of the lower extremity are the following:
Posterior tibial	The **posterior tibial vein** is formed by the union of the **medial** and **lateral plantar veins** behind the medial malleolus. It ascends deep in the muscle at the back of the leg, receives blood from the **peroneal vein,** and unites with the anterior tibial vein just below the knee.
Anterior tibial	The **anterior tibial vein** is an upward continuation of the **dorsalis pedis** veins in the foot. It runs between the tibia and fibula and unites with the posterior tibial to form the popliteal vein.
Popliteal	The **popliteal vein,** located just behind the knee, receives blood from the anterior and posterior tibials and the small saphenous vein.
Femoral	The **femoral vein** is an upward continuation of the popliteal just above the knee. The femorals run up the posterior of the leg and drain deep structures of the thighs. After receiving the great saphenous veins in the region of the groin, they continue as the right and left **external iliac veins.** The right and left **internal iliac veins** receive blood from the pelvic wall and viscera, external genitals, buttocks, and medial aspect of the thigh. The right and left **common iliac veins** are formed by the union of the internal and external iliacs. The common iliacs unite to form the inferior vena cava.

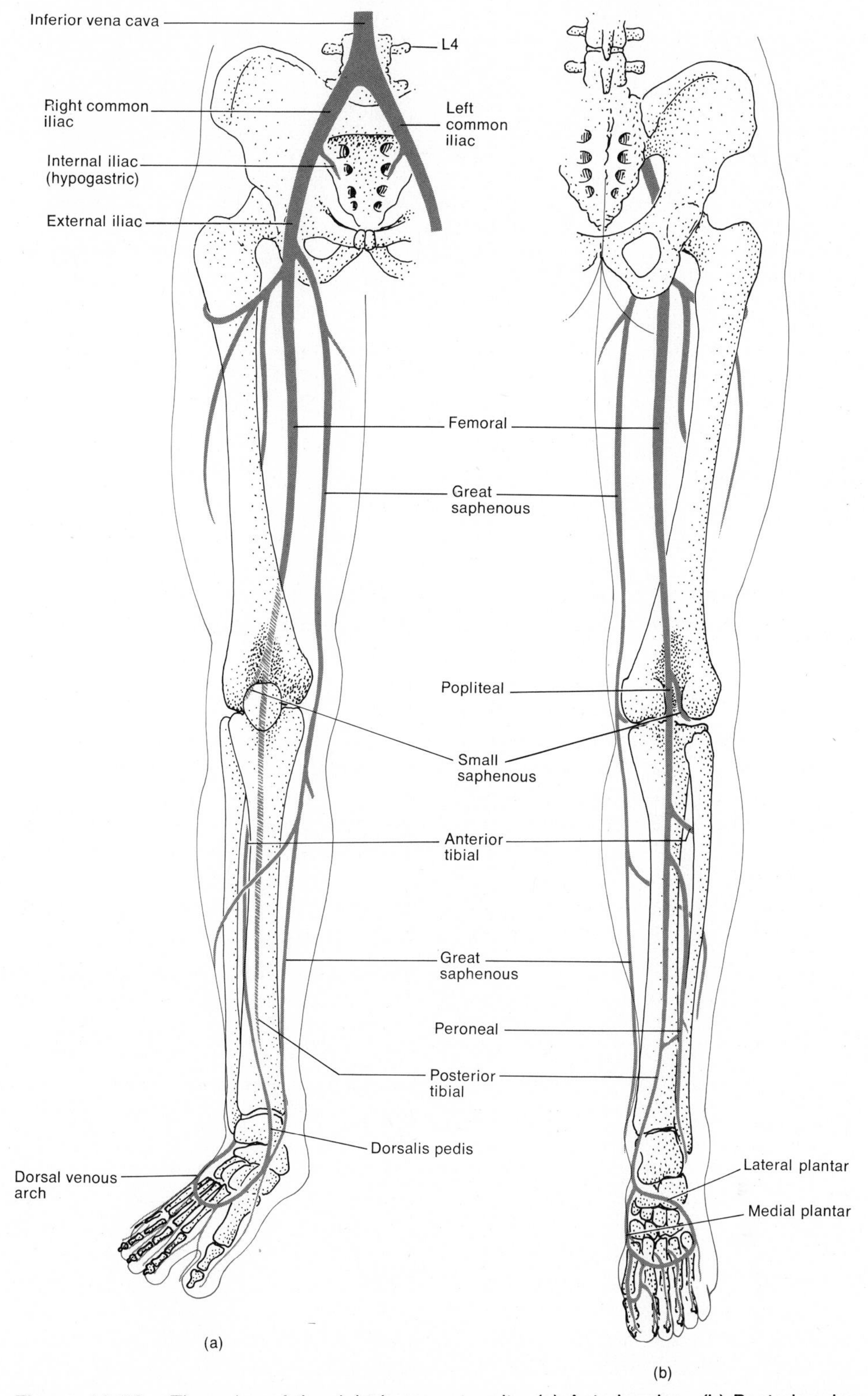

Figure 11–20. The veins of the right lower extremity. (a) Anterior view. (b) Posterior view.

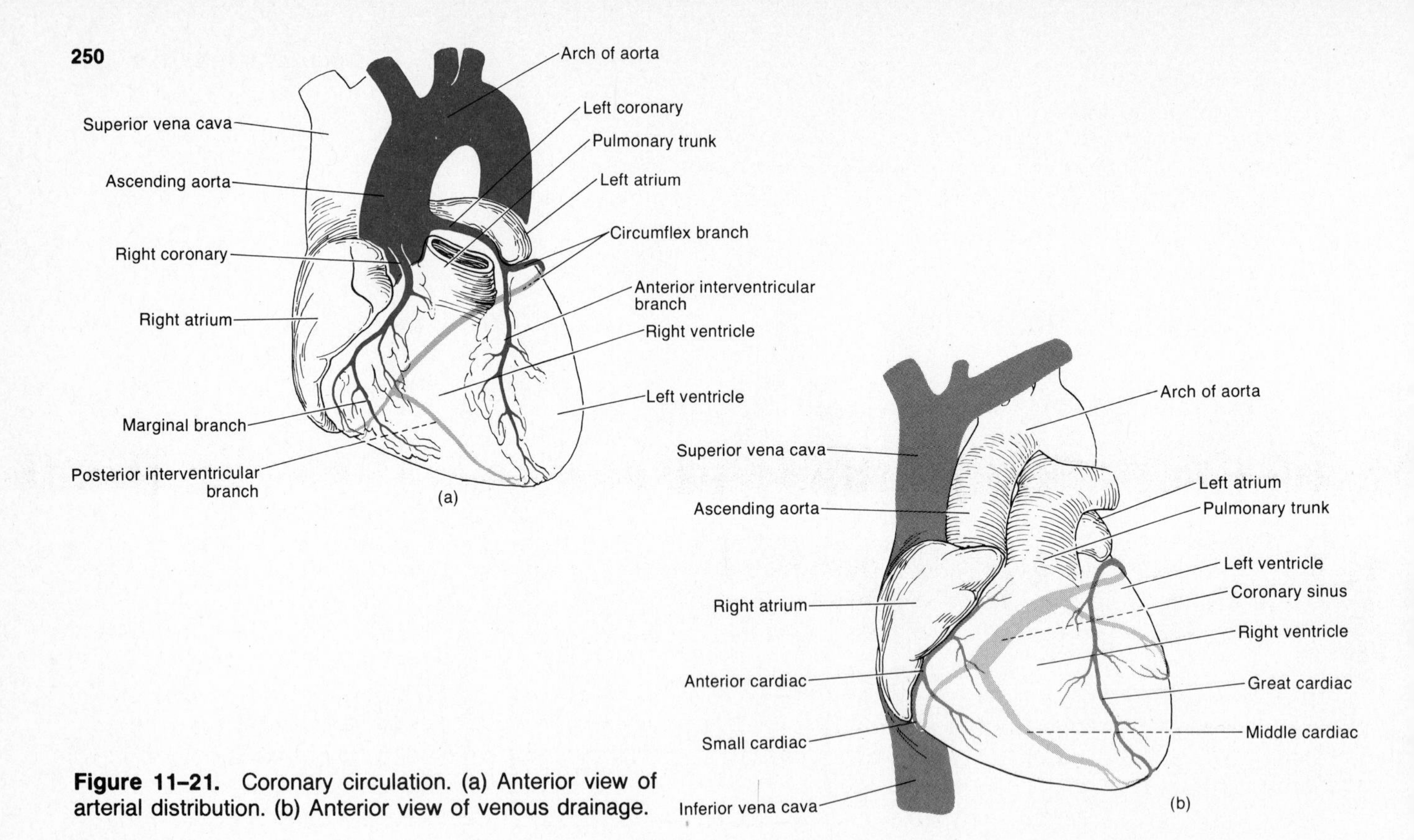

Figure 11–21. Coronary circulation. (a) Anterior view of arterial distribution. (b) Anterior view of venous drainage.

Coronary circulation

You may have assumed that the blood collected in the chambers of the heart supplies the walls of the heart with oxygen and nutrients. In that case, you will be surprised to learn that the walls of the heart, like any other tissue, including larger blood vessels, have their own blood vessels. Nutrients could not possibly diffuse through all the layers of cells that make up the heart tissue. And the blood in the right chambers of the heart would never supply enough oxygen. The flow of blood through the numerous vessels that pierce the myocardium is called the **coronary circulation.** It is a specialized part of the systemic circulation (Figure 11–21). The vessels that serve the myocardium include the *left coronary artery,* which originates as a branch of the ascending aorta. This artery runs under the left atrium and divides into the anterior descending and circumflex branches. The *anterior interventricular branch* follows the anterior interventricular sulcus and supplies oxygenated blood to the walls of both ventricles. The *circumflex branch* distributes oxygenated blood to the walls of the left ventricle and left atrium. The *right coronary artery* also originates as a branch of the ascending aorta. It runs under the right atrium and divides into the posterior descending and marginal branches. The *posterior interventricular branch* follows the posterior interventricular sulcus and supplies the walls of the two ventricles with oxygenated blood. The *marginal branch* transports oxygenated blood to the myocardium of the right ventricle and right atrium. The left ventricle receives the most abundant blood supply because of the enormous work it must do.

As blood passes through the arterial system of the heart, it delivers oxygen and nutrients and collects carbon dioxide and wastes. Most of the deoxygenated blood, which carries the carbon dioxide and wastes, is collected by a large vein, the *coronary sinus,* which empties into the right atrium (See Figure 11–5). The principal tributaries of the coronary sinus are the *great cardiac vein,* which drains the anterior aspect of the heart, and the *middle cardiac vein,* which drains the posterior aspect of the heart.

The majority of "heart problems" result from some foul-up in the coronary circulation. If a reduced oxygen supply weakens the cells, but does not actually kill them, the condition is called **ischemia.** *Angina pectoris* is ischemia of the myocardium. The name comes from the area in which the pain is felt. Remember that pain impulses originating from most visceral muscles are referred to an area on the surface of the body. Angina pectoris occurs when coronary circulation is somewhat reduced for some reason. Stress, which produces constriction of vessel walls, is a common cause. Equally common is strenuous exercise after a heavy meal. When any quantity of food is dumped into the stomach, the body increases blood flow to the digestive tract. The digestive glands can then receive enough oxygen for their increased activities, and the digested food can be quickly absorbed into the bloodstream. As a consequence, some blood is diverted away from other organs, including the heart. Exercise, however, increases heart muscle activity and thus increases its need for oxygen. Thus, doing heavy work while food is in the stomach can lead to oxygen deficiency in the myocardium. Angina pectoris weakens the heart muscle, but it does not produce a full-scale heart attack. The simple remedy of taking nitroglycerin, a drug that dilates vessels and thereby increases the area of blood flow, brings coronary circulation back to normal and stops the pain of angina. Because repeated attacks of angina can weaken the heart and lead to serious heart trouble, angina patients are told to avoid activities and stresses that bring on the attacks.

A much more serious problem is *myocardial infarction,* commonly called a "coronary" or "heart attack." **Infarction** means death of an area of tissue because of a drastically reduced or completely interrupted blood supply. Myocardial infarction results from a thrombus or embolus in one of the coronary arteries. The tissue on the far side of the obstruction dies, and the heart muscle loses at least some of its strength. The aftereffects depend partly on the size and location of the infarcted, or dead, area which is replaced by scar tissue.

Hepatic portal circulation

Blood coming into the liver is derived from two sources. The hepatic artery delivers oxygenated blood from the systemic circulation, and the hepatic portal vein delivers deoxygenated blood from the digestive organs. The term **hepatic portal circulation** refers to this flow of blood from the digestive organs to the liver before returning to the heart (Figure 11–22). Hepatic portal blood is rich with substances that have been absorbed from the digestive tract. One of the roles of the liver is to monitor these substances before they are passed into the general circulation. For example, the liver stores nutrients such as glucose. It modifies other digested substances so that they may be more easily used by cells. And it detoxifies harmful substances that have been absorbed by the digestive tract.

The hepatic portal system of veins includes veins that drain blood from the pancreas, spleen, stomach, intestines, and gallbladder, and transport it to the portal vein of the liver. The *hepatic portal vein* is formed by the union of the superior mesenteric

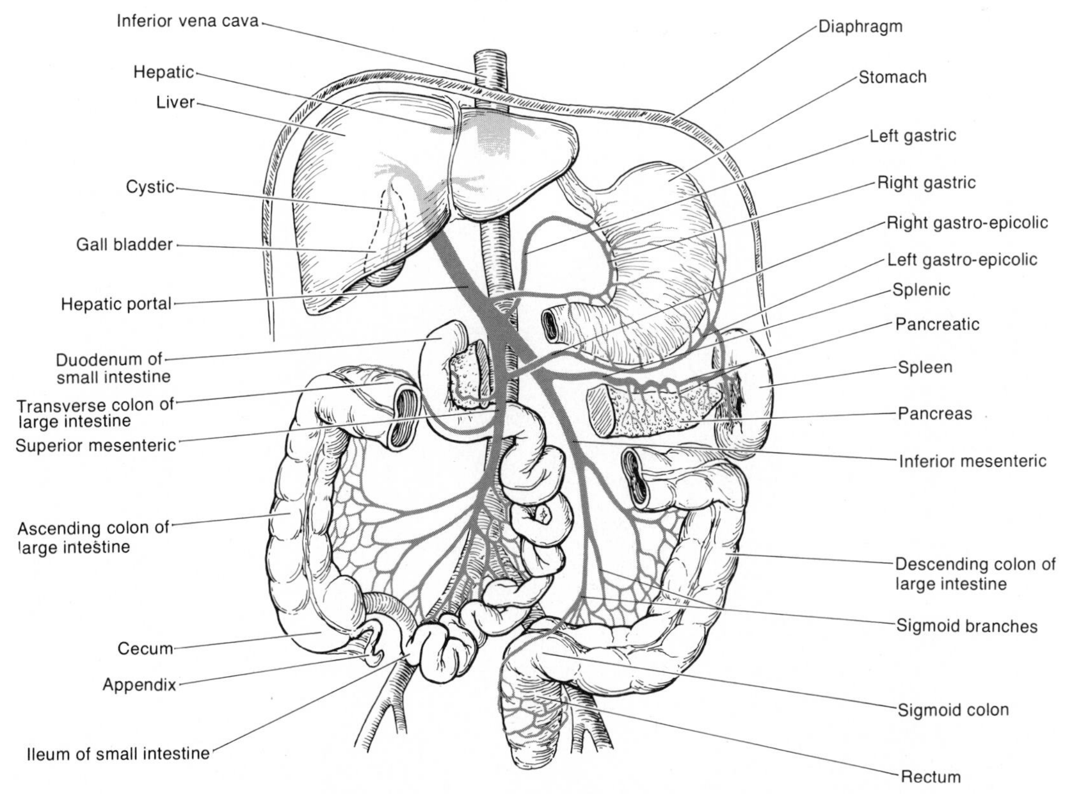

Figure 11–22. Hepatic portal circulation.

and splenic veins. The *superior mesenteric vein* drains blood from the small intestine and portions of the large intestine and stomach. The *splenic vein* drains the spleen and receives tributaries from the stomach, pancreas, and colon. The tributaries coming from the stomach are the *coronary, pyloric,* and *gastroepiploic veins.* The *pancreatic veins* come from the pancreas, and the *inferior mesenteric veins* come from portions of the colon. Before the hepatic portal vein enters the liver, it receives the *cystic vein* from the gallbladder. Ultimately, the blood leaves the liver through the *hepatic veins,* which enter the inferior vena cava.

Pulmonary circulation

The flow of deoxygenated blood from the right ventricle to the lungs and the return of oxygenated blood from the lungs to the left atrium is called the **pulmonary circulation** (Figure 11–23). The *pulmonary trunk* emerges from the right ventricle, passes upward, backward, and to the left. It then divides into two branches. These are the right pulmonary artery, which runs to the right lung, and the left pulmonary artery, which goes to the left lung. Upon entering the lungs, the branches divide and subdivide. They grow smaller in size and ultimately form capillaries around the alveoli in the lungs. Carbon dioxide is passed from the blood into the alveoli to be breathed out of the lungs. Oxygen breathed in by the lungs is passed from the alveoli into the blood. The capillaries then unite. They grow larger in size and become veins. Eventually, two *pulmonary veins* exit from each lung and transport the oxygenated blood to the left atrium. The pulmonary veins are the only postnatal veins that carry oxygenated blood. Contraction of the left ventricle then sends the blood into the systemic circulation.

Fetal circulation

The fetal circulatory system is necessarily different from postnatal circulation because the lungs and digestive tract of a fetus are nonfunctional. Instead, the fetus derives its oxygen and nutrients and eliminates its carbon dioxide and wastes through the maternal blood. This exchange of materials between fetus and mother constitutes the **fetal circulation** (Figure 11–24).

The exchange of materials between fetal and maternal circulation occurs through a structure called the *placenta.* It is attached to the navel of the fetus by the umbilical cord, and it communicates with the mother through countless small blood vessels that emerge from the walls of the uterus. The umbilical cord contains blood vessels that branch into capillaries within the placenta. Wastes from the fetal blood diffuse out of the capillaries, into the

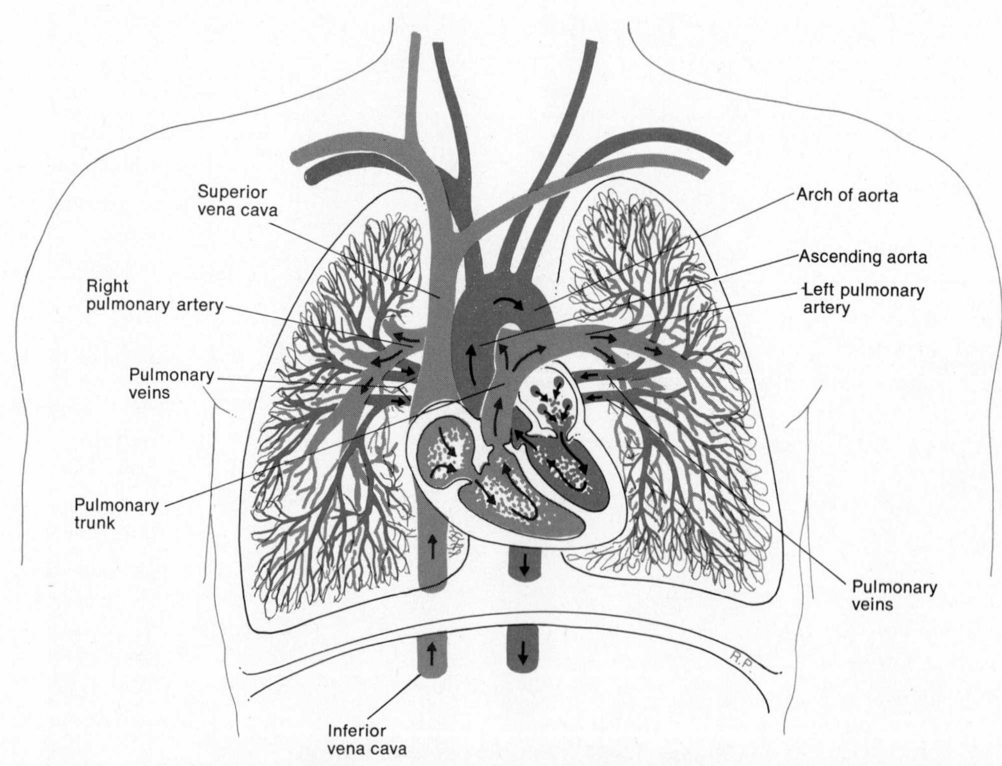

Figure 11–23. Pulmonary circulation.

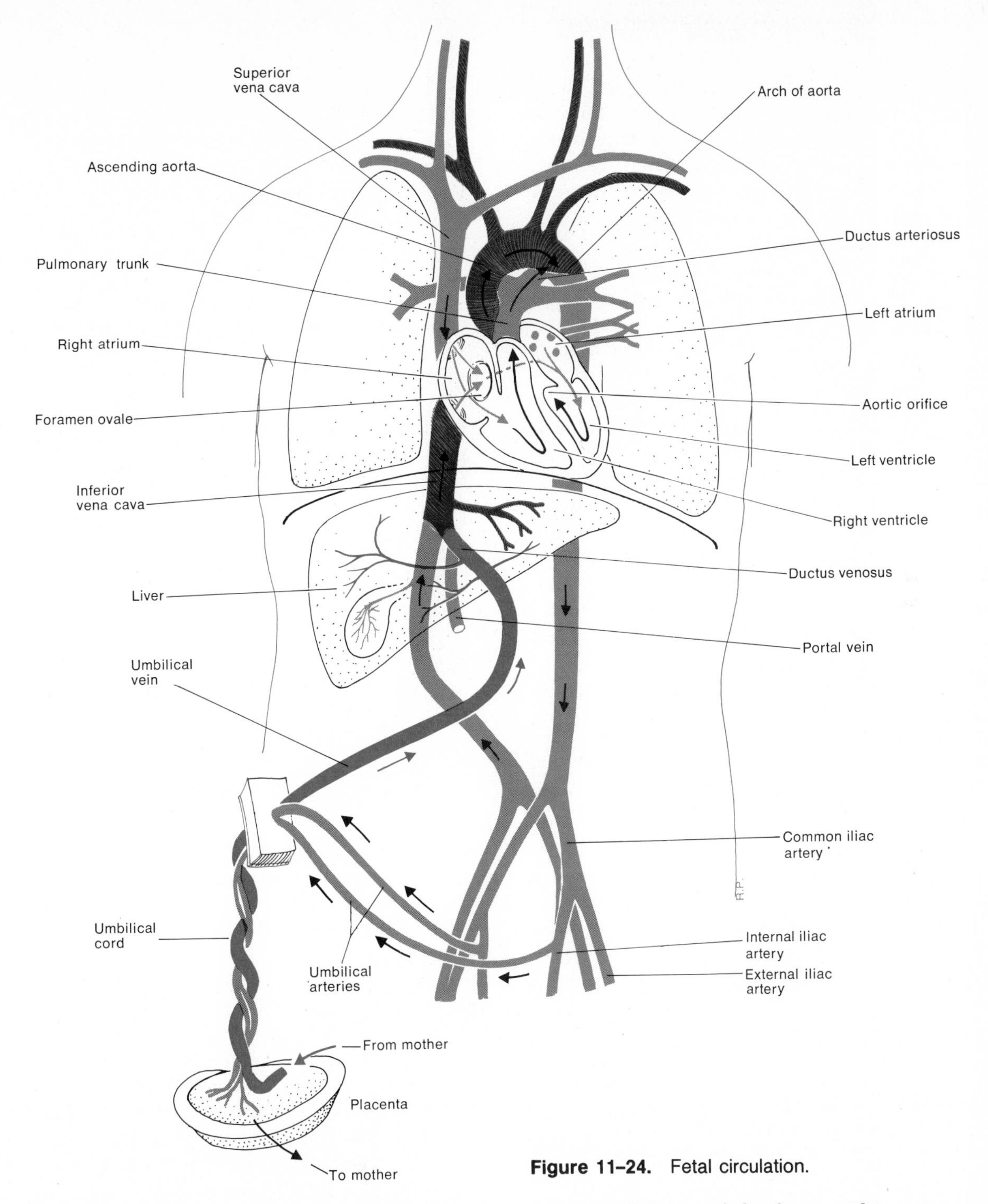

Figure 11–24. Fetal circulation.

interstitial fluid of the placenta, and finally into blood vessels belonging to the mother. Nutrients travel the opposite route, going from the maternal blood vessels to the interstitial fluid of the placenta to the fetal capillaries. There is no mixing of maternal and fetal blood since all exchanges occur by diffusion through capillaries.

Blood leaves the fetus by two *umbilical arteries.* These arteries are branches of the internal iliac arteries. The umbilical arteries are included in the umbilical cord. At the placenta, the blood picks up oxygen and nutrients and eliminates carbon dioxide and wastes. The oxygenated blood enters the fetus by way of the *umbilical vein.* This vein ascends to the liver of the fetus where it divides into two branches. Some of the blood flows through the branch that joins the hepatic portal vein and enters the liver. Since the fetal liver does not function completely, it needs only a small amount of blood. Most of the blood, therefore, flows into the second branch, called the *ductus venosus.* The ductus venosus connects with the inferior vena cava.

Circulation through other portions of the fetus is not unlike the postnatal circulation. Deoxygenated blood returning from the lower regions is mingled with oxygenated blood from the ductus venosus in the inferior vena cava. This mixed blood then enters the right atrium. The circulation of blood through the upper portion of the fetus is also very similar to the postnatal flow. Deoxygenated blood returning from the upper regions of the fetus is collected by the superior vena cava, and it also passes into the right atrium.

However, unlike postnatal circulation, most of the blood does not pass into the right ventricle and go to the lungs since the fetal lungs do not operate. In the fetus, an opening called the *foramen ovale* exists in the septum between the right and left atria. A valve in the inferior vena cava directs most of the blood through the foramen ovale so that it may be sent directly into the systemic circulation. The blood that does descend into the right ventricle is pumped into the pulmonary trunk, but only a small amount of this blood actually reaches the lungs. Most of the blood in the pulmonary trunk is sent through the *ductus arteriosus.* This is a small vessel connecting the pulmonary trunk with the aorta, and it enables another portion of the blood in excess of nutrient requirements for the lungs to bypass the fetal lungs. The blood in the aorta is carried to all parts of the fetus through its systemic branches. When the common iliac arteries branch into the external and internal iliacs, part of the blood flows into the internal iliacs. It then goes to the umbilical arteries and back to the placenta for another exchange of materials. Note that the only vessel which carries fully oxygenated blood is the umbilical vein.

At birth, when lung, digestive, and liver functions are established, the special structures of fetal circulation are no longer needed. Their postnatal fates are

1. The umbilical arteries atrophy to become the lateral umbilical ligaments.
2. The umbilical vein becomes the round ligament of the liver.
3. The placenta is shed by the mother as the "afterbirth."
4. The ductus venosus becomes the ligamentum venosum, a fibrous cord in the liver.
5. The foramen ovale normally closes shortly after birth to become the fossa ovalis, a depression in the interatrial septum (see Figure 11–5).
6. The ductus arteriosus closes, atrophies, and becomes the ligamentum arteriosum.

Usually the ductus arteriosus closes shortly after birth. When it fails to close or closes imperfectly, some of the blood shuttles uselessly back and forth between the heart and lungs. This condition is called *patent ductus arteriosus.* A surgical procedure easily remedies this problem.

In addition to the heart and the veins, arteries, and capillaries that route blood through the body, the circulatory system includes the structures of the lymph vascular system. Next, we shall examine the vessels, glands, and three organs of the lymph vascular system.

THE LYMPH VASCULAR SYSTEM

Lymph vessels, a series of small glands called nodes, and three organs – the tonsils, the thymus, and the spleen – make up the **lymph vascular system.** The primary function of the lymph vascular system is to drain, from the tissue spaces, protein-containing fluid which escapes from the blood capillaries. Such proteins cannot be directly reabsorbed. Other functions of the lymph vascular system are to clean and return cellular wastes to the blood, to produce agranular leucocytes, and to develop immunities.

Lymphatic vessels

Lymphatic vessels originate as tubes that begin in spaces between cells. The tubes are called **lymph capillaries.** (See Figure 11–3.) Lymph capillaries, originate in most parts of the body. They are, however, slightly larger and more permeable than blood capillaries. Just as blood capillaries converge to form venules and veins, lymph capillaries unite to form larger and larger lymph vessels called **lymphatics** (Figure 11–25). Lymphatics resemble veins in structure except that they have thinner walls, more valves, and contain lymph nodes at various intervals. Ultimately, lymphatics converge into two main channels – the thoracic duct and the right lymphatic duct.

The *thoracic duct,* or left lymphatic duct, begins as a dilation located in front of the second lumbar vertebra. This dilation is called the *cisterna chyli.* The thoracic duct receives lymph from the left side of the head, neck, and chest, the left upper extremity, and the entire body below the ribs. It then empties the lymph into the left subclavian vein. A pair of valves at this junction prevents the passage of venous blood into the thoracic duct. The *right lymphatic duct* drains lymph from the upper right side of the body and empties it into the right subclavian vein at the junction of the right subclavian and jugular veins.

Lymphangiography is a procedure by which lymphatic vessels and lymph organs are filled with an opaque substance in order to be filmed. Such a film is called a *lymphangiogram.* Lymphangiograms are useful in detecting edema and carcinomas and in localizing lymph nodes for either surgical or radiotherapeutic treatment. A normal lymphangiogram of lymphatic vessels and a few nodes in the upper thighs and pelvis is shown in Figure 11–26.

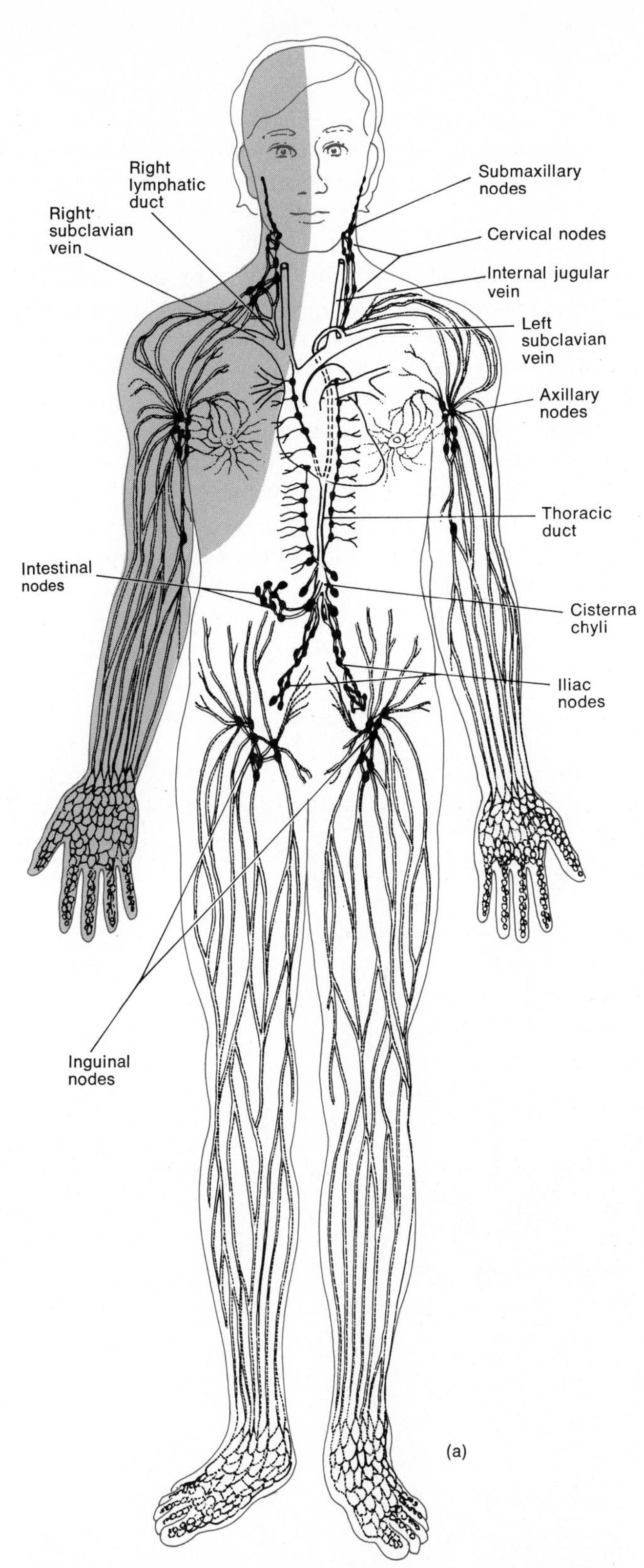

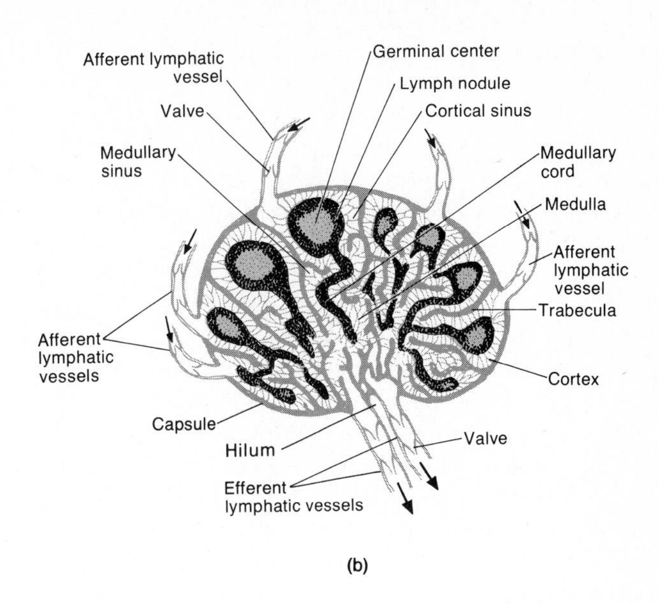

Figure 11–25. The lymph vascular system. (a) Location of the principal lymphatics and lymph nodes. The colored area indicates those portions of the body drained by the right lymphatic duct. All other areas of the body are drained by the thoracic duct. (b) Enlargement of a lymph node showing internal structure and path taken by circulating lymph (arrows).

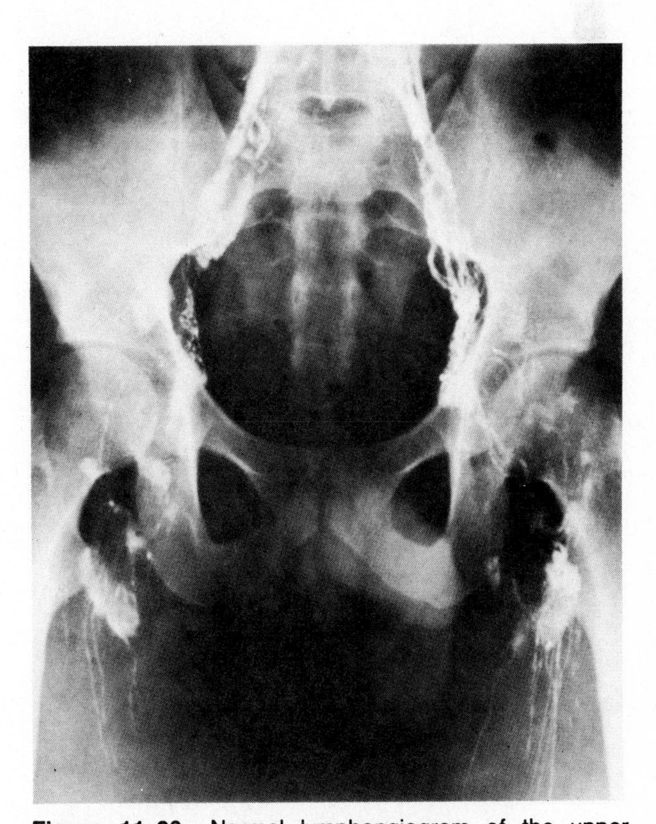

Figure 11–26. Normal lymphangiogram of the upper thighs and pelvis. Can you identify the lymphatics and lymph nodes? (Courtesy of Lester W. Paul and John H. Juhl, *The Essentials of Roentgen Interpretation,* 3d ed., Harper & Row, Publishers, Inc., New York, 1972.)

Lymph nodes

The oval- or bean-shaped structures located along the length of lymphatics are called **lymph nodes** or **lymph glands** (Figure 11–25b). They range from 1 to 25 mm (0.04 to 1 in.) in length. Structurally, a lymph node contains a slight depression on one side called a *hilum,* where the blood vessels enter and leave the node. Each node is covered by a *capsule* of fibrous connective tissue that partially extends into the node. The capsular extensions are called *trabeculae.* The capsule, trabeculae, and hilum constitute the stroma (framework) of a lymph node. The parenchyma of a lymph node is specialized into two regions. The outer *cortex* contains densely packed lymphocytes arranged in masses called *lymph nodules.* The nodules often contain lighter-staining central areas, the *germinal centers,* where lymphocytes are produced. The inner region of a lymph node is called the *medulla.* In the medulla, the lymphocytes are arranged in strands called *medullary cords.*

The circulation of lymph through a node involves afferent lymphatic vessels, sinuses within the node, and efferent lymphatic vessels. *Afferent lymphatic vessels* enter the convex surface of the node at several points. They contain valves which open toward the node so that the circulation of lymph through the afferent lymphatic vessels is into the lymph node. Once inside the lymph node, the lymph enters the sinuses of the node, which are a series of irregular channels. Lymph from the afferent lymphatic vessels enters the *cortical sinuses* under the capsule. From here, the lymph circulates to the *medullary sinuses,* which are between the medullary cords. From these sinuses, the lymph circulates into the *efferent lymphatic vessels.* These vessels are located at the hilum of the lymph node. The efferent vessels are wider and fewer in number than the afferent vessels and contain valves that

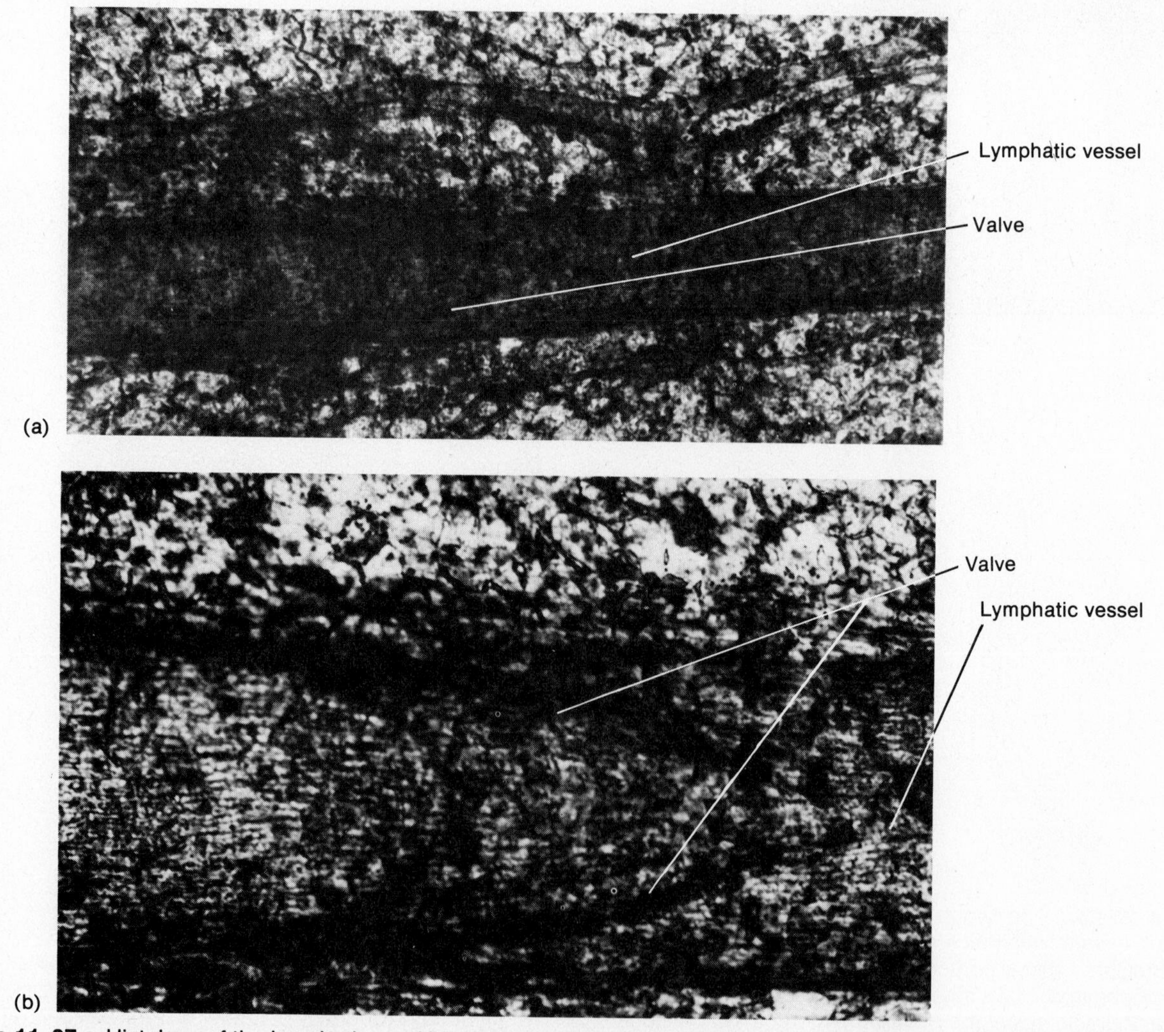

Figure 11–27. Histology of the lymphatics and lymph nodes. (a) Longitudinal section of a lymphatic vessel at a magnification of 15X. (b) Enlarged aspect of a valve at a magnification of 50X.

open away from the lymph node. Thus, they convey lymph out of the node.

As the lymph circulates through the nodes, it is processed by the reticuloendothelial cells that line the sinuses. Reticuloendothelial cells are multifunctional. They can phagocytose microorganisms and cellular debris. They may develop into monocytes and lymphocytes. And they may develop directly into plasma cells and release antibodies. Thus, the lymph is filtered, and it picks up agranular leucocytes and antibodies. At times, however, the number of entering microbes is so great that the node itself may become infected. It then becomes enlarged and tender.

Histological features of lymphatics and lymph nodes are shown in Figure 11–27.

The majority of lymph nodes appear in groups or chains in certain areas of the body (Figure 11–25a). Among the larger and clinically important groups are the following: (1) The *cervical lymph nodes*, which are located by the sternocleidomastoid muscles, drain lymph from the head and neck. (2) The *submaxillary lymph nodes*, located in the floor of the mouth, drain the nose, lips, and teeth. (3) The *axillary lymph nodes* in the underarm and chest region drain the skin and muscles of the chest, including the breasts. (4) The *inguinal nodes* in the groin area drain the external genitals and the lower extremities.

Tonsils

Tonsils are basically masses of lymphoid tissue embedded in mucous membrane (see Figure 20–4). The *pharyngeal tonsils* are located on the posterior wall of the passageway between the nose and throat. The *palatine tonsils* are situated on either side of the mouth cavity near the soft palate. These are the ones commonly removed by a tonsillectomy. The *lingual tonsil* is located at the base of the

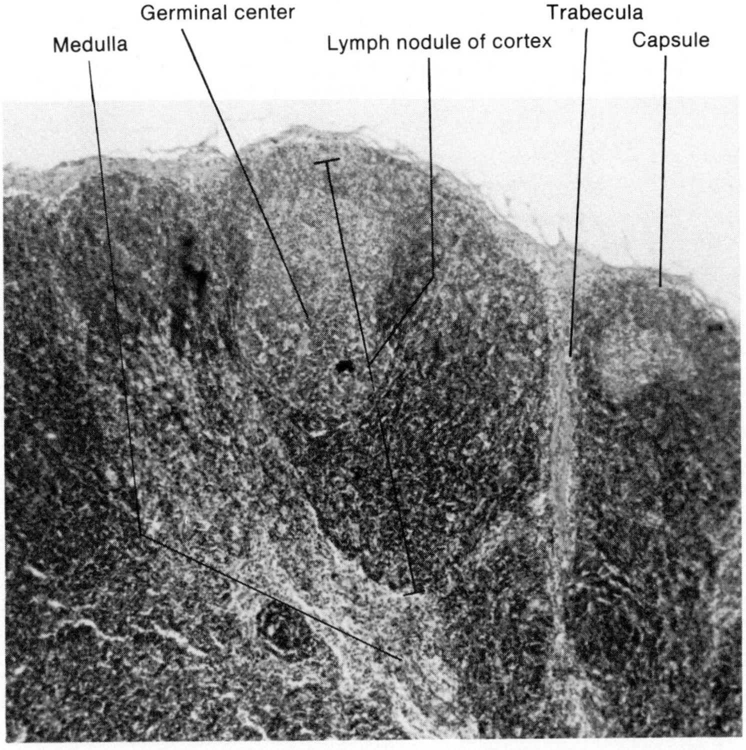

Germinal center
Medulla
Lymph nodule of cortex
Trabecula
Capsule
(c)

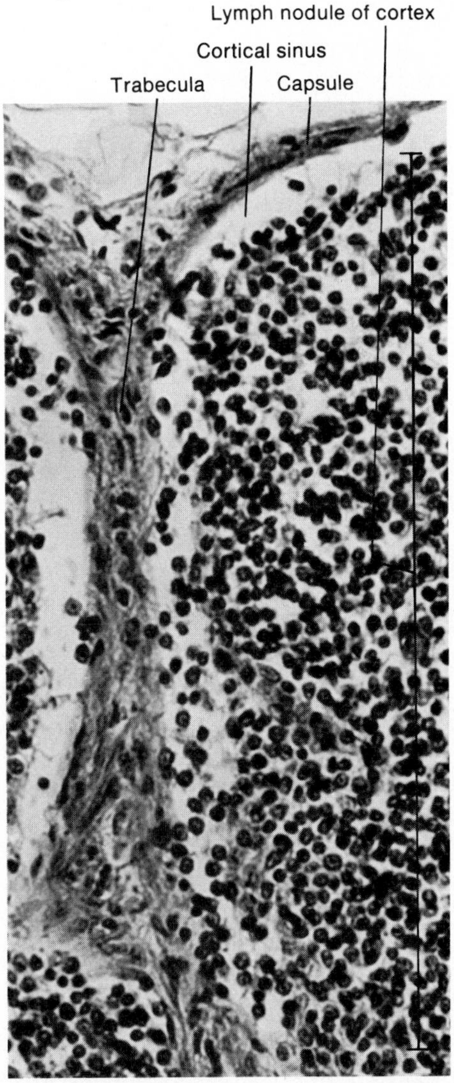

Lymph nodule of cortex
Cortical sinus
Trabecula
Capsule
(d)

Figure 11–27 (cont.). (c) Section through a lymph node at a magnification of 15X. (d) Enlarged aspect of a lymph node at a magnification of 90X. (Courtesy of Victor B. Eichler, Wichita State University.)

tongue and may also have to be removed by a tonsillectomy. The tonsils are supplied with reticuloendothelial cells. The tonsils function in filtering lymph and producing lymphocytes.

Spleen

The oval-shaped **spleen** is the largest mass of lymphatic tissue in the body, measuring about 12 cm (5 in.) in length. It is situated in the left hypochondriac region inferior to the diaphragm and posterolateral to the stomach (see Figure 1–4). The spleen is surrounded by a capsule consisting of fibroelastic tissue and scattered smooth muscle. The capsule, in turn, is covered by a serous membrane, the peritoneum. Like lymph nodes, the spleen contains trabeculae and a hilum. The capsule, trabeculae, and hilum constitute the stroma of the spleen.

The parenchyma of the spleen consists of two different kinds of tissue called white pulp and red pulp. *White pulp* is essentially lymphoid tissue arranged around arteries. The clusters of lymphocytes surrounding the arteries at intervals and the expansions are referred to as *splenic nodules* or *Malpighian corpuscles.* The *red pulp* consists of venous sinuses filled with blood and cords of splenic tissue called *splenic* or *Billroth's cords.* Veins are closely associated with the red pulp.

The splenic artery and vein and the efferent lymphatics pass through the hilum. The spleen functions in the phagocytosis of bacteria and worn-out red blood cells and platelets. It also produces lymphocytes, monocytes, and plasma cells. In addition, the spleen stores and releases blood in case of hemorrhage. The release seems to be purely sympathetic. The sympathetic impulses cause the smooth muscle of the spleen to contract.

Thymus gland

The **thymus gland** is a bilobed mass of lymphatic tissue in the upper thoracic cavity. It is found along the trachea behind the sternum. (See Figure 17–15.) The thymus is relatively large in children. It reaches its maximum size at puberty and then undergoes involution. Eventually, it is replaced by fat and connective tissue. The thymus, like the pancreas, is an organ that specializes in servicing two body systems. The pancreas serves both the endocrine and digestive systems. The thymus produces hormones that belong to the endocrine system, but it also performs a vital function as part of the lymphatic system. This function is related to immunity.

Lymph circulation

The flow of lymph is from tissue spaces to the large lymphatic ducts to the subclavian veins. The flow is maintained primarily by the milking action of muscle tissue. Skeletal muscle contractions compress lymph vessels and force lymph toward the subclavian veins. Moreover, lymph vessels, like veins, contain valves, and the valves ensure the movement of lymph toward the subclavian veins. Another factor that maintains lymph flow is respiratory movements. These movements create a pressure gradient (difference) between the two ends of the lymphatic vascular system. As a result of respiratory movements, the pressure in the thoracic region is zero or negative. Because of the pressure gradient, lymph flows from the tissue spaces, where the pressure is higher, toward the thoracic region, where it is lower.

Under normal circumstances, the amount of interstitial fluid remains fairly constant. The excessive accumulation of lymph in tissue spaces is referred to as **edema.** Several conditions may bring about edema. One cause is an obstruction to the flow of lymph in the lymphatic channels somewhere between the lymphatic capillaries and the subclavian veins. Infected lymph nodes may bring about this condition. Other causes are excessive lymph formation and increased permeability of blood capillary walls. A rise in capillary blood pressure, in which interstitial fluid is formed at a faster rate than it is passed into lymphatics, also may result in edema.

APPLICATIONS TO HEALTH

Blood disorders

The various blood disorders that can arise affect differing portions of the blood. With anemia, for example, the patient may have an abnormally low number of red blood cells, whereas with mononucleosis white blood cells are affected. There are numerous blood disorders, and all may have wide-ranging effects.

Anemia

Anemia is a sign rather than a diagnosis. Many different kinds of anemia exist, and all are characterized by less than normal numbers of erythrocytes or by a below normal hemoglobin content. These conditions lead to fatigue and intolerance to cold, both of which are related to lack of oxygen needed for energy and heat production, and paleness due to low hemoglobin content. But diagnosis cannot be made nor can treatment begin until the cause of the anemia is discovered.

HEMORRHAGIC ANEMIA. An excessive loss of erythrocytes through bleeding is called **hemorrhagic anemia.** Common causes are large wounds, stomach ulcers, and excessive menstrual bleeding (hypermenorrhea). If bleeding is extraordinarily heavy, the anemia is termed acute. Excessive loss of blood fluid may endanger the individual's life. Slow, prolonged bleeding is more apt to produce a chronic anemia whose chief symptom is fatigue.

HEMOLYTIC ANEMIA. The term **hemolytic anemia** comes from the word *hemolysis,* the rupturing of erythrocyte cell membranes. The cell is destroyed, and its hemoglobin pours out into the plasma. A characteristic sign includes distortions in the shapes of the erythrocytes that are progressing toward hemolysis. There may also be a sharp increase in the number of reticulocytes or immature red blood cells that normally remain in marrow. The reticulocytes increase since the destruction of red blood cells stimulates hematopoiesis. Agents that may cause hemolytic anemia are parasites, toxins, and antibodies from incompatible blood, such as occurs with an Rh^- mother and Rh^+ fetus. Erythroblastosis fetalis of the newborn is an example of a hemolytic anemia.

APLASTIC ANEMIA. Destruction or inhibition of the red bone marrow results in **aplastic anemia.** Typically, the marrow is replaced by fatty tissue, fibrous tissue, or tumor cells. Toxins and certain medications are causes. Many of the medications inhibit the enzymes involved in hemopoiesis.

SICKLE CELL ANEMIA. The erythrocytes of a person with **sickle cell anemia** manufacture an abnormal kind of hemoglobin. When the erythrocyte gives up its oxygen to the interstitial fluid, its hemoglobin tends to lose its integrity in places of low oxygen tension and forms long, stiff, rodlike structures that bend the erythrocyte into a sickle shape which gives the disorder its name. The sickled cells rupture easily. And even though hematopoiesis is stimulated by the loss of the cells, it cannot keep pace with the hemolysis. The individual consequently suffers from a hemolytic anemia that reduces the amount of oxygen which can be supplied to his tissues. Prolonged oxygen reduction may eventually cause extensive tissue damage. Furthermore, because of the shape of the sickled cells, they tend to get stuck in blood vessels–a situation that can cut off blood supply to an organ altogether.

Polycythemia

The term **polycythemia** refers to a condition characterized by an abnormal increase in the number of red blood cells. Increases of 2 to 3 million cells per cubic millimeter are considered to be polycythemic. The disorder is harmful because the thickness of the blood (viscosity) is greatly increased due to the extra red blood cells, and viscosity contributes to a tendency to thrombosis and hemorrhage. It also causes a rise in blood pressure. The tendency for thrombosis to develop results from too many red blood cells piling up as they try to enter smaller vessels. The tendency to hemorrhage is because hyperemia (unusually large amount of blood in an organ part) is observed in all organs.

There are two basic types of polycythemia: *primary* and *secondary.* The primary type is characterized by an overactivity of the red bone marrow and by an enlarged liver and spleen. Its cause is unknown. The secondary type is the result of a lack of oxygen in the arteries of people suffering from chronic cardiac or pulmonary disease. Other causes include kidney tumors or cysts, which could increase secretion of the hormone aldosterone resulting in increasing blood volume; liver cancer, which could interfere with the normal function of the liver in disposing of millions of old red blood cells daily; and very high altitudes.

The percentage of the blood that is made up of red blood cells is called the *hematocrit.* It is determined by centrifuging blood and noting the ratio of red blood cells to plasma. The average hematocrit is 45 percent. This means that in 100 ml of blood there are 45 ml of cells and 55 ml of plasma. Anemic blood may have a hematocrit of 15 percent, whereas polycythemic blood may have a hematocrit of 65 percent.

Disorders involving white blood cells

Two well-known disorders affecting white blood cells are infectious mononucleosis and leukemia.

INFECTIOUS MONONUCLEOSIS. **Infectious mononucleosis** is a contagious disease of viral origin that occurs mainly in children and young adults. The trademark of the disease is an elevated white count with an abnormally high percentage of lymphocytes and mononucleocytes–hence, the name mononucleosis. As mentioned earlier in the chapter, an increase in the number of monocytes usually indicates a chronic infection. The various signs and symptoms include slight fever, sore throat, brilliant red throat and soft palate, stiff neck, cough, and malaise. The spleen may enlarge; and secondary complications involving the liver, heart, kidneys, and nervous system may develop.

There is no cure for mononucleosis, and treatment consists of watching for and treating any complications. Usually the disease runs its course in a few weeks, and the individual usually suffers no permanent ill effects.

LEUKEMIA. This disorder is also called "cancer of the blood." **Leukemia** is an uncontrolled, greatly accelerated production of white cells. Many of the cells fail to reach maturity. As with most cancers, the symptoms and the cause of death do not result so much from the cancer cells themselves as from the interference of the cancer cells with normal body processes. The accumulation of cells leads to abnormalities in organ functions. For example, the anemia and bleeding problems commonly seen in leukemia result from the "crowding out" of normal bone marrow cells. This interferes with the normal production of red blood cells and platelets. The most common cause of death from leukemia is internal hemorrhaging, especially cerebral hemorrhage that destroys the vital centers in the brain. The second most frequent cause of death is uncontrolled infection. This happens because there is a lack of mature or normal white blood cells available to fight infection.

Therapy may temporarily stop the pathologic process. The abnormal accumulation of leucocytes may be reduced or even eliminated by using x-ray and antileukemic drugs. Partial or complete remissions may be induced, with some lasting as long as 15 years.

Cardiovascular disorders

Diseases of the heart and blood vessels are the biggest single killers in the developed world. These diseases account for approximately 53 percent of all deaths. A recent comparison indicates that cardiovascular disease kills more people than cancer, accidents, pneumonia, influenza, and diabetes combined. Some of the cardiovascular problems involve aneurysms, atherosclerosis, hypertension, and various heart disorders.

Aneurysm

A blood-filled sac formed by an outpouching in an arterial or venous wall is called an **aneurysm.** Aneurysms may occur in any major blood vessel of the body and include the following types:

1. Berry, which is a small aneurysm of a cerebral artery. If it ruptures, it may cause a hemorrhage below the dura mater (Figure 11–28a). Hemorrhaging is one cause of stroke.
2. Ventricular, which is a focal dilatation of a ventricle of the heart (Figure 11–28b).
3. Aortic, which is a focal dilation of the aorta (Figure 11–28c).

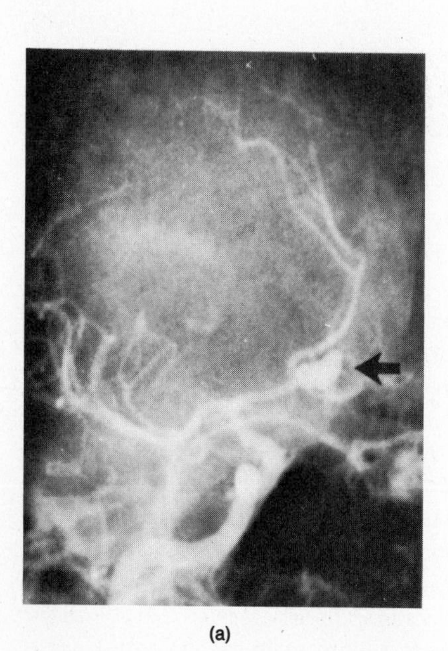

(a)

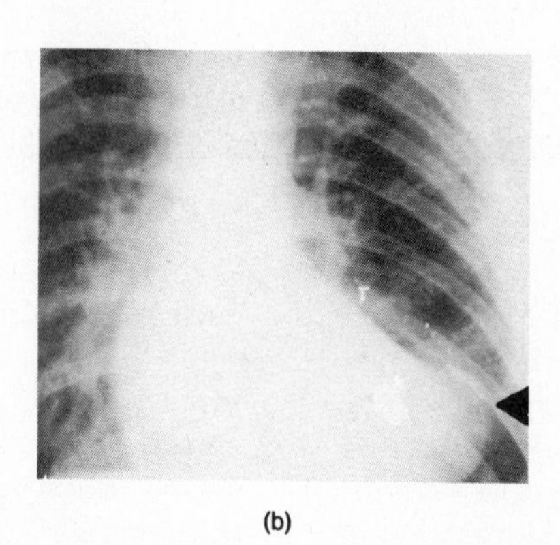

(b)

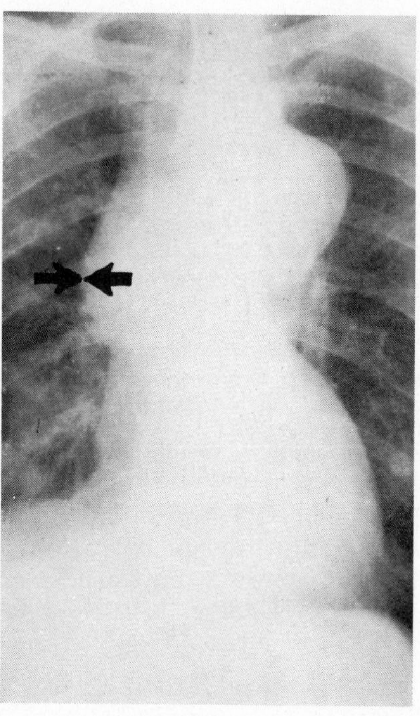

(c)

Figure 11–28. Aneurysms. (a) Aneurysm of the anterior cerebral artery. (b) Ventricular aneurysm. (c) Aortic aneurysm. (Courtesy of Lester W. Paul and John H. Juhl, *The Essentials of Roentgen Interpretation,* 3d ed., Harper & Row, Publishers, Inc., New York, 1972.)

Atherosclerosis

A hardening of the arteries is described by the term **atherosclerosis.** Atherosclerosis is responsible for the most important and prevalent of all clinical complications. In this disorder, the inner layer of the artery becomes thickened with soft fatty deposits, called *plaques.* The plaque looks like a pearly gray or yellow mound of tissue on the inside of the blood vessel wall. It usually consists of a core of lipid (mainly cholesterol) covered by a cap of fibrous (scar) tissue. As the plaques increase in size, they may impede or cut off blood flow in affected arteries. This causes damage to the tissues supplied by these arteries. An additional danger is that the lipid core of the plaques may be washed into the bloodstream. There, it could become an embolus and obstruct small arteries and capillaries quite a distance away from the original site of formation. A third possibility is that the plaque will provide a roughened surface for clot formation.

Atherosclerosis is generally a slow, progressive disease. It may start in childhood, and its development may produce absolutely no symptoms for 20 to 40 years or longer. Even if it reaches the advanced stages, the individual may feel no symptoms, and the condition may be discovered only at postmortem examination. Diagnosis during life is made possible by injecting radiopaque substances into the blood and then taking x-rays of the arteries. This technique is called *angiography* or *arteriography.* The film is called an *arteriogram* (Figure 11–29).

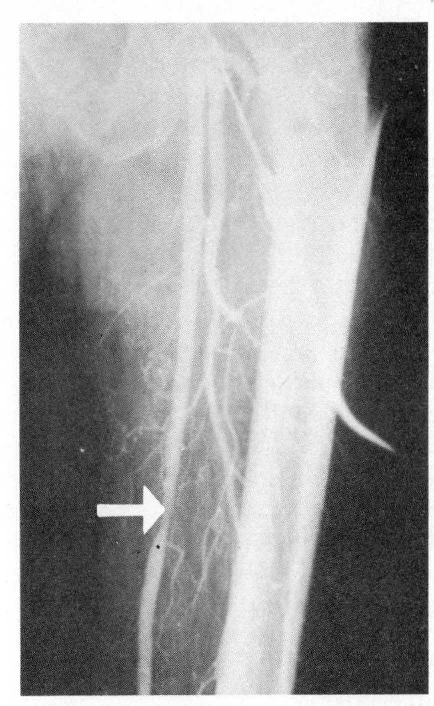

Figure 11–29. Atherosclerosis. Femoral arteriogram showing an atherosclerotic plaque (arrow) in the middle third of the thigh. (Courtesy of Lester W. Paul and John H. Juhl, *The Essentials of Roentgen Interpretation,* 3d ed., Harper & Row, Publishers, Inc., New York, 1972.)

Animal experiments have given us considerable scientific information about the plaques. They begin as yellowish fatty streaks of lipids that appear under the tunica intima. It is possible to produce the streaks in many animals by feeding them a diet that is high in fat and cholesterol. This raises the blood lipid levels—a condition called *hyperlipidemia. Hyper* means over or above, whereas *lipo* means fat. Hyperlipidemia is an important factor in increasing the risk of atherosclerosis. Patients with high blood levels of cholesterol should be identified and treated with appropriate diet and drug therapy.

Hypertension

Hypertension, or high blood pressure, is the commonest of the diseases affecting the heart and blood vessels. Statistics from a recent National Health Survey indicate that hypertension afflicts at least 17 million American adults and perhaps as many as 22 million.

Primary hypertension, or essential hypertension, is a persistently elevated blood pressure that cannot be attributed to any particular organic cause. Specifically, the diastolic pressure continually exceeds 95 mm Hg. Approximately 85 percent of all hypertension cases fit this definition. The other 15 percent is called *secondary hypertension.* Secondary hypertension is caused by disorders such as atherosclerosis, kidney disease, and adrenal hypersecretion. Atherosclerosis increases blood pressure by reducing the elasticity of the arterial walls and by narrowing the space through which the blood can flow. Both kidney diseases and obstruction of blood flow to the kidney may cause the kidney to release an enzyme called renin into the blood. This enzyme catalyzes the formation of angiotensin from a plasma protein. Angiotensin is a powerful blood-vessel constrictor. It is the most potent agent known for raising blood pressure. Aldosteronism, the hypersecretion of aldosterone, may also cause an increase in blood pressure. Aldosterone is the adrenal cortex hormone that promotes the retention of salt and water by the kidneys. It thus tends to increase plasma volume. Pheochromocytoma is a benign tumor of the adrenal medulla. It produces and releases into the blood large quantities of norepinephrine and epinephrine. These hormones also raise blood pressure by stimulating the heart and constricting blood vessels.

High blood pressure is of considerable concern because of the harm it can do to certain body organs such as the heart, brain, and kidneys if it remains uncontrolled for long periods. The heart is most commonly affected by high blood pressure. When pressure is high, the heart uses more energy in pumping. Because of the increased effort, the heart muscle thickens, and the heart becomes enlarged. The heart also needs more oxygen. If it cannot meet the demands put on it, angina pectoris or even myocardial infarction may occur. Continued high blood pressure may produce a cerebral vascular accident, or "stroke." In this case, severe strain has been imposed on the cerebral arteries that supply the brain. These arteries are usually less protected by the surrounding tissues than are the major arteries in other parts of the body. As a result, one or more of these weakened cerebral arteries may finally rupture, and a brain hemorrhage follows.

The kidney is another prime target of hypertension. The principal site of damage is in the arterioles that supply this vital organ. The continual high blood pressure pushing against the walls of the arterioles causes them to thicken, thus narrowing the lumen. The blood supply to the kidney is, thereby, gradually reduced. In response, the kidney may secrete renin, which raises the blood pressure even higher and complicates the problem. The reduced blood flow to the kidney cells may eventually lead to the death of the cells.

At present, the causes of primary hypertension are unknown. Medical science cannot cure it. However, almost all cases of hypertension, whether mild or very severe, can be controlled by a variety of effective drugs that reduce elevated blood pressure.

Heart disorders

It is estimated that one in every five persons who reaches 60 will have a **heart attack.** And it is also estimated that one in every four persons between 30 and 60 has the potential to be stricken. Heart disease is epidemic in this country, despite the fact that some of the causes can be foreseen and prevented.

RISK FACTORS. The Framingham, Massachusetts, Heart Study, which began in 1950 and is still going on, is the longest and most famous study ever made of the susceptibility of a community to heart disease. Approximately 13,000 people in the town have participated in the investigation by receiving examinations every 2 years since the study began. The results of this research indicate that people who develop combinations of certain risk factors eventually have heart attacks. These factors are high cholesterol blood level, high blood pressure, cigarette smoking, overweight, lack of exercise, and diabetes.

The first five risk factors – high cholesterol blood level that leads to atherosclerosis, high blood pressure, cigarette smoking, overweight, and lack of exercise – all contribute to increasing the work load of the heart. The effects of high cholesterol blood level and hypertension have already been discussed. Cigarette smoking increases the work load through the effects of nicotine, which stimulates the adrenal gland to oversecrete aldosterone, epinephrine, and norepinephrine – powerful vasoconstrictors. Overweight people develop miles of extra capillaries to nourish their unwanted fat tissue. This means that the heart has to work harder to pump the blood through more vessels. Lack of exercise means that venous return gets less help from contracting skeletal muscles. In addition, regular exercise strengthens the smooth muscle of blood vessels and enables them to assist general circulation more efficiently. Exercise also increases cardiac efficiency and output.

People with three, four, or more risk factors form an especially high-risk group. The incidence of serious heart attacks in this high-risk group is far greater than in groups that have no risk factor or only one. The people who are most apt to develop atherosclerosis and who, consequently, run the highest risk of all have three risk factors: high cholesterol, hypertension, and cigarette smoking. Other researchers list emotional stress as an important risk factor. But there is still controversy among medical people as to the relative importance of this factor.

The risk factors make a person more susceptible to heart trouble because they strain the heart or increase the likelihood that its oxygen supply will be shut off at some time. Generally, the immediate cause of the heart trouble is one of the following: (1) failure of the heart's blood supply, (2) faulty heart architecture, or (3) failure of the heart's conductivity. Of these three reasons, the first two are far more common than the third.

FAILURE OF BLOOD SUPPLY. Angina pectoris and myocardial infarction result from insufficient oxygen supply to the myocardium. Coronary artery disease takes about one in twelve of all Americans who die between the ages of 25 and 34. It claims almost one in four of all those who die between 35 and 44. It has been reported that 50 to 65 percent of all sudden deaths are due to coronary heart disease.

At least half of the deaths from myocardial infarction occur before the patient reaches the hospital. These early deaths could result from an irregular heart rhythm, which is called an *arrhythmia.* Sometimes this progresses to the stage called *cardiac arrest* or ventricular fibrillation, in which the heart stops functioning. An arrhythmia is an abnormal, irregular rhythm change of the heart, caused by disturbances in the conduction system. This abnormal rhythm of the heartbeat could result in cardiac arrest because in this condition the heart is not capable of supplying the oxygen demands of the body. Serious arrhythmias can be controlled, and the normal heart rhythm can be reestablished, if they are detected and treated early enough. Coronary care units have reduced hospital mortality rates from acute myocardial infarctions by about 30 to 20 percent or less by preventing or controlling serious arrhythmias.

FAULTY ARCHITECTURE. Less than 1 percent of all new babies have a **congenital,** or **inborn, heart defect.** Even so, the total number in this country each year is estimated to be 30,000 to 40,000. Some of these infants may be able to live quite healthy and long lives without any need for repairing their hearts. But sometimes an inborn heart defect is so severe that an infant lives only a few hours. One of the more common of these defects is patent ductus arteriosus. The seriousness of this condition is that the connection between the aorta and the pulmonary artery remains open instead of closing completely after birth. This results in aortic blood flowing into the lower-pressure pulmonary trunk, thus increasing the pulmonary trunk blood pressure. This increases considerably the work of both ventricles and overworks the heart.

Another common group of congenital problems are the septal defects. A **septal defect** is an opening in the septum that separates the interior of the heart into a left and right side. *Atrial septal defect* is a hole caused by the failure of the fetal foramen ovale to close off the two atria from one another. Because pressure in the right atrium is low, atrial septal defect generally allows a good deal of blood to flow from the left atrium to the right. This results in an overload of the pulmonary circulation, producing fatigability, increased respiratory infections, and growth failure, if it occurs early in life because the systemic circulation may be deprived of a considerable portion of the blood destined for the organs and tissues of the body. *Ventricular septal defect* is caused by an abnormal development of the interventricular septum. Deoxygenated blood subsequently gets mixed with the oxygenated blood that is pumped into the systemic circulation. Consequently, the victim suffers *cyanosis,* a blue or dark purple discoloration of the skin. Cyanosis results from insufficient oxygen in the blood. It occurs whenever deoxygenated blood reaches the cells because of heart defect, lung defect, or suffocation. Septal openings can now be sewn shut or covered with synthetic patches.

A third defect is **valvular stenosis.** It is a narrowing, or *stenosis,* of one of the valves regulating blood flow inside the heart. Narrowing may occur in the valve itself, most commonly in the mitral valve, from rheumatic heart disease or the aortic valve from sclerosis or rheumatic fever. Or it may occur in an area near a valve. The seriousness of all types of stenoses stems from the fact that they all place a severe work load on the heart by making it work harder to push the blood through the abnormally narrow valve openings. As a result of mitral stenosis, blood pressure is increased and angina pectoris and heart failure may accompany the progress of this disorder. The majority of stenosed valves are totally replaced with artificial valves developed in recent years.

The last congenital defect that we shall discuss is tetralogy of Fallot. **Tetralogy of Fallot** is a combination of four defects causing a "blue baby." These are: a ventricular septal opening, an aorta that emerges from both ventricles instead of solely from the left ventricle, and, as a result of these, a stenosed pulmonary semilunar valve and an enlarged right ventricle (Figure 11–30). Because of the ventricular septal defect, both oxygenated and unoxygenated blood are mixed in the ventricles. However, the tissues of the body are much more starved for oxygen than are those of a child with simple ventricular septal defect. Because the aorta also emerges from the right ventricle and the pulmonary artery is stenosed, very little blood ever gets to the lungs and pulmonary circulation is bypassed almost completely. Today it is possible to completely cure cases of tetralogy of Fallot when the patient is of proper age and condition. Open-heart operations are performed in which the narrowed pulmonary valve is cut open and the septal defect is sealed with a Dacron patch.

PROBLEMS ARISING FROM FAULTY CONDUCTION. As noted earlier, the term arrhythmia refers to any variation in the rate, rhythm, or synchrony of the heart. It arises when electrical impulses through the

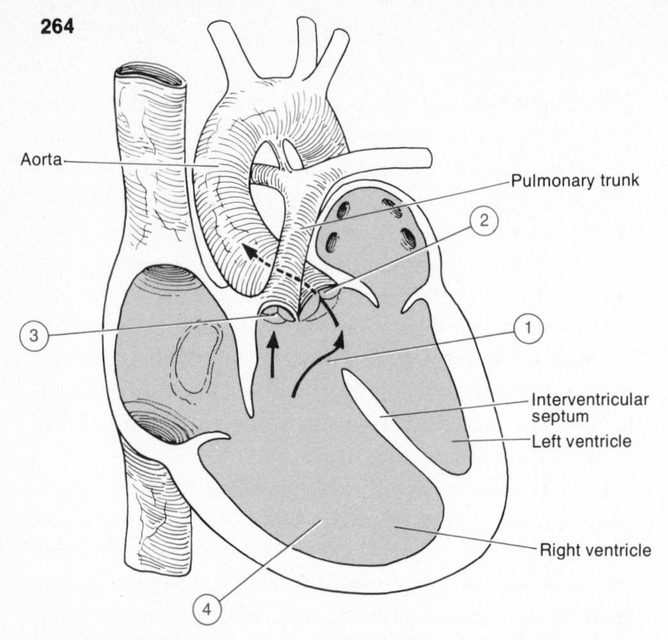

Figure 11–30. Tetralogy of Fallot. The four abnormalities associated with this condition are indicated by numbers. (1) Opening in the interventricular septum. (2) Origin of the aorta in both ventricles. (3) Stenosed pulmonary semilunar valve. (4) Enlarged right ventricle.

heart are blocked at critical points in the conduction system. One such arrhythmia is called a **heart block.** Perhaps the most common blockage is in the atrioventricular node, which conducts impulses from the atria to the ventricles. This disturbance is called *atrioventricular (AV) block.* And it usually indicates a myocardial infarction, atherosclerosis, rheumatic heart disease, diphtheria, or syphilis. In a first-degree AV block, which can be detected only by the use of an electrocardiograph, the transmission of impulses from the atria to the ventricles is delayed. Here, the P-R interval is greater than it should be. In a second-degree AV block, every second impulse fails to reach the ventricles so that the ventricular rate is about one-half that of the atrial rate. When ventricular contraction does not occur (dropped beat), oxygenated blood is not pumped efficiently to all parts of the body. The patient may feel faint and dizzy or may collapse if there are many dropped ventricular beats. In a third-degree or complete AV block, impulses reach the ventricle

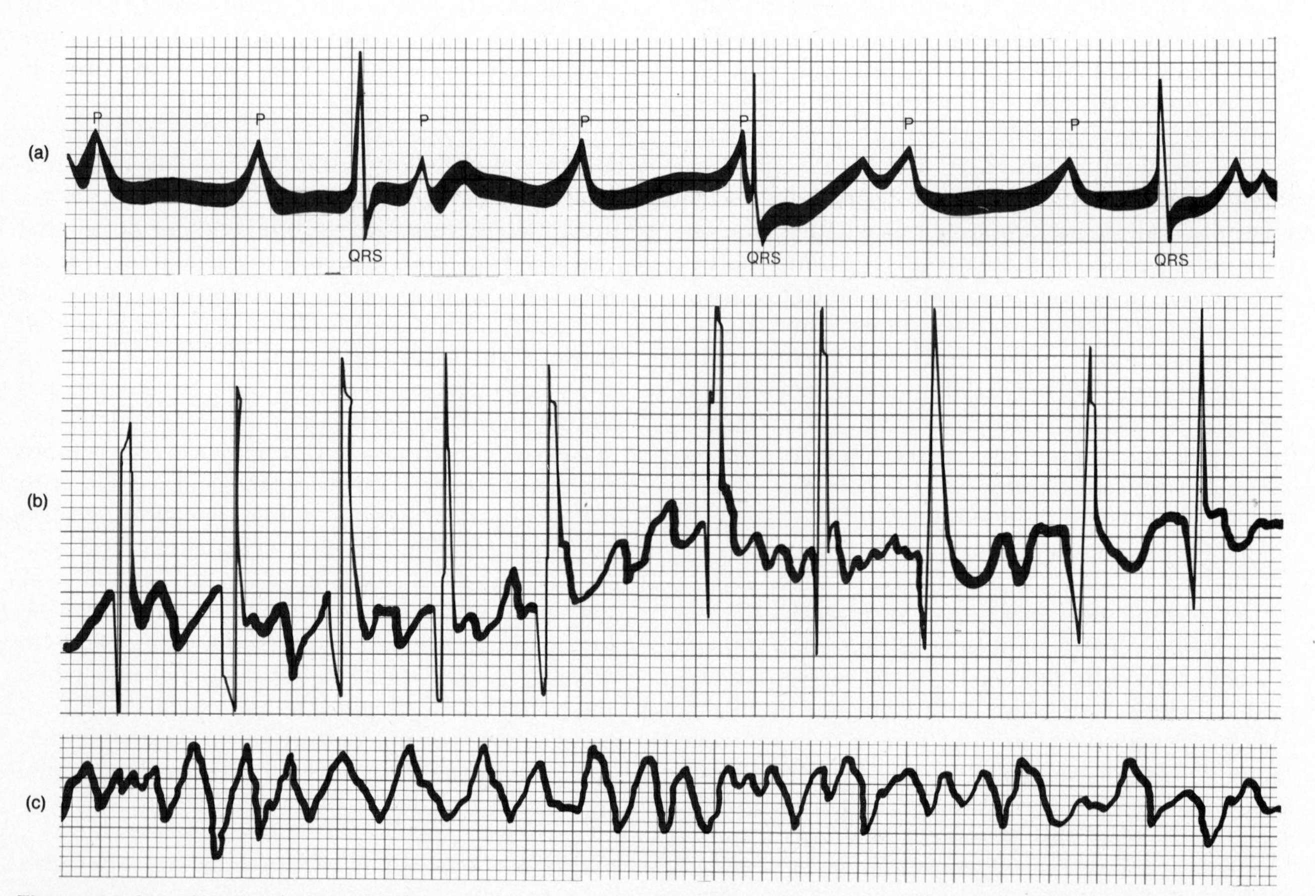

Figure 11–31. Abnormal electrocardiograms. (a) Complete heart block. There is no fixed ratio between atrial contractions (P waves) and ventricular contractions (QRS waves). (b) Atrial fibrillation. There is no regular atrial contraction and, therefore, no P wave. Since the ventricles contract irregularly and independently, the QRS wave appears at irregular intervals. (c) Ventricular fibrillation. In general, there is no rhythm of any kind.

at irregular intervals, and some never reach it at all. The result is that atrial and ventricular rates are out of synchronization (Figure 11–31a). The ventricles may go into systole at any time. This could occur when the atria are in systole or just before the atria go into systole. Or the ventricles may take a rest for a few cardiac cycles.

With complete AV block, many patients may have vertigo, unconsciousness, or convulsions. These symptoms result from a decreased cardiac output with consequent diminished cerebral blood flow and cerebral hypoxia or lack of sufficient oxygen. Among the causes of AV block are excessive stimulation by the vagus nerves that depresses conductivity of the junctional fibers, destruction of the AV bundle as a result of coronary infarct, atherosclerosis, myocarditis, or depression caused by various drugs. Other heart blocks include *intraatrial (IA) block, interventricular (IV) block,* and *bundle branch block (BBB).* In the latter condition, the ventricles do not contract together because of the delay in the impulse in the blocked branch.

RHYTHMS INDICATING HEART TROUBLE. Two rhythms that indicate heart trouble are atrial flutter and fibrillation. In **atrial flutter** the atrial rhythm averages between 240 and 360 beats per minute. The condition is essentially very rapid atrial contractions accompanied by a second-degree AV block. It is typically indicative of severe damage to heart muscle. Atrial flutter usually becomes fibrillation after a few days or weeks. **Atrial fibrillation** is an asychronous contraction of the atrial muscles that causes the atria to contract irregularly and still faster. An electrocardiogram of atrial fibrillation is shown in Figure 11–31b. Atrial flutter and fibrillation occur in myocardial infarction, acute and chronic rheumatic heart disease, and hyperthyroidism. Atrial fibrillation results in complete uncoordination of atrial contraction so that atrial pumping ceases altogether. When the muscle fibrillates, the muscle fibers of the atrium quiver individually instead of contracting together. The quivering cancels out the pumping function of the atrium. In a strong heart, atrial fibrillation reduces the pumping effectiveness of the heart by 25 to 30 percent.

Ventricular fibrillation is another kind of rhythm that indicates heart trouble. It is characterized by asynchronous, irregular, haphazard ventricular muscle contractions. The rate may be rapid or slow. The impulse travels to the different parts of the ventricles at different rates. Thus, part of the ventricle may be contracting, while other parts are still unstimulated. Ventricular contraction becomes ineffective and circulatory failure and death occur immediately unless the arrhythmia is reversed quickly (Figure 11–31c). Ventricular fibrillation may be caused by coronary occlusion. It sometimes occurs during surgical procedures on the heart or pericardium. And it may be the cause of death in electrocution.

We will now direct our attention to a study of the nervous system. First we will consider the basic structure and functions of nervous tissue.

Key terms associated with the blood vascular and lymphatic systems

Adenitis (*adeno* = gland) Enlarged, tender, and inflamed lymph nodes resulting from an infection.

Angiocardiography (*angio* = vessel; *cardio* = heart; *graph* = write, record) Recording of an image of the chambers of the heart by x-ray revealed by the direct injection of radiopaque dyes.

Arteriography Recording of an image of arteries by x-ray revealed by the direct injection of dyes.

Bradycardia (*brady* = slow) Slow heartbeat or pulse.

Cardiac arrest Complete stoppage of the heartbeat.

Cardiomegaly (*megalo* = great, large) Heart enlargement.

Corpuscles Cellular elements in the blood such as red and white cells.

Cyanosis Slightly bluish, dark purple skin coloration due to oxygen deficiency in systemic blood.

Defibrillator A mechanical device for applying electrical shock to the heart to terminate abnormal cardiac rhythms.

Elephantiasis Great enlargement of a limb (especially lower limbs) and scrotum, resulting from obstruction of lymph glands or vessels; caused by a tiny parasitic worm.

Embolus Any foreign or abnormal particle that is transported through the blood. It may be solid, liquid, or gaseous. Obstruction occurs if embolus becomes stuck in small vessels.

Epistaxis A nosebleed.

Fractionated blood (*fract* = break) Blood that has been separated into its components. Only the part of the blood needed by the patient is given.

Blood cells

Packed red cells Erythrocytes separated from citrated, whole human blood plasma, which has been identified with the donor's blood group and Rh type. May be used in the treatment of any type of severe anemia, except that produced by excessive bleeding. In the latter case, whole blood is given in order to bring the fluid volume up to normal.

Platelet concentrates Platelets obtained from freshly drawn whole blood. Used for platelet-deficiency disorders such as hemophilia.

Hem, Hemo, Hema, Hemato (*heme* = iron) Various combining forms meaning blood.

Hematoma (hemangioma) (*haemo, hemato* = blood; *oma* = tumor) Leakage of blood from a vessel, which clots to form a solid mass or swelling in any tissue.

Hemolysis (laking) A swelling and subsequent rupture of erythrocytes with the liberation of hemoglobin into the surrounding fluid.

Hemorrhage (*rrhage* = bursting forth) Bleeding, either internal (from blood vessels into tissues) or external (from blood vessels directly to surface of body).

Hypersplenism Abnormal splenic activity involving highly increased blood cell destruction.

Lymphadenectomy (*ectomy* = cutting out, removal of) Removal of a lymph node.

Lymphadenopathy (*patho* = disease, suffering) Enlarged, sometimes tender lymph glands.

Lymphangioma A benign tumor of the lymph vessels.

Lymphangitis Inflammation of the lymphatic vessels.

Lymphedema Accumulation of lymph fluid producing subcutaneous tissue swelling.

Lymphoma Any tumor composed of lymph tissue. Malignancy of reticuloendothelial cells of lymph nodes is called Hodgkin's disease.

Lymphostasis (*stasis* = halt) A lymph flow stoppage.

Murmurs A heart sound produced by blood passing through a valve or an opening in the heart (septal defect). One of the primary means of clinically diagnosing heart disease.

Occlusion The closure or obstruction of the lumen of a structure, such as the lumen of a blood vessel.

Phlebitis Inflammation of a vein.

Plasmolysis (crenation) (*plas* = mold, shape, form; *lysis* = dissolve) Shrinking of the cytoplasm in any cell (in this case, erythrocytes) producing knobbed, starry shapes. Results from a loss of water by osmosis, as when blood is mixed with a 5 percent salt solution.

Septicemia (*sep* = decay, putrefaction; *emia* = condition of blood) Toxins or disease-causing bacteria in the blood. Also called "blood poisoning."

Tachycardia (*tachy* = rapid, fast) Rapid heart rate.

Thrombophlebitis (*thrombo* = clot) Inflammation of a vein with clot formation.

Thrombus (*thrombo* = clot, lump) A blood clot that is lodged in a vessel.

Transfusion The transfer of whole blood, individual blood components (red blood cells only or plasma only), or bone marrow directly into the blood stream of a recipient.

Direct (immediate) Transfer of blood directly from one person to another without exposing the blood to air.

Indirect (mediate) Transfer of blood from a donor to a flask or other container and then to the recipient. Permits blood to be stored for an emergency. Also permits breaking down blood into its components so that patient receives only a needed part of the blood.

Exchange Removing and discarding blood from the recipient while simultaneously replacing it with donor blood. This method is used for erythroblastosis fetalis and for poisoning.

Reciprocal Blood is transferred from a person who has recovered from a contagious infection into the vessels of a patient suffering with the same infection. An equal amount of blood is returned from the patient to the well person. This method allows the patient to receive antibody-bearing lymphocytes from the recovered person.

Venesection Opening of a vein for withdrawal of blood.

Whole blood Blood containing all formed elements, plasma, and plasma solutes in natural concentration.

Citrated whole blood Whole blood protected from coagulation by a citrate.

Heparinized whole blood Whole blood in a heparin solution to prevent coagulation.

Chapter summary in outline

BLOOD

1. The principal functions of blood are the transportation of O_2 and CO_2, nutrients and wastes, and hormones and enzymes. It regulates pH, normal body temperature, and water content of cells; and protects against disease. Blood consists of plasma and formed elements.

Formed elements

1. Formed elements are erythrocytes, leucocytes, and thrombocytes.
2. Erythrocytes, or red blood cells, are biconcave discs without nuclei that contain hemoglobin. Erythrocyte formation is called erythropoiesis and occurs in adult red marrow of certain bones.
3. Leucocytes, or white blood cells, are nucleated cells. Two principal types are granular (neutrophils, eosinophils, and basophils) and agranular (lymphocytes and monocytes).
4. One function of leucocytes, especially neutrophils, is to combat inflammation and infection through phagocytosis.
5. In response to the presence of foreign proteins called antigens, lymphocytes are changed into plasma cells. Plasma cells produce antibodies, which cover antigens and render them harmless. This is called the antigen-antibody response and is important in combating infection and providing immunities.
6. Thrombocytes, or platelets, are disc-shaped structures without nuclei. They initiate clotting.

Plasma

1. The liquid portion of blood, called plasma, consists of 92 percent water and 8 percent solutes. Important solutes include proteins (albumins, globulins, and fibrinogen), foods, enzymes and hormones, gases, and electrolytes.

INTERSTITIAL FLUID AND LYMPH

1. Interstitial fluid bathes body cells, whereas lymph is found in lymphatic vessels.
2. These fluids are basically similar in chemical composition. They differ chemically from plasma in that both contain less protein, a variable number of leucocytes, and no platelets or erythrocytes.

HEART

1. The parietal pericardium, consisting of an outer fibrous layer and an inner serous layer, encloses the heart.
2. The wall of the heart has three layers called the epicardium, myocardium, and endocardium. The chambers include two upper atria and two lower ventricles.
3. All valves of the heart prevent the back flow of blood. The blood flows through the heart from the superior and inferior venae cavae, to the right atrium, through the tricuspid valve to the right ventricle, through the pulmonary artery to the lungs, through the pulmonary veins to the left atrium, through the bicuspid valve to the left ventricle, and out through the aorta.
4. The conduction system consists of tissue specialized for impulse conduction. Components are the sinu-atrial node (pacemaker), atrioventricular node, bundle of His, bundle branches, and Purkinje fibers.
5. The record of electrical changes during cardiac cycle is referred to as an electrocardiogram (ECG). Normal ECG consists of a P wave (spread of impulse from SA node over atria), QRS wave (spread of impulse through ventricles), and T wave (ventricular repolarization).
6. The first heart sound (lubb) represents the closing of the atrioventricular valves. The second sound (dupp) represents the closing of the semilunar valves.
7. Surface landmarks are useful for identifying the position of the borders and valves of the heart.

BLOOD VESSELS

1. Arteries carry blood away from the heart. They are stronger and thicker than veins, consisting of a tunica interna, tunica media, and tunica externa.
2. Many arteries anastomose, which means that the distal ends of two or more vessels unite. An alternate blood route from an anastomosis is called collateral circulation.
3. Capillaries are microscopic blood vessels through which materials are exchanged between blood and interstitial fluid. They unite to form venules, which in turn form veins to carry blood back to the heart.
4. Veins have less elastic tissue and smooth muscle than arteries, and they contain valves to prevent back flow of blood.
5. Weak valves can lead to varicose veins or hemorrhoids.

CIRCULATORY ROUTES

1. The systemic circulation takes oxygenated blood from the left ventricle through the aorta to all parts of the body except the lungs. It includes the coronary and hepatic portal circulations.
2. The coronary circulation takes oxygenated blood through the arterial system of the myocardium. Deoxygenated blood returns to the right atrium via the coronary sinus. Complications of this system are angina pectoris and myocardial infarction.
3. The hepatic portal circulation takes blood from the veins of the pancreas, spleen, stomach, intestines, and gallbladder to the portal vein of the liver. It enables the liver to utilize nutrients and detoxify harmful substances in the blood.
4. The pulmonary circulation takes deoxygenated blood from the right ventricle to the lungs and returns oxygenated blood from the lungs to the left atrium. It allows blood to be oxygenated for systemic circulation.
5. The fetal circulation involves the exchange of materials between the fetus and mother through the placenta.

LYMPH VASCULAR SYSTEM

1. This system consists of lymph vessels, lymph, lymph nodes, and lymph organs. Its primary function is to drain, from tissue spaces, protein-containing fluid which escapes from blood capillaries.
2. Lymphatic vessels are similar in structure to veins. All lymphatics deliver lymph to either the thoracic duct or right lymphatic duct.
3. Lymph nodes are oval-shaped structures located along lymphatics. Lymph passing through the nodes is filtered, and it picks up antibodies and agranular leucocytes.
4. Lymph organs that filter lymph and add white blood cells and antibodies are the tonsils, spleen, and thymus gland.
5. Lymph circulation is from tissue spaces to large lymphatic ducts to the subclavian veins.

APPLICATIONS TO HEALTH

Blood disorders

1. Anemia is indicated by a decreased erythrocyte count or hemoglobin deficiency. Kinds of anemia include hemorrhagic, hemolytic, aplastic, and sickle cell anemia.
2. Polycythemia is an abnormal increase in the number of erythrocytes.
3. Infectious mononucleosis is characterized by an elevated white cell count, especially the monocytes. The cause is unknown.
4. Leukemia is the uncontrolled production of white blood cells that interferes with normal clotting and vital body activities.

Cardiovascular disorders

1. An aneurysm is a sac formed by an outpocketing of a portion of an arterial or venous wall.
2. Atherosclerosis is the hardening of the arteries caused by formation of plaques.
3. Hypertension is high blood pressure. Primary hypertension cannot be linked to a specific organic cause. Secondary hypertension may be caused by atherosclerosis, kidney disorders, excessive aldosterone secretion, and tumors.
4. Heart disorders related to inadequate blood supply are angina pectoris and myocardial infarction.
5. Congenital heart defects include patent ductus arteriosus, septal defects, valvular stenosis, and tetralogy of Fallot.
6. Heart conditions relative to conduction problems include heart blocks, atrial flutter and fibrillation, and ventricular fibrillation.

Review questions and problems

1. Distinguish between the blood vascular system and lymph vascular system.
2. Define the principal physical characteristics of blood. List the functions of blood and their relationship to other systems of the body.
3. Distinguish between plasma and formed elements. Where are the formed elements produced?
4. Describe the microscopic appearance of erythrocytes. What is the essential function of erythrocytes?
5. What is a reticulocyte count? What is its diagnostic significance?
6. Describe the classification of leucocytes. What are their functions?
7. What is the importance of diapedesis and phagocytosis in fighting bacterial invasion?
8. What is a differential count? What is its significance?
9. Distinguish between leucocytosis and leucopenia.
10. Describe the antigen-antibody response. How is the response protective?
11. What are the major chemicals in plasma? What do they do?
12. What is the difference between plasma and serum?
13. Compare interstitial fluid and lymph with regard to location, chemical composition, and function.
14. Describe the location of the heart in the mediastinum. Distinguish the subdivisions of the parietal pericardium. What is the purpose of this structure?
15. Compare the three portions of the wall of the heart. Define atria and ventricles. What vessels enter or exit the atria and ventricles?
16. Discuss the principal kinds of valves in the heart and how they operate.
17. Describe the components of the conduction system of the heart.
18. What is an electrocardiogram? Describe the importance of the deflection waves of an ECG.
19. Describe the various sounds of the heart and the importance of each.
20. Prepare a diagram to illustrate the surface landmarks used to locate the borders and valves of the heart.
21. Describe the structural and functional differences among arteries, capillaries, and veins.
22. Discuss the importance of the elasticity and contractility of arteries. What is an anastomosis?
23. Define varicose veins and hemorrhoids.
24. What is meant by a circulatory route? Define systemic circulation.
25. By means of a diagram, indicate the major divisions of the aorta, their principal arterial branches, and the regions supplied.

26. Trace a drop of blood from the arch of the aorta through its systemic circulatory route and back to the heart again. Remember that the major branches of the arch are the brachiocephalic artery, left common carotid artery, and left subclavian artery. In your answer, be sure to indicate which veins return the blood to the heart.
27. What is the circle of Willis? Why is it important?
28. What are visceral branches of an artery? Parietal branches?
29. What major organs are supplied by branches of the thoracic aorta? How is blood returned from these organs to the heart?
30. What organs are supplied by the celiac artery, superior mesenteric, renal, inferior mesenteric, inferior phrenic, and middle sacral? How is blood returned from these organs to the heart?
31. Trace a drop of blood from the common iliac arteries through their branches to the respective organs and back to the heart again.
32. What is a deep vein? A superficial vein? Define a venous sinus in relation to blood vessels. What are the three major groups of systemic veins?
33. Describe the route of blood in the coronary circulation. Distinguish between angina pectoris and myocardial infarction.
34. What is hepatic portal circulation? Describe the route by means of a diagram. Why is this route important?
35. Define pulmonary circulation. Prepare a diagram to indicate the route. What is the purpose of the route?
36. Discuss in detail the anatomy and physiology of fetal circulation. Be sure to indicate the function of the umbilical arteries, umbilical vein, ductus venosus, foramen ovale, and ductus arteriosus.
37. What is the fate of the special structures involved in fetal circulation once postnatal circulation is established?
38. Describe the cause and treatment of patent ductus arteriosus.
39. Identify the components of the lymph vascular system. What is its function?
40. How do lymphatic vessels originate? Compare veins and lymphatics with regard to structure.
41. Construct a diagram to indicate the role of the thoracic duct and right lymphatic duct in draining lymph from different regions of the body.
42. What is a lymphangiogram? What is its diagnostic value?
43. Describe the structure of a lymph node. What functions do lymph nodes serve?
44. Identify four groups of clinically important lymph nodes.
45. Compare the functions of the tonsils, spleen, and thymus gland as lymphatic organs.
46. Define anemia. Contrast the causes of hemorrhagic, hemolytic, aplastic, and sickle cell anemias.
47. What is leukemia, and what is the cause of some of its symptoms?
48. Define hematocrit. Compare the hematocrits of polycythemic and anemic blood.
49. What is infectious mononucleosis?
50. What is an aneurysm? Distinguish three types on the basis of location.
51. Discuss the causes, symptoms, and diagnosis of atherosclerosis.
52. Compare primary and secondary hypertension with regard to cause.
53. Describe the risk factors involved in heart disease.
54. Distinguish among the following congenital heart disorders: patent ductus arteriosus, septal defects, valvular stenosis, and tetralogy of Fallot.
55. Define an arrhythmia. What is a heart block? Distinguish the various kinds of heart block.
56. Describe atrial flutter and atrial fibrillation. What is ventricular fibrillation?
57. Refer to the glossary of key terms associated with the blood vascular and lymphatic systems. Be sure that you can define each term listed.

Selected readings

Berne, R. M., and N. L. Matthew. *Cardiovascular Physiology.* St. Louis: Mosby, 1972.

Cunningham's Textbook of Anatomy. 11th ed., edited by G. J. Romanes. London: Oxford University Press, 1972. Pp. 837–967.

Dawson, Helen L. *Basic Human Anatomy.* 2d ed. New York: Appleton-Century-Crofts, 1974. Pp. 134–164.

Guyton, Arthur C. *Textbook of Medical Physiology.* 4th ed. Philadelphia: W. B. Saunders, 1971.

Ham, Arthur W. *Histology.* 7th ed. Philadelphia: J. B. Lippincott, 1974.

Leeson, Thomas S., and C. Roland Leeson. *Histology.* 2d ed. Philadelphia: W. B. Saunders, 1970.

Netter, Frank H. *Heart.* CIBA Collection of Medical Illustrations, Vol. 5. Summit, NJ: CIBA Pharmaceutical Co., 1969.

Rapaport, S. I. *Introduction to Hematology.* New York: Harper and Row, 1971.

Simmons, A. *Basic Hematology.* Springfield, IL: C. C. Thomas, 1973.

CHAPTER 12

NERVOUS TISSUE

STUDENT OBJECTIVES

After you have read this chapter, you should be able to:

1. Describe the function of the nervous system in maintaining homeostasis
2. Classify the organs of the nervous system into central and peripheral divisions
3. Contrast the histological characteristics of neuroglia and neurons
4. Categorize neurons on the basis of structure and function
5. Define a nerve impulse and describe its importance in the body

The **nervous system** is the control tower and communications network of the body. In human beings it performs three broad functions. First, it stimulates movements that are vital to life as well as movements that simply make life easier and more enjoyable. Second, it shares responsibility with the endocrine system for the maintenance of homeostasis. And, third, it allows us to express uniquely human traits. Human life simply cannot exist without a functioning nervous system. For instance, skeletal and most smooth muscle cells cannot contract until they are stimulated by a nerve impulse. If the intercostal muscles and diaphragm do not contract, we cannot breathe. If the smooth muscles of the digestive system do not contract, food cannot be pushed through the esophagus, stomach, and intestines. And if the digestive glands are not stimulated to release their secretions, food cannot be digested. It is obvious, then, that our cells cannot receive nutrients unless the digestive system is connected to a functioning nervous system. But suppose our muscles and glands could stimulate themselves. Even then, we could not live very long without our nervous system. This is because of the second great function of the nervous system—keeping the body in homeostasis. Recall that homeostasis is the maintenance of a constant internal environment. The nervous system *senses* changes that occur inside the body and in the outside environment. It then interprets these changes and may initiate a course of action. This property is called *integration.* After deciding which action to take, it elicits a *response* by sending impulses to the appropriate muscles and glands. The third broad function of the human nervous system is to provide the uniquely human pleasures of thinking and acting upon our thoughts and feelings.

ORGANIZATION OF THE NERVOUS SYSTEM

For convenience of study, the entire nervous system may be divided into two principal portions: (1) the central nervous system and (2) the peripheral nervous system (Figure 12–1). The **central nervous system (CNS)** is the control center for the entire system and consists of the brain and the spinal cord. All body sensations must be relayed to the central nervous system if they are to be felt and acted upon. All the impulses that stimulate muscles to contract and glands to secrete must pass from the central system.

The **peripheral nervous system (PNS)** contains all remaining nervous tissue—that is, the nerves which connect the central nervous system to all other parts of the body. The peripheral nervous system is frequently divided into a somatic nervous

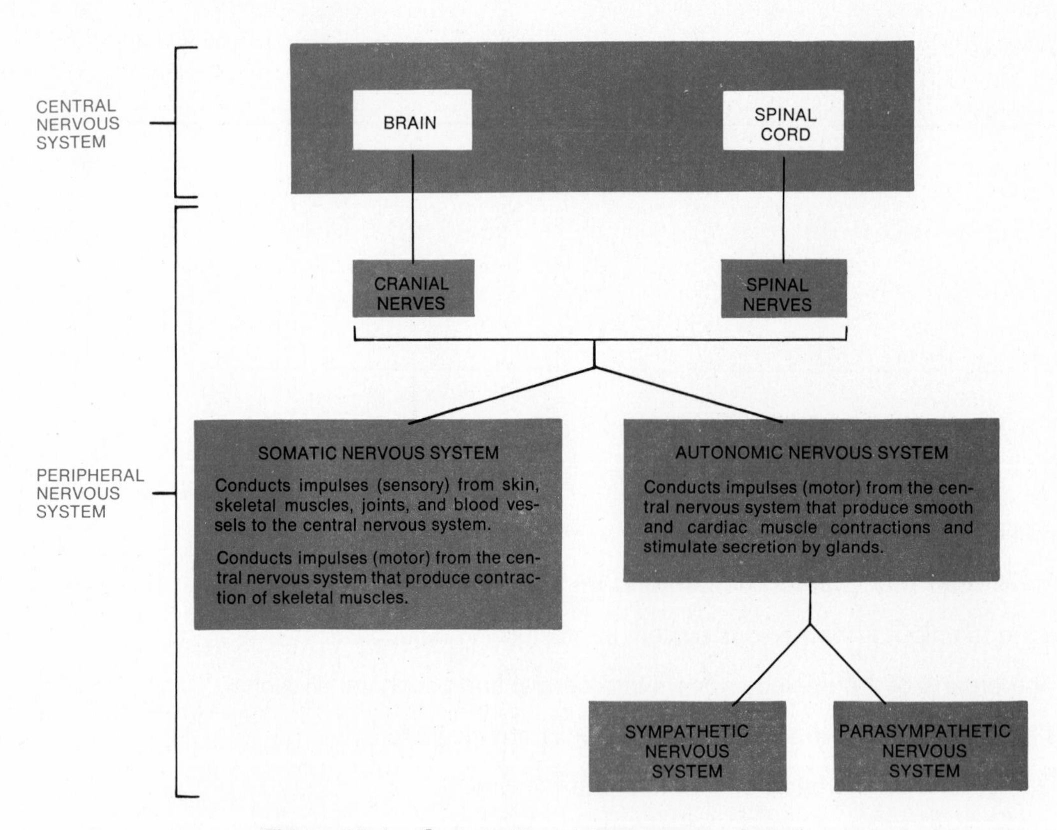

Figure 12–1. Organization of the nervous system.

system (SNS) and an autonomic nervous system (ANS) on the basis of function.

The *somatic nervous system* (*soma,* meaning body) contains sensory nerves that conduct impulses from the skin, skeletal muscles, joints, viscera, and blood vessels to the central nervous system. The somatic nervous system also contains motor nerves that conduct impulses from the central nervous system to skeletal muscles. Since the somatic nervous system produces movement only in skeletal muscles, it is under our conscious control and is therefore said to be voluntary.

The *autonomic nervous system,* by contrast, contains motor nerves that convey impulses from the central nervous system to smooth muscle, cardiac muscle, and glands. The autonomic system produces responses only in involuntary muscles and glands. For this reason, the autonomic system is usually considered to be involuntary. With very few exceptions, the viscera receive nerve fibers from the two divisions of the autonomic nervous system: the sympathetic division and the parasympathetic division. The particular functions of the sympathetic and parasympathetic divisions will be considered in Chapter 15.

HISTOLOGY OF THE NERVOUS SYSTEM

Despite the organizational complexity of the nervous system, it consists of only two principal kinds of cells. The first of these, the neurons or nerve cells, make up the nervous tissue that forms the structural and functional portion of the system. Neurons are highly specialized for impulse conduction and are responsible for all the special attributes of the nervous system, such as thinking, controlling muscle activity, and regulating glands. The second type of cell, the neuroglia, forms a special kind of connective tissue component of the nervous system. Neuroglia generally perform the less specialized activity of binding together nervous tissue, forming myelin, and phagocytosis. They do not transmit impulses.

Neuroglia

The connective tissue cells of the nervous system are called **neuroglia** or **glial cells.** The combining form *neuro* means nerve, and *glia* means glue. Nervous tissue lacks the compactness of muscle and rigidity of bone and must be heavily supported by thick connective tissue. Many of the glial cells form a supporting network by twining around the nerve cells in the brain and spinal cord. Other glial cells bind nervous tissue to other supporting structures and attach the neurons to their blood vessels. A few of the glial cells also serve very specialized functions. For example, many nerve fibers are coated with a thick, fatty sheath that is produced by a particular type of neuroglia. Certain small glial cells are phagocytotic. They protect the central nervous system from disease by engulfing invading microbes and clearing away debris. All these duties of the neural connective tissue are divided among several different kinds of glial cells, which are classified according to their size and shape. Neuroglia are of special clinical interest because they are a very common source of tumors of the nervous system. Exhibit 12–1 lists the cells and summarizes their functions.

Neurons

Nerve cells, called **neurons,** are responsible for the conduction of impulses from one part of the body to another.

Structure

A neuron consists of three structurally and functionally distinct portions: (1) the cell body, (2) dendrites, and (3) an axon (Figure 12–2a). The **cell body** or **perikaryon** contains a well-defined nucleus and nucleolus surrounded by a granular cytoplasm. Within the cytoplasm are found typical organelles such as mitochondria and Golgi apparatus. Also found within the cytoplasm are structures characteristic of neurons. These are Nissl bodies and neurofibrils. The *Nissl bodies* consist of orderly arrangements of granular (rough) ER and free ribosomes. The function of Nissl bodies is protein synthesis. The newly synthesized proteins pass from the perikaryon into the neuronal processes, mainly the axon, at the rate of about 1 mm (0.0394 in.)/day. These proteins replace those lost during metabolism and are used for growth of neurons and regeneration of peripheral nerve fibers. The details of regeneration are discussed in Chapter 13. *Neurofibrils* are long, thin fibrils. Neurofibrils are believed to be composed of protein threads and may assume a function in support.

The cytoplasmic processes of neurons differ according to the direction in which they conduct impulses. On the basis of this difference, the processes are of two kinds: dendrites and axons. **Dendrites** are short, highly branched extensions of the cytoplasm of the cell body. A neuron usually has several main dendrites. Dendrites typically contain Nissl bodies and mitochondria. The function of dendrites is to conduct an impulse toward the cell body. The second type of cytoplasmic process, called an **axon** or

Exhibit 12–1. NEUROGLIA

TYPE	DESCRIPTION	MICROSCOPIC APPEARANCE	FUNCTION
Astrocytes (*astro* = star; *cyte* = cell)	Star-shaped cells with numerous processes		Twine around nerve cells to form a supporting network in the brain and spinal cord; attach neurons to their blood vessels
Oligodendrocytes (*oligo* = few; *dendro* = tree)	Resemble astrocytes in some ways, but the processes are fewer and shorter		Give support by forming semirigid connective tissue rows between neurons in brain and spinal cord; produce a thick, fatty sheath called the myelin sheath on neurons of the central nervous system
Microglia (*micro* = small)	Small cells with few processes; normally stationary; if nervous tissue is damaged, they may migrate to injured area		Engulf and destroy microbes and cellular debris

axis cylinder, is a single, highly specialized, and relatively long process that conducts impulses away from the cell body to another neuron or to another tissue.

An axon originates from the cell body as a small conical elevation called the *axon hillock.* An axon contains mitochondria and neurofibrils but does not contain Nissl bodies. Its cytoplasm, called *axoplasm,* is surrounded by a plasma membrane known as the *axolemma.* Axons vary in length from a few millimeters (1 mm = 0.0394 in.) within the brain to a meter (3.28 ft) or more between the spinal cord and toes. Along the course of an axon, there may be one or more side branches called *axon collaterals.* The axon and its collaterals each terminate by branching into many fine filaments called *telodendria.*

The term **nerve fiber** is applied to an axon and certain coverings or sheaths. We will first consider the structure of nerve fibers of the peripheral nervous system. Figure 12–2b shows a cross section of a nerve fiber of the peripheral nervous system. Many axons, especially large, peripheral axons, are surrounded by a white, lipid, segmented covering called the *myelin sheath.* Axons containing such a covering are called *myelinated,* whereas those that lack it are called *unmyelinated.* Myelin is responsi-

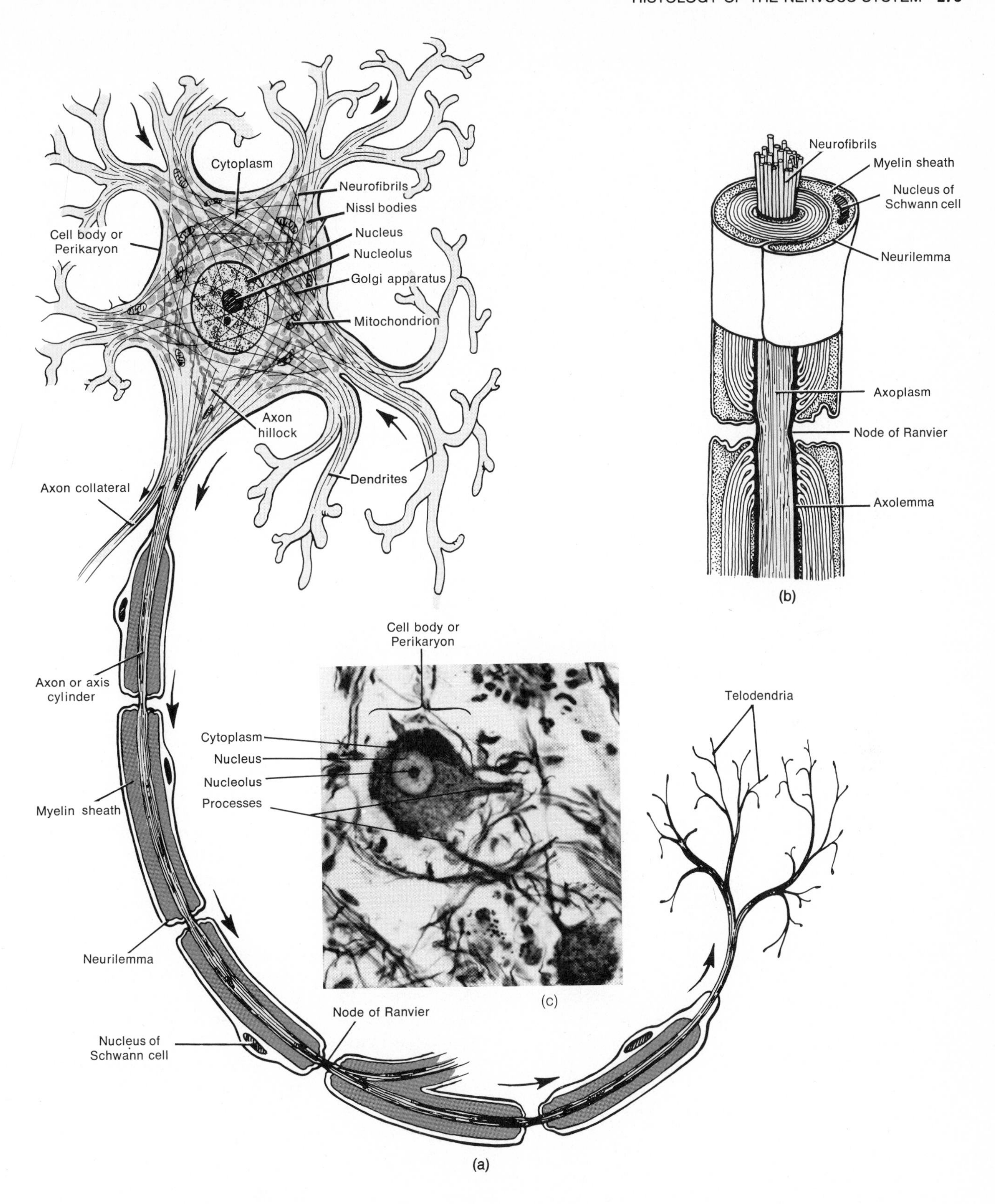

Figure 12–2. Structure of a neuron. (a) Shown in an entire multipolar neuron. The arrows indicate the direction in which a nerve impulse passes. (b) Cross section through a myelinated fiber. (c) Photomicrograph of a multipolar neuron from a sympathetic ganglion at a magnification of 640X. (Courtesy of Edward J. Reith, from *Atlas of Descriptive Histology,* by Edward J. Reith and Michael H. Ross, Harper & Row, Publishers, Inc., New York, 1970.)

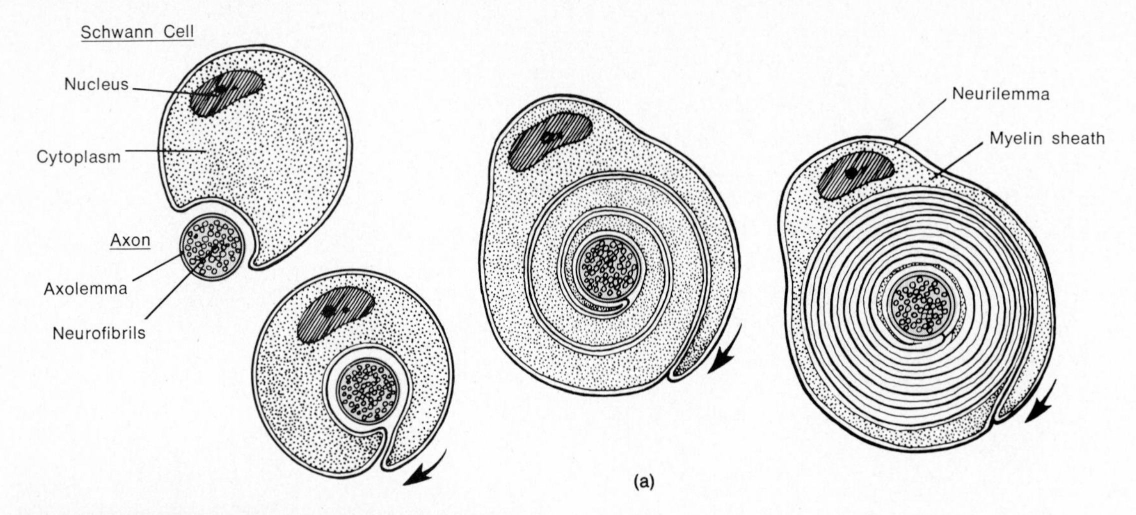

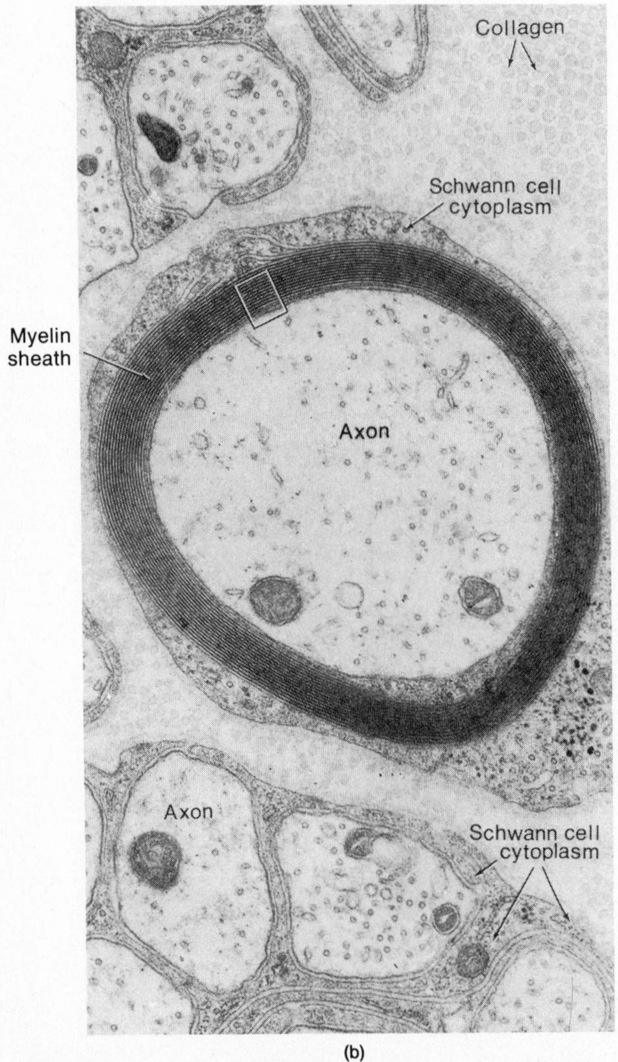

Figure 12–3. Myelin sheath. (a) Stages in the formation of a myelin sheath by a Schwann cell. (b) Electron micrograph of a nerve in cross section showing a myelinated nerve axon and several unmyelinated nerve axons at a magnification of 12,000X. (Courtesy of William Bloom and Don W. Fawcett, *A Textbook of Histology,* W. B. Saunders, Philadelphia, 1968.)

ble for the color of white matter of nerves and the brain and spinal cord. The myelin sheath is produced by flattened cells, called *Schwann cells,* located along the axon. They are neuroglial cells. In this process, a Schwann cell encircles the axon until its ends meet and overlap (Figure 12–3). The cell then winds around the axon several times and, in doing so, the cytoplasm and nucleus are pushed to the outside layer. The inner layer, consisting of several layers of Schwann cell membrane, is the myelin sheath. The function of the myelin sheath is to insulate and maintain the axon. The *neurilemma* or *sheath of Schwann* is the peripheral nucleated cytoplasmic layer of the Schwann cell that encloses the myelin sheath. The neurilemma is found only around fibers of the peripheral nervous system. Its function is to assist in the regeneration of injured axons. Between the segments of the myelin sheath are unmyelinated gaps called *nodes of Ranvier.* Unmyelinated fibers are also enclosed by Schwann cells, but without multiple wrappings.

Nerve fibers of the central nervous system may be myelinated or unmyelinated. Myelination for central nervous system axons is accomplished by oligodendrocytes in the same manner as Schwann cells myelinate peripheral nervous system axons. Myelinated axons of the central nervous system also contain nodes of Ranvier. A principal difference between myelinated fibers of the central nervous system and the peripheral nervous system is that fibers of the central nervous system do not contain a neurilemma.

Kinds of neurons

The many different kinds of neurons found in the body may be classified according to two criteria: structure and function. The structural classification is based upon the number of processes extending from the cell body. For example, **multipolar neurons** have several dendrites and one axon. Most neurons in the brain and spinal cord are of this type. **Bipolar neurons** have one dendrite and one axon and are found in the retina of the eye, the inner ear, taste buds, and the olfactory area. The third structural type of neuron is called a **unipolar neuron.** This has only one process extending from the cell body. The single process divides into a central branch, which functions as an axon, and a peripheral branch, which is an axon in structure but functions as a dendrite. Unipolar neurons originate in the embryo as bipolar neurons. During development, the axon and dendrite fuse into a single process.

The functional classification of neurons is based upon the direction in which they carry impulses. **Sensory neurons,** called **afferent neurons,** carry impulses from receptors in the skin and sense organs to the brain and spinal cord. Sensory neurons are usually unipolar. **Motor neurons,** called **efferent neurons,** convey impulses from the brain and spinal cord to effectors, which may be either muscles or glands. Other neurons, called **association, internuncial,** or **connecting neurons,** carry impulses from sensory neurons to motor neurons and are located inside the brain and spinal cord.

As you will see later, the processes of afferent and efferent neurons are arranged into bundles called nerves. Since nerves lie outside the central nervous system, they belong to the peripheral nervous system. The functional components of nerves are the nerve fibers, and the fibers may be grouped according to the following scheme:

1. *General somatic afferent.* These fibers belong to the somatic portion of the peripheral nervous system. Their function is to conduct impulses from the skin, skeletal muscles, and joints to the central nervous system.
2. *General somatic efferent.* Again, these fibers belong to the somatic portion of the peripheral nervous system. They function by conducting impulses from the central nervous system to skeletal muscles. Impulses over these fibers cause the contraction of skeletal muscles.
3. *General visceral afferent.* These fibers also belong to the somatic portion of the peripheral nervous system. They convey impulses from the viscera and blood vessels to the central nervous system.
4. *General visceral efferent.* Unlike the previously mentioned fibers, general visceral efferent fibers belong to the autonomic nervous system. Appropriately, they are also called *autonomic fibers.* Their function is to convey impulses from the central nervous system to smooth muscle, cardiac muscle, and glands. Impulses passing over these fibers cause contractions of smooth and cardiac muscle and secretion by glands.

Before you move on to the next section, examine Figure 12–1 again. Using your knowledge about the classification of nerve fibers just discussed, see if you can determine where each type of fiber fits into the scheme outlined for the organization of the nervous system.

Varieties of neurons

Although all neurons conform to the general plan previously described, there are considerable differences in structure. For example, cell bodies range in size from 5 μm for the smallest cells to 135 μm for large motor neurons. Also, the pattern of dendritic branching is varied and distinctive for neurons in different parts of the body. Moreover, the axons of very small neurons are only a fraction of a millimeter in length and lack a myelin sheath. Axons of large neurons are up to a meter in length in some cases and are usually enclosed in a myelin sheath. A few patterns of diversity are shown in Figure 12–4.

In Figure 12–4a, note the structure of a typical afferent (sensory) neuron. Compare it to the typical efferent (motor) neuron shown in Figure 12–2a. What structural differences do you observe? A few examples of association neurons are also shown in Figure 12–4. These include a *stellate cell* (Figure 12–4b), *cell of Martinotti* (Figure 12–4c), and a *horizontal cell of Cajal* (Figure 12–4d), all of which are found in the cerebral cortex, the outer layer of the cerebrum of the brain. Also shown in Figure 12–4e is a *granule cell,* an association neuron in the cortex of the cerebellum of the brain.

PHYSIOLOGY OF NERVOUS TISSUE

One of the most striking features of nervous tissue is its ability to send electrical messages called nerve impulses. Although it is beyond the scope of this text to describe the details of a nerve impulse, certain concepts must be understood in order to know how the nervous system works. Very simply, a **nerve impulse** is a wave of negativity that travels along the surface of the membrane of a neuron. Among other things, a nerve impulse depends upon the movement of ions between interstitial fluid and the inside of a neuron. For a nerve impulse to begin, a stimulus of adequate strength must be applied to

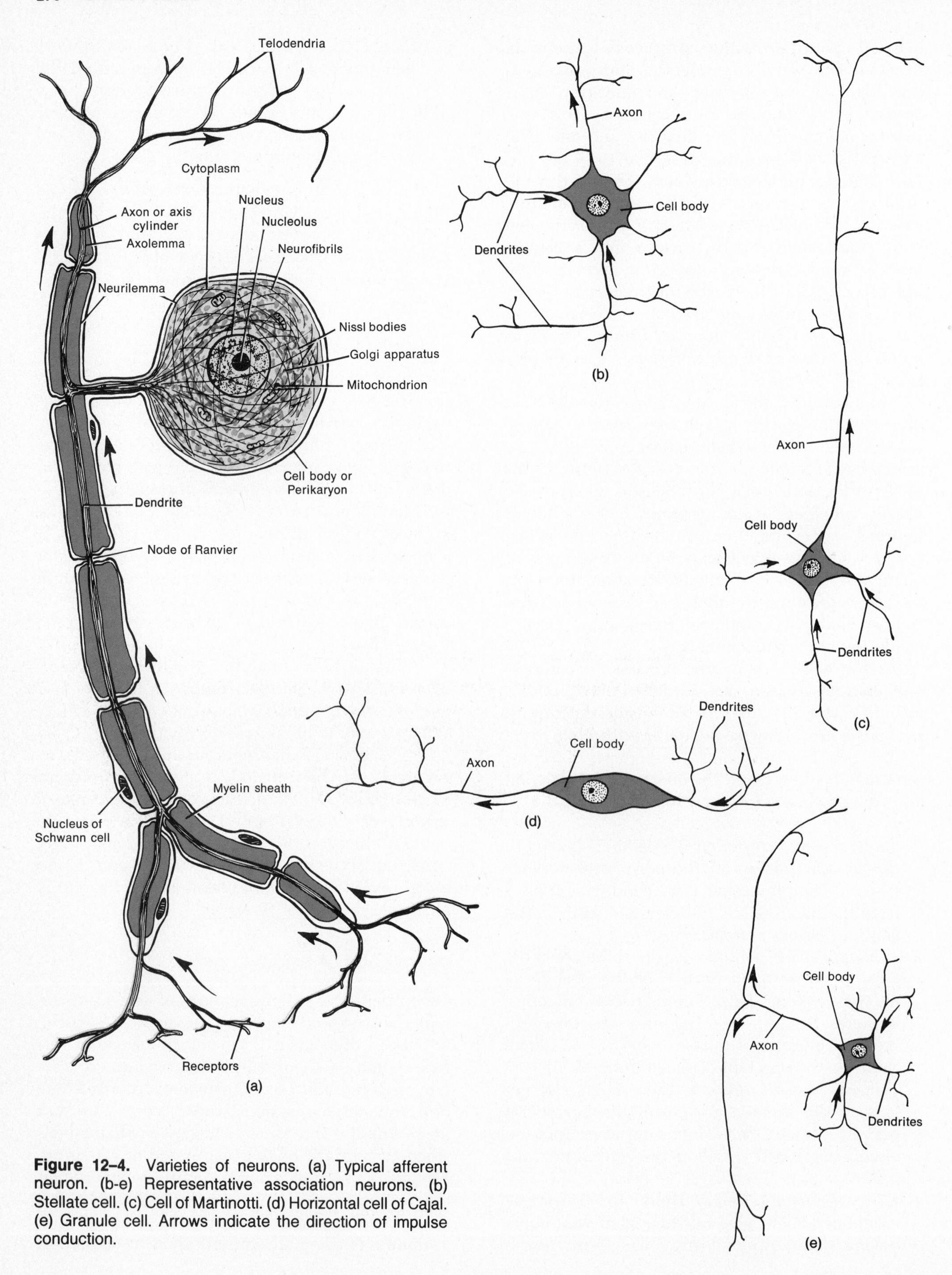

Figure 12–4. Varieties of neurons. (a) Typical afferent neuron. (b-e) Representative association neurons. (b) Stellate cell. (c) Cell of Martinotti. (d) Horizontal cell of Cajal. (e) Granule cell. Arrows indicate the direction of impulse conduction.

the neuron. A stimulus is a change in the environment of sufficient strength to initiate an impulse. The ability of a neuron to respond to a stimulus and convert it into an impulse is known as **irritability.**

The nerve impulse is the most rapid way that the body can respond to environmental changes. It provides the quickest means for achieving homeostasis, the ability of the body to maintain normalcy. The speed of a nerve impulse is determined by the size, type, and physiological condition of the nerve fiber. For example, myelinated fibers with the largest diameters can transmit impulses at speeds up to about 100 m (328 ft)/second. Unmyelinated fibers with the smallest diameters can transmit impulses at the rate of about 0.5 m (1.5 ft)/second.

In addition to irritability, neurons are also capable of **conductivity,** the ability to transmit an impulse to another neuron or another tissue, such as a muscle or a gland. The junction between two neurons is called a **synapse,** and it is across the synapse that the impulses are conducted from one neuron to another. A neuron located before the synapse is known as a *presynaptic neuron,* and one located after the synapse is referred to as a *postsynaptic neuron.* Axons of presynaptic neurons release chemical transmitters into the synapse. Some transmitters permit the impulse to pass to the postsynaptic neuron or other tissue of the body. Other transmitters inhibit the impulse.

Impulse conduction at a synapse is one-way. The direction is from the presynaptic neuron to the postsynaptic neuron. In other words, impulses must move forward over their pathways. They cannot back up into another presynaptic neuron. Such a mechanism is very important in preventing impulse conduction along wrong pathways. Imagine the result if impulses transmitted along the motor neurons that move your hand could move back and stimulate the sensory neuron which relays information about heat. You would feel heat, cry in pain, and go through all the emotions of being burned every time you simply wanted to move your hand.

We will now examine how nervous tissue is organized into structural and functional units of the nervous system. We will begin this discussion with a study of the spinal cord and spinal nerves.

Chapter summary in outline

ORGANIZATION

1. The nervous system controls and integrates all body activities by sensing changes, interpreting them, and reacting to them.
2. The entire nervous system is divided into a central nervous system (CNS) and a peripheral nervous system (PNS).
3. The central nervous system consists of the brain and spinal cord.
4. The peripheral nervous system consists of the somatic nervous system (SNS) and the autonomic nervous system (ANS).

HISTOLOGY

1. Neuroglia are connective tissue cells that support, attach neurons to blood vessels, produce myelin, and carry on phagocytosis.
2. Neurons, or nerve cells, consist of a perikaryon (cell body), dendrites, and axons. The cell body contains specialized Nissl bodies and neurofibrils.
3. Dendrites are cytoplasmic processes that conduct impulses toward the cell body.
4. Axons are single, relatively long processes that conduct impulses away from the cell body or to another neuron or tissue.
5. An axon together with certain coverings or sheaths is termed a nerve fiber.
6. Many axons contain a myelin sheath, produced by Schwann cells. The myelin sheath insulates and maintains the axon.
7. The neurilemma is the peripheral nucleated cytoplasmic layer of the Schwann cell that encloses the myelin sheath. It is found only around fibers of the peripheral nervous system. Its function is to assist in regeneration.
8. On the basis of structure, neurons are classified as multipolar, bipolar, and unipolar.
9. On the basis of function, neurons are classified as sensory (afferent), which transmit impulses toward the central nervous system; association (internuncial or connecting), which transmit impulses from sensory to motor neurons; and motor (efferent), which transmit impulses from the brain and spinal cord to effectors.
10. Nerve fibers are grouped as general somatic afferent, general somatic efferent, general visceral afferent, and general visceral efferent.

PHYSIOLOGY

1. The ability of a neuron to respond to a stimulus and convert it into a nerve impulse is called irritability.
2. A nerve impulse is a wave of negativity that travels along the surface of the membrane of a neuron.
3. The speed of a nerve impulse is determined by the size, type, and physiological condition of the nerve fiber.
4. Conductivity is the ability of a neuron to transmit a nerve impulse to another neuron or another tissue.
5. The junction between neurons is called a synapse. Impulse conduction across a synapse requires the release of chemical transmitters by presynaptic neurons.
6. Impulse conduction at a synapse is one-way.

Review questions and problems

1. What is the overall function of the nervous system?
2. Distinguish between the central and peripheral nervous system.
3. What are neuroglia? List their principal functions.
4. Define a neuron. Diagram a neuron, label it, and list the function next to each labeled part.
5. What is the myelin sheath? How is it formed?
6. Define the neurilemma. Why is it important?
7. Discuss the structural classification of neurons. Give an example of each.
8. Describe the functional classification of neurons.
9. Distinguish among the following kinds of fibers: general somatic afferent, general somatic efferent, general visceral afferent, and general visceral efferent.
10. What structural differences exist between a typical afferent and a typical efferent neuron?
11. Define a nerve impulse.
12. Compare irritability and conductivity.
13. What factors determine the rate of impulse conduction?
14. What is a synapse? How does an impulse cross a synapse?

Selected readings

Dawson, Helen L. *Basic Human Anatomy.* 2d ed. New York: Appleton-Century-Crofts, 1974. Pp. 175–181.

Ham, Arthur W. *Histology.* 7th ed. Philadelphia: J. B. Lippincott, 1974. Chap. 21.

Leeson, Thomas S., and C. Roland Leeson. *Histology.* 2d ed. Philadelphia: W. B. Saunders, 1970. Chap. 10.

CHAPTER 13

THE SPINAL CORD AND THE SPINAL NERVES

STUDENT OBJECTIVES

After you have read this chapter, you should be able to:

1. Define white matter, gray matter, nerve, ganglion, tract, nucleus, and horn
2. Describe the gross anatomical features of the spinal cord
3. Explain how the spinal cord is protected by the meninges
4. Define a spinal puncture
5. Describe the structure of the spinal cord in cross section
6. Explain the functions of the spinal cord as a conduction pathway and a reflex center
7. List the location, origin, termination, and function of the principal ascending and descending tracts of the spinal cord
8. Describe the components of a reflex arc
9. Compare the mechanism of a stretch reflex, flexor reflex, and crossed extensor reflex
10. Define a spinal nerve
11. Describe the composition and coverings of a spinal nerve
12. Identify the names of the 31 pairs of spinal nerves
13. Explain how a spinal nerve branches upon leaving the intervertebral foramen
14. Define a plexus
15. Explain the composition and distribution of the cervical, brachial, lumbar, sacral, and coccygeal plexuses
16. Define an intercostal nerve
17. Define a dermatome and its clinical importance
18. Describe spinal cord injury and list the immediate and long-range effects
19. Describe the conditions necessary for peripheral nerve regeneration

We will now examine how nervous tissue is put together to form structural and functional units of the nervous system. In this chapter, our main concern will be a study of the structure and function of the spinal cord and the nerves that originate from it called spinal nerves.

A study of the spinal cord, spinal nerves, or any other part of the nervous system must first consider how neurons are grouped.

THE GROUPING OF NEURAL TISSUE

Neurons are not strung out in a maze of separate, crisscrossing, tangled fibers. Rather, the axons of a group of neurons are usually found neatly bundled together and headed in the same direction. The cell bodies and dendrites of such a bundle of neurons are usually collected together in specific areas. The term **white matter** refers to aggregations of myelinated axons from many neurons. The fatty substance, myelin, has a whitish color that gives white matter its name. The gray-colored areas of the nervous system are called, obviously enough, **gray matter.** They contain either nerve cell bodies and dendrites or bundles of unmyelinated axons.

A **nerve** is a bundle of fibers located outside the central nervous system. Since the dendrites of somatic afferent neurons and axons of somatic efferent neurons of the peripheral nervous system are myelinated, most nerves are white matter. Nerve cell bodies that lie outside the central nervous system are generally grouped together with other nerve cell bodies to form **ganglia.** The term *ganglion* means knot. Ganglia, since they are made up prinicipally of unmyelinated nerve cell bodies, are masses of gray matter.

A **tract** is a bundle of fibers located inside the central nervous system. Tracts may run long distances up and down the spinal cord. Short tracts exist inside the brain and connect parts of the brain with each other and with the spinal cord. The chief tracts that conduct impulses up the cord are concerned with sensory impulses and are called *ascending tracts.* By contrast, bundles of fibers that carry impulses down the cord are motor tracts and are called *descending tracts.* The major tracts consist of myelinated fibers, and they are therefore white matter. A **nucleus** is a mass of nerve cell bodies and dendrites inside the central nervous system. It consists of gray matter. **Horns** are the chief areas of gray matter in the spinal cord. For purposes of study and discussion, the white matter, gray matter, ganglia, tracts, and nuclei of the body's nervous system are categorized into two major divisions: the central nervous system (CNS) and the peripheral nervous system (PNS). The central nervous system consists of the brain and spinal cord. It is the control center for the entire nervous system. The peripheral nervous system consists of all nervous tissue that connects the CNS to all other parts of the body, that is, the periphery. You might wish to review the overall organization of the nervous system in Figure 12–1 before moving on.

THE SPINAL CORD

General features

The **spinal cord** is a cylindrical structure that is slightly flattened anteriorly and posteriorly. It begins as a continuation of the medulla oblongata, the inferior part of the brain stem, and extends from the foramen magnum of the occipital bone to about the level of the second lumbar vertebra (Figure 13–1). The length of an adult spinal cord ranges from 42 to 45 cm (16 to 18 in.). The diameter of the cord varies at different levels.

When the cord is viewed externally, two conspicuous enlargements can be seen. The superior enlargement, the *cervical enlargement,* extends from the fourth cervical to the first thoracic vertebra (Figure 13–1). Nerves that supply the upper extremities arise from the cervical enlargement. The inferior enlargement, called the *lumbar enlargement,* extends from the ninth to twelfth thoracic vertebra. Nerves that supply the lower extremities arise from the lumbar enlargement.

Below the lumbar enlargement, the spinal cord tapers to a conical portion known as the *conus medullaris.* The conus medullaris lies at about the level of the first or second lumbar vertebra. Arising from the conus medullaris is the *filum terminale,* a nonnervous fibrous tissue of the spinal cord that extends down to the coccyx (Figure 13–1). The filum terminale consists mostly of pia mater, the innermost of three membranes that cover and protect the spinal cord and brain. Some of the nerves that arise from the lower portion of the cord do not leave the vertebral column immediately. Instead, they run farther down in the vertebral canal and look like wisps of coarse hair flowing from the end of the cord. They are appropriately named the *cauda equina,* which means horse's tail.

The spinal cord is viewed as a series of 31 segments, each giving rise to a pair of spinal nerves. In the future, when we refer to a *spinal segment,* we are speaking of a region of the spinal cord from which a given pair of spinal nerves arises. If you

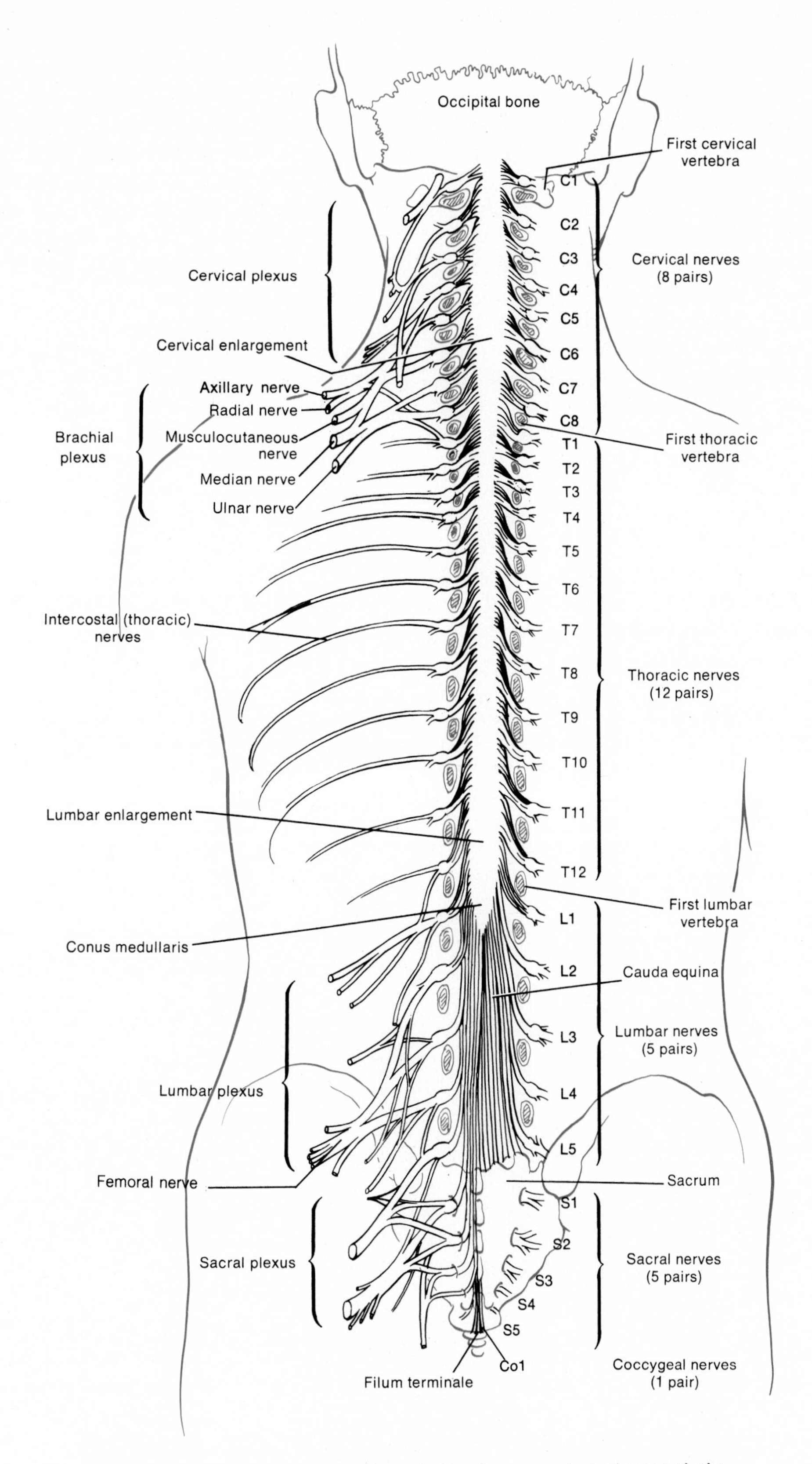

Figure 13-1. The spinal cord and spinal nerves viewed posteriorly.

examine the cross section of the spinal cord in Figure 13–3, you will see that the cord is partially divided into right and left sides by two grooves. One of these, the *anterior median fissure,* is a deep, wide groove on the anterior (ventral) surface. The other groove is the *posterior median sulcus,* a shallower, narrow groove on the posterior (dorsal) surface.

Protection and coverings

The spinal cord is located within the vertebral canal of the vertebral column. The vertebral canal, as you may recall, is formed by the vertebral foramina arranged one on top of the other. Since the wall of the vertebral canal is essentially a ring of bone surrounding the spinal cord, the cord is well protected within the canal. A certain degree of protection is also provided by the vertebral ligaments.

The cord receives further protection from several coverings around it plus a cushion of cerebrospinal fluid. At this point, we will consider the coverings only. The coverings are collectively called *meninges,* and they run continuously around the spinal cord and brain. Those associated specifically with the cord are frequently referred to as spinal meninges (Figure 13–2). The outer spinal meninx is called the *dura mater* (or tough mother) and forms a tube from the level of the second sacral vertebra, where it is fused with the filum terminale, to the foramen magnum, where it is continuous with the dura mater of the brain. The dura mater is composed of tough, fibrous connective tissue. Between the dura mater and the wall of the vertebral canal is the *epidural space* which is filled with fat, connective tissue, and blood vessels. It serves as padding around the cord. The middle spinal meninx is called the *arachnoid* (or spider layer). It is a very delicate fibrous membrane that forms a tube inside the dura mater. It is also continuous with the arachnoid of the brain. Between the dura mater and the arachnoid is a space called the *subdural space,* which contains some serous fluid. The inner meninx is known as the *pia mater* (or delicate mother). It is a transparent fibrous membrane that forms a tube around and adheres to the surface of the cord and the brain. It contains blood vessels. The space between the arachnoid and the pia mater is the *subarachnoid space,* which contains cere-

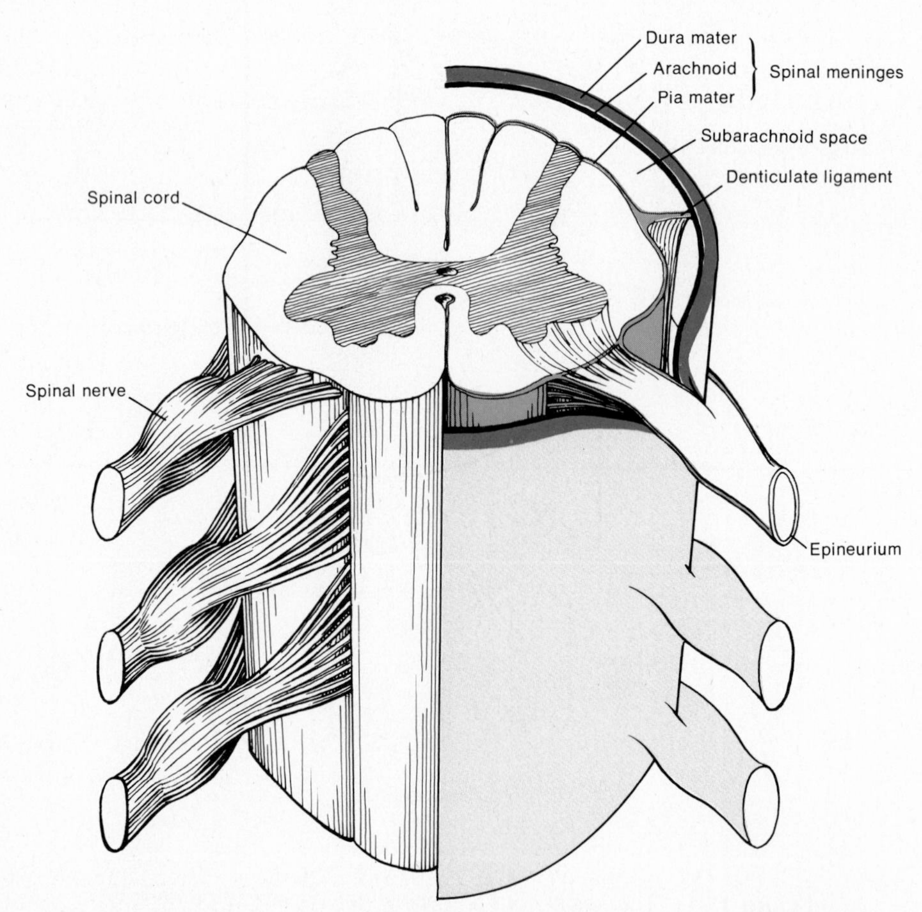

Figure 13-2. The location of the spinal meninges as seen in a cross section of the spinal cord.

brospinal fluid. All three spinal meninges cover the spinal nerves as they exit the spinal column through the intervertebral foramina. The spinal cord is suspended in the middle of its dural sheath by membranous extensions of the pia mater. These extensions, called the *denticulate ligaments,* are attached on either side of the length of the spinal cord between the ventral and dorsal nerve roots. The ligaments afford protection for the spinal cord against shock and sudden displacement.

Inflammation of the meninges is known as *meningitis.* If only the dura mater becomes inflamed, the condition is specifically called *pachymeningitis.* Inflammation of the arachnoid and/or pia mater is referred to as *leptomeningitis.* When a person has meningitis, it is usually leptomeningitis.

Spinal puncture

The removal of cerebrospinal fluid from the subarachnoid space in the lumbar region of the spinal cord is a *spinal (lumbar) puncture* or *tap.* The procedure is normally done between the third and fourth lumbar vertebrae. The spinous process of the fourth lumbar vertebra is easily located by drawing a line across the highest points of the iliac crests. This line will pass right through the spinous process of the vertebra. If you stop to think about it, the site of a lumbar puncture poses no danger to the spinal cord when the needle is inserted to withdraw fluid. The spinal cord ends at or above the second lumbar vertebra. If the patient lies on his or her side and brings the knees and chest together, the vertebrae separate enough so that the needle can be conveniently inserted. Lumbar punctures are not only used to withdraw fluid for diagnostic purposes, but they are also used to introduce antibiotics (as in the case of meningitis) and to administer anesthesia as a spinal block for surgery below the thorax or for childbirth.

Structure in cross section

The spinal cord consists of both gray and white matter. Looking at the cross section of the cord in Figure 13–3, you can see that the gray matter lies in an area that is shaped like a letter H or a butterfly. The gray matter consists primarily of nerve cell bodies and unmyelinated axons and dendrites of association and motor neurons. The white matter surrounds the gray matter and consists of bundles of myelinated axons of motor and sensory neurons.

If you examine the gray matter carefully, you will notice a cross bar of the H called the *gray commissure.* It connects the upright portions of the H. In the center of the gray commissure is a small space

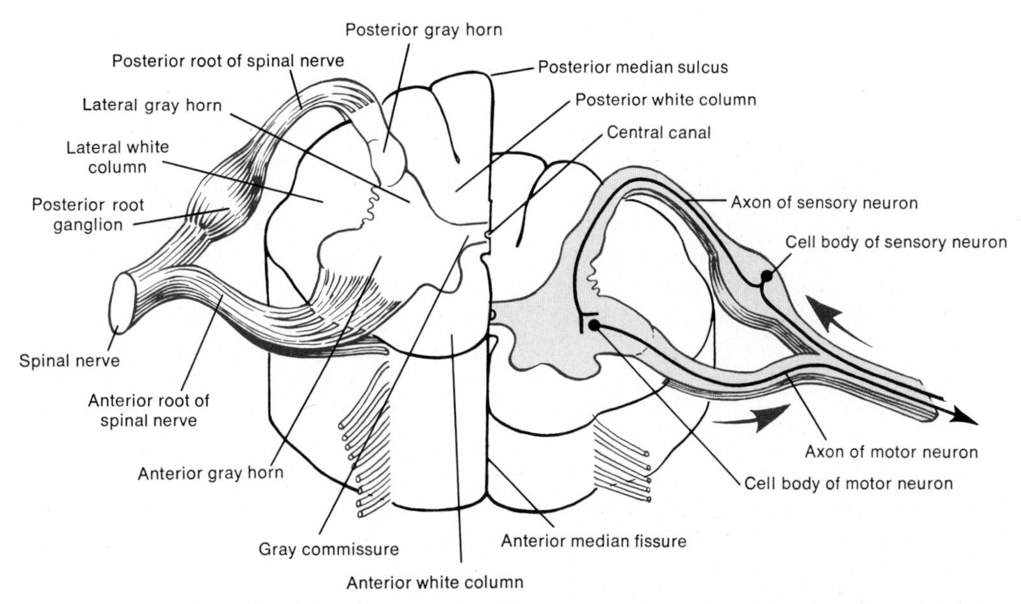

Figure 13-3. The organization of gray and white matter in the spinal cord as seen in a cross section of the spinal cord. The right side of the figure has been sectioned at a lower level than the left side so that you can see what is inside the posterior root ganglion, posterior root of the spinal nerve, anterior root of the spinal nerve, and the spinal nerve.

called the *central canal,* which runs the length of the cord and is continuous with the fourth ventricle of the medulla. It contains cerebrospinal fluid. On the basis of location, the upright portions of the H are further subdivided into regions. Those closer to the front of the cord are called *anterior (ventral) gray horns.* They represent the motor part of the gray matter. The regions closer to the back of the cord are referred to as *posterior (dorsal) gray horns.* They represent the sensory part of the gray matter. The regions between the anterior and posterior gray horns are designated as *lateral gray horns.* They represent the origin of preganglionic fibers of the autonomic nervous system which are discussed in Chapter 15. The lateral gray horns are most prominent in the thoracic and upper lumbar segments of the cord.

The gray matter of the cord also contains several nuclei which serve as relay stations for impulses and origins for certain nerves. Nuclei, as you may recall, are clusters of nerve cell bodies and dendrites inside the spinal cord and brain. The names, locations, and functions of the nuclei in the gray matter of the spinal cord will be considered later as they apply to a specific activity of the nervous system.

The white matter on each side of the cord, like the gray, is also organized into regions on the basis of location. The anterior and posterior gray horns divide the white matter into three broad areas. These are an *anterior (ventral) white column (funiculus),* a *posterior (dorsal) white column (funiculus),* and a *lateral white column (funiculus).* Each column or funiculus, in turn, consists of several distinct bundles of myelinated fibers that run the length of the cord. These bundles are called *fasciculi* or *tracts.* The long *ascending tracts* consist of sensory axons that conduct impulses entering the cord upward to the brain. The long *descending tracts* consist of motor axons that conduct impulses downward from the brain to the spinal cord. Thus, the ascending tracts are sensory tracts and the descending tracts are motor tracts. Still other shorter tracts contain ascending and descending axons that convey impulses from one level of the cord to another.

Functions

One of the principal functions of the spinal cord is to convey sensory impulses from the periphery to the brain and to conduct motor impulses from the brain to the periphery. A second principal function is related to reflexes, a topic we will consider shortly.

Conduction pathway

The vital function of conveying sensory and motor information is carried out by the ascending and descending tracts of the cord. The names of the tracts are generally descriptive enough to indicate the column (funiculus) in which the tract travels, where the cell bodies of the tract originate, where the axons of the tract terminate, and the direction of impulse conduction within the tract. For example, the anterior spinothalamic tract is located in the anterior column, it originates in the spinal cord, it terminates in the thalamus (a part of the brain), and it is an ascending (sensory) tract because it conveys impulses from the cord upward to the brain. As another example, consider the lateral corticospinal tract. It is located in the lateral column, it originates in the cerebral cortex (outer surface of cerebrum), it terminates in the spinal cord, and it is a descending (motor) tract because it conveys impulses from the brain downward to the cord.

The locations of some of the more important ascending and descending tracts in their respective columns are diagramed in Figure 13–4. You might also want to refer to Exhibit 13–1, which is a summary of some of the more important tracts. The tracts are listed by name, column through which they pass, origin, termination, and function. At this time you should familiarize yourself with these tracts.

Reflex center

The second principal function of the spinal cord deals with reflexes. To understand what reflexes mean, we will first have to examine the general structure of spinal nerves. Spinal nerves are the lines of communication between the spinal cord tracts and the periphery. Examination of the cross section of the spinal cord in Figure 13–3 reveals that each pair of spinal nerves is connected to a segment of the cord by two points of attachment called roots. The **posterior** or **dorsal (sensory) root** contains sensory nerve fibers only and conducts impulses from the periphery to the spinal cord. These fibers extend into the posterior (dorsal) gray horn. Each dorsal root also has a swelling, the **posterior** or **dorsal (sensory) root ganglion,** which contains the cell bodies of the sensory neurons from the periphery. The other point of attachment of a spinal nerve to the cord is called the **anterior** or **ventral (motor) root.** It contains motor nerve fibers only and conducts impulses from the spinal cord to the periphery. The cell bodies of the motor neurons are

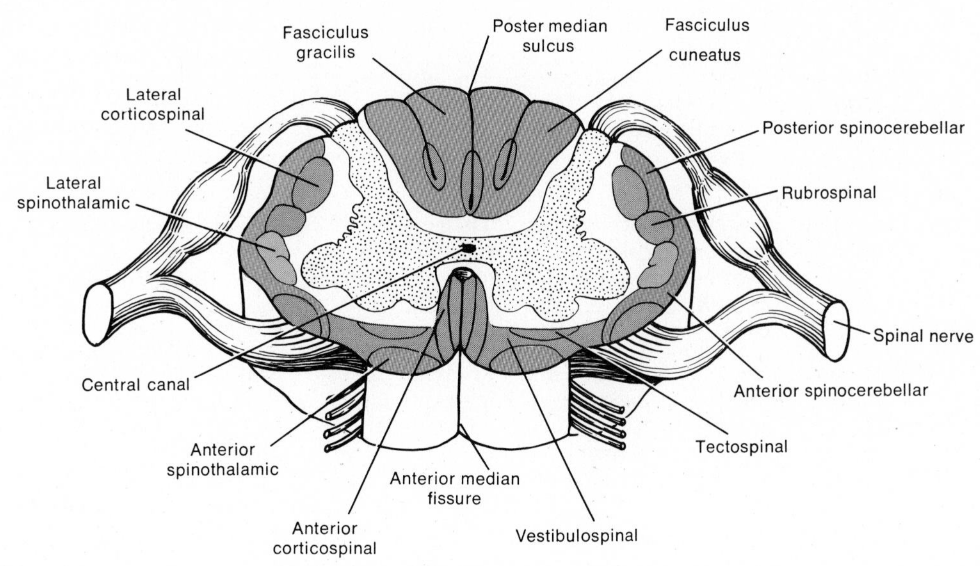

Figure 13-4. Selected tracts of the spinal cord. Ascending (sensory) tracts are indicated in color; descending (motor) tracts are shown in gray.

located in the gray matter of the cord. If the motor impulse goes to a skeletal muscle, the cell bodies are in the anterior (ventral) gray horn. If, instead, the impulse goes to smooth muscle, cardiac muscle, or a gland through the autonomic nervous system, the cell bodies are found in the lateral gray horn.

The reflex arc

The path that an impulse takes from its origin in the dendrites or cell body of a neuron in one part of the body to its termination somewhere else in the body is called a *conduction pathway.* All conduction pathways consist of circuits of neurons arranged in various patterns. One type of pathway is known as a **reflex arc,** the functional unit of the nervous system. A reflex arc contains two or more neurons over which impulses are conducted from a receptor to the brain or spinal cord and back to an effector. The basic components of a reflex arc are as follows:

1. The **receptor** is the distal end of a dendrite of a sensory neuron. Its role in the reflex arc is to respond to a change (stimulus) in the internal or external environment by initiating a nerve impulse.
2. The **sensory neuron** passes the impulse generated by the receptor to the central nervous system.
3. The **center** is a region, usually within the central nervous system, where an incoming sensory impulse turns into an outgoing motor impulse. In the center, the impulse may be inhibited, transmitted, or rerouted. The center of the majority of reflex arcs also contains an association neuron between the sensory neuron and the motor neuron leading to a muscle or a gland. However, in the center of some reflex arcs, the sensory neuron delivers the impulse directly to the motor neuron.
4. The **motor neuron** transmits the impulse from the sensory or association neuron in the center to the organ of the body that will respond.
5. The **effector** is the organ of the body, either a muscle or a gland, that responds to the motor impulse. This response is called a *reflex action* or, just simply, a *reflex.*

Reflexes that are mediated (carried out) by the spinal cord alone are called *spinal reflexes.* In addition, reflexes that result in the contraction of skeletal muscles are known as *somatic reflexes.* Those that bring about the contraction of smooth or cardiac muscle or secretion by glands are referred to as *visceral (autonomic) reflexes.* Our concern at this point is to examine a few somatic spinal reflexes. Among these are the stretch reflex, the flexor reflex, and the crossed extensor reflex.

Exhibit 13–1. SELECTED ASCENDING AND DESCENDING TRACTS OF THE SPINAL CORD

TRACT	LOCATION (COLUMN)	ORIGIN	TERMINATION	FUNCTION
A. Ascending tracts				
1. Anterior (ventral) spinothalamic	Anterior (ventral) column	Posterior (dorsal) gray horn on one side of the cord but crosses to the opposite side of the cord	Thalamus; impulse is eventually conveyed to cerebral cortex	Conveys sensations for crude touch and pressure from one side of the body to the opposite side of the thalamus. Eventually sensations reach cerebral cortex.
2. Lateral spinothalamic	Lateral column	Posterior (dorsal) gray horn on one side of the cord but crosses to the opposite side of the cord	Thalamus; impulse is eventually conveyed to cerebral cortex	Conveys sensations for pain and temperature from one side of the body to the opposite side of the thalamus. Eventually sensations reach cerebral cortex.
3. Fasciculus gracilis and fasciculus cuneatus	Posterior (dorsal) column	Nucleus gracilis and nucleus cuneatus on one side of the medulla but crosses to the opposite side of the medulla	Medulla; impulse is eventually conveyed to cerebral cortex	Convey sensations from one side of the body to the opposite side of the medulla for fine touch; two-point discrimination (ability to distinguish that two points on the skin are touched even though they are close together); proprioception (awareness of the precise position of body parts and their direction of movement); sterognosis (ability to recognize the size, shape, and texture of an object); weight discrimination (ability to assess the weight of an object); and vibrations. Eventually sensations reach cerebral cortex.
4. Posterior (dorsal) spinocerebellar	Posterior (dorsal) column	Posterior (dorsal) gray horn on the same side of the cord	Cerebellum	Conveys sensations from one side of the body to the same side of the cerebellum for subconscious proprioception
5. Anterior (ventral) spinocerebellar	Anterior (ventral) column	Anterior (ventral) gray horn on one side of the cord but crosses to the opposite side of the cord	Cerebellum	Conveys sensations from one side of the body to the opposite side of the cerebellum for subconscious proprioception
B. Descending tracts				
1. Lateral corticospinal	Lateral column	Cerebral cortex on one side of the brain but crosses in the base of the medulla to the opposite side of the cord	Anterior (ventral) gray horn	Conveys motor impulses from one side of the cortex to skeletal muscles on the opposite side of the body that coordinate very precise, discrete movements
2. Anterior (ventral) corticospinal	Anterior (ventral) column	Cerebral cortex on one side of the brain to the same side of the cord	Anterior (ventral) gray horn	Conveys motor impulses from one side of the cortex to skeletal muscles on the same side of the body that coordinate very precise, discrete movements
3. Rubrospinal	Lateral column	Midbrain (red nucleus) but crosses to the opposite side of the cord	Anterior (ventral) gray horn	Conveys motor impulses from one side of the medulla to skeletal muscles on the opposite side of the body that are concerned with muscle tone and posture
4. Tectospinal	Anterior (ventral) column	Midbrain but crosses to opposite side of the cord	Anterior (ventral) gray horn	Conveys motor impulses from one side of the midbrain to skeletal muscles on the opposite side of the body that control movements of the head in response to auditory, visual, and cutaneous stimuli
5. Vestibulospinal	Anterior (ventral) column	Medulla on one side of the brain to the same side of the cord	Anterior (ventral) gray horn	Conveys motor impulses from one side of the medulla to skeletal muscles on the same side of the body that regulate body tone in response to movements of the head (equilibrium)

Stretch reflex

The **stretch reflex** is based upon a *two-neuron* or *monosynaptic reflex arc.* This means that only two neurons are involved and there is only one synapse in the pathway (Figure 13–5). This type of reflex results in the contraction of a muscle when it is stretched. Slight stretching of a muscle stimulates receptors in the muscle called *neuromuscular spindles.* The spindles monitor any changes in the length of the muscle. Once the spindle is stimulated, an impulse is sent along a sensory neuron to the spinal cord. The sensory neuron enters the posterior root of a spinal nerve and synapses with a motor neuron in the anterior gray horn. The impulse crosses the synapse and travels along the motor neuron. The motor neuron leaves the cord through the anterior root of the spinal nerve and terminates in a skeletal muscle. Once the impulse travels along the motor neuron and reaches the muscle, the stretched muscle contracts. As a result of this reflex, the stretch is counteracted by contraction of the muscle. Since the sensory impulse enters the spinal cord on the same side that the motor impulse leaves the cord, we say that the reflex arc is an *ipsilateral reflex arc.* All monosynaptic reflex arcs are ipsilateral. If you stop to think about it, the stretch reflex is very important in helping you to maintain muscle tone.

The stretch reflex forms the basis for several tests that are used in a neurological examination. One such reflex is the *knee jerk* or *patellar reflex.* This reflex is illustrated in Figure 13–5. For example, when a physician taps your patellar ligament (stimulus), neuromuscular spindles in the quadriceps femoris muscle attached to the tendon send the sensory impulse to the spinal cord. The returning motor impulse causes contraction of the muscle. The response is extension of the leg at the knee, or a knee jerk. We shall consider the operation and meaning of other clinically important reflexes later. Let us now consider the flexor reflex and crossed extensor reflex.

Flexor reflex and crossed extensor reflex

Reflexes other than stretch reflexes involve one or more association neurons in addition to the sensory and motor neuron. Thus, they are referred to as *polysynaptic reflex arcs.* One example of a reflex based on a polysynaptic reflex arc is the **flexor reflex** (Figure 13–6). Let us assume that you accidentally step on a thumbtack. As a result of the painful stimulus, you immediately withdraw your foot. What has happened? A sensory neuron transmits the impulse from the receptor to the spinal cord. The impulse is picked up by an association neuron and then conveyed to a motor neuron. The motor neuron stimulates the muscles of your foot and you withdraw your foot. As you can see, a flexor reflex is protective. It helps you to move a limb of the body away from the harmful stimulus.

Notice again, that the stretch reflex just described is also ipsilateral. The incoming and outgoing impulses are on the same side of the spinal cord.

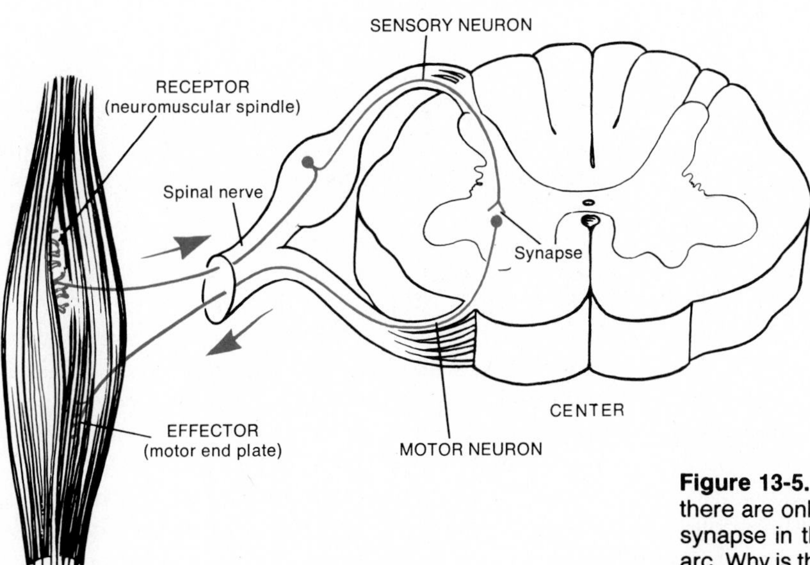

Figure 13-5. Stretch reflex. Notice that in a stretch reflex there are only two neurons involved, and there is only one synapse in the pathway. Thus it is a monosynaptic reflex arc. Why is the reflex arc shown referred to as an ipsilateral reflex arc?

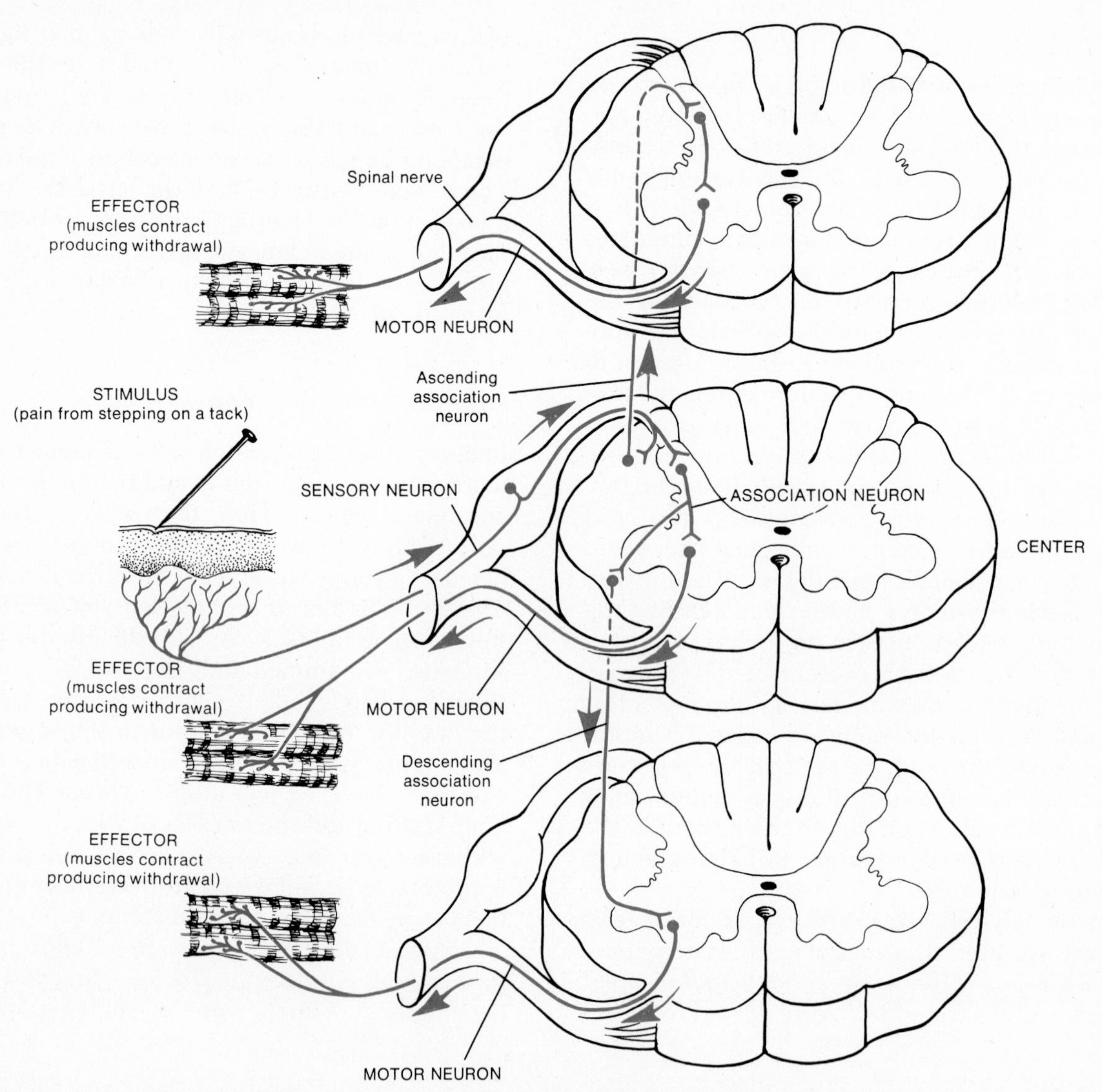

Figure 13-6. Flexor reflex. This reflex arc is a polysynaptic ipsilateral reflex arc because it involves more than one synapse. This is because it contains association neurons as well as sensory and motor neurons. Why is the reflex arc shown also an intersegmental reflex arc?

But, the stretch reflex also illustrates another feature of reflex arcs that you should know about. In the monosynaptic stretch reflex previously described, the returning motor impulse affected only the quadriceps muscle of the thigh. When you withdraw your entire lower or upper extremity from a painful or irritating stimulus, more than one muscle is involved. Therefore, several motor fibers are returning impulses to several lower or upper extremity muscles at the same time. In other words, a single sensory impulse is causing several motor responses. This kind of reflex arc, in which a single sensory neuron splits into ascending and descending branches, each forming a synapse with association neurons at different segments of the cord, is called an *intersegmental reflex arc.* Because of intersegmental reflex arcs, a single sensory neuron can activate several motor neurons and thereby bring about the stimulation of more than one effector.

The story does not end when you withdraw an extremity from a harmful stimulus. Something else may happen. When you withdraw your foot, you temporarily lose your balance as your body weight shifts to the unaffected lower extremity. What is your body response? Well, you do whatever you can to regain your balance so that you do not fall. This means motor impulses are also sent to your unaffected lower extremity and both upper extremities. The motor impulses that travel to your unaffected lower extremity cause extension at the knee so that you can place your entire body weight on the foot.

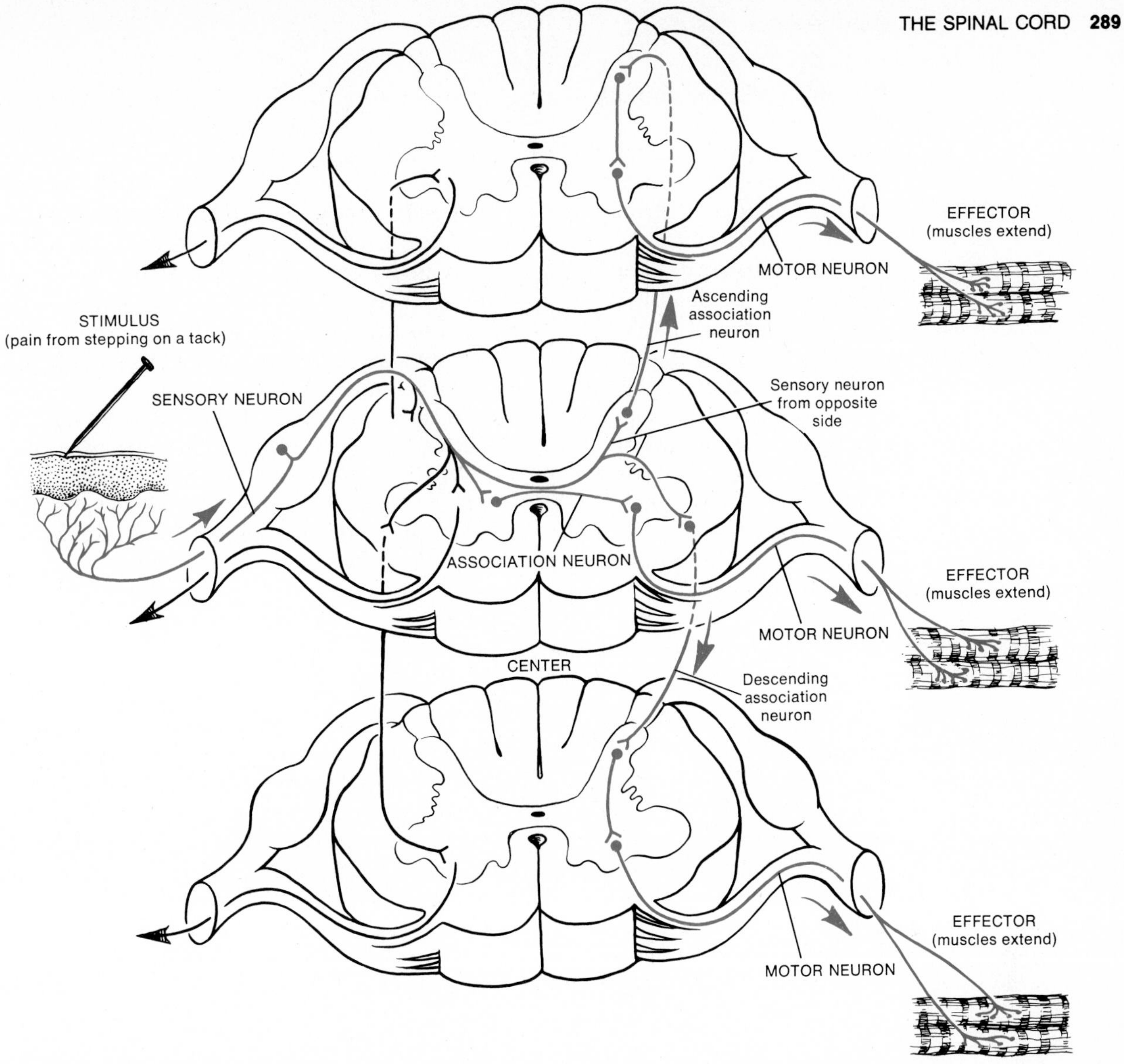

Figure 13-7. Crossed extensor reflex. Although the flexor reflex is shown on the left of the diagram so that you can correlate it with the crossed extensor reflex on the left, concentrate your attention on the crossed extensor reflex. Why is the crossed extensor reflex classified as a contralateral reflex arc?

The impulses get there by crossing in the spinal cord. Look at Figure 13–7. The incoming sensory impulse not only fires the flexor reflex that causes you to withdraw, it also fires an extensor reflex. The incoming sensory impulse crosses to the other side of the cord through another association neuron at that level and several levels above and below the point of sensory stimulation. From these levels, the motor neurons cause extension of the knee, thus permitting you to maintain balance. Unlike the flexor reflex, which passes over an ipsilateral reflex arc, the extensor reflex passes over a *contralateral reflex arc,* that is, one in which the impulse enters one side of the spinal cord and exits on the opposite side. The reflex just described in which extension of the muscles in one limb occurs as a result of flexion of the muscles of the opposite limb is simply called a **crossed extensor reflex.**

The flexor reflex and crossed extensor reflex illustrate another very important feature of many reflexes. This is referred to as *reciprocal inhibition.* Reciprocal inhibition means that when a reflex excites a muscle to cause its contraction, it also inhibits another muscle to allow its extension. Excitation and inhibition occur at the same time. For example, in the flexor reflex, when the flexor muscles of your lower extremity are contracting, the extensor muscles of the same extremity are being extended. If both sets of muscles contracted at the same time, you would not be able to flex your limb because both sets of muscles would pull on the limb bones at the same time. But, because of reciprocal inhibition, one set of muscles contracts while the other set is being extended.

In the crossed extensor reflex, reciprocal inhibition also occurs. While you are flexing the muscles of the limb that has been stimulated by the thumbtack, the muscles of your other limb are producing extension to help you keep your balance. Reciprocal inhibition is very important in coordinating body movements. You may recall from Chapter 10 that skeletal muscles act in groups rather than alone and that each muscle in the group has a specific role in bringing about the movement. In flexing the forearm at the elbow, there are a prime mover, an antagonist, and a synergist involved. The prime mover (biceps) contracts to bring about flexion, the antagonist (triceps) extends to give way to the action of the prime mover, and the synergist (deltoid) helps the prime mover to perform its role more efficiently.

We will now turn our attention to the other main theme of this chapter, the spinal nerves.

SPINAL NERVES

Composition and coverings

It has already been indicated that a **spinal nerve** has two points of attachment to the cord, a posterior root and an anterior root. A short distance from the spinal cord the roots unite to form a spinal nerve. Since the posterior root contains sensory fibers and the anterior root contains motor fibers, a spinal nerve is referred to as a *mixed nerve*. The posterior root ganglion contains cell bodies of sensory neurons. The posterior and anterior roots unite to form the spinal nerve at the intervertebral foramen.

If you examine the cross section of a spinal nerve in Figure 13–8, you can see that the nerve contains many fibers that are surrounded by several different coverings. The individual fibers, whether myelinated or unmyelinated, are wrapped in a connective tissue covering called the *endoneurium*. Groups of fibers are arranged in bundles called fascicles, and each bundle is wrapped in another connective tissue covering called the *perineurium*. Finally, the outermost covering around the entire nerve is called the *epineurium*. The spinal meninges fuse with the epineurium as the nerve exits from the vertebral canal.

Names

The 31 pairs of spinal nerves are named and numbered according to the region and level of the spinal cord from which they emerge (see Figure 13–1). The first cervical pair emerges between the atlas and the occipital bone. All other spinal nerves leave the backbone from the intervertebral foramina between adjoining vertebrae. There are 8 pairs of cervical nerves, 12 pairs of thoracic nerves, 5 pairs of lumbar nerves, 5 pairs of sacral nerves, and 1 pair of coccygeal nerves. During fetal life, the spinal cord and vertebral column grow at different rates, the cord growing more slowly. As a result of this

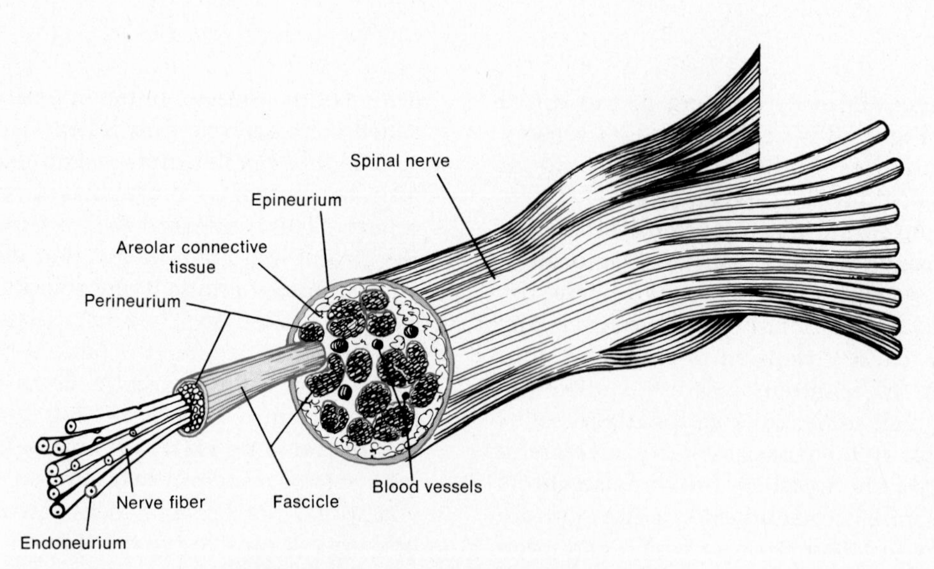

Figure 13-8. Coverings of a spinal nerve.

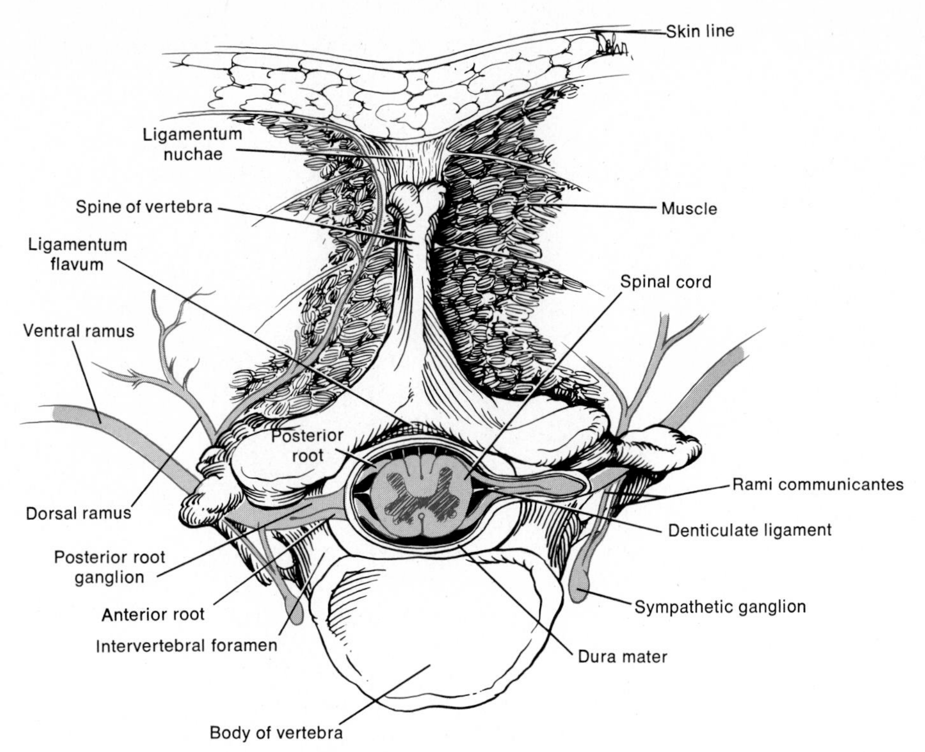

Figure 13-9. Branches of a typical spinal nerve.

uneven growth, not all the spinal cord segments are in line with their corresponding vertebrae. Remember that the spinal cord terminates near the first or second lumbar vertebra. Thus, the lower lumbar, sacral, and coccygeal nerves must descend more and more to reach their foramina before emerging from the vertebral column. This arrangement constitutes the cauda equina.

Branches

Shortly after a spinal nerve leaves its intervertebral foramen, it divides into several branches (Figure 13–9). These branches are known as rami. The *dorsal ramus* innervates (supplies) the muscles and skin of the dorsal surface of the back. The *ventral ramus* of a spinal nerve innervates all the structures of the extremities and the lateral and ventral trunk. Except for thoracic nerves T2 to T12, the ventral rami of other spinal nerves enter into the formation of plexuses before supplying a part of the body. These will be described shortly. In addition to dorsal and ventral rami, spinal nerves also give off a *meningeal branch.* This branch supplies the vertebrae, vertebral ligaments, blood vessels of the spinal cord, and the meninges. Other branches of a spinal nerve are the *rami communicantes.* Since the communicantes are part of the autonomic nervous system, we will wait until Chapter 15 before discussing them.

Plexuses

The ventral rami of spinal nerves, except for T2 to T12, do not go directly to the structures of the body they supply. Instead, they form networks with adjacent nerves on both sides of the body. Such networks are called **plexuses,** meaning braids. The principal plexuses are named the cervical plexus, the brachial plexus, the lumbar plexus, and the sacral plexus. Emerging from the plexuses are nerves bearing names that are often descriptive of the general regions they supply or the course they take. Each of these nerves, in turn, may have several branches named for the specific structures they innervate.

The cervical plexus

The **cervical plexus** is formed by the ventral rami of the first four cervical nerves (C1 to C4) with contributions from C5. It is located in both sides of the neck alongside the first four cervical vertebrae (Figure 13–10). The *roots* of the plexus indicated in

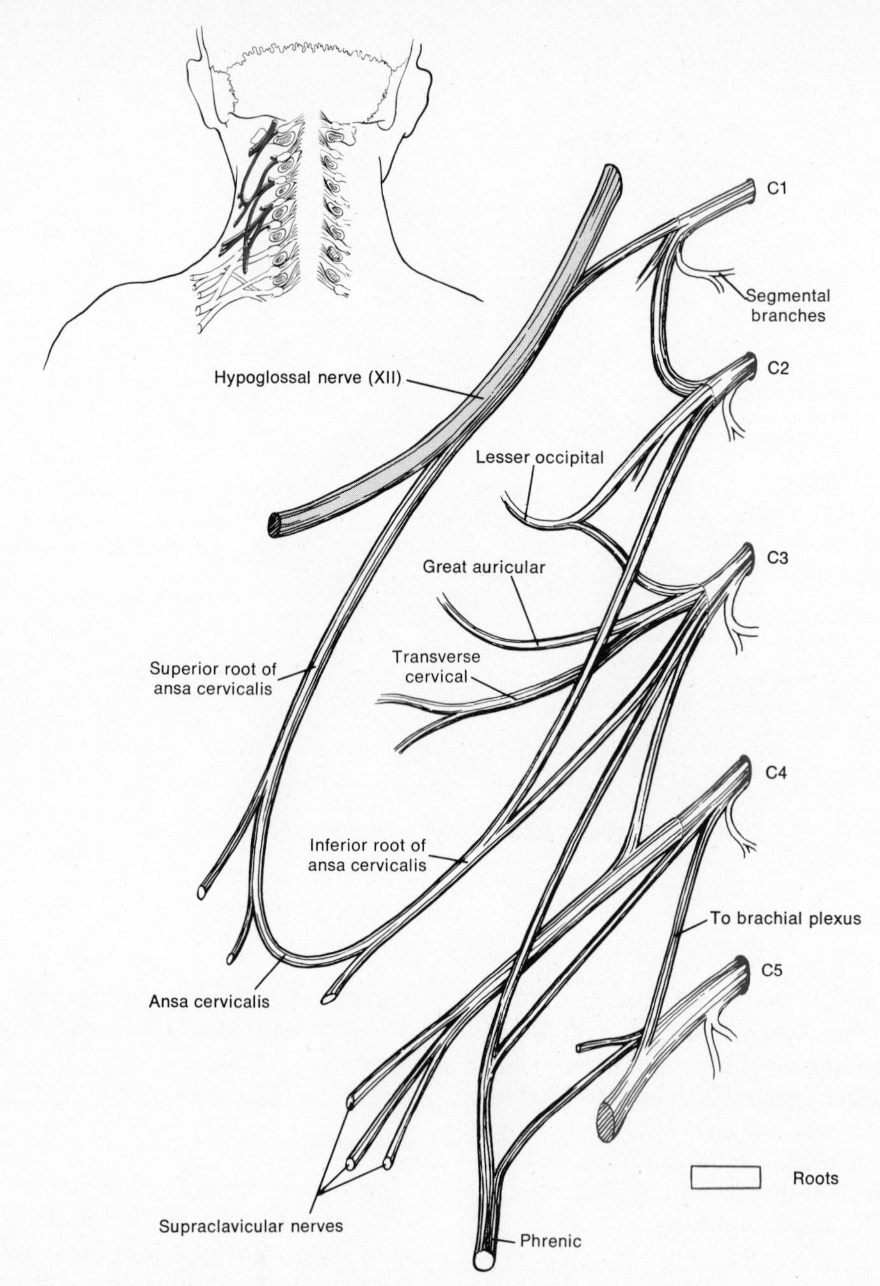

Figure 13-10. The cervical plexus. Be sure to consult Exhibit 13-2 so that you can determine the distribution of each of the nerves of the plexus.

Exhibit 13–2. THE CERVICAL PLEXUS

NERVE	ORIGIN	DISTRIBUTION
Superficial or cutaneous branches		
Lesser occipital	C2–C3	Skin of the scalp behind and above ear
Greater auricular	C2–C3	Skin in front, below, and over ear
Transverse cervical	C2–C3	Skin over anterior aspect of neck
Supraclaviculars	C3–C4	Skin over upper portion of chest and shoulder
Deep or largely motor branches		
Ansa cervicalis	This nerve is divided into a superior ramus and an inferior ramus.	
Superior root	C1–C2	Infrahyoid, thyrohyoid, and geniohyoid muscles of neck
Inferior root	C3–C4	Omohyoid, sternohyoid, and sternothyroid muscles of neck
Phrenic	C3–C5	Diaphragm between chest and abdomen
Segmental branches	C1–C5	Prevertebral (deep) muscles of neck, levator scapulae, sternocleidomastoid, and trapezius muscles

Figure 13-11. The brachial plexus. Consult Exhibit 13-3 so that you can determine the distribution of each of the nerves of the plexus.

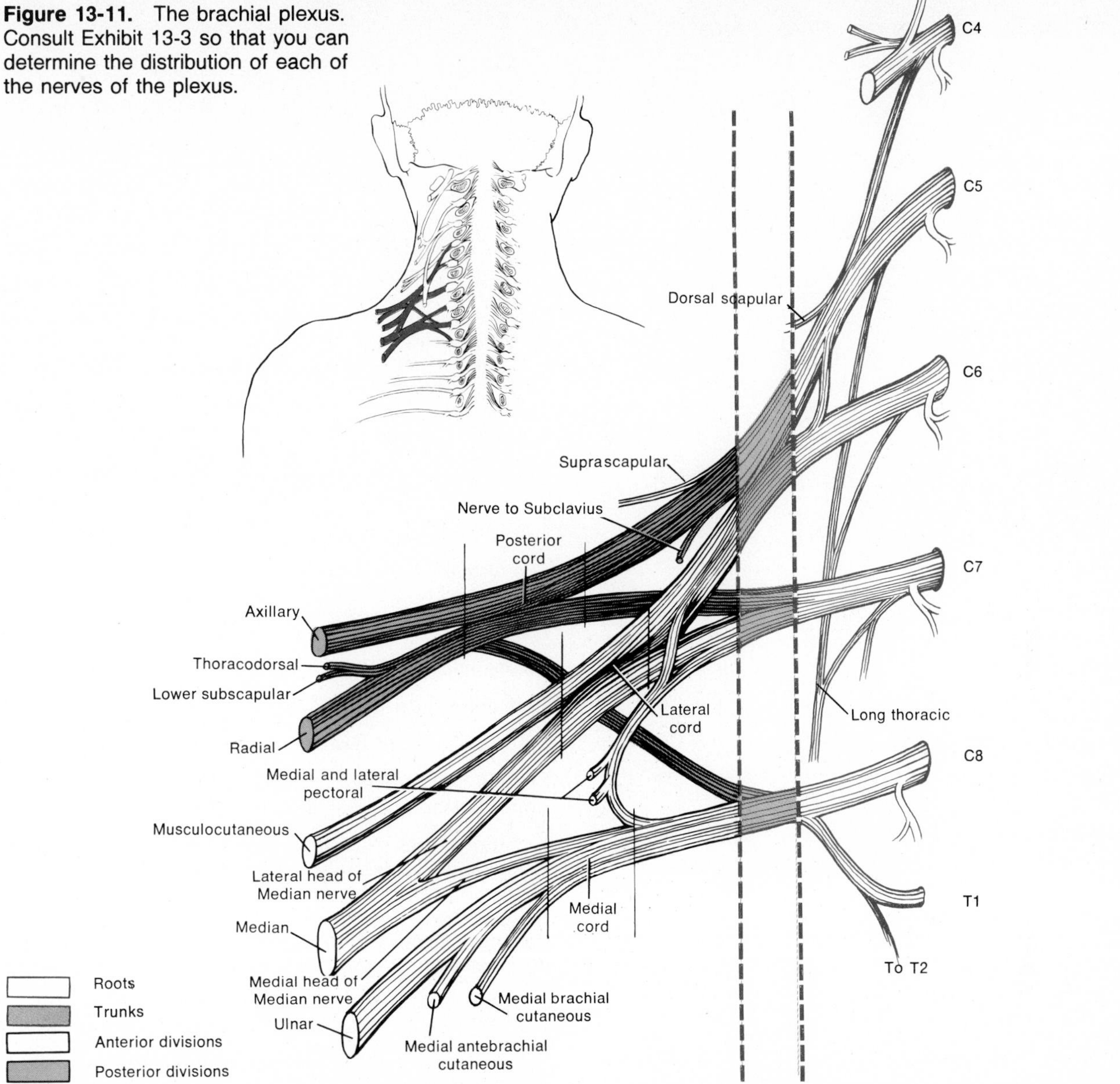

the diagram are simply continuations of the ventral rami. The cervical plexus supplies the skin and muscles of the head, neck, and upper part of the shoulders. Branches of the cervical plexus also connect with cranial nerves XI (accessory) and XII (hypoglossal). A very important nerve that arises from the cervical plexus is the phrenic nerve. It supplies motor fibers to the diaphragm. Severing or diseases of the spinal cord above the level of the origin of the phrenic nerve results in paralysis of the diaphragm and death since the phrenic nerve no longer sends impulses to the diaphragm. Contractions of the diaphragm are essential for breathing.

A summary of the nerves and distributions of the cervical plexus is provided in Exhibit 13–2. The relationship of the cervical plexus to the other plexuses is shown in Figure 13–1.

The brachial plexus

The **brachial plexus** is formed by the ventral rami of spinal nerves C5 to C8 and T1 with contributions from C4 and T2. On either side of the last four cervical and first thoracic vertebrae, the brachial plexus extends downward and laterally, passes over the first rib behind the clavicle, and then enters the axilla (Figure 13–11). The brachial plexus constitutes the entire nerve supply for the upper extremities, as well as a number of neck and shoulder muscles.

The *roots* of the brachial plexus, like those of the cervical plexus, are continuations of the ventral rami of the spinal nerves. The roots of C5 and C6 unite to form the *upper trunk,* C7 becomes the *middle trunk,* and C8 to T1 form the *lower trunk.* Each trunk, in turn, divides into an *anterior division* and a *posterior division.* The divisions then unite to form cords. The *posterior cord* is formed by the union of the posterior divisions of the upper and middle trunks, and the *medial cord* is formed as a continuation of the anterior division of the lower trunk. The peripheral nerves arise from the cords.

Exhibit 13–3. THE BRACHIAL PLEXUS

NERVE	ORIGIN	DISTRIBUTION
Root nerves		
Dorsal scapular	C5	Levator scapulae, rhomboideus major, and rhomboideus minor muscles
Long thoracic	C5–C7	Serratus anterior muscle
Trunk nerves		
Nerve to subclavius	C5	Subclavius muscle
Suprascapular	C5–C6	Supraspinatus and infraspinatus muscles
Lateral cord nerves		
Musculocutaneous	C5–C7	Coracobrachialis, biceps, and brachialis muscles
Median (lateral head)	C5–C7	See distribution for Median (medial head)
Lateral pectoral	C5–C7	Pectoralis major muscle
Posterior cord nerves		
Upper subscapular	C5–C6	Subscapularis muscle
Thoracodorsal	C6–C8	Latissimus dorsi muscle
Lower subscapular	C5–C6	Subscapularis and teres major muscles
Axillary (circumflex)	C5–C6	Deltoid and teres minor muscles; skin over deltoid and upper posterior aspect of arm
Radial	C5–C8, T1	Extensor muscles of the arm and forearm (e.g., triceps, brachioradialis, extensor carpi radialis longus, extensor digitorum, extensor carpi ulnaris, extensor indicis); skin of the posterior arm and forearm, lateral two-thirds of the dorsum of hand, and fingers over the proximal and middle phalanges
Medial cord nerves		
Medial pectoral	C8–T1	Pectoralis major and pectoralis minor muscles
Medial brachial cutaneous	C8–T1	Skin of the medial and posterior aspects of the lower third of the arm
Medial antebrachial cutaneous	C8–T1	Skin of the medial and posterior aspects of the forearm
Median (medial head)	C5–C8, T1	The medial and lateral heads of the median nerve form the median nerve. It is distributed to the flexors of the forearm (e.g., pronator teres, flexor carpi radialis, flexor digitorum superficialis) except the flexor carpi ulnaris and flexor digitorum profundus; skin of the lateral two-thirds of the palm of the hand and fingers
Ulnar	C7–C8, T1	Flexor carpi ulnaris and flexor digitorum profundus muscles; skin of the medial side of the hand, little finger, and medial half of the ring finger
Other cutaneous distributions		
Intercostobrachial	Second intercostal nerve	Skin over medial side of arm
Upper lateral brachial cutaneous	Axillary	Skin over deltoid muscle and down to elbow
Posterior brachial cutaneous	Radial	Skin over posterior aspect of arm
Lower lateral brachial cutaneous	Radial	Skin over lateral aspect of elbow
Lateral antebrachial cutaneous	Musculocutaneous	Skin over lateral aspect of forearm
Posterior antebrachial cutaneous	Musculocutaneous	Skin over posterior aspect of forearm

You may recall from Chapter 10, that in giving a deltoid intramuscular injection, precautions are taken not to damage the radial nerve. This is one of the nerves that arises from the brachial plexus and supplies the muscles on the posterior aspect of the arm and forearm. Radial nerve damage is indicated by *wrist drop,* which is an inability to extend the hand at the wrist. The median nerve is another nerve arising from the brachial plexus. It supplies most of the muscles of the anterior forearm and some of the muscles in the palm. Damage to it is indicated by inability to pronate the forearm, flex the wrist properly, and flex the fingers and thumb properly. Still another important nerve of the brachial plexus is the ulnar nerve. It supplies the anteromedial muscles of the forearm and most of the muscles of the palm. If it is damaged, there is an inability to flex and adduct the wrist and the fingers cannot be spread easily.

A summary of the nerves and distributions of the brachial plexus is found in Exhibit 13–3. The relationship of the brachial plexus to the other plexuses is shown in Figure 13–1.

The lumbar plexus

The **lumbar plexus** is formed by the ventral rami of spinal nerves L1 to L4. It differs from the brachial plexus in that there is no intricate interlacing of fibers. It also consists of *roots* and an *anterior* and *posterior division.* On either side of the first four lumbar vertebrae, the lumbar plexus passes obliquely outward behind the psoas major muscle (posterior division) and anterior to the quadratus lumborum muscle (anterior division) and then gives rise to its peripheral nerves (Figure 13–12). The lumbar plexus supplies the anterior abdominal wall, external genitals, and part of the lower extremity.

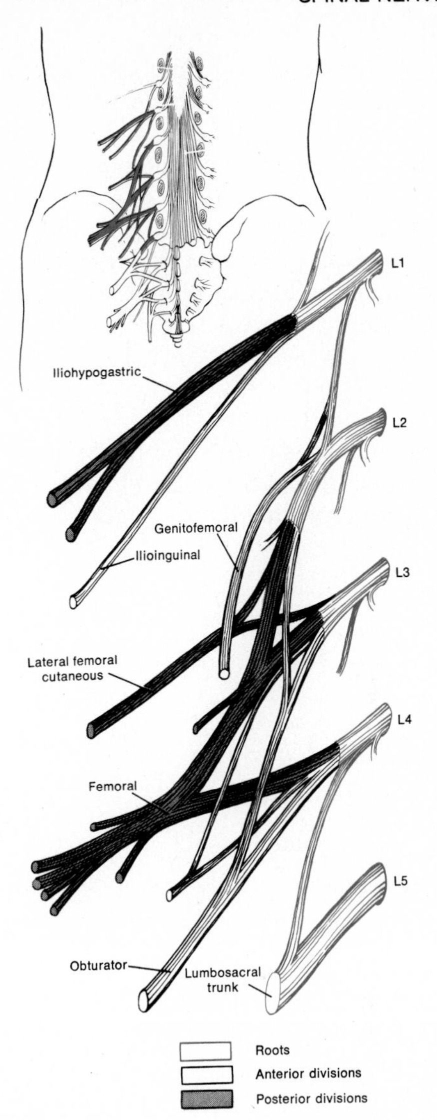

Figure 13-12. The lumbar plexus. Be sure to consult Exhibit 13-4 so that you can determine the distribution of each of the nerves of the plexus.

Exhibit 13–4. THE LUMBAR PLEXUS

NERVE	ORIGIN	DISTRIBUTION
Iliohypogastric	T12–L1	Muscles of the anterolateral abdominal wall (rectus abdominus, external oblique, internal oblique, transversus abdominis); skin of the lower abdomen and buttock
Ilioinguinal	L1	Muscles of the anterolateral abdominal wall as indicated above; skin of the upper medial aspect of thigh, root of penis and scrotum in the male, and labia majora and mons pubis in the female
Genitofemoral	L1–L2	Cremaster muscle; skin over the middle of the anterior surface of thigh, scrotum in the male, and labia majora in the female
Lateral femoral cutaneous	L2–L3	Skin over the lateral aspect of the thigh and buttock
Femoral	L2–L4	Flexor muscles of the leg (e.g., psoas major, iliacus, pectineus, rectus femoris, vastus lateralis, vastus medialis, vastus intermedius, sartorius); skin on front and over medial aspect of thigh and medial side of leg and foot
Obturator	L2–L4	Adductor muscles of leg (e.g., obturator externus, pectineus, adductor longus, adductor brevis, adductor magnus, gracilis); skin over medial aspect of thigh
Saphenous	L2–L4	Skin over medial aspect of leg and foot

The largest nerve arising from the lumbar plexus is the femoral nerve. Injury to the nerve is indicated by an inability to extend the leg and by loss of sensation in the skin over the anteromedial aspect of the thigh.

A summary of the nerves and distributions of the lumbar plexus is presented in Exhibit 13–4. The relationship of the lumbar plexus to the other plexuses is shown in Figure 13–1.

The sacral plexus

The **sacral plexus** is formed by the ventral rami of spinal nerves L4 to L5 and S1 to S4. It is situated largely in front of the sacrum (Figure 13–13). Like the lumbar plexus, it also contains *roots* and an *anterior* and *posterior division.* The sacral plexus supplies the buttocks, perineum, and lower extremities. The largest nerve arising from the sacral plexus, and, in fact, the largest nerve in the body, is the sciatic nerve. This nerve may be injured because of a slipped disc, dislocated hip, pressure from the uterus during pregnancy, or an improperly given gluteal intramuscular injection. Recall again from Chapter 10 that the sciatic nerve is avoided by giving an intramuscular injection into the upper, outer quadrant of the buttock. The sciatic nerve, and its branches, supplies the entire musculature of the leg and foot. Damage to them results in *foot drop,* an inability to dorsiflex the foot.

Exhibit 13–5. THE SACRAL PLEXUS

NERVE	ORIGIN	DISTRIBUTION
Superior gluteal	L4–L5, S1	Gluteus minimus and gluteus medius muscles
Inferior gluteal	L5–S2	Gluteus maximus muscle
Nerve to piriformis	S1–S2	Piriformis muscle
Nerve to quadratus femoris	L4–L5, S1	Inferior gemellus and quadratus femoris muscles
Nerve to obturator internus	L5–S2	Superior gemellus and obturator internus muscles
Perforating cutaneous	S2–S3	Skin over lower medial aspect of buttock
Posterior cutaneous	S1–S3	Skin over anal region, upper posterior aspect of thigh, upper part of calf, scrotum in the male, and labia majora in the female
Sciatic	L4–S3	The sciatic is actually two nerves, the tibial and common peroneal, bound together by a common sheath of connective tissue. It splits into its two divisions, usually at the knee. (See below for the distributions.) As the sciatic nerve descends through the thigh, it sends branches to the hamstring muscles (biceps femoris, semitendinosus, semimembranosus) and adductor magnus.
Tibial (medial popliteal)	L4–S3	Gastrocnemius, plantaris, soleus, popliteus, tibialis posterior, flexor digitorum, and hallucis muscles. Branches of the tibial nerve in the foot are the medial plantar nerve and the lateral plantar nerve.
Medial plantar		Abductor hallucis, flexor digitorum brevis, and flexor hallucis muscles; skin over the medial two-thirds of the plantar surface of the foot
Lateral plantar		Remaining muscles of the foot not supplied by the medial plantar nerve; skin over the lateral third of the plantar surface of the foot
Common peroneal (lateral popliteal)	L4–S2	This nerve divides into a superficial peroneal and a deep peroneal branch. The distributions of each are described below.
Superficial peroneal		Peroneus longus and peroneus brevis muscles; skin over distal third of anterior aspect of leg and dorsum of foot
Deep peroneal		Tibialis anterior, extensor hallucis longus, peroneus tertius, and extensor digitorum brevis muscles; skin over great and second toes
Pudendal	S2–S4	Muscles of the perineum; skin of penis and scrotum in the male and clitoris, labia majora, labia minora, and lower vagina in the female

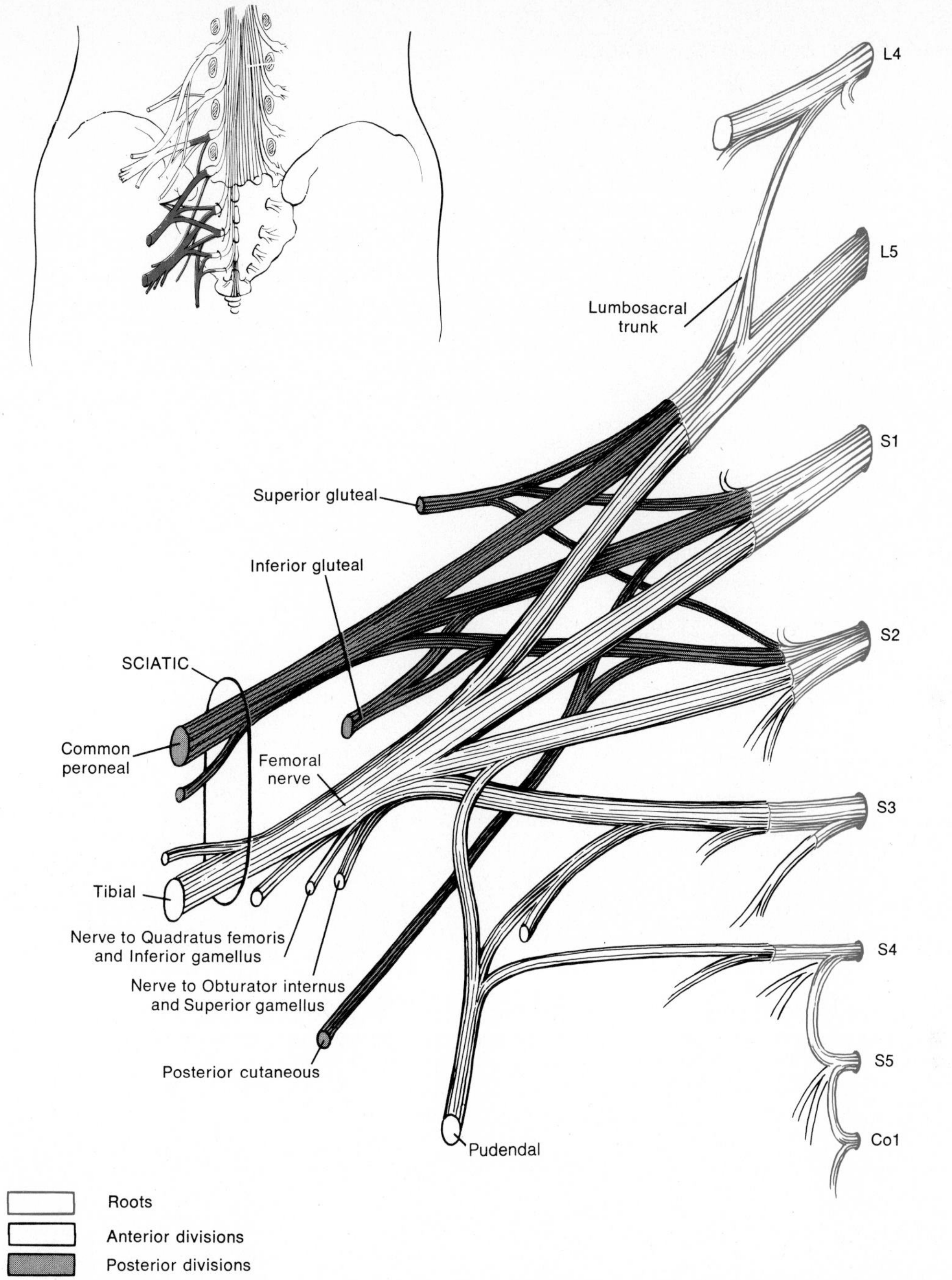

Figure 13-13. The sacral plexus. Be sure to consult Exhibit 13-5 so that you can determine the distribution of each of the nerves of the plexus.

A summary of the nerves and distributions of the sacral plexus is given in Exhibit 13–5. The relationship of the sacral plexus to the other plexuses is shown in Figure 13–1.

The coccygeal plexus

The **coccygeal plexus** is formed by the ventral rami of spinal nerves S4 to S5 and Co1. It distributes sensory fibers to the skin over the coccyx.

Intercostal (thoracic) nerves

It was noted earlier that spinal nerves T2 to T12 do not enter into the formation of plexuses. These nerves are called **intercostal** or **thoracic** nerves. Instead of forming a plexus, they are distributed directly to the structures they supply. After leaving the vertebral foramina, nerves T3 to T6 pass in the costal grooves of the ribs and are distributed to the intercostal muscles and skin of the anterior and lateral chest wall. Nerves T7 to T12 supply the abdominal muscles and overlying skin. T2 supplies the skin of the axilla and posteromedial aspect of the arm.

From the discussion of spinal nerves thus far, it should be obvious that specific spinal nerves supply fairly well-defined areas of the skin. We will now turn our attention to the relationship between spinal nerves and the skin.

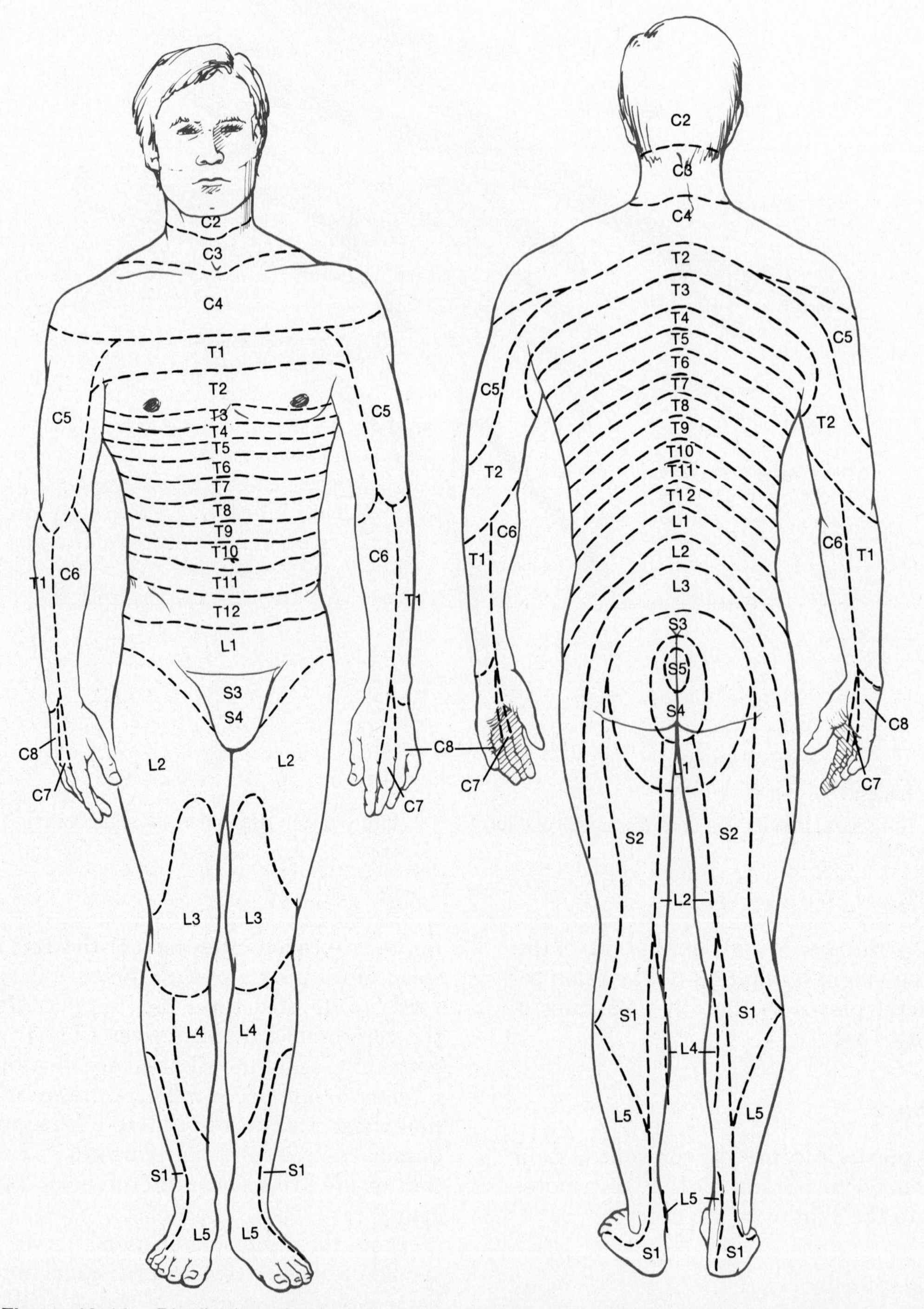

Figure 13-14. Distribution of spinal nerves to dermatomes. (a) Anterior view. (b) Posterior view.

Dermatomes

The skin over the entire body is supplied segmentally by spinal nerves. This means that the spinal nerves innervate specific, constant segments of the skin. With the exception of spinal nerve C1, all other spinal nerves supply branches to the skin. The skin segment supplied by a given spinal nerve is called a **dermatome** (Figure 13–14). In the neck and trunk, the dermatomes form consecutive bands of skin. Also, in the trunk, there is an overlap of adjacent dermatome nerve supply. Thus, there is little loss of sensation if only a single nerve supply to a dermatome is interrupted. Most of the skin of the face and scalp is supplied by cranial nerve V (trigeminal).

Knowledge of the location of dermatomes has important clinical applications. For example, if a physician knows that a particular dermatome is associated with a particular spinal nerve, it is possible to determine which segment of the spinal cord or spinal nerve is malfunctioning. If a dermatome is stimulated by the physician and the sensation is not picked up by the patient, it can be assumed that the nerve supplying the dermatome is involved.

Another topic we would like to consider at this point is spinal cord injury. By now, you should have enough information to help you understand some of the consequences of a damaged spinal cord.

APPLICATIONS TO HEALTH

Spinal cord injury

The spinal cord may be damaged by fracture or dislocation of the vertebrae enclosing it or by wounds, such as those from bullets or shell fragments. All can result in **transection,** that is, partial or complete severing of the spinal cord. Complete transection means that all ascending and descending pathways are cut. It results in loss of all sensations and voluntary muscular movements below the level of transection. In fact, individuals with complete cervical transections close to the base of the skull usually die of asphyxiation before treatment can be administered. This happens because impulses from the phrenic nerves to the breathing muscles are interrupted. If the upper cervical cord is partially transected, both the arms and legs are paralyzed, and the patient is classified as *quadriplegic.* Partial transection between the cervical and lumbar enlargements results in paralysis of the legs only, and the patient is classified as *paraplegic.*

In the case of partial transection, there is an initial period of **spinal shock** lasting from a few days to several weeks. During this period, all reflex activity is abolished, a condition called *areflexia.* However, in time, there is a return of reflex activity. For example, the first reflex to return is the knee jerk. Its reappearance may take several days. Next, the flexor reflex appears. This may take up to several months. Then the crossed extensor reflex appears. Visceral reflexes such as erection and ejaculation are also affected by transection. For instance, bladder and bowel function are no longer under voluntary control.

We will now conclude our discussion of the spinal cord and spinal nerves with a description of some of the factors related to the degeneration and regeneration of peripheral nerves.

Peripheral nerve damage

Unlike the cells of epithelial tissue, neurons have very limited powers for regeneration. Before or around the time of birth, the cell bodies of most developing nerve cells lose their mitotic apparatus and thus their ability to reproduce. This means that if the cell body of a neuron is destroyed, the neuron cannot be replaced by other reproducing cells in the tissue. The function of the neuron is permanently lost. However, axons that have a neurilemma can be repaired, as long as the cell body is intact and fibers are in association with Schwann cells. Most nerves that lie outside the brain and spinal cord consist of axons and dendrites that are covered with a neurilemma. A person who injures a nerve in his upper extremity, for example, has a good chance of regaining nerve function. Axons that lie within the brain and spinal cord do not have a neurilemma. This means that any injury to the brain or spinal cord is permanent.

When there is damage to an axon (or to dendrites of somatic afferent neurons), there are usually changes in the cell body and always changes in the axon portion distal to the site of damage. These changes associated with the cell body are referred to as the axon reaction, whereas those associated with the distal portion of the axon are called wallerian degeneration. The axon reaction occurs in essentially the same way, whether the damaged fiber is in the central or peripheral nervous system. The wallerian reaction, however, is different depending on whether the fiber is central or peripheral. We will consider the axon reaction first.

Axon reaction

When there is damage to an axon of a central or peripheral neuron, certain structural changes occur within the cell body. As noted above, these are collectively called the axon reaction. One of the most significant features of the axon reaction occurs from 24 to 48 hours following damage. The Nissl bodies, arranged in an orderly fashion in an uninjured cell body, are broken down into finely granular masses. This alteration is called *chromatolysis.* It begins between the axon hillock and nucleus but spreads throughout the cell body. As a result of chromatolysis, the cell body swells and the swelling reaches its maximum between 10 and 20 days following injury. Chromatolysis results in a loss of ribosomes by the rough endoplasmic reticulum and an increase in the number of free ribosomes. Following chromatolysis, there are signs of recovery in the cell body. There is an acceleration of RNA and protein synthesis, which favors regeneration of the axon. Recovery often takes several months and involves the restoration of normal levels of RNA and proteins and the Nissl bodies to their usual, uninjured patterns. Another sign of the axon reaction is the off-center position of the nucleus in the cell body. This makes it possible to readily identify the cell bodies of damaged fibers through a microscope.

Wallerian degeneration

Accelerated protein synthesis is required for the repair of the damaged axon. The proteins synthesized pass from the cell body into the axon at about the rate of 1 mm (0.039 in.)/day. The proteins assist in regenerating the damaged axon. The part of the axon distal to the damage becomes slightly swollen and then breaks up into fragments by the third to fifth day. The myelin sheath around the axon also undergoes degeneration. Degeneration of the distal portion of the axon and myelin sheath is called **wallerian degeneration.** Following degeneration, there is phagocytosis of the remains.

Even though there is degeneration of the axon and myelin sheath, the neurilemma of the Schwann cells remains. The Schwann cells in the distal portion of the axon multiply by mitosis and grow toward the proximal portion of the axon. This growth results in the formation of a tube that serves to unite the distal and proximal portions of the axon. This "tunnel" provides a means for new axons to grow from the proximal area, across the injured area, and into the distal area so that the severed portions of the axon can be united.

Regeneration

During the first few days following damage, regenerating axons begin to invade the tube formed by the Schwann cells. Axons from the proximal area grow at the rate of about 1.5 mm (0.059 in.)/day across the area of damage, find their way into the distal portion, and unite with the axons in the distal area. Thus, sensory and motor connections are reestablished. In time, a new myelin sheath is also produced as a result of the activity of the Schwann cells.

We will now direct our attention to some of the important features related to the brain and cranial nerves.

Chapter summary in outline

THE GROUPING OF NEURAL TISSUE

1. White matter is an aggregation of myelinated axons.
2. Gray matter is a collection of nerve cell bodies and dendrites or unmyelinated axons.
3. A nerve is a bundle of fibers outside the central nervous system.
4. A ganglion is a collection of cell bodies outside the central nervous system.
5. A bundle of fibers of similar function inside the central nervous system forms a tract.
6. A mass of nerve cell bodies and dendrites inside the brain forms a nucleus.
7. A horn is an area of gray matter in the spinal cord.

THE SPINAL CORD

General features

1. The spinal cord begins as a continuation of the medulla oblongata and terminates at about the second lumbar vertebra.
2. It contains a cervical and lumbar enlargement, which serve as points of origin for nerves to the extremities.
3. The tapered portion of the spinal cord is the conus medullaris, from which arises the filum terminale and cauda equina.
4. The spinal cord is partially divided into right and left sides by the anterior median fissure and posterior median sulcus.

Protection and coverings

1. The spinal cord is protected by the wall of the vertebral canal, meninges, and cerebrospinal fluid.
2. The spinal meninges from outside to inside are the dura mater, arachnoid, and pia mater. Inflammation of the meninges is called meningitis.

3. The removal of cerebrospinal fluid from the subarachnoid space in the lumbar region of the cord is called a spinal puncture.

Structure
1. Parts of the spinal cord observed in cross section are the gray commissure; central canal; anterior, posterior, and lateral gray horns; anterior, posterior, and lateral white columns; and ascending and descending tracts.
2. The spinal cord conveys sensory and motor information by way of the ascending and descending tracts, respectively.
3. In serving as a reflex center, the posterior root, posterior root ganglion, and anterior root convey an impulse.

Reflex arc
1. The reflex arc is the functional unit of the nervous system.
2. Components of a reflex arc include the receptor, sensory neuron, center, motor neuron, and effector.
3. Important somatic spinal reflexes include the stretch reflex, the flexor reflex, and the crossed extensor reflex.
4. Reflex arcs may be classified as ipsilateral, monosynaptic, polysynaptic, intersegmental, and contralateral.

SPINAL NERVES
Composition and coverings
1. Spinal nerves are attached to the spinal cord by means of a posterior root and an anterior root. All spinal nerves are mixed.
2. Spinal nerves are covered by endoneurium, perineurium, and epineurium.

Names and branches
1. The 31 pairs of spinal nerves are named and numbered according to the region and level of the spinal cord from which they emerge.
2. Branches of a spinal nerve include the dorsal ramus, ventral ramus, meningeal branch, and rami communicantes.

Plexuses
1. The ventral rami of spinal nerves, except for T2 to T12, form networks of nerves called plexuses.
2. Emerging from the plexuses are nerves bearing names that are often descriptive of the general regions they supply or the course they take.
3. The principal plexuses are called the cervical, brachial, lumbar, sacral, and coccygeal plexuses.
4. Nerves that do not form plexuses are called intercostal nerves (T2 to T12). They are distributed directly to the structures they supply.

Dermatomes
1. With the exception of spinal nerve C1, all spinal nerves supply branches to the skin.
2. The skin segment supplied by a given spinal nerve is called a dermatome.
3. Knowledge of dermatomes helps a physician to determine which segment of the spinal cord or spinal nerve is malfunctioning.

APPLICATIONS TO HEALTH
1. Complete or partial severing of the spinal cord is called transection. It may result in quadriplegia or paraplegia. Partial transection is followed by a period of loss of reflex activity called areflexia.
2. Axons with a neurilemma are capable of regeneration. The regeneration is preceded by the axon reaction and wallerian degeneration.

Review questions and problems

1. Define the following terms: white matter, gray matter, nerve, ganglion, tract, nucleus, and horn.
2. Describe the location of the spinal cord. What are the cervical and lumbar enlargements?
3. Define conus medullaris, filum terminale, and cauda equina. What is a spinal segment? How is the spinal cord partially divided into a right and left side?
4. Describe the bony covering of the spinal cord.
5. Explain the location and composition of the spinal meninges. Describe the location of the epidural, subdural, and subarachnoid spaces. What are the denticulate ligaments?
6. Define meningitis. Distinguish between pachymeningitis and leptomeningitis.
7. What is a spinal puncture? Give several purposes served by a spinal puncture.
8. Based upon your knowledge of the structure of the spinal cord in cross section, define the following: gray commissure, central canal, anterior gray horn, lateral gray horn, posterior gray horn, anterior white column, lateral white column, posterior white column, ascending tract, and descending tract.
9. Describe the function of the spinal cord as a conduction pathway.
10. Using Exhibit 13–1 as a guide, be sure that you can list the location, origin, termination, and function of the principal ascending and descending tracts.
11. Describe how the spinal cord serves as a reflex center.
12. What is a reflex arc? List and define the components of a reflex arc.
13. Describe the mechanism of a stretch reflex, a flexor reflex, and a crossed extensor reflex.
14. Define the following terms: monosynaptic reflex arc, ipsilateral reflex arc, polysynaptic reflex arc, intersegmental reflex arc, contralateral reflex arc, and reciprocal inhibition.
15. Define a spinal nerve. Why are all spinal nerves classified as mixed nerves?
16. Describe how a spinal nerve is attached to the spinal cord.
17. Explain how a spinal nerve is enveloped by its several different coverings.
18. How are spinal nerves named and numbered?
19. Describe the branches and innervations of a typical spinal nerve.
20. What is a plexus?
21. Explain the location, origin, nerves, and distributions of the following plexuses: cervical, brachial, lumbar, and sacral.
22. What are intercostal nerves?
23. Define a dermatome. Why is a knowledge of dermatomes important?
24. What is transection? Distinguish between quadriplegia and paraplegia.
25. What is meant by spinal shock?
26. Explain the conditions necessary for peripheral nerve repair.
27. Distinguish between the axon reaction and wallerian degeneration.

Selected readings

Barr, M. L. *The Human Nervous System.* New York: Harper and Row, 1972.

Cunningham's Textbook of Anatomy. 11th ed., edited by G. J. Romanes. London: Oxford University Press, 1972. Pp. 592–599, 734–775.

Dawson, Helen L. *Basic Human Anatomy.* 2d ed. New York: Appleton-Century-Crofts, 1974. Pp. 191–194, 204–214.

Everett, N. B. *Functional Neuroanatomy.* 6th ed. Philadelphia: Lea and Febiger, 1971.

Gray, Henry. *Anatomy of the Human Body.* 29th ed., edited by Charles Mayo Goss. Philadelphia: Lea and Febiger, 1973. Pp. 792–802.

Netter, Frank H. *Nervous System.* CIBA Collection of Medical Illustrations, Vol. 1. Summit, NJ: CIBA Pharmaceutical Co., 1958.

Truex, Raymond C., and Malcolm B. Carpenter. *Human Neuroanatomy.* 6th ed. Baltimore: Williams and Wilkins, 1969.

CHAPTER 14

THE BRAIN AND THE CRANIAL NERVES

STUDENT OBJECTIVES

After you have read this chapter, you should be able to:

1. Identify the principal areas of the brain
2. Describe the location of the cranial meninges
3. Explain the formation and circulation of cerebrospinal fluid
4. Describe the blood supply to the brain and the concept of the blood-brain barrier
5. Compare the components of the brain stem with regard to structure and function
6. Identify the structure and functions of the diencephalon
7. Describe the anatomical characteristics and functions of the cerebellum
8. Identify the structural features of the cerebrum
9. Describe the lobes, tracts, and basal ganglia of the cerebrum
10. Describe the structure and functions of the limbic system
11. Compare the motor, sensory, and association areas of the cerebrum
12. Describe the principle of the electroencephalograph and its significance in the diagnosis of certain disorders
13. Define a cranial nerve
14. Identify the 12 pairs of cranial nerves by name, number, type, location, and function
15. Explain the effects of injury on cranial nerves

Attention will now be turned to a discussion of the brain and the 12 pairs of cranial nerves. We will consider how the brain is protected, what its principal parts are, how the parts function, how the brain is related to the spinal cord, and how the brain is related to the paired cranial nerves.

BRAIN

Protection and coverings

The **brain** of an average adult is one of the largest organs of the body, weighing about 1,300 g (3 lb). By looking at Figure 14–1, you will see that the brain resembles a rather stocky mushroom. It is divided into four principal areas: the brain stem, diencephalon, cerebellum, and cerebrum. The **brain stem,** the stalk of the mushroom, consists of the medulla oblongata, pons varolii, and midbrain. The lower end of the brain stem is a continuation of the spinal cord. The **diencephalon** consists primarily of the thalamus and hypothalamus, and over it spreads the cap of the mushroom, the large cerebrum. The **cerebrum** constitutes about seven-eighths of the total weight of the brain and occupies most of the skull. Inferior to the cerebrum and posterior to the brain stem is the **cerebellum.**

The brain is protected by the cranial bones. These have already been described and may be reviewed in Chapter 6. Like the spinal cord, the brain is also protected by meninges. The *cranial meninges* surround the brain and are continuous with the spinal meninges. The cranial meninges have the same basic structure and bear the same names as the spinal meninges, namely, the outermost *dura mater,* middle *arachnoid,* and innermost *pia mater* (see Figure 14–2). The cranial dura mater consists of two layers. The thicker, outer layer (endosteal layer) lightly adheres to the cranial bones and serves as a periosteum. The thinner, inner layer (meningeal layer) contains a mesothelial layer on its smooth surface. The spinal dura mater corresponds with the meningeal layer of the cranial dura mater.

Cerebrospinal fluid

The brain, as well as the rest of the central nervous system, is further protected against injury by a substance called **cerebrospinal fluid.** This fluid circulates through the subarachnoid space around the brain and cord and through the ventricles of the brain. The subarachnoid space is the area between the arachnoid and pia mater. The **ventricles** are cavities within the brain that communicate with each other, the central canal of the spinal cord, and the subarachnoid space. The two *lateral ventricles* are located one in each hemisphere (side) of the cerebrum under the corpus callosum (Figure 14–2). The *third ventricle* is a slit between and inferior to the right and left halves of the thalamus and between the lateral ventricles. Each lateral ventricle communicates with the third ventricle by a narrow, oval opening, the *interventricular foramen* or *foramen of Monro.* The *fourth ventricle* lies in the brain stem inferior to the cerebellum. It communicates with the third ventricle via the *cerebral aqueduct (aqueduct of Sylvius)* which passes through the midbrain. The roof of the fourth ventricle has three openings: a *median aperture (foramen of Magendie)* and two *lateral apertures (foramina of Luschka).* Through these openings, the fourth ventricle also communicates with the subarachnoid space of the brain and cord.

Before we describe the formation and flow of cerebrospinal fluid, carefully examine the sagittal section of the brain shown in Figure 14–2 and be sure that you can locate each of the structures just described.

The entire central nervous system holds about 125 ml (2 oz) of cerebrospinal fluid. It is a clear, colorless fluid of watery consistency. Chemically, it contains proteins, glucose, urea, and salts. It also contains some white blood cells. The fluid serves as a shock absorber for the central nervous system. It also circulates nutritive substances filtered from the blood. Cerebrospinal fluid is formed primarily by filtration from networks of capillaries, called **choroid plexuses,** located in the ventricles (Figure 14–2). The fluid formed in the choroid plexuses of the lateral ventricles circulates through the interventricular foramen to the third ventricle, where more fluid is added by the choroid plexus in the third ventricle. It then flows through the cerebral aqueduct to the fourth ventricle. Here, further additions occur from the choroid plexus in the fourth ventricle. The fluid then circulates through the apertures of the fourth ventricle into the subarachnoid space around the back of the brain. It also passes downward to the subarachnoid space around the posterior surface of the cord, up the anterior surface of the cord, and around the anterior part of the brain. From here, it is gradually reabsorbed into veins. Some cerebrospinal fluid may be formed by ependymal (neuroglial) cells in the central canal of the spinal cord. This small quantity of fluid ascends to reach the fourth ventricle. Most of the fluid is absorbed into the superior sagittal sinus (see Figure 11–17). The absorption actually occurs through **arachnoid villi,** which are fingerlike projections of the arachnoid that push into the superior sagittal sinus. Normally, cerebrospinal fluid is absorbed as rapidly as it is formed. Each 24 hours, the body

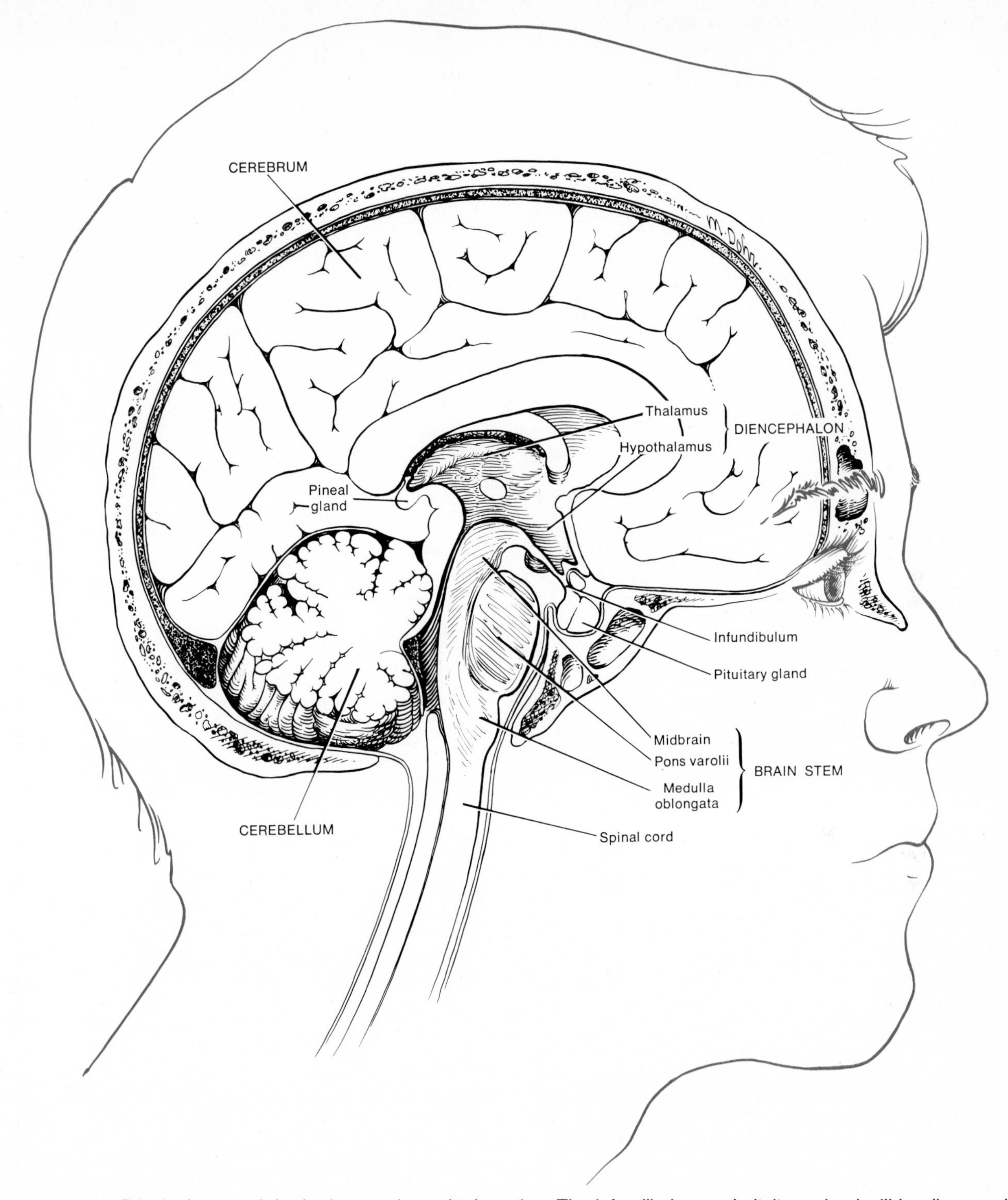

Figure 14-1. Principal parts of the brain seen in sagittal section. The infundibulum and pituitary gland will be discussed in conjunction with the nervous system.

forms and absorbs about 400 ml (7 oz) of the fluid. Now examine the three-dimensional view of the ventricles in Figure 14–3. This will help to give you a better understanding of the path taken by cerebrospinal fluid.

If an obstruction, such as a tumor, arises in the brain and interferes with the drainage of the fluid from the ventricles into the subarachnoid space, large amounts of fluid start building up in the ventricles. Fluid pressure inside the brain increases, and, if the fontanels have not yet closed, the head bulges to relieve the pressure. This condition is called **internal hydrocephalus.** The term *hydro* means water, and *cephalo* means head. If an obstruction interferes with drainage somewhere in the subarachnoid space and cerebrospinal fluid accumulates inside the space, the condition is referred to as **external hydrocephalus.**

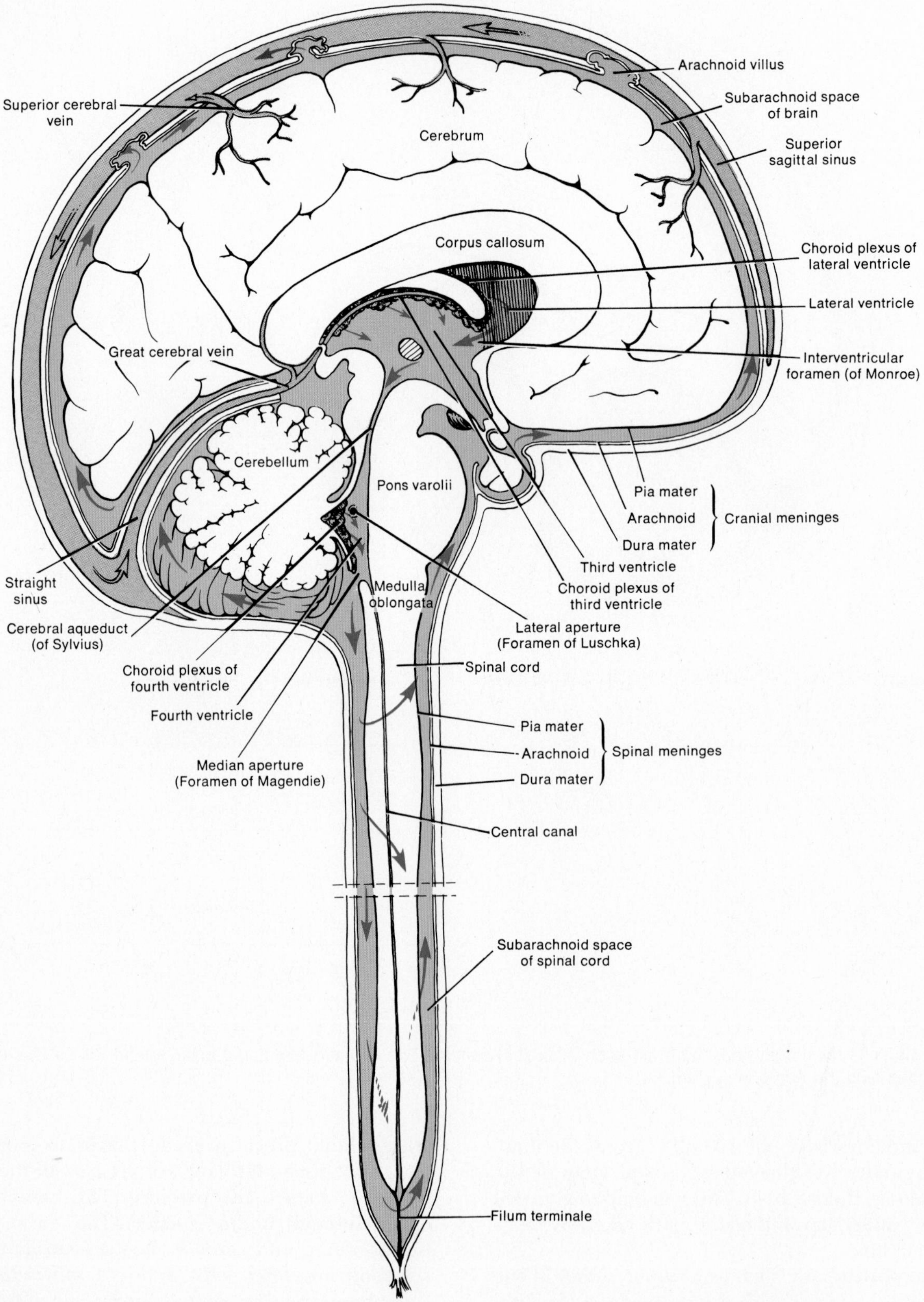

Figure 14-2. The brain and meninges seen in sagittal section. The direction of flow of cerebrospinal fluid is indicated by colored arrows.

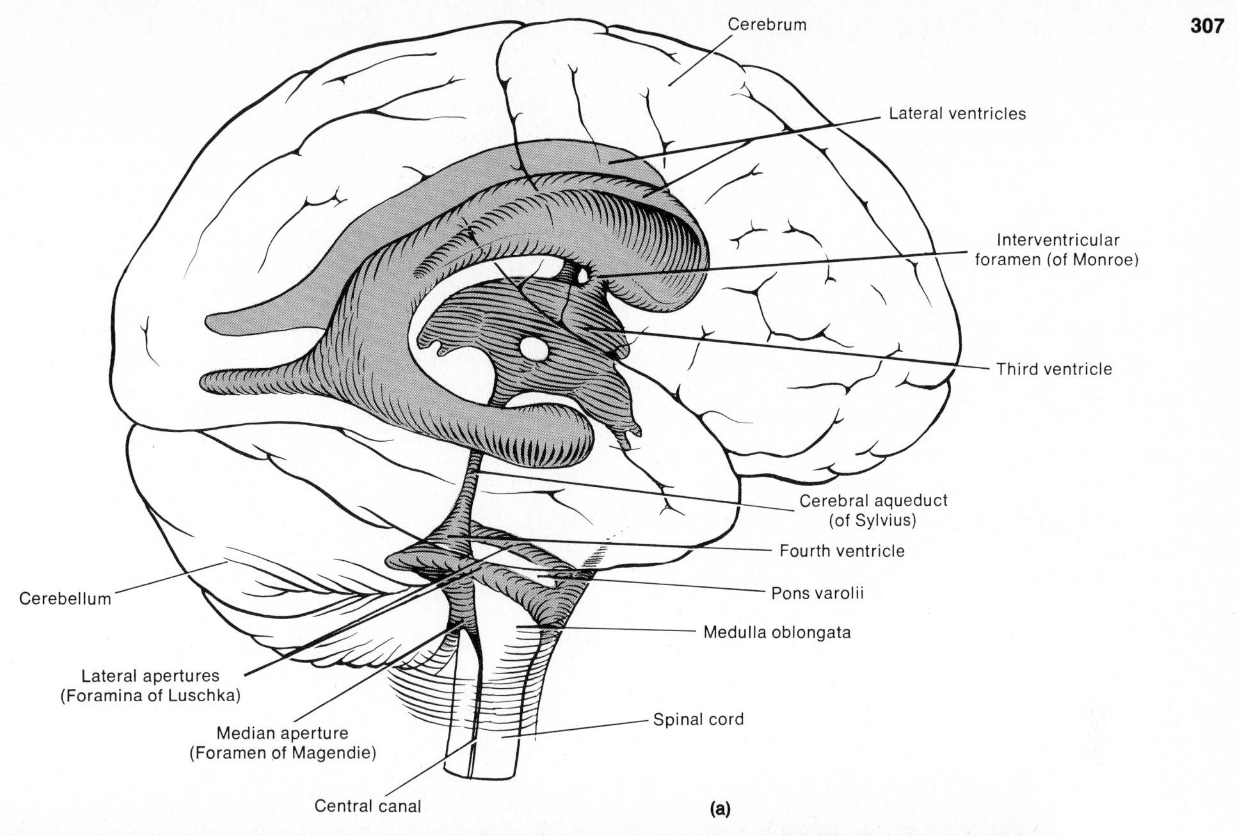

(a)

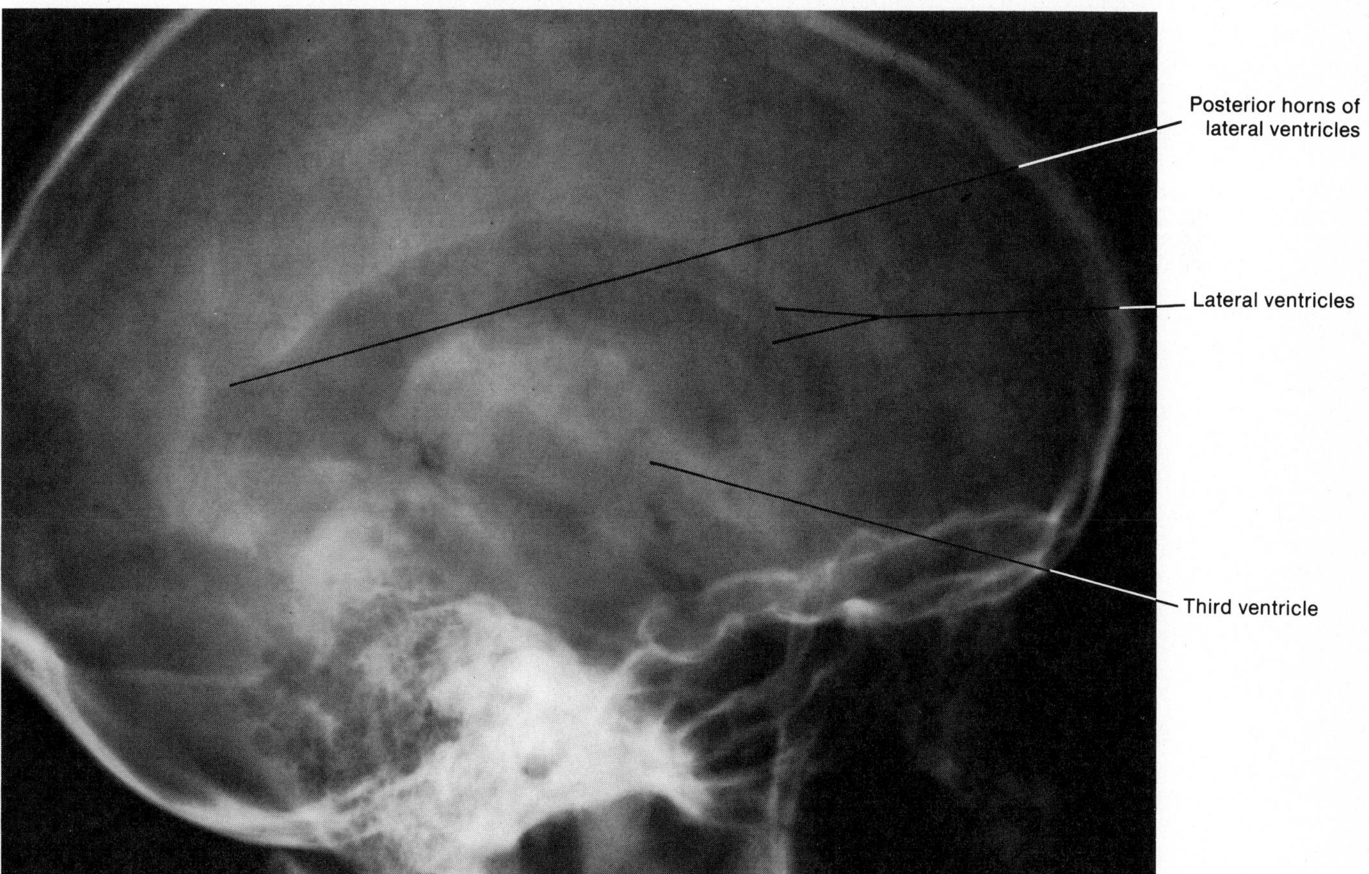

(b)

Figure 14-3. Ventricles of the brain and their interconnections. (a) Diagrammatic lateral projection. (b) Roentgenogram, lateral view. (Courtesy of John C. Bennett, St. Mary's Hospital, San Francisco.)

Before we start a discussion of the parts of the brain and their functions, we will first examine the blood supply to the brain.

Blood supply to the brain

The names and locations of the blood vessels that supply the brain have already been discussed in detail in Exhibit 11–4 and Figure 11–13. At this point we only want to concentrate on the importance of blood flow to the brain. The brain is well supplied with blood vessels, which bring it oxygen and nutrients. Although the brain actually consumes less oxygen than most other organs of the body, it must receive its allotment of oxygen continuously. If the blood supply to the brain is interrupted for only a few moments, unconsciousness may result. An interruption that lasts a minute or two can weaken the brain cells by starving them of oxygen. If the cells are totally deprived of oxygen for about 4 minutes, many of them are permanently injured. Sometimes babies are cut off from the oxygen supply of their mothers' blood before they leave the birth canal and can breathe. Many of these children are born dead or suffer throughout their lives from brain damage that may result in mental retardation, epilepsy, and paralysis.

Blood supplying the brain also contains glucose, the principal source of energy for brain cells. Because carbohydrate storage in the brain is limited, the blood must also transport a continuous supply of glucose. If blood entering the brain has a low glucose level, conditions such as mental confusion, dizziness, convulsions, and even loss of consciousness may occur.

Substances such as glucose, oxygen, and certain ions pass rapidly from the circulating blood into brain cells. Other substances such as creatinine, urea, chloride, insulin, and sucrose pass quite slowly. There are even substances, such as proteins and most antibiotics, that do not pass at all from the blood into brain cells. The differential rates of passage of certain materials from the blood into the brain suggest a concept called the **blood-brain barrier.** Electron micrograph studies of the capillaries of the brain reveal that they do differ from other capillaries of the body. For example, brain capillaries either lack or contain fewer pores than other body capillaries. Brain capillaries are also constructed of more densely packed cells. Moreover, brain capillaries are surrounded by large numbers of glial cells. Finally, brain capillaries are surrounded by a continuous basement membrane. All these features form an effective barrier to the passage of certain materials. Thus, substances that move through the barrier are either very small molecules or require the assistance of a carrier molecule to move through by active transport. The actual function of the blood-brain barrier is not known. However, it is speculated that the barrier may protect brain cells from potentially harmful substances.

Now we can examine the principal parts of the brain and their functions. We will begin with the brain stem, which is composed of the medulla oblongata, pons varolii, and midbrain (mesencephalon).

The brain stem

Medulla oblongata

The **medulla oblongata,** or simply **medulla,** is a continuation of the upper portion of the spinal cord and forms the inferior part of the brain stem. Its position in relation to the other parts of the brain may be noted in Figure 14–1. The medulla lies just superior to the level of the foramen magnum and extends upward to the inferior portion of the pons varolii. The medulla measures only 3 cm (about 1 in.) in length. Within the medulla is the fourth ventricle, an expansion of the central canal of the spinal cord.

In addition to containing the fourth ventricle, the medulla also contains all ascending and descending tracts that communicate between the spinal cord and various parts of the brain. These tracts constitute the white matter of the medulla. Some of the tracts undergo crossing as they pass through the medulla. Let us see how this crossing occurs and what it means.

On the ventral side of the medulla are two roughly triangular structures called *pyramids.* These can be observed in Figure 14–4. The pyramids are composed of the largest motor tracts that run from the outer region of the cerebrum (cerebral cortex) to the spinal cord. Just above the junction of the medulla with the cord, most of the fibers in the left pyramid cross to the right side, and most of the fibers in the right pyramid cross to the left. This crossing is referred to as **decussation of pyramids.** The adaptive value, if any, of this phenomenon is unknown. The principal motor fibers that undergo decussation belong to the lateral corticospinal tract. This tract originates in the cerebral cortex and runs down to the medulla. Here, the fibers cross in the pyramids and descend in the lateral columns of the spinal cord. From here the fibers terminate in skeletal muscles. As a result of the crossing, fibers that originate in the left cerebral cortex stimulate muscles on the right side of the

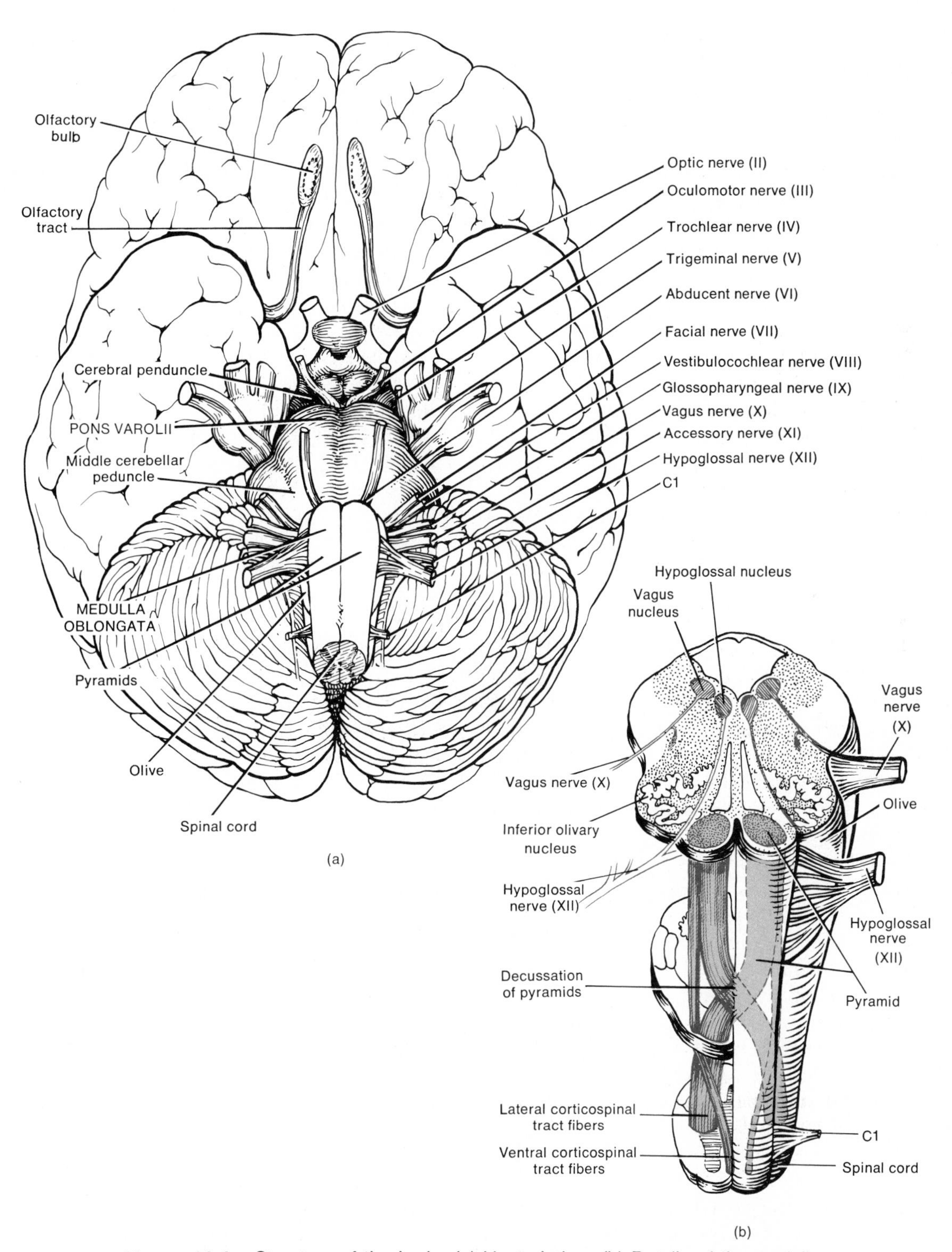

Figure 14-4. Structure of the brain. (a) Ventral view. (b) Details of the medulla.

body, and fibers that originate in the right cerebral cortex stimulate muscles on the left side of the body. The phenomenon of decussation explains why motor areas of one side of the cerebral cortex control muscular movements on the opposite side of the body.

The dorsal side of the medulla contains two pairs of prominent nuclei. These are the right and left *nucleus gracilis* and *nucleus cuneatus.* These nuclei receive sensory fibers from some ascending tracts (right and left fasciculus gracilis and fasciculus cuneatus) of the spinal cord and relay the sensory information to the opposite side of the medulla. This information is conveyed to the sensory areas of the cerebral cortex. Nearly all sensory impulses received on one side of the body cross in the medulla or spinal cord and are registered on the opposite side of the cerebral cortex.

In addition to its function as a conduction pathway for motor and sensory impulses between the brain and spinal cord, the medulla also contains an area of dispersed gray matter containing some white fibers. This region is called the **reticular formation.** Actually, portions of the reticular formation are found in the spinal cord, pons, midbrain, and diencephalon. The reticular formation functions in consciousness and arousal. As far as the medulla is concerned, within its portion of the reticular system are nuclei that contain three vital reflex centers. These are the *cardiac center,* which regulates heartbeat; the *respiratory center,* which adjusts the rate and depth of breathing; and the *vasoconstrictor center,* which regulates the diameter of blood vessels. Other centers in the medulla coordinate swallowing, vomiting, coughing, sneezing, hiccoughing, and blinking reflexes.

The medulla also contains the nuclei of origin of four pairs of cranial nerves. These are the glossopharyngeal nerves (IX), which relay impulses related to swallowing, salivation, and taste; the vagus nerves (X), which relay impulses to and from many thoracic and abdominal viscera; the accessory nerves (XI), which convey impulses related to head and shoulder movements; and the hypoglossal nerves (XII), which convey impulses that involve tongue movements.

On each lateral surface of the medulla is an oval projection called the *olive.* The olive contains an inferior olivary nucleus and two accessory olivary nuclei. The nuclei are connected to the cerebellum by fibers and will be discussed in conjunction with the cerebellum.

In view of the many vital activities controlled by the medulla, it is not surprising that a hard blow to the base of the skull can be fatal. Nonfatal medullary injury may be indicated by cranial nerve malfunctions on the same side of the body as the area of medullary injury (cranial nerves do not decussate), paralysis and loss of sensation on the opposite side of the body, and irregularities in respiratory control.

Pons varolii

The next area of the brain stem we will consider is the **pons varolii** or **pons.** Its relationship to other parts of the brain can be seen in Figures 14–1 and 14–4a. The pons, which means bridge, lies directly above the medulla and anterior to the cerebellum. It measures about 2.5 cm (1 in.) in length. Like the medulla, the pons consists of white fibers with nuclei scattered throughout. As the name implies, one of the chief functions of the pons is to serve as a bridge connecting the spinal cord with the brain and parts of the brain with each other. These connections are achieved by fibers that run in two principal directions. The transverse fibers connect with the cerebellum through the middle cerebellar peduncles of the cerebellum. The longitudinal fibers of the pons belong to the motor and sensory tracts that connect the spinal cord or medulla with the upper parts of the brain stem.

The nuclei for certain paired cranial nerves are also contained in the pons. These include the trigeminal nerves (V), which relay impulses for chewing and for sensations of the head and face; the abducent nerves (VI), which regulate some eyeball movements; the facial nerves (VII), which conduct impulses related to taste, salivation, and facial expression; and the vestibulocochlear nerves (VIII), which are concerned with hearing and equilibrium.

Another important nucleus in the reticular formation of the pons is the *pneumotaxic center.* Together with the respiratory center in the medulla, it helps to control respiration.

Midbrain

The last part of the brain stem we will consider is the **midbrain** or **mesencephalon.** It extends from the pons to the lower portion of the cerebrum (Figure 14–1). The midbrain is about 2.5 cm (1 in.) in length. The cerebral aqueduct passes through the midbrain and serves to connect the third ventricle above with the fourth ventricle below.

The ventral portion of the midbrain, called the *basilar portion,* contains a pair of fiber bundles referred to as *cerebral peduncles.* The cerebral peduncles contain many motor fibers that convey impulses from the cerebral cortex to the pons and spinal cord. They also contain sensory fibers that pass from the spinal cord to the thalamus. The cerebral peduncles constitute the main connection for tracts between upper parts of the brain and lower parts of the brain and spinal cord.

The dorsal portion of the midbrain is called the *tegmental portion.* It contains four rounded eminences, the *corpora quadrigemina.* Two of the eminences are known as the *superior colliculi.* These serve as a reflex center for movements of the eyeballs and head in response to visual and other stimuli. The other two eminences are referred to as the *inferior colliculi.* They serve as reflex centers for movements of the head and trunk in response to auditory stimuli. The tegmental portion also contains the *substantia nigra,* a large, heavily pigmented nucleus near the cerebral peduncles.

A very important nucleus in the reticular formation of the midbrain is the *red nucleus.* Fibers from the cerebellum and cerebral cortex terminate in the red nucleus. The red nucleus is also the origin of cell bodies of the descending rubrospinal tract. Other nuclei in the midbrain are associated with cranial nerves. These include the oculomotor nerves (III), which mediate some movements of the eyeballs and changes in the size of the pupils and the shape of the lenses; and the trochlear nerves (IV), which conduct impulses that move the eyeballs.

Before leaving the discussion of the brain stem, I would like to call your attention to a structure called the *medial lemniscus,* which is common to the medulla, pons, and midbrain. The medial lemniscus is a band of white fibers containing axons that convey impulses for fine touch, proprioception, and vibrations from the spinal cord to the cerebral cortex.

We will now consider some of the important structural and functional aspects of the diencephalon.

The diencephalon

As indicated earlier, the **diencephalon** consists principally of the thalamus and hypothalamus. The relationship of these structures to the rest of the brain may be noted in Figure 14–1.

Thalamus

The **thalamus** is a large, oval structure located above the midbrain (Figure 14–5a). It consists of two masses of gray matter covered by a thin layer of white matter. It measures about 3 cm (1 in.) in length and constitutes four-fifths of the diencephalon. The thalamus contains numerous nuclei that are organized into masses (Figure 14–5b). The nuclei in the thalamus serve as relay stations for all sensory impulses, except smell, to the cerebral cortex. These include the *medial geniculate nuclei* (hearing), the *lateral geniculate nuclei* (vision), the *ventral posterior nuclei* (general sensations and taste), the *ventral lateral nuclei* (voluntary motor actions), and *ventral anterior nuclei* (voluntary motor actions and arousal). The thalamus is the principal relay station for sensory impulses that reach the cerebral cortex from the spinal cord, brain stem, cerebellum, and parts of the cerebrum.

The thalamus also functions as an interpretation center. That is, some sensory signals that enter the thalamus are interpreted there. For example, at the thalamic level, one can have conscious recognition of pain and temperature and some awareness of crude touch and pressure. A final feature of the thalamus we can note at this point is that it also contains a *reticular nucleus* in its reticular formation and an *anterior nucleus* in the floor of the lateral ventricle.

Hypothalamus

The **hypothalamus** is a very small portion of the diencephalon that weighs only about 4 g ($^1/_7$ oz). Its relationship to other parts of the brain is shown in Figures 14–1 and 14–5a. The hypothalamus forms the floor and part of the lateral walls of the third ventricle. The hypothalamus is protected by the sphenoid bone and indirectly by the sella turcica. Despite its relatively small size, nuclei in the hypothalamus control many body activities, most of which are related to controlling homeostasis. A summary of the more important functions of the hypothalamus is listed below.

1. The hypothalamus controls and integrates the autonomic nervous system, which stimulates smooth muscle, regulates the rate of contraction of cardiac muscle, and controls the secretions of many of the body's glands. Through the autonomic system, the hypothalamus is the chief regulator of visceral activities. For instance, it regulates the heartbeat, the movement of food through the digestive tract, and contraction of the urinary bladder.
2. The hypothalamus is involved in the reception and interpretation of sensory impulses from the viscera.
3. The hypothalamus is the principal intermediary between the nervous system and endocrine system – the two great control systems of the body. The hypothalamus lies just above the pituitary,

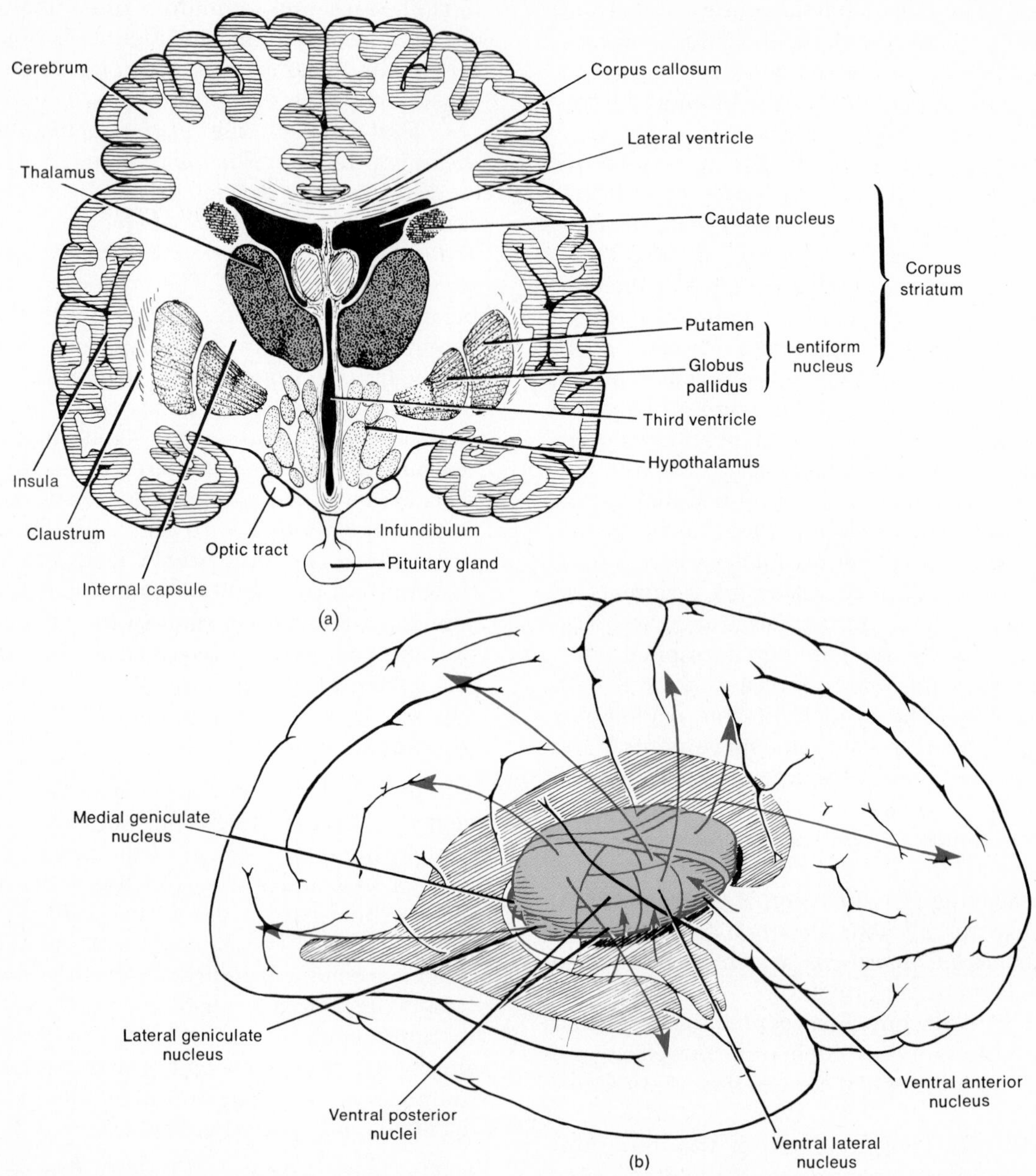

Figure 14-5. The thalamus. (a) Frontal section showing the thalamus and associated structures. (b) Right lateral view of the thalamic nuclei.

which is the major endocrine gland. When the hypothalamus detects certain changes in the body, it releases chemicals that stimulate the anterior pituitary gland. The anterior pituitary then releases hormones that determine the fate of carbohydrates, fats, and proteins, regulate the concentrations of some ions, and stimulate the sex organs. The hypothalamus also produces two hormones, which will be described in Chapter 17.

4. The hypothalamus is the center for the mind-over-body phenomenon. And, believe it or not, the mind can have a great deal of influence over how the body functions. When the cerebral cortex interprets strong emotions, it often sends impulses over tracts that connect the cortex with the hypothalamus. The hypothalamus then sends impulses over the autonomic system and releases chemicals that stimulate the anterior pituitary gland. The result can be a wide range of changes in body activities. For instance, when you panic, impulses go out from the hypothalamus to stimulate your heart to beat faster. Likewise, continued psychological stress can produce long-term abnormalities in body function that can make a person quite ill. These are the so-called psychosomatic disorders. Psychosomatic disorders are real and not imaginary.
5. The hypothalamus may be the area of the brain that feels rage and aggression.

6. It controls normal body temperature. Certain cells of the hypothalamus serve as a thermostat —a mechanism sensitive to changes in temperature. If blood flowing through the hypothalamus is above normal temperature, the hypothalamus sends impulses over the autonomic system to stimulate activities that promote heat loss. Heat can be lost through relaxation of the smooth muscle in the blood vessels and by sweating. Conversely, if the temperature of the blood is below normal, the hypothalamus sends out impulses that promote heat retention by the body. Heat can be retained through the contraction of cutaneous blood vessels, cessation of sweating, and shivering.
7. The hypothalamus regulates the amount of food intake through two centers. The *feeding center* is stimulated by hunger sensations from an empty stomach. When enough food has been ingested, the *satiety center* becomes stimulated and sends out impulses that inhibit the feeding center.
8. The hypothalamus also contains a *thirst center*. Certain cells in the hypothalamus become stimulated when the water content of the blood is low. The stimulated cells produce the sensation of thirst in the hypothalamus.
9. The hypothalamus serves as one of the centers that maintains the waking state and sleep patterns.

We will now turn our attention to the third principal part of the brain, the cerebellum.

The cerebellum

The **cerebellum** is the second largest portion of the brain and occupies the inferior and posterior aspects of the cranial cavity. Specifically, it is below the posterior portion of the cerebrum and is separated from it by the *transverse fissure*. (See Figure 14–1.) The cerebellum is also separated from the cerebrum by an extension of the cranial dura mater called the *tentorium cerebelli*. The cerebellum is shaped somewhat like a butterfly. The central constricted area is called the *vermis*, which means worm-shaped, and the lateral "wings" are referred to as *hemispheres* (Figure 14–6). Between the hemispheres is another extension of the cranial dura mater, the *falx cerebelli*.

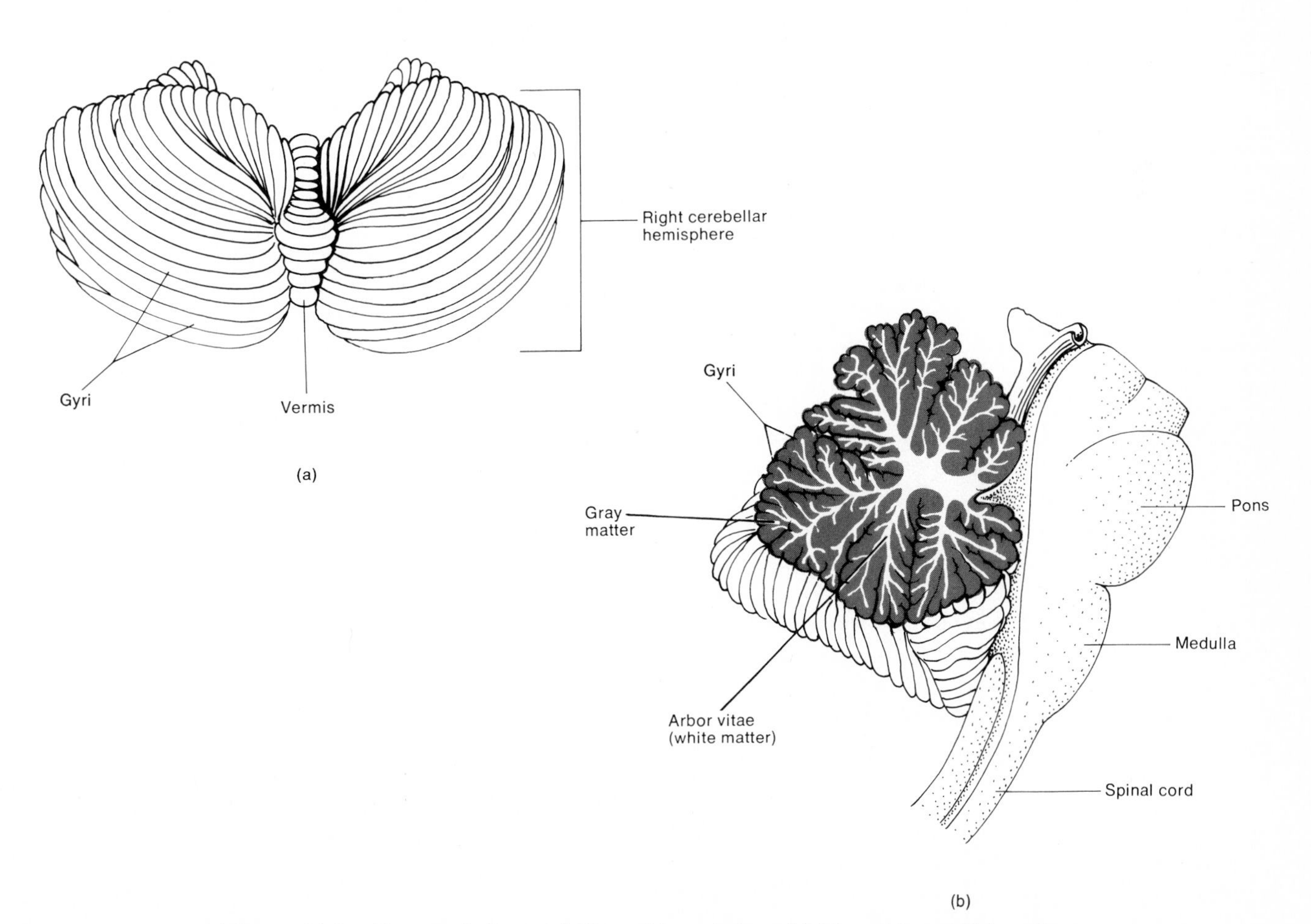

Figure 14-6. The cerebellum. (a) Viewed from below. (b) Viewed in sagittal section.

The surface of the cerebellum, called the *cortex*, consists of gray matter that is thrown into a series of slender, parallel ridges called *gyri*. These gyri are less prominent than those to be seen on the cerebral cortex. Beneath the gray matter are white matter tracts *(arbor vitae)* that resemble branches of a tree. Deep within the white matter are masses of gray matter, the *cerebellar nuclei*.

The cerebellum is attached to the brain stem by three paired bundles of fibers called *cerebellar peduncles*. These are as follows:

1. *Inferior cerebellar peduncles*, which connect the cerebellum with the medulla at the base of the brain stem and with the spinal cord
2. *Middle cerebellar peduncles*, which connect the cerebellum with the pons
3. *Superior cerebellar peduncles*, which connect the cerebellum with the midbrain

The cerebellum is a motor area of the brain that produces certain unconscious movements in the skeletal muscles. These movements are required (1) for coordination, (2) for maintenance of posture, and (3) for keeping the body balanced. The cerebellar peduncles are the fiber tracts that allow the cerebellum to carry out its functions.

Let us now see how the cerebellum produces coordinated movement. Motor areas of the cerebral cortex voluntarily initiate muscle contraction. Once the movement has begun, the sensory areas of the cortex receive impulses from nerves in the joints. The impulses provide information about the extent of muscle contraction and the amount of joint movement. The term *proprioception* is applied to this sense of the position of one body part relative to another. The cerebral cortex uses the proprioceptive sensations to decide what muscles need to contract next and how much contraction is needed in order to continue moving in the desired direction. The cerebral cortex sends its decision over tracts to the pons and midbrain, which relay the impulses over the middle and superior cerebellar peduncles to the cerebellum. The cerebellum is now ready to act. It sends subconscious motor impulses over the inferior cerebellar peduncles to the medulla and spinal cord. The impulses pass out of the spinal cord and over nerves that stimulate the prime movers and synergists to contract and that inhibit the contraction of the antagonists. The result is smooth, coordinated movement. A well-functioning cerebellum is particularly essential for skilled, delicate movements such as playing the piano or threading a needle.

The cerebellum also transmits impulses that control postural muscles. In other words, the cerebellum is required for maintaining normal muscle tone. Finally, the cerebellum maintains body equilibrium. The inner ear contains structures that sense balance. Information such as whether the body is leaning to the left or to the right or about to tip over frontward is probably transmitted from the inner ear to the cerebellum. The cerebellum then discharges impulses that bring about the contraction of the muscles necessary for maintaining equilibrium.

Damage to the cerebellum through trauma or disease is characterized by certain symptoms involving skeletal muscles. For example, there may be muscle uncoordination, called *ataxia*. A sign of ataxia is the inability to accurately touch an indicated point on the body without looking. For instance, blindfolded people with ataxia cannot touch the tip of their nose with a finger because they cannot coordinate movement with their sense of where a body part is located. Another sign of ataxia is a change in the speech pattern due to incoordination of speech muscles. Cerebellar damage may also result in disturbances of gait in which the subject staggers or cannot coordinate normal walking movements.

Within the gray matter of the cerebellum are four pairs of cerebellar nuclei. Fibers from the nuclei are connected to the midbrain via the superior cerebellar peduncles, to the pons through the middle cerebellar peduncles, and to the lower brain stem and spinal cord by the inferior cerebellar peduncles. One of the more prominent pairs of nuclei is the *dentate nuclei*. These nuclei are connected to the cerebral cortex by tracts so that motor impulses between the cerebral cortex and cerebellum may influence each other.

The final portion of the brain we will consider is the cerebrum.

The cerebrum

Supported on the brain stem and forming the bulk of the brain is the **cerebrum** (Figure 14–1). The surface of the cerebrum is composed of gray matter [2 to 4 mm (0.08 to 0.16 in.) thick] and is referred to as the *cerebral cortex*. The word *cortex* means rind or bark. The cortex, containing millions and millions of cells, consists of six layers of nerve cell bodies. Underneath the cortex lies the cerebral white matter.

During embryonic development when there is a rapid increase in brain size, the gray matter of the cortex enlarges out of all proportion to the underlying white matter. As a result, the cortical region rolls and folds upon itself. The upfolds are called *gyri* or *convolutions* (Figure 14–7a). The deep downfolds are referred to as *fissures*, whereas the shallow

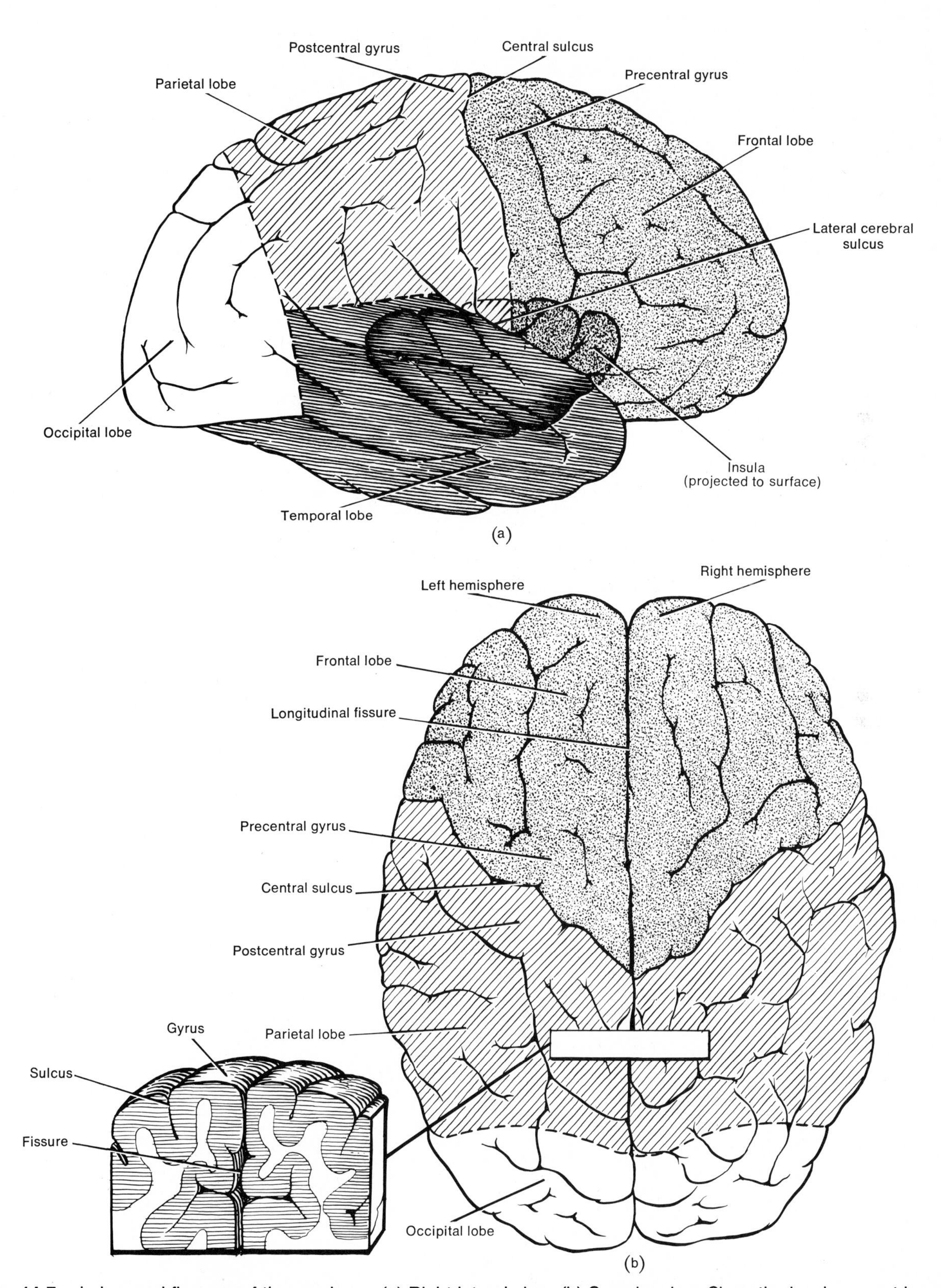

Figure 14-7. Lobes and fissures of the cerebrum. (a) Right lateral view. (b) Superior view. Since the insula cannot be seen externally, it has been projected to the surface. It can be seen in Figure 14-5a. The insert in (b) indicates the relative differences between a gyrus, sulcus, and fissure.

downfolds are termed *sulci.* The most prominent fissure, the *longitudinal fissure,* almost completely separates the cerebrum into right and left halves, or *hemispheres* (Figure 14–7b). The hemispheres, however, are connected internally by a large bundle of transverse fibers composed of white matter and called the *corpus callosum* (Figure 14–1). Between the hemispheres is an extension of the cranial dura mater called the *falx cerebri.*

Lobes

Each cerebral hemisphere is further subdivided into four lobes by other sulci or fissures (Figure 14–7). The *central sulcus,* or *fissure of Rolando,* separates the *frontal lobe* from the *parietal lobe.* A very important gyrus, the precentral gyrus, is located immediately in front of the central sulcus. The *lateral cerebral sulcus,* or *fissure of Sylvius,* separates the *frontal lobe* from the *temporal lobe.* The *parietooccipital sulcus* separates the *parietal lobe* from the *occipital lobe.* Another prominent fissure, the *transverse fissure,* separates the cerebrum from the cerebellum. The frontal lobe, parietal lobe, temporal lobe, and occipital lobe are named after the bones that cover them. A fifth part of the cerebrum, the *insula (island of Reil),* lies deep within the lateral cerebral fissure, under the parietal, frontal, and temporal lobes. It cannot be seen in an external view of the brain.

White matter

The white matter underlying the cortex consists of myelinated axons running in three principal directions:

1. *Association fibers* connect and transmit impulses between gyri within the same hemisphere.
2. *Commissural fibers* transmit impulses from the gyri in one hemisphere to the gyri in the other hemisphere; the names of three important groups of commissural fibers include the *corpus collosum, anterior commissure,* and *posterior commissure.*
3. *Projection fibers* form ascending and descending tracts that transmit impulses from the cerebrum to other parts of the brain and spinal cord.

Basal ganglia

The **basal ganglia** or **cerebral nuclei** are paired masses of gray matter within each cerebral hemisphere (Figure 14–8). The largest of the basal ganglia of each hemisphere is the *corpus striatum.*

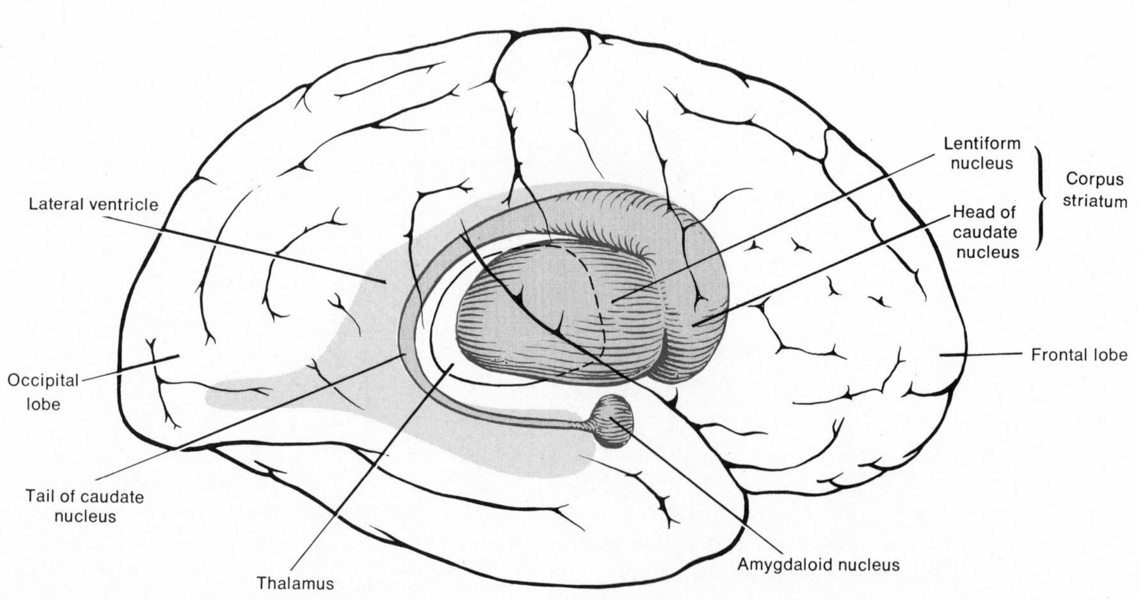

Figure 14-8. Basal ganglia. In this right lateral view of the cerebrum, the basal ganglia have been projected to the surface. Refer to Figure 14-5a to note the positions of the basal ganglia in the frontal section of the cerebrum.

It consists of the *caudate nucleus* and the *lentiform nucleus.* The lentiform nucleus, in turn, is subdivided into a lateral portion called the *putamen* and a medial portion called the *globus pallidus.* If you examine the coronal section of the brain in Figure 14–5a, you can see the two divisions of the lentiform nucleus and a structure called the *internal capsule.* It is a white matter group of sensory and motor projection tracts that connect the cerebral cortex with the brain stem and spinal cord. The portion of the internal capsule passing between the lentiform nucleus and the caudate nucleus and between the lentiform nucleus and thalamus is sometimes included as part of the corpus striatum.

Other structures frequently considered part of the basal ganglia are the claustrum and amygdaloid nucleus. The *claustrum* is a thin sheet of gray matter lateral to the putamen. The *amygdaloid nucleus* is located at the end of the tail of the caudate nucleus. Some authorities also consider the *substantia nigra,* the *subthalamic nucleus,* and the *red nucleus* to be part of the basal ganglia. The substantia nigra is a large motor nucleus in the midbrain, and its role in Parkinson's disease will be described later. The subthalamic nucleus lies against the internal capsule. Its major connection is with the globus pallidus. A lesion in the nucleus results in a motor disturbance on the opposite side of the body called *hemiballismus,* which is characterized by involuntary movements coming on suddenly with great force and rapidity. The movements are purposeless and generally of the withdrawal type, although they may be jerky. The spontaneous movements affect the proximal portions of the extremities most severely, especially the arms.

The basal ganglia are interconnected by many fibers. They are also connected to the cerebral cortex, thalamus, and hypothalamus. The caudate nucleus and the putamen control large unconscious movements of the skeletal muscles. An example of this is swinging the arms while walking. Such gross movements are also consciously controlled by the cerebral cortex. The globus pallidus is concerned with the regulation of muscle tone required for specific body movements. For example, if you wished to perform a very specific function with one of your hands, you might first position your body appropriately and then tense the muscles of the upper arm. Damage to the nuclei results in abnormal body movements, such as uncontrollable shaking, called tremor, and involuntary movements of skeletal muscle. Moreover, destruction of a substantial portion of the caudate nucleus almost totally paralyzes the opposite part of the body. The caudate nucleus is an area often affected by stroke.

The limbic system

Certain components of the cerebral hemispheres and diencephalon constitute the **limbic system.** It includes the following regions of gray matter:

1. *Limbic lobe.* This is formed by two gyri of the cerebral hemisphere, the cingulate gyrus and the hippocampal gyrus.
2. *Hippocampus.* This is an extension of the hippocampal gyrus that extends into the floor of the lateral ventricle.
3. *Amygdaloid nucleus.* This is located at the end of the tail of the caudate nucleus.
4. *Hypothalamus.* The regions of the hypothalamus that form part of the limbic system are the perifornical nuclei.
5. *Anterior nucleus of the thalamus.* This nucleus is located in the floor of the lateral ventricle.

In general terms, the limbic system functions in (1) emotional aspects of behavior related to survival, together with visceral responses accompanying these emotions, and (2) memory. Although behavior is a function of the entire nervous system, the limbic system controls most of the involuntary aspects of behavior. Experiments conducted on the limbic system of monkeys and other animals provide some indication of its function. For example, stimulation of the amygdaloid nucleus transmits impulses through the hypothalamus that alter blood pressure, heart rate, digestive activities, urination, defecation, movements of the pupil, and secretion of various hormones by the anterior pituitary gland. Stimulation of the amygdaloid nucleus also sends signals to the lower brain stem resulting in changes in muscle tone, posture, and movements related to eating (licking, chewing, and swallowing). Stimulation of other regions of the amygdaloid nucleus results in sexual excitement, such as erection, copulation, ejaculation, ovulation, and premature labor. From such experiments, it is concluded that the amygdaloid nucleus assumes a major role in controlling the overall pattern of behavior.

Other experiments have shown that the limbic system is associated with pleasure and pain. When certain areas of the limbic system of the hypothalamus, thalamus, and midbrain are stimulated, animals' reactions indicate they are experiencing intense punishment, as though they are feeling severe pain. When other areas are stimulated, the animals' reactions indicate they are experiencing extreme pleasure. In still other studies, stimulation of the perifornical nuclei of the hypothalamus result in a behavioral pattern called *rage.* This pattern consists of the animal developing a defensive posture, extending its claws, raising its tail, hissing, spitting, growling, dilating its pupils, and opening its eyes wide. Stimulating other areas of the limbic system results in an opposite behavioral pattern that consists of docility, tameness, and affection.

We can conclude from the foregoing discussion that the limbic system assumes a primary function in emotions such as pain, pleasure, anger, rage, fear, sorrow, sexual feelings, docility, and affection. Because of these functions, the limbic system is sometimes called the *visceral* or *emotional brain.*

We will now turn our attention to the functional areas of the cerebral cortex to understand how the cortex carries out its many vital functions.

Functional areas of the cerebral cortex

The functions of the cerebrum are numerous and complex. In a very general way, the cerebral cortex is divided into motor, sensory, and association areas. The **motor areas** are the regions that govern muscular movement. The **sensory areas** are concerned with the interpretation of sensory impulses. And the **association areas** are concerned with emotional and intellectual processes. We will first consider the sensory areas.

SENSORY AREAS. The *general sensory area* or *somesthetic area* is found directly behind the central sulcus of the cerebrum on the postcentral gyrus. It extends from the longitudinal fissure on the top of the cerebrum to the lateral cerebral sulcus. You will note in Figure 14–9a that the general sensory area is designated by the areas numbered 3, 1, and 2.* The general sensory area receives sensations from cutaneous, muscular, and visceral receptors in various parts of the body, and each specific point of the general sensory area receives sensations from specific parts of the body. Essentially the entire body is spatially represented in the general sensory area. The portion of the sensory area receiving stimuli from body parts is not dependent on the size of the part, but rather on the number of receptors. For example, a greater portion of the sensory area receives impulses from the lips than from the thorax. The major function of the general sensory area is to localize exactly the points of the body where the sensations originate. The thalamus is capable of localizing sensations in a general way. That is, the thalamus can recognize a sensation from a large area of the body such as the leg. But, it cannot distinguish the specific part of the leg from which the sensation originates. This must be performed by the general sensory area of the cortex.

* These numbers, as well as most of the others shown, are based upon K. Brodmann's cytoarchitectural map of the cerebral cortex. His map, first published in 1909, is an attempt to establish a basis for structural and functional correlations.

Posterior to the general sensory area is the *somesthetic association area.* It corresponds to the areas numbered 5 and 7 in Figure 14–9a. The somesthetic association area receives input from the thalamus, other lower portions of the brain, and the general sensory area. The function of the somesthetic association area is to integrate and interpret sensations. For example, this area permits you to determine the exact shape and texture of an object without looking at it. It also enables you to determine the orientation of one object to another as they are felt. The somesthetic association area also permits you to sense the relationship of one part of the body to another. Another role assumed by the somesthetic association area is the storage of memories of past sensory experiences. Thus, you can compare sensations with previous experiences.

Other sensory areas of the cortex include the following:

1. *Primary visual area* (area 17). This is located on the medial surface of the occipital lobe. It receives sensory impulses from the eyes. It interprets basic visual sensations such as the shape and color of an object.
2. *Visual association area* (areas 18 and 19). This region is located in the occipital lobe and receives sensory signals from the primary visual area and the thalamus. It functions in more complex aspects of vision. For example, it relates present to past visual experiences with recognition of what is seen and appreciation of its significance.
3. *Primary auditory area* (areas 41 and 42). This area is found in the upper part of the temporal lobe near the lateral cerebral sulcus. It interprets the basic characteristics of sound such as pitch and rhythm.
4. *Auditory association area* (area 22). This area is inferior to the primary auditory area. One of its functions is to interpret the meaning of sound, that is, to determine if the sound is speech, music, or a noise. Another of its functions is to interpret the meaning of speech by translating words into thoughts.
5. *Primary gustatory area* (area 43). This area is found at the base of the postcentral gyrus above the lateral cerebral sulcus. Its role is to interpret sensations related to taste.
6. *Primary olfactory area.* This area is in the temporal lobe and interprets sensations related to smell.

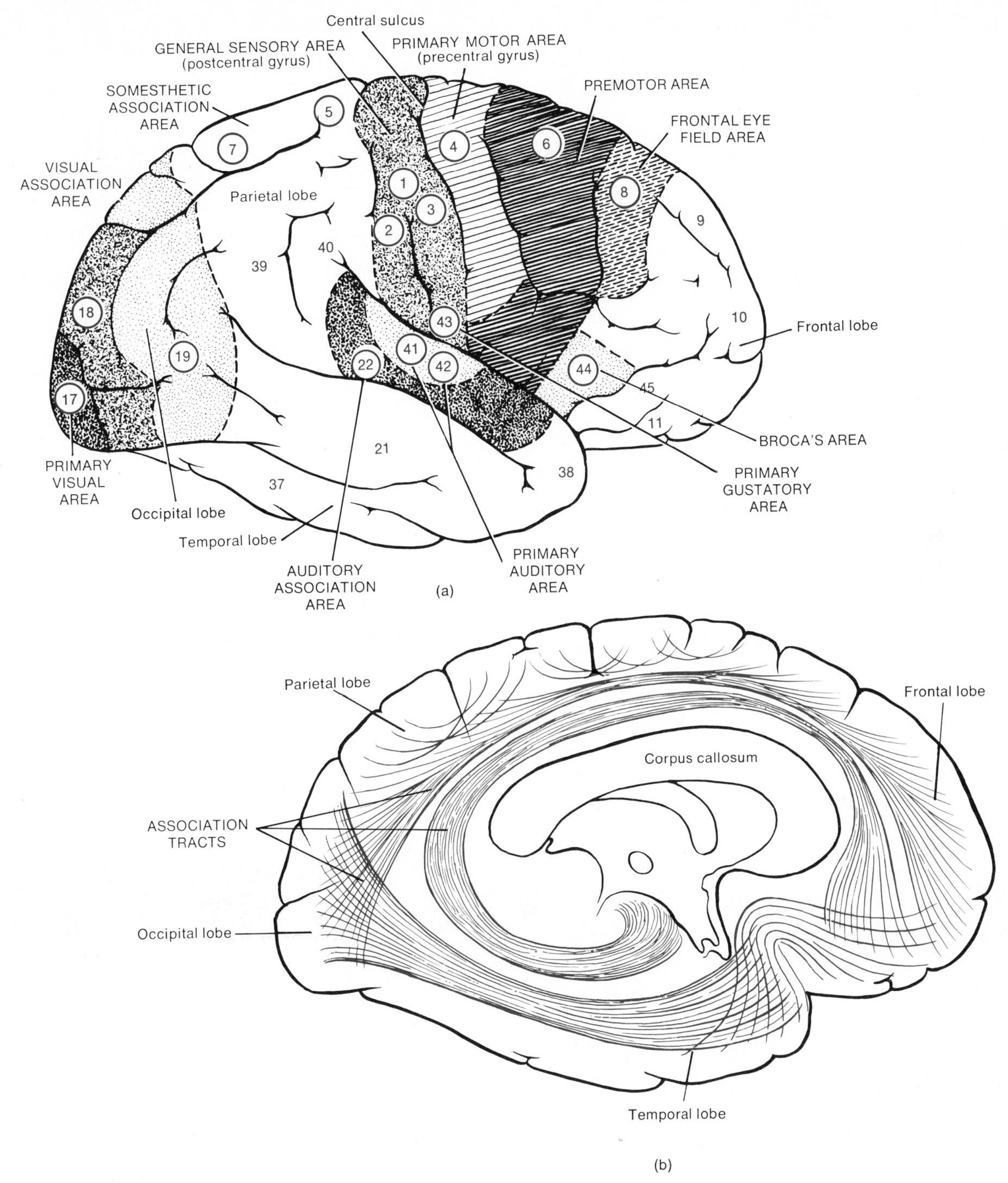

Figure 14-9. Functional areas of the cerebrum. (a) The lateral view indicates the sensory and motor areas. Although the right hemisphere is illustrated, Broca's area is in the left hemisphere of most people. (b) The sagittal section shows the association tracts.

7. *Gnostic area.* This is also called the *common integrative area* and is located between the somesthetic, visual, and auditory association areas. The gnostic area receives impulses from these areas, as well as from the taste and smell areas, the thalamus, and lower portions of the brain stem. The function of the gnostic area is to integrate all thoughts from the different sensory areas so that a common thought can be drawn from the various sensory inputs. Once this is done, it sends signals to the other parts of the brain to cause the appropriate response to the sensory signal.

Let us now turn attention to the motor areas of the cerebral cortex.

MOTOR AREAS. The *primary motor area* is located mostly in the precentral gyrus of the frontal lobe (Figure 14–9a). This region is also designated as area number 4. Like the general sensory area, the primary motor area consists of specific points that control specific muscles or groups of muscles. Stimulation of a specific point of the primary motor area results in a muscular contraction, usually on the opposite side of the body.

The *premotor area* (area 6) is in front of the motor area. It is concerned with learned motor activities of a complex and sequential nature. It sends out impulses that cause a specific group of muscles to contract in a specific sequence. A good example of this is writing. Another is playing a musical instrument. The premotor area, in other words, controls skilled movements.

The *frontal eye field area,* or area 8, is sometimes included in the premotor area. This area controls voluntary scanning movements of the eyes, like those used when looking up a word in a dictionary or in an index of a textbook.

The *language areas* are also important parts of the motor cortex. The use of language for communication is a major characteristic that sets humans apart from other organisms. The ability to speak depends on how you interpret speech and translate thoughts into speech. You communicate with other humans by speech or written words. When you listen to someone speaking, sounds are relayed to the primary auditory area of the cortex. The sounds are then interpreted as words in the auditory association area. The words are interpreted as thoughts in the gnostic area. In a similar manner, written words are interpreted by the visual association area and converted into thoughts by the gnostic area. Thus, you can translate speech or written words into thoughts.

The translation of thoughts into speech involves *Broca's area* or the *motor speech area,* designated as area 44. From this area, a sequence of signals is sent to the premotor area regions that control the muscles of the larynx, throat, and mouth. The impulses from the premotor area to the muscles result in specific, coordinated contractions that enable you to speak. Simultaneously, impulses are sent from Broca's area to the primary motor area. From here, impulses reach your breathing muscles to regulate the proper flow of air past the vocal cords. The coordinated contractions of your speech and breathing muscles enable you to translate your thoughts into speech.

Broca's area is usually located in the left cerebral hemisphere of most individuals, regardless of whether they are left-handed or right-handed. Injury to the sensory or motor speech areas results in *aphasia,* which is an inability to speak; *agraphia,* an inability to write; *word deafness,* an inability to understand spoken words; or *word blindness,* an inability to understand written words.

ASSOCIATION AREAS. The *association areas* of the cortex are made up of association tracts that connect motor and sensory areas (see Figure 14–9b). The association region of the cortex occupies the greater portion of the lateral surfaces of the occipital, parietal, and temporal lobes and the frontal lobes in front of the motor areas. The association areas are concerned with memory, emotions, reasoning, will, judgment, personality traits, and intelligence.

Brain waves: electroencephalogram

Brain cells have the capacity to generate electrical activity as a result of literally millions of action potentials of individual neurons. These electrical potentials are called **brain waves** and indicate activity of the cerebral cortex. Brain waves pass easily through the skull, and they can be detected by sensors called electrodes. A record of such waves is called an **electroencephalogram,** or **EEG.** An EEG is obtained by placing electrodes on the head of an individual and amplifying the waves by using an instrument called an electroencephalograph. As indicated in Figure 14–10, four kinds of waves are produced by normal individuals:

1. *Alpha waves.* These rhythmic waves occur at about a frequency of 10 to 12 cycles/second. They are found in the EEGs of almost all normal individuals when they are awake and in the resting state. These waves disappear entirely during sleep.
2. *Beta waves.* The frequency of these waves is between 15 to 60 cycles/second. Beta waves generally appear when the nervous system is active, that is, during periods of sensory input and mental activity.
3. *Theta waves.* These waves have frequencies of 5 to 8 cycles/second. Theta waves are normal in children and also occur in the EEGs of adults undergoing a great deal of emotional stress.
4. *Delta waves.* The frequency of these waves is between 1 to 5 cycles/second. Delta waves appear during sleep. They are normal in an awake infant. But, when they are produced by an awake adult, they indicate certain types of brain damage.

Distinct EEG patterns appear in certain abnormalities. In fact, the EEG is used clinically in the diagnosis of epilepsy, infectious diseases, tumors,

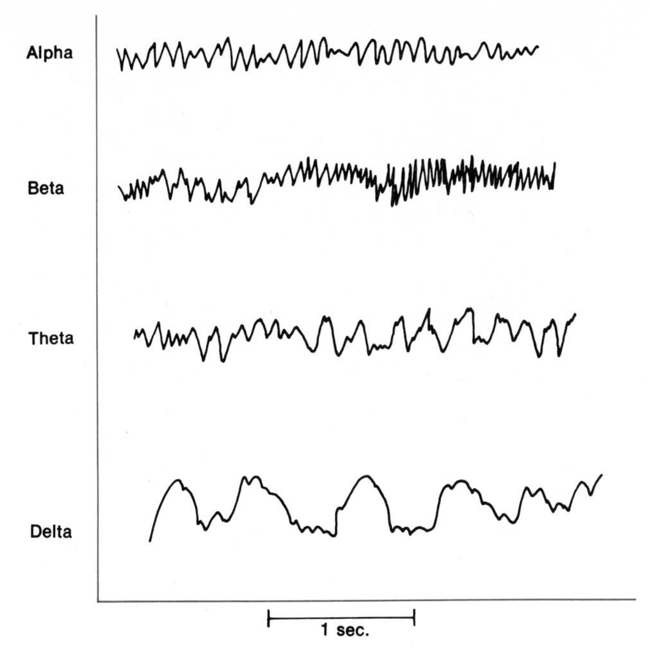

Figure 14-10. Kinds of waves recorded in an electroencephalograph.

trauma, and hematomas. Electroencephalograms are also used to provide information regarding sleep and wakefulness.

Just as we noted the relationship of spinal nerves to the spinal cord in the last chapter, we will now examine how the 12 pairs of cranial nerves are related to the brain.

CRANIAL NERVES

Of the 12 pairs of **cranial nerves,** 10 pairs originate from the brain stem, but all 12 pairs leave the skull through foramina in the base of the skull.

The cranial nerves are designated in two ways, with Roman numerals and with names. The Roman numerals indicate the order in which the nerves arise from the brain (front to back), whereas the names indicate the distribution or function of the nerves. Some of the cranial nerves are referred to as *mixed nerves.* That is, they contain both sensory and motor fibers. Other cranial nerves contain sensory fibers only. The cell bodies of the sensory fibers are located in ganglia outside the brain. The cell bodies of motor fibers lie in nuclei within the brain.

As we discuss the cranial nerves, you will notice that some of the motor fibers control subconscious movements, whereas the somatic nervous system has been defined as a conscious system. The reason for this seeming contradiction is that some of the fibers of the autonomic system leave the brain bundled together with the somatic fibers of the cranial nerves. Damage to a cranial nerve can easily include damage to the autonomic fibers that travel with it. Therefore, subconscious functions transmitted by the autonomic fibers are described along with the conscious functions of the somatic fibers of the cranial nerves.

Although the discussion of cranial nerves that follows is presented in the singular, be aware that the cranial nerves are paired structures.

I. Olfactory

The **olfactory nerve** is entirely sensory and conveys impulses related to smell. It arises in the olfactory mucosa of the nasal cavity as bipolar neurons (see Figure 16–4). The dendrites and cell bodies of these neurons are generally limited to the mucosa covering the superior nasal conchae and the adjacent nasal septum. Axons from the neurons pass through the cribriform plate of the ethmoid bone and synapse with other olfactory neurons in the *olfactory bulb,* an extension of the brain lying above the cribriform plate. The axons of these neurons make up the *olfactory tract.* The fibers from the tract terminate in the primary olfactory area in the cerebral cortex.

II. Optic

The **optic nerve** is also entirely sensory and conveys impulses related to vision. The dendrites and cell bodies arise in the rods and cones of the retina of the eye. Axons from the retina enter the optic foramina where the two optic nerves unite to form the *optic chiasma* (see Figure 16–11). In the optic chiasma, fibers from the medial half of each retina cross to the opposite side; those from the lateral half remain on the same side. From the optic chiasma, the fibers pass posteriorly to the *optic tracts.* From the optic tracts, the majority of fibers terminate in a nucleus of the thalamus. There, they synapse with neurons that pass to the visual areas of the cerebral cortex. Some of the fibers from the optic chiasma terminate in the superior colliculi of the midbrain. They synapse with neurons whose fibers terminate in the nuclei that convey impulses to the oculomotor (III), trochlear (IV), and abducent (VI) nerves, nerves that control the extrinsic and intrinsic muscles of the eyeball. Through this relay, there are widespread motor responses to light stimuli.

III. Oculomotor

The **oculomotor nerve** is a mixed cranial nerve. It originates from neurons in a nucleus in the ventral portion of the midbrain. It runs forward, divides into a superior and inferior branch, and passes through the superior orbital fissure in the orbit. The superior branch is distributed to the superior rectus (an extrinsic eyeball muscle) and the levator palpebrae superioris (the muscle of the upper eyelid). The inferior branch is distributed to the medial rectus, inferior rectus, and inferior oblique muscles, all extrinsic eyeball muscles. These distributions to the levator palpebrae superioris and extrinsic eyeball muscles constitute the motor portion of the oculomotor nerve. Through these distributions, impulses are sent that control movements of the eyeball and upper eyelid.

The inferior branch of the oculomotor nerve also gives off a branch to the *ciliary ganglion,* a relay station of the autonomic nervous system, that connects a nucleus in the midbrain with the intrinsic eyeball muscles. These intrinsic muscles include the ciliary muscle of the eyeball and the sphincter muscle of the iris. Through the ciliary ganglion, the oculomotor nerve controls the smooth muscle (ciliary muscle) that is responsible for accommodation of the lens for near vision and the smooth muscle (sphincter muscle of iris) that is responsible for constriction of the pupil of the eyeball.

The sensory portion of the oculomotor nerve consists of afferent fibers from proprioceptors in the eyeball muscles supplied by the nerve to the midbrain. These fibers convey impulses that are related to muscle sense (proprioception).

IV. Trochlear

The **trochlear nerve** is a mixed cranial nerve. It is the smallest of all cranial nerves. The motor portion originates in a nucleus in the midbrain, and axons from the nucleus pass through the superior orbital fissure of the orbit. The motor fibers innervate the superior oblique muscle of the eyeball, another extrinsic eyeball muscle. It controls movement of the eyeball.

The sensory portion of the trochlear nerve consists of afferent fibers that run from proprioceptors in the superior oblique muscle to the nucleus of the nerve in the midbrain. The sensory portion is responsible for muscle sense.

V. Trigeminal

The **trigeminal nerve** is a mixed cranial nerve and is the largest of the cranial nerves. As indicated by its name, the trigeminal nerve has three sensory branches: ophthalmic, maxillary, and mandibular. The trigeminal nerve contains two roots on the ventral lateral surface of the pons. The large sensory root has a swelling called the *semilunar (gasserian) ganglion* located in the petrous portion of the temporal bone. From this ganglion, the *ophthalmic branch* enters the orbit via the superior orbital fissure, the *maxillary branch* enters the foramen rotundum, and the *mandibular branch* pierces the foramen ovale. The smaller motor root originates in a nucleus in the pons. The motor fibers join the mandibular branch and supply the muscles of mastication. These motor fibers, which control chewing movements, constitute the motor portion of the trigeminal nerve.

The sensory portion of the trigeminal nerve delivers impulses related to touch, pain, and temperature and consists of the ophthalmic, maxillary, and mandibular branches. The ophthalmic branch receives sensory fibers from the skin over the upper eyelid, eyeball, lacrimal glands, the upper part of the nasal cavity, the side of the nose, forehead, and anterior half of the scalp. The maxillary branch receives sensory fibers from the mucosa of the nose, palate, parts of the pharynx, upper teeth, upper lip, cheek, and lower eyelid. The mandibular branch transmits sensory fibers from the anterior two-thirds of the tongue, lower teeth, skin over the mandible and side of the head in front of the ear, and the mucosa of the floor of the mouth. Sensory fibers from the three branches of the trigeminal nerve enter the semilunar ganglion and terminate in a nucleus in the pons.

VI. Abducent

The **abducent nerve** is a mixed cranial nerve. It originates from a nucleus in the pons. The motor fibers extend from the nucleus to the lateral rectus muscle of the eyeball, an extrinsic eyeball muscle. Impulses over the fibers bring about movement of the eyeball.

The sensory fibers run from proprioceptors in the lateral rectus muscle to the pons and mediate muscle sense. The abducent nerve reaches the lateral rectus muscle through the superior orbital fissure of the orbit.

VII. Facial

The **facial nerve** is a mixed nerve. Its motor fibers originate from a nucleus in the pons and enter the petrous portion of the temporal bone. The motor fibers are distributed to facial and scalp muscles. Impulses over these fibers cause contraction of the muscles of facial expression. Some motor fibers are also distributed to the sublingual and submandibular glands.

The sensory fibers extend from the taste buds of the anterior two-thirds of the tongue to the *genicu-*

late ganglion, a swelling on the facial nerve. From here, the fibers pass to a nucleus in the medulla. In the medulla, the fibers synapse with neurons that pass to the gustatory area of the cerebral cortex. The sensory portion of the facial nerve conveys sensations related to taste.

VIII. Vestibulocochlear

The **vestibulocochlear nerve** is a sensory cranial nerve. It consists of two branches: the cochlear (auditory) branch and the vestibular branch. The *cochlear branch,* which conveys impulses associated with hearing, arises in the spiral organ of Corti in the cochlea of the internal ear (see Figure 16–13c). The cell bodies of the cochlear branch are in the *spiral ganglion* in the cochlea. From here, the axons terminate in a nucleus in the lower pons. Ultimately, the fibers synapse with neurons that relay the impulses to the auditory areas of the cerebral cortex.

The *vestibular branch* arises in the semicircular canals in the inner ear. Fibers from the semicircular canals run to the *vestibular ganglion,* where the cell bodies are contained. The cell bodies of the fibers in the ganglion synapse with fibers that extend to a nucleus in the lower pons. Some of the fibers also enter the cerebellum. The vestibular branch transmits impulses related to equilibrium.

IX. Glossopharyngeal

The **glossopharyngeal nerve** is a mixed cranial nerve. The motor fibers of this nerve originate in a nucleus in the medulla. The nerve leaves the skull through the jugular foramen. The motor fibers are distributed to the swallowing muscles of the pharynx and the parotid gland to mediate swallowing movements and the secretion of saliva.

The sensory fibers of the glossopharyngeal nerve supply the pharynx and taste buds of the posterior one-third of the tongue. Some sensory fibers also come from receptors in the carotid sinus, which, as you will see later, assumes an important role in regulating blood pressure. The sensory fibers of the glossopharyngeal nerve terminate in a nucleus in the medulla.

X. Vagus

The **vagus nerve** is a mixed cranial nerve and is widely distributed from the head and neck into the thorax and abdomen. Its motor fibers originate in a nucleus in the medulla and terminate in the muscles of the pharynx, larynx, respiratory passageways, heart, esophagus, stomach, small intestine, most of the large intestine, and gallbladder. Impulses along the motor fibers bring about visceral, cardiac, and skeletal muscle movement.

Sensory fibers of the vagus nerve supply essentially the same structures as the motor fibers. They convey impulses for various sensations from the pharynx, larynx, and viscera. The fibers terminate in the medulla and pons.

XI. Accessory

The **accessory nerve** (formerly the spinal accessory nerve) is a mixed cranial nerve. It differs from all other cranial nerves in that it originates from both the brain stem and the spinal cord. The *bulbar (medullary) portion* originates from nuclei in the medulla, passes through the jugular foramen, and supplies the voluntary muscles of the pharynx, larynx, and soft palate that are used in swallowing. The *spinal portion* originates in the anterior gray horn of the first five segments of the cervical portion of the spinal cord. The fibers from the segments join, enter the foramen magnum, and pass out through the jugular foramen along with the bulbar portion. The spinal portion conveys motor impulses to the sternocleidomastoid and trapezius muscles so that movements of the head can be accomplished.

The sensory fibers run from proprioceptors in the muscles supplied by its motor neurons and end in upper cervical posterior root ganglia. They conduct impulses for proprioception.

XII. Hypoglossal

The **hypoglossal nerve** is a mixed nerve. The motor fibers originate in a nucleus in the medulla, pass through the hypoglossal canal, and supply the muscles of the tongue. These fibers conduct impulses related to speech and swallowing.

The sensory portion of the hypoglossal nerve consists of fibers from proprioceptors in the tongue muscles to the medulla. The sensory fibers conduct impulses for muscle sense.

A summary of the cranial nerves is presented in Exhibit 14–1. It contains a listing of the cranial nerves by number, name, type, location, and function and comments. The cranial nerves are illustrated in Figure 14–11.

We will now conclude our discussion of the nervous system by considering some of the pertinent structural features of the autonomic nervous system, as well as some of its functional aspects.

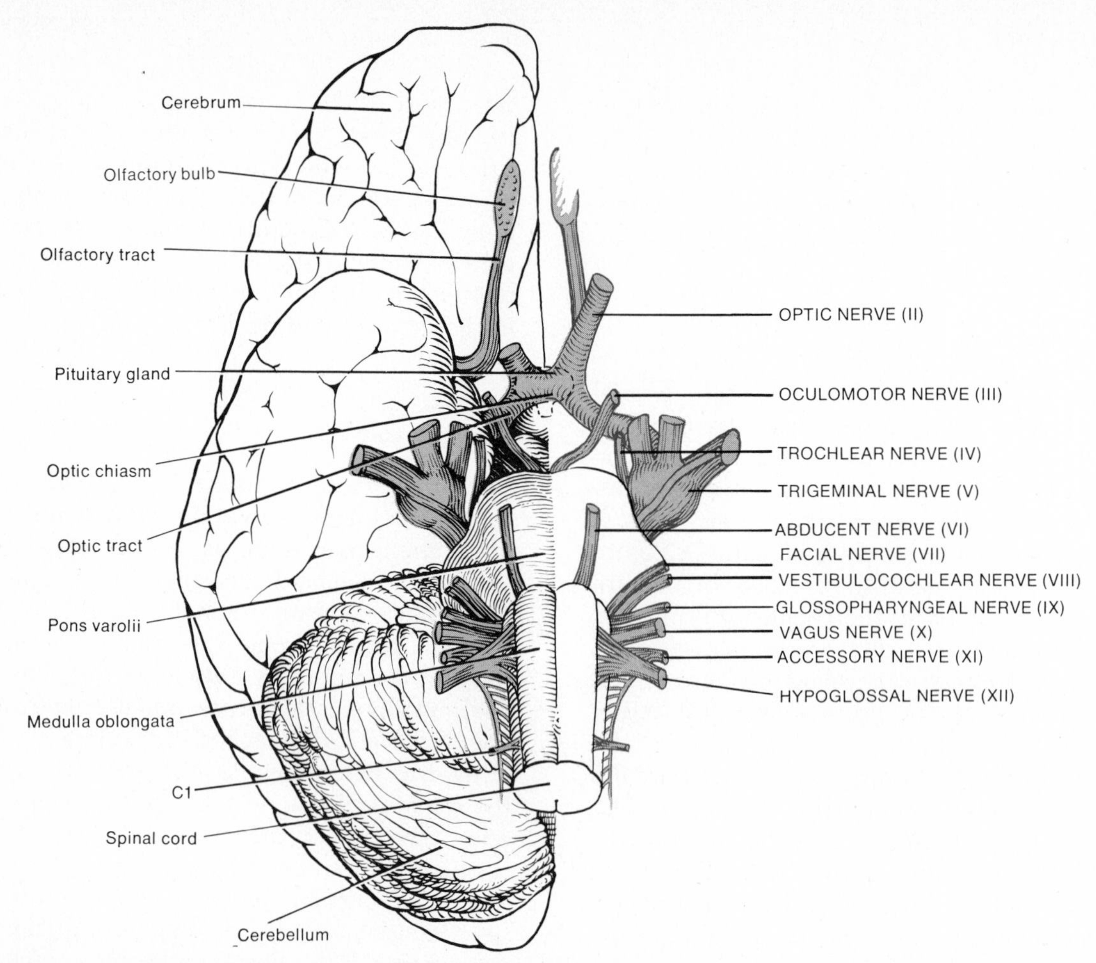

Figure 14-11. The cranial nerves. The olfactory nerve (I) cannot be seen in this view. It arises in the olfactory mucosa, and its fibers enter the olfactory bulb from below and are torn when the brain is removed from the skull.

Exhibit 14–1. SUMMARY OF CRANIAL NERVES (see Figure 14–11)

NUMBER, NAME, AND (TYPE)	LOCATION	FUNCTION AND COMMENTS
I. Olfactory (sensory)	Arises in olfactory mucosa, passes through olfactory bulb and olfactory tract, and terminates in the olfactory area of the cerebral cortex.	Smell. Fractures of the cribriform plate of the ethmoid may result in the loss of the sense of smell *(anosmia).*
II. Optic (sensory)	Arises in the retina of the eye, forms the optic chiasma, passes through the optic tracts, a nucleus in the thalamus, and terminates in the visual areas of the cerebral cortex.	Vision. Fractures and growths in the orbit and diseases of the nervous system may affect the optic nerve resulting in visual field defects and loss of visual acuity.
III. Oculomotor (mixed)	The motor portion originates in the midbrain and is distributed to the levator palpebrae superioris of the upper eyelid, four extrinsic eyeball muscles (superior rectus, medial rectus, inferior rectus, and inferior oblique), the ciliary muscle of the eyeball, and the sphincter muscle of the iris.	Movement of the eyelid and eyeball, accommodation of the lens for near vision, and constriction of the pupil.
	The sensory portion consists of afferent fibers from proprioceptors in the eyeball muscles supplied to the midbrain.	Muscle sense (proprioception).

NUMBER, NAME, AND (TYPE)	LOCATION	FUNCTION AND COMMENTS
III. Oculomotor (mixed) *continued*		Complete paralysis of the oculomotor, trochlear, or abducent nerves is indicated by squints *(strabismus)* and seeing two objects instead of one *(diplopia).* In oculomotor paralysis, the affected eye looks downward and outward, the upper eyelid droops *(ptosis),* the pupil is dilated and fixed, and there is a loss of accommodation for near vision.
IV. Trochlear (mixed)	The motor portion originates in the midbrain and is distributed to the superior oblique muscle.	Movement of the eyeball.
	The sensory portion consists of afferent fibers from proprioceptors in the superior oblique muscle to the midbrain.	Muscle sense (proprioception).
		In trochlear nerve paralysis, the individual may carry his head tilted to the affected side.
V. Trigeminal (mixed)	The motor portion originates in the pons and terminates in the muscles of mastication.	Chewing. Injury results in paralysis of the muscles of mastication.
	The sensory portion consists of three branches: *ophthalmic*—contains sensory fibers from upper eyelid, eyeball, nasal cavity, side of nose, forehead, and anterior half of scalp; *maxillary*—contains sensory fibers from palate, pharynx, upper teeth, upper lip, cheek, and lower eyelid; *mandibular*—contains sensory fibers from anterior two-thirds of tongue, lower teeth, skin over mandible and side of head in front of ear. The three branches terminate in the pons.	Conveys sensations for touch, pain, and temperature from the head, face, and pharynx. Injury results in a loss of sensation of touch and temperature.
		Injury to the motor portion results in paralysis of the muscles of mastication. Injury to a sensory division results in a loss of sensation of touch and temperature.
		Neuralgia (pain) of one or more branches of the trigeminal nerve is called *tic douloureux* or *trigeminal neuralgia.*
VI. Abducent (mixed)	The motor portion originates in the pons and is distributed to the lateral rectus muscle.	Movement of the eyeball.
	The sensory portion consists of afferent fibers from proprioceptors in the lateral rectus to the pons.	Muscle sense (proprioception).
		In abducent nerve paralysis, the affected eye looks inward.
VII. Facial (mixed)	The motor portion originates in the pons and is distributed to facial and scalp muscles and the sublingual and submandibular glands.	Facial expression and the secretion of saliva. Injury produces paralysis of the muscles of facial expression and reduced salivary secretion.
	The sensory portion arises from taste buds on the anterior two-thirds of the tongue, passes through the geniculate ganglion, a nucleus in the medulla, and terminates in the gustatory area of the cerebral cortex.	Taste. Injury produces loss of taste.

Exhibit 14–1. SUMMARY OF CRANIAL NERVES (cont.)

NUMBER, NAME, AND (TYPE)	LOCATION	FUNCTION AND COMMENTS
VIII. Vestibulocochlear (sensory)	Consists of a cochlear and a vestibular branch. The cochlear branch arises in the spiral organ of Corti, forms the spiral ganglion, passes through a nucleus in the pons, and terminates in the auditory areas of the cerebral cortex. The vestibular branch arises in the semicircular canals, forms the vestibular ganglion, and terminates in a nucleus in the pons.	The cochlear branch conveys impulses associated with hearing; the vestibular branch conveys impulses associated with equilibrium. Injury to the cochlear branch may cause ringing or buzzing sounds or deafness. The commonest symptoms of damage to the vestibular branch are giddiness and loss of balance.
IX. Glossopharyngeal (mixed)	The motor portion originates in the medulla and is distributed to the swallowing muscles of the pharynx and the parotid gland.	Swallowing movements and the secretion of saliva. Injury results in pain during swallowing and reduced secretion of saliva.
	The sensory portion arises from taste buds on the posterior one-third of the tongue, and from the carotid sinus and terminates in the medulla.	Taste and regulation of blood pressure. Injury results in loss of sensations in the throat and loss of taste from posterior part of tongue.
X. Vagus (mixed)	The motor portion originates in the medulla and terminates in the muscles of the pharynx, larynx, respiratory passageways, esophagus, heart, stomach, small intestine, most of large intestine, and gallbladder.	Visceral muscle movement.
	The sensory portion arises from the same structures supplied by the motor fibers and terminates in the medulla and pons.	Sensations from organs supplied.
		Severing of both nerves in the upper body interferes with swallowing, paralyzes vocal cords, and interrupts sensations from many organs. Injury to both nerves in the abdominal area has little effect since the abdominal organs are also supplied by autonomic fibers that arise from the spinal cord.
XI. Accessory (mixed)	The motor portion consists of a bulbar portion and a spinal portion. The bulbar portion originates from the medulla and supplies voluntary muscles of the pharynx, larynx, and soft palate. The spinal portion originates from the anterior gray horn of the first five cervical segments of the spinal cord and supplies the sternocleidomastoid and trapezius muscles.	The bulbar portion mediates swallowing movements; the spinal portion mediates movements of the head. Injury to the spinal portion results in an inability to turn the head or raise the shoulders.
	The sensory portion consists of afferent fibers from proprioceptors in the muscles supplied by the motor components.	Muscle sense (proprioception).
XII. Hypoglossal (mixed)	The motor portion originates in the medulla and supplies the muscles of the tongue.	Movement of the tongue during speech and swallowing.
	The sensory portion consists of fibers from proprioceptors in the tongue muscles that terminate in the medulla.	Muscle sense (proprioception).
		Injury results in difficulty in chewing, speaking, and swallowing. The tongue, when protruded, curls toward the affected side and the affected side becomes atrophied, shrunken, and deeply furrowed.

Chapter summary in outline

BRAIN

1. The brain consists of four principal areas: brain stem (medulla, pons, midbrain), diencephalon (thalamus and hypothalamus), cerebellum, and cerebrum.
2. The brain is protected by the cranial meninges and cerebrospinal fluid.
3. Cerebrospinal fluid is formed in the choroid plexuses and circulates through the subarachnoid space, ventricles, and central canal. Most of the fluid is absorbed by the arachnoid villi of the superior sagittal sinus.
4. The blood supply to the brain is via the circle of Willis. The blood continuously delivers oxygen and glucose to brain cells. The differential rates of passage of certain materials from the blood into the brain is based upon a blood-brain barrier.

Brain Stem

1. The medulla oblongata is continuous with the upper part of the spinal cord. Within it, the decussation of pyramids occurs. The medulla contains nuclei that are reflex centers for heartbeat, respiration, and vasoconstriction. The medulla also contains the nuclei of origin of cranial nerves IX to XII.
2. The pons varolii is superior to the medulla. It serves as a bridge between the spinal cord and other parts of the brain and contains the nuclei of origin of cranial nerves V to VIII.
3. The midbrain connects the pons and cerebellum with the cerebrum. It conveys motor impulses from the cerebrum to the cerebellum and spinal cord and conveys sensory impulses from the spinal cord to the thalamus. It also coordinates auditory and visual stimuli and contains the nuclei of origin of cranial nerves III and IV.

Diencephalon

1. Consists of the thalamus and hypothalamus.
2. The thalamus is superior to the midbrain and contains nuclei that serve as relay stations for all sensory impulses, except smell, to the cerebral cortex. The thalamus also registers conscious recognition of pain and temperature and some awareness of crude touch and pressure.
3. The hypothalamus is inferior to the thalamus. It controls the autonomic nervous systems, connects the nervous and endocrine system, controls body temperature, food and fluid intake, the waking state, and sleep.

Cerebellum

1. The cerebellum occupies the inferior and posterior aspects of the cranial cavity. It consists of two hemispheres and a central constricted vermis.
2. It is attached to the brain stem by three pairs of cerebellar peduncles.
3. The cerebellum functions in the coordination of skeletal muscles, the maintenance of posture, and keeping the body balanced.

Cerebrum

1. The cerebrum is the largest part of the brain. Its cortex contains convolutions, fissures, and sulci.
2. The cerebral lobes are named the frontal, parietal, temporal, and occipital.
3. The white matter is under the cortex and consists of myelinated axons running in three principal directions.
4. The basal ganglia are paired masses of gray matter in the cerebral hemispheres. They help to control muscular movements.
5. The limbic system is found in the cerebral hemispheres and diencephalon. It functions in emotional aspects of behavior and memory.
6. The motor areas of the cerebral cortex are the regions that govern muscular movement. The sensory areas are concerned with the interpretation of sensory impulses. The association areas are concerned with emotional and intellectual processes.
7. Brain waves generated by the cerebral cortex are recorded as an EEG. They may be used to diagnose epilepsy, infections, and tumors.

CRANIAL NERVES

1. The 12 pairs of cranial nerves leave the skull through foramina in the base of the skull. Some are sensory and some are mixed.
2. They are named primarily on the basis of distribution and are numbered on the basis of order of origin.
3. Pertinent data concerning the cranial nerves is summarized in Exhibit 14–1.

Review questions and problems

1. Identify the four principal parts of the brain and the components of each, where applicable.
2. Describe the location of the cranial meninges.
3. Where is cerebrospinal fluid formed? Describe its circulation. Where is cerebrospinal fluid absorbed?
4. Distinguish between internal and external hydrocephalus.
5. Describe the blood supply to the brain. Explain the importance of oxygen and glucose to brain cells.
6. What is the blood-brain barrier? Is it of any advantage?
7. Describe the location and structure of the medulla. Define decussation of pyramids. Why is it important?
8. List the principal functions of the medulla.
9. Describe the location and structure of the pons. What are its functions?
10. Describe the location and structure of the midbrain. What are some of its functions?
11. Describe the location and structure of the thalamus. List some of the functions of the thalamus.
12. Where is the hypothalamus located? Explain some of the major functions of the hypothalamus.
13. Describe the location of the cerebellum. List the principal parts of the cerebellum.
14. Describe the relationship of the dural extensions to the cerebellum.
15. What are cerebellar peduncles? List and explain the function of each.
16. Explain the functions of the cerebellum. What is ataxia?
17. Where is the cerebrum located? Describe the cortex, convolutions, fissures, and sulci of the cerebrum.
18. List and locate the lobes of the cerebrum. How are they separated from each other? What is the insula?
19. Describe the organization of cerebral white matter. Be sure to indicate the function of each group of fibers.

20. What are basal ganglia? Name the important basal ganglia and list the function of each.
21. Describe the effects of damage on the basal ganglia.
22. Define the limbic system. Explain several of its functions.
23. What is meant by a sensory area of the cerebral cortex? List, locate, and give the function of each sensory area.
24. What is meant by a motor area of the cerebral cortex? List, locate, and give the function of each motor area.
25. What is an association area of the cerebral cortex? What are its functions?
26. Define an electroencephalogram. List the types of waves recorded on an EEG and explain the importance of each. What is the diagnostic value of an EEG?
27. Define a cranial nerve. How are cranial nerves named and numbered? Distinguish between a mixed and a sensory cranial nerve.
28. For each of the 12 pairs of cranial nerves, list *(a)* its name, number, and type; *(b)* its location; and *(c)* its function. In addition, list the effects of damage, where applicable.

Selected readings

Barr, M. L. *The Human Nervous System.* New York: Harper and Row, 1972.

Cunningham's Textbook of Anatomy. 11th ed., edited by G. J. Romanes. London: Oxford University Press, 1972. Pp. 599–734.

Dawson, Helen L. *Basic Human Anatomy.* 2d ed. New York: Appleton-Century-Crofts, 1974. Pp. 185–191, 198–204.

Gray, Henry. *Anatomy of the Human Body.* 29th ed., edited by Charles Mayo Goss. Philadelphia: Lea and Febiger, 1973. Pp. 802–876.

Netter, Frank H. *Nervous System.* CIBA Collection of Medical Illustrations, Vol. 1. Summit, NJ: CIBA Pharmaceutical Co., 1958.

Truex, Raymond C., and Malcolm B. Carpenter. *Human Neuroanatomy.* 6th ed. Baltimore: Williams and Wilkins, 1969.

CHAPTER 15

THE AUTONOMIC NERVOUS SYSTEM

STUDENT OBJECTIVES

After you have read this chapter, you should be able to:

1. Define the autonomic nervous system
2. Compare the structural and functional differences between the somatic and autonomic nervous systems
3. Compare the sympathetic and parasympathetic divisions of the autonomic nervous system in terms of structure
4. Explain the classification of autonomic fibers on the basis of chemical transmitters released
5. Describe how the autonomic nervous system functions
6. Explain the mechanism of a visceral autonomic reflex and describe the components of a visceral autonomic reflex arc
7. Explain the clinical symptoms of disorders of the nervous system, including poliomyelitis, syphilis, cerebral palsy, Parkinsonism, epilepsy, multiple sclerosis, stroke, and tumors
8. Define key terms associated with the nervous system

The portion of the nervous system that regulates the activities of smooth muscle, cardiac muscle, and glands is the **autonomic nervous system.** Structurally, the system consists of visceral efferent neurons organized into nerves, ganglia, and plexuses. Functionally, it usually operates without any conscious control. Physiologists originally thought that the autonomic system functioned autonomously, without any control from the central nervous system—hence its name, the autonomic nervous system. In truth, however, the autonomic system is neither structurally nor functionally independent of the central nervous system. It is regulated by centers in the brain, in particular by the cerebral cortex, the hypothalamus, and the medulla oblongata. However, the autonomic nervous system does differ from the somatic nervous system in some ways. For convenience of study the two are separated.

COMPARISON OF SOMATIC AND AUTONOMIC SYSTEMS

Whereas the efferent portion of the somatic system produces conscious movement in skeletal muscles, the autonomic nervous system regulates visceral activities. And it generally does so involuntarily and automatically. Examples of visceral activities that are regulated by the autonomic nervous system are changes in the size of the pupil of the eye, accommodation for near vision, dilatation and constriction of blood vessels, adjustment of the rate and force of the heartbeat, emptying of the urinary bladder by contraction of its smooth muscle, movements of the gastrointestinal tract, formation of gooseflesh, and secretion by most glands. Note again that these activities usually lie beyond conscious control. They are automatic activities.

Also unlike the somatic system, the autonomic nervous system is entirely motor. All its axons are efferent fibers, which transmit impulses from the central nervous system to visceral effectors. Autonomic fibers are called **visceral efferent fibers. Visceral effectors** include cardiac muscle, smooth muscle, and glandular epithelium. It should not be inferred that there are no afferent (sensory) impulses from visceral effectors, however. Visceral sensations pass over visceral afferent fibers that structurally and functionally belong to the somatic nervous system. Some of the functions of these fibers were described with the cranial and spinal nerves. However, the hypothalamus, which largely controls the autonomic nervous system, also receives impulses from the visceral sensory fibers.

The somatic and autonomic systems also differ in the way they bring about effector relaxation. The autonomic system consists of two principal divisions: the **sympathetic** and the **parasympathetic.** Many organs innervated by the autonomic system receive visceral efferent neurons from both components of the autonomic system—one set from the sympathetic division, another from the parasympathetic division. In general, impulses transmitted by the fibers of one division stimulate the organ to start or increase activity, whereas impulses from the other division bring about a decrease or halt in the activity of the organ. In the somatic nervous system, only one kind of motor neuron innervates an organ, which is always a skeletal muscle. When the somatic neurons stimulate the cells of the muscle, the muscle becomes active. When the neuron ceases to stimulate the organ, contraction stops altogether. Skeletal muscle cells of each motor unit contract only when they are stimulated by their motor neuron. When the impulse stops, contraction stops.

STRUCTURE OF THE AUTONOMIC NERVOUS SYSTEM

The autonomic nervous system is divided into the **sympathetic** and **parasympathetic** divisions. These are also referred to as the **thoracolumbar** and **craniosacral** divisions, respectively. Let us see what this means by discussing the general features applicable to both divisions.

Visceral efferent neurons

Autonomic visceral efferent pathways always consist of two neurons. One runs from the central nervous system to a ganglion. The other runs directly from the ganglion to the effector.

The first of these visceral efferent neurons in an autonomic pathway is called a **preganglionic neuron** (Figure 15–1). Preganglionic neurons have their cell bodies in the brain or spinal cord. Their myelinated axons, called **preganglionic fibers,** pass out of the central nervous system as part of a cranial or spinal nerve. At some point, they leave their somatic nerves and run to autonomic ganglia where they synapse with the dendrites or cell bodies of postganglionic neurons.

Postganglionic neurons, the second visceral efferent neurons in an autonomic pathway, lie entirely outside the central nervous system (see Figure 15–1). Their cell bodies and dendrites (if they have dendrites) are located in the autonomic ganglia, where they synapse with the preganglionic fibers. The axons of postganglionic neurons are called **post-**

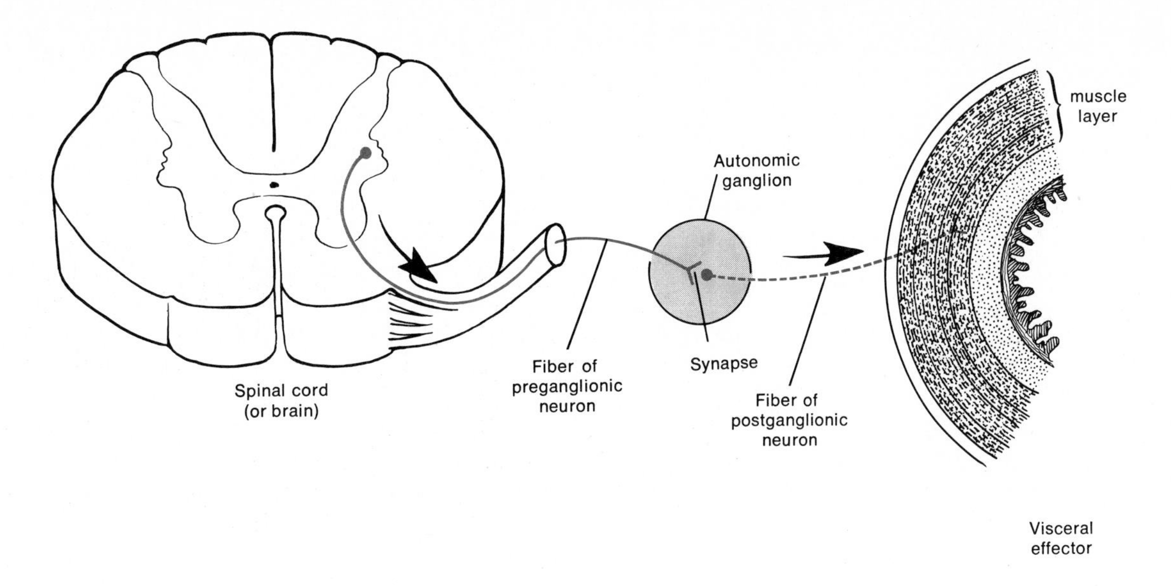

Figure 15-1. Relationship between preganglionic and postganglionic neurons of the autonomic nervous system.

ganglionic fibers. Postganglionic fibers are nonmyelinated, and they terminate in visceral effectors.

If we review this organization, then, it can be seen that preganglionic neurons convey efferent impulses from the central nervous system to autonomic ganglia. Postganglionic neurons relay the impulses from the autonomic ganglia to visceral effectors.

Preganglionic neurons

In the sympathetic division, the preganglionic neurons have their cell bodies in the lateral gray horns of the thoracic segment and first two lumbar segments of the spinal cord (Figure 15–2). It is for this reason that the sympathetic division is also called the **thoracolumbar division** and the fibers of the sympathetic preganglionic neurons are known as the **thoracolumbar outflow.**

The cell bodies of the preganglionic neurons of the parasympathetic division are located in nuclei in the brain stem and in the lateral gray horns of the second through fourth sacral segments of the spinal cord (Figure 15–2). Thus the synonymous term **craniosacral division.** The fibers of the parasympathetic preganglionic neurons are referred to as the **craniosacral outflow.**

Autonomic ganglia

Autonomic pathways also include **autonomic ganglia,** where synapses between visceral efferent neurons occur. Autonomic ganglia differ from posterior root ganglia in that the latter contain cell bodies of sensory neurons and no synapses occur within them. The autonomic ganglia may be divided into three general groups (Figure 15–3). The *sympathetic trunk* or *vertebral chain ganglia* are a series of ganglia that lie in a horizontal row on either side of the veretebral column, extending from the base of the skull to the coccyx. They receive preganglionic fibers only from the thoracolumbar (sympathetic) division (Figure 15–2).

A second kind of ganglion of the autonomic nervous system is called a *prevertebral* or *collateral ganglion* (Figure 15–3). The ganglia of this group lie anterior to the spinal column and close to the large abdominal arteries from which their names are derived. Examples of prevertebral ganglia so named are the celiac ganglion, on either side of the celiac artery just below the diaphragm; the superior mesenteric ganglion, near the beginning of the superior mesenteric artery in the upper abdomen; and the inferior mesenteric ganglion, located near the beginning of the inferior mesenteric artery in the middle of the abdomen (Figure 15–2). Prevertebral ganglia receive preganglionic fibers from the thoracolumbar (sympathetic) division.

The third kind of autonomic ganglion is referred to as a *terminal* or *intramural ganglion.* The ganglia of this group are located at the end of a visceral efferent pathway very close to visceral effectors or within the walls of visceral effectors. Terminal ganglia receive preganglionic fibers from the

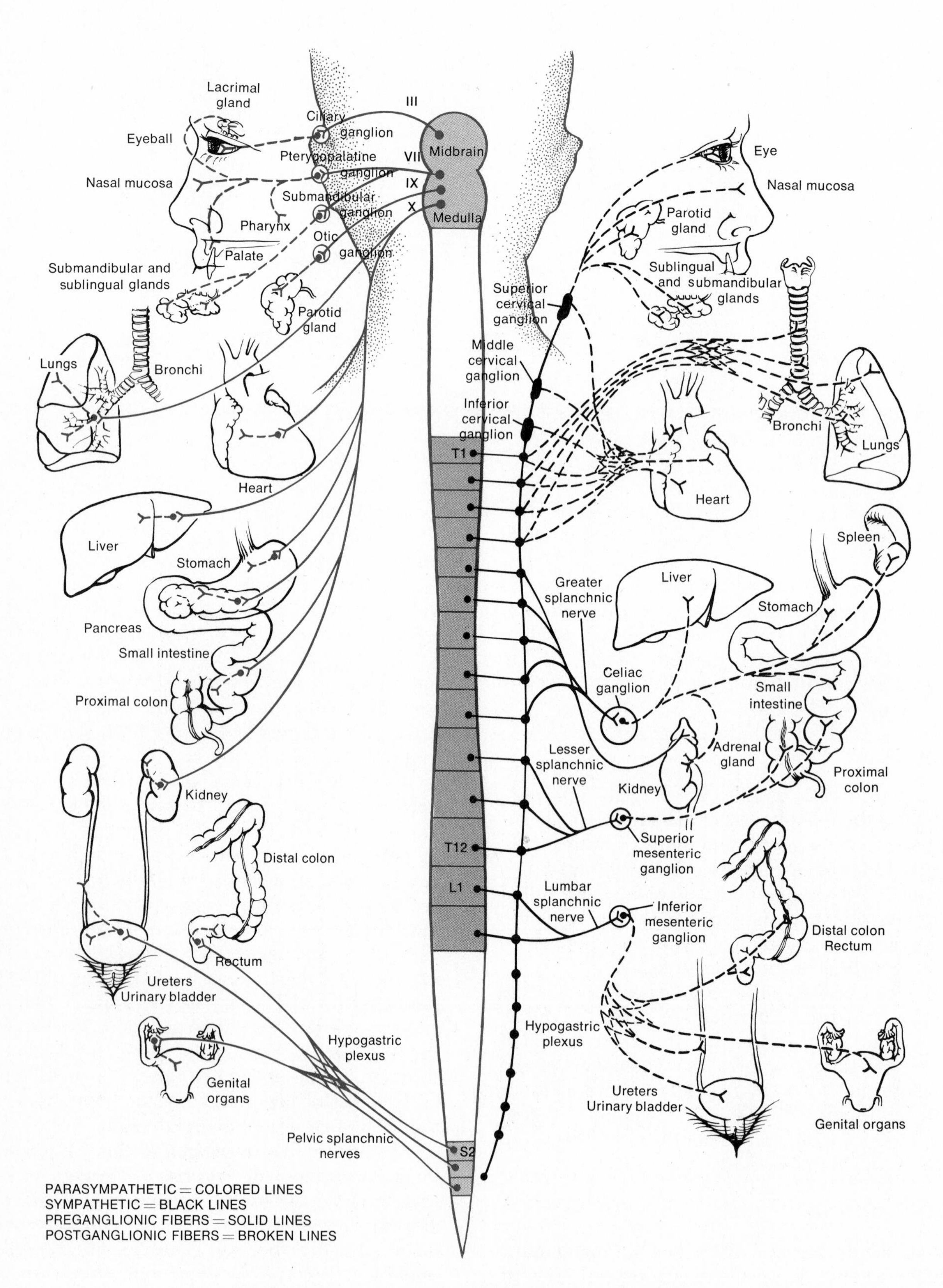

Figure 15-2. Structure of the autonomic nervous system.

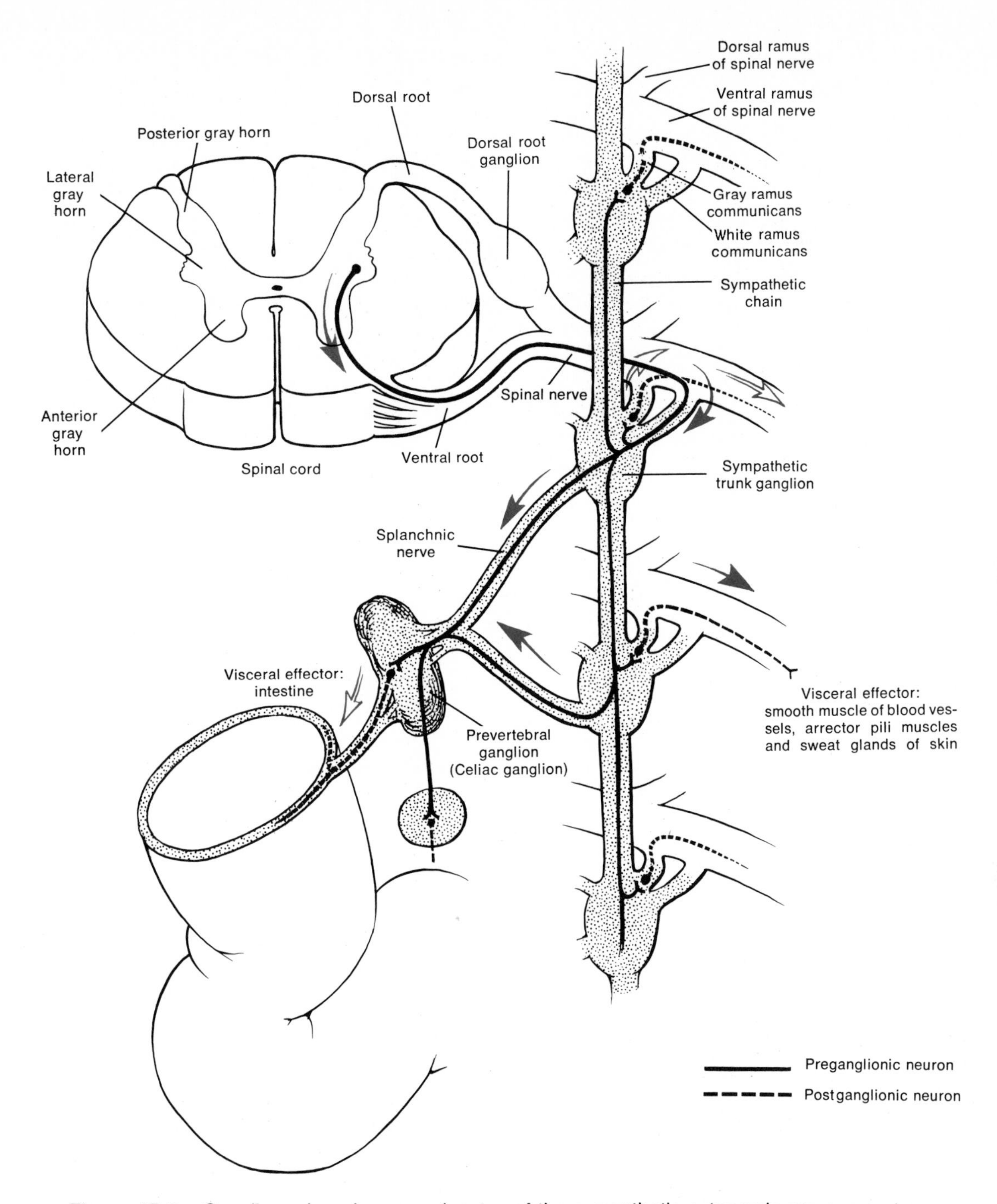

Figure 15-3. Ganglia and rami communicantes of the sympathetic autonomic nervous system.

craniosacral (parasympathetic) division. The preganglionic fibers do not pass through sympathetic trunk ganglia (Figure 15–2).

In addition to autonomic ganglia, the autonomic nervous system also contains **autonomic plexuses.** Slender nerve filaments from ganglia containing postganglionic nerve cell bodies arranged in a branching network constitute an autonomic plexus. Several will be described shortly.

Postganglionic neurons

Axons from preganglionic neurons of the sympathetic division pass to ganglia of the sympathetic trunk. They can either synapse in the sympathetic chain ganglia with postganglionic sympathetics or they can continue, without synapsing, through the chain ganglia to end at a prevertebral ganglion where synapses with the postganglionic sympathetics can take place. Each sympathetic preganglionic fiber synapses with several postganglionic fibers in the ganglion and the postganglionic fibers pass to several visceral effectors. You will see the importance of this shortly. Upon exiting their respective ganglia, the postsynaptic fibers innervate their visceral effectors.

Axons from preganglionic neurons of the parasympathetic division pass to terminal ganglia, near or within a visceral effector. Within the ganglion, the presynaptic neuron usually synapses with only four or five postsynaptic neurons to a single visceral effector. The meaning of this will also be explained shortly. Upon exiting their respective ganglia, the postsynaptic fibers supply their visceral effectors.

With this background in mind, we can now examine some specific structural features of the sympathetic and parasympathetic divisions of the autonomic nervous system.

Sympathetic division of the autonomic nervous system

As noted earlier, the preganglionic fibers of the sympathetic division have their cell bodies located in the lateral gray horn of the spinal cord in the thoracic and first two lumbar segments (Figure 15–2). The preganglionic fibers are myelinated and leave the spinal cord through the ventral root of a spinal nerve along with the somatic motor fibers emerging at the same segmental levels. The preganglionic fibers then branch away from the somatic fibers, enter the ventral ramus of the spinal nerve, and then pass on to the nearest sympathetic trunk ganglion on the same side. In their course from the ventral rami to the sympathetic trunk, the sympathetic preganglionic fibers are contained in structures referred to as **white rami communicantes** (Figure 15–3). They are so-called because they contain myelinated fibers. Only thoracic and upper lumbar nerves have white rami communicantes. The white rami communicantes connect the ventral ramus of the spinal nerve with the ganglia of the sympathetic trunk.

When a preganglionic fiber of a white ramus communicans enters the sympathetic trunk, it may terminate (synapse) in several ways. Some fibers synapse in the first ganglion at the level of entry. Others pass up or down the sympathetic trunk for a variable distance to form the fibers on which the ganglia are strung. These fibers, known as *sympathetic chains* (Figure 15–3), may not synapse until they reach a ganglion in the cervical or sacral area. Some of the postganglionic fibers leaving the sympathetic trunk ganglia pass directly to visceral effectors of the head, neck, chest, and abdomen. Most, however, rejoin the spinal nerves before supplying peripheral visceral effectors such as sweat glands and the smooth muscle in blood vessels and around hair follicles. The **gray ramus communicans** is the structure containing the postganglionic fiber that runs from the ganglion of the sympathetic trunk to the spinal nerve (Figure 15–3). The term gray refers to the fact that the fiber is unmyelinated. All spinal nerves have gray rami communicantes. Gray rami communicantes outnumber the white rami since there is a gray ramus leading to each of the 31 pairs of spinal nerves.

In most cases, a preganglionic fiber terminates by synapsing with a large number of postganglionic cells in a ganglion, usually 20 or more. Very often, the postganglionic fibers terminate in widely separated organs of the body. Thus, an impulse that starts in a single preganglionic neuron may affect several visceral effectors. For this reason, most sympathetic responses have widespread effects on the body.

The sympathetic trunks are two in number, situated anterolaterally to the spinal cord, one on either side. Each consists of a series of ganglia arranged more or less segmentally. The divisions of the sympathetic trunk are named on the basis of location. Typically, there are 22 ganglia in each chain: 3 cervical, 11 thoracic, 4 lumbar, and 4 sacral. Although the trunk extends downward from the neck, thorax, and abdomen to the coccyx, it receives preganglionic fibers only from the thoracic and lumbar segments of the spinal cord (Figure 15–2).

The cervical portion of each sympathetic trunk is located in the neck anterior to the prevertebral muscles. It is subdivided into a superior, middle, and inferior ganglion (Figure 15–2). The *superior cervical ganglion* is behind the internal carotid artery on a level with the second or third cervical vertebra. Postganglionic fibers leaving the ganglion are distributed to the eye, nasal mucosa, and the submandibular, sublingual, and parotid salivary glands. The *middle cervical ganglion* is situated at the level of the sixth cervical vertebra. Postganglionic fibers from it innervate the heart. The *inferior cervical ganglion* is located near the first rib. It also transmits postganglionic fibers that supply the heart.

The thoracic portion of each sympathetic trunk usually consists of 11 segmentally arranged ganglia. They are situated in the upper part of the thorax ventral to the necks of the corresponding ribs. This portion of the sympathetic trunk receives most of the sympathetic preganglionic fibers. Postganglionic fibers from the thoracic sympathetic trunk innervate the heart, lungs, bronchi, and other thoracic viscera.

The lumbar portion of each sympathetic trunk is found on either side of the corresponding lumbar vertebrae. The sacral portion of the sympathetic trunk lies in the pelvic cavity on the medial side of the sacral foramina.

It should be emphasized again that postganglionic neurons whose cell bodies are in the ganglia of the sympathetic trunk typically send their fibers back to the spinal nerves as part of the gray rami communicantes. Each spinal nerve has one. Once the postganglionic fibers from the sympathetic trunk have entered a spinal nerve via the gray rami communicantes, they are distributed to blood vessels, smooth muscle, and sweat glands. Since most sympathetic ganglia are located a considerable distance from the visceral effectors they supply, postganglionic sympathetic fibers to the visceral effectors are relatively long.

Before leaving the discussion of the sympathetic division, it should be noted that some preganglionic fibers pass through the sympathetic trunk without terminating in the trunk. Beyond the trunk, they form nerves known as splanchnic nerves (Figure 15–2). The greater splanchnic nerve is formed by the union of branches from the sixth to tenth thoracic ganglia; the lesser splanchnic nerve is derived from the ninth or tenth thoracic ganglia; the lowest splanchnic nerve is formed by branches of the last thoracic ganglion; and the lumbar splanchnic nerve is formed by branches of the first two lumbar ganglia. After passing through the trunk of ganglia, the splanchnic nerves terminate in the *celiac* or *"solar" plexus.* Within the plexus, the preganglionic fibers synapse in ganglia with postganglionic cell bodies. These ganglia are prevertebral ganglia. The greater splanchnic nerve passes to the celiac ganglion of the celiac plexus. From here, postganglionic fibers are distributed to the stomach, spleen, liver, kidney, and small intestine. The lesser splanchnic nerve passes through the celiac plexus to the superior mesenteric ganglion of the superior mesenteric plexus. Postganglionic fibers from the ganglion innervate the small intestine and colon. The lumbar splanchnic nerve enters the inferior mesenteric plexus. Within the plexus, the preganglionic fibers synapse with postganglionic fibers in the inferior mesenteric ganglion. These fibers pass through the hypogastric plexus and supply the distal colon and rectum, urinary bladder, and genital organs. As noted earlier, the postganglionic fibers leaving the prevertebral ganglia follow the course of various arteries to abdominal and pelvic visceral effectors.

Let us now examine some specific structural features of the parasympathetic division of the autonomic nervous system.

Parasympathetic division of the autonomic nervous system

The preganglionic cell bodies of the parasympathetic division are found in nuclei in the brain stem and the lateral gray horn of the second through fourth sacral segments of the spinal cord (Figure 15–2). Their fibers emerge either as a component of a cranial nerve or part of the ventral root of a spinal nerve. The **cranial parasympathetic outflow** consists of preganglionic fibers that leave the brain stem by way of the oculomotor nerves (III), facial nerves (VII), glossopharyngeal nerves (IX), and vagus nerves (X). The **sacral parasympathetic outflow** consists of preganglionic fibers that leave the ventral roots of the second through fourth sacral nerves. The preganglionic fibers of both the cranial and sacral outflows end in terminal ganglia where they synapse with postganglionic neurons. We will first look at the cranial outflow.

The four pairs of cranial parasympathetic ganglia innervate structures in the head and are located very close to the organs they innervate (Figure 15–2). The *ciliary ganglion* is near the back of an orbit lateral to each optic nerve. Preganglionic fibers pass with the oculomotor nerve to the ciliary ganglion. Postganglionic fibers from the ganglion innervate the eyeball. Each *pterygopalatine ganglion* is situated lateral to a sphenopalatine foramen. It receives preganglionic fibers from the facial nerve and transmits postganglionic fibers to the nasal mucosa, palate, pharynx, and lacrimal gland. Each *submandibular ganglion* is found near the duct of a submandibular salivary gland. It receives preganglionic fibers from the facial nerve and transmits postganglionic fibers that innervate the submandibular and sublingual salivary glands. The *otic ganglia* are situated just below each foramen ovale. The otic ganglion receives preganglionic fibers from the glossopharyngeal nerve and transmits postganglionic fibers that innervate the parotid salivary gland. Ganglia associated with the cranial outflow are classified as terminal ganglia. Note that since the terminal ganglia are very close to their visceral effectors, postganglionic parasympathetic fibers are relatively short. This is in contrast to postganglionic sympathetic fibers which are relatively long.

The last component of the cranial outflow, the preganglionic fibers that leave the brain via the vagus nerves, has the most extensive distribution of the parasympathetic fibers (Figure 15–2). Each vagus nerve enters into the formation of several plexuses in the thorax and abdomen. As it passes through the thorax, it sends fibers to the *superficial cardiac plexus* in the arch of the aorta and the *deep cervical plexus* anterior to the branching of the trachea. These plexuses contain terminal ganglia, and the postganglionic parasympathetic fibers emerging from them supply the heart. Also in the thorax is the *pulmonary plexus*, in front and behind the roots of the lungs and within the lungs themselves. It receives preganglionic fibers from the vagus and transmits postganglionic parasympathetic fibers to the lungs and bronchi. Other plexuses associated with the vagus nerve will be described in later chapters in conjunction with the appropriate thoracic, abdominal, and pelvic viscera. Just be aware of the fact that postganglionic fibers from these plexuses innervate viscera such as the liver, pancreas, stomach, kidney, small intestine, and part of the colon.

The sacral parasympathetic outflow consists of preganglionic fibers from the ventral roots of the second through fourth sacral nerves. Collectively, they form the *pelvic splanchnic nerves* (see Figure 15–2). They pass into the hypogastric plexus. From ganglia within the plexus, parasympathetic postganglionic fibers are distributed to the colon, ureters, urinary bladder, and reproductive organs.

A very brief comparison of some of the salient structural differences between the sympathetic and parasympathetic divisions is presented in Exhibit 15–1.

With this knowledge of the structure of the autonomic nervous system, we can now take a look at some of its functions.

PHYSIOLOGY OF THE AUTONOMIC SYSTEM

Chemical transmitters

Autonomic fibers, like other axons of the body, release chemical transmitters at synapses and at points of contact between autonomic fibers and visceral effectors. These points are called **neuroeffector junctions.** On the basis of the particular chemical transmitter produced, autonomic fibers may be classified as either cholinergic or adrenergic. **Cholinergic fibers** release *acetylcholine* and include the following: (1) all sympathetic and parasympathetic preganglionic axons, (2) parasympathetic postganglionic axons, and (3) some sympathetic postganglionic axons. The cholinergic sympathetic postganglionic axons include those to sweat glands and those to blood vessels in skeletal muscles and the external genitalia. Since acetylcholine is quickly inactivated by the enzyme cholinesterase, the effects of cholinergic fibers are short-lived and are not widespread. **Adrenergic fibers** produce the chemical transmitter *norepinephrine,* also called *sympathin.* Most sympathetic postganglionic axons are adrenergic. Since norepinephrine is inactivated much more slowly than acetylcholine and norepinephrine may enter the blood stream, the effects of sympathetic stimulation are longer lasting and more widespread than parasympathetic stimulation.

Activities of the autonomic system

Most visceral effectors have dual innervation. That is, they receive fibers from both the sympathetic and parasympathetic divisions. In these cases, impulses from one division stimulate the organ to initiate or increase its activities, whereas impulses from the other division decrease or inhibit the organ's activities. The stimulating division may be either the sympathetic or the parasympathetic, depending on the organ involved. For example, sympathetic impulses increase heart activity, but parasympathetic impulses decrease heart activity. On the other hand, parasympathetic impulses increase digestive activities, whereas sympathetic impulses inhibit them. A summary of the activities of the autonomic system is presented in Exhibit 15–2.

The parasympathetic division is primarily concerned with activities that restore and conserve body energy. It is a rest-repose system. For instance, under normal body conditions, parasympathetic impulses to the digestive glands and the smooth muscle of the digestive system dominate over sympathetic impulses. This situation allows energy-supplying foods to be digested and absorbed by the body.

The sympathetic division, by contrast, is primarily concerned with processes involving the expenditure of energy. When the body is in homeostasis, the main function of the sympathetic division is to counteract the parasympathetic effects just enough to carry out normal processes requiring energy. However, during a time of extreme stress, the sympathetic dominates the parasympathetic. For example, when people are confronted with a dangerous situation, their bodies become very alert, and they sometimes perform feats of unusual strength. This is because fear stimulates the sympathetic division. Activation of the sympathetic division sets into operation a series of physiological responses collectively called the *fight-or-flight* response. It produces the following effects: (1) the pupils of the eyes dilate; (2) the heart rate increases; (3) the blood vessels of the skin and viscera constrict; and (4) the remainder of the blood vessels dilate. This causes a rise in blood pressure and a faster flow of blood into the dilated blood vessels of skeletal muscles, cardiac muscle, the lungs, and brain – organs that are involved in fighting off the

Exhibit 15–1. COMPARISONS BETWEEN SYMPATHETIC AND PARASYMPATHETIC DIVISIONS OF THE AUTONOMIC NERVOUS SYSTEM

SYMPATHETIC	PARASYMPATHETIC
Forms thoracolumbar outflow	Forms craniosacral outflow
Contains sympathetic trunk and prevertebral ganglia	Contains terminal ganglia
Ganglia are distant from visceral effectors	Ganglia are near or within visceral effectors
Each preganglionic fiber synapses with many postganglionic neurons that pass to many visceral effectors	Each preganglionic fiber usually synapses with four or five postganglionic neurons that pass to a single visceral effector
Distributed throughout the body	Distribution limited primarily to head and viscera of thorax, abdomen, and pelvis

Exhibit 15–2. ACTIVITIES OF THE AUTONOMIC NERVOUS SYSTEM

VISCERAL EFFECTOR	EFFECT OF SYMPATHETIC STIMULATION	EFFECT OF PARASYMPATHETIC STIMULATION
Eye		
Iris	Contracts the dilator muscle of the iris and brings about dilatation of the pupil	Contracts sphincter muscle of the iris and brings about constriction of the pupil
Ciliary muscle	Relaxes the ciliary muscle and accommodates the lens for far vision	Contracts the ciliary muscle and accommodates the lens for near vision
Glands		
Sweat	Stimulates secretion	No innervation
Lacrimal (tear)	No innervation	Normal or excessive secretion
Salivary.	Decreases secretion and saliva is mucus-rich	Increases secretion of thin, watery saliva
Gastric	No known effect	Secretion stimulated
Intestinal	No known effect	Secretion stimulated
Adrenal medulla	Promotes epinephrine and norepinephrine secretion	No innervation
Lungs (bronchial tubes)	Dilatation	Constriction
Heart	Increases rate and strength of contraction; dilates coronary vessels that supply blood to heart muscle cells	Decreases rate and strength of contraction; constricts coronary vessels
Blood vessels		
Skin	Constriction	No innervation for most
Skeletal muscle	Dilatation	No innervation
Visceral organs (except heart and lungs)	Constriction	No innervation for most
Liver	Promotes glycogenolysis; decreases bile secretion	Promotes glycogenesis; increases bile secretion
Stomach	Decreases motility	Increases motility
Intestines	Decreases motility	Increases motility
Kidney	Constriction that results in decreased urine volume	No effect
Pancreas	Inhibits secretion	Promotes secretion
Spleen	Contraction and discharge of stored blood into general circulation	No innervation
Urinary bladder	Relaxes muscular wall	Contracts muscular wall
Arrector pili of hair follicles	Contraction results in erection of hairs ("goose pimples")	No innervation
Uterus	Inhibits contraction if nonpregnant; stimulates contraction if pregnant	Minimal effect
Sex organs	Vasoconstriction of ductus deferens, seminal vesicle, prostate; results in ejaculation	Vasodilation and erection

danger. Rapid breathing occurs as the bronchioles dilate to allow movement of air in and out of the lungs at a faster rate. Blood sugar level rises as liver glycogen is converted to glucose to supply the additional energy needs of the body. The sympathetic division also stimulates the medulla of the adrenal gland to produce epinephrine and norepinephrine, hormones that intensify and prolong the sympathetic effects noted above. During this period of stress, the sympathetic effects inhibit other processes that are not essential for meeting the situation. For example, muscular movements of the gastrointestinal tract and digestive secretions are slowed down or even stopped.

Visceral autonomic reflexes

When we first began the discussion of the autonomic nervous system, it was indicated that afferent (sensory) neurons do participate in the functioning of the autonomic nervous system. In Chapter 13, we analyzed some somatic reflexes, that is, reflexes that result in skeletal muscle contractions. To understand the relationship of afferent neurons to the autonomic nervous system, we will now discuss a visceral autonomic reflex. A **visceral autonomic reflex** adjusts the activity of a visceral effector. In other words, it results in the contraction of smooth or cardiac muscle or secretion by a gland. Such reflexes assume a key role in activities such as regulating heart action, blood pressure, respiration, digestion, defecation, and bladder functions.

A visceral autonomic reflex arc consists of the following components:

1. *Receptor.* The receptor is the distal end of an afferent neuron in an exteroceptor or enteroceptor.
2. *Afferent neuron.* This neuron, either a somatic afferent or visceral afferent neuron, conducts the sensory impulse to the spinal cord or brain.
3. *Association neurons.* Neurons in the central nervous system.
4. *Visceral efferent preganglionic neuron.* In the thoracic and abdominal regions, this neuron is in the lateral gray horn of the spinal cord. The axon passes through the ventral root of the spinal nerve, the white ramus communicans, and enters a sympathetic trunk or prevertebral ganglion where it synapses with a postganglionic neuron. In the cranial and sacral regions, the visceral efferent preganglionic axon passes to a terminal ganglion where it synapses with a postganglionic neuron. The role of the visceral efferent preganglionic neuron is to convey a motor impulse from the brain or spinal cord to an autonomic ganglion.
5. *Visceral efferent postganglionic neuron.* This neuron conducts a motor impulse from a visceral efferent preganglionic neuron to the visceral effector.
6. *Visceral effector.* As noted previously, a visceral effector is smooth muscle, cardiac muscle, or a gland of the body.

The basic difference between a somatic reflex arc and a visceral autonomic reflex arc is that in a somatic reflex, only one efferent neuron is involved. In a visceral autonomic reflex arc, two efferent neurons are involved.

Visceral sensations do not always reach the cerebral cortex. Therefore, they do not reach conscious levels. Instead, they remain at subconscious levels. For example, under normal conditions, you are not aware of muscular contractions of the digestive organs, heartbeat, changes in the diameter of blood vessels, and pupil dilation and constriction. When your body is making adjustments in such visceral activities, they are handled by visceral reflex arcs whose centers are in the spinal cord or lower regions of the brain. Among such centers are the cardiac, respiratory, vasomotor, swallowing, and vomiting centers in the medulla and the temperature-control center in the hypothalamus. Thus, stimuli delivered by visceral afferent neurons synapse in these centers, and the returning motor impulses conducted by visceral efferent neurons bring about an adjustment in the visceral effector without conscious recognition. The sensations are interpreted and acted upon subconsciously. Some visceral sensations that do give rise to conscious recognition are hunger, nausea, and fullness of the urinary bladder and rectum.

We will now conclude our discussion of the nervous system by considering a few applications to health.

APPLICATIONS TO HEALTH

Many disorders can affect the various parts of the nervous system. Some of the diseases are known to be caused by viruses or bacteria. Other conditions are caused by damage to the nervous system during birth. The origins of many conditions, however, are unknown. Here, we shall discuss the origins and describe the symptoms of some common nervous system disorders.

Poliomyelitis

Poliomyelitis, also known as **infantile paralysis,** is a viral infection that is most common during childhood. The onset of the disease is marked by fever, severe headache, a stiff neck and back, deep muscle pain and weakness, and loss of some somatic reflexes. The virus may spread by means of the blood and respiratory passages to the central nervous system where it destroys the motor nerve cell bodies, specifically those in the anterior horns of the spinal cord and in the nuclei of the cranial nerves. The injury to the spinal gray matter gives the disease its name—*polio,* meaning gray matter, and *myel,* meaning spinal cord. Destruction of the anterior horns produces paralysis of one or more limbs. Poliomyelitis can cause death from respiratory or heart failure, if the virus invades the brain cells of the vital medullary centers. In recent years, an immunization against the disease has been used. However, nearly 100 percent immunization is achieved only if the child is immunized shortly after birth.

Syphilis

Syphilis is a venereal disease caused by the *Treponema pallidum* bacterium. Venereal diseases are infectious disorders that can be spread through sexual contact of any sort. The disease process of syphilis goes through several stages: primary, secondary, latent, and sometimes tertiary. During the primary stage, the chief symptom is an open sore, called a chancre, at the point of contact. The chancre eventually heals. About 6 weeks later, a range of generalized symptoms, such as a skin rash, fever, and aches in the joints and muscles, ushers in the secondary stage. At this stage, syphilis can usually be treated with antibiotics. However, even if the individuals do not undergo treatment, their symptoms will eventually disappear. Within a few years, they will usually cease to be infectious. The signs and symptoms of the disease disappear, but a blood test would show positive results. During this later "symptomless" period, called the latent stage, the bacteria may invade and slowly destroy any of the body organs. This is why untreated syphilis is so dangerous. When symptoms of organ degeneration appear, the disease is said to be in its tertiary stage. If the syphilis bacteria attack the organs of the nervous system, the tertiary stage is called *neurosyphilis.* Neurosyphilis may take any one of several forms, depending on the tissue involved. For instance, about 2 years after the onset of the disease, the bacteria may attack the meninges, producing meningitis. Or the blood vessels that supply the brain may become infected. In this case, symptoms depend on the parts of the brain that are destroyed by oxygen and glucose starvation. Over the years, the bacteria may spread through the nerve cells of the brain. As one nerve cell after another is destroyed, patients experience corresponding symptoms. Cerebellar damage is manifested by uncoordinated movements. For instance, they may have trouble with skilled activities such as writing. As the motor areas become extensively damaged, victims may be unable to control urine and bowel movements. Eventually, they may be bedridden, without even the motor control necessary to feed themselves. Damage to the cerebral cortex produces memory loss and personality changes that range from irritability to confusion to hallucinations.

A common form of neurosyphilis is *tabes dorsalis,* a progressive degeneration of the posterior columns of the spinal cord. The sensory ganglia and the sensory nerve roots are also affected. Tabes dorsalis forms an interesting contrast with polio. The polio virus attacks the anterior columns and destroys motor neurons. Polio victims are unable to move the affected muscles voluntarily, but they retain all their sensory functions. Persons with tabes dorsalis suffer from just the opposite problem. They retain motor control but lose many of their sensory functions. They often have tingling or numbness in their limbs and trunk. Normally, receptors in the joints are able to tell the central nervous system how much a joint is flexed and where one part of the body is in relation to another part. This information is necessary for coordinating movement and maintaining posture and balance. When the sensory nerve roots are destroyed, however, this information cannot pass from the receptors to the brain. Consequently, persons with tabes dorsalis must use their eyes in order to carry out motor activities successfully. They have trouble walking in the dark because they do not know whether their legs are flexed or where their feet are. In lighted areas, they may walk with a characteristic shuffle, which consists of jerking the knee up and then letting the leg extend abruptly to the ground.

Syphilis can be treated with antibiotics during the primary, secondary, and latent periods. Some forms of neurosyphilis can also be successfully treated; however, the prognosis for tabes dorsalis is very poor. Unfortunately, not everyone with syphilis shows noticeable symptoms during the first two stages of the disease. But syphilis usually can be diagnosed through a blood test. The importance of these blood tests and follow-up treatment cannot be overemphasized.

Cerebral palsy

The term **cerebral palsy** refers to a group of motor disorders caused by damage to the motor areas of the brain during fetal life, birth, or infancy. One cause is infection of the mother with German measles during the first 3 months of pregnancy. During early pregnancy, certain cells in the fetus are dividing and changing in order to lay down the basic structures of the brain. These cells can be abnormally changed by toxin from the measles virus. Radiation during fetal life, temporary oxygen starvation during birth, and hydrocephalus during infancy can also damage brain cells.

Cases of cerebral palsy are categorized into three groups depending on whether the cortex, the basal ganglia of the cerebrum, or the cerebellum is affected most severely. Most cerebral palsy victims have at least some damage in all three areas. The location and extent of motor damage determine the symptoms. For instance, the cerebral palsy victim may have a contorted face caused by partial facial paralysis. If the tongue is paralyzed, the individual may be able to make only guttural sounds. Extensive damage to the cerebellum causes very uncoordinated movements. Although cerebral palsy refers only to motor damage, sensory and association areas of the brain may be affected as well. The person may be deaf or partially blind. About 70 percent of cerebral palsy victims appear to be mentally retarded. However, the apparent mental slowness is often due to the person's inability to speak or hear well. These people are often much brighter than they seem.

Cerebral palsy is not a progressive disease. In other words, it does not get worse as time elapses. Once the damage is done, however, it is irreversible.

Parkinsonism

This disorder, also called **Parkinson's disease,** is a progressive malfunction of the basal ganglia of the cerebrum. Recall that the basal ganglia regulate unconscious contractions of skeletal muscles that aid activities desired by the motor areas of the cerebral cortex. Examples of movement produced by the basal ganglia are swinging the arms when walking and making facial expressions when talking. In Parkinsonism, the basal ganglia produce useless skeletal movements that often interfere with voluntary movement. For instance, the muscles of the upper extremities may alternately contract and relax so that the patient's hands shake. This type of shaking is called *tremor.* Other muscles may contract continuously and make the involved part of the body rigid. *Rigidity* of the facial muscles gives the face a masklike appearance. The expression is characterized by a wide-eyed, unblinking stare and a slightly open mouth with saliva drooling from the corners. Vision, hearing, and intelligence are unaffected by the disorder, indicating that Parkinsonism does not attack the cerebral cortex.

Parkinsonism seems to be caused by a malfunction at the neuron synapses. The motor neurons of the basal ganglia release the chemical transmitter acetylcholine. In normal people, the basal ganglia also produce a synaptic transmitter called dopamine, which quickly inactivates the acetylcholine and prevents continuous conduction across the synapse. People with Parkinsonism do not manufacture enough dopamine in their bodies. As a result, stimulated basal ganglia neurons do not easily stop conducting impulses. Unfortunately, injections of dopamine are useless because the drug is not able to diffuse from the blood into the brain because of the blood-brain barrier. However, a few years ago researchers developed a drug called levodopa, which is very similar to dopamine. Levodopa is able to diffuse into the brain, where it is converted by a chemical reaction to dopamine.

Epilepsy

Epilepsy is a disorder characterized by short, recurrent, periodic attacks of motor, sensory, and/or psychological malfunction. The attacks, called **epileptic seizures,** are brought on by abnormal and irregular discharges of electricity by millions of neurons in the brain. The discharges stimulate many of the neurons to send impulses over their conduction pathways. As a result, a person undergoing an attack may contract skeletal muscles involuntarily. Lights, noise, or smells may be sensed when the eyes, ears, and nose actually have not been stimulated. The electrical discharges may also inhibit certain brain centers. For instance, the waking center in the brain may be depressed so that the person loses consciousness.

Many different types of epileptic seizures exist. The particular type of seizure depends on the area of the brain that is electrically stimulated and whether the stimulation is restricted to a small area or spreads throughout the brain. *Grand mal* seizures are brought on by a burst of electrical discharges that travel throughout the motor areas and spread to the areas of consciousness in the brain. The person loses consciousness, has spasms of voluntary muscles, and may also lose urinary and bowel control. Sensory and intellectual areas may also be involved. For instance, just as the attack begins, the person may sense a peculiar taste in the mouth or see flashes of light or have olfactory hallucinations. The unconsciousness and motor activity last a few minutes. Then the muscles relax, and the person awakens. Afterward, the individual may be mentally confused for a short period of time. Studies with EEGs show that grand mal attacks are charac-

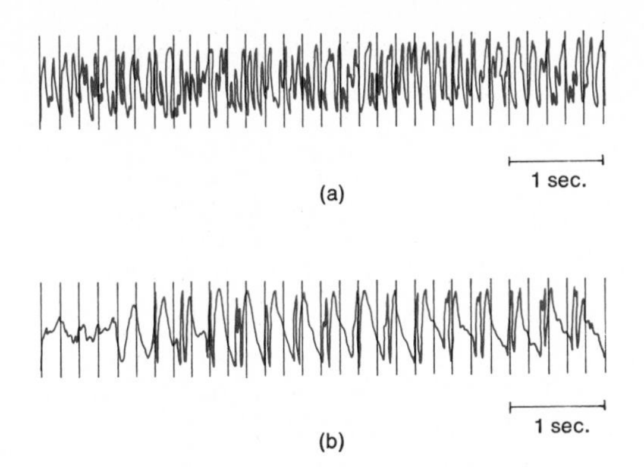

Figure 15-4. Electroencephalographs. (a) Grand mal seizure. (b) Petit mal seizure.

terized by rapid brain waves occurring at the rate of 25 to 30 per second (Figure 15–4). The normal adult rate is 10 waves/second.

Many epileptics suffer from electrical discharges that are restricted to one or several relatively small areas of the brain. An example is the *petit mal* form, which apparently involves the thalamus and hypothalamus. Petit mal seizures are characterized by an abnormally slow brain wave pattern occurring at the rate of 3 waves/second (see Figure 15–4). The person may lose contact with the environment for about 5 to 30 seconds but does not undergo the embarrassing loss of motor control that is typical of a grand mal seizure. The victim merely seems to be daydreaming. A few people experience several hundred petit mal seizures each day. For them, the chief problems are a loss of productivity in school or work and periodic inattentiveness while driving a car.

Some epileptics experience motor seizures that are restricted to the precentral motor area of one cerebral hemisphere. These attacks consist of spasms that pass up or down one side of the body. People who suffer from sensory seizures may see lights or distorted objects if the discharge occurs in the occipital lobe. They may hear voices or a roaring in their ears if the discharge is located in the temporal lobe. Or, they may taste something unpleasant if the discharge is in the parietal lobe. People undergoing attacks of localized motor or sensory disturbances may or may not lose consciousness. A form of epilepsy that is sometimes confused with mental illness is *psychomotor epilepsy*. The electrical outburst occurs in the temporal lobe, where it causes the person to lose contact with reality. It may spread to some of the motor areas and produce mild spasms in some of the voluntary muscles. These persons may stare into space and involuntarily smack their lips or clap their hands during an attack. If the motor areas are not involved, they may simply walk aimlessly. When they come back to reality, they are surprised to find themselves in a strange or different place.

The causes of epilepsy are varied. Many conditions can cause nerve cells to produce periodic bursts of impulses. Head injuries, tumors and abscesses of the brain, and childhood infections, such as mumps, whooping cough, and measles, are some of the causes. Epilepsy may also be idiopathic, that is, have no demonstrable cause. It should be noted, however, that epilepsy almost never affects intelligence. If frequent severe seizures are allowed to occur over a long period of time, some cerebral damage may occasionally result. However, damage can be prevented by controlling the seizures with drug therapy.

Epileptic seizures can be eliminated or alleviated by drugs that make neurons more difficult to stimulate. Many of these drugs change the permeability of the neuron cell membrane so that it does not depolarize as easily.

Multiple sclerosis

This disorder causes a progressive destruction of the myelin sheaths of neurons in the central nervous system. The sheaths deteriorate to *scleroses,* which are hardened scars or plaques, in multiple regions. Hence, the disorder is given the name **multiple sclerosis.** The destruction of the myelin sheaths interferes with the transmission of impulses from one neuron to another, literally "short-circuiting" conduction pathways. Multiple sclerosis is one of the most common disorders of the central nervous system. Usually the first symptoms occur between the ages of 20 and 40. The early symptoms are generally produced by the formation of just a few plaques and are, consequently, mild. For instance, plaque formation in the cerebellum may produce some incoordination in one hand. The patient complains that his or her handwriting has become sloppy. A short-circuiting of some pathways in the corticospinal tract may partially paralyze the leg muscles so that the patient drags a foot when walking. Following a period of remission during which the symptoms disappear for a while, a new series of plaques develop, and the victim suffers a second attack. One attack follows another over the years. Each time the plaques form, some of the neurons are damaged by the hardening of their sheaths. Other neurons are uninjured by their plaques. The result is a progressive loss of function interspersed with remission periods during which the undamaged neurons regain their ability to transmit impulses.

The symptoms of multiple sclerosis depend on the areas of the central nervous system most heavily infested with plaques. Sclerosis of the white matter of the spinal cord is common. As the sheaths of the neurons in the corticospinal tract deteriorate, the patient loses the ability to contract skeletal muscles. Damage to the ascending tracts produces numbness and short-circuits information about the position of body parts and the flexion of joints. Damage to either set of tracts also destroys spinal cord reflexes. Very often the white matter of the brain stem and cerebellum deteriorates. In this case, the patient loses muscular coordination, cannot control movement of the eyeballs, and pronounces words slowly and hesitatingly. Damage to the tracts in the cerebral white matter can short-circuit impulses running to and from the cerebral cortex. In such cases, the person is unable to carry out motor activities initiated in the cortex or cannot express himself or herself in speech. Incoming sensory impulses, such as those for vision, may be interrupted.

As the disease progresses, most voluntary motor control is eventually lost, and the patient becomes bedridden. Death occurs anywhere from 7 to 30 years after the first symptoms appear. The usual cause of death is a severe infection indirectly brought on by the loss of motor activity. For instance, the involuntary muscles that form the walls of the bladder often become paralyzed. When these muscles are unable to contract, the bladder cannot squeeze urine to the outside of the body. A tube may be inserted so that the urine can drain out. But without the squeezing action of the bladder walls, the bladder never totally empties. Some of the urine then stagnates, forming a good environment for bacteria. The patient undergoes bouts of bladder infection that easily spreads to the kidney and destroys the kidney cells.

The etiology of multiple sclerosis is unknown. Many researchers suspect that the disease is caused by a common virus that does not affect the myelin sheaths of most people. Other possible causes are metal poisons, accidental injury to the central nervous system, and diseases of the blood vessels that supply the central nervous system.

Cerebral vascular accidents (CVA)

By far the most common type of brain disorder is **stroke,** also called **cerebral apoplexy** and **cerebrovascular accident.** Stroke is the destruction of brain tissue or infarction resulting from any of several disorders in the vessels that supply the brain. Common causes of stroke are intracerebral hemorrhage, embolism, and atherosclerosis of the cerebral arteries. An *intracerebral hemorrhage* is a rupture of one of the vessels within the pia mater or brain. Blood oozes into the brain and can damage neurons by increasing intracranial fluid pressure in all or part of the brain. An *embolus* is a blood clot or bit of foreign material, most often debris from an inflammation, that becomes lodged in an artery and blocks circulation. *Atherosclerosis* is the formation of plaques on the walls of the arteries. The plaques may slow down circulation by partially closing off the vessel. Both emboli and atherosclerosis cause brain damage by reducing or shutting off the supply of oxygen and glucose needed by brain cells.

Many elderly people suffer "little strokes," which are short periods of reduced blood supply. One cause is atherosclerosis. Another is arteriosclerosis, or hardening of the arteries, that occurs with aging. Damage is generally either undetectable or very mild. During these mild strokes the person may have a short blackout, blurred vision, or dizziness, and does not realize that anything serious has occurred. However, a stroke can also cause sudden, massive damage. Severe stroke causes about 21 percent of all deaths from cardiovascular disease. If the person recovers, he or she may suffer partial paralysis, speech difficulty, mental disturbances, severe headaches, visual disturbances, deafness, and dizziness.

The type of malfunction following a stroke depends on the parts of the brain that were injured. Vascular disorders are more common after the age of 40, and thus stroke is primarily a disease of the middle-aged and elderly.

Tumors of the nervous system

Brain tumors are generally divided into different classes that present a variety of symptoms, depending on their location and the extent of impairment. Classes include tumors of the skull, meninges, neuroglia, cranial nerves, pituitary, and pineal body. These tumors may be congenital, or they may occur at any age. They are most common, however, in early adult or middle life. Males and females are affected about equally. Malignant tumors of the neuroglia are called *gliomas,* and the most common of these are *astrocytomas.* Another common malignant tumor is the *medulloblastoma,* which occurs in the cerebellum of children and young adults. A *meningioma* is generally a benign tumor that arises in the meninges.

Symptoms arise when the tumor damages neurons by invading them directly, by putting pressure on other parts of the brain as the tumor enlarges, and by obstructing the flow of cerebrospinal fluid.

To determine whether a patient has a brain tumor, the physician first tests the patient's reflexes, coordination, muscular strength, and ability to interpret stimuli correctly, and the functioning of certain involuntary muscles such as the muscles that control dilation of the pupil of the eye. In this way, the physician can determine what parts of the nervous system seem to be diseased. An EEG may be given to determine whether the brain is involved. If the cause of malfunction is a brain tumor, the tumor may, finally, be located through various techniques that provide pictures of the brain and its blood vessels.

We will now consider structures associated with the special senses.

Key terms associated with the nervous system

Analgesia (*an* = without; *algia* = painful condition) Insensibility to pain.
Anesthesia (*esthesia* = feeling) Loss of feeling.
Aphasia (*a* = without; *phasis* = speech; *ia* = condition) Diminished or complete loss of ability to comprehend and/or express spoken or written words, due to injury or disease of the brain centers; most common cause is stroke.
Bacterial meningitis Acute inflammation of the meninges caused by bacteria.
Coma Abnormally deep unconsciousness with an absence of voluntary response to stimuli; varying degrees of reflex activity remain. May be due to illness or to an injury.
Neuralgia (*neur* = nerve) Attacks of pain along the entire course or branch of a peripheral sensory nerve; one common type involves one or more branches of the trigeminal nerve and is called trigeminal neuralgia (tic douloureux).
Neuritis Inflammation of a nerve; can result from irritation to the nerve produced by trauma, bone fractures, nutritional deficiency (usually thiamine), poisons such as carbon monoxide and carbon tetrachloride, heavy metals such as lead, and some drugs. Neuritis of the facial nerve that results in paralysis of facial muscles is called Bell's palsy.
Paralysis Diminished or total loss of motor function resulting from damage to nervous tissue or a muscle.
Sciatica Severe pain along the sciatic nerve and its branches. Usually due to rupture of an intervertebral disc or to osteoarthritis of the lower spinal column: the disc or arthritic joint puts pressure on the nerve root supplying the sciatic nerve and thereby causes the pain.
Shingles Acute inflammation caused by a virus that attacks sensory cell bodies of dorsal root ganglia. Inflammation spreads peripherally along a spinal nerve and infiltrates dermis and epidermis over the nerve, producing a characteristic line of skin blisters.
Spastic (*spas* = draw or pull) Resembling spasms or convulsions.
Spina bifida (*bifid* = into two parts) An abnormality in one or many vertebral arches. The arches fail to fuse during embryonic development so that part of the spinal cord may be exposed.
Torpor Abnormal inactivity or no response to normal stimuli.
Viral encephalitis Acute inflammation of the brain caused by any of a number of viruses.

Chapter summary in outline

AUTONOMIC NERVOUS SYSTEM

1. The autonomic nervous system regulates the activities of smooth muscle, cardiac muscle, and glands.
2. It usually operates without conscious control.
3. Unlike the somatic nervous system, the autonomic nervous system is entirely motor.

Structure

1. Autonomic visceral efferent pathways consist of a preganglionic neuron and a postganglionic neuron.
2. The preganglionic neurons of the sympathetic division originate in the thoracic and first two lumbar segments of the cord, thus the division is also called the thoracolumbar division.
3. The preganglionic neurons of the parasympathetic division originate in the brain stem and second through fourth sacral segments of the cord, thus the division is also called the craniosacral division.
4. Autonomic ganglia are of three kinds: sympathetic trunk (sympathetic) on either side of the spinal cord; prevertebral (sympathetic) are anterior to the spinal cord near large abdominal vessels; and terminal (parasympathetic) near or within visceral effectors.
5. Postganglionic sympathetic fibers usually pass to several visceral effectors.
6. Postganglionic parasympathetic fibers usually pass to only a single visceral effector.
7. Sympathetic preganglionic fibers are contained in white rami communicantes that connect the ventral ramus of a spinal nerve with sympathetic trunk ganglia.
8. Sympathetic postganglionic fibers are contained in gray rami communicantes that run from sympathetic trunk ganglia to a spinal nerve.

Chemical transmitters

1. Autonomic fibers release chemical transmitters at synapses and neuroeffector junctions.
2. Adrenergic fibers (most sympathetic postganglionic) release norepinephrine, and cholinergic fibers (all others) release acetylcholine.

Activities

1. Most visceral effectors receive dual innervation. One division usually stimulates, and the other usually inhibits a visceral effector.
2. The parasympathetic division is primarily concerned with activities that restore and conserve body energy.
3. The sympathetic division is primarily concerned with processes involving the expenditure of energy.
4. Activation of the sympathetic division sets into operation the fight-or-flight response.

Visceral reflexes

1. A visceral autonomic reflex is one that adjusts the activity of a visceral effector.
2. A visceral autonomic reflex arc consists of a receptor, an afferent neuron, association neurons, a visceral efferent preganglionic neuron, a visceral efferent postganglionic neuron, and a visceral effector.

APPLICATIONS TO HEALTH

1. Poliomyelitis is a viral infection that results in paralysis.
2. Syphilis is caused by the bacterium *Treponema pallidum* and may result in blindness, memory defects, abnormal behavior, and loss of sensory functions in trunk and limbs.
3. Cerebral palsy includes a group of central nervous system disorders that primarily involve the cerebral cortex, cerebellum, and basal ganglia. The disorders damage motor centers.
4. Parkinsonism is a malfunction of the basal ganglia of the cerebrum caused by insufficient dopamine.
5. With epilepsy, the victim experiences convulsive seizures. It results from irregular electrical discharges of brain cells and may be diagnosed by an EEG.
6. Multiple sclerosis is the destruction of myelin sheaths of the neurons of the central nervous system. Impulse transmission is interrupted.
7. Cerebral vascular accidents are also called strokes. Brain tissue is destroyed due to hemorrhage, embolism, and atherosclerosis.
8. Tumors may involve the neuroglia (gliomas), the cerebellum (medulloblastomas), and the meninges (meningiomas).

Review questions and problems

1. Define the autonomic nervous system.
2. Contrast the structural and functional differences between the autonomic nervous system and the somatic nervous system.
3. Describe the organization of the autonomic nervous system into sympathetic and parasympathetic divisions.
4. Define a visceral efferent neuron. Distinguish between a preganglionic neuron and a postganglionic neuron.
5. Explain the origin of preganglionic neurons of the sympathetic and parasympathetic divisions of the autonomic nervous system.
6. What is an autonomic ganglion? Describe the location and function of the three types of autonomic ganglia.
7. Explain the termination of postganglionic neurons of the sympathetic and parasympathetic divisions of the autonomic nervous system.
8. What are white rami communicantes? What are gray rami communicantes?
9. Compare the significant structural features of the sympathetic and parasympathetic divisions of the autonomic nervous system.
10. Describe the classification of autonomic fibers into cholinergic and adrenergic.
11. What is meant by dual innervation? Give several examples. Explain its importance to the body.
12. What is the primary function of the parasympathetic division of the autonomic nervous system? Explain.
13. What is the primary function of the sympathetic division of the autonomic nervous system? Explain.
14. Describe the physiological activities involved in the fight-or-flight response.
15. Define a visceral autonomic reflex. How does it differ from a somatic reflex?
16. Describe the components, and their functions, of a visceral autonomic reflex arc.
17. Define poliomyelitis. What are some of its principal symptoms?
18. How does syphilis involve the nervous system?
19. What is cerebral palsy? Describe some of its symptoms.
20. Define Parkinsonism. Explain its symptoms and treatment.
21. Contrast the various kinds of epilepsy with regard to clinical symptoms. What is the diagnostic usefulness of the EEG for epilepsy?
22. Define multiple sclerosis and list a few of its outstanding symptoms.
23. What is a stroke? What are some common causes?
24. Contrast the kinds of tumors that affect the nervous system.
25. Refer to the glossary of key terms at the end of the chapter. Be sure that you can define each term.

Selected readings

Cunningham's Textbook of Anatomy. 11th ed., edited by G. J. Romanes. London: Oxford University Press, 1972. Pp. 775–793.

Dawson, Helen L. *Basic Human Anatomy.* 2d ed. New York: Appleton-Century-Crofts, 1974. Pp. 214–218.

Gray, Henry. *Anatomy of the Human Body.* 29th ed., edited by Charles Mayo Goss. Philadelphia: Lea and Febiger, 1973. Pp. 1006–1034.

Netter, Frank H. *Nervous System.* CIBA Collection of Medical Illustrations, Vol. 1. Summit, NJ: CIBA Pharmaceutical Co., 1958.

Truex, Raymond C., and Malcolm B. Carpenter. *Human Neuroanatomy.* 6th ed. Baltimore: Williams and Wilkins, 1969.

CHAPTER 16

SENSORY STRUCTURES

STUDENT OBJECTIVES

After you have read this chapter, you should be able to:

1. Define a sensation and list the four prerequisites necessary for its transmission
2. Define projection, adaptation, afterimages, and modality as characteristics of sensations
3. Compare the location and function of exteroceptors, visceroceptors, and proprioceptors
4. Describe the distribution of cutaneous receptors by interpreting the results of the two-point discrimination test
5. List the location and function of the receptors for touch, pressure, cold, heat, pain, and proprioception
6. Distinguish among somatic, visceral, referred, and phantom pain
7. Locate the receptors for olfaction and describe the neural pathway for smell
8. Identify the gustatory receptors and describe the neural pathway for taste
9. Describe the structure and physiology of the accessory visual organs
10. List the structural divisions of the eyeball
11. Discuss retinal image formation by describing refraction, accommodation, constriction of the pupil, convergence, and inverted image formation
12. Define emmetropia, myopia, hypermetropia, and astigmatism
13. Describe the afferent pathway of light impulses to the brain
14. Define the anatomical subdivisions of the ear
15. List or describe the principal events involved in hearing
16. Identify the receptor organs for equilibrium
17. Discuss the receptor organs' roles in the maintenance of dynamic and static equilibrium
18. Contrast the causes and symptoms of cataracts, glaucoma, conjunctivitis, trachoma, Ménière's disease, and impacted cerumen
19. Define key terms associated with the sense organs

Your ability to "sense" stimuli is vital to your survival. If you could not "sense" pain, you would probably not remove your hand quickly from a hot pot handle. You would not be alerted to the fact that your appendix was about to burst, your heart was malfunctioning, you were developing a cavity in a tooth, or an ulcer was forming in your stomach. In all these cases a great deal of damage could be done to your body before you did something about the situation. If you could not see, you might run the risk of physical injury whenever you faced an unfamiliar obstacle. If you could not smell, you might breathe in a harmful gas. If you could not hear, you might not notice verbal warnings, a car honking, or an aggressor sneaking up. If you could not taste, you might swallow some kind of poisonous material. In short, if you could not "sense" your environment and make the necessary homeostatic adjustments, you could not survive on your own.

As vital as the ability to "sense" stimuli is to your survival, it also provides us with many pleasures. Think of the pleasures involved in listening to your favorite music, enjoying the taste of your favorite food, smelling a flower, or seeing a ballet.

SENSATIONS

Definition

In its broadest context, the term **sensation** refers to a state of awareness of external or internal conditions of the body. In order for a sensation to occur, four prerequisites must be met: (1) A *stimulus*, or change in the environment, capable of initiating a response by the nervous system must be present. (2) A *receptor* or sense organ must pick up the stimulus and convert it to a nervous impulse. (3) The impulse must be *conducted* along a nervous pathway from the receptor or sense organ to the brain. (4) A region of the brain must *translate* the impulse into a sensation. A **sense receptor** or **sense organ** may be viewed as specialized nervous tissue that exhibits a high degree of sensitivity to internal or external conditions.

A receptor may be quite simple. For example, it might be just the dendrites of a single neuron that pick up the sensation of pain in the skin. Or, it may be a complex organ, such the the eye, that contains highly specialized neurons, epithelium, and connective tissues. Regardless of complexity, though, all sense receptors contain the dendrites of sensory neurons. The dendrites occur either alone or in close association with specialized cells belonging to other types of tissues. Receptors exhibit a very high degree of excitability, and their threshold stimulus is low. Except for receptors associated with pain, each is specialized for receiving a particular kind of stimulus.

The majority of sensory impulses are conducted to the sensory areas of the cerebral cortex because it is only in this region of the body that a stimulus can produce conscious feeling. Sensory impulses that terminate in the spinal cord or brain stem can initiate motor activities, but they can never produce conscious sensations. Once a stimulus is received by a receptor and converted into an impulse, the impulse is conducted along an afferent pathway that enters either the spinal cord or the brain.

Before studying the receptors, you will want to know something about the general characteristics of sensations and their basic classifications.

Characteristics

Although it may seem contrary to personal experience, conscious sensations occur in the cortical regions of the brain. In other words, you see, hear, and feel pain in the brain. You seem to see with your eyes, hear with your ears, and feel pain in an injured part of your body only because the cortex interprets the sensation as coming from the stimulated sense receptor. The term **projection** describes this process by which the brain refers sensations to their point of stimulation.

A second characteristic of many sensations is **adaptation.** According to this phenomenon, a sensation may disappear even though a stimulus is still being applied. For example, when you first get into a tub of hot water, you might feel an intense burning sensation. But, after a brief period of time, the sensation decreases to one of comfortable warmth, even though the stimulus (hot water) is still present. Other examples of adaptation include placing a ring on your finger, putting on your shoes or hat, and sitting on a chair. Initially, you are conscious of the sensations involved, but they are lost soon thereafter.

Sensations may also be characterized by **afterimages.** That is, some sensations tend to persist even though the stimulus has been removed. This phenomenon is just the opposite of adaptation. One common example of afterimage occurs when you look at a bright light and then look away or close your eyes. You still see the light for several seconds or minutes afterward.

A fourth characteristic of sensations is **modality,** which refers to the specific kind of sensation felt. The sensation may be one of pain, pressure, touch, body position, equilibrium, hearing, vision, smell, or taste. In other words, the distinct properties by which one sensation may be distinguished from another is its modality. As sensory signals enter the brain, they are sorted according to their modality and transferred to the appropriate region where a specific sensation is produced.

Classification

One convenient method of classifying sensations is to categorize them according to the location of the receptor. On this basis, receptors may be classified as exteroceptors, visceroceptors, and proprioceptors. **Exteroceptors** provide information about the external environment. They pick up stimuli outside the body and transmit sensations of hearing, sight, touch, pressure, temperature, and pain on the skin. Exteroceptors are located near the surface of the body.

Visceroceptors or **enteroceptors** pick up information about the internal environment. These sensations arise from within the body and may be felt as pain, taste, fatigue, hunger, thirst, and nausea. Visceroceptors are located in blood vessels and viscera.

Proprioceptors allow us to feel position and movement. Such sensations give us information about muscle tension, the position and tension of our joints, and equilibrium. These receptors provide information about body position and movements.

Sensations may also be classified according to the simplicity or complexity of the receptor and the neural pathway involved. *General senses* are those that involve a relatively simple receptor and neural pathway. In addition, the receptors for general sensations are numerous and are found throughout widespread areas of the body. Examples include cutaneous sensations, such as touch, pressure, heat, cold, and pain. *Special senses,* by contrast, involve rather complex receptors and neural pathways. The receptors for each special sense are found in only one or two specific areas of the body. Among the special senses are smell, taste, sight, and hearing. These will be discussed later in the chapter.

GENERAL SENSES

Your skin contains the receptor organs for many general senses. Receptor organs are also located in your muscles, tendons, joints, subcutaneous tissue, and viscera.

Cutaneous sensations

Touch, pressure, cold, heat, and pain are known as the **cutaneous sensations.** The receptors for these sensations are located in the skin, connective tissue, and the ends of the gastrointestinal tract. Inasmuch as the sensation of pain is not limited to cutaneous receptors, we shall consider pain under a separate heading. The cutaneous receptors are randomly distributed over the body surface so that some parts of the skin are densely populated with receptors and other parts contain only a few, widely separated ones. Areas of the body that have few cutaneous receptors are relatively insensitive, whereas those regions containing large numbers of cutaneous receptors are very sensitive. This can be demonstrated by using the *two-point discrimination test* for touch (Figure 16–1). In this test, a compass is applied to the skin, and the distance in millimeters between the two points of the compass is varied. The subject then indicates when two points are felt and when only one is felt.

The compass may be placed on the tip of the tongue, an area where receptors are very densely packed. The distance between the two points can then be narrowed to 1.4 mm (0.06 in.). At this distance, the points are able to stimulate two different receptors, and the subject feels touched by two objects. However, if the distance is decreased below 1.4 mm, the subject feels only one point, even though both points are touching the tongue. This is because the points are so close together that they reach only one receptor. The compass can then be placed on the back of the neck, where receptors are relatively few and far between. In this case, the subject feels two distinctly different points only if the distance between them is 36.2 mm (1.43 in.) or more.

The results of this test indicate that the more sensitive the area, the closer the compass points may be placed and still be felt separately. The following order for receptors illustrated, from greatest sensitivity to least, has been established from the test: tip of tongue, tip of finger, side of nose, back of hand, and back of neck.

Cutaneous receptors, have relatively simple structures. Basically, they consist of the dendrites of one or several sensory neurons that may or may not be enclosed in a capsule of epithelial or connective tissue. Impulses received by cutaneous touch receptors pass along somatic afferent neurons in spinal and cranial nerves, through the thalamus, to the general sensory area of the parietal lobe of the cortex.

Fine touch

Cutaneous receptors for **fine** or **light touch** include Meissner's corpuscles, Merkel's discs, and root hair plexuses (Figure 16–2). *Meissner's corpuscles* are egg-shaped receptors containing a mass of dendrites enclosed by epithelium. They are found in the papillae of the skin and enable us to detect when two points on the skin are being touched. Meissner's corpuscles are most numerous in the fingertips, palms of the hand, and soles of the feet. They are also abundant in the eyelids, tip of the tongue, lips, nipples, clitoris, and tip of the penis. *Merkel's discs* are receptors for touch that consist of disclike formations of dendrites attached to deeper layers of epidermal cells. They are distributed in many of the same locations as Meissner's corpuscles. *Root hair plexuses* are dendrites arranged in networks around the roots of hairs. If a hair shaft is moved, the dendrites are stimulated. Since the root hair plexuses are not surrounded by any supportive or protective structures, they are called *free,* or *naked, nerve endings.*

Deep pressure

Sensations of **deep pressure** are longer lasting and have less variation in intensity than do sensations of touch. Moreover, whereas light touch is felt in a small, "pinprick" area, deep pressure is felt over a much larger area. The deep pressure receptors are oval-shaped structures called *Pacinian corpuscles* (Figure 16–2). They are composed of a capsule resembling an onion made from layers of connec-

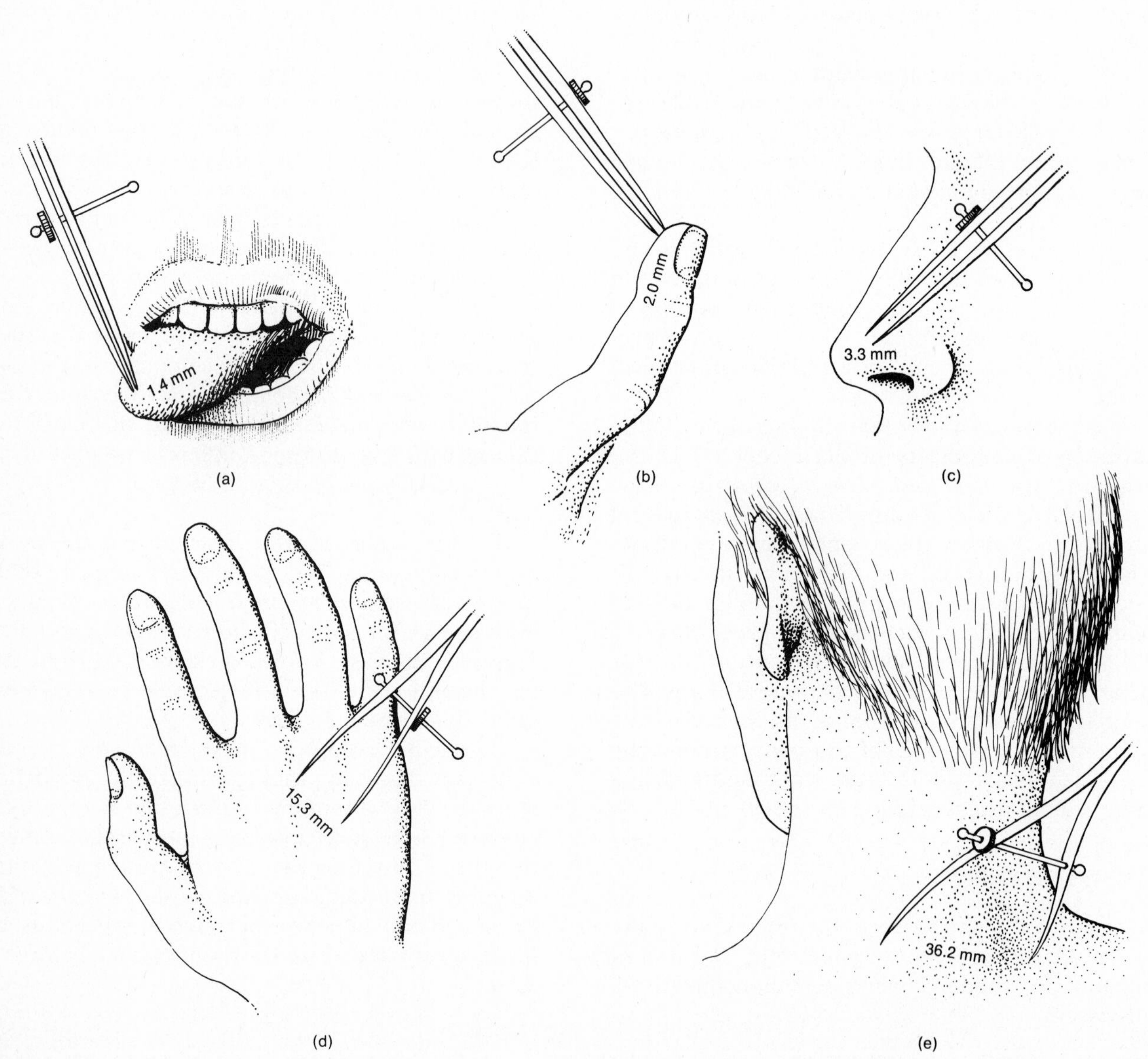

Figure 16–1. Two-point discrimination test for touch. (a) Tip of tongue. (b) Tip of finger. (c) Side of nose. (d) Back of hand. (e) Back of neck.

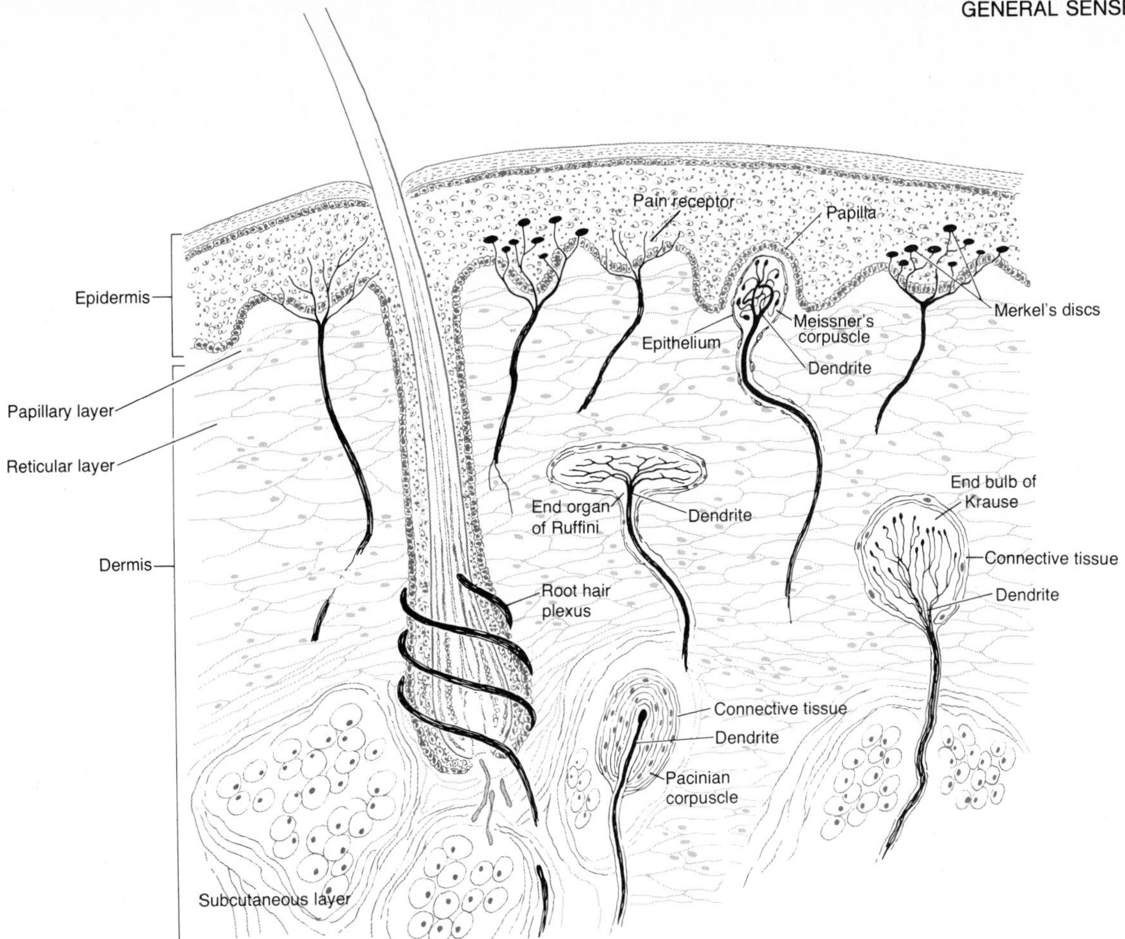

Figure 16–2. Structure and location of cutaneous receptors.

tive tissue enclosing dendrites. Pacinian corpuscles are found in the deep subcutaneous tissues that lie under mucous membranes, in serous membranes of the abdominal cavity, around joints and tendons, and in some viscera.

Cold

The cutaneous receptors for the sensation of **cold** are called *end bulbs of Krause* (Figure 16–2). The commonest form of these receptors is an oval-shaped connective tissue capsule containing dendrites. They are widely distributed in the dermis and subcutaneous connective tissue and are also located in the cornea of the eye, the tip of the tongue, and external genitals.

Heat

The cutaneous receptors for **heat** are referred to as *end organs of Ruffini* (Figure 16–2). The end organs of Ruffini are deeply embedded in the dermis and are less abundant than cold receptors.

Pain sensations

The receptors for **pain** are simply the branching ends of the dendrites of certain sensory neurons. Pain receptors are found in practically every tissue of the body and adapt only slightly or not at all. They may be stimulated by any type of stimulus. Excessive stimulation of any sense organ causes pain. For example, when stimuli for other sensations such as touch, pressure, heat, and cold reach a certain threshold, they stimulate pain receptors as well. Additional stimuli for pain receptors include excessive distention or dilation of an organ, prolonged muscular contractions, muscle spasms, inadequate blood flow to an organ, or the presence of certain chemical substances. Pain receptors, because of their sensitivity to all stimuli, have a general protective function in that they inform us of changes that could be potentially dangerous to health or life. Adaptation to pain does not readily occur. This is rather important since pain indicates disorder or disease. If we became used to it and ignored it, irreparable damage could result.

Sensory impulses for pain are conducted to the central nervous system through spinal and cranial nerves. The lateral spinothalamic tracts of the cord relay impulses to the thalamus. From here, the impulses may be relayed to the postcentral convolution of the parietal lobe. Recognition of the kind and intensity of most pain is ultimately localized in the cerebral cortex. Some pain discrimination occurs at subcortical levels.

In general, pain may be divided into two types: somatic and visceral. **Somatic pain** arises from stimulation of the skin receptors. In this case, it is called *superficial somatic pain.* It may also arise from stimulation of receptors in skeletal muscles, joints, tendons, and fascia, in which case it is called *deep somatic pain.* **Visceral pain** results from stimulation of receptors in the viscera.

In most cases of somatic pain and in some instances of visceral pain, the cortex accurately projects the pain back to the stimulated area. For example, if you burn your finger, you feel the pain in your finger. Also, an individual with inflammation of the lining of the pleural cavity experiences pain in the affected area. In most instances of visceral pain, however, the sensation of pain is not projected back to the point of stimulation. Rather, the pain may be felt in or just under the skin that overlies the stimulated organ. Or, the pain may be felt in a surface area of the body that is quite far removed from the stimulated organ. This phenomenon is called **referred pain.** In general, the area to which the pain is referred and the visceral organ involved receive their innervation from the same segment of the spinal cord. Consider the following example:

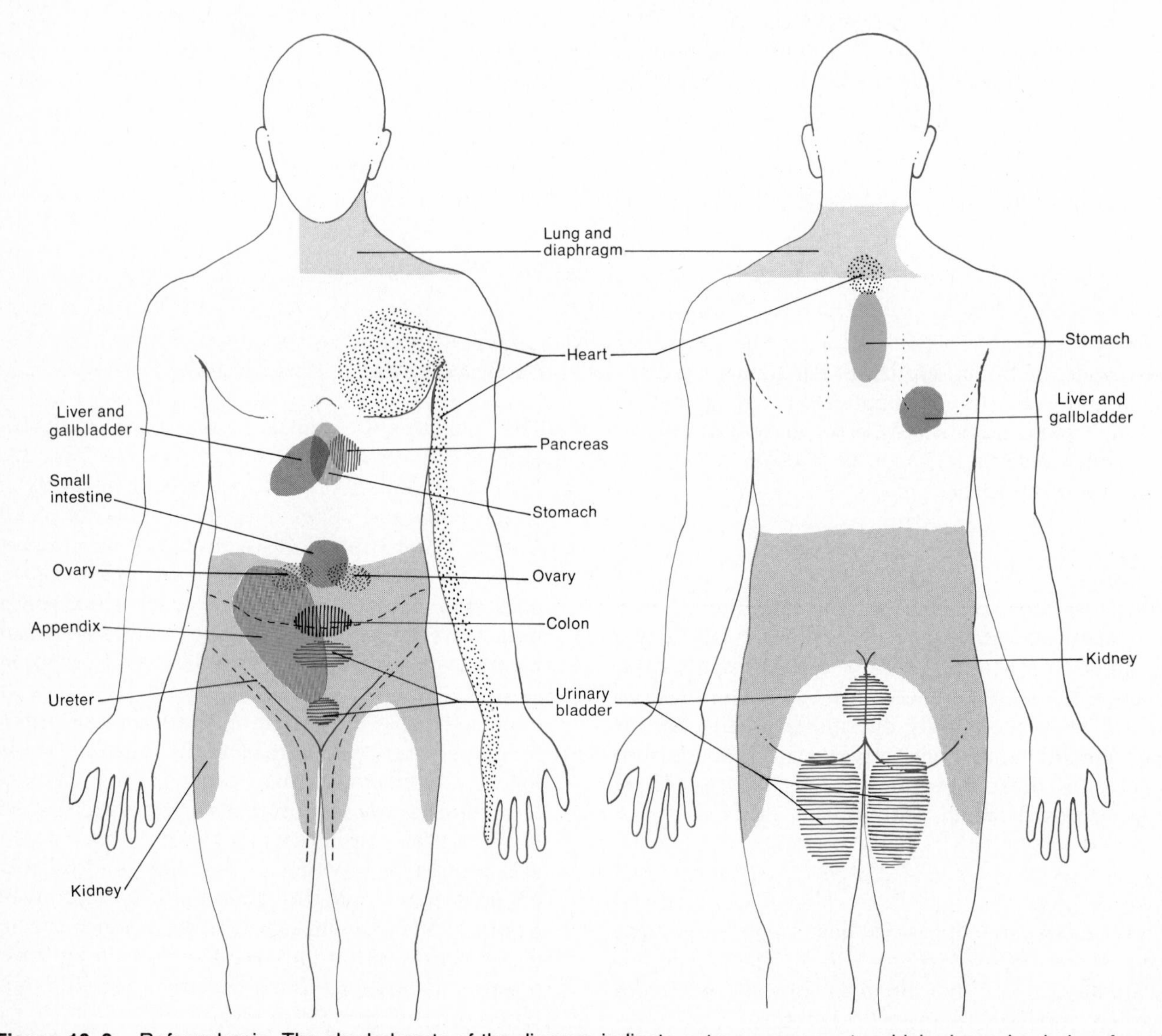

Figure 16–3. Referred pain. The shaded parts of the diagram indicate cutaneous areas to which visceral pain is referred.

Afferent fibers from the heart enter spinal cord segments T1 to T4 as do afferent fibers from the skin over the heart and the skin over the medial surface of the left arm. Thus, the pain of a heart attack is typically felt in the skin over the heart and down the left arm. Figure 16–3 illustrates cutaneous regions to which visceral pain may be referred.

A kind of pain frequently experienced by amputees is called **phantom pain.** In this instance, the person experiences pain in a limb or part of a limb after it has been amputated. Here is how phantom pain occurs. Let us say that a foot has been amputated. A sensory nerve that originally terminated in the foot is severed during the operation but repairs itself and returns to function within the remaining leg. From past experience the brain has always projected stimulation of the neuron back to the foot. So when the distal end of this neuron is now stimulated, the brain continues to project the sensation back to the missing part. Thus, even though the foot has been amputated, the patient still "feels" pain in his toes.

Proprioceptive sensations

An awareness of the activities of muscles, tendons, and joints is provided by the **proprioceptive** or **kinesthetic sense.** It informs us of the degree to which muscles are contracted and of the amount of tension that is consequently created in the tendons. The proprioceptive sense enables us to recognize the location and rate of movement of one part of the body in relation to other parts. It also allows us to estimate weight and to determine the muscular work necessary to perform a task. With the proprioceptive sense, we can judge the position and movements of our limbs without using our eyes when we walk, type, play a musical instrument, or dress in the dark.

Proprioceptive receptors are located in muscles, tendons, and joints. The receptors for proprioception are of three types. The *joint kinesthetic receptors* are located in the capsules of joints. These receptors provide feedback information on the degree and rate of angulation (change of position) of a joint. The other two receptors for proprioception, neuromuscular spindles and Golgi tendon organs, provide feedback information from muscles. *Neuromuscular spindles* consist of nerve filaments that are wrapped around specialized muscle fibers. They are located in almost all skeletal muscles and are found in large numbers in the muscles of the extremities. Neuromuscular spindles provide feedback information on the degree of stretch of a muscle. The information is relayed to the central nervous system to assist in the coordination and efficiency of muscle contraction. Neuromuscular spindles are involved in the stretch and extensor reflexes that were described in Chapter 13. *Golgi tendon organs* are also proprioceptive receptors that provide information on skeletal muscles. These are located at the junction of the muscle and tendon and function by sensing the tension applied to a tendon. The degree of tension informs the central nervous system of the degree of contraction of the muscle.

Proprioceptors adapt only slightly. This is beneficial since the brain must be apprised of the status of different parts of the body at all times so that adjustments can be made to ensure coordination.

The afferent pathway for muscle sense consists of impulses sent from proprioceptors into cranial and spinal nerves. Impulses for conscious proprioception travel in ascending tracts in the cord, where they are relayed to the thalamus and cerebral cortex. The sensation is registered in the general sensory area in the parietal lobe of the cortex posterior to the central fissure. Proprioceptive impulses that result in reflex action travel to the cerebellum through spinocerebellar tracts.

We have already noted in Chapter 13 that sensory fibers terminating in the spinal cord can bring about spinal reflexes. These reflexes do not require immediate action by the brain in order to occur. Sensory fibers that terminate in the lower brain stem can bring about much more complex motor reactions than simple spinal reflexes. When sensory information reaches the lower brain stem, it causes subconscious motor reactions. Sensory signals that reach the thalamus can be localized crudely in the body. In fact, at the thalamic level, sensations are distinguished on the basis of their modality. That is, the sensations are sorted out by type. When sensory information reaches the cerebral cortex, we experience very precise localization. It is at this level that memories of previous sensory information are stored and perception of the sensation occurs on the basis of past experiences.

SPECIAL SENSES

In contrast to the general senses, the special senses of smell, taste, sight, hearing, and equilibrium have receptor organs that are highly complex. The sense of smell is the least specialized, as opposed to the sense of sight, which is the most specialized. Like the general senses, however, the special senses allow us to detect changes in our environment. First, you will read about the anatomy of each special sense organ.

Olfactory sensations

The receptors for the **olfactory sense** are located in the nasal epithelium in the superior portion of the nasal cavity on either side of the nasal septum (Figure 16–4a). The nasal epithelium consists of two principal kinds of cells: supporting and olfactory (Figure 16–4b, c). The *supporting cells* are columnar epithelial cells of the mucous membrane lining the nose. Olfactory glands in the mucosa keep the mucous membrane moist. The *olfactory cells* are bipolar neurons whose cell bodies lie between the supporting cells. The distal (free) end of each olfactory cell contains six to eight dendrites, called *olfactory hairs.* The unmyelinated axons of the olfactory cells unite to form the *olfactory nerves,* which pass through foramina in the cribriform plate of the ethmoid bone. The olfactory nerves terminate in paired masses of gray matter called the *olfactory bulbs.* The olfactory bulbs lie beneath the frontal lobes of the cerebrum on either side of the crista galli of the ethmoid bone. The first synapse of the olfactory neural pathway occurs in the olfactory bulbs between the axons of the olfactory nerves and the dendrites of neurons inside the olfactory bulbs. Axons of these neurons run posteriorly to form the *olfactory tract.* From here, impulses are conveyed to the olfactory portion of the cortex. In the cortex, the impulses are interpreted as odor and give rise to the sensation of smell.

The mechanism by which the stimulus for smell is converted to a nerve impulse is explained by

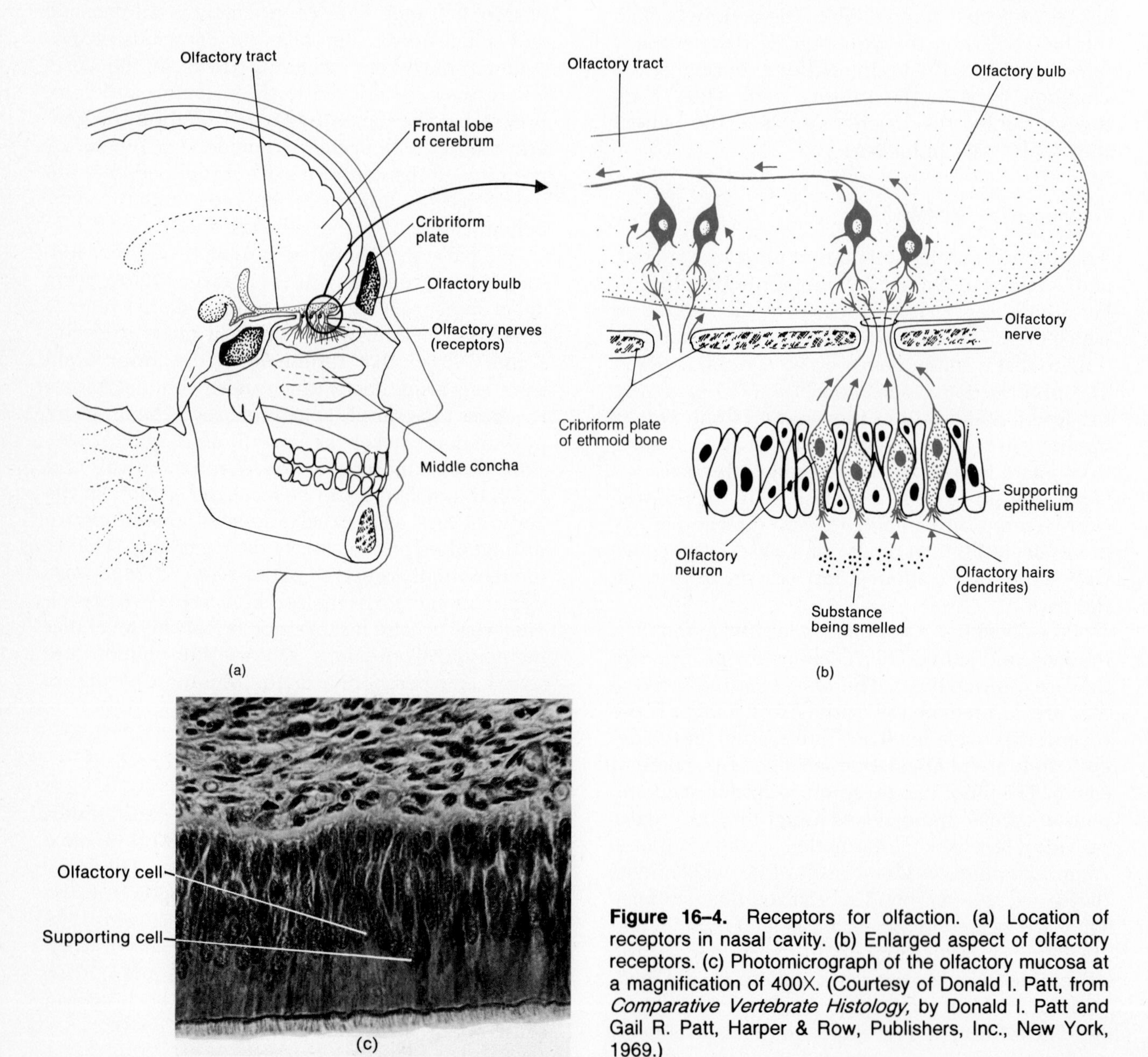

Figure 16–4. Receptors for olfaction. (a) Location of receptors in nasal cavity. (b) Enlarged aspect of olfactory receptors. (c) Photomicrograph of the olfactory mucosa at a magnification of 400X. (Courtesy of Donald I. Patt, from *Comparative Vertebrate Histology,* by Donald I. Patt and Gail R. Patt, Harper & Row, Publishers, Inc., New York, 1969.)

three widely accepted theories. One theory holds that substances capable of producing odors emit gaseous particles. On entering the nasal cavities, these particles become dissolved in the mucus of the nasal membrane. This fluid then acts chemically on the olfactory hairs to create a nerve impulse. According to the second theory, radiant energy given off by the molecules of the stimulating substance is the stimulus rather than the molecules themselves. The third theory purports that substances detected by smell are usually soluble in fat. Since the membrane of an olfactory hair is largely fat, it is assumed that molecules of substances to be smelled are dissolved in the membrane where they initiate a nerve impulse.

The sensation of smell happens quickly, but adaptation to odors also occurs rapidly. For this reason, we become accustomed to some odors and are also able to endure unpleasant ones. Rapid adaptation also accounts for the failure of a person to detect a gas that accumulates slowly in a room. The cortex stores memories of odors quite well. Once you have smelled a substance, you generally recognize its odor if you smell it again.

The supporting cells of the nasal epithelium and tear ducts are innervated by branches of the trigeminal nerve (V). The nerve receives stimuli of pain, cold, heat, tickling, and pressure. Olfactory stimuli, such as pepper, onions, ammonia, ether, and chloroform, are irritating and may cause tearing because they stimulate the receptors of the trigeminal nerve as well as the olfactory neurons.

Gustatory sensations

The receptors for **gustatory sensations,** or sensations of taste, are located in the taste buds (Figure 16–5b). Taste buds are most numerous on the tongue, but they are also found on the soft palate and in the throat. The *taste buds* are oval-shaped bodies consisting of two kinds of cells. The *supporting cells* are specialized epithelium that forms a capsule. Inside each capsule are 4 to 20 *gustatory cells,* which are the sensory neurons for taste. Each gustatory cell contains a dendrite that projects to the external surface through an opening in the taste bud called the *taste pore.* Gustatory cells make contact with taste stimuli through the taste pore.

Taste buds are found in some connective tissue elevations on the tongue called **papillae** (Figure 16–5a). The papillae give the upper surface of the tongue its characteristic rough appearance. *Circumvallate papillae,* the largest type, are circular and form an inverted V-shaped row at the posterior portion of the tongue. *Fungiform* (mushroom-shaped) *papillae* are knoblike elevations and are found primarily on the tip and sides of the tongue. All circumvallate and most fungiform papillae contain taste buds. *Filiform papillae* are threadlike structures that cover the anterior two-thirds of the tongue. (See Figure 20–5.)

In order for gustatory cells to be stimulated, the substances we taste must be in solution in the saliva so that they can enter the taste pores in the taste buds. Despite the many substances we taste, there are basically only four taste sensations: sour, salt, bitter, and sweet. Each taste is due to a different response to different chemicals. Certain regions of the tongue react more strongly than other regions to particular taste sensations. For example, although the tip of the tongue reacts to all four taste sensations, it is highly sensitive to sweet and salty substances. The posterior portion of the tongue is highly sensitive to bitter substances, and the lateral edges of the tongue are more sensitive to sour substances. (See Figure 20–5.)

The cranial nerves that supply afferent fibers to taste buds are the facial nerve (VII), which supplies the anterior two-thirds of the tongue; the glossopharyngeal (IX), which supplies the posterior one-third of the tongue; and the vagus (X), which supplies the epiglottis area of the throat. Taste impulses are conveyed from the gustatory cells in taste buds to the three nerves just cited. From these, the impulses enter the medulla, pass through the thalamus, and terminate in the parietal lobe of the cortex.

Visual sensations

The structures related to vision are the eyeball, which is the receptor organ for visual sensations, the optic nerve, the brain, and a number of accessory structures.

Accessory structures of the eye

Among the accessory organs are the eyebrows, eyelids, eyelashes, and the lacrimal apparatus, which produces tears (Figure 16–6). The *eyebrows* form a transverse arch at the junction of the upper eyelid and forehead. Structurally, they resemble the hairy scalp. The skin of the eyebrows is richly supplied with sebaceous glands. The hairs are generally coarse and directed laterally. Deep to the skin of the eyebrows are the fibers of the orbicularis oculi muscles. The eyebrows protect the eyeballs from falling objects, prevent perspiration from getting into the eyes, and shade the eyes from the direct rays of the sun.

The *eyelids* or *palpebrae,* upper and lower, consist of dense folds of skin. They shade the eyes during sleep, protect the eyes from excessive light rays and foreign objects, and spread lubricating secre-

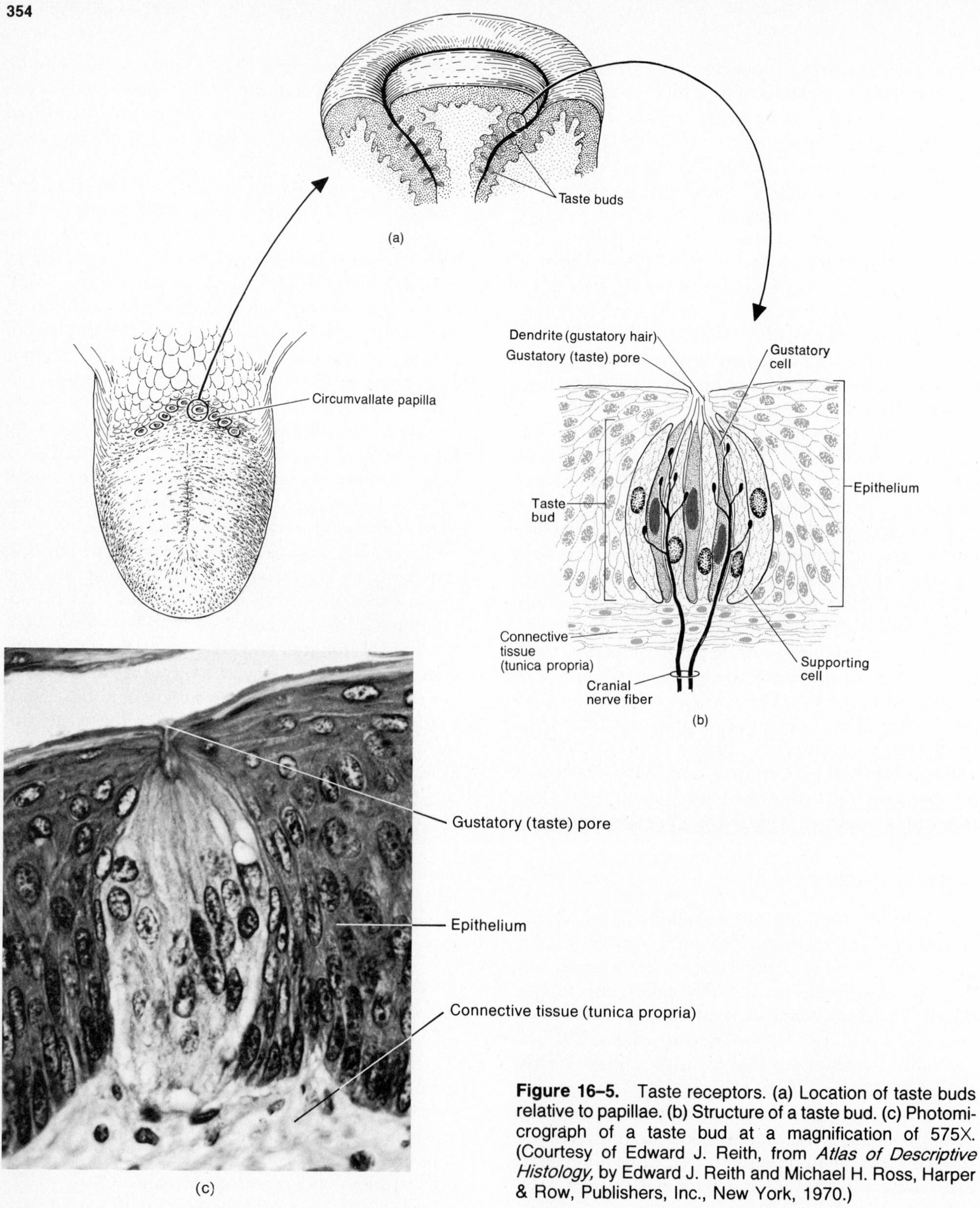

Figure 16–5. Taste receptors. (a) Location of taste buds relative to papillae. (b) Structure of a taste bud. (c) Photomicrograph of a taste bud at a magnification of 575X. (Courtesy of Edward J. Reith, from *Atlas of Descriptive Histology,* by Edward J. Reith and Michael H. Ross, Harper & Row, Publishers, Inc., New York, 1970.)

tions over the surface of the eyeballs. The upper eyelid is more movable than the lower and has attached to it a special levator muscle known as the *levator palpebrae superioris.* The space between the upper and lower eyelids that exposes the eyeball is called the *palpebral fissure.* Its angles are known as the *lateral canthus,* which is narrower and closer to the temporal bones, and the *medial canthus,* which is broader and nearer the nasal bones. In the medial canthus, there is a small reddish elevation, the *lacrimal caruncle,* containing sebaceous and sudoriferous glands. A whitish material secreted by the caruncle collects in the medial canthus.

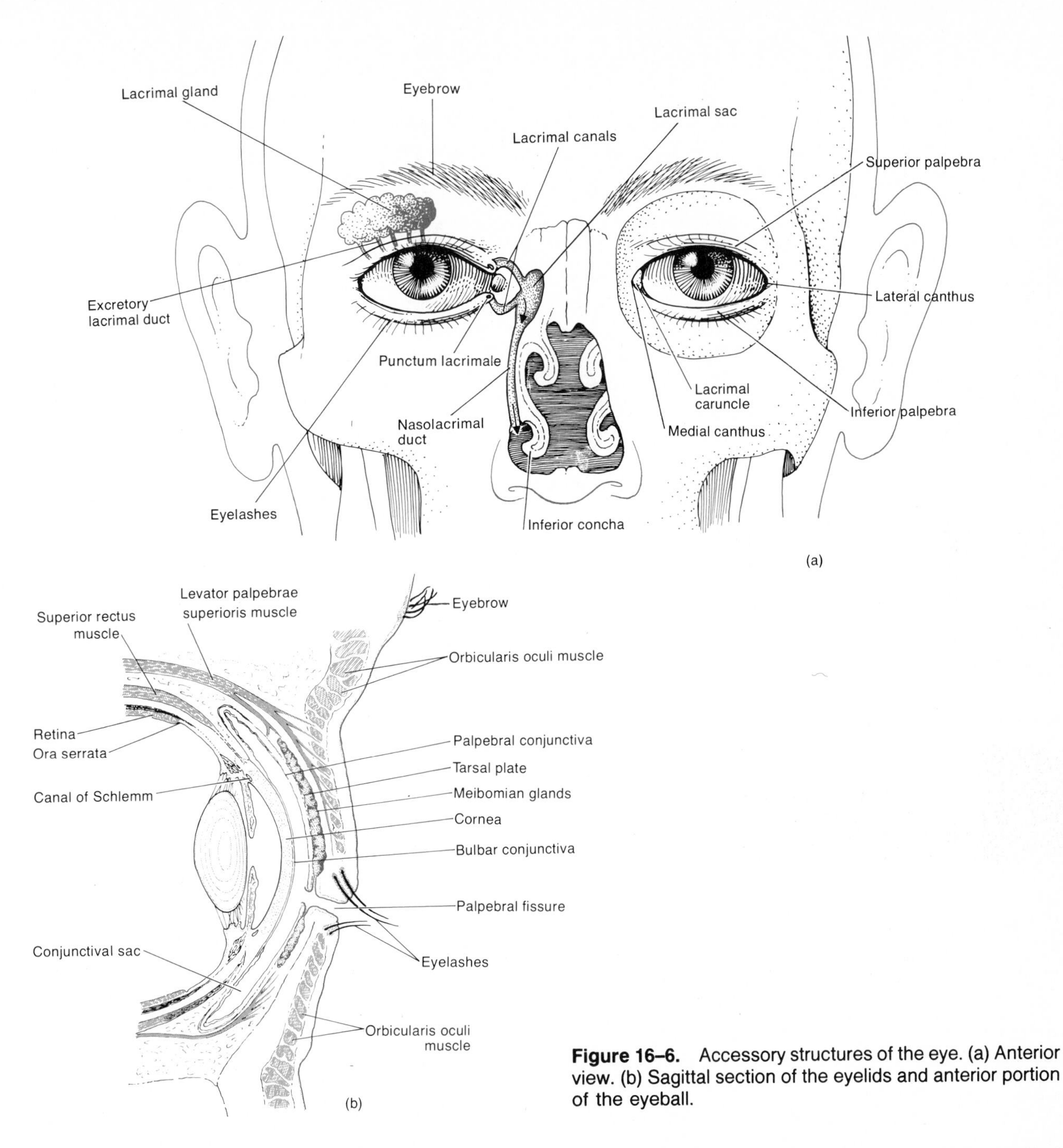

Figure 16–6. Accessory structures of the eye. (a) Anterior view. (b) Sagittal section of the eyelids and anterior portion of the eyeball.

From superficial to deep, each eyelid consists of epidermis, dermis, subcutaneous areolar tissue, fibers of the orbicularis oculi muscle, tarsal plate, tarsal glands, and conjunctiva. The *tarsal plate* forms the border of the eyelid and contains the roots of the eyelashes. It is a thickened fold of connective tissue. Embedded in grooves on the deep surface of each tarsal plate is a row of elongated tarsal glands known as *Meibomian glands.* These are modified sebaceous glands, and their oily secretion helps to keep the eyelids from adhering to each other. Infection of these glands produces a *chalazion,* or *cyst,* on the eyelid. Sebaceous glands at the base of the hair follicles of the eyelashes called *glands of Zeis* pour a lubricating fluid into the follicles. Infection of these glands is called a *sty.* The *conjunctiva* is a thin mucous membrane called the *palpebral conjunctiva* when it lines the inner aspect of the eyelids and called the *bulbar conjunctiva* when it is reflected from the eyelids onto the eyeball to the periphery of the cornea. Inflammation of this membrane is called *conjunctivitis,* or *pinkeye.*

Projecting from the border of each eyelid, anterior to the Meibomian glands, is a row of short, thick hairs, the *eyelashes.* In the upper lid, they are relatively long and turn upward, whereas in the lower lid, they are relatively short and turn downward.

The *lacrimal apparatus* is a term used for a group of structures that manufactures and drains away tears. These structures are the lacrimal glands, the excretory lacrimal ducts, the lacrimal canals, the lacrimal sacs, and the nasolacrimal ducts. A *lacrimal gland* is a compound tubuloacinar gland and is located at the superior lateral portion of both orbits. Each is about the size and shape of an almond. Leading from the lacrimal glands are 6 to 12 *excretory lacrimal ducts* that empty lacrimal fluid, or tears, onto the surface of the conjunctiva of the upper lid. From here, the lacrimal fluid passes medially and enters two small openings *(puncta lacrimalia)* that appear as two small dots, one in each papilla of the eyelid, at the medial canthus of the eye. The lacrimal secretion then passes into two ducts, the *lacrimal canals,* and is next conveyed into the lacrimal sac. The *lacrimal sac* is the superior expanded portion of the nasolacrimal duct, a canal that transports the lacrimal secretion into the inferior meatus of the nose.

The *lacrimal secretion* is a watery solution containing salts, some mucus, and a bactericidal enzyme called lysozyme. It cleans, lubricates, and moistens the external surface of the eyeball. After being secreted by the lacrimal glands, it is spread over the surface of the eyeball by the blinking action of the eyelids. Normally, the secretion is carried away by evaporation or by passing into the lacrimal canals as fast as it is produced. If, however, an irritating substance makes contact with the conjunctiva, the lacrimal glands are stimulated to oversecrete. Tears then accumulate more rapidly than they can be carried away. This is a very important protective mechanism since the tears dilute and wash away the irritating substance. "Watery" eyes also occur when an inflammation of the nasal mucosa, such as a cold, obstructs the nasolacrimal ducts so that drainage of the tears is blocked.

Structure of the eyeball

The adult **eyeball** measures about 2.5 cm (1 in.) in diameter. Of its total surface area, only the anterior one-sixth is exposed. The remainder is recessed and protected by the orbit into which it fits. Anatomically, the eyeball can be divided into three principal layers: (1) fibrous tunic, (2) vascular tunic, and (3) retina (Figure 16–7a-c).

The **fibrous tunic** is the outer coat of the eyeball. It can be divided into two regions, the posterior sclera and the anterior cornea. The *sclera,* called the "white of the eye," is a white coat of fibrous tissue that covers all the eyeball except the anterior colored portion. The sclera gives shape to the eyeball and affords protection for its inner parts. Its posterior surface is pierced by the optic nerve. The anterior portion of the fibrous tunic is called the *cornea.* It is a nonvascular, transparent fibrous coat that covers the iris, the colored part of the eye. Like the sclera, the cornea is composed of fibrous tissue. The outer surface of the cornea contains an epithelial layer that is continuous with the epithelium of the bulbar conjunctiva. As you will see shortly, an unequal curvature of the surface of the cornea can cause blurred vision, or astigmatism. At the junction of the sclera and cornea is a structure referred to as the *canal of Schlemm.*

The **vascular tunic** is the middle layer of the eyeball and is composed of three portions: the posterior choroid, the anterior ciliary body, and iris. Collectively, these three structures are called the *uvea.* The *choroid* is a thin, dark brown membrane that lines most of the internal surface of the sclera. It contains numerous blood vessels and a large amount of pigment. The choroid absorbs light rays so they are not reflected back out of the eyeball. Through its blood supply, it also maintains the nutrition of the retina. The optic nerve also pierces the choroid at the back of the eyeball. The anterior portion of the choroid becomes the *ciliary body.* It is the thickest portion of the vascular tunic (Figure 16–7a-c). It extends from the *ora serrata* of the retina (inner tunic) to a point just behind the sclerocorneal junction. The ora serrata is simply the jagged margin of the retina. This second division of the vascular tunic contains the *ciliary muscle*—a smooth muscle that alters the shape of the lens for near or far vision. The *iris* is the third portion of the vascular tunic. It consists of circular and radial smooth muscle fibers arranged to form a doughnut-shaped structure. The black hole in the center of the iris is the *pupil,* the area through which light enters the eyeball. The iris is suspended between the cornea and the lens and is attached at its outer margin to the ciliary body. One of the principal functions of the iris is to regulate the amount of light entering the eyeball. For example, when the eye is stimulated by bright light, the circular muscles of the iris contract and decrease the size of the pupil. When the eye must adjust to dim light, the radial muscles of the iris contract and increase the size of the pupil. Smooth muscles inside the eyeball, such as the ciliary, radial, and circular muscles, originate and insert within the eyeball and adjust the eye internally for vision. They are referred to as *intrinsic eye muscles.* Skeletal muscles, such as the recti and oblique muscles, originate in the orbit and insert on the outside surface of the eyeball. (See Figure 10–6.) They move the eyeball in various directions and are called *extrinsic eye muscles.*

The third and inner coat of the eye, the **retina,** lies only in the posterior portion of the eye. Its primary function is image formation. It consists of a nervous tissue layer and pigmented layer (Figure 16–7a-c). The outer pigmented layer is composed of epithelial cells lying in contact with the choroid. The inner nervous layer contains three zones of neurons. These three zones, named in the order in which they conduct impulses, are the photoreceptor neurons, bipolar neurons, and ganglion neurons.

The dendrites of the photoreceptor neurons are called rods and cones because of their respective shapes. They are visual receptors highly specialized for stimulation by light rays. **Rods** are specialized for vision in dim light. They also allow us to discriminate between different shades of dark and light and permit us to see shapes and movement. **Cones,** by contrast, are specialized for color vision and for sharpness of vision, called *visual acuity.* Cones are stimulated only by bright light. This is why we cannot see color by moonlight. It is estimated that there are 7 million cones and somewhere between 10 and 20 times as many rods. Cones are most densely concentrated in the *central fovea,* a small depression in the center of the macula lutea. (See Figure 16–7a.) The *macula lutea,* or yellow spot, is situated in the exact center of the retina. The fovea is the area of sharpest vision because of the high concentration of cones. Rods are absent from the fovea and macula, but they increase in density toward the periphery of the retina.

When impulses for sight have passed through the photoreceptor neurons, they are conducted across synapses to the bipolar neurons in the intermediate zone of the nervous layer of the retina. From here, the impulses are passed to the ganglion neurons.

The axons of the ganglion neurons extend posteriorly to a small area of the retina called the *optic disc,* or *blind spot.* This region contains openings through which the fibers of the ganglion neurons exit as the optic nerve. Since this area contains neither rods nor cones, and only nerve fibers, no image is formed on it. Thus, it is called the blind spot. (See Figure 16–7a.)

In addition to the fibrous tunic, vascular tunic, and retina, the eyeball itself contains the lens, just behind the pupil and iris (Figure 16–7a-c). The *lens* is constructed of numerous layers of protein fibers arranged like the layers of an onion. Normally, the lens is perfectly transparent. It is enclosed by a clear plastic capsule and held in position by the *suspensory ligament.* An opacity of the lens is known as a *cataract.*

The interior part of the eyeball contains a large cavity that is divided into two smaller cavities (see Figure 16–7a). These are called the anterior cavity and the posterior cavity. They are separated from each other by the lens. The *anterior cavity,* in turn, has two subdivisions referred to as the anterior chamber and the posterior chamber. The *anterior chamber* lies behind the cornea and in front of the iris. The *posterior chamber* lies behind the iris and in front of the suspensory ligaments and lens. The anterior cavity is filled with a clear, watery fluid called the *aqueous humor.* The fluid is believed to be secreted in the posterior chamber by the epithelium of the ciliary bodies behind the iris. From the posterior chamber, the fluid permeates the posterior cavity and then passes forward between the iris and the lens, through the pupil into the anterior chamber. From the anterior chamber, the aqueous humor, which is continually produced, is drained off into the *canal of Schlemm* and then into the blood. The pressure in the eye, called *intraocular pressure,* is produced mainly by the aqueous humor. The intraocular pressure keeps the retina smoothly applied to the choroid so that the retina may form clear images. Normal intraocular pressure (about 24 mm Hg) is maintained by the drainage of the aqueous humor through the canal of Schlemm and into the blood. Abnormal elevation of intraocular pressure, called *glaucoma,* results in degeneration of the retina and blindness. In addition to maintaining normal intraocular pressure, the aqueous humor is also the principal link between the circulatory system and the lens and cornea. Neither the lens nor the cornea has blood vessles.

The second, and larger, cavity of the eyeball is the *posterior cavity.* It lies between the lens and the retina and contains a soft, jellylike substance called the *vitreous humor.* This substance contributes to intraocular pressure, helps to prevent the eyeball from collapsing, and holds the retina flush against the internal portions of the eyeball. However, the vitreous humor, unlike the aqueous humor, does not undergo constant replacement. It is formed during embryonic life and is not replaced thereafter.

Before you read the next section, refer to Exhibit 1–3, which contains some surface features of the

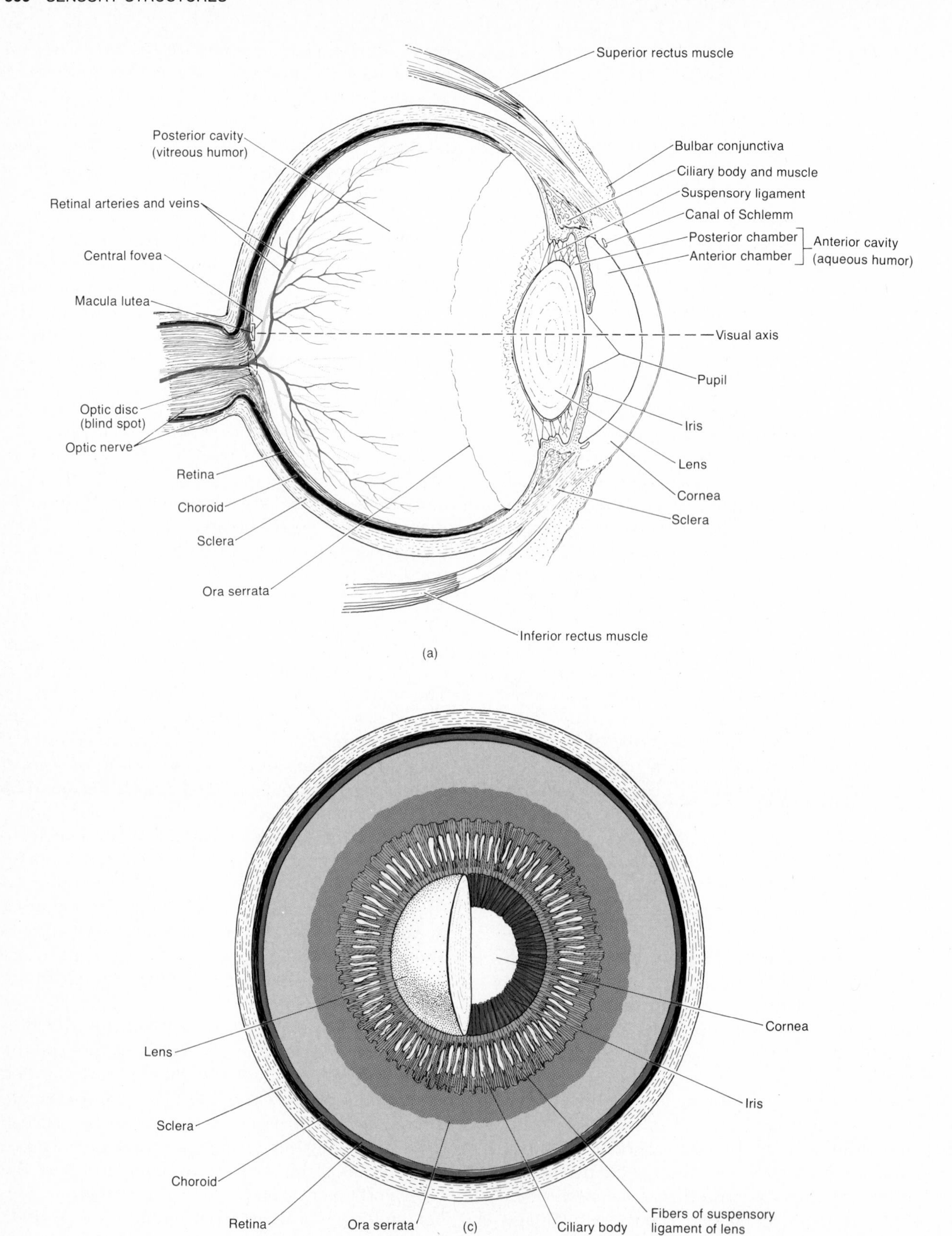

Figure 16–7. Structure of the eyeball. (a) Gross anatomy in sagittal section. (b) Section through the anterior part of the eyeball at the sclerocorneal junction. (c) Anterior portion of the eyeball seen from behind. (d) Diagram of the microscopic structure of the retina. (e) Photomicrograph of the posterior wall of the eyeball, showing the layers of the retina at a magnification of 100X. (Courtesy of Donald I. Patt, from *Comparative Vertebrate Histology,* by Donald I. Patt and Gail R. Patt, Harper & Row, Publishers, Inc., New York, 1969.)

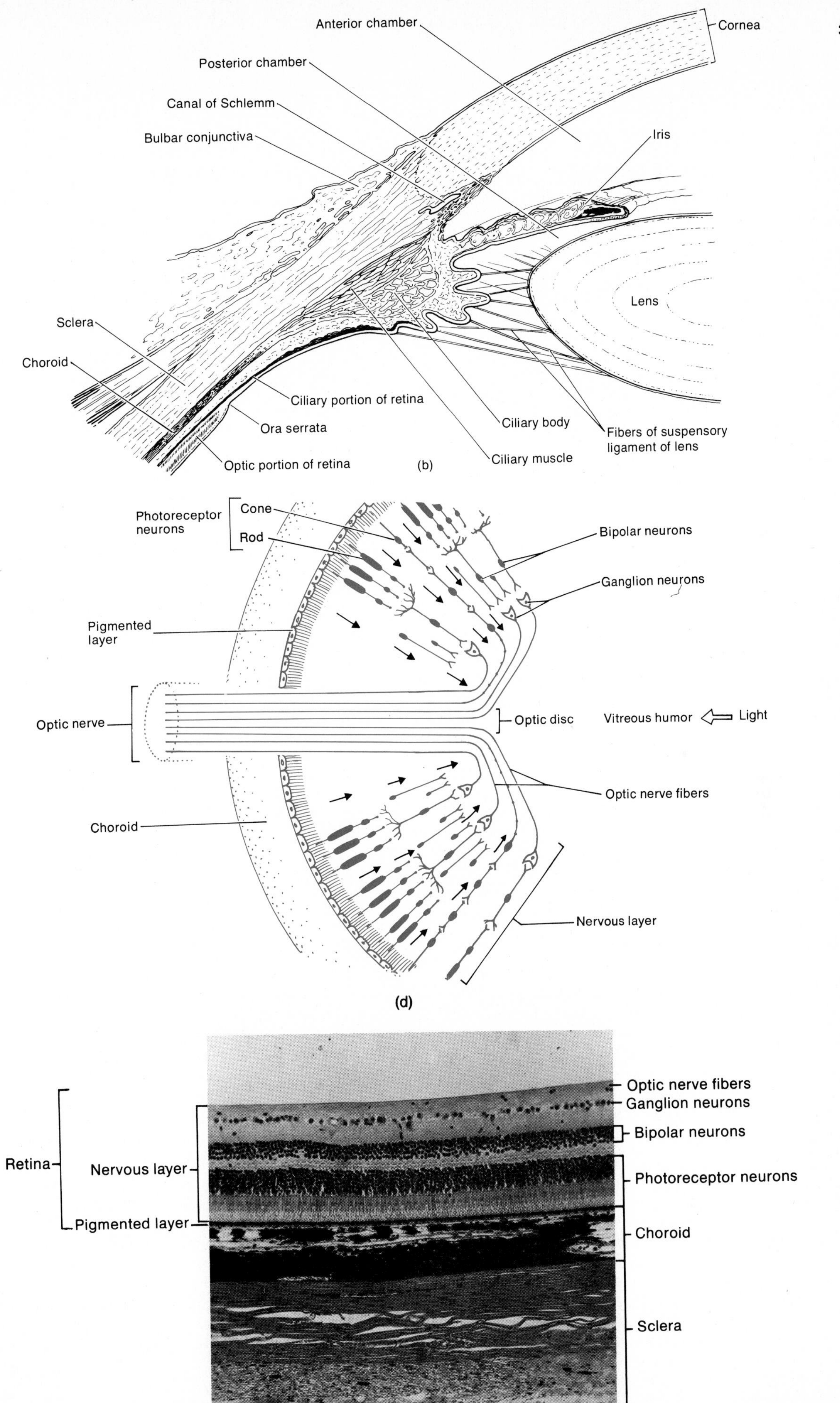

Anterior chamber
Cornea
Posterior chamber
Canal of Schlemm
Bulbar conjunctiva
Iris
Lens
Sclera
Choroid
Ciliary portion of retina
Ora serrata
Optic portion of retina
Ciliary body
Fibers of suspensory ligament of lens
Ciliary muscle
(b)
Photoreceptor neurons
Cone
Rod
Bipolar neurons
Ganglion neurons
Pigmented layer
Optic nerve
Optic disc
Vitreous humor
Light
Optic nerve fibers
Choroid
Nervous layer
(d)
Optic nerve fibers
Ganglion neurons
Bipolar neurons
Photoreceptor neurons
Choroid
Sclera
Retina
Nervous layer
Pigmented layer
(e)

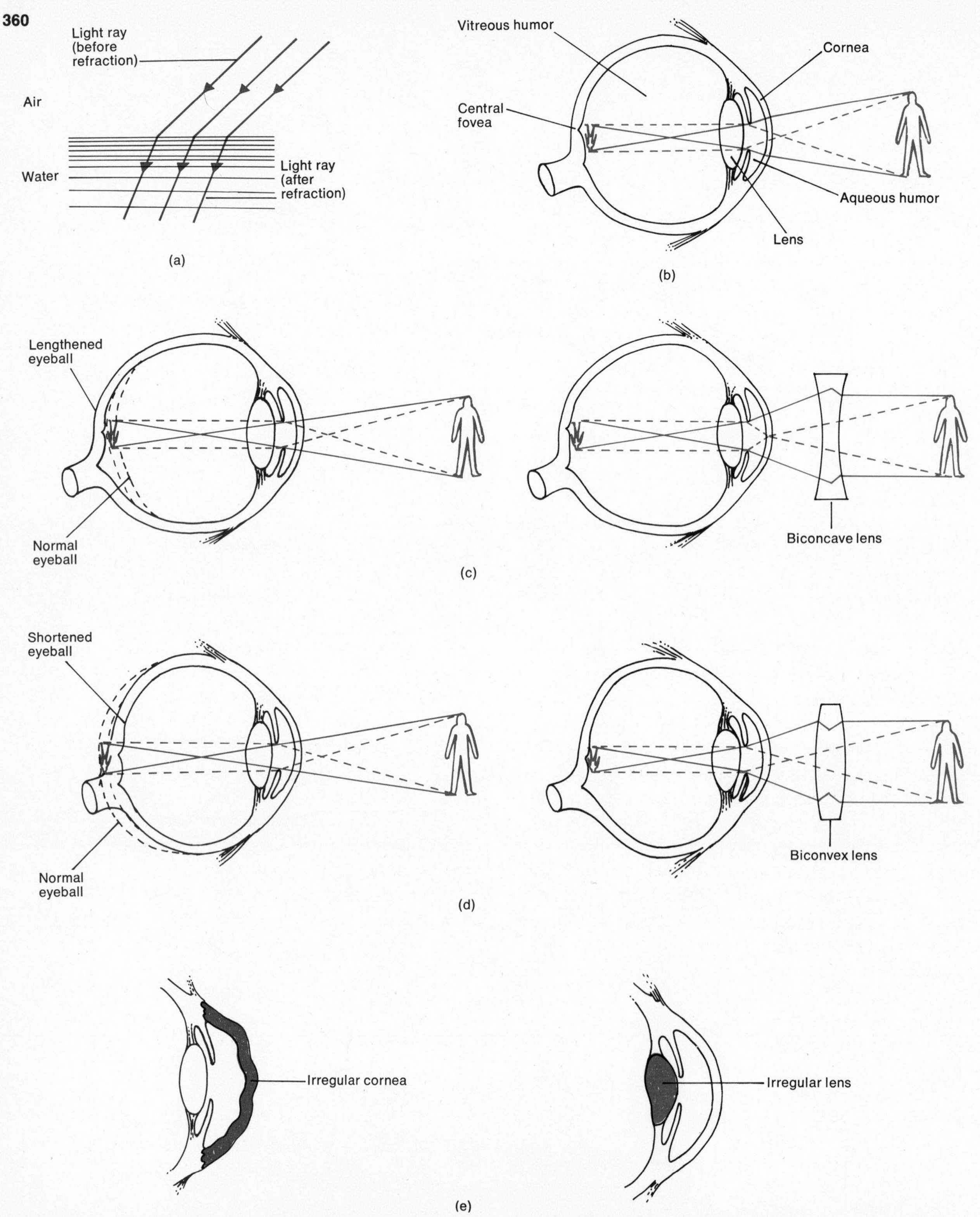

Figure 16–8. Normal and abnormal refraction in the eyeball. (a) Refraction of light rays passing from air into water. (b) In the normal or emmetropic eye, light rays from an object are bent sufficiently by the four refracting media and converged on the central fovea. A clear image is formed. (c) In the myopic eye, the image is focused in front of the retina. The condition may result from an elongated eyeball or a thickened lens. Correction is by use of a concave lens. (d) In the hypermetropic eye, the image is focused behind the retina. The condition may be the result of the eyeball being too short or the lens being too thin. Correction is by a convex lens. (e) Astigmatism. This is a condition in which the curvature of the cornea or lens is uneven. As a result, horizontal and vertical rays are focused at two different points on the retina. Suitable glasses correct the refraction of an astigmatic eye. On the left, astigmatism resulting from an irregular cornea. On the right, astigmatism resulting from an irregular lens. The image is not focused on the area of sharpest vision of the retina. This results in blurred or distorted vision.

eye. See how many structures you can find and define the function of each.

Physiology of vision

Before light can reach the rods and cones of the retina, it must first pass through the cornea, aqueous humor, pupil, lens, and vitreous humor. Moreover, for vision to occur, light reaching the rods and cones must form an image on the retina. The resulting nerve impulses must then be conducted to the visual areas of the cerebral cortex. In discussing the physiology of vision, let us first consider retinal image formation.

RETINAL IMAGE FORMATION. The formation of an image on the retina requires four basic processes, all concerned with focusing light rays. These basic processes are (1) refraction of light rays, (2) accommodation of the lens, (3) constriction of the pupil, and (4) convergence of the eyes.

Refraction and accommodation. When light rays traveling through a transparent medium (such as air) pass into a second transparent medium with a different density (such as water), the rays bend at the surface of the two media. This is called *refraction* (Figure 16–8a). The eye has four such media of refraction: the cornea, aqueous humor, lens, and vitreous humor (Figure 16–8b). Light rays entering the eye from the air are refracted at the following points: (1) the anterior surface of the cornea as they pass from the lighter air into the denser cornea; (2) the posterior surface of the cornea as they pass into the less dense aqueous humor; (3) the anterior surface of the lens as they pass from the aqueous humor into the denser lens; and (4) the posterior surface of the lens as they pass from the lens into the less dense vitreous humor.

When an object is 6m (20 ft) or more away from the viewer, the light rays that are reflected from the object are nearly parallel to one another. The degree of refraction that takes place at each surface in the eye is very precise. Because of this, the parallel rays are sufficiently bent to fall exactly on the central fovea, where vision is sharpest. However, light rays that are reflected from close-by objects are divergent rather than parallel. As a result, they must be refracted toward each other to a greater extent. This necessary change in refraction is brought about by the lens of the eye.

If the surfaces of a lens curve outward, as in a biconvex lens, the lens will refract the rays toward each other so that they eventually intersect (Figure 16–8d). The more the lens curves outward, the more acutely it bends the rays toward each other. Conversely, when the surfaces of a lens curve inward, as in a biconcave lens, the rays bend away from each other (Figure 16–8c). The lens of the eye is biconvex. Furthermore, it has the unique ability to change the focusing power of the eye by becoming moderately curved at one moment and greatly curved the next. When the eye is focusing on a close object, the lens curves greatly in order to bend the rays toward the central fovea. This increase in the curvature of the lens is called *accommodation* (Figure 16–9). During accommodation, the ciliary muscle contracts, pulling the ciliary body and choroid forward toward the lens. This releases the tension on the lens and suspensory ligament. Due to its elasticity, the lens shortens, thickens, and bulges. In near vision, the ciliary muscle is contracted, and the lens is bulging. In far vision, the ciliary muscle is relaxed, and the lens is flatter. With aging, the lens loses elasticity and, therefore, its ability to accommodate.

The normal eye, referred to as an *emmetropic eye*, can sufficiently refract light rays from an object 6 m (20 ft) away to focus a clear object on the retina.

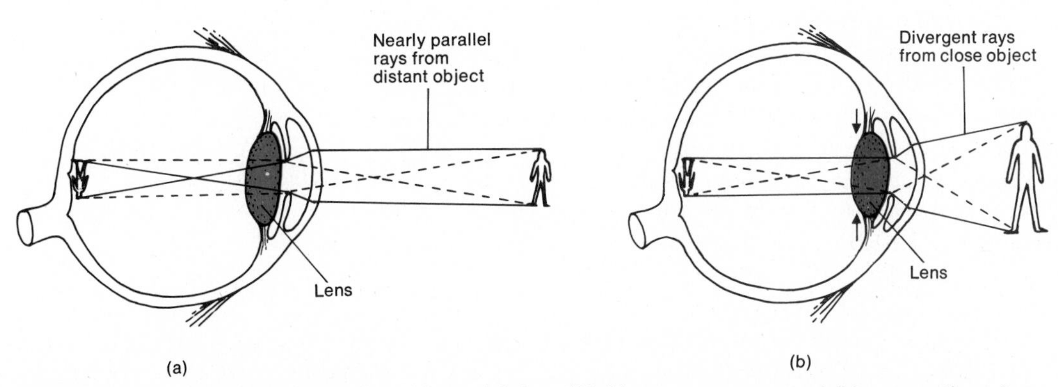

Figure 16–9. Accommodation for objects (a) 6 m (20 ft) or more away and (b) nearer than 6 m.

Many individuals, however, do not have this ability because of abnormalities related to improper refraction. Among these abnormalities are *myopia* (nearsightedness), *hypermetropia* (farsightedness), and *astigmatism* (irregularities in the surface of the lens or cornea). The conditions are illustrated and explained in Figure 16–8c, e. Why do you think nearsightedness can be corrected with glasses containing biconcave lenses? How would you correct farsightedness?

Constriction of the pupil. The muscles of the iris also assume a function in the formation of clear retinal images. Part of the accommodation mechanism consists of the contraction of the circular muscle fibers of the iris to constrict the pupil. Constricting the pupil means narrowing the diameter of the hole through which light enters the eye. This occurs simultaneously with accommodation of the lens and prevents light rays from entering the eye through the periphery of the lens. Light rays entering at the periphery would not be brought to focus on the retina and would result in blurred vision. The pupil, as noted earlier, also constricts in bright light to protect the retina from sudden or intense stimulation.

Convergence. When birds use their eyes, they see a set of objects off to the left through one eye and an entirely different set of objects off to the right through the other eye. This characteristic effectively doubles their field of vision and allows them to detect a predator sneaking up from behind. In human beings, both eyes focus on only one set of objects–a characteristic called *single binocular vision.*

Single binocular vision occurs when light rays from an object are directed toward corresponding points on the two retinas. When we stare straight ahead at a distant object, the incoming light rays are aimed directly at both pupils and are refracted to identical spots on the retinas of both eyes. But as we move close to the object, our eyes must rotate medially–that is, become "crossed"–in order for the light rays from the object to hit the same points on both retinas. The term *convergence* refers to this medial movement of the two eyeballs so that they are both directed toward the object being viewed. The nearer the object, the greater the degree of convergence necessary to maintain single binocular vision. Convergence is brought about by the coordinated action of the extrinsic eye muscles.

Inverted image. Images are actually focused upside down on the retina. They also undergo mirror reversal. That is, light reflected from the right side of an object hits the left side of the retina and vice versa. Note in Figure 16–8b that reflected light from the top of the object crosses light from the bottom of the object and strikes the retina below the central fovea. Reflected light from the bottom of the object crosses light from the top of the object and strikes the retina above the central fovea. The reason why we do not see a topsy-turvy world is that the brain learns early in life to coordinate visual images with the exact locations of objects. The brain stores memories of reaching and touching objects and automatically turns visual images right-side-up and right-side-around.

STIMULATION OF PHOTORECEPTORS. After an image is formed on the retina by refraction, accommodation, constriction of the pupil, and convergence, light impulses must be converted into nerve impulses by the rods and cones. The exact mechanism by which light acts as a stimulus to initiate nerve impulses that result in the sensation of sight is not entirely clear. The following data, however, are known. Rods contain a reddish purple pigmented compound called *rhodopsin,* or *visual purple.* This substance consists of the protein scotopsin plus retinene, a derivative of vitamin A. When light rays strike a rod, rhodopsin rapidly breaks down, and this chemical breakdown stimulates impulse conduction by the rods (Figure 16–10).

Rhodopsin is highly light sensitive–so much so that even the light rays from the moon or from a

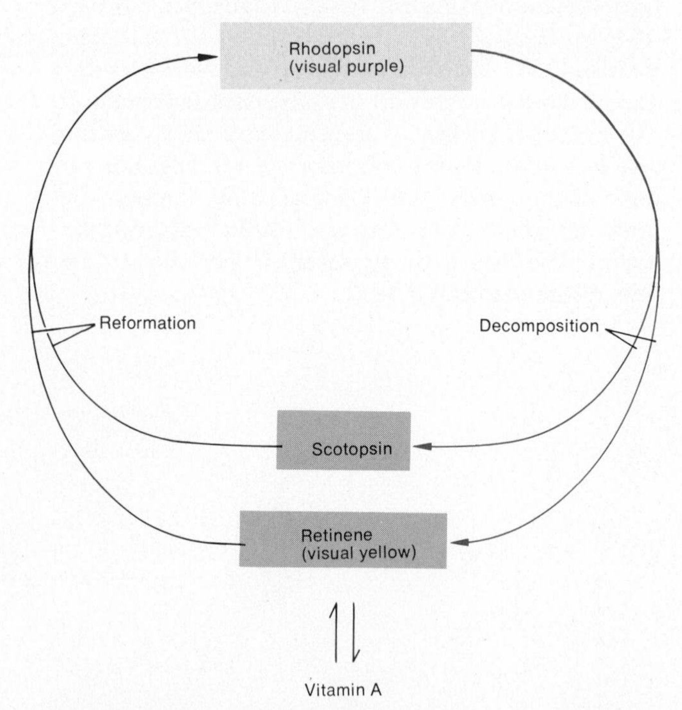

Figure 16–10. The rhodopsin cycle. Color indicates breakdown reactions in light. Black indicates reformation of rhodopsin in darkness.

candle will break down some of it and thereby allow us to see. The rods, then, are uniquely specialized for night vision. However, they are of only limited help for daylight vision. This is because rhodopsin breaks down extremely quickly in bright light and is rebuilt slowly. In bright light, the rhodopsin is destroyed faster than it can be manufactured. In dim light, production is able to keep pace with a slower rate of breakdown. These characteristics of rhodopsin are responsible for the experience of having to adjust to a dark room after walking in from the sunshine. The period of adjustment is the time it takes for the completely destroyed rhodopsin to reform. The adjustment period is normal. *Night blindness* is the lack of normal night vision following the adjustment period. Night blindness is not considered normal. It is most often caused by vitamin A deficiency.

Cones, which are the receptors for daylight and color, contain photosensitive chemicals that require brighter light for their breakdown. Unlike rhodopsin, the photosensitive chemicals of the cones reform quickly. It is believed that there are three types of cones and that each contains a different visual pigment. Each pigment has a different maximum absorption of light of a different color so that each responds best to light of a given wavelength (color). One type of cone responds best to red light, the second to green light, and the third to blue light. Just as an artist can obtain any color by mixing specific amounts of the primary colors, it is believed that cones can perceive any color by differential stimulation. Stimulation of a given rod by two or more colors may produce any combination of colors.

AFFERENT PATHWAY TO THE BRAIN. From the rods and cones, impulses are transmitted through bipolar neurons to ganglion cells. The cell bodies of the ganglion cells lie in the retina and their axons leave the eyeball via the optic nerve (Figure 16–11). The axons pass through the *optic chiasma,* a crossing point of the optic nerves. Some fibers cross to the opposite side. Others remain uncrossed. Upon passing through the optic chiasma, the fibers, now part of the *optic tract,* enter the brain and terminate in the thalamus. Here the fibers synapse with third-order neurons whose axons pass to the visual centers located in the occipital lobes of the cerebral cortex.

Analysis of the afferent pathway to the brain reveals that the visual field of each eye is divided into two regions. These are referred to as the *medial,* or *nasal, half* and the *lateral,* or *temporal, half.* For each eye, light rays from an object in the nasal half of the visual field fall on the temporal half of the retina. Light rays from an object in the temporal half of the vision field fall on the nasal half of the retina. Note that in the optic chiasma nerve fibers from the nasal halves of the retinas cross and continue on to the thalamus. Also note that nerve fibers from the temporal halves of the retinas do not cross but continue directly on to the thalamus. As a result, the visual center in the cortex of the right occipital lobe "sees" the left side of an object via impulses from the temporal half of the retina of the right eye and the nasal half of the retina of the left eye. The cortex of the left occipital lobe interprets visual sensations from the right side of an object via impulses from the nasal half of the right eye and the temporal half of the left eye. Blind spots in the field of vision may indicate a brain tumor along one of the afferent pathways. For instance, a symptom of tumor in the right optic tract might be an inability to see the left side of a normal field of vision without moving the eyeball.

Auditory sensations and equilibrium

In addition to containing receptors for sound waves, the **ear** also contains receptors for equilibrium. After reviewing the anatomy of the ear, we shall turn to the physiology of hearing and equilibrium. Anatomically, the ear is subdivided into three principal regions: (1) the external or outer ear, (2) the middle ear, and (3) the internal or inner ear.

The external or outer ear

The *external* or *outer ear* is structurally designed to collect sound waves and then direct them inward (Figure 16–12a). It consists of the pinna, the external auditory canal, and the tympanic membrane, also called the eardrum. The *pinna,* or *auricle,* is a trumpet-shaped flap of elastic cartilage covered by thick skin that has relatively few cutaneous receptors. The rim of the pinna is called the helix, and the inferior portion is referred to as the lobe. (See Exhibit 1–3 for additional surface features of the external ear.) The pinna is attached to the head by ligaments and muscles. The *external auditory canal* or *meatus* is a tube, about 2.5 cm (1 in.) in length, that lies in the external auditory meatus of the temporal bone. It leads from the pinna to the eardrum. The walls of the canal consist of bone lined with cartilage that is continuous with the cartilage of the pinna. The cartilage in the external auditory canal is covered with thin, highly sensitive skin. Near the exterior opening, the canal contains a few hairs and specialized sebaceous glands called

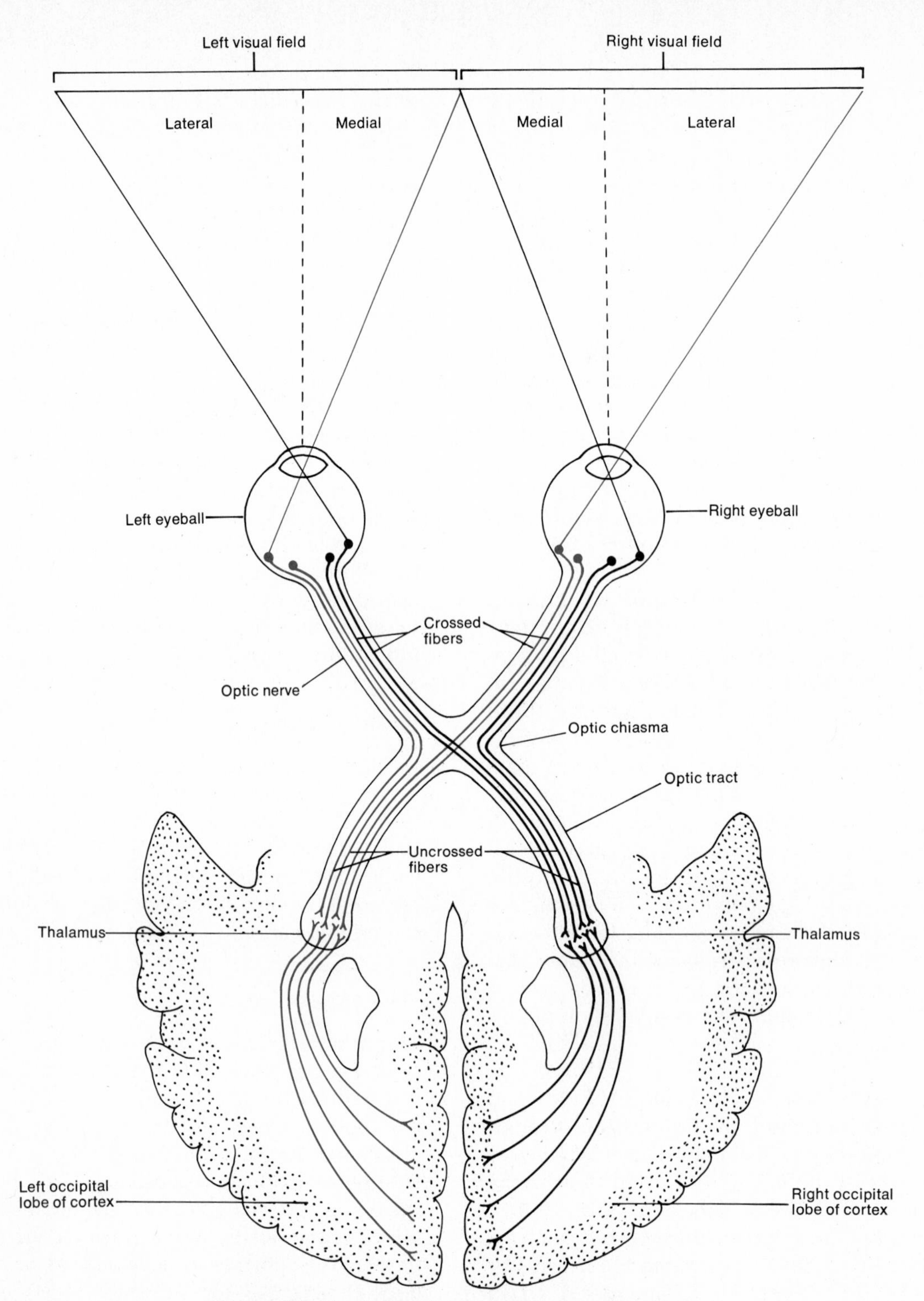

Figure 16–11. Afferent pathway for visual impulses.

ceruminous glands, which secrete *cerumen* (earwax.) The combination of hairs and cerumen prevents foreign objects from entering the ear. The *tympanic membrane* or *eardrum* is a thin, semitransparent partition of fibrous connective tissue between the external auditory meatus and the middle ear. Its external surface is concave and is covered with skin. Its internal surface is convex and is covered with a mucous membrane.

The middle ear

Also called the *tympanic cavity,* the *middle ear* is a small, epithelial-lined, air-filled cavity hollowed out of the temporal bone (Figure 16–12a, b). The cavity is separated from the external ear by the eardrum and from the internal ear by a very thin bony partition that contains two small openings, called the oval window and the round window. The

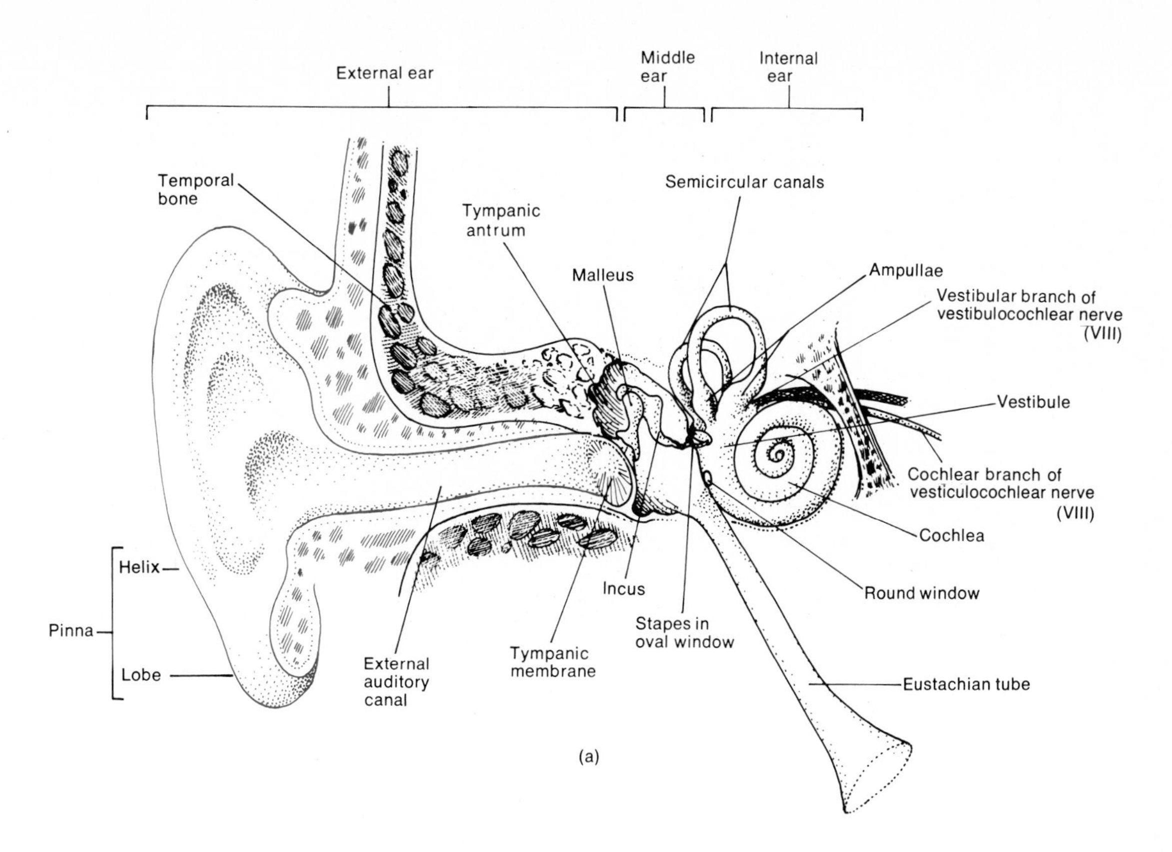

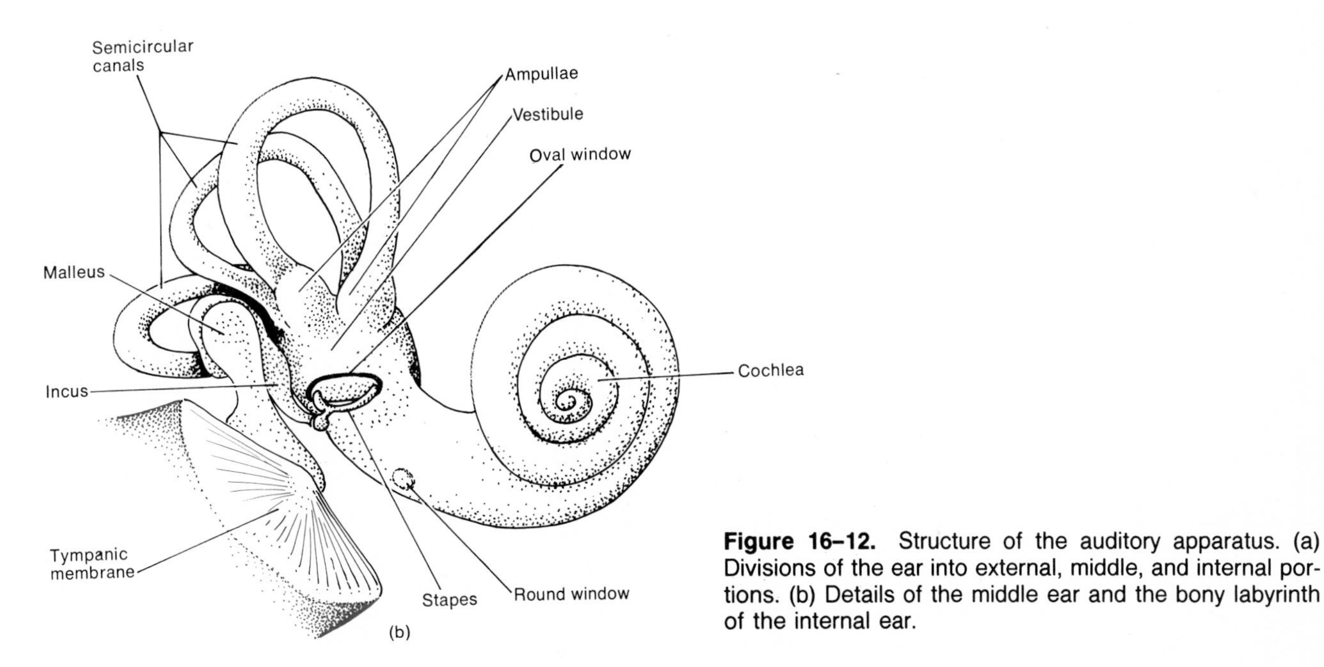

Figure 16–12. Structure of the auditory apparatus. (a) Divisions of the ear into external, middle, and internal portions. (b) Details of the middle ear and the bony labyrinth of the internal ear.

posterior wall of the cavity communicates with the mastoid cells of the temporal bone through a chamber called the *tympanic antrum.* This anatomical fact explains why a middle ear infection may spread to the temporal bone, causing mastoiditis, or even spread to the brain. The anterior wall of the cavity contains an opening that leads into the *Eustachian tube,* also called the internal auditory tube. The Eustachian tube connects the middle ear with the nose and nasopharynx of the throat. Through this passageway, infections may travel from the throat and nose to the ear. The function of the tube is to equalize air pressure on both sides of the tympanic membrane. Abrupt changes in external or internal air pressure might otherwise cause the eardrum to rupture. Since the tube opens during swallowing and yawning, these activities allow atmospheric air to enter or leave the middle ear until the internal pressure equals the external pressure. Any sudden pressure changes against the eardrum may be equalized by deliberately swallowing.

Extending across the middle ear are three exceedingly small bones referred to as **auditory ossicles.** These are called the malleus, incus, and stapes. According to their shapes, they are commonly named the hammer, anvil, and stirrup, respectively. The "handle" of the **malleus** is attached to the internal surface of the tympanic membrane. Its head articulates with the base of the incus. The **incus** is the intermediate bone in the series and articulates with the stapes. The base of the **stapes** fits into a small opening between the middle and inner ear called the *fenestra vestibuli,* or *oval window.* Directly below the oval window is another opening, the *fenestra cochlea,* or *round window.* This opening, which also separates the middle and inner ears, is enclosed by a membrane called the secondary tympanic membrane. The auditory ossicles are attached to the tympanic membrane, to each other, and to the oval window by means of ligaments and muscles.

The internal or inner ear

The *internal* or *inner ear* is also called the *labyrinth* (Figure 16–13a). because of its complicated series of canals. Structurally, it consists of two main divisions: (1) a bony labyrinth and (2) a membranous labyrinth that fits within the bony labyrinth. The *bony labyrinth* is a series of cavities within the petrous portion of the temporal bone. It can be divided into three areas named on the basis of shape. These areas are the vestibule, cochlea, and semicircular canals. The bony labyrinth is lined with periosteum and contains a fluid called the *perilymph.* This fluid surrounds the *membranous labyrinth,* a series of sacs and tubes lying inside and having the same general form as the bony labyrinth. Epithelium lines the membranous labyrinth, and it contains a fluid called the *endolymph.*

The *vestibule* constitutes the oval central portion of the bony labyrinth. The membranous labyrinth within the vestibule consists of two sacs called the *utricle* and *saccule.* These sacs are connected to each other by a small duct.

Projecting upward and posteriorly from the vestibule are the three bony *semicircular canals.* Each of the semicircular canals is arranged at approximately right angles to the other two. On the basis of their positions, they are called the superior, posterior, and lateral canals. One end of each canal enlarges into a swelling called the *ampulla.* Inside the bony semicircular canals lie portions of the membranous labyrinth, the *semicircular ducts* or *membranous semicircular canals.* These structures are almost identical in shape to the bony semicircular canals and communicate with the utricle of the vestibule.

Lying in front of the vestibule is the *cochlea,* so designated because of its resemblance to a snail's shell. The cochlea consists of a bony spiral canal that makes about 2¾ turns around a central bony core called the *modiolus.* A cross section through the cochlea shows that the canal is divided by partitions into three separate channels resembling the letter Y lying on its side (Figure 16–13b). The stem of the Y is a bony shelf that protrudes into the canal. The wings of the Y are composed of the bony labyrinth. The channel above the bony parition is called the *scala vestibuli,* and the channel below, the *scala tympani.* The cochlea adjoins the wall of the vestibule, into which the scala vestibuli opens. The scala tympani terminates at the round window. The perilymph of the vestibule is continuous with that of the scala vestibuli. The third channel (between the wings of the Y) is the membranous labyrinth, the *cochlear duct.* The cochlear duct is separated from the scala vestibuli by the *vestibular membrane,* also called *Reissner's membrane.* It is separated from the scala tympani by the *basilar membrane.* Resting on the basilar membrane is the *spiral organ,* or *organ of Corti,* the organ of hearing (Figure 16–13c). The organ of Corti is a series of epithelial cells on the inner surface of the basilar membrane. It consists of a number of supporting cells and hair cells, which are the receptors for auditory sensations. The inner hair cells are medially placed in a single row and extend the entire length of the cochlea. The outer hair cells are arranged in several rows throughout the cochlea. The hair cells have long hairlike processes at their free ends which extend into the endolymph of the cochlear duct. The basal ends of the hair cells are in contact with fibers of the cochlear branch of the vestibulocochlear nerve (VIII). Projecting over and in contact with the hair cells of the organ of Corti is the *tectorial membrane,* a very delicate and flexible gelatinous membrane.

Physiology of hearing

Sound waves result from the alternate compression and decompression of air. They originate from a vibrating object and travel through air in much the same way that waves travel over the surface of water. The events involved in the physiology of hearing sound waves are listed below. While reading this succession of events, you will want to make constant reference to Figure 16–14.

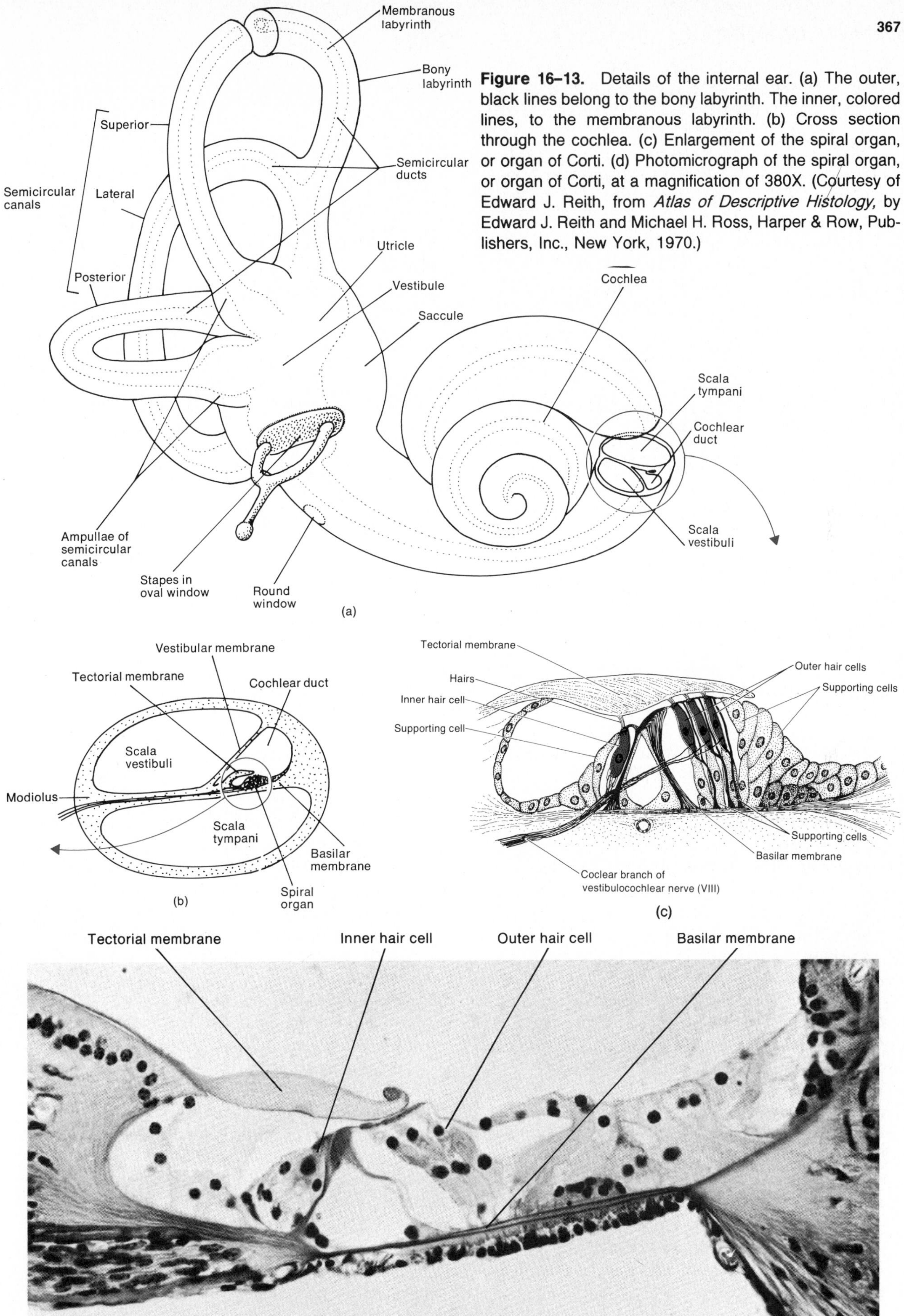

Figure 16–13. Details of the internal ear. (a) The outer, black lines belong to the bony labyrinth. The inner, colored lines, to the membranous labyrinth. (b) Cross section through the cochlea. (c) Enlargement of the spiral organ, or organ of Corti. (d) Photomicrograph of the spiral organ, or organ of Corti, at a magnification of 380X. (Courtesy of Edward J. Reith, from *Atlas of Descriptive Histology,* by Edward J. Reith and Michael H. Ross, Harper & Row, Publishers, Inc., New York, 1970.)

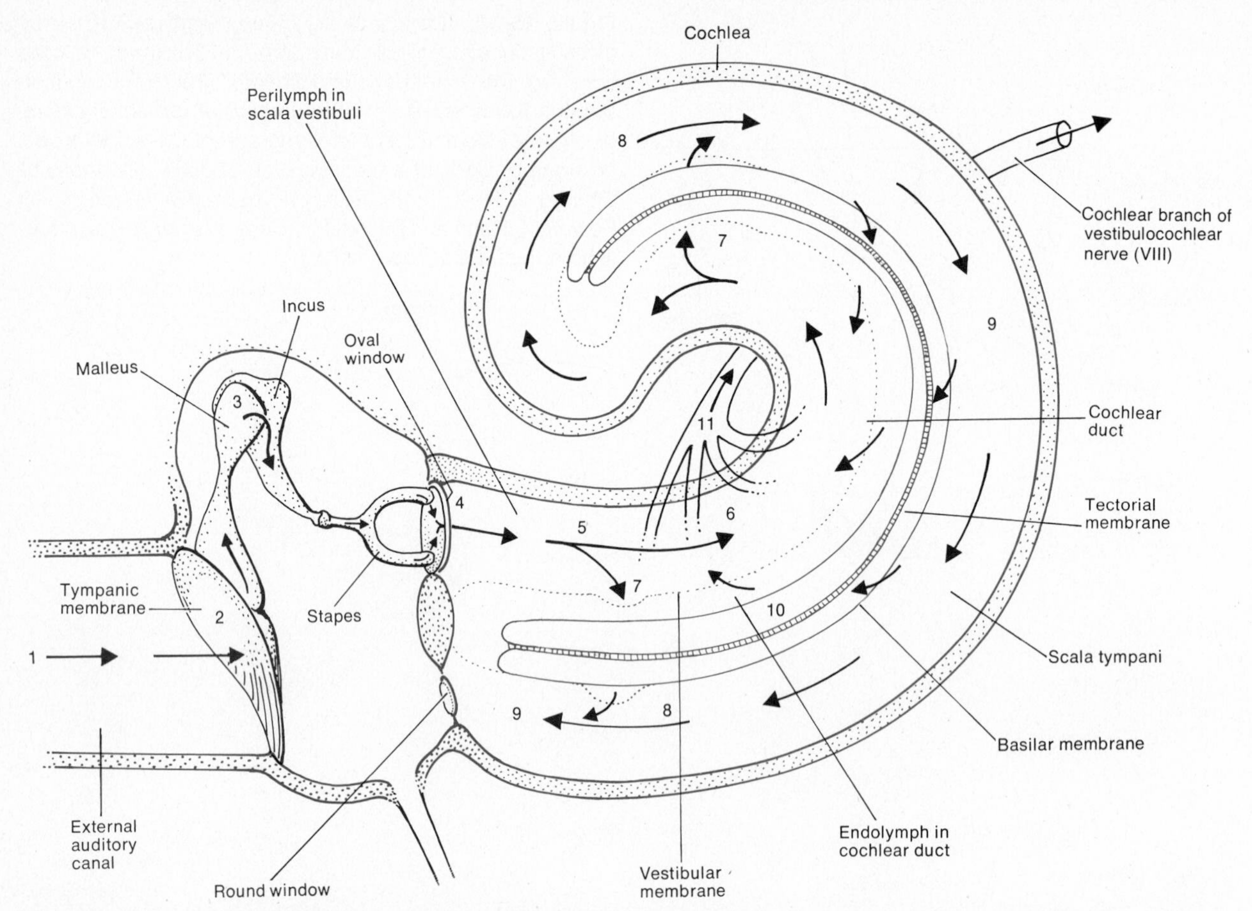

Figure 16–14. The physiology of hearing. Follow the text very carefully to understand the meaning of the numbers.

1. Sound waves that reach the ear are directed by the pinna into the external auditory canal.
2. When the waves strike the tympanic membrane, the alternate compression and decompression of the air cause the membrane to vibrate.
3. The central area of the tympanic membrane is connected to the malleus, which also starts to vibrate. The vibration is then picked up by the incus, which transmits the vibration to the stapes.
4. As the stapes moves back and forth, it pushes the oval window in and out.
5. The movement of the oval window sets up waves in the perilymph.
6. As the window bulges inward, it pushes the perilymph of the scala vestibuli up into the cochlea.
7. This pressure pushes the vestibular membrane inward and increases the pressure of the endolymph inside the cochlear duct.
8. The basilar membrane gives under the pressure and bulges out into the scala tympani.
9. The sudden pressure in the scala tympani pushes the perilymph toward the round window, causing it to bulge back into the middle ear. Conversely, as the sound wave subsides, the stapes moves backward, and the procedure is reversed. That is, the fluid moves in the opposite direction along the same pathway, and the basilar membrane bulges into the cochlear duct.
10. When the basilar membrane vibrates, the hair cells of the organ of Corti are moved against the tectorial membrane. In some unknown manner, the movement of the hairs stimulates the dendrites of neurons at their base, and sound waves are converted into nerve impulses.
11. The impulses are then passed on to the cochlear branch of the vestibulocochlear nerve and the medulla. Here, some impulses cross to the opposite side and finally travel to the auditory area of the temporal lobe of the cerebral cortex.

It has been demonstrated that if sound waves passed directly to the oval window without first

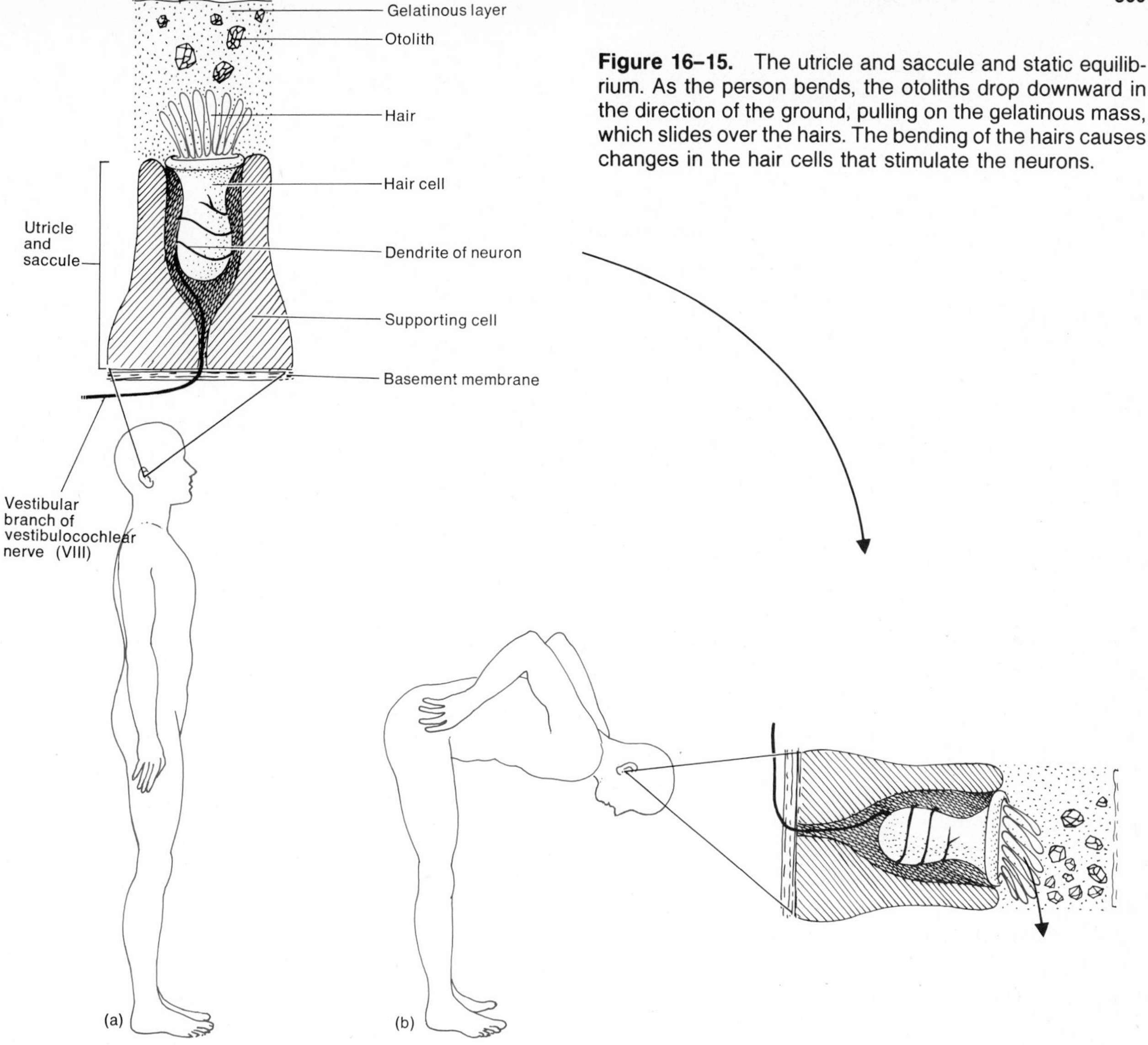

Figure 16–15. The utricle and saccule and static equilibrium. As the person bends, the otoliths drop downward in the direction of the ground, pulling on the gelatinous mass, which slides over the hairs. The bending of the hairs causes changes in the hair cells that stimulate the neurons.

passing through the tympanic membrane and auditory bones, hearing would be inadequate. This is because a minimal amount of sound energy is required to transmit sound waves through the perilymph of the cochlea. Since the tympanic membrane has a surface area about 22 times larger than that of the oval window, it can collect about 22 times more sound energy. This energy is sufficient to transmit sound waves through the perilymph.

Physiology of equilibrium

The term *equilibrium* has two meanings. One kind of equilibrium, called *static equilibrium,* refers to the orientation of the body (mainly the head) relative to the ground. The second kind of equilibrium, called *dynamic equilibrium,* is the maintenance of the position of the body (mainly the head) in response to sudden movements or to a change in the rate or direction of movement. The receptor organs for equilibrium are the saccule, utricle, and semicircular ducts.

The utricle and saccule each contain within their walls sensory hair cells that project into the cavity of the membranous labyrinth (Figure 16–15). The hairs are coated with a gelatinous layer in which particles of calcium carbonate, called *otoliths,* are embedded. When the head tips downward, the otoliths slide with gravity in the direction of the ground. As the particles move, they exert a downward pull on the gelatinous mass, which in turn exerts a downward pull on the hairs and makes them bend. The movement of the hairs stimulates the dendrites at the bases of the hair cells. The impulse is then transmitted to the temporal lobe of the brain through the vestibular branch of the vestibulocochlear nerve. The utricle and saccule are considered to be sense organs of static equilibrium. They provide information regarding the

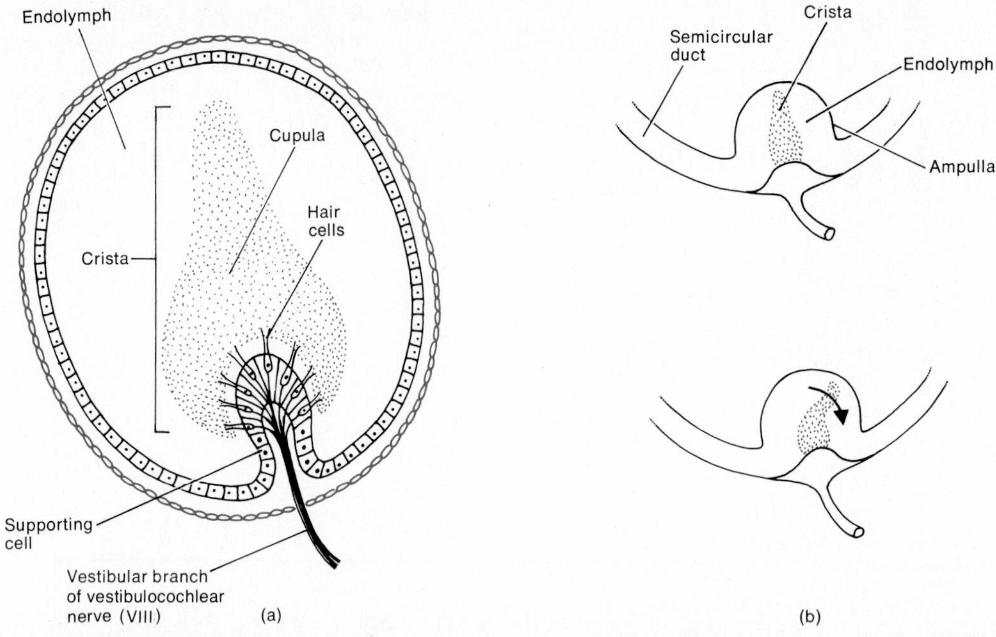

Figure 16–16. The semicircular ducts and dynamic equilibrium. (a) Enlarged aspect of a crista. (b) Cupula at rest (above) and in movement (below). When the endolymph in the semicircular duct moves, the cupula is displaced. The impulse is picked up by the vestibular branch of the vestibulocochlear nerve and relayed to the brain.

orientation of the head in space and are essential for the maintenance of posture.

Let us now consider the role of the semicircular ducts in maintaining dynamic equilibrium. The three semicircular ducts are positioned at right angles to each other in three planes: frontal (the superior duct), sagittal (the posterior duct), and lateral (the lateral duct). This positioning permits correction of an imbalance in three planes. In the ampulla, the dilated portion of each duct, there is a small elevation called the *crista* (Figure 16–16a). Each crista is composed of a group of hair cells covered by a mass of gelatinous material called the *cupula.* When the head moves, the endolymph in the semicircular ducts flows over the hairs and bends them as water in a stream bends the plant life growing at its bottom. The movement of the hairs stimulates sensory neurons, and the impulses pass over the vestibular branch of the vestibulocochlear nerve (Figure 16–16b). The impulses then reach the temporal lobe of the cerebrum. Impulses are then sent to the muscles that must contract in order to maintain body balance in the new position.

APPLICATIONS TO HEALTH

The sense organs can be altered or damaged by numerous disorders. The causes of disorder can range from congenital origins to bacterial infections to the effects of old age. Here, we shall discuss only a few of the more common disorders of the eyes and ears.

Eye disorders

Cataract

Disorders most commonly resulting in blindness are those that primarily affect elderly people. The most prevalent one is **cataract** formation. This disorder causes the lens or its capsule to lose its transparency (Figure 16–17). Cataracts can occur at any age, but we shall discuss only the type that develops with old age. Quite often as a person gets older the cells in the lenses of the eyes degenerate and are replaced with nontransparent fibrous protein. Or the lenses may start to manufacture nontransparent protein. The main symptom of cataract is a progressive, painless loss of vision. The degree of loss depends upon the location and extent of the opacity. If vision loss is very gradual, frequent changes in glasses may help to maintain useful vision for a while. This is because the changes initially affect the density of the lens and thus change the refraction of light rays. Eventually, though, the changes may be so extensive that light rays are blocked out altogether. At this point, surgery is indicated. Essentially, the surgical procedure consists of removing the opaque lens and substituting an artificial lens by means of eyeglasses.

Glaucoma

The second most common cause of blindness, especially in the elderly, is **glaucoma.** This disorder is characterized by an abnormally high pressure of the fluid inside the eyeball. The aqueous humor does not return into the bloodstream through the canal of Schlemm as quickly as it is formed (see Figure 16–7a). The fluid accumulates and puts pressure on the neurons of the retina. If the pressure continues over a long period of time, it destroys the neurons and brings about blindness. It can affect a person of any age, but 95 percent of the victims are past 40 years of age. Glaucoma affects the eyesight of more than 1 million people in this country.

Treatment is through drugs or surgery. Drug treatment involves reducing the pressure in the eye by giving cholinergic compounds that cause the sphincter muscles of the iris and ciliary body to contract. These contractions prevent the obstruction of the anterior chamber and the canal of Schlemm. This results in improved fluid outflow from the eye. Surgery consists of the removal of a small piece of the iris (iridectomy) to allow drainage and lessen the interior pressure.

Conjunctivitis (pinkeye)

Many different kinds of eye inflammations exist. But the most common type is **conjunctivitis,** an inflammation of the membrane that lines the insides of the eyelids and covers the cornea.

Conjunctivitis can be caused by microorganisms – most often the pneumococci or staphylococci bacteria. In such cases, the inflammation is very contagious. It can also be caused by a number of irritants, in which case the inflammation is not contagious. The irritants include dust, smoke, wind, air pollution, and excessive glare from intense light on water, sand, or snow. The condition may be acute or chronic. The epidemic type in children is extremely contagious, but normally it is not serious.

Trachoma

This is a chronic contagious conjunctivitis caused by an organism called the TRIC agent. The organism has characteristics of both viruses and bacteria. **Trachoma** is characterized by many granulations or fleshy projections on the eyelids. These projections can irritate and inflame the cornea if untreated and reduce vision. The disease produces an excessive growth of subconjunctival tissue and the invasion of blood vessels into the upper half of the front of the cornea. The disease progresses until it covers the entire cornea, bringing about a loss of vision because of corneal opacity. Antibiotics, such as tetracycline and the sulfa drugs, kill the organisms that cause trachoma and have reduced the seriousness of this infection.

Ear disorders

Ménière's disease

An important cause of deafness and loss of equilibrium in adults is **Ménière's disease** of the inner ear. It is a disturbance or malfunction of any part of the

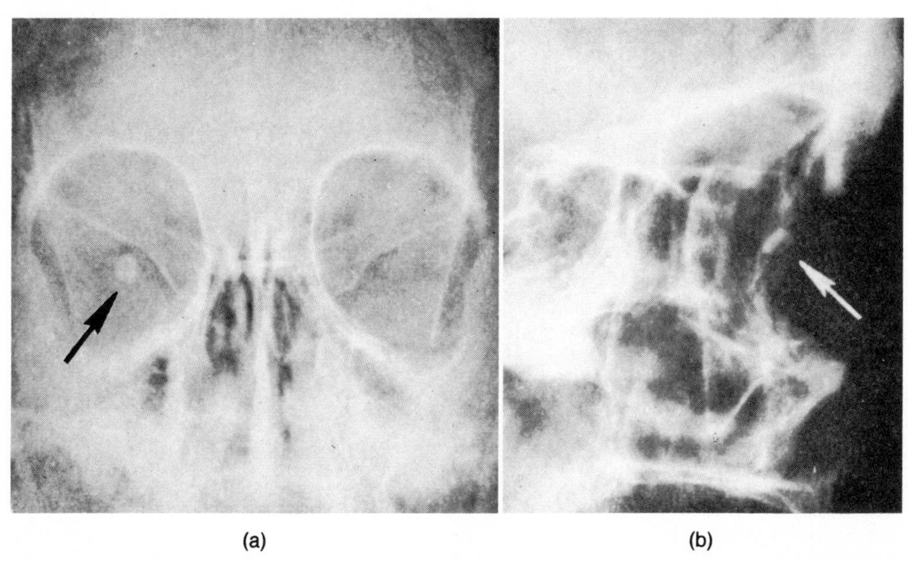

Figure 16–17. Roentgenogram of a cataract (arrow) in the lens of the right eye. (a) Anterior view. (b) Right lateral view. (Courtesy of Lester W. Paul and John H. Juhl, *The Essentials of Roentgen Interpretation,* Harper & Row, Publishers, Inc., New York, 1972.)

inner ear. It can be the result of many causes. These include (1) infection of the middle ear; (2) trauma from brain concussion producing hemorrhage or a splitting of the labyrinth; (3) cardiovascular diseases, such as atherosclerosis and blood vessel disturbances; (4) congenital malformation of the labyrinth; (5) an excessive formation of endolymph; and (6) an allergy. The last two causes can both produce an increase in pressure in the cochlear duct and vestibular system. This, in turn, causes a progressive atrophy of the hair cells of the cochlear or semicircular ducts.

If the cochlear duct is injured, typical symptoms are hissing, roaring, or ringing in the ears and/or deafness. If the semicircular ducts are involved, the person has a feeling of dizziness or motion sickness, often producing nausea and vomiting. The dizzy spells can be mild or severe, lasting from a few minutes to several days. Ménière's disease is a chronic disorder. It affects both sexes equally and usually begins in late middle life. In severe cases, surgical treatment should be considered. Treatment with ultrasound, which consists of beaming ultrasound waves into the labyrinth, selectively destroys the semicircular ducts (equilibrium) while preserving the cochlear duct (hearing). This has produced excellent results. Another successful surgical procedure involves electrical coagulation of the labyrinth, which means changing the fluid to a jelly or solid. Of all the symptoms of Ménière's disease, the tendency of the patient to fall in the direction of the affected ear is probably the most incapacitating and potentially dangerous.

Impacted cerumen

Some people produce an abnormal amount of cerumen, or earwax, in the external auditory canal of the ear. Here, it becomes impacted, or pressed so firmly together that it becomes immovable, and prevents sound waves from reaching the tympanic membrane. The treatment for **impacted cerumen** is usually periodic ear irrigation or removal of wax with a blunt instrument.

Key terms associated with the sense organs

Achromatopsia (*a* = without; *chrom* = color) Complete color blindness.

Ametropia (*ametro* = disproportionate; *ops* = eye; *ia* = condition) Refractive defect of the eye resulting in an inability to focus images properly on the retina.

Blepharitis (*blepharo* = eyelid) An eyelid inflammation.

Eustachitis Eustachian tube infection or inflammation.

Keratitis (*kerato* = cornea) An inflammation or infection of the cornea.

Keratoplasty (*plasty* = mold, shape, form) Corneal graft or transplant; an opaque cornea is removed and replaced with a normal transparent cornea to restore vision.

Labyrinthitis Inner ear or labyrinth inflammation.

Myringitis (*myringa* = eardrum) Inflammation of the eardrum; also called tympanitis.

Nystagmus A constant, rapid involuntary eyeball movement, possibly caused by a disease of the central nervous system.

Otalgia (*oto* = ear; *algia* = pain) Earache.

Otitis Inflammation of the ear.

Presbyopia (*presby* = old) Inability to focus clearly on nearby objects due to loss of elasticity of the crystalline lens. Loss usually caused by aging.

Ptosis (*ptosis* = fall) Falling or drooping of the eyelid. (This expression is also used for the slipping of any organ below its normal position.)

Retinoblastoma (*blast* = bud; *oma* = tumor) A relatively common tumor arising from immature retinal cells and accounting for 2 percent of childhood malignancies.

Strabismus An eye muscle disorder, commonly called "crossed eyes." The eyeballs do not move in unison. May be caused by lack of coordination of the extrinsic eye muscles.

Tinnitus A ringing in the ears.

Chapter summary in outline

SENSATIONS

Definition

1. Awareness of conditions and changes in these conditions inside and outside the body is sensation.
2. The prerequisites are receiving a stimulus, converting it into an impulse, conducting the impulse to the brain, and translating the impulse into a sensation.
3. A receptor picks up a stimulus for a sensation.

Characteristics

1. Projection occurs when the brain refers a sensation to the point of stimulation.
2. Adaptation is the loss of sensation even though the stimulus is still applied.
3. An afterimage is the persistence of a sensation even though the stimulus is removed.
4. The modality is the property by which one sensation is distinguished from another.

Classification

1. Exteroceptors receive stimuli from the external environment.
2. Visceroceptors receive stimuli from blood vessels and viscera.

3. Proprioceptors receive stimuli from muscles, tendons, and joints for body position and movement.

GENERAL SENSES

Cutaneous sensations

1. Touch, pressure, cold, and heat sensations from the skin, connective tissue, and gastrointestinal tract are called the cutaneous sensations.
2. Receptors for these sensations are Meissner's corpuscles, Merkel's discs, root hair plexuses, Pacinian corpuscles, end bulbs of Krause, and end organs of Ruffini.

Pain sensations

1. Receptors are found in almost every body tissue.
2. Two general kinds of pain, recognized in the parietal lobe of the cortex, are somatic and visceral.
3. Referred pain is felt in the skin near or away from the organ sending pain impulses.
4. With phantom pain, a person "feels" pain in a limb that has been amputated.

Proprioceptive sensations

1. Receptors, found in muscles, tendons, and joints, inform us of muscle tone, movement of body parts, and body position.
2. This sense is called proprioception.
3. The receptors involved are joint kinesthetic receptors, neuromuscular spindles, and Golgi tendon organs.

SPECIAL SENSES

Olfactory sensations

1. Receptor cells in the nasal epithelium send impulses to the olfactory bulbs, olfactory tracts, and cortex.

Gustatory sensations

1. Receptors in the taste buds send impulses to the cranial nerves, thalamus, and cortex.

Visual sensations

1. Accessory structures of the eyes include the eyebrows, eyelids, eyelashes, and the lacrimal apparatus.
2. The eye is constructed of three coats: *(a)* fibrous tunic (sclera and cornea); *(b)* vascular tunic (choroid, ciliary body, and iris); and *(c)* retina, which contains rods and cones.
3. The anterior cavity contains aqueous humor, and the posterior cavity contains vitreous humor.
4. Retinal image formation involves refraction of light, accommodation of lens, constriction of pupil, convergence, and inverted image formation.
5. Improper refraction may result from myopia (nearsightedness), hypermetropia (farsightedness), and astigmatism (corneal or lens abnormalities).
6. Rods and cones convert light rays into visual nerve impulses; rhodopsin is necessary for the conversion.
7. Impulses from rods and cones are conveyed through retina to optic nerve, optic chiasma, optic tract, thalamus, and cortex.

Auditory sensations and equilibrium

1. The ear consists of three anatomical subdivisions: *(a)* the external or outer ear (pinna, external auditory canal, and tympanic membrane); *(b)* the middle ear (Eustachian tube, ossicles, oval window, and round window); and *(c)* the internal or inner ear (bony labyrinth and membranous labyrinth). The internal ear contains the organ of Corti, the organ of hearing.
2. Sound waves are caused by the alternate compression and decompression of air.
3. Waves enter the external auditory canal, strike the tympanic membrane, pass through the ossicles, strike the oval window, set up waves in the perilymph, strike the vestibular membrane and scala tympani, increase pressure in the endolymph, strike the basilar membrane, and stimulate hairs on the spiral organ. A sound impulse is then initiated.
4. Static equilibrium is the relationship of the body relative to the pull of gravity. The utricle and saccule are the sense organs of static equilibrium.
5. Dynamic equilibrium is equilibrium in response to movement of the body. The semicircular ducts are the sense organs of dynamic equilibrium.

APPLICATIONS TO HEALTH

1. Cataract is the loss of transparency of the lens or capsule.
2. Glaucoma is abnormally high intraocular pressure, which destroys neurons of the retina.
3. Conjunctivitis is an inflammation of the conjunctiva.
4. Trachoma is a chronic, contagious inflammation of the conjunctiva.
5. Ménière's disease is the malfunction of the inner ear that may cause deafness and loss of equilibrium.
6. Impacted cerumen is an abnormal amount of earwax in the external auditory canal.

Review questions and problems

1. Define a sensation and a sense receptor. What prerequisities are necessary for the perception of a sensation?
2. Describe the following characteristics of a sensation: projection, adaptation, afterimage, and modality.
3. Can you think of any examples of adaptation not discussed in the text?
4. Compare the location and function of exteroceptors, visceroceptors, and proprioceptors.
5. Distinguish between a general sense and a special sense.
6. What is a cutaneous sensation? How are cutaneous receptors distributed over the body? Relate your response to the two-point discrimination test.
7. For each of the following cutaneous sensations, describe the receptor involved in terms of structure, function, and location: touch, pressure, cold, and heat.
8. How do cutaneous sensations help you to survive?
9. Why are pain receptors important? Differentiate somatic pain, visceral pain, referred pain, and phantom pain.
10. What is proprioception?
11. Describe the structure, location, and function of the proprioceptive receptors.
12. Discuss the origin and path of an impulse that results in smelling.
13. How are papillae related to taste buds? Describe the structure and location of the papillae. Discuss how an impulse for taste travels from a taste bud to the brain.
14. Describe the structure and importance of the following accessory structures of the eye: eyelids, eyelashes, eyebrows, and lacrimal apparatus.
15. By means of a labeled diagram, indicate the principal anatomical structures of the eye. How is the retina adapted to its function?
16. Distinguish a sty from a chalazion.

17. How do extrinsic eye muscles differ from intrinsic eye muscles?
18. Describe the location and contents of the chambers of the eye. What is intraocular pressure? How is the canal of Schlemm related to this pressure?
19. Explain how each of the following events is related to the physiology of vision: *(a)* refraction of light, *(b)* accommodation, *(c)* constriction of the pupil, *(d)* convergence, and *(e)* inverted image formation.
20. Distinguish emmetropia, myopia, hypermetropia, and astigmatism by means of a diagram.
21. How is a light stimulus converted into an impulse?
22. What is night blindness? What causes it?
23. Describe the path of a visual impulse from the optic nerve to the brain.
24. Define visual field. Relate the visual field to image formation on the retina.
25. Diagram the principal parts of the outer, middle, and inner ear. Describe the function of each part labeled.
26. Explain the events involved in the transmission of sound from the pinna to the spiral organ.
27. What is the afferent pathway for sound impulses from the vestibulocochlear nerve to the brain?
28. Compare the function of the semicircular ducts in maintaining dynamic equilibirum with the role of the saccule and utricle in maintaining static equilibrium.
29. Define each of the following: cataract, glaucoma, conjunctivitis, trachoma, Ménière's disease, and impacted cerumen.
30. Refer to the key terms associated with the sense organs. Be sure that you can define each term listed.

Selected readings

Basmajian, John V. *Primary Anatomy.* 6th ed. Baltimore: Williams and Wilkins, 1970. Chaps. 12, 13.

Cunningham's Textbook of Anatomy. 11th ed., edited by G. J. Romanes. London: Oxford University Press, 1972. Pp. 803–836.

Dawson, Helen L. *Basic Human Anatomy.* 2d ed. New York: Appleton-Century-Crofts, 1974. Pp. 329–348.

Gray, Henry. *Anatomy of the Human Body.* 29th ed., edited by Charles Mayo Goss. Philadelphia: Lea and Febiger, 1973. Chap. 13.

CHAPTER 17

THE ENDOCRINE SYSTEM

STUDENT OBJECTIVES

After you have read this chapter, you should be able to:

1. Define an endocrine gland
2. Identify the relationship between an endocrine gland and a target organ
3. Explain the role of cyclic AMP in bringing about hormonal responses
4. Describe the location and blood supply of the anterior pituitary
5. List the hormones of the anterior pituitary, their target organs, and their functions
6. Describe the location and nerve supply of the posterior pituitary
7. Describe the source of hormones stored by the posterior pituitary, their target organs, and their functions
8. Describe the location and histology of the thyroid gland
9. Explain the physiological effects of the thyroid secretions
10. Describe the location and histology of the parathyroid glands
11. Explain the physiological effects of the parathyroid hormone
12. Identify the location and histology of the adrenal glands
13. Contrast the adrenal cortex and adrenal medulla with respect to hormones secreted and the effects of the hormones
14. Explain the correlation between the adrenal medulla and the sympathetic nervous system
15. Describe the location and histology of the pancreas
16. Explain the physiological effects of the pancreatic hormones
17. Identify the location, structure, and functions of the pineal gland
18. Identify the location, structure, and functions of the thymus gland
19. Describe pituitary dwarfism, Simmond's disease, giantism, and acromegaly as disorders of the anterior pituitary
20. Describe diabetes insipidus as a malfunctioning of the posterior pituitary
21. Explain cretinism, myxedema, exophthalmic goiter, and simple goiter as dysfunctions of the thyroid gland
22. Compare the effects of hypo- and hypersecretions of adrenocortical hormones
23. Describe diabetes mellitus and hyperinsulinism and dysfunctions of the pancreas
24. Define key terms associated with the endocrine system

In previous chapters, you learned how the nervous system controls the body through electrical impulses that are delivered over neurons. Now it is time to look at the body's other control system, the endocrine system. The endocrine organs affect bodily activities by releasing chemical messengers, called hormones, into the bloodstream. Obviously, bodily activities would become counterproductive and ineffective if the two great control systems were to pull in opposite directions. The nervous and endocrine systems, therefore, coordinate their activities like an interlocking supersystem. Certain parts of the nervous system routinely stimulate or inhibit the release of hormones. The hormones, in turn, are quite capable of stimulating or inhibiting the flow of particular nerve impulses. Like the proverbial horse and carriage, the two systems go together. In this chapter, you will study the various endocrine glands, the organs that make up the body's means of chemical control.

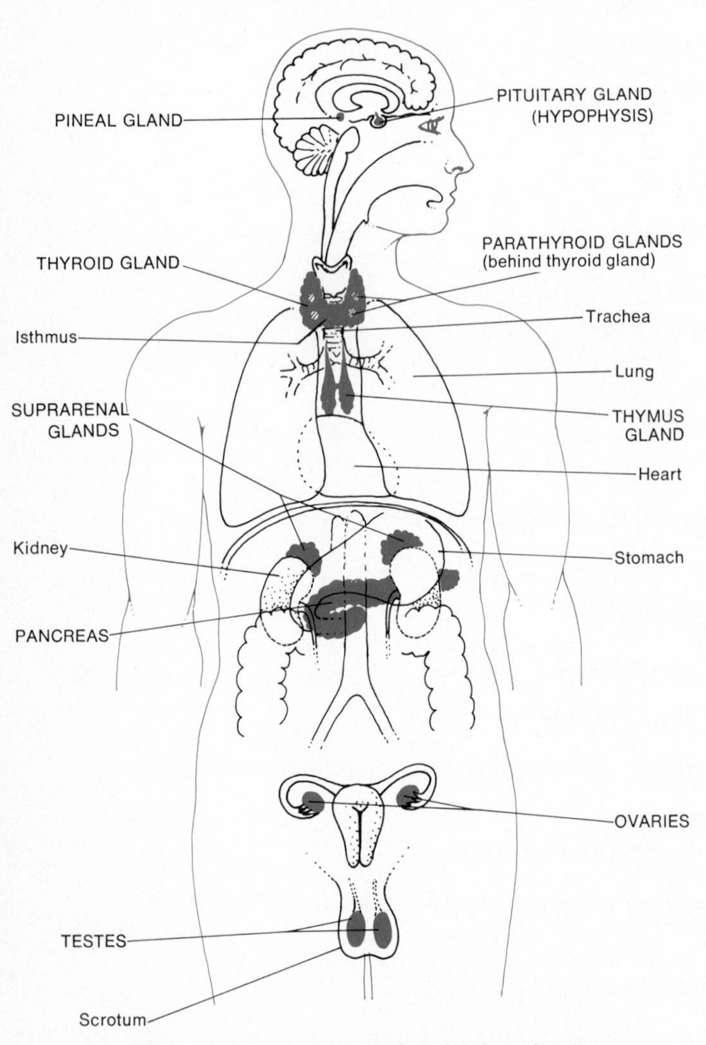

Figure 17–1. Location of endocrine glands.

ENDOCRINE GLANDS

The body contains two different kinds of glands: exocrine and endocrine. **Exocrine glands,** mentioned in Chapter 3, secrete their products into ducts. The ducts then carry the secretions into body cavities or to the external surface of the body. So far, we have talked only about exocrine glands. They include sweat, sebaceous, mucous, and digestive glands. **Endocrine glands,** by contrast, secrete their products into the extracellular space around the secretory cells. Since they secrete internally, the term *endo,* meaning within, is used. The secretion passes through the capillaries and into the blood. Since they have no ducts, endocrine glands are also referred to as *ductless glands.* The endocrine glands of the body are the pituitary, thyroid, parathyroids, adrenals, pancreas, ovaries, testes, pineal, and thymus. The placenta or "afterbirth" is, in some ways, a temporary endocrine gland. Moreover, although the kidneys, stomach, and small intestine produce hormones, they will be treated in later chapters as parts of the systems to which they belong. The endocrine glands make up the **endocrine system.** The location of many organs of the endocrine system is illustrated in Figure 17–1.

The secretions of endocrine glands are called **hormones,** the term *hormone* meaning to set in motion. A hormone may be a protein, an amine, or a steroid. Amines, like proteins, contain carbon, hydrogen, and nitrogen. Unlike proteins, they lack oxygen, and they do not contain peptide bonds. A steroid is a type of lipid. The one thing that all hormones have in common–whether protein, amine, or steroid–is maintaining homeostasis by changing the physiological activities of cells. A hormone may stimulate changes in the cells of an organ or in groups of organs. These are called the *target organs* of a hormone. Or, the hormone may directly affect the activities of all the cells in the body.

The general effects of most hormones are fairly well known. For example, we know that the hormones insulin and glucagon regulate blood sugar level; the hormone prolactin stimulates milk secretion by the mammary glands; the thyroid hormone, thyroxin, helps to control metabolic rate; and progesterone helps to prepare the uterus for implantation. However, the manner in which hormones affect different cells of the body is not entirely clear.

In recent years, physiologists have learned that one particular chemical substance seems to be involved in many reactions involving hormones. This substance is **cyclic AMP.** Cyclic AMP is very closely related to ATP, the principal energy-storing chemical in cells. Current speculation is that a hormone circulating in the blood reaches a target cell and brings a specific message for that cell. In order to give the message to the cell, the hormone must attach to a specific receptor site on the plasma membrane (Figure 17–2). This attachment is probably like an enzyme and substrate fitting together. In this way, the hormone message is given to the plasma membrane of the target cell. In some manner not fully understood, a series of chemical changes occurs in the membrane, and the result is that some of the ATP inside the cell is changed into cyclic AMP. The cyclic AMP functions as a second messenger inside the cell and performs a specific function according to the message indicated by the hormone.

Within the cell, cyclic AMP gears up the appropriate chemicals to get a specific job done. In a liver or skeletal muscle cell, cyclic AMP would stimulate the change of glucose into glycogen to lower blood sugar level according to a message from insulin, but would raise blood sugar level by stimulating the changing of glycogen according to a message from glucagon. In a cell of the thyroid gland, cyclic AMP would stimulate the cell to secrete its hormone called thyroxin in response to a message received from another hormone called thyroid-stimulating hormone. In other words, cyclic AMP induces a specific target cell to perform a specific function based upon the message it receives from the hormone that attaches to the membrane. As you have probably guessed, different target cells will respond only to specific hormones as they circulate through the blood. In this way, different target cells accomplish different functions. This is a very efficient way for all your body cells to cooperate to help you maintain homeostasis. It is also a very efficient way to make sure that a particular target cell does not do the wrong thing. Keep in mind that hormones are not enzymes; they stimulate enzymatic activity.

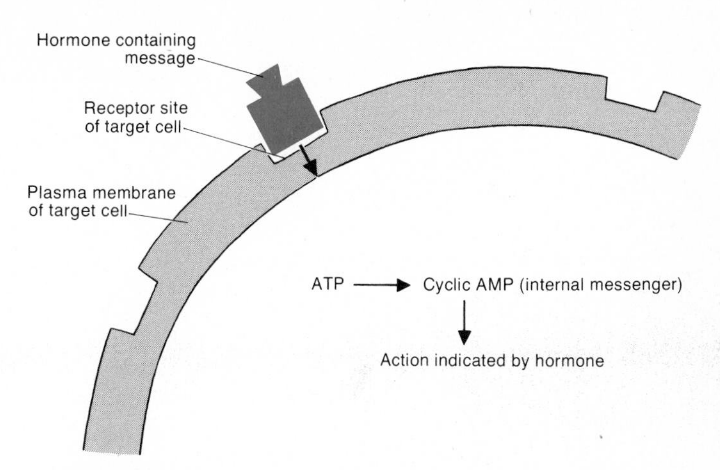

Figure 17–2. Proposed relationship between a hormone, target cell, and cyclic AMP.

PITUITARY GLAND

The hormones of the **pituitary gland,** also called the **hypophysis,** regulate so many body activities that the pituitary has been nicknamed the "master gland." Surprisingly, the hypophysis is a rather small round structure measuring about 1.3 cm (0.5 in.) in diameter and weighing about 0.5 g. It lies in the sella turcica of the sphenoid bone and is attached to the hypothalamus of the brain via a stalk-like structure. This structure is called the *infundibulum.* (See Figure 14–1.)

The pituitary is divided structurally and functionally into an anterior lobe and a posterior lobe, both of which are connected to the hypothalamus. The *anterior lobe* contains many glandular epithelium cells and forms the glandular part of the pituitary. A system of blood vessels connects the anterior lobe with the hypothalamus. The *posterior lobe* contains axons of neurons, which form the neural part of the pituitary. Other nerve fibers connect the neurohypophysis directly with the hypothalamus.

The hypophysis is innervated by axons of the hypothalamic neurons via the infundibulum and by unmyelinated postganglionic axons from the superior cervical sympathetic ganglion. The glossopharyngeal nerve may contribute some parasympathetic fibers.

The adenohypophysis

The anterior lobe of the pituitary is also called the **adenohypophysis.** It releases hormones that regulate a whole range of bodily activities, from growth to reproduction. However, the release of these hormones is either stimulated or inhibited by chemical secretions that come from the hypothalamus of the brain. Such chemicals are called *releasing factors.* This is one hookup between the nervous system and the endocrine system. The hypothalamic secretions are delivered to the adenohypophysis in the following way. The blood supply to the adenohypophysis and infundibulum is derived from several *superior hypophyseal arteries.* These arteries are branches of the internal carotid and posterior communicating arteries (Figure 17–3). The superior hypophyseal arteries form a network or plexus of capillaries, the *primary plexus,* in the infundibulum near the inferior portion of the hypothalamus. Releasing factors from the hypothalamus diffuse into this plexus. This plexus drains into veins,

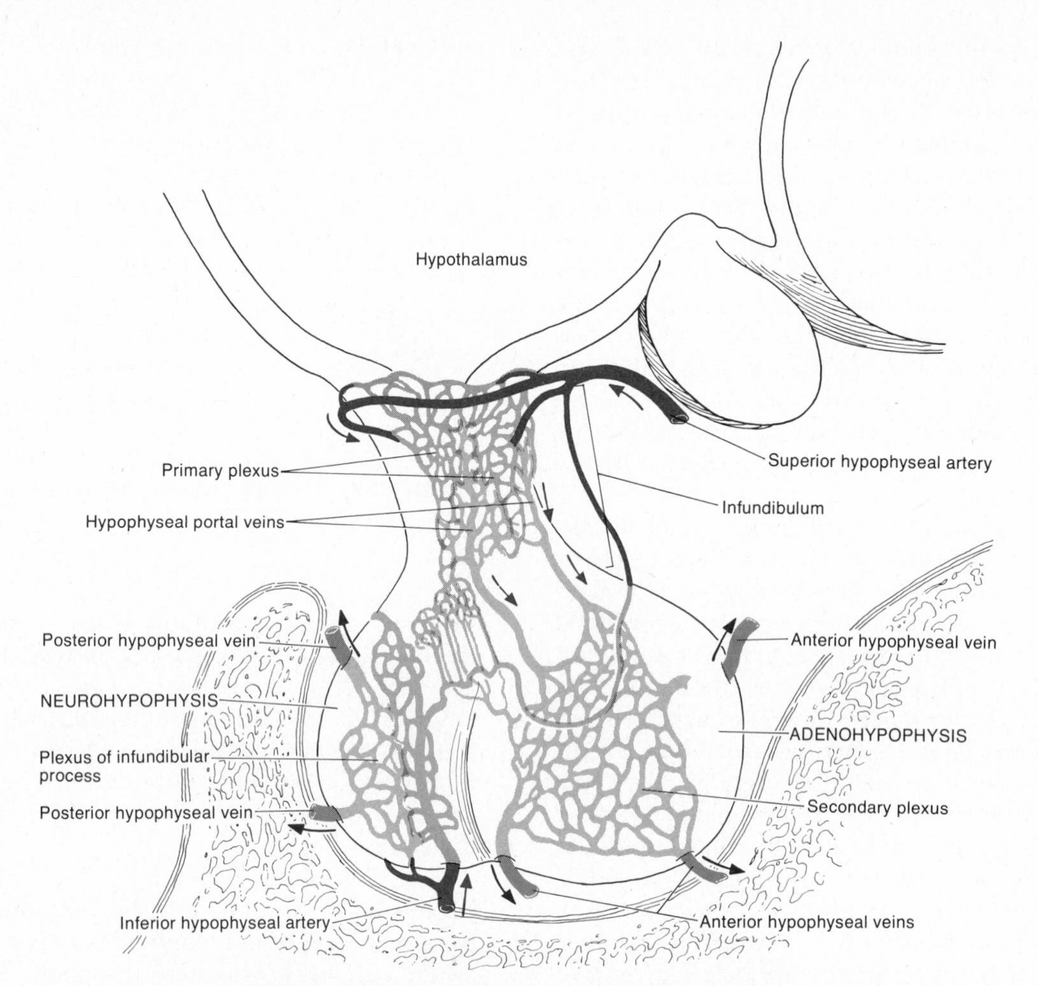

Figure 17–3. Blood supply of the pituitary gland.

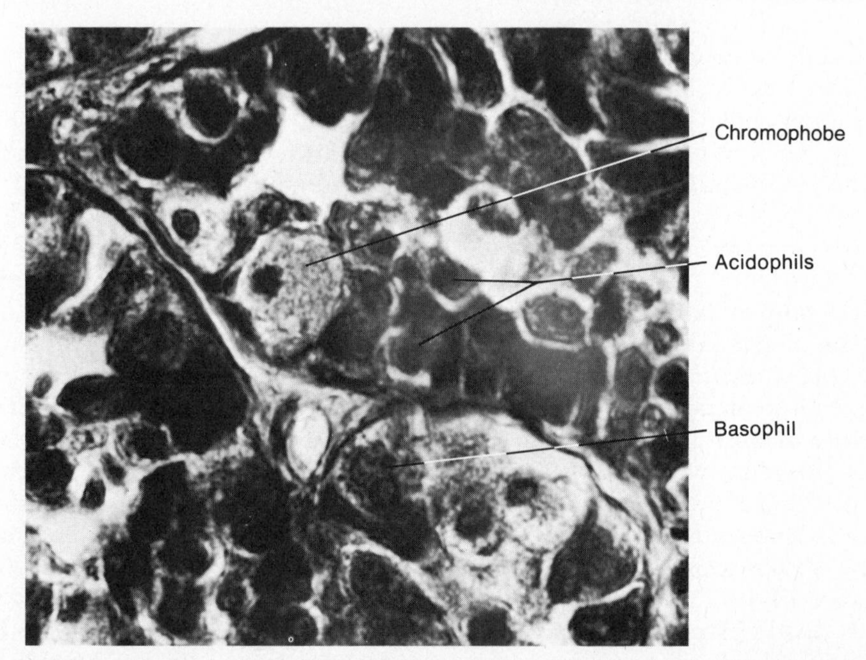

Figure 17–4. Histology of the adenohypophysis. The cytoplasm of the acidophils stains with acid dyes such as eosin. The cytoplasm of the basophils stains with basic dyes or hemtoxylin. The chromophobes stain poorly. The cells are shown at a magnification of 160X. (Courtesy of Victor B. Eichler, Wichita State University.)

known as the *long hypophyseal portal veins,* that pass down the infundibulum. At its inferior portion, the veins form a *secondary plexus* in the adenohypophysis. From this plexus, hormones of the adenohypophysis pass into the anterior hypophyseal veins for distribution to target organs.

When the anterior lobe receives the proper chemical stimulation from the hypothalamus, its glandular cells secrete any one of six hormones. The glandular cells, themselves, are called acidophils, basophils, or chromophobes, depending on the way their cytoplasm reacts to laboratory stains (Figure 17–4). The acidophils, which stain pink, secrete two hormones: human growth hormone, which controls bodily growth; and prolactin, which initiates milk secretion by the mammary glands. Basophils stain darkly and release the other four hormones. These are thyroid-stimulating hormone, which controls the thyroid gland; adrenocorticotrophic hormone, which regulates the cortical regions of the adrenal glands; follicle-stimulating hormone, which stimulates the production of egg and sperm in the reproductive organs; and luteinizing hormone, which stimulates other sexual and reproductive activities. The chromophobes may also be involved in the secretion of the adrenocorticotrophic hormone.

Except for the growth hormone, all the secretions are referred to as *trophic hormones,* which means that their target organs are other endocrine glands. Prolactin, follicle-stimulating hormone, and luteinizing hormone are also called *gonadotrophic hormones* because they regulate the functions of the gonads. The gonads (ovaries and testes) are the endocrine glands that produce sex hormones.

Human growth hormone (HGH)

The *human growth hormone (HGH)* is also referred to as *somatotropin* and as the *somatotrophic hormone (STH).* The word root *soma* means body, and *troph* means nourishment. Its principal function is to act upon the hard and soft tissues of the body to increase their rate of growth and to maintain their size once growth is attained. HGH causes cells to grow and multiply by increasing the rate at which amino acids enter cells and are built up into proteins. The building processes are called *anabolism.* Thus, HGH is considered to be a hormone of protein anabolism. HGH has two other functions as well. It causes cells to switch from burning carbohydrates to burning fats for energy. For example, it stimulates adipose tissue to release some of its fat. And it stimulates the other cells of the body to break down the released fat molecules. When chemical bonds are broken, energy is released. Since energy-releasing processes are referred to as *catabolism,* we can say that HGH promotes fat catabolism. At the same time, HGH accelerates the rate at which glycogen stored in the liver is converted into glucose and released into the blood. However, since the cells are using fats for energy, they do not consume much glucose. The end result is an increase in the level of blood sugar, and long continued excessive amounts of the hormone may lead to diabetes. This is called the *diabetogenic effect.* Another hormone of the body, insulin, decreases blood sugar level. In this regard, HGH and insulin are antagonistic. The former is *hyperglycemic,* which means that it increases blood sugar level, and the latter is *hypoglycemic,* which means that it decreases blood sugar level.

Thyroid-stimulating hormone (TSH)

This hormone is also called *thyrotropin* and *TSH.* Its function is to stimulate the synthesis and secretion of the hormones produced by the thyroid gland.

Adrenocorticotrophic hormone (ACTH)

This hormone, also called *adrenocorticotropin* and *ACTH,* has a dual function. Its trophic function is to control the production and secretion of the adrenal cortex hormones. The adrenal glands will be discussed later in this chapter. Its effects on the body other than on the adrenal cortex are to increase fat mobilization, bring about hyperglycemia, and help provide resistance to stress.

Follicle-stimulating hormone (FSH)

In the female, *follicle-stimulating hormone,* or *FSH,* is transported by the blood to the ovaries where it stimulates the development of ova each month. FSH also stimulates cells in the ovaries to secrete estrogens, or female sex hormones. In the male, FSH stimulates the testes to produce sperm and to secrete testosterone, a male sex hormone.

Luteinizing hormone (LH or ICSH)

The *luteinizing hormone* is called *luteotropin* and *LH* in the female and *interstitial cell stimulating hormone (ICSH)* in the male. LH assumes a role in the development of an ovum following the action of FSH. Namely, it stimulates the ovary to release the developed ovum and prepares the uterus for implantation of a fertilized ovum. It also stimulates the secretion of progesterone (another female sex hormone) and readies the mammary glands for milk secretion. In the male, ICSH stimulates the interstitial cells in the testes to develop and secrete testosterone.

Prolactin

Prolactin, or *lactogenic hormone,* together with other hormones, initiates and maintains milk secretion by the mammary glands.

The neurohypophysis

In a strict sense, the posterior lobe, or **neurohypophysis,** is not an endocrine gland since it does not synthesize hormones. The posterior lobe consists of supporting cells called pituicytes, which are similar in appearance to the neuroglia of the nervous system. It also contains neuron fibers that establish an important connection between the hypothalamus and neurohypophysis (Figure 17–5). The cell bodies of the neurons originate in nuclei in the hypothalamus. The fibers project from the hypothalamus, form the *hypothalamic-hypophyseal tract,* and terminate on blood capillaries in the neurohypophysis. The cell bodies of the neurons produce two hormones, oxytocin and the antidiuretic hormone. Following their production, the hormones are transported in the neuron fibers into the neurohypophysis and are stored in the axon terminals resting on the capillaries. Later, when the hypothalamus is properly stimulated, it sends impulses over the neurons. The impulses cause the release of the hormones from the axon terminals into the blood.

The blood supply to the neurohypophysis is from the *inferior hypophyseal arteries,* derived from the internal carotid arteries. Within the neurohypophysis, the inferior hypophyseal arteries form a plexus of capillaries called the *plexus of the infundibular process.* From this plexus blood is drained into the *posterior hypophyseal veins.*

Oxytocin

This hormone stimulates the contraction of the smooth muscle cells in the pregnant uterus and the contractile cells around the ducts of the mammary glands. It is released in large quantities just prior to giving birth. When labor begins, the uterus and vagina distend. This distention initiates afferent impulses to the hypothalamus. These impulses stimulate the secretion of more oxytocin by the hypothalamus. The oxytocin migrates along the nerve

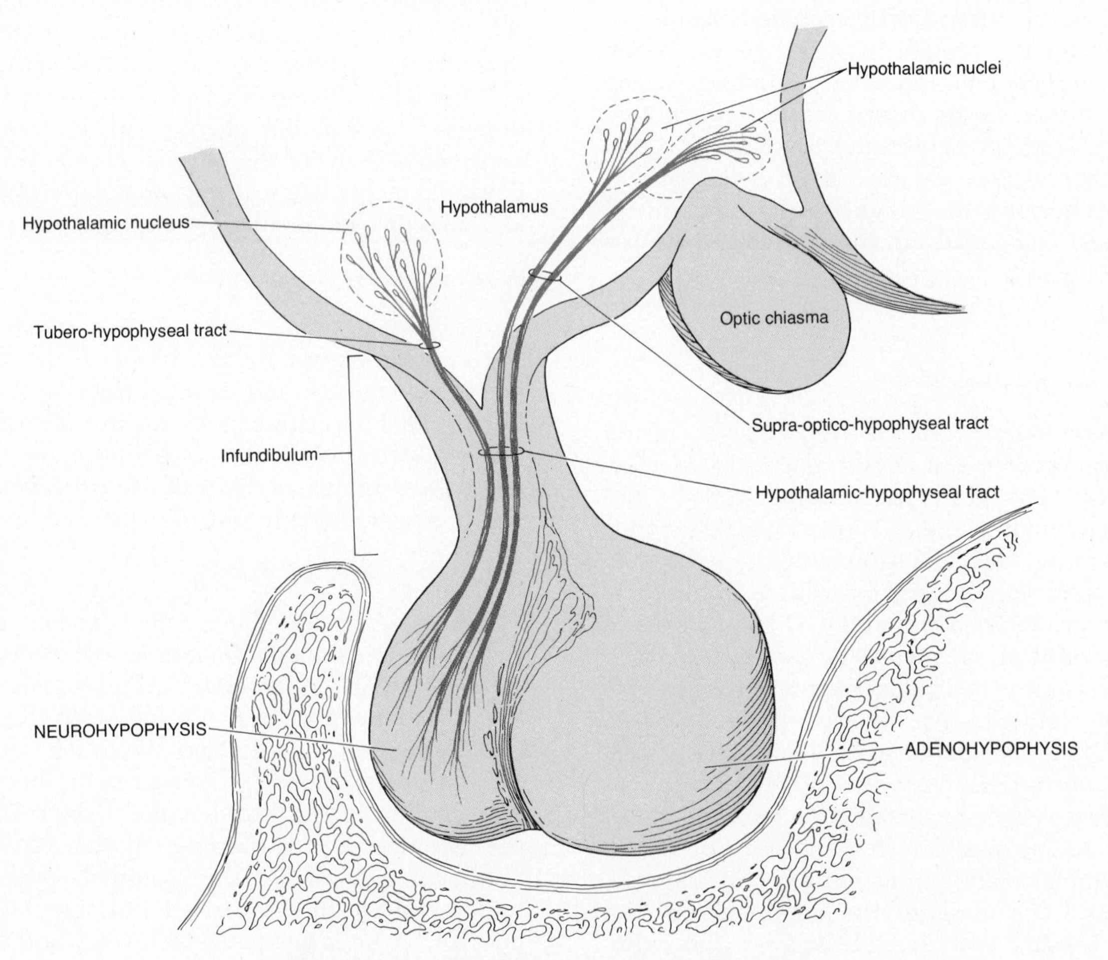

Figure 17–5. The hypothalamic-hypophyseal tract.

fibers of the hypothalamus to the neurohypophysis. The impulses also cause the neurohypophysis to release oxytocin into the blood. It is then carried by the blood to the uterus to reinforce uterine contractions. The effect of oxytocin on milk ejection is as follows. Milk formed by the glandular cells of the breasts is stored until the baby begins active suckling. From about 30 seconds to 1 minute after nursing begins, the baby receives no milk. During this latent period, nerve impulses from the nipple are transmitted to the hypothalamus. The hypothalamus sends impulses down the neurosecretory neurons which release oxytocin from their axonic ends in the neurohypophysis. Oxytocin then flows from the neurohypophysis via the blood to the breasts, where it stimulates smooth muscle cells to contract and eject milk out of the mammary glands.

Vasopressin-ADH or antidiuretic hormone

This hormone has two principal physiological activities. One of these is to cause a rise in blood pressure by bringing about constriction of arterioles. This effect is noted if large quantities of the purified hormone are injected. Only in rare instances, however, does the body secrete enough hormone to significantly affect blood pressure. Thus the name vasopressin. The more important physiological activity of vasopressin-ADH is its effect on urine volume. For this reason, the hormone is simply referred to as the antidiuretic hormone or ADH. ADH causes the kidneys to remove water from newly forming urine and return it to the bloodstream. Since water is the chief constituent of urine, ADH decreases urine volume. In the absence of ADH, urine output may be increased 10 times. An *antidiuretic* is any chemical substance that prevents excessive urine production.

THYROID GLAND

The endocrine organ located just below the larynx is called the **thyroid gland.** The two lateral lobes lie one on either side of the trachea and are connected by a mass of tissue called an *isthmus* that lies in front of the trachea just below the cricoid cartilage (Figure 17–6). The pyramidal lobe extends upward. The gland weighs about 25 g (almost 1 oz) and has a very rich blood supply, receiving about 80 to 120

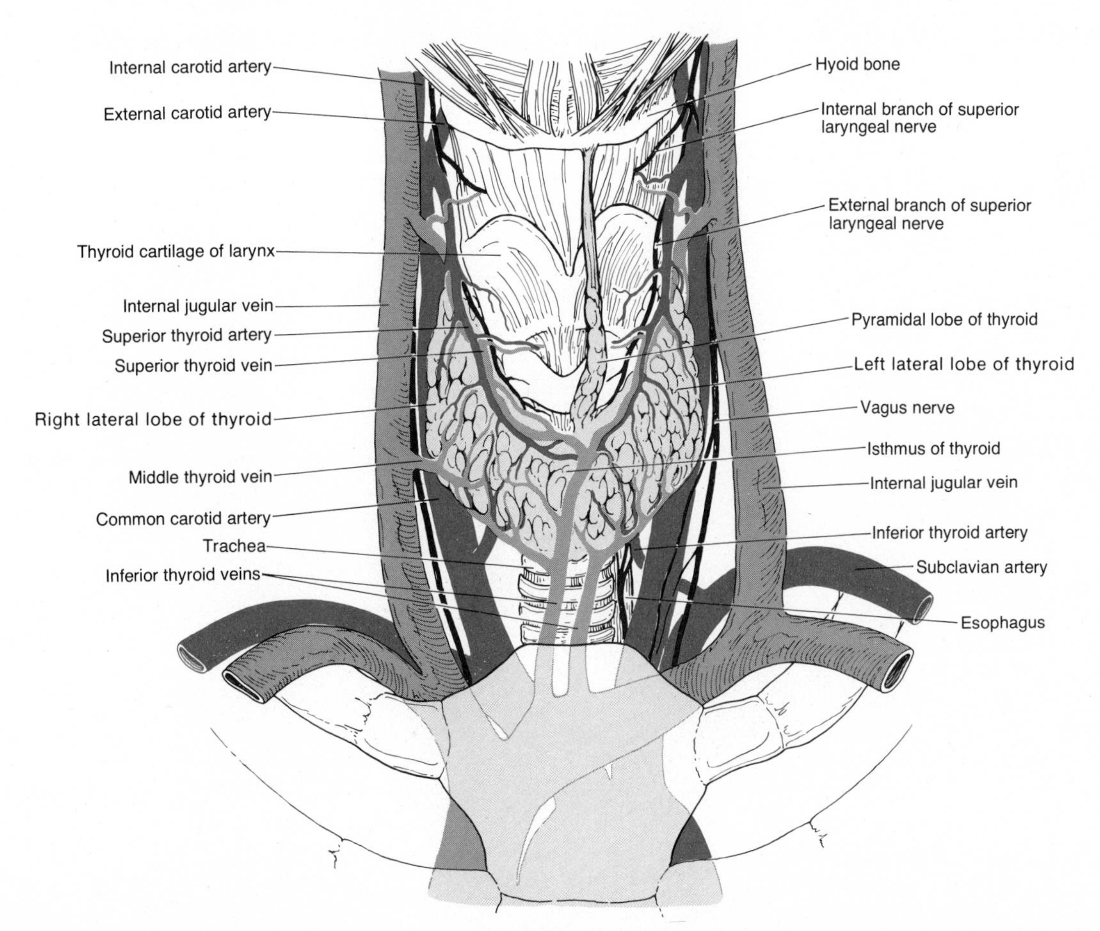

Figure 17–6. Location and blood supply of the thyroid gland seen in anterior view. The pyramidal lobe of the thyroid is not constant and, when present, it may be attached to the hyoid bone by a muscle.

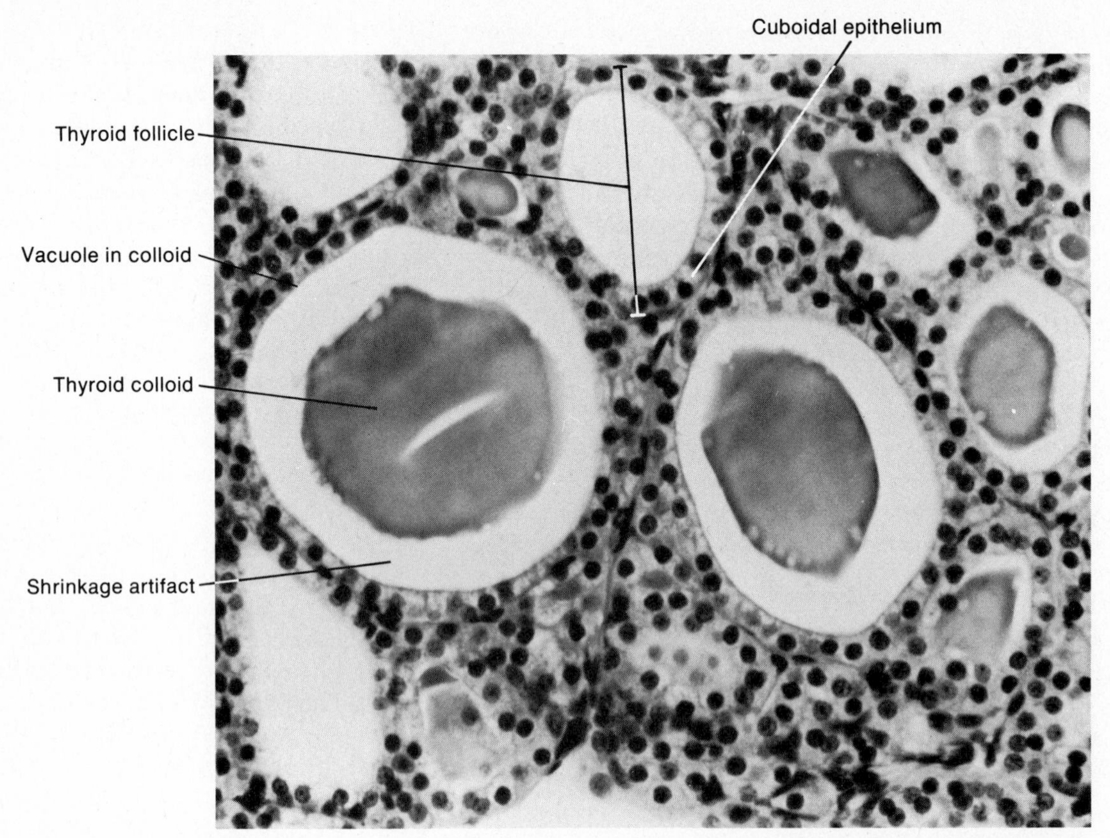

Figure 17–7. Histology of the thyroid gland at a magnification of 90X. (Courtesy of Victor B. Eichler, Wichita State University.)

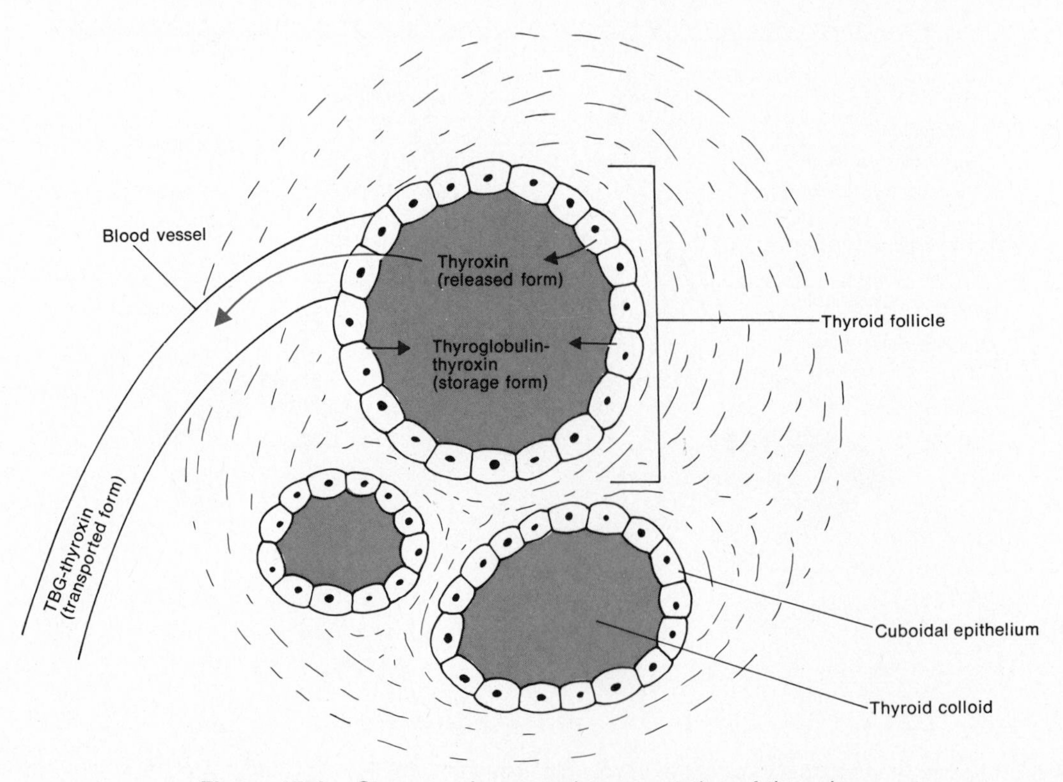

Figure 17–8. Storage, release, and transportation of thyroxin.

ml blood/minute. The main blood supply is from the superior thyroid artery, a branch of the external carotid artery, and the inferior thyroid artery, a branch of the subclavian artery. The veins draining the thyroid are the superior and middle thyroid veins that pass into the internal jugular veins and the inferior thyroid veins that join the brachiocephalic veins. Histologically, the thyroid gland is composed of spherical-shaped sacs called *thyroid follicles* (Figure 17–7). The walls of each follicle consist of cells that reach the surface of the lumen of the follicle (principal cells) and cells that do not reach the lumen (parafollicular cells). The principal cells manufacture the hormones thyroxin (T4 or tetraiodothyronine) and triiodothyronine (T3). Thyroxin is considered to be the major hormone produced by the principal cells. The parafollicular cells produce the hormone thyrocalcitonin. The interior of each thyroid follicle is filled with *thyroid colloid,* a stored form of the thyroid hormones.

The nerve supply of the thyroid consists of postganglionic fibers from the superior and middle cervical sympathetic ganglia. Preganglionic fibers from the ganglia are derived from the second through seventh thoracic segments of the spinal cord.

Physiology of the thyroid

One of the unique features of the thyroid gland is its ability to store its hormones and release them in a steady flow over a long period of time. For example, the principal hormone, *thyroxin,* is synthesized from iodine and an amino acid called tyrosine. Synthesis usually occurs on a fairly continuous basis. However, if the body has no immediate need for the hormone, it combines with *thyroglobulin,* a protein secreted by the follicle cells, and is stored in the thyroid colloid (Figure 17–8). When demand for thyroxin occurs, the hormone splits apart from the thyroglobulin and is released into the blood. There, it combines with a plasma protein called *thyroxin-binding globulin* or *TBG.* The thyroxin is released from TBG as it enters tissue cells.

The iodine in thyroxin and thyrocalcitonin that is bound to TBG is called *protein-bound iodine* or *PBI.* Under normal circumstances, the amount of PBI in the blood is fairly constant—4 to 8 μg PBI/100 ml blood. This amount can easily be measured. For this reason, PBI is a good index of thyroid hormone secretion and is often used as a tool to diagnose suspected thyroid malfunction.

Thyroxin action and control

Thyroxin is produced by the principal cells of the thyroid gland. The major function of thyroxin is to control the rate of metabolism by regulating the catabolic or energy-releasing processes and the anabolic or building-up processes. Thyroxin increases the rate at which carbohydrates are burned. And it stimulates cells to break down proteins for energy instead of using them for building processes. At the same time, thyroxin decreases the breakdown of fats. The overall effect, though, is to increase catabolism. It thus produces energy and raises body temperature as heat energy is given off. This is called the *calorigenic effect.* Thyroxin is also an important factor in the regulation of tissue growth and in the development of tissues. It works with the human growth hormone to accelerate body growth. Hyposecretion of thyroxin during the early years of life causes some organs to fail to develop. Finally, thyroxin acts as a diuretic, increases the reactivity of the nervous system, and increases the heart rate.

Thyrocalcitonin action and control

The thyroid hormone produced by the parafollicular cells of the thyroid gland is *thyrocalcitonin.* It is involved in the homeostasis of blood calcium level. As you remember, bones are continually remolded during adult life. Part of this process consists of the breakdown of osseous tissue and the release of calcium into the blood. The other part of the process is the deposition of calcium in the bones and the subsequent laying down of new ossified tissue. Thyrocalcitonin lowers the amount of calcium in the blood by inhibiting bone breakdown and by accelerating the absorption of calcium by the bones. If thyrocalcitonin is administered to a person with normal blood calcium levels, it causes *hypocalcemia,* or low blood calcium level. If thyrocalcitonin is given to an individual with *hypercalcemia* (high blood calcium level), the level returns to normal.

PARATHYROIDS

Embedded on the posterior surfaces of the lateral lobes of the thyroid are small, round masses of tissue called the **parathyroid glands.** Typically, two parathyroids, superior and inferior, are attached to each lateral thyroid lobe. They measure about 3 to 8 mm (0.1 to 0.3 in.) in length, 2 to 5 mm (0.07 to 0.2 in.) in width, and 0.5 to 2 mm (0.02 to 0.07 in.) in thickness. The combined weight of the four glands is between 0.05 and 3 g (Figure 17–9).

The parathyroids are abundantly supplied with blood from branches of the superior and inferior thyroid arteries. Blood is drained by the superior and middle thyroid veins which drain into the internal jugular vein and by the inferior thyroid veins

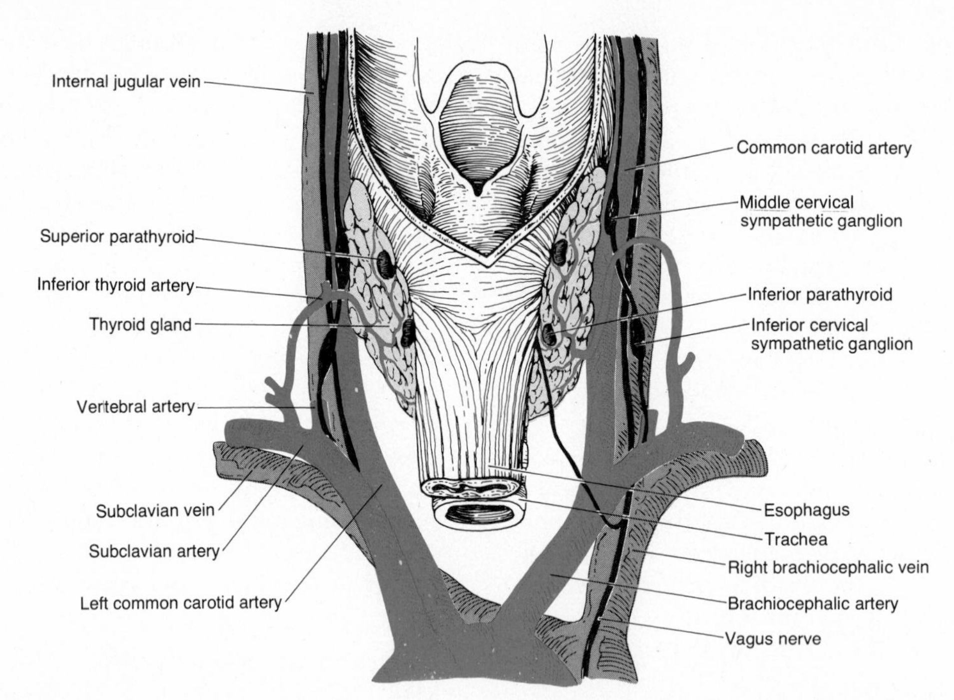

Figure 17–9. Location and blood supply of the parathyroid glands seen in posterior view.

which empty into the brachiocephalic or internal jugular veins. The nerve supply of the parathyroids is derived from the cervical sympathetic ganglia and the pharyngeal branch of the vagus. Histologically, the parathyroids contain two kinds of epithelial cells (Figure 17–10). The first kind, called *principal* or *chief cells,* is believed to be the major synthesizer of the parathyroid hormone called the parathyroid hormone. Some researchers believe that the other kind of cell, called an *oxyphil cell,* synthesizes a reserve capacity of hormone.

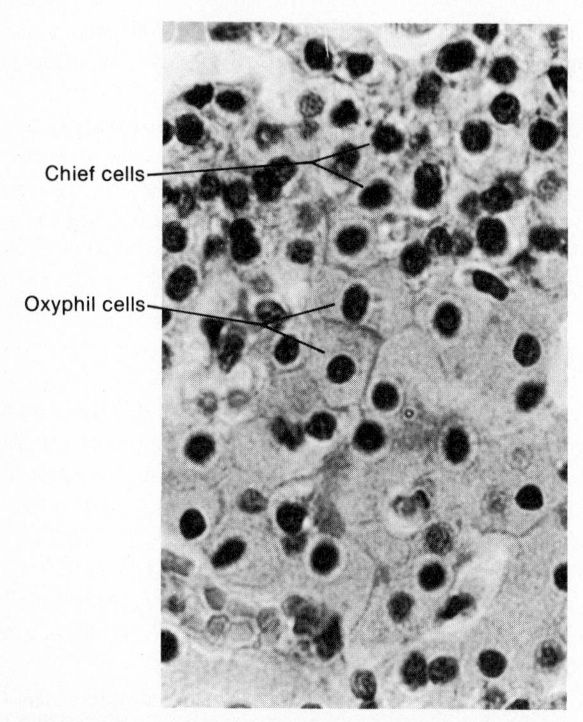

Figure 17–10. Histology of the parathyroid glands at a magnification of 50X. (Courtesy of Victor B. Eichler, Wichita State University.)

Physiology of the parathyroids

Parathyroid hormone (PTH) or *parathormone* controls the homeostasis of ions in the blood, especially the homeostasis of calcium and phosphate ions. First, if adequate amounts of vitamin D are present, PTH increases the rate at which calcium is absorbed from the intestine into the blood. Second, PTH increases the number of osteoclasts, or bone-destroying cells. As a result, bone tissue is broken down, and calcium and phosphate ions are released into the blood. Recall that thyrocalcitonin secreted by the thyroid has the opposite effect. Finally, PTH produces two changes in the kidneys: (1) it increases the rate at which the kidneys remove calcium ions from the urine and return them to the blood; (2) it accelerates the transportation of phosphate ions from the blood into the urine for elimination. More phosphate is lost through the urine than

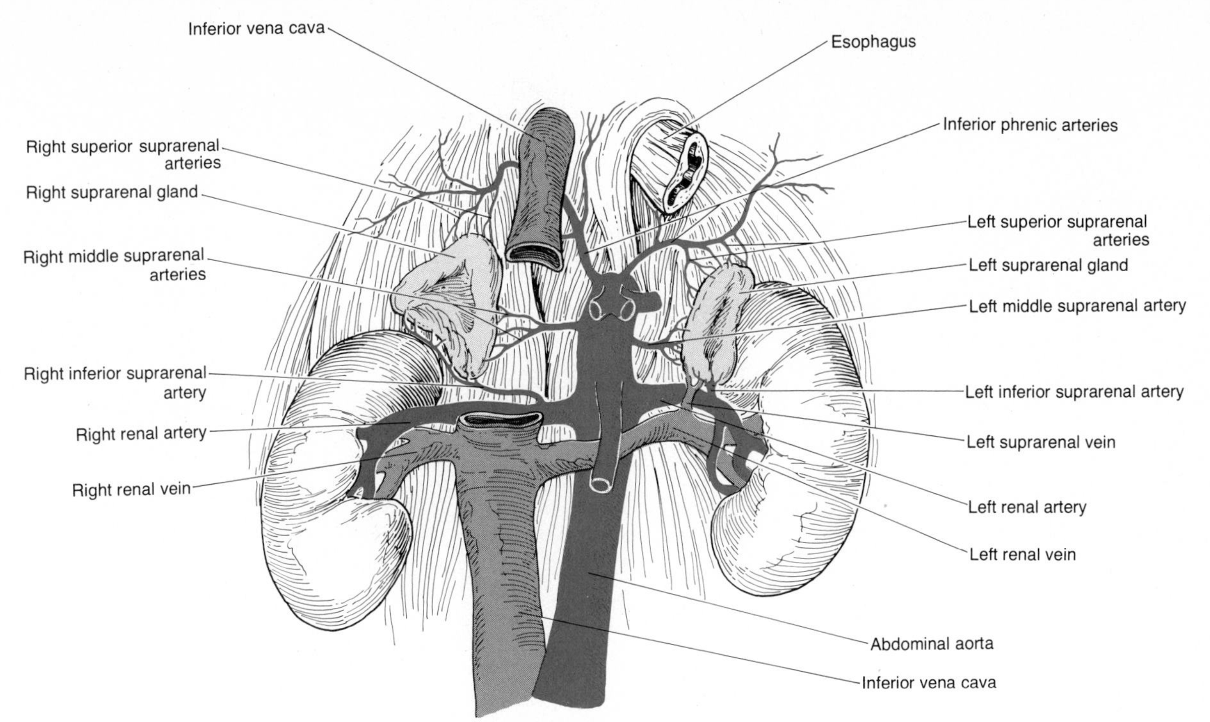

Figure 17–11. Location and blood supply of the adrenal glands.

is gained from the bones. The overall effect of PTH, then, is to decrease blood phosphate level and increase blood calcium level. As far as blood calcium level is concerned, PTH and thyrocalcitonin are antagonists.

THE ADRENALS

The body has two **adrenal** or **suprarenal glands** superior to each kidney (Figure 17–11). Each adrenal gland is structurally and functionally differentiated into two sections: the outer *adrenal cortex,* which makes up the bulk of the gland, and the inner *adrenal medulla* (Figure 17–12a). Covering the gland is a thick layer of fatty connective tissue and an outer, thin fibrous capsule.

The average dimensions of the adult adrenal gland are about 50 mm (2 in.) in length, 30 mm (1.1 in.) in width, and 10 mm (0.4 in.) in thickness. The combined weight of the glands varies between 10 and 20 g (1/3 to 2/3 oz). The adrenals, like the thyroid, are among the most vascular organs of the body. The main arteries of supply are the three suprarenal arteries arising from the inferior phrenic artery, the aorta, and the renal arteries. The suprarenal vein of the right adrenal gland drains into the inferior vena cava, whereas the suprarenal vein of the left adrenal gland empties into the left renal vein.

The principal nerve supply to the adrenal glands is from preganglionic fibers from the splanchnic nerves and from the celiac and associated sympathetic plexuses. These myelinated fibers end on the secretory cells of the gland found in a region called the medulla.

The adrenal cortex

Histologically, the cortex is subdivided into three zones (Figure 17–12c). Each zone has a different cellular arrangement and secretes different hormones. The outer zone, directly underneath the connective tissue covering, is referred to as the *zona glomerulosa.* Its cells are arranged in arched loops or round balls, and they secrete a group of hormones called mineralocorticoids. The middle zone, or *zona fasciculata,* is the widest of the three zones and consists of cells arranged in long, straight cords. The zona fasciculata secretes glucocorticoid hormones. The inner zone, the *zona reticularis,* contains cords of cells that branch freely. This zone synthesizes sex hormones, chiefly male hormones called androgens.

Mineralocorticoids

These hormones help control electrolyte homeostasis, particularly the concentrations of sodium and potassium. Although the adrenal cortex secretes at least three different hormones classified as mineralocorticoids, one of these hormones is responsible for about 95 percent of the mineralocorticoid activity. The name of this hormone is *aldosterone.*

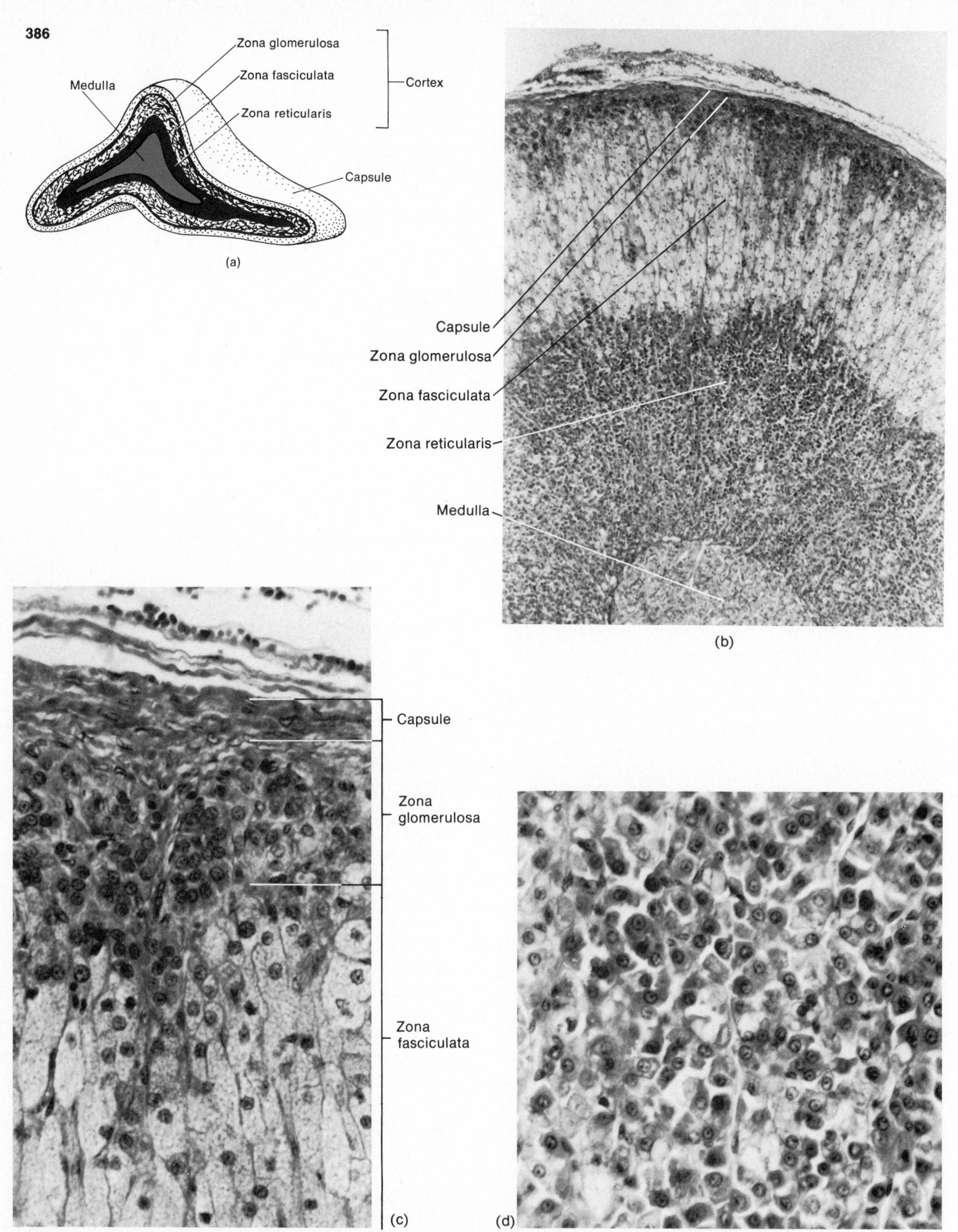

Figure 17–12. Subdivisions and histology of the adrenal gland. (a) Diagrammatic representation. (b) Photomicrograph at a magnification of 15X. (c) Histology of the capsule, zona glomerulosa, and zona fasciculata at a magnification of 90X. (d) Histology of the zona reticularis at a magnification of 90X. (Courtesy of Victor B. Eichler, Wichita State University.)

Aldosterone acts on the tubule cells in the kidneys and causes them to increase their reabsorption of sodium, with the result that sodium ions are removed from the urine and returned to the blood. In this manner, aldosterone prevents rapid depletion of sodium from the body. On the other hand, aldosterone decreases reabsorption of potassium. Large amounts of potassium are moved from the blood into the urine.

Glucocorticoids

The *glucocorticoids* are a group of hormones that are largely concerned with normal metabolism and the ability of the body to resist stress. Three examples of glucocorticoids are *hydrocortisone (cortisol), corticosterone,* and *cortisone.* Of the three, hydrocortisone is the most abundant. The glucocorticoids have the following effects on the body:

1. Glucocorticoids work with other hormones in promoting normal metabolism. Their role is to make sure that enough energy is provided. They increase the rate at which amino acids are removed from cells and transported to the liver. The amino acids may be synthesized into new proteins, such as the enzymes that are needed for the metabolic reactions. Or, if the body's reserves of glycogen and fat are low, the liver may convert the amino acids to glucose. Glucocorticoids also promote the breakdown of carbohydrates to glucose. Both processes increase blood sugar level. Glucocorticoids are therefore hyperglycemic. In addition, the hormones encourage the movement of fats from storage depots to all the cells, where they are catabolized for energy.
2. Glucocorticoids work in many ways to provide resistance to stress. One of the more obvious ways is their hyperglycemic effect. A sudden increase in available glucose makes the body more alert. Additional glucose gives the body energy for combating a range of stressors, such as fright, temperature extremes, high altitude, bleeding, and infection. Glucocorticoids also make the blood vessels more sensitive to vessel-constricting chemicals. They thereby raise blood pressure. This is advantageous if the stressor happens to be blood loss, which causes a drop in blood pressure.
3. Glucocorticoids decrease the blood vessel dilatation and edema associated with inflammations. They are thus anti-inflammatories. Unfortunately, they also decrease connective-tissue regeneration and are thereby responsible for slow wound healing.

Gonadocorticoids

The adrenal cortices secrete both male and female *gonadocorticoids,* or *sex hormones.* But the amount of sex hormones secreted by the adrenals is usually so small that it is insignificant. The exception is hypersecretion—an abnormality that will be described shortly.

The adrenal medulla

The adrenal medulla consists of hormone-producing cells, called *chromaffin cells,* which surround large blood-containing sinuses (see Figure 17–12c). Chromaffin cells develop from the same source as the postganglionic cells of the sympathetic division of the nervous system. They are directly innervated by preganglionic cells of the sympathetic division of the autonomic nervous system and may be regarded as postganglionic cells that are specialized to secrete. In all other visceral effectors, preganglionic sympathetic fibers first synapse with postganglionic neurons before innervating the effector. In the adrenal medulla, however, the preganglionic fibers pass directly into the chromaffin cells of the gland. The secretion of hormones from the chromaffin cells is directly controlled by the autonomic nervous system, and innervation by the preganglionic fibers allows the gland to respond very rapidly to a stimulus.

The two principal hormones synthesized by the adrenal medulla are epinephrine and norepinephrine. Epinephrine constitutes about 80 percent of the total secretion of the gland and is more potent in its action than norepinephrine. Both hormones are *sympathomimetic.* That is, they produce effects similar to those brought about by the sympathetic division of the autonomic nervous system. And, to a large extent, they are responsible for the fight-or-flight response. Like the glucocorticoids of the adrenal cortices, these hormones help the body resist stress situations. However, unlike the cortical hormones, the medullary hormones are not essential for life. Under stress conditions, impulses received by the hypothalamus are conveyed to sympathetic preganglionic neurons, which cause the chromaffin cells to increase their output of epinephrine and norepinephrine. Epinephrine increases blood pressure by increasing the heart rate and by constricting the blood vessels. It accelerates the rate of respiration, dilates respiratory passageways, decreases the rate of digestion, increases the efficiency of muscular contractions, increases blood sugar level, and stimulates cellular metabolism.

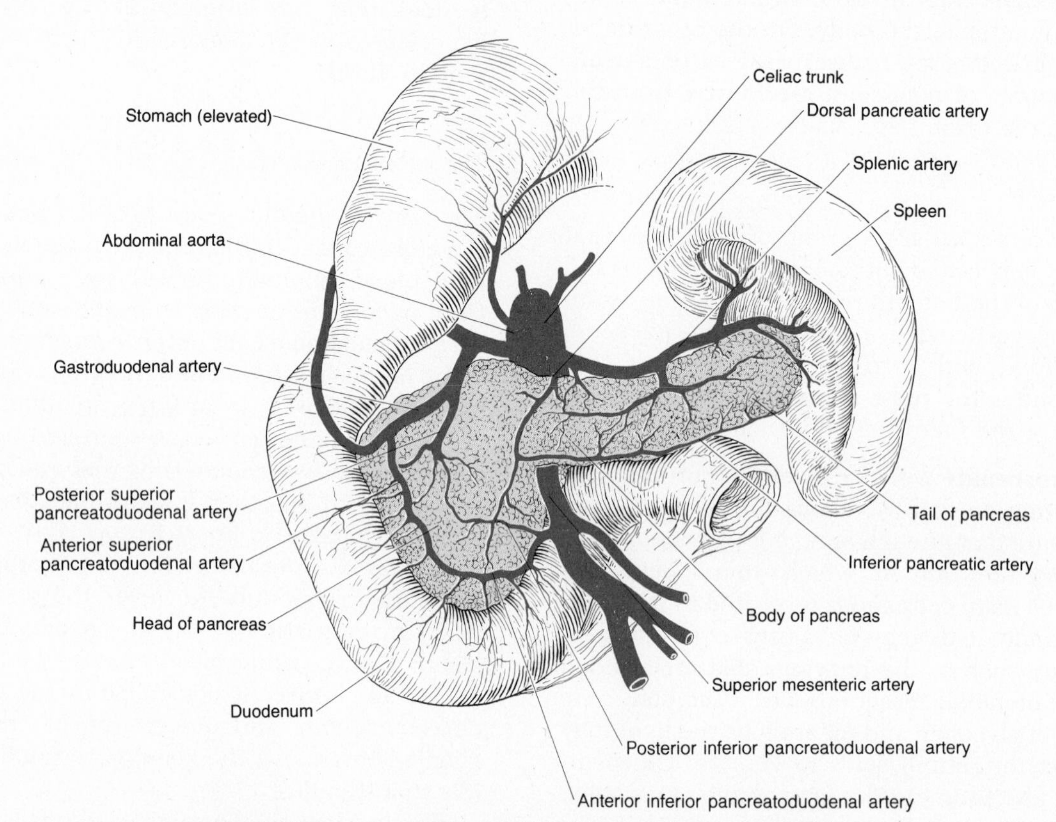

Figure 17–13. Location and blood supply of the pancreas.

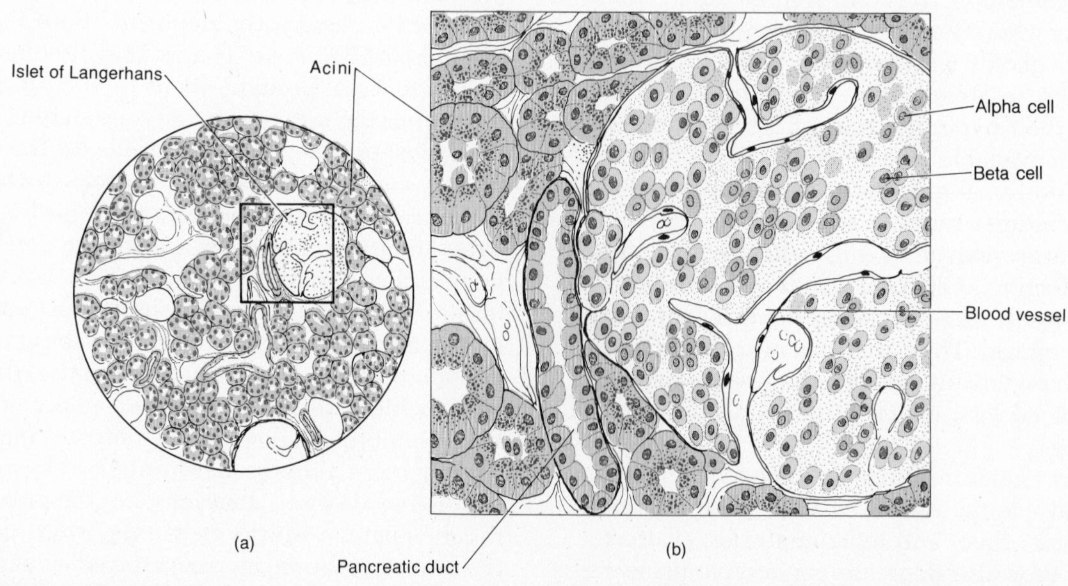

Figure 17–14. Histology of the pancreas. (a) View of two islets of Langerhans (cells that produce hormones) and several acini (cells that produce digestive enzymes). (b) Enlarged aspect of a portion of an islet of Langerhans and several acini.

PANCREAS

Because of its functions, the **pancreas** can be classified as both an endocrine and an exocrine gland. Since the exocrine functions of the gland will be discussed in the chapter on the digestive system, we shall treat only its endocrine functions at this point. The pancreas is a flattened organ located posterior and slightly inferior to the stomach (Figure 17–13). The adult pancreas consists of a head, body, and tail. Its average length is 12 to 15 cm (5 to 6 in.), and its average weight is about 85 g (3 oz). The arterial supply of the pancreas is from the superior and inferior pancreatoduodenal arteries and from the splenic and superior mesenteric veins. The nerves to the pancreas are branches of the celiac plexus. The glandular portion of the pancreas is innervated by the craniosacral division of the autonomic nervous system, whereas the blood vessels of the pancreas are innervated by the thoracolumbar division of the autonomic nervous system. The endocrine portion of the pancreas consists of clusters of cells called *islets of Langerhans* (Figure 17–14). Two kinds of cells are found in these clusters: (1) *alpha cells,* which constitute about 25 percent of the islet cells and secrete the hormone glucagon; and (2) *beta cells,* which constitute about 75 percent of the islet cells and secrete the hormone insulin. The islets are surrounded by blood capillaries and by the cells that form the exocrine part of the gland.

Physiology of the pancreas

The endocrine secretions of the pancreas–glucagon and insulin–are concerned with regulation of the blood sugar level. Let us now examine how this takes place.

Glucagon

The product of the alpha cells is *glucagon,* a hormone whose principal physiological activity is to increase the blood glucose level. Glucagon does this by accelerating the conversion of liver glycogen into glucose. The liver then releases the glucose into the blood, and the blood sugar level rises. If for some reason the alpha cells secrete glucagon continuously, hyperglycemia may result.

Insulin

The beta cells of the islets produce a hormone called *insulin.* This hormone increases the build-up of proteins in cells. But its chief physiological action is opposite that of glucagon. Insulin decreases blood sugar level. This is accomplished in two ways. First, insulin accelerates the transport of glucose from the blood into body cells, especially into the cells of the skeletal muscles. Second, insulin accelerates the conversion of glucose into glycogen.

OVARIES AND TESTES

The female gonads, called the **ovaries,** are paired oval-shaped bodies located in the pelvic cavity (see Figure 17–1). The ovaries produce female sex hormones that are responsible for the development and maintenance of the female sexual characteristics. Along with the gonadotrophic hormones of the pituitary, the sex hormones also regulate the menstrual cycle, maintain pregnancy, and ready the mammary glands for lactation. The male has two oval-shaped glands, called **testes,** that lie in the scrotum. The testes produce the male sex hormones that stimulate the development and maintenance of the male sexual characteristics. In Chapter 18, more will be said about the sex hormones and the anatomy of the testes and ovaries.

PINEAL GLAND

The cone-shaped gland located in the roof of the third ventricle is known as the **pineal gland,** or **epiphysis cerebri** (see Figure 17–1). The gland is about 5 to 8 mm (0.2 to 0.3 in.) long and 9 mm wide. It weighs about 0.2 g. It is covered by a capsule formed by the pia mater. It consists of masses of parenchymal and glial cells. Around the cells are scattered preganglionic sympathetic fibers. The pineal gland starts to degenerate at about age 7, and in the adult it is largely fibrous tissue. The posterior cerebral artery supplies the pineal with blood, and the great cerebral vein drains it.

Although many anatomical facts concerning the pineal gland have been known for years, its physiology is still somewhat obscure. One hormone secreted by the pineal gland is *melatonin,* which appears to affect the secretion of hormones by the ovaries. It has been known for years that light stimulates the sexual endocrine glands. Researchers have also discovered that blood levels of melatonin are low during the day and high at night. Putting these observations together, some investigators now believe that melatonin inhibits the activities of the ovaries. During daylight hours, light entering the eye stimulates neurons to transmit impulses to the pineal that inhibit melatonin secretion. Without melatonin interference, the ovaries are free to step up their hormone production. But at night, the pineal gland is able to release melatonin, and ovarian function is slowed down. One of the functions of the pineal gland might very well be regulation of the activities of the sexual endocrine glands, particularly the menstrual cycle.

Some evidence also exists that the pineal secretes a second hormone called *adrenoglomerulotropin.* This hormone may stimulate the adrenal cortex to secrete aldosterone. Still other functions attributed to the pineal gland are the secretion of a growth-inhibiting factor and the secretion of a hormone called *serotonin* that is involved in normal brain physiology.

THYMUS GLAND

Usually a bilobed organ, the **thymus gland** is located in the upper mediastinum posterior to the sternum and between the lungs (Figure 17–15). The gland is conspicuous in the infant, and during puberty it reaches its absolute maximum size, when it weighs between 30 and 40 g (1 to 1.3 oz). After puberty, the thymic tissue, which consists primarily of lymphocytes, is replaced by fat. By the time the person reaches maturity, the gland has atrophied. The thymus secretes a hormone that sensitizes the body to produce antibodies. The arterial supply is derived mainly from the internal mammary and inferior thyroid vessels. The veins that drain the thymus are the internal mammary, brachiocephalic, and thyroid veins. Postganglionic sympathetic and parasympathetic fibers supply the gland. The thymus is believed to secrete a hormone that enables the lymphocytes and plasma cells to produce antibodies for the defense of the body.

APPLICATIONS TO HEALTH

Disorders of the pituitary

Disorders of the endocrine system, in general, are based upon under- or overproduction of hormones. The term **hyposecretion** describes an underproduction, whereas the term **hypersecretion** means an oversecretion. The anterior pituitary gland produces many hormones. All these hormones, with the exception of the human growth hormone, directly control the activities of other endocrine glands. It is hardly surprising, then, that hypo- or hypersecretion of an anterior pituitary hormone produces widespread and complicated abnormalities.

Among the clinically interesting disorders related to the adenohypophysis are those involving the human growth hormone. Human growth hormone builds up cells, particularly those of bone tissue. If the hormone is hyposecreted during the growth years, bone growth is slow, and the epiphyseal plates close before normal height is reached. This is the condition called **pituitary dwarfism.** Other organs of the body also fail to grow, and the pituitary dwarf is childlike in many physical respects. Treating the condition requires administration of human growth hormone during childhood, before the epiphyseal plates close.

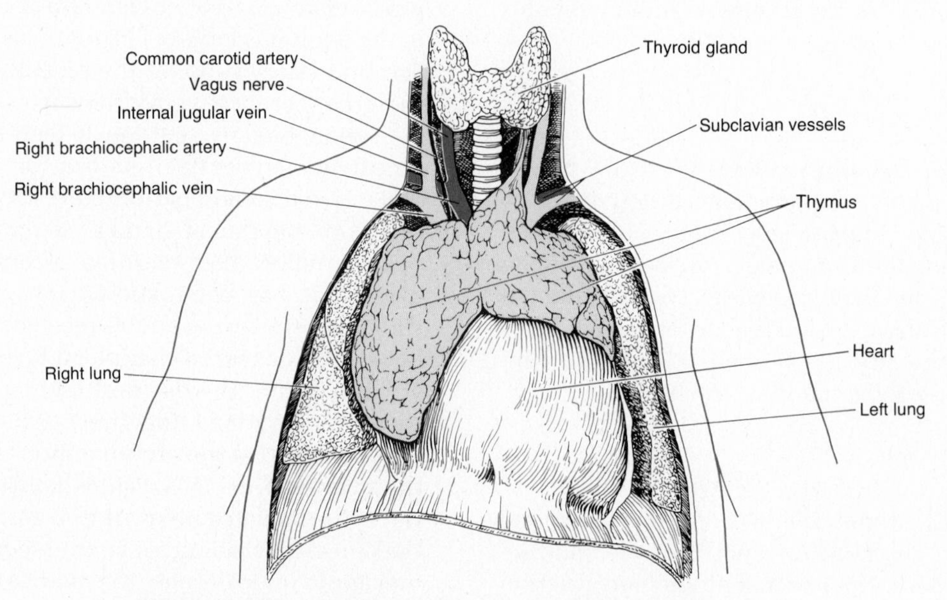

Figure 17–15. Location of the thymus gland in a young child.

If secretion of human growth hormone is normal during childhood, but lower than normal during adult life, a rare condition called **pituitary cachexia (Simmond's disease)** occurs. The tissues of a person with Simmond's disease waste away, or **atrophy.** The victim becomes quite thin and shows signs of premature aging. For instance, as the connective tissue degenerates, it loses its elasticity, and the skin hangs and becomes wrinkled. The atrophy occurs because the person is not receiving enough human growth hormone to stimulate the protein-building activities that are required for replacing cells and cell parts.

Hypersecretion of the human growth hormone produces completely different disorders. For example, hyperactivity during childhood years results in **giantism,** which is an abnormal increase in the length of long bones. Hypersecretion during adulthood is called **acromegaly.** Acromegaly cannot produce further lengthening of the long bones because the epiphyseal plates are already closed. Instead, the bones of the hands, feet, cheeks, and jaws thicken. Other tissues also grow. For instance, the eyelids, lips, tongue, and nose enlarge, the skin thickens and furrows, especially on the forehead and soles of the feet.

The principal abnormality associated with dysfunction of the neurohypophysis is **diabetes** *(overflow)* **insipidus** *(tasteless).* This disorder should not be confused with diabetes mellitus *(sugar),* which is a disorder of the pancreas and is characterized by sugar in the urine. Diabetes insipidus is the result of a hyposecretion of ADH, usually caused by damage to the neurohypophysis or to the hypothalamus. Symptoms of the disorder include excretion of large amounts of urine and subsequent thirst. Diabetes insipidus is treated by administering ADH.

Disorders of the thyroid

Hyposecretion of thyroxin during the growth years results in a condition called **cretinism.** Two outstanding clinical symptoms of the cretin are dwarfism and mental retardation. The first is caused by failure of the skeleton to grow and mature. The second is caused by failure of the brain to develop fully. Recall that one of the functions of thyroid hormone is to control tissue growth and development. Cretins also exhibit retarded sexual development and a yellowish skin color. Flat pads of fat develop, giving the cretin the characteristic round face and thick nose; a large, thick, protruding tongue; and protruding abdomen. Because the energy-producing metabolic reactions are so slow, the cretin has a low body temperature and general lethargy. Carbohydrates are stored rather than utilized. Heart rate is also slow. If the condition is diagnosed early, the symptoms are reversible following administration of the thyroid hormone.

Hypothyroidism during the adult years produces a disorder called **myxedema.** The name refers to the fact that thyroxin is a diuretic. Lack of thyroxin causes the body to retain water. And one of the hallmarks of myxedema is an edema that causes the facial tissues to swell and look puffy. Another symptom caused by the retention of water is an increase in blood volume that frequently causes high blood pressure. Like the cretin, the person with myxedema also suffers from slow heart rate, low body temperature, muscular weakness, general lethargy, and a tendency to gain weight easily. The long-term combination of a slow heart rate and high blood pressure may overwork the heart muscles, causing the heart to enlarge. Because the brain has already reached maturity, the person with myxedema does not experience mental retardation. However, in moderately severe cases, nerve reactivity may be dulled so that the person lacks mental alertness. Myxedema occurs eight times more frequently in females than in males. Its symptoms are abolished by the administration of thyroxin.

Hypersecretion of thyroxin gives rise to a condition called **exophthalmic goiter.** This disease, like myxedema, is also more frequent in females, affecting eight females to every one male. One of its primary symptoms is an enlarged thyroid, called a *goiter,* which may be two to three times its original size. The two other characteristic symptoms are an edema behind the eyeball, which causes the eye to protrude (**exophthalmos**), and an abnormally high metabolic rate. The high metabolic rate produces a range of effects that are generally opposite to those of myxedema. The person has an increased pulse. The body temperature is high, and the skin is warm, moist, and flushed. Weight loss occurs, and the person is usually full of "nervous" energy. The thyroxin also increases the responsiveness of the nervous system. Thus a person with exophthalmic goiter may become irritable and may exhibit tremors of the fingers when they are extended. The usual methods for treating hyperthyroidism are administering drugs that suppress thyroxin synthesis or surgically removing a part of the gland.

The term **goiter** simply means an enlargement of the thyroid gland. It is a symptom of many thyroid disorders. It may also occur if the gland does not receive enough iodine to produce sufficient thyroxin for the body's needs. The follicular cells then enlarge in a futile attempt to produce more thyroxin, and they secrete large quantities of colloid.

This is called *simple goiter.* Simple goiter is most often caused by a lower-than-average amount of iodine in the diet. However, it may also develop if iodine intake is not increased during certain conditions that put a high demand on the body for thyroxin. Such conditions are frequent exposure to cold and high fat and protein diets.

Disorders of the parathyroids

A normal amount of calcium in the extracellular fluid is necessary to maintain the resting state of neurons. A deficiency of calcium caused by **hypoparathyroidism** causes neurons to depolarize without the usual stimulus. As a result, nervous impulses increase and result in muscle twitches, spasms, and convulsions. This condition is called **tetany.** The effects of hypocalcemic tetany are observed in the **Trousseau** and **Chvostek signs.** Trousseau sign is observed when the binding of a blood pressure cuff around the upper arm produces contraction of the fingers and inability to open the hand. The Chvostek sign is a contracture of the facial muscles elicited by tapping the facial nerves at the angle of the jaw. Hypoparathyroidism results from surgical removal of the parathyroids or from parathyroid damage caused by parathyroid disease, infection, hemorrhage, or mechanical injury.

Hyperparathyroidism causes demineralization of bone. This condition is called **osteitis fibrosa cystica** because the areas of destroyed bone tissue are replaced by cavities that fill with fibrous tissue. The bones thus become deformed and are highly susceptible to fracture. Hyperparathyroidism is usually caused by a tumor in the parathyroids.

Disorders of the adrenals

Hypersecretion of the mineralocorticoid aldosterone results in a decrease in the body's potassium concentration. As you remember, potassium movement is involved in the transmission of nerve impulses. Consequently, if potassium depletion is great enough, neurons cannot depolarize and muscular paralysis results. Hypersecretion also brings about excessive retention of sodium and water. The water increases the volume of the blood and causes high blood pressure. It also increases the volume of the interstitial fluid, producing edema.

Disorders associated with glucocorticoids include Addison's disease and Cushing's syndrome. Hyposecretion of glucocorticoids results in the condition called **Addison's disease.** Clinical symptoms include hypoglycemia, which leads to muscular weakness, mental lethargy, and weight loss. In addition, increased potassium blood levels and decreased sodium blood levels lead to low blood pressure and dehydration. **Cushing's syndrome** is a hypersecretion of glucocorticoids, especially hydrocortisone and cortisone. The condition is characterized by the redistribution of fat. This results in spindly legs accompanied by a characteristic "moon face," "buffalo hump" on the back, and pendulous abdomen. The facial skin is flushed, and the skin covering the abdomen develops stretch marks. The individual also bruises easily, and wound healing is poor.

The **adrenogenital syndrome** results from overproduction of sex hormones, particularly the male androgens, by the adrenal cortex. Hypersecretion in male infants and young male children results in an enlarged penis. In young boys, it also causes premature development of male sexual characteristics. Hypersecretion in adult males is characterized by overgrowth of body hair, enlargement of the penis, and increased sexual drive. Hypersecretion in young girls results in premature sexual development. Hypersecretion in both girls and women usually produces a receding hairline, baldness, an increase in body hair, deepening of the voice, muscular arms and legs, small breasts, and an enlarged clitoris.

Tumors of the chromaffin cells of the adrenal medulla, called **pheochromocytomas,** cause hypersecretion of the medullary hormones. The oversecretion causes high blood pressure, high levels of sugar in the blood and urine, an elevated basal metabolic rate, nervousness, and sweating. Since the medullary hormones create the same effects as does sympathetic nervous stimulation, hypersecretion puts the individual into a prolonged version of the fight-or-flight response. Needless to say, this eventually wears out the body, and the individual eventually suffers from general weakness.

Endocrine disorders of the pancreas

Hyposecretion of insulin results in a number of clinical symptoms referred to as **diabetes mellitus.** Typically an inherited disease, diabetes mellitus is caused by the destruction or malfunction of the beta cells. Among the symptoms are hyperglycemia and excretion of glucose in the urine as hyperglycemia increases. There is also an inability to reabsorb water, resulting in increased urine production, dehydration, loss of sodium, and thirst. Although the cells need glucose for energy-releasing reactions, glucose cannot enter the cells without the help of insulin. The cells start breaking down large quantities of fats and proteins into glucose.

When the fats are decomposed, organic acids called ketone bodies are formed as side products. Excessive decomposition of fats produces more ketone bodies than the body can neutralize through its buffer systems. As a result, the blood pH falls. This form of acidosis is called **ketosis.** The catabolism of stored fats and proteins also causes weight loss. As lipids are transported by the blood from storage depots to hungry cells, lipid particles are deposited on the walls of blood vessels. The deposition leads to atherosclerosis and a multitude of circulatory problems.

Hyperinsulinism is much rarer than hyposecretion and is generally the result of a malignant tumor in an islet. The principal symptom is a decreased blood glucose level, which stimulates the secretion of epinephrine, glucagon, and the growth hormone. As a consequence, anxiety, sweating, tremor, increased heart rate, and weakness occur. Moreover, brain cells do not have enough glucose to function efficiently. This leads to mental disorientation, convulsions, unconsciousness, shock, and eventual death as the vital centers in the medulla are affected.

Key terms associated with the endocrine system

Acromegaly (*acro* = extremity; *megaly* = large) Oversecretion of the pituitary growth during adulthood resulting in enlarged bones of the hands, feet, and face.

Addison's disease Caused by hyposecretion of glucocorticoids by the adrenal cortex; amino acids are not catabolized, and blood sugar is low.

Aldosteronism A disorder caused by hypersecretion of adrenal mineralocorticoids; potassium depletion occurs, sometimes causing paralysis; sodium and water are retained, causing high blood pressure and edema.

Cretinism Hyposecretion of the thyroid, resulting in dwarfism and mental retardation.

Cushing's syndrome A disease of hypersecretion of the adrenal glucocorticoids; amino acid and fat catabolism increase; fat is redeposited in face and trunk.

Diabetes insipidus A disorder caused by hyposecretion of the antidiuretic hormone (ADH) of the neurohypophysis; basic clinical symptoms are high urine production and thirst.

Diabetes mellitus (*meli* = sugar) A chronic hereditary disease characterized by high blood sugar and due to an absolute or relative insufficiency of insulin; symptoms may appear at any age.

Feminizing adenoma Malignant tumors of the adrenal gland that secrete abnormally high amounts of female sex hormones and produce female secondary sexual characteristics in the male.

Goiter (*gutter* = throat) An enlargement of the thyroid gland with typical swelling of the front of the neck; *simple goiter* most often caused by insufficient iodine, and person suffers symptoms of thyroxin hyposecretion. *Exophthalmic goiter* (Graves' disease) caused by overactivity of thyroid gland; body's metabolic reactions increased. *Exophthalmos* means protrusion of eyeballs.

Hyperplasia Excessive development of any tissue.

Hypoplasia The defective development of any tissue.

Myxedema Hyposecretion of the thyroid resulting in puffy features due to edema.

Neuroblastoma Malignant tumor arising from the adrenal medulla associated with metastases to bones.

Pituitary cachexia (Simmond's disease) Hyposecretion of pituitary growth hormone during adult years; reduced protein anabolism, among other metabolic abnormalities, causes wasting away of tissue.

Pituitary dwarfism Unusual shortness and general underdevelopment of all organs caused by hyposecretion of the human growth hormone of the anterior pituitary gland during childhood.

Pituitary giantism A hypersecretion of the human growth hormone of the anterior pituitary gland during childhood resulting in abnormal growth, particularly of long bones.

Thyroid storm An aggravation of all symptoms of hyperthyroidism resulting from trauma, surgery, unusual emotional stress, or labor.

Virilism Masculinization.

Virilizing adenoma Malignant tumors of the adrenal gland that secrete high amounts of male sex hormones and produce male secondary sexual characteristics in the female.

Chapter summary in outline

ENDOCRINE GLANDS

1. Exocrine glands (sweat, sebaceous, digestive) secrete their products through ducts into body cavities or onto body surfaces.
2. Endocrine glands are ductless and secrete hormones into the blood.
3. Hormones are proteins, amines, or steroids that change the physiological activities of cells in order to maintain homeostasis. Many hormones exert their effects through cyclic AMP.
4. Organs that exhibit changes in response to hormones are called target organs.

PITUITARY GLAND

1. This gland is differentiated into the adenohypophysis (the anterior lobe and glandular portion) and the neurohypophysis (the posterior lobe and nervous portion).
2. The adenohypophysis secretes trophic hormones and gonadotrophic hormones. These hormones are regulated by releasing factors.
3. The blood supply to the adenohypophysis is from the superior hypophyseal arteries.
4. Hormones of the adenohypophysis are: *(a)* human growth hormone (regulates growth); *(b)* thyroid-stimulating hormone (regulates activities of thyroid); *(c)* adrenocorticotrophic hormone (regulates adrenal cortex); *(d)* follicle-stimulating hormone (regulates ovaries and testes); *(e)* luteinizing hormone (regulates female and male reproductive activities); and *(f)* prolactin (initiates milk secretion).
5. The neural connection between the hypothalamus and neurohypophysis is via the hypothalamic-hypophyseal tract.

6. Hormones of the neurohypophysis are oxytocin (stimulates contraction of uterus and ejection of milk) and ADH (stimulates arteriole constriction and water reabsorption by the kidneys).

THYROID GLAND
1. The gland synthesizes thyroxin and triiodothyronine, which control the rate of metabolism by increasing the catabolism of carbohydrates and proteins and serve as a diuretic.
2. Thyrocalcitonin regulates the homeostasis of blood calcium.

PARATHYROIDS
1. Parathyroid hormone regulates the homeostasis of calcium and phosphate.

ADRENALS
1. These glands consist of an outer cortex and inner medulla.
2. Cortical secretions are mineralocorticoids (regulate sodium reabsorption and potassium excretion); glucocorticoids (normal metabolism and resistance to stress); and gonadocorticoids (male and female sex hormones).
3. Medullary secretions are epinephrine and norepinephrine, which produce effects similar to sympathetic responses.

PANCREAS
1. Alpha cells of the pancreas secrete glucagon (increases blood glucose level), and beta cells secrete insulin (decreases blood glucose level).

OVARIES AND TESTES
1. Ovaries are located in the pelvic cavity and produce sex hormones related to development and maintenance of female sexual characteristics.
2. Testes lie inside the scrotum and produce sex hormones related to the development and maintenance of male sexual characteristics.

PINEAL GLAND
1. This gland secretes melatonin (possibly regulates menstrual cycle), adrenoglomerulotropin (may stimulate adrenal cortex), and serotonin (involved in normal brain physiology).

THYMUS GLAND
1. This gland is believed to secrete a hormone related to antibody production.

Review questions and problems

1. Distinguish between an endocrine gland and an exocrine gland. What is the relationship between an endocrine gland and a target organ?
2. What is a hormone? Distinguish between trophic and gonadotrophic hormones. How do hormones exert their effects through cyclic AMP?
3. In what respect is the pituitary gland actually two separate glands? Describe the histology of the adenohypophysis. Why does the anterior lobe of the gland have such an abundant blood supply?
4. What hormones are produced by the adenohypophysis, and what are their functions?
5. Relate the importance of neurohumors to secretions of the adenohypophysis.
6. Discuss the histology of the neurohypophysis and the function and regulation of the hormones produced by the neurohypophysis.
7. Describe the structure and importance of the hypothalamic-hypophyseal tract.
8. Describe the location and histology of the thyroid gland.
9. List the functions of the thyroid hormones.
10. How is thyroxin synthesized, stored, and secreted?
11. Where are the parathyroids located? What is their histology? What are the functions of the parathyroid hormone?
12. Compare the adrenal cortex and adrenal medulla with regard to location and histology.
13. Describe the function of the hormones produced by the adrenal cortex.
14. What relationship does the adrenal medulla have to the autonomic nervous system? What is the action of adrenal medullary hormones?
15. Describe the location of the pancreas and the histology of the islets of Langerhans. What are the actions of glucagon and insulin?
16. Where is the pineal gland located? What are its assumed functions?
17. Describe the location of the thymus gland. What is its proposed function?
18. Distinguish between hyposecretion and hypersecretion. What are the principal clinical symptoms of pituitary dwarfism, Simmond's disease, giantism, and acromegaly?
19. In diabetes insipidus, why does the patient exhibit high urine production and thirst?
20. What clinical symptoms are present in cretinism, myxedema, exophthalmic goiter, and simple goiter? Relate these symptoms to the normal activity of thyroxin.
21. Distinguish between the cause and symptoms of tetany and the cause and symptoms of osteitis fibrosa cystica.
22. What are the effects of hypersecretion of aldosterone? Describe Addison's disease, Cushing's syndrome, and the adrenogenital syndrome. What is a pheochromocytoma?
23. What are the principal effects of hypoinsulinism and hyperinsulinism?

Selected readings

Basmajian, John V. *Primary Anatomy.* 6th ed. Baltimore: Williams and Wilkins, 1970. Chap. 14.

Cunningham's Textbook of Anatomy. 11th ed., edited by G. J. Romanes. London: Oxford University Press, 1972. Pp. 557–579.

Gray, Henry. *Anatomy of the Human Body.* 29th ed., edited by Charles Mayo Goss. Philadelphia: Lea and Febiger, 1973. Chap. 18.

Netter, Frank H. *Endocrine System and Selected Metabolic Diseases.* CIBA Collection of Medical Illustrations, Vol. 4. Summit, NJ: CIBA Pharmaceutical Co., 1965.

Turner, C. D., and J. T. Bagnara. *General Endocrinology.* 5th ed. Philadelphia: W. B. Saunders, 1971.

CHAPTER 18

THE REPRODUCTIVE SYSTEMS

STUDENT OBJECTIVES

After you have read this chapter, you should be able to:

1. Define reproduction and classify the organs of reproduction by function
2. Describe the structure and function of the scrotum
3. Explain the structure, histology, and functions of the testes
4. Describe the structure of a spermatozoan
5. Explain the physiological effects of testosterone
6. Describe the straight tubules and rete testis as components of the duct system of the testes
7. Describe the location, structure, histology, and functions of the epididymis
8. Explain the structure and functions of the ductus deferens
9. Describe the procedure employed in a vasectomy
10. Describe the structure and function of the ejaculatory duct
11. List and describe the three subdivisions of the male urethra
12. Explain the location and functions of the seminal vesicles, prostate gland, and bulbourethral glands, the accessory reproductive glands
13. Describe the composition of semen
14. Explain the structure and functions of the penis
15. Describe the location, histology, and functions of the ovaries
16. Explain the location, structure, histology, and functions of the uterine tubes
17. Describe the location and ligamentous attachments of the uterus
18. Explain the histology and blood supply of the uterus
19. List and describe the physiological effects of estrogens and progesterone
20. Describe the location, structure, and functions of the vagina
21. List and explain the components of the vulva and note the function of each component
22. Describe the anatomical landmarks of the perineum
23. Explain the structure and histology of the mammary glands
24. Describe the symptoms of venereal diseases, prostate disorders, impotence, sterility, menstrual disorders, ovarian cysts, and breast and cervical cancer
25. Define key terms associated with the reproductive systems

In its broadest sense, **reproduction** may be viewed as the self-perpetuation of genetic molecules, that is, molecules which determine the characteristics of all living forms. Reproduction is the mechanism by which the thread of life is sustained. It is the process by which a single cell duplicates its genetic material, allowing an organism to grow and to repair itself, and in this sense, reproduction enables the individual organism to maintain its own life. But reproduction is also the process by which genetic material is passed from generation to generation. In this regard, reproduction maintains the life of the species.

The male and female reproductive systems are organized into several types of organs that may be grouped on the basis of their general functions. For example, the testes and ovaries, also called gonads, function in the production of sperm cells and ova, respectively. Sperm cells and ova are collectively called gametes. The gonads also secrete hormones. Other reproductive organs that constitute the ducts transport, receive, and store gametes. Still other reproductive organs, called accessory glands, produce materials that support gametes.

THE MALE REPRODUCTIVE SYSTEM

The organs of the male reproductive system (Figure 18–1) are the testes, or male gonads, which produce sperm, a number of ducts which either store or transport sperm to the exterior, accessory glands that add secretions comprising the semen, and several supporting structures, including the penis.

Scrotum

The **scrotum** is a pouching of the abdominal wall consisting of loose skin and superficial fascia (see Figure 18–1). It is the supporting structure for the primary male reproductive organs. Externally, it looks like a single pouch of skin separated into lateral portions by a medial ridge called the *raphe.* Internally, it is divided by a septum into two sacs, each of which contains a single testis. The septum

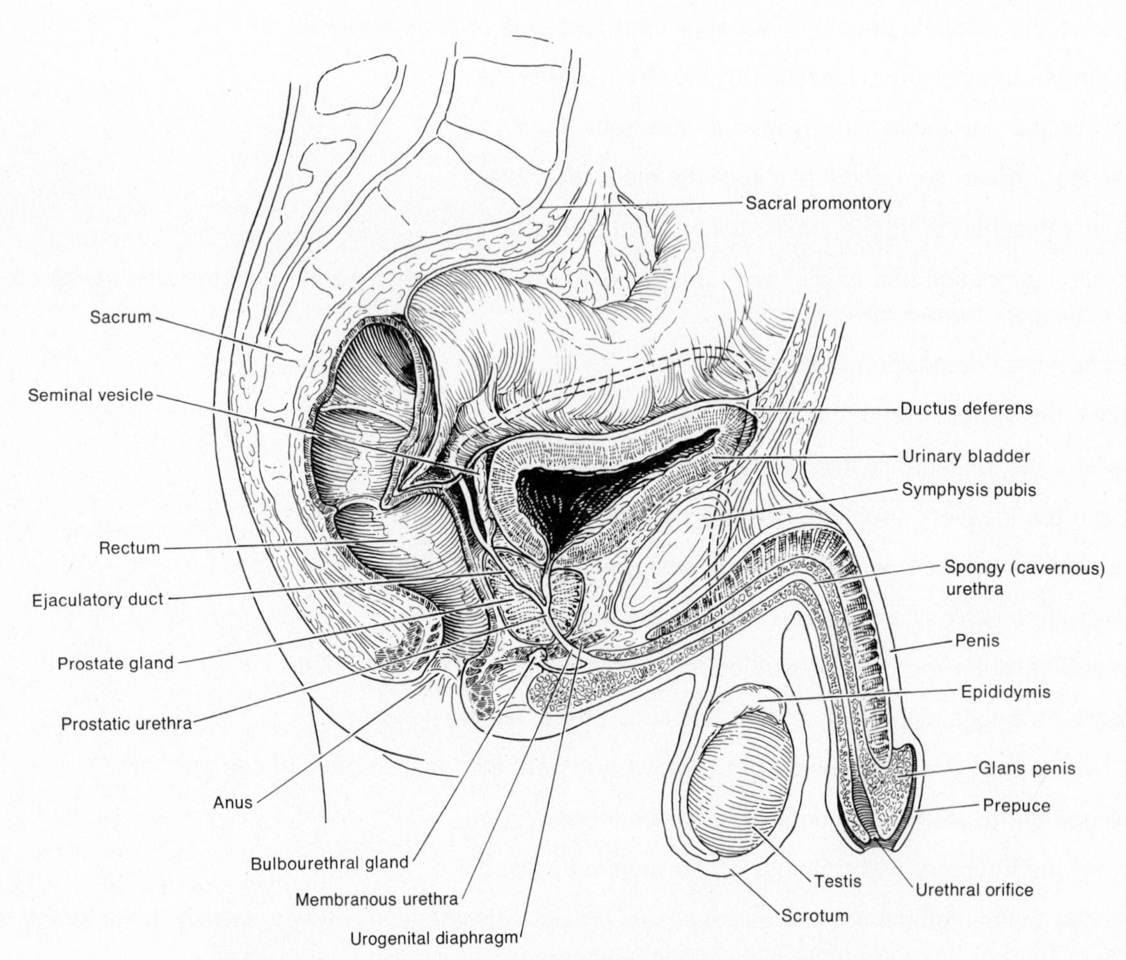

Figure 18–1. Male organs of reproduction seen in sagittal view.

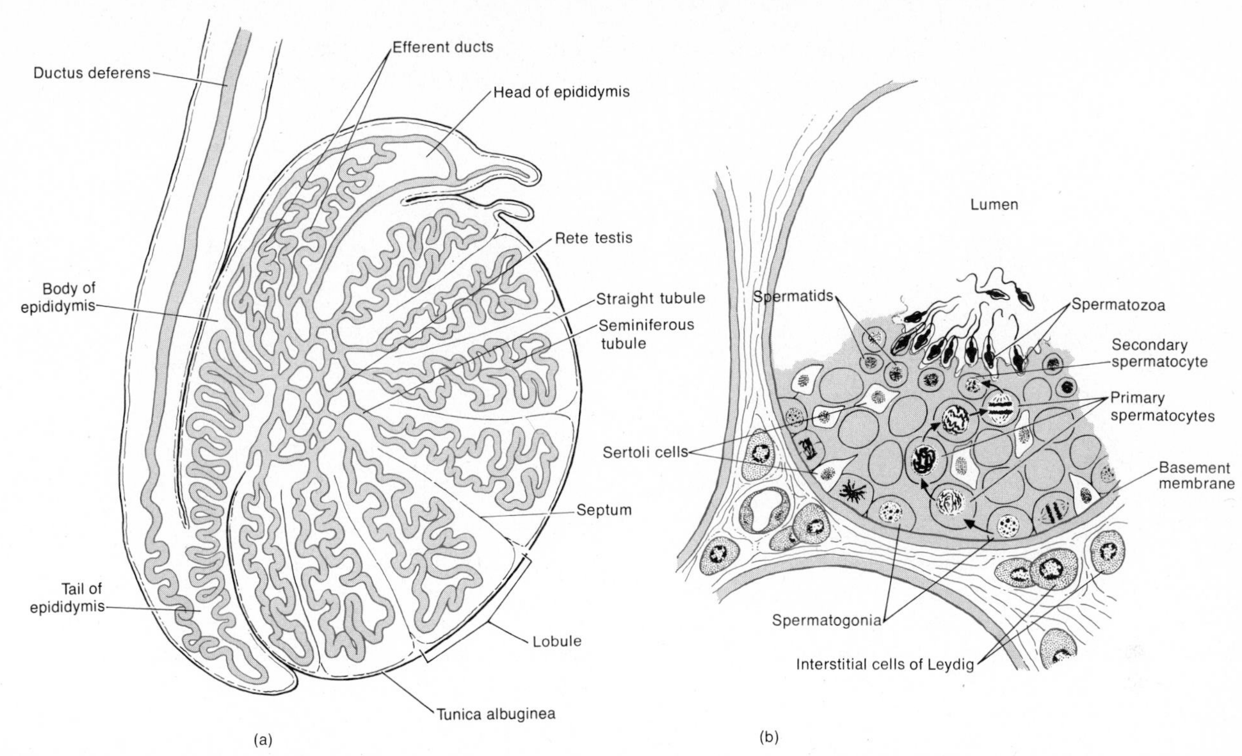

Figure 18–2. The testis. (a) Sectional view of a testis showing its system of tubes. (b) Cross section of a seminiferous tubule showing the stages of spermatogenesis.

consists of superficial fascia and bundles of smooth muscle fibers, the *dartos.* The dartos is also found in the subcutaneous tissue of the scrotum. The testes are the organs that produce sperm. Sperm production and survival requires a temperature that is lower than body temperature. Because the scrotum is isolated from the body cavities, it is not as warm as the rest of the abdominal cavity. Thus, it supplies an environment about 3°F below body temperature. Exposure to cold causes contraction of the smooth muscle fibers, moving the testes closer to the abdomen where they can absorb body heat. This causes the skin of the scrotum to appear more wrinkled. Exposure to warmth reverses the process. Another muscle in the spermatic cord, the cremaster muscle, also elevates the testes upon exposure to cold. The blood supply of the scrotum is derived from the internal pudendal branch of the internal iliac, the cremasteric branch of the inferior epigastric artery, and the external pudendal artery from the femoral artery. The scrotal veins follow the arteries. The scrotal nerves are derived from the pudendal, posterior cutaneous of the thigh, and ilioinguinal nerves.

The testes

The **testes** are paired oval-shaped glands measuring about 5 cm (2 in.) in length and 2.5 cm (1 in.) in diameter. They average between 10 and 15 g (Figure 18–2a). During most of fetal life they lie in the pelvic cavity, but about 2 months prior to birth they descend into the scrotum. When the testes do not descend, the condition is referred to as *cryptorchidism.* Cryptorchidism results in sterility because the sperm cells are destroyed by the higher body temperature of the pelvic cavity. Undescended testes can be placed in the scrotum by administering hormones or by surgical means prior to puberty without ill effects.

The testes are covered by a dense layer of white fibrous tissue, the *tunica albuginea,* that extends inward and divides each testis into a series of internal compartments called *lobules.* Each lobule contains one to three tightly coiled tubules, the *seminiferous tubules,* that produce the sperm by a process called *spermatogenesis.* A cross section through a seminiferous tubule reveals that the tubule is packed with sperm cells in various stages

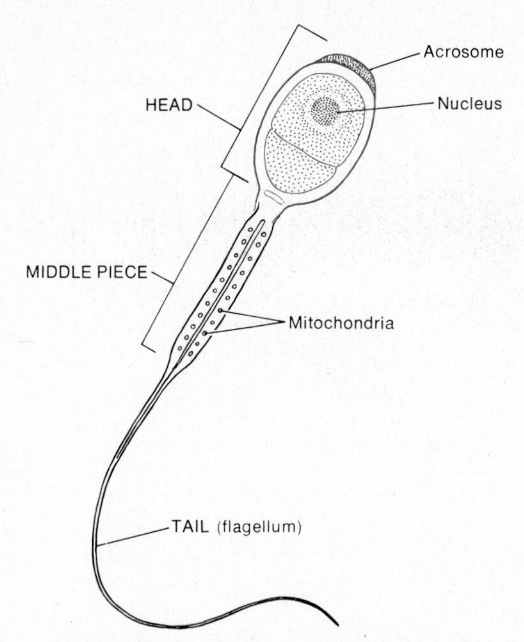

Figure 18–3. Parts of a spermatozoan.

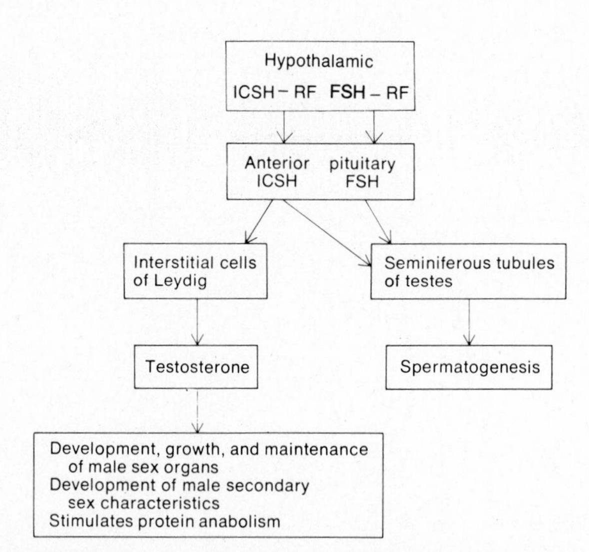

Figure 18–4. The secretion and physiological effects of testosterone.

of development (Figure 18–2b). The most immature cells, the *spermatogonia,* are located against the basement membrane. Moving toward the lumen in the center of the tube, one can see layers of progressively more mature cells. In order of advancing maturity, these are primary spermatocytes, secondary spermatocytes, and spermatids. By the time a **sperm cell,** or **spermatozoan,** has reached full maturity, it is in the lumen of the tubule and begins to be moved through a series of ducts. Embedded between the developing sperm cells in the tubules are *Sertoli cells* that produce secretions for the supplying of nutrients to the spermatozoa. Between the seminiferous tubules are clusters of *interstitial cells of Leydig.* These cells secrete the male hormone testosterone.

The testes are supplied by the testicular arteries, which arise from the aorta immediately below the origin of the renal arteries. Blood from the testes is drained on the right side by the testicular vein which enters the inferior vena cava and on the left side by the testicular vein which enters the left renal vein. The nerve supply to the testes is from the testicular plexus which contains vagal parasympathetic fibers and sympathetic fibers from the tenth thoracic segment of the spinal cord.

The functions of the testes are to produce sperm and testosterone.

Spermatozoa

Spermatozoa, once ejaculated, have a life expectancy of about 24 hours. A spermatozoan is highly adapted for reaching and penetrating a female ovum. It is composed of a head, a middle piece, and a tail (Figure 18–3). The head contains the nuclear material and the acrosome, which contains chemicals that effect penetration of the sperm cell into the ovum. Numerous mitochondria are contained in the middle piece. Mitochondria carry on the catabolism that provides energy for locomotion. The tail, a typical flagellum, propels the sperm along its way.

Testosterone

Secretions of the anterior pituitary gland assume a major role in the developmental changes associated with puberty. The anterior pituitary starts to secrete and release gonadotrophic hormones (FSH and ICSH) at the onset of puberty. Their release is controlled by releasing factors from the hypothalamus known as FSH-RF and ICSH-RF. These are the follicle-stimulating-hormone releasing factor and the interstitial-cell-stimulating-hormone releasing factor, respectively. Once secreted, the gonadotrophic hormones have profound effects on male reproductive organs. We shall examine the effects of gonadotrophic hormones on female reproductive organs later.

At the onset of puberty, the follice-stimulating hormone (FSH) is released by the anterior pituitary. It acts on the seminiferous tubules to initiate spermatogenesis (Figure 18–4). However, spermatogenesis is not completed under the influence of FSH alone. Interstitial-cell-stimulating hormone

(ICSH) is also secreted by the anterior pituitary at the onset of puberty. This hormone also acts on the seminiferous tubules and further assists the tubules to develop mature sperm, but the chief function of ICSH is to stimulate the interstitial cells of Leydig to secrete testosterone.

Testosterone has a number of effects on the body. It controls the development, growth, and maintenance of the male sex organs; the development of male secondary sex characteristics (hair pattern, muscular development, voice changes); bone growth; protein anabolism; normal sexual behavior; and sperm production.

Ducts

When the sperm mature, they are moved through the seminiferous tubules to the **straight tubules** (see Figure 18–2a). The straight tubules form a network of ducts in the center of the testis called the **rete testis.** Some of the cells lining the rete testis possess cilia that probably help to push the sperm along. The sperm are next transported out of the testis through a series of coiled **efferent ducts** that empty into a single tube called the epididymis. At this point, the sperm are morphologically mature.

The epididymis

The two **epididymides** are highly coiled tubes, measuring about 6 m (20 ft) in length and 1 mm in diameter, that lie tightly packed within the scrotum. Each epididymis attaches to its testis, where it receives the efferent ducts, and then it descends along the posterior side of the testis, makes a loop, and ascends (see Figures 18–1 and 18–2). The superior portion of the epididymis is known as the *head,* the portion posterior to the testis is called the *body,* and the inferior portion is referred to as the *tail.* The epididymis is lined with pseudostratified columnar epithelium, and its wall contains smooth muscle. The free surfaces of the columnar cells contain long, branching microvilli called *stereocilia* (Figure 18–5). Functionally, the epididymis is the site of sperm maturation, it stores spermatozoa in anticipation of ejaculation, and it propels spermatozoa toward the urethra during ejaculation. Propulsion of the sperm is accomplished by peristaltic contractions of the smooth muscle.

The ductus deferens

The terminal portion of the tail of the epididymis is less coiled and considerably thicker. At this point it is referred to as the **ductus (vas) deferens** or **seminal duct** (see Figure 18–2a). The ductus deferens, about 45 cm long (18 in.), ascends along the

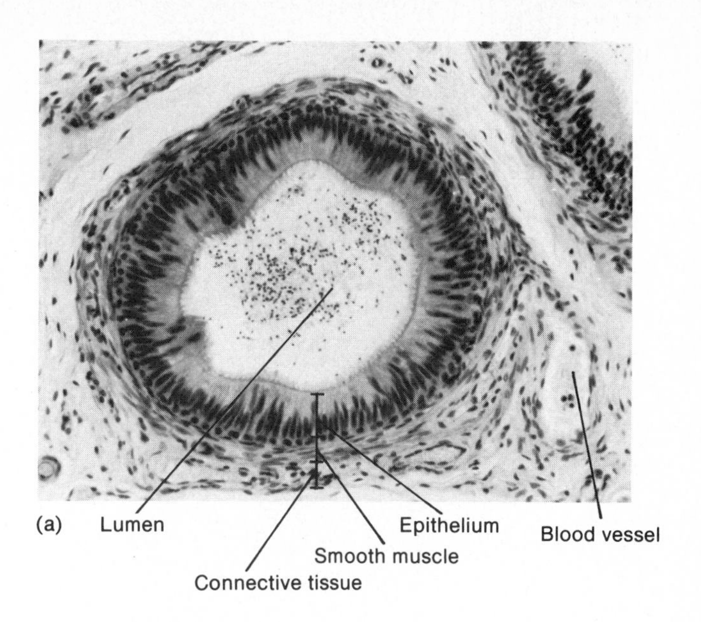

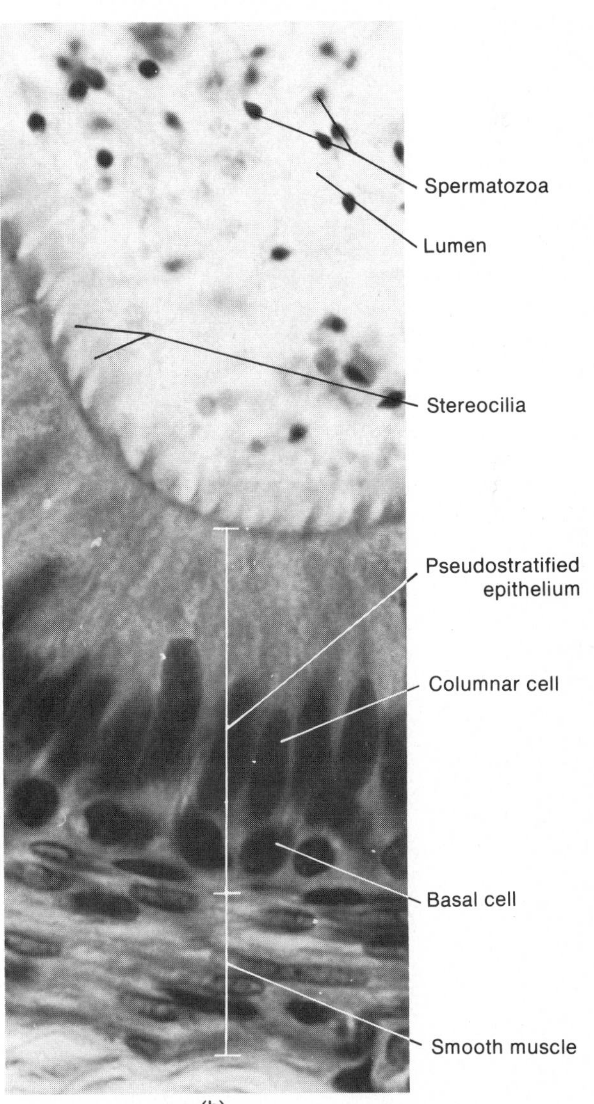

Figure 18–5. Histology of the epididymis. (a) Cross section through the epididymis at a magnification of 30X. (b) Enlarged aspect of the epithelium at a magnification of 200X. (Courtesy of Victor B. Eichler, Wichita State University.)

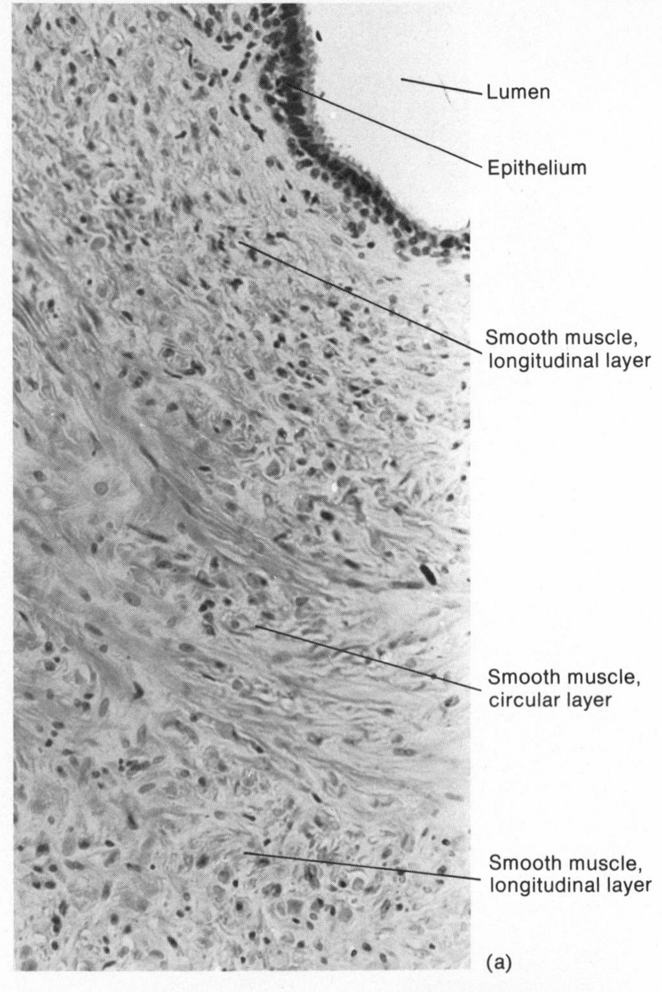

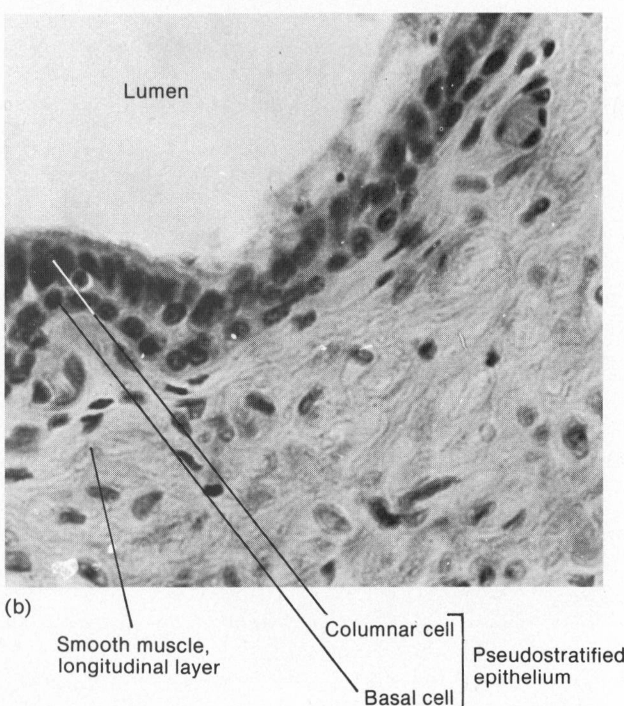

Figure 18–6. Histology of the ductus deferens. (a) Cross section of a portion of the ductus deferens at a magnification of 40X. (b) Enlarged aspect of the epithelium and muscularis at a magnification of 100X. (Courtesy of Victor B. Eichler, Wichita State University.)

posterior border of the testis, penetrates the inguinal canal, and enters the abdomen, where it loops over the top and down the posterior surface of the bladder. The dilated terminal portion of the ductus deferens is known as the *ampulla* (see Figure 18–1). Histologically, the ductus deferens is lined with pseudostratified epithelium and contains a heavy coat of three layers of muscle (Figure 18–6). Peristaltic contractions of the muscular coat propel the spermatozoa toward the urethra during ejaculation.

Traveling with the ductus deferens are the testicular artery, autonomic nerves, veins that drain the testes, lymphatics, and a small circular band of skeletal muscle called the *cremaster muscle.* These structures together constitute the **spermatic cord,** a supporting structure of the male reproductive system. The cremaster muscle elevates the testes during sexual stimulation and exposure to cold. The spermatic cord passes through the *inguinal canal,* a slitlike passageway in the anterior abdominal wall just superior to the medial half of the inguinal ligament. The area of the inguinal canal and spermatic cord represents a weak spot in the abdominal wall, and it is frequently the site of a *hernia.* Very simply, a hernia is a rupture or separation of a portion of the abdominal wall resulting in the protrusion of a part of a viscera.

One method of sterilization of males is called *vasectomy,* in which a portion of each ductus deferens is removed. In the procedure, an incision is made in the scrotum, the ducts are located, and each is tied in two places. Then the portion between the ties is excised. Although sperm production continues in the testes, the sperm cannot reach the exterior since the ducts are cut.

The ejaculatory duct

Posterior to the urinary bladder, each ductus deferens joins its **ejaculatory duct** (Figure 18–7). Each duct is about 2 cm (1 in.) long. Both ejaculatory ducts eject spermatozoa into the prostatic urethra. The urethra is the terminal duct of the system, serving as a common passageway for both spermatozoa and urine.

Urethra

In the male, the **urethra** passes through the prostate gland, the urogenital diaphragm, and the penis. It measures about 20 cm (8 in.) in length and is subdivided into three parts (see Figures 18–1 and 18–7). The *prostatic portion* is about 2 to 3 cm (1 in.) long and passes through the prostate gland. It continues inferiorly as the membranous portion as it passes

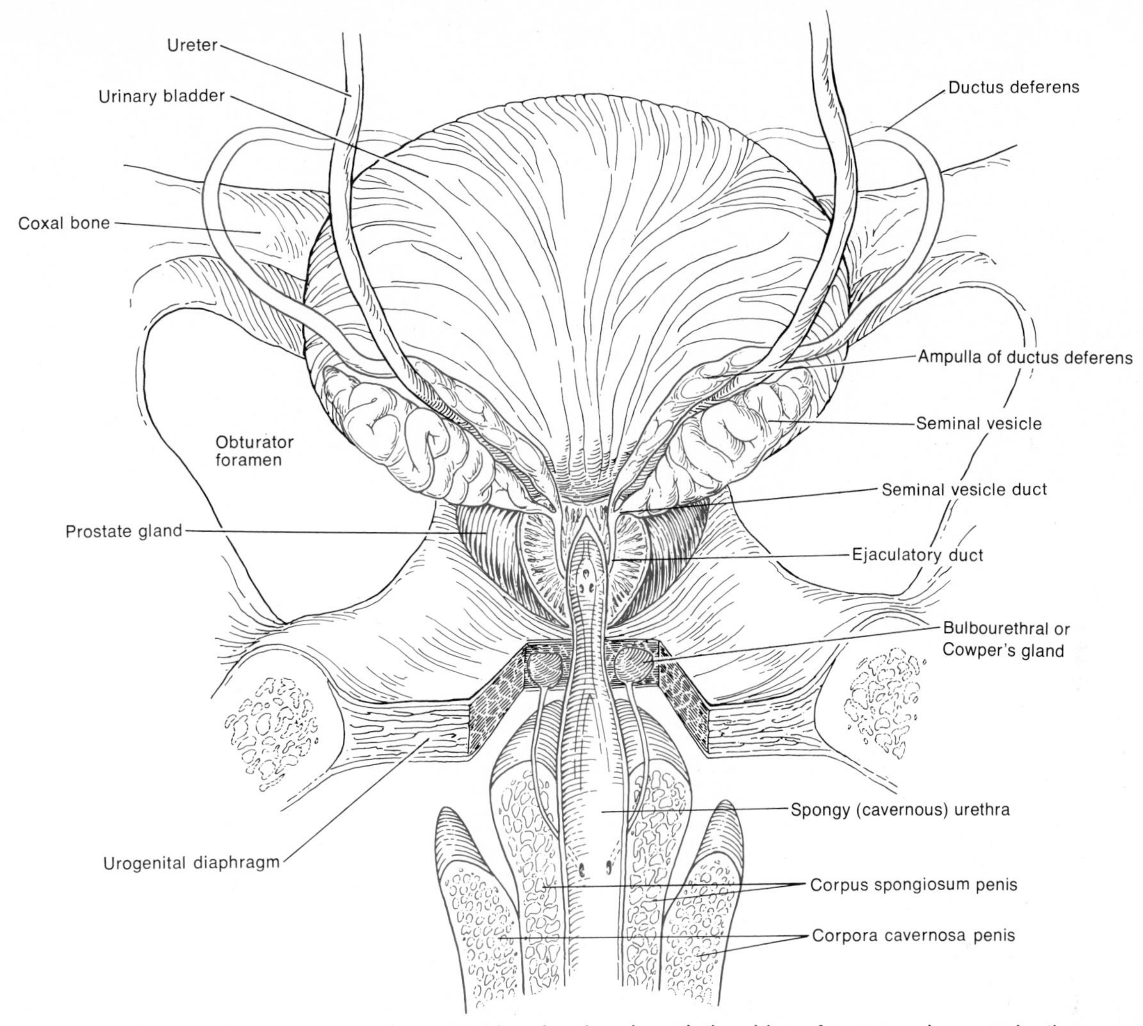

Figure 18–7. Posterior view of the urinary bladder showing the relationships of some male reproductive organs.

through the urogenital diaphragm, a muscular partition between the two ischiopubic rami. The *membranous portion* is about 1 cm (½ in.) in length. After passing through the urogenital diaphragm, it is known as the *spongy (cavernous) portion* of the urethra. This portion is about 15 cm (6 in.) long. The spongy urethra enters the bulb of the penis and terminates at the external urethral orifice.

Accessory glands

Whereas the ducts of the male reproductive system store and transport sperm cells, the **accessory glands** secrete the liquid portion of semen. The first of the accessory glands we shall consider are the paired **seminal vesicles** (Figure 18–7). These glands are convoluted pouchlike structures, about 5 cm (2 in.) in length, lying posterior to and at the base of the urinary bladder, in front of the rectum. They secrete the alkaline viscous component of semen rich in the sugar fructose and pass it into the ejaculatory duct. The seminal vesicles contribute about 60 percent of the volume of semen.

The **prostate gland** is a single, doughnut-shaped gland about the size of a chestnut (see Figure 18–7). It measures about 3.7 cm (1½ in.) from side to side, 2.5 cm (1 in.) from front to back, and 3.5 cm (1½ in.) in height. It is inferior to the urinary bladder and surrounds the upper portion of the urethra. The prostate secretes an alkaline fluid that constitutes about 13 to 33 percent of the semen into the prostatic urethra. In older men, the prostate sometimes enlarges to the point where it compresses the urethra and obstructs urine flow. At this stage, surgical removal of part of or the entire gland (prostatectomy) usually is indicated. The prostate gland is also a common tumor site in older males.

The paired **bulbourethral** or **Cowper's glands** are about the size and shape of peas. They are located beneath the prostate on either side of the

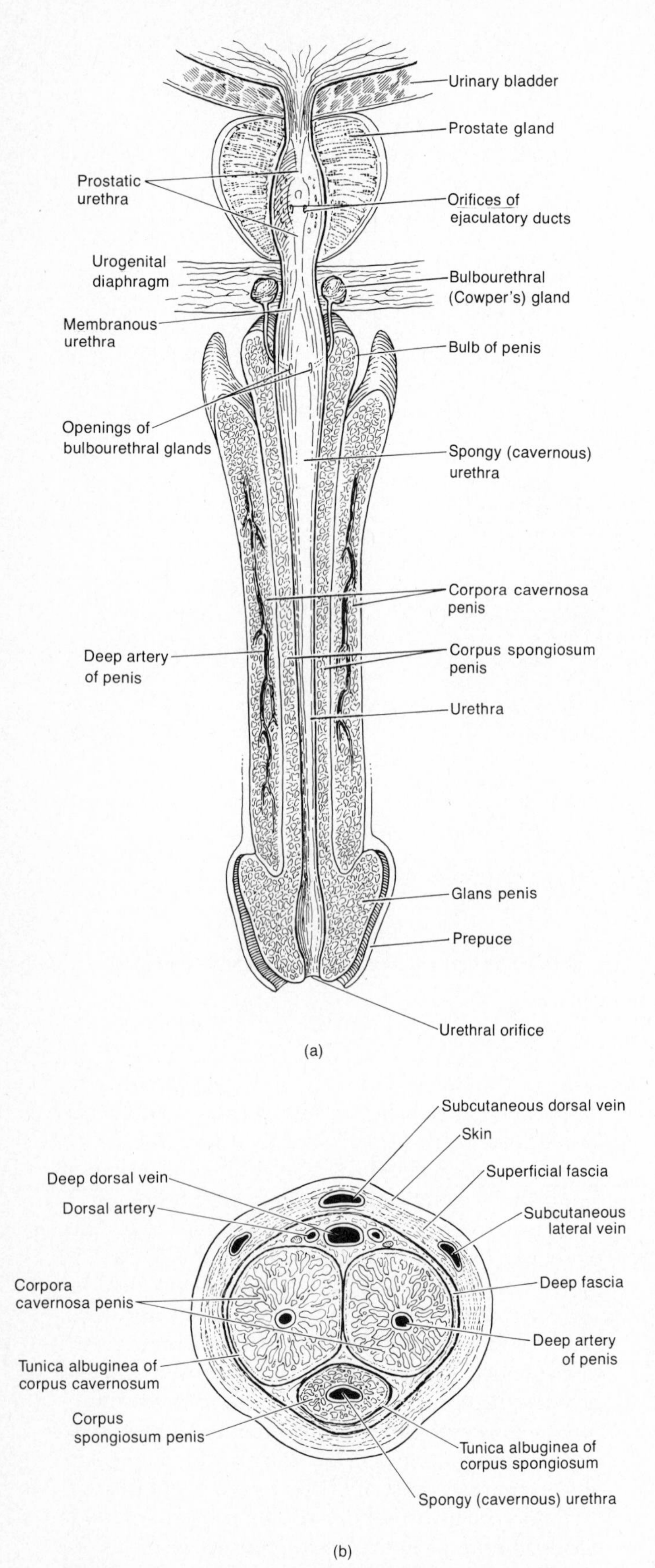

Figure 18–8. Internal structure of the penis. (a) View from the floor of the penis. (b) Cross section.

urethra (Figure 18–7). Like the prostate, the Cowper's glands secrete an alkaline fluid; their ducts open into the spongy urethra.

Semen

In our discussion of the male reproductive organs we have mentioned a fluid called semen. **Semen,** or **seminal fluid,** is a mixture of sperm and the secretions of the seminal vesicles, the prostate gland, and the bulbourethral glands. The average volume of semen for each ejaculation is 3 to 4 ml, and the average range of spermatozoa ejaculated is between 300 and 400 million. When the number of spermatozoa falls below approximately 100 million, the male is likely to be physiologically sterile. Even though only a single spermatozoan fertilizes an ovum, it is hypothesized that fertilization requires the combined action of a tremendous number of spermatozoa. The ovum is enclosed by cells that form a barrier between it and the sperm. An enzyme called hyaluronidase is secreted by the acrosomes of sperm. Hyaluronidase is believed to dissolve intercellular materials of the cells covering the ovum, giving the sperm a passageway into the ovum. Apparently, vast numbers of sperm are required to secrete an effective amount of the enzyme.

Semen has a pH range of 7.35 to 7.50; that is, it is slightly alkaline. The prostatic secretion gives semen a milky appearance, and fluids from the seminal vesicles and bulbourethral glands give it a mucoid consistency. Semen provides spermatozoa with a transportation medium and nutrients and acts as a buffer to neutralize the acid environment of the female reproductive system and the male urethra from urine. Semen also activates sperm with enzymes after ejaculation.

Penis

The **penis** is used to introduce spermatozoa into the female vagina (Figure 18–8). The distal end of the penis is a slightly enlarged region called the *glans,* which means shaped like an acorn. Covering the glans is the loosely fitting skin, called the *prepuce* or *foreskin,* that is removed during circumcision. Internally, the penis is composed of three cylindrical masses of tissue bound together by fibrous tissue. The two dorsally located masses are called the *corpora cavernosa penis,* and the smaller ventral mass, the *corpus spongiosum penis,* contains the spongy urethra. All three masses of tissue are spongelike and contain venous sinuses. Under the influence of sexual excitation, the arteries supplying the penis dilate, and large quantities of blood enter the venous sinuses. Expansion of these spaces compresses the veins draining the penis so most entering blood is retained. These vascular changes result in an *erection.* The penis returns to its flaccid state when the arteries constrict and pressure on the veins is relieved. Erection prevents urination by the male during sexual excitation and ejaculation. During ejaculation, the smooth muscle sphincter at the base of the urinary bladder is closed due to the higher pressure in the urethra caused by expansion of the corpus spongiosum penis. Thus, urine is not expelled during ejaculation and semen does not enter the urinary bladder.

The penis has a very rich blood supply from the internal pudendal artery and the femoral artery. The veins drain into corresponding vessels. The sensory nerves to the penis are branches from the pudendal and ilioinguinal nerves. The corpora have a sympathetic and parasympathetic supply. As a result of parasympathetic stimulation, the blood vessels dilate and the muscles contract. The result is that blood is trapped within the penis and causes an erection. The musculature of the penis is supplied by the pudendal nerve. The muscles include the bulbocavernosus muscle, which overlies the bulb of the penis; the ischiocavernosus muscles on either side of the penis; and the superficial transverse perineus muscles, on either side of the bulb of the penis (see Figure 18–19a).

THE FEMALE REPRODUCTIVE SYSTEM

The female organs of reproduction (Figure 18–9) include the ovaries, which produce ova; the uterine tubes, which transport the ova to the uterus (or womb); the vagina; and external organs that constitute the vulva or pudendum. The mammary glands, or breasts, also are considered part of the female reproductive system.

Ovaries

The **ovaries,** or female gonads, are paired glands resembling almonds (with their shells on) in size and shape. They are about 3 cm (1.2 in.) long, 2 cm (0.8 in.) wide, and about 1 cm (½ in.) thick. They are positioned in the upper pelvic cavity, one on each side of the uterus (see Figure 18–9). The ovaries are maintained in position by a series of ligaments (Figure 18–10). They are suspended by a part

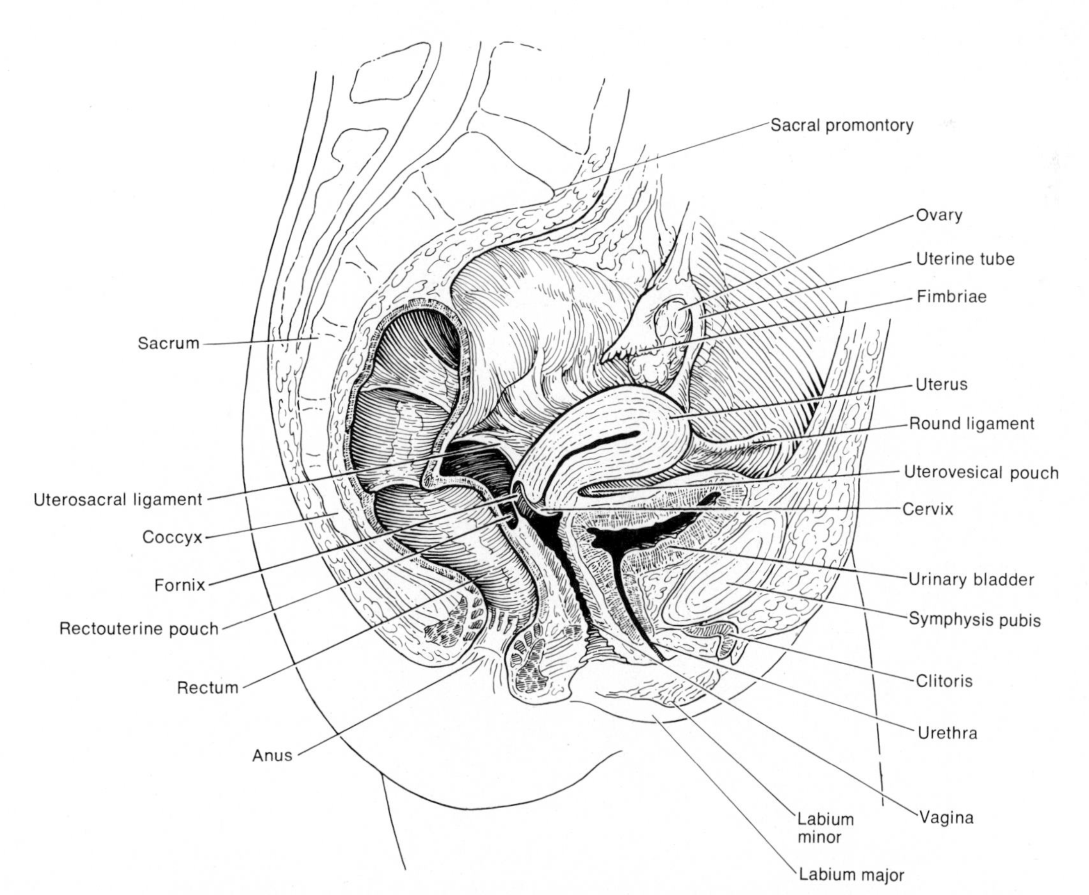

Figure 18–9. Female organs of reproduction seen in sagittal section.

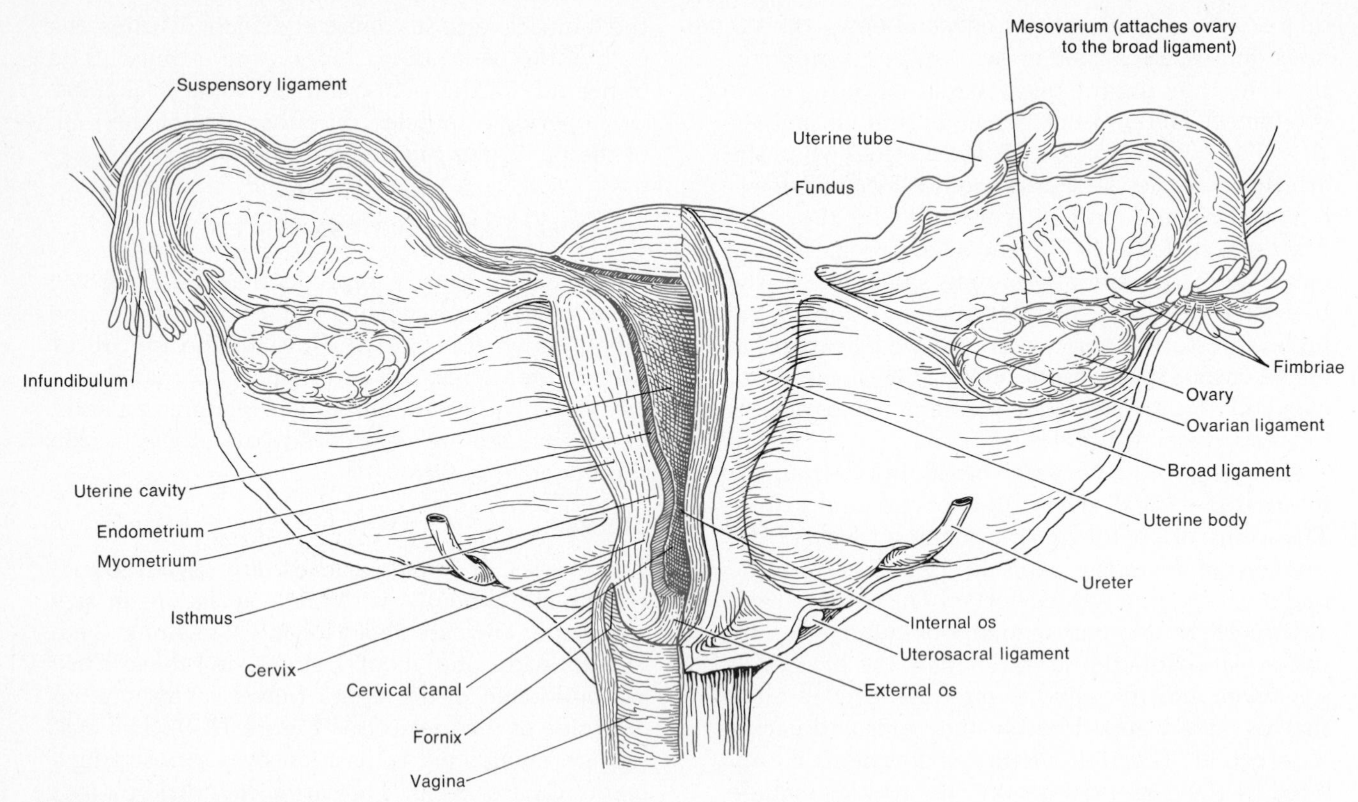

Figure 18–10. The uterus and associated female reproductive structures. The left side of the figure has been sectioned to show internal structures.

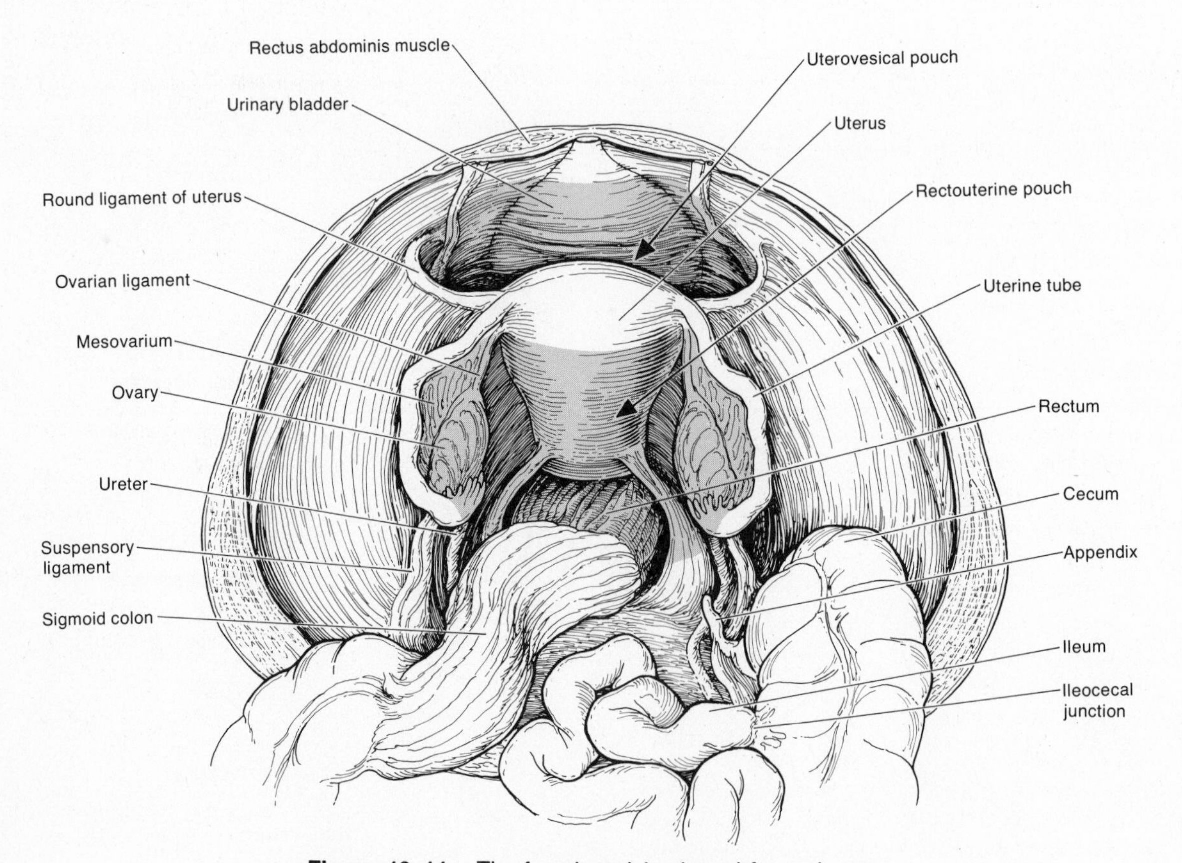

Figure 18–11. The female pelvis viewed from above.

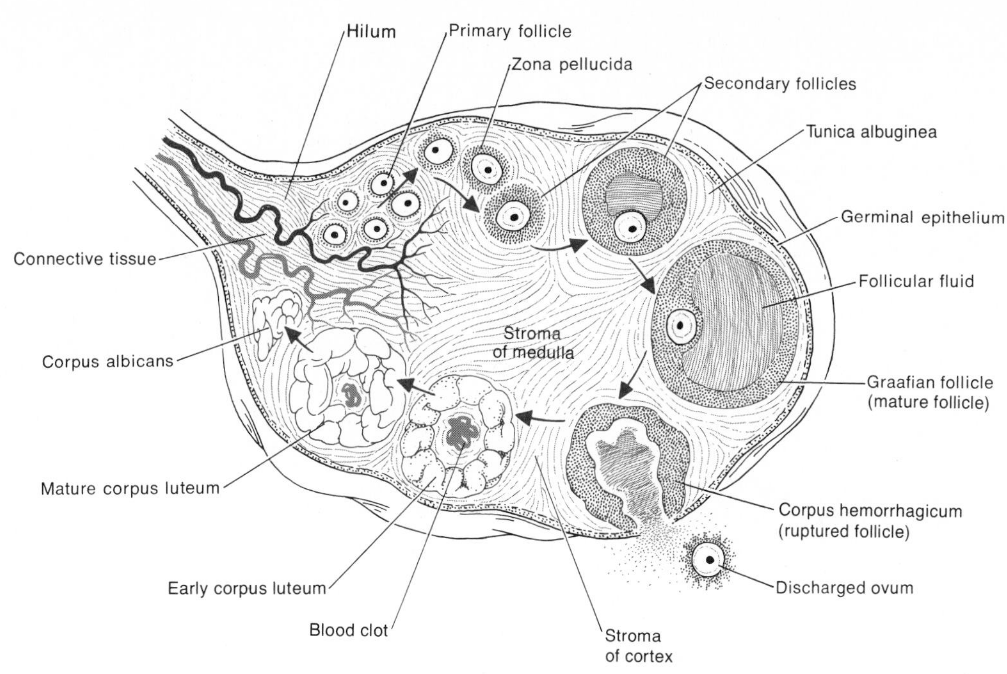

Figure 18–12. Parts of the ovary seen in sectional view. The arrows indicate the sequence of developmental stages that occur as part of the ovarian cycle.

of the broad ligament of the uterus, a fold of peritoneum called the *mesovarium;* anchored to the uterus by the *ovarian ligament;* and attached to the pelvic wall by the *suspensory ligament.* These ligaments can also be seen in Figure 18–11, the pelvic organs viewed from above. Each ovary also contains a *hilus,* the point of entrance for blood vessels and nerves.

In microscopic view it can be seen that each ovary consists of the following parts (Figure 18–12):

1. *Germinal epithelium.* This is a layer of simple cuboidal epithelium which covers the free surface of the ovary.
2. *Tunica albuginea.* This is a capsule of collagenous connective tissue immediately deep to the germinal epithelium.
3. *Stroma.* This is a region of connective tissue deep to the tunica albuginea. The stroma is composed of an outer, more dense layer called the *cortex* and an inner, looser layer known as the *medulla.* The cortex contains ovarian follicles.
4. *Ovarian follicles.* These are ova and their surrounding tissues in various stages of development.
5. *Graafian follicle.* This is a mature ovum and its surrounding tissues.
6. *Corpus luteum.* This is a glandular body developed from a Graafian follicle after extrusion of an ovum (ovulation). The corpus luteum produces the hormone progesterone.

The ovarian blood supply is furnished by the ovarian arteries and branches of the uterine arteries. The ovaries are drained by the ovarian veins. On the right side, they drain into the inferior vena cava, and on the left side, they drain into the renal vein. Sympathetic and parasympathetic nerve fibers to the ovaries are said to terminate on the blood vessels and not enter the substance of the ovaries.

The ovaries produce mature ova, discharge mature ova (ovulation), and secrete the female sexual hormones. The ovaries are analogous to the testes of the male.

Uterine tubes

The female body contains two **uterine,** or **fallopian, tubes,** which transport the ova from the ovaries to the uterus (see Figure 18–10). Measuring about 10 cm (4 in.) long, the tubes are positioned between the folds of the broad ligaments of the uterus. The funnel-shaped open end of each tube, called the *infundibulum,* lies very close to the ovary but is not attached to it and is surrounded by a fringe of fingerlike projections called *fimbriae.* From the infundibulum the tube extends inward and downward and attaches to the upper side of the uterus.

Histologically, the uterine tubes are composed of three layers (Figure 18–13). The internal *mucosa* contains ciliated columnar cells and secretory cells,

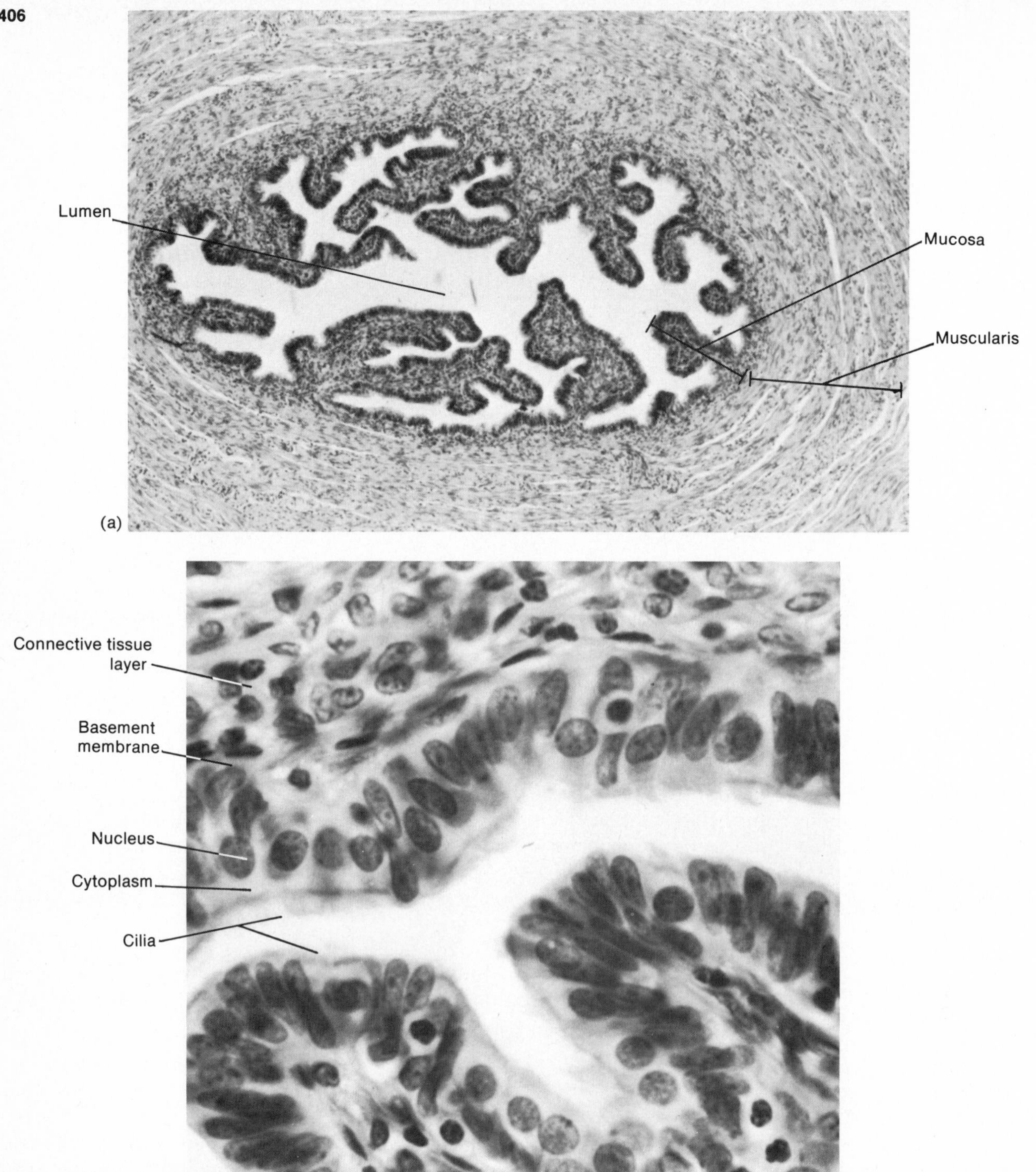

Figure 18–13. Histology of the uterine tube. (a) Cross section through the uterine tube at a magnification of 20X. (b) Enlarged aspect of the epithelium at a magnification of 200X. (Courtesy of Victor B. Eichler, Wichita State University.)

which are believed to aid the nutrition of the ovum. The middle layer, the *muscularis,* is composed of a thick circular region of smooth muscle and an outer, thin, longitudinal region of smooth muscle. Wavelike contractions of the muscularis help to move the ovum down into the uterus. The outer layer of the uterine tubes is a *serous membrane.*

About once a month a mature ovum is released to the surface of the ovary near the infundibulum of the uterine tube, a process called **ovulation.** The ovum adheres to the surface of the ovary until it is swept off by the ciliary action of the epithelium of the infundibulum which sweeps over the ovary. The ovum is then moved along by ciliary action supplemented by the wavelike contractions of the muscularis of the uterine tube. Under normal cir-

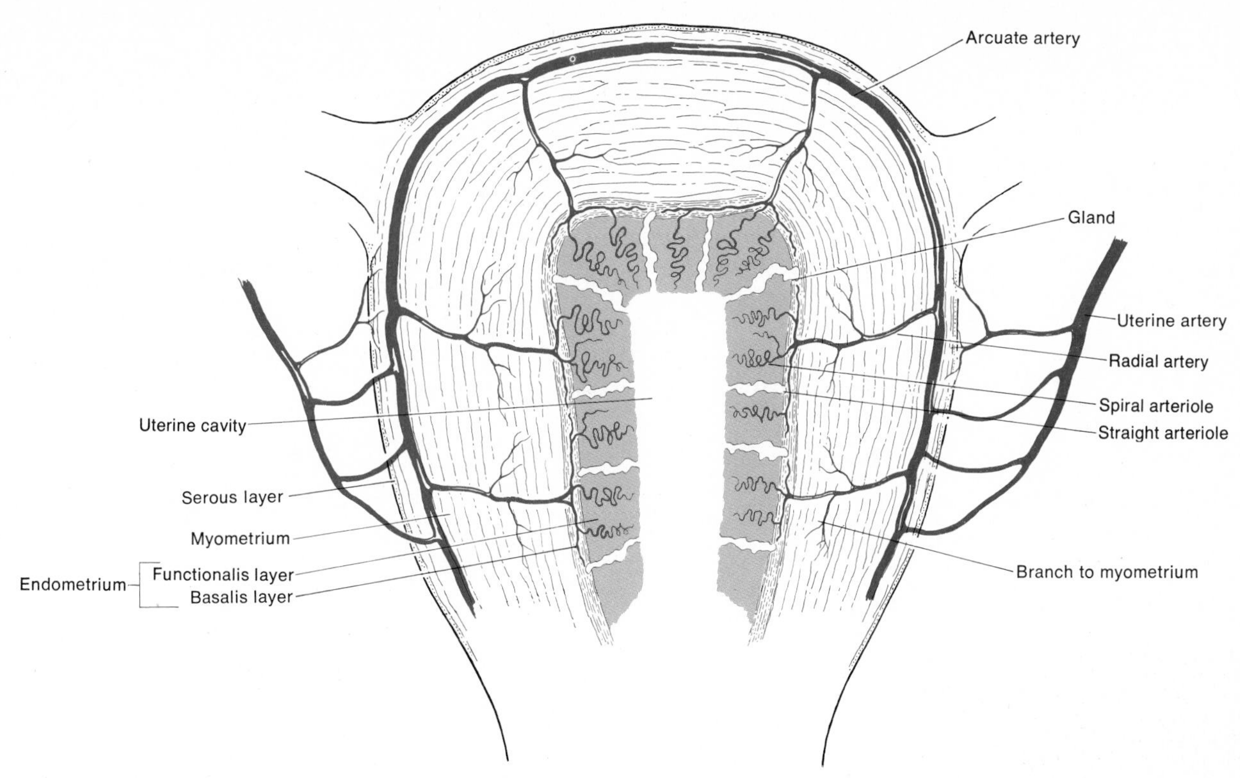

Figure 18–15. Blood supply to the uterus.

with puberty in the female (Figure 18–16). The release of the gonadotrophic hormones is controlled by releasing factors from the hypothalamus called the follicle-stimulating-hormone releasing factor (FSH-RF) and the luteinizing-hormone releasing factor (LH-RF). FSH stimulates the initial development of the ovarian follicles and the secretion of estrogens by the follicles. Another anterior pituitary hormone, the luteinizing hormone (LH), stimulates the further development of ovarian follicles, brings about ovulation, and stimulates progesterone production by ovarian cells. The female sex hormones, estrogens and progesterone, affect the body in different ways. **Estrogens** have four main functions. First is the development and maintenance of female reproductive organs, especially the endometrium, secondary sex characteristics, and the breasts. The secondary sex characteristics include fat distribution to the breasts, abdomen, and hips, voice pitch, and hair pattern. Second, they control fluid and electrolyte balance. Third, they increase protein anabolism. Fourth, they help cause an increase in the female sex drive. High levels of estrogens in the blood inhibit the secretion of FSH by the anterior pituitary gland. This inhibition provides the basis for the action of one kind of contraceptive pill. **Progesterone,** the other female sex hormone, works with estrogens to prepare the endometrium for implantation and to prepare the mammary glands for milk secretion.

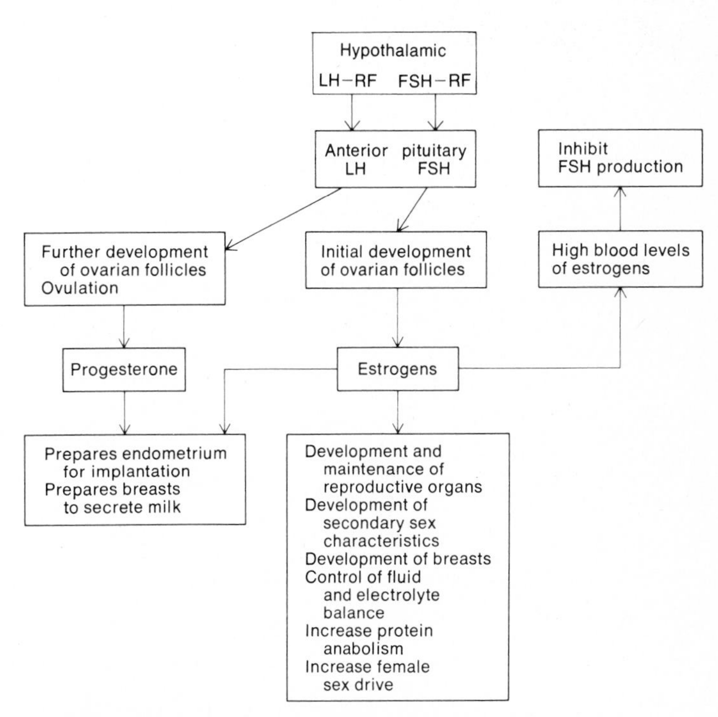

Figure 18–16. Physiological effects of estrogens and progesterone.

Vagina

We now shall continue our discussion of the female reproductive organs, with consideration of the vagina. The **vagina** serves as a passageway for the menstrual flow; as the receptacle for the penis during coitus, or sexual intercourse; and as the lower portion of the birth canal. It is a muscular, tubular organ lined with mucous membrane, measures about 10 cm (4 in.) in length (Figures 18–9 and 18–10), and is situated between the bladder and the rectum. It is directed upward and backward where it attaches to the uterus. A recess, called the *fornix,* surrounds the vaginal attachment to the cervix. The dorsal recess, called the posterior fornix, is larger than the ventral and two lateral fornices. The fornices make possible the use of contraceptive diaphragms. The mucosa of the vagina consists of stratified squamous epithelium and connective tissue that lies in a series of transverse folds, the *rugae,* and is capable of a good deal of distention. The muscularis is composed of smooth muscle that can stretch considerably (Figure 18–17). This distention is important because the vagina receives the penis

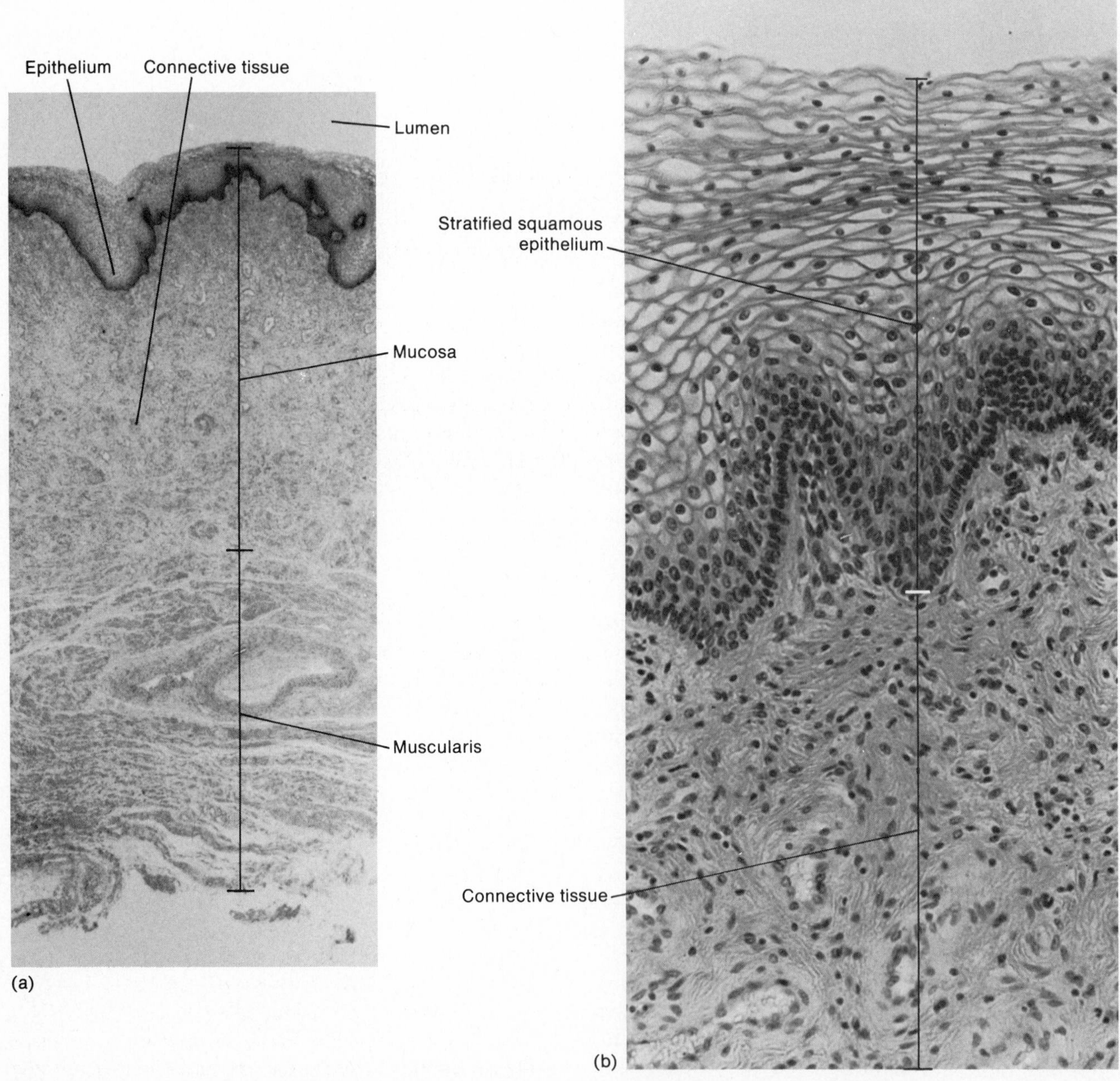

Figure 18–17. Histology of the vagina. (a) Section of the wall of the vagina at a magnification of 6X. (b) Enlarged aspect of the mucosa at a magnification of 50X. (Courtesy of Victor B. Eichler, Wichita State University.)

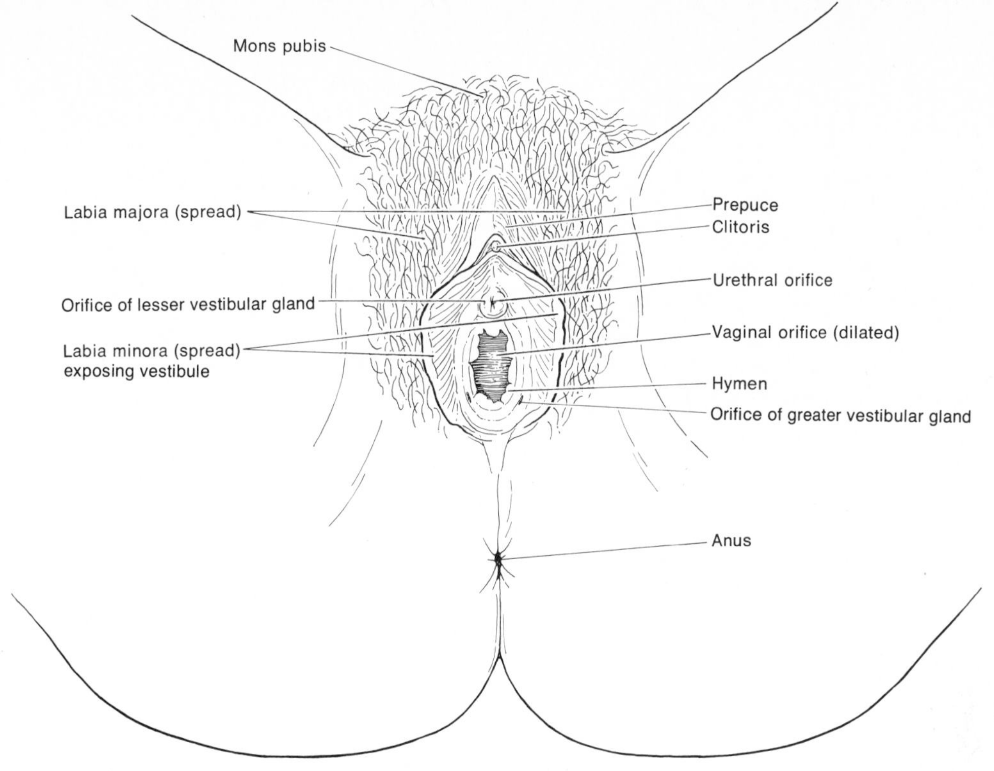

Figure 18–18. The vulva.

during sexual intercourse and serves as the lower portion of the birth canal. At the lower end of the vaginal opening *(vaginal orifice)* is a thin fold of vascularized mucous membrane called the *hymen,* which forms a border around the orifice, partially closing it (Figure 18–18). Sometimes the hymen completely covers the orifice, a condition called *imperforate hymen,* and surgery is required to open the orifice to permit the discharge of the menstrual flow. The mucosa of the vagina contains large amounts of glycogen that, upon decomposition, produce organic acids. These acids create a low pH environment in the vagina, a situation that retards microbial growth. However, the acidity is also injurious to sperm cells. For this reason, the buffering action of semen is important. Semen neutralizes the acidity of the vagina to ensure survival of the sperm.

Vulva

The term **vulva** or **pudendum** is a collective designation for the external genitalia of the female (Figure 18–18).

The *mons pubis (veneris)* is an elevation of adipose tissue covered by coarse pubic hair situated over the symphysis pubis. It lies in front of the vaginal and urethral openings. From the mons pubis, two longitudinal folds of skin, the *labia majora,* extend downward and backward. The labia majora, the female homologue of the scrotum, contain an abundance of adipose tissue and sebaceous and sweat glands; they are covered by hair on their upper outer surfaces. Medial to the labia majora are two folds of skin called the *labia minora.* Unlike the labia majora, the labia minora are devoid of hair and have relatively few sweat glands. They do, however, contain numerous sebaceous glands.

The *clitoris* is a small, cylindrical mass of erectile tissue and nerves. It is located just behind the junction of the labia minora. A layer of skin called the *prepuce,* or *foreskin,* is formed at the point where the labia minora unite and covers the body of the clitoris. The exposed portion of the clitoris is referred to as the *glans.* The clitoris is homologous to the penis of the male in that it is capable of enlargement upon tactile stimulation and assumes a role in sexual excitement of the female.

The cleft between the labia minora is called the *vestibule.* Within the vestibule are the hymen, vaginal orifice, urethral orifice, and the openings of several ducts. The *vaginal orifice* occupies the greater portion of the vestibule and is bordered by the hymen. In front of the vaginal orifice and behind the clitoris is the *urethral orifice.* Behind and to either side of the urethral orifice are the openings of the ducts of the *lesser vestibular* or *Skene's glands.* These glands secrete mucus and are the female homologue of the prostate gland. On either side of the vaginal orifice itself are two small glands, the *greater vestibular* or *Bartholin's glands.* These glands open by a duct into the space between the hymen and labia minora and produce a mucoid secretion that serves to supplement lubrication during sexual intercourse. Whereas the lesser vestibular glands are homologous to the male prostate, the greater vestibular glands are homologous to the male bulbourethral or Cowper's glands.

Perineum

The **perineum** is the diamond-shaped area at the lower end of the trunk between the thighs and buttocks of both males and females (Figure 18–19). It is surrounded anteriorly by the symphysis pubis, laterally by the ischial tuberosities, and posteriorly by the coccyx. A transverse line drawn between the ischial tuberosities divides the perineum into an anterior *urogenital triangle* that contains the external genitalia and a posterior *anal triangle* that contains the anus. In the female, the region between the vagina and anus is known as the clinical perineum. If the vagina is too small to accommodate the head of an emerging fetus, the skin and underlying tissue of the clinical perineum tears. To avoid this, a small incision, called an *episiotomy,* is made in the perineal skin just prior to delivery.

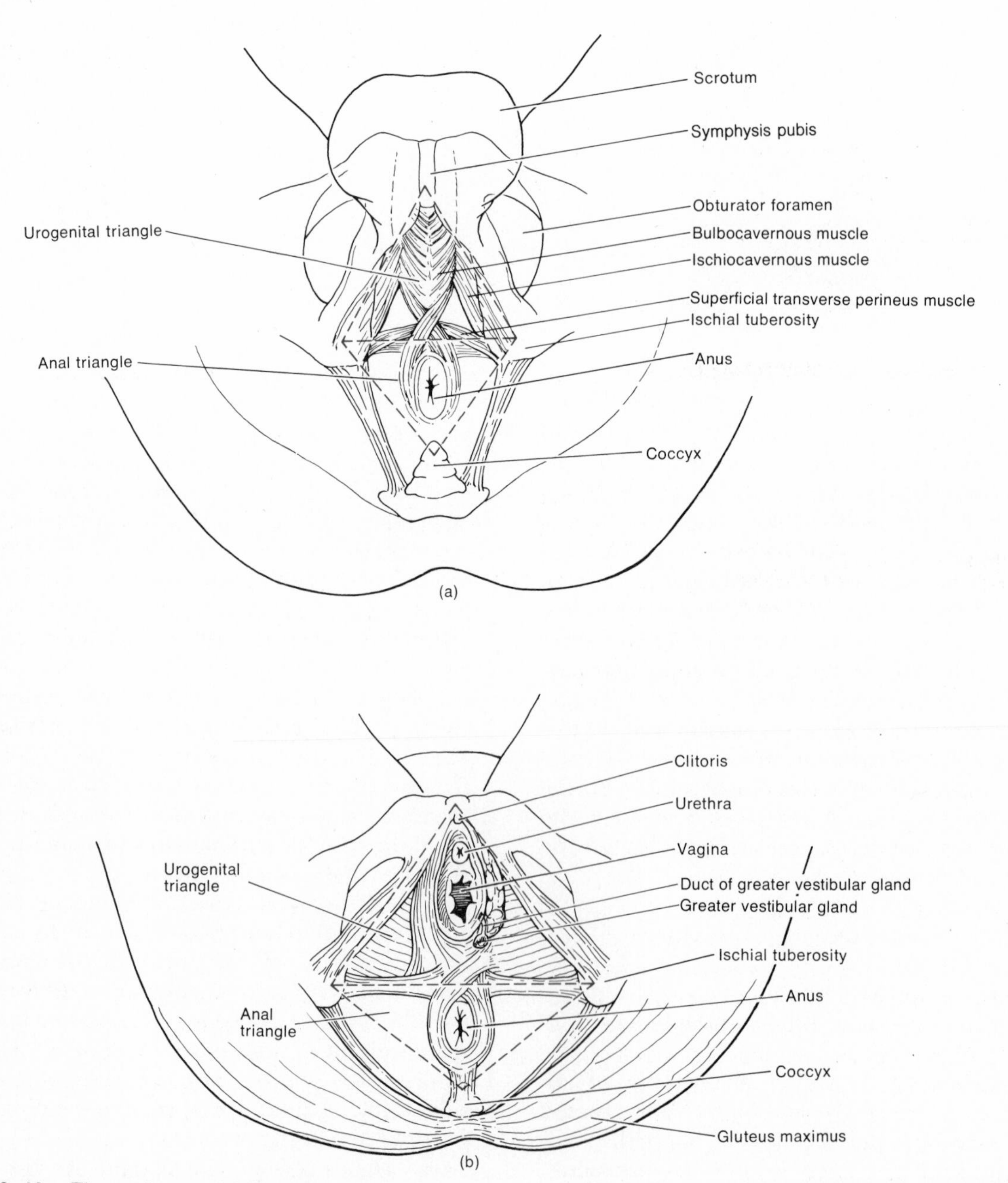

Figure 18–19. The perineum. (a) Borders seen in the male. (b) Dissected view of some regions of the female perineum.

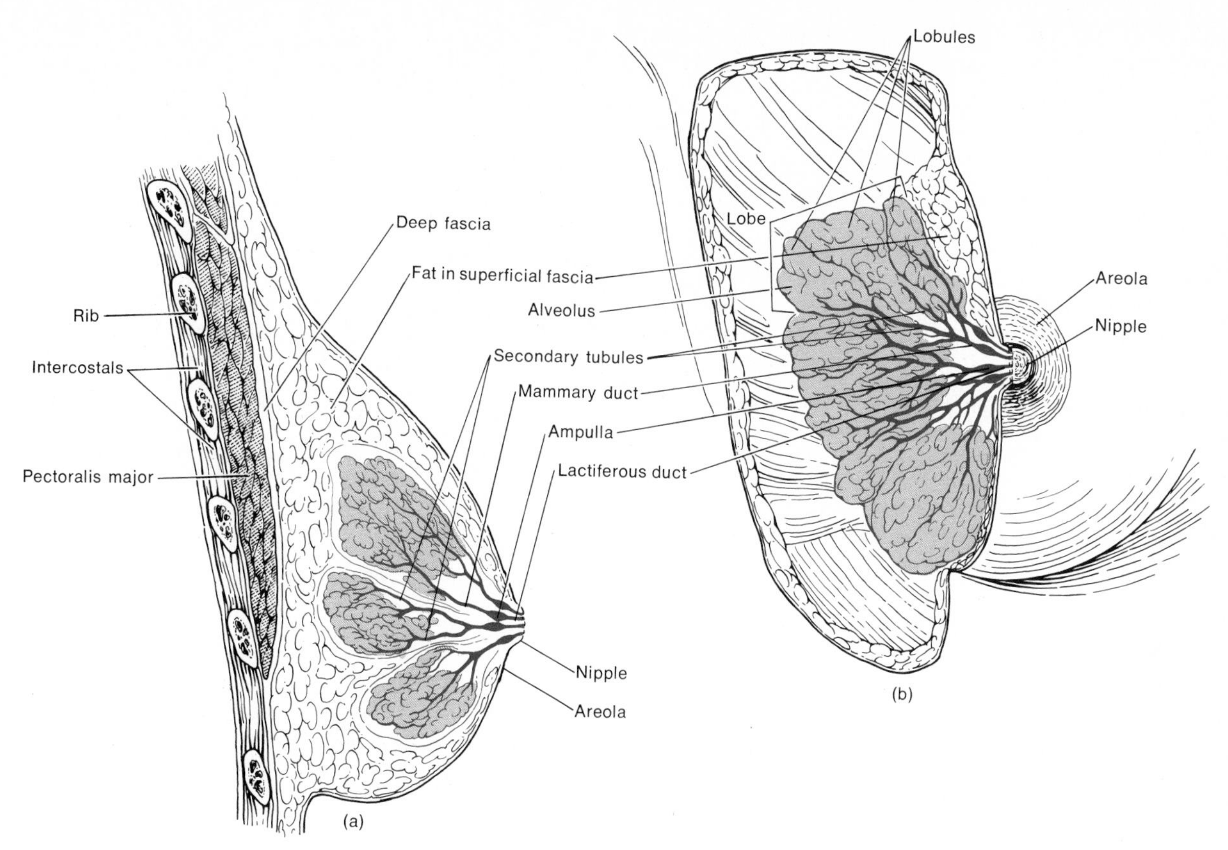

Figure 18–20. The mammary glands. (a) Sagittal section. (b) Front view partially sectioned.

Mammary glands

The **mammary glands** are branched tubuloalveolar glands that lie over the pectoralis major muscles and are attached to them by a layer of connective tissue (Figure 18–20). Internally, each mammary gland consists of 15 to 20 *lobes,* or compartments, separated by adipose tissue. The amount of adipose tissue present is the principal determinant of the size of the breasts. However, the size of the breasts has nothing to do with the amount of milk produced. Within each lobe are several smaller compartments, called *lobules,* that are composed of connective tissue in which milk-secreting cells referred to as *alveoli* are embedded (Figure 18–21). Between the lobules are located strands of connective tissue called the *suspensory ligaments of Cooper.* These ligaments run between the skin and deep fascia and serve to support the breast. Alveoli are arranged in grapelike clusters. They convey the milk into a series of *secondary tubules.* From here, the milk passes into the *mammary ducts.* As the mammary ducts approach the nipple, expanded sinuses called *ampullae,* where milk may be stored, are present. The ampullae continue as *lactiferous ducts* that terminate in the *nipple.* Each lactiferous duct conveys milk from one of the lobes to the exterior, although some may join before reaching the surface. The circular pigmented area of skin surrounding the nipple is called the *areola.* It appears rough because it contains modified sebaceous glands.

At birth, both male and female mammary glands are undeveloped and appear as slight elevations on the chest. With the onset of puberty, the female breasts begin to develop—the mammary ducts elongate, extensive fat deposition occurs, and the areola and nipple grow and become pigmented. These changes are correlated with an increased output of estrogen by the ovary. Further mammary development occurs at sexual maturity, with the onset of ovulation and the formation of the corpus luteum. During adolescence, lobules are formed and fat deposition continues, increasing the size of the glands. Although these changes are associated with estrogen and progesterone secretion by the ovaries, ovarian secretion is ultimately controlled by FSH.

The essential function of the mammary glands is milk secretion or lactation.

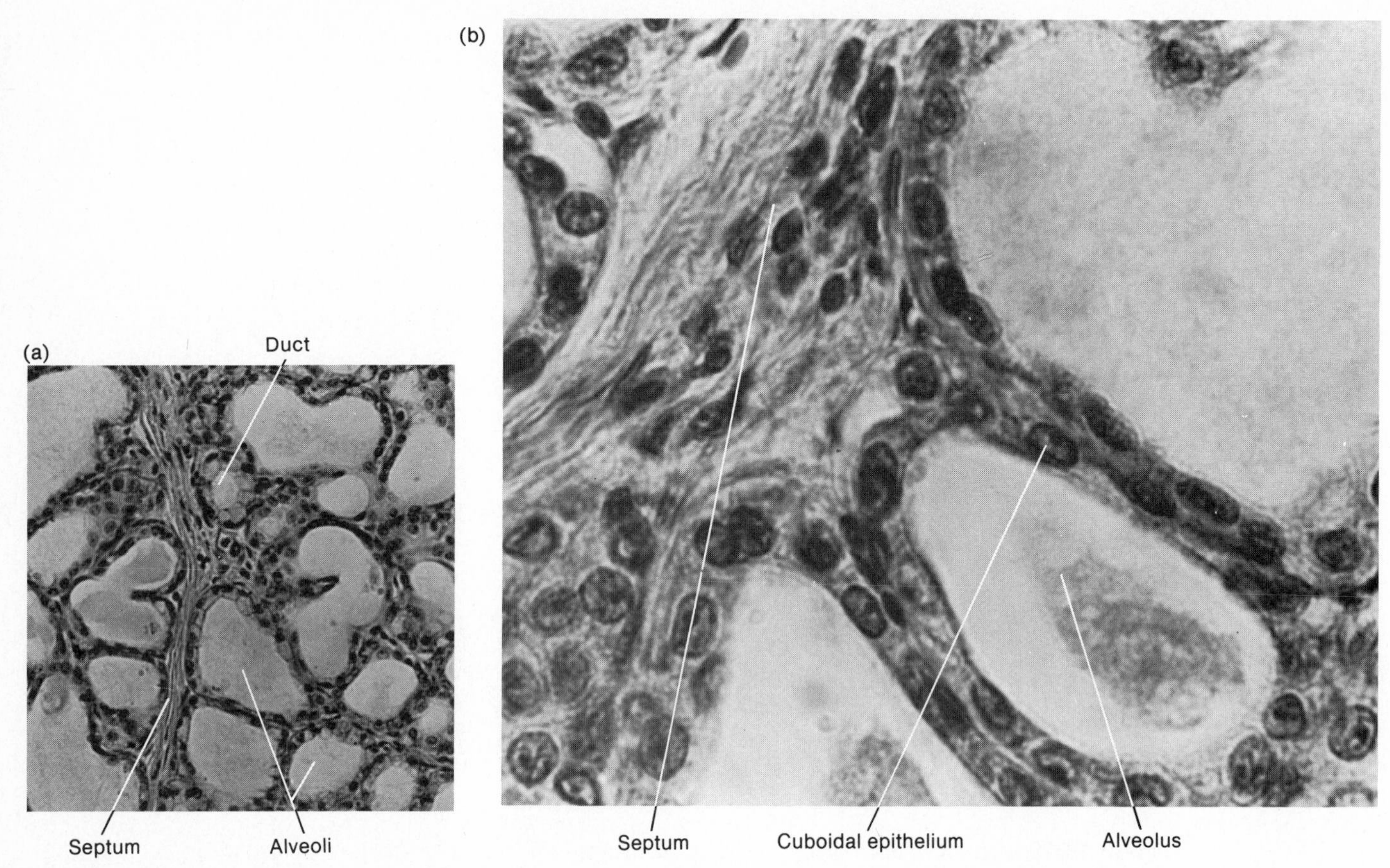

Figure 18–21. Histology of a lactating mammary gland. (a) Section showing numerous alveoli at a magnification of 40X. (b) Enlarged aspect at a magnification of 200X. (Courtesy of Victor B. Eichler, Wichita State University.)

APPLICATIONS TO HEALTH

In discussing the disorders that occur in the reproductive systems, we shall consider venereal diseases first. Then we shall look at some common diseases involving the male and female reproductive tracts, respectively.

Venereal diseases

The term venereal comes from Venus, the goddess of love. **Venereal diseases** represent a group of infectious diseases that are spread primarily through sexual intercourse. With the exception of the common cold, venereal diseases are ranked as the number 1 communicable diseases in the United States. Gonorrhea and syphilis are the two most common venereal diseases.

Gonorrhea, more commonly known as "clap," is an infectious disease that primarily affects the mucous membrane of the urogenital tract, the rectum, and occasionally the eyes. The disease is caused by the bacterium *Neisseria gonorrhoeae.* Males usually suffer inflammation of the urethra, with pus and painful urination. Fibrosis sometimes occurs in an advanced stage of gonorrhea, causing stricture of the urethra. There also may be involvement of the epididymis and prostate gland. In females, infection may occur in the urethra, vagina, and cervix, and there may be a discharge of pus. If the uterine tubes become involved, sterility and pelvic inflammation may result. Females commonly can harbor the disease and be asymptomatic.

Discharges from the involved mucous membranes are the source of infection, and the bacteria are transmitted by direct contact, usually sexual. The bacteria may be transmitted to the eyes of a newborn when the baby passes through the birth canal. Initially, the infant develops conjunctivitis. But later, other structures of the eye may become involved, and blindness may result. Administration of a 1 percent silver nitrate solution or penicillin in the eyes of the infant is very effective in preventing their infection. Penicillin is also the drug of choice for the treatment of gonorrhea in adults.

Syphilis is an infectious disease caused by the bacterium *Treponema pallidum.* It also is acquired through sexual contact. The early stages of the disease primarily affect the organs that are most likely to have made sexual contact–the genital organs, the mouth, and the rectum. The point where the bacteria enter the body is marked by a lesion called a *chancre.* In males, it usually occurs on the penis, and in females, it usually occurs in the vagina or cervix. The chancre heals without scarring. Following the initial infection, the bacteria enter the bloodstream and are spread throughout the body. In some individuals, an active secondary stage of the disease occurs, characterized by lesions of the skin and mucous membranes and often fever. The signs of the secondary stage also go away without medical treatment. During the next several years, the disease progresses without symptoms and is said to be in a latent phase. When symptoms again appear, anywhere from 5 to 40 years after the initial infection, the person is said to be in the tertiary stage of the disease. Tertiary syphilis may involve the circulatory system, skin, bones, viscera, and the nervous system. The effects of the syphilis bacterium on the nervous system have already been discussed in Chapter 15.

Male disorders

Among the common male reproductive disorders are those involving the prostate gland and sexual functions.

Prostate disorders

The prostate gland is susceptible to infection, enlargement, and benign and malignant tumors. Because the prostate surrounds the urethra, any of these disorders can cause obstruction to the flow of urine. Prolonged obstruction also may result in serious changes in the bladder, ureters, and kidneys.

Acute and chronic infections of the prostate gland are common in postpubescent males, many times in association with inflammation of the urethra. In **acute prostatitis** the prostate gland becomes swollen and very tender. Appropriate antibiotic therapy, bed rest, and above-normal fluid intake are effective in treatment.

Chronic prostatitis is one of the most common chronic infections in men of the middle and later years. On examination, the prostate gland feels enlarged, soft, and extremely tender. The surface outline is irregular and may be hard. This disease frequently produces no symptoms, but the prostate is believed to harbor infectious microorganisms responsible for some allergic conditions, arthritis, and inflammations of nerves (neuritis), muscles (myositis), and the iris (iritis).

An **enlarged prostate** gland occurs in approximately one-third of all males over 60 years of age. The enlarged gland is from two to four times larger than normal. The cause is unknown, and the enlarged condition usually can be detected by rectal examination.

Tumors of the male reproductive system primarily involve the prostate gland. Both benign and malignant growths are common in elderly men. Both types of tumors put pressure on the urethra, making urination painful and difficult. At times, the excessive back pressure destroys kidney tissue and gives rise to an increased susceptibility to infection. Therefore, even if the tumor is benign, surgery is indicated to remove the prostate or parts of it if the tumor is obstructive and perpetuates urinary tract infections.

Male sexual function abnormalities

Included here are those disorders that prevent satisfactory performance of the sex act or interfere with male fertility. **Impotence** is the inability of an adult male to attain or hold an erection long enough for normal intercourse. Impotence could be the result of physical abnormalities of the penis; systemic disorders such as syphilis; vascular disturbances; neurological disorders; or psychic factors such as fear of causing pregnancy, fear of venereal disease, religious inhibitions, and emotional immaturity. Impotence does not mean infertility.

Infertility, or **sterility,** is an inability to fertilize the ovum and does not imply impotence. Male fertility requires viable spermatozoa, adequate production of spermatozoa by the testes, unobstructed transportation of sperm through the seminal tract, and satisfactory deposition within the vagina. The tubules of the testes are sensitive to many factors–x-rays, infections, toxins, malnutrition, and others–that may cause degenerative changes and produce male sterility.

If inadequate spermatozoa production is suspected, a sperm analysis should be performed. Analysis includes measuring the volume of semen, counting the number of sperm per milliliter, evaluating sperm motility at 4 hours after ejaculation, and determining the percentage of abnormal sperm forms (not to exceed 20 percent).

Female disorders

Common disorders of the female reproductive system include menstrual abnormalities, ovarian cysts, leukorrhea, infertility, breast tumors, and cervical cancer.

Abnormalities of menstruation

Disorders of the female reproductive system frequently include menstrual disorders. This is hardly surprising because proper menstruation reflects not only the health of the uterus but the health of the glands that control it, that is, the ovaries and the pituitary gland.

Amenorrhea is the absence of menstruation in a woman. If the woman has never menstruated, the condition is called *primary amenorrhea.* Primary amenorrhea can be caused by endocrine disorders, most often in the pituitary gland and hypothalamus, or by genetically caused abnormal development of the ovaries or uterus. *Secondary amenorrhea* is cessation of uterine bleeding in women who have previously menstruated. The first cause considered is pregnancy. If that is ruled out, various endocrine disturbances are considered.

Dysmenorrhea is painful menstruation caused by contractions of the uterine muscles. A primary cause is believed to be low levels of progesterone. Recall that progesterone prevents uterine contractions. It can also be caused by pelvic inflammatory disease, uterine tumors, cystic ovaries, or congenital defects.

Abnormal uterine bleeding includes menstruation of excessive duration and/or excessive amount, too-frequent menstruation, intermenstrual bleeding, and postmenopausal bleeding. These abnormalities may be caused by disordered hormonal regulation, emotional factors, and systemic diseases.

Ovarian cysts

Ovarian cysts are tumors of the ovary that contain fluid. Follicular cysts may occur in the ovaries of elderly women, in ovaries that have inflammatory diseases, and in menstruating females. They have thin walls and contain a serous albuminous material. Cysts may also arise from the corpus luteum or the endometrium. *Endometriosis* is a painful disorder characterized by endometrial tissue or cysts in abnormal locations, such as in the uterine tubes, ovaries, vagina, peritoneum, or any other place in the body outside the uterus.

Leukorrhea

Leukorrhea is a nonbloody vaginal discharge that may occur at any age and affects most women at some time. It is not a disease; it is a symptom of infection or congestion of some portion of the reproductive tract. It may be a normal discharge in some women. If it is evidence of an infection, it may be caused by a protozoan microorganism called *Trichomonas vaginalis,* a yeast, a virus, or a bacterium.

Female infertility

Female infertility, or the inability to conceive, occurs in about 10 percent of married females in the United States. Once it is established that ovulation occurs regularly, the reproductive tract is examined for functional and anatomical disorders to determine the possibility of union of the sperm and the ovum in the oviduct.

Diseases of the breasts

The breasts are highly susceptible to cysts and tumors. Men are also susceptible to breast tumor, but certain breast cancers are 100 times more common in women than in men. Usually these growths can be detected early by the woman who inspects and palpates her breasts regularly. To **palpate** means to feel or examine by touch. Unfortunately, so few women practice periodic self-examination that many growths are discovered by accident and often too late for proper treatment.

In the female, the benign *fibroadenoma* is a common tumor of the breast. It occurs most frequently in young women. Fibroadenomas have a firm rubbery consistency and are easily moved about within the mammary tissue. The usual treatment is excision of the growth. The breast itself is not removed.

Breast cancer has the highest fatality rate of all cancers affecting women, but it is rare in men. In the female, breast cancer is rarely seen before age 30, and its occurrence rises rapidly after menopause.

Breast cancer is generally not painful until it becomes quite advanced, so often it is not discovered early, or if it is noted, it is ignored. Any lump, be it ever so small, should be reported to a doctor at once. If there is no evidence of *metastasis* (the spread of cancer cells from one part of the body to another or from one organ to another), the treatment of choice is a *modified* or *radical mastectomy.* A radical mastectomy involves removal of the affected breast, along with the underlying pectoral muscles and the axillary lymph nodes. Metastasis of cancerous cells is usually through the lymphatics or blood. Radiation treatments may follow the surgery to ensure the destruction of any remaining stray cancer cells.

Cervical cancer

Another common disorder of the female reproductive system is cancer of the uterine cervix. It ranks third in frequency after breast and skin cancers. **Cervical cancer** starts with a change in the shape of the cervical cells called *cervical dysplasia.* Cervical dysplasia is not a cancer in itself, but the abnormal cells tend to become malignant.

Early diagnosis of cancer of the uterus is accomplished by the *Papanicolaou test,* or "Pap" smear. In this generally painless procedure, a few cells from the vaginal fornix (that part of the vagina surrounding the cervix) and the cervix are removed with a swab and examined microscopically. Malignant cells have a characteristic appearance and are indicative of an early stage of cancer, even before any symptoms occur. Estimates indicate that the "Pap" smear is more than 90 percent reliable in detecting cancer of the cervix. Treatment of cervical cancer may involve complete or partial removal of the uterus, called a *hysterectomy,* or radiation treatments.

Key terms associated with the reproductive systems

Castration Excision of the testes or ovaries.

Copulation Sexual intercourse. Coitus refers to sexual intercourse between human beings.

Hermaphroditism Presence of both male and female sex organs in one individual.

Hysterectomy Removal of the uterus.

Impotence Inability for sexual intercourse in the man.

Menarche Beginning of menstruation.

Menopause Cessation of menstruation.

Salpingitis Inflammation or infection of the uterine tube.

Vaginitis Inflammation of the vagina.

Chapter summary in outline

MALE REPRODUCTIVE SYSTEM

1. The scrotum supports and protects the testes and provides an appropriate temperature for the production and survival of spermatozoa.
2. The major functions of the testes are sperm production and the secretion of testosterone.
3. FSH and ICSH maintain the growth and development of the male reproductive organs.
4. Spermatozoa are conveyed from the testes to the exterior through the seminiferous tubules, straight tubules, rete testis, efferent ducts, epididymis, ductus deferens, ejaculatory duct, and urethra.
5. The male urethra is divided into prostatic, membranous, and spongy portions.
6. The seminal vesicles, prostate, and Cowper's glands secrete the liquid portion of semen. They are accessory reproductive glands.
7. Semen is a mixture of sperm and secreted liquids.
8. The penis serves as the organ of copulation.

FEMALE REPRODUCTIVE SYSTEM

1. The ovaries produce ova and secrete estrogens and progesterone.
2. The uterine tubes convey ova from the ovaries to the uterus and are the sites of fertilization.
3. The normal position of the uterus is maintained by a series of ligaments.
4. The uterus is associated with menstruation, implantation of a fertilized ovum, development of the fetus, and labor.
5. FSH and LH control the ovarian cycle. Estrogens and progesterone control the menstrual cycle.
6. The vagina serves as a passageway for the menstrual flow, as the lower portion of the birth canal, and as the receptacle for the penis.
7. The vulva is a collective designation for the external genitalia of the female.
8. The perineum is a diamond-shaped area at the lower end of the trunk between the thighs and buttocks.
9. The mammary glands function in the secretion of milk.

APPLICATIONS TO HEALTH

1. Venereal diseases are a group of infectious diseases spread primarily through sexual intercourse.
2. Conditions that affect the prostate are prostatitis, enlarged prostate, and tumors.
3. Impotence is the inability of the male to attain or hold an erection long enough for intercourse.
4. Infertility is the inability of a male's sperm to fertilize an ovum.
5. Menstrual disorders include amenorrhea, dysmenorrhea, and abnormal bleeding.
6. Ovarian cysts are tumors that contain fluid.
7. Leukorrhea is a nonbloody vaginal discharge that may be caused by an infection.
8. The mammary glands are susceptible to benign fibroadenomas and malignant tumors. The removal of a malignant breast, pectoral muscles, and lymph nodes is called a radical mastectomy.
9. Cervical cancer can be diagnosed by a "Pap" test. Complete or partial removal of the uterus is called a hysterectomy.

Review questions and problems

1. Define reproduction. List the male and female organs of reproduction.
2. Describe the function of the scrotum in protecting the testes from temperature fluctuations. What is cryptorchidism?
3. Describe the internal structure of a testis. Where are the sperm cells made?
4. Describe the principal features of spermatogenesis.
5. Identify the principal parts of a spermatozoan. List the function of each.
6. Explain the effects of FSH and ICSH on the male reproductive system.
7. Describe the physiological effects of testosterone.
8. Describe the location, structure, and histology of the epididymis, ductus deferens, and ejaculatory duct.
9. Trace the course of a sperm cell through the male system of ducts from the seminiferous tubules to the urethra.
10. Explain the location of the three subdivisions of the male urethra.
11. What is the spermatic cord?
12. Briefly explain the locations and functions of the seminal vesicles, prostate gland, and Cowper's glands.
13. What is seminal fluid? What is its function?
14. Describe the structure of the penis.
15. How is the penis structurally adapted as an organ of copulation?
16. Describe the histology of the ovaries.
17. How are the ovaries held in position in the pelvic cavity? What is ovulation?
18. What is the function of the uterine tubes? Explain in terms of their histology. Define an ectopic pregnancy.
19. Diagram the principal parts of the uterus.
20. Describe the arrangement of ligaments that hold the uterus in its normal position. Explain the two major malpositions of the uterus.
21. Explain the histology of the uterus.
22. Discuss the blood supply to the uterus. Why is an abundant blood supply important?
23. List the physiological effects of estrogens and progesterone.
24. What is the function of the vagina? Describe its structure.
25. List the parts of the vulva and the functions of each part.
26. What is the perineum? Define episiotomy.
27. Describe the structure of the mammary glands. How are the breasts supported?
28. Describe the passage of milk from the areolar cells of the mammary gland to the nipple.
29. Define venereal disease. Distinguish between gonorrhea and syphilis.
30. Describe several disorders that affect the prostate gland.
31. Distinguish between impotence and infertility.
32. What are some of the causes of amenorrhea, dysmenorrhea, and abnormal uterine bleeding?
33. What are ovarian cysts? Define endometriosis.
34. What are some possible causes of leukorrhea?
35. Distinguish between a fibroadenoma and a malignant tumor of the breast.
36. What is a radical mastectomy?
37. What is a "Pap" smear? What is a hysterectomy?
38. Refer to the glossary of key terms at the end of the chapter. Be sure that you can define each term.

Selected readings

Arey, L. B. *Developmental Anatomy.* 7th ed. Philadelphia: W. B. Saunders, 1974.

Dawson, Helen L. *Basic Human Anatomy.* 2d ed. New York: Appleton-Century-Crofts, 1974. Chap. 12.

Gray, Henry. *Anatomy of the Human Body.* 29th ed., edited by Charles Mayo Goss. Philadelphia: Lea and Febiger, 1973. Pp. 1299–1337.

Netter, Frank H. *Reproductive System.* CIBA Collection of Medical Illustrations, Vol. 2. Summit, NJ: CIBA Pharmaceutical Co., 1965.

Patten, B. M. *Human Embryology.* 3d ed. New York: McGraw-Hill, 1968.

CHAPTER 19

THE RESPIRATORY SYSTEM

STUDENT OBJECTIVES

After you have read this chapter, you should be able to:

1. Identify the organs of the respiratory system
2. Compare the structure of the external and internal nose
3. Contrast the functions of the external and internal nose in filtering, warming, and moistening air
4. Differentiate between the three regions of the pharynx and describe their roles in respiration
5. Identify the anatomical features of the larynx related to respiration and voice production
6. Describe the tubes that form the bronchial tree with regard to structure and location
7. Contrast tracheotomy and intubation as alternate methods for clearing air passageways
8. Identify the coverings of the lungs and the gross anatomical features of the lungs
9. Describe the structure of a bronchopulmonary segment and a lobule of the lung
10. Describe the role of alveoli in the diffusion of respiratory gases
11. List the sequence of pressure changes involved in inspiration and expiration
12. Compare the volumes and capacities of air exchanged in respiration
13. Describe the parts of the nervous system that control respiration
14. Compare the roles of the Hering-Breuer reflex and the pneumotaxic center in controlling respiration
15. Define coughing, sneezing, sighing, yawning, crying, laughing, and hiccoughing as modified respiratory movements
16. Define hay fever, bronchial asthma, emphysema, pneumonia, tuberculosis, and hyaline membrane disease as disorders of the respiratory system
17. Describe the effects of pollutants on the epithelium of the respiratory system
18. Define key terms associated with the respiratory system

Cells need a continuous supply of oxygen to carry out the activities that are vital to their survival. Many of these activities release quantities of carbon dioxide. Since an excessive amount of carbon dioxide is poisonous to cells, the gas must be eliminated quickly and efficiently. The two systems that are designed to supply oxygen and eliminate carbon dioxide are the circulatory system and the respiratory system. The **respiratory system** consists of organs that exchange gases between the atmosphere and blood. These organs are the nose, pharynx, larynx, trachea, bronchi, and lungs (Figure 19–1). In turn, the blood transports gases between the lungs and the cells. The overall exchange of gases between the atmosphere, the blood, and the cells is called **respiration.** Both the respiratory and circulatory systems participate equally in respiration. Failure of either system has the same effect on the body: rapid death of cells from oxygen starvation and disruption of homeostasis.

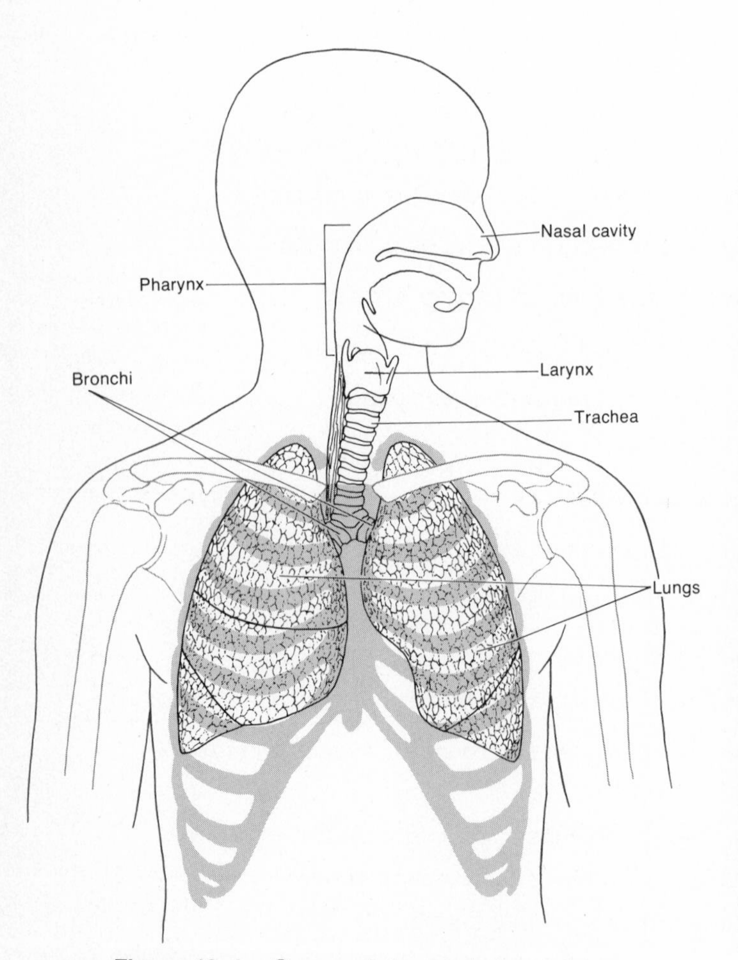

Figure 19–1. Organs of the respiratory system.

THE RESPIRATORY ORGANS

Each of the respiratory organs will now be described in some detail.

Nose

The **nose** has an external portion jutting out from the face and an internal portion lying hidden inside the skull. Externally, the nose consists of a supporting framework of bone and cartilage covered with skin and lined with mucous membrane. The bridge of the nose is formed by the nasal bones, which hold it in a fixed position. Because it has a framework of pliable cartilage, the rest of the external nose is quite flexible. On the undersurface of the external nose are two openings called the *nostrils,* or *external nares* (Figure 19–2). You may review the surface anatomy of the nose by referring back to Exhibit 1–3.

The internal region of the nose is a large cavity within the skull that lies below the cranium and above the mouth. Anteriorly, the internal nose merges with the external nose, and posteriorly it communicates with the throat (pharynx) through two openings called the *internal nares.* Four paranasal sinuses (frontal, sphenoid, maxillary, and ethmoid) and the nasolacrimal ducts also open into the internal nose. The lateral walls of the internal nose are formed by the ethmoid, maxillae, and inferior conchae bones. The ethmoid forms the roof, and the floor is formed by the palatine bones and the maxilla of the hard palate. Occasionally the palatine and maxillary bones fail to fuse during embryonic life, and a child is born with a crack in the bony wall that separates the internal nose from the mouth. This condition is called *cleft palate.*

The inside of both the external and internal nose is divided into right and left *nasal cavities* by a vertical partition called the *nasal septum.* Cartilage is the primary material making up the anterior portion of the septum. The remainder is formed by the vomer and the perpendicular plate of the ethmoid. (See Figure 6–7.) The anterior portions of the nasal cavities, which are just inside the nostrils, are called the *vestibules.* The vestibules are surrounded by cartilage as opposed to bone of the upper nasal cavity. The interior structures of the nose are specialized for three functions. First, incoming air is warmed, moistened, and filtered. Second, olfactory stimuli are received, and third, large hollow resonating chambers are provided for speech sounds. These three functions are accomplished in the following manner. When air enters the nostrils, it passes first through the vestibule. The vestibule is

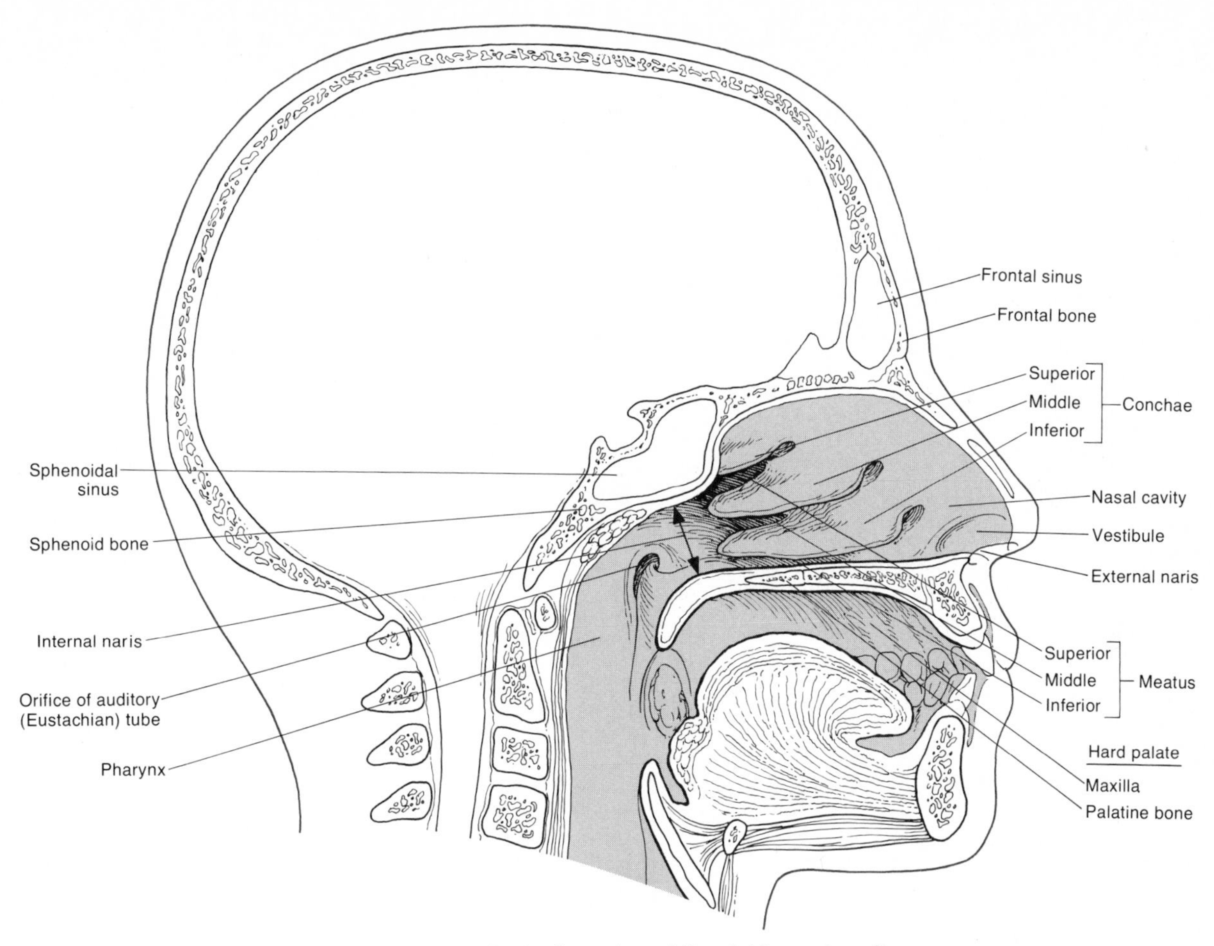

Figure 19–2. Sagittal section of the right nasal cavity.

lined by skin containing coarse hairs that filter out large dust particles. The air then passes into the rest of the cavity. Three shelves formed by projections of the superior, middle, and inferior conchae or turbinates extend out of the lateral wall of the cavity. The conchae, almost reaching the septum, subdivide each nasal cavity into a series of groovelike passageways called *meati.* These are called the *superior, middle,* and *inferior meati.* Mucous membrane lines the cavity and its shelves. The olfactory receptors lie in the membrane lining the upper portion of the cavity, also called the olfactory region (see Figure 16–4). Below the olfactory region, the membrane contains pseudostratified ciliated columnar cells with many goblet cells and capillaries. As the air whirls around the turbinates and meati, it is warmed by the capillaries. Mucus secreted by the goblet cells moistens the air and traps dust particles. Drainage from the lacrimal ducts, and perhaps secretions from the paranasal sinuses, also help to moisten the air. The cilia move the resulting mucus-dust packages along to the throat so that they can be eliminated from the body. As the air passes through the top of the cavity, chemicals in the air may stimulate the olfactory receptors.

The arterial supply to the nasal cavity is principally from the sphenopalatine branch of the maxillary artery. The remainder is supplied by the ophthalmic artery. The veins of the nasal cavity drain into the sphenopalatine vein, the facial vein, and the ophthalmic vein.

The nerve supply of the nasal cavity consists of olfactory cells in the olfactory epithelium associated with the olfactory nerve (see Figure 16–4) and the nerves of general sensation. These nerves are branches of the ophthalmic division of the trigeminal nerve and the maxillary division of the trigeminal nerve.

Pharynx

The **pharynx,** or throat, is a tube about 13 cm (5 in.) long that starts at the internal nares and runs partway down the neck (Figure 19–3). It lies just in back of the nasal cavity and oral cavity and just in front

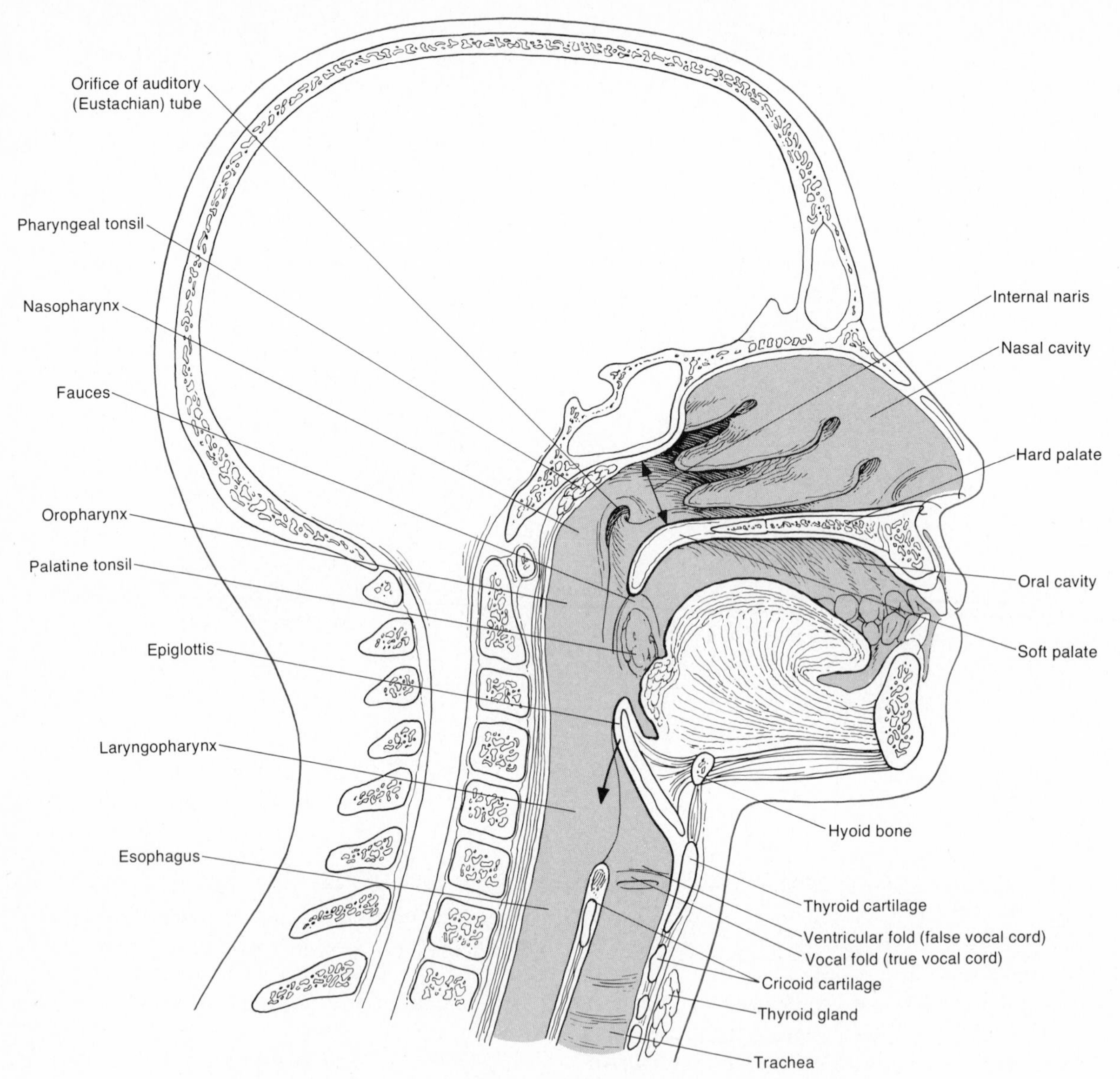

Figure 19–3. Sagittal section of the head, neck, and upper chest.

of the cervical vertebrae. Its walls are composed of skeletal muscles. The interior of the walls is lined with mucous membrane. As you might expect from such a structurally simple tube, the functions of the pharynx are limited to serving as a passageway for air and food and providing a resonating chamber for speech sounds.

The uppermost portion of the pharynx is called the *nasopharynx*. This part lies behind the internal nasal cavity and extends over the soft palate. There are four openings in its walls: two internal nares plus two openings that lead into the auditory or Eustachian tubes. The posterior wall of the nasopharynx also contains the pharyngeal tonsils, or adenoids. Through the internal nares the nasopharynx exchanges air with the nasal cavities and receives the packages of dust-laden mucus. The nasopharynx has a lining of pseudostratified ciliated epithelium. Cilia in the walls of the nasopharynx move the mucus down toward the mouth. The nasopharynx also exchanges small amounts of air with the auditory canal so that the air pressure inside the middle ear equals the pressure of the atmospheric air flowing through the nose and pharynx.

The second portion of the pharynx, the *oropharynx*, lies behind the oral cavity and extends from the soft palate down to the hyoid bone. It receives only one opening, the *fauces*, or opening from the mouth. It is lined by stratified squamous epithelium. This portion of the pharynx is both respiratory and digestive in function since it is a common passageway for both air and food. Two pairs of tonsils, the palatine tonsils and the lingual tonsils, are found in the oropharynx. The lingual tonsils lie at the base of the tongue and are illustrated in Figure 20–4.

The lowest portion of the pharynx is called the *laryngopharynx*. The laryngopharynx extends downward from the hyoid bone and empties into the esophagus (food tube) posteriorly and into the larynx (voice box) anteriorly. Like the oropharynx, the laryngopharynx is both respiratory and digestive in function and is lined by stratified squamous epithelium.

The arteries of the pharynx are the ascending pharyngeal, the ascending palatine branch of the facial, the descending palatine and pharyngeal branches of the maxillary, and the muscular branches of the superior thyroid artery. The veins of the pharynx drain into the pterygoid plexus and the internal jugular vein.

Most of the muscles of the pharynx are innervated by the pharyngeal plexus. This plexus is formed by the pharyngeal branches of the glossopharyngeal and vagal nerves and the superior cervical sympathetic ganglion.

Larynx

The **larynx,** or voice box, is a short passageway that connects the pharynx with the trachea. It lies in the midline of the neck anterior to the fourth through sixth cervical vertebra. The walls of the larynx are supported by nine pieces of cartilage. Three are single and three are paired. The three single pieces are the large thyroid cartilage and the smaller epiglottic and cricoid cartilage (see Figure 19–3). Of the paired cartilages, the arytenoid cartilages are the most important. The paired corniculate and cuneiform cartilages are of lesser significance. The *thyroid cartilage,* or Adam's apple, consists of two fused plates that form the anterior wall of the larynx and give it its triangular shape (Figure 19–4a and b). In males the thyroid cartilage is bigger than it is in females.

The *epiglottis* is a large, leaf-shaped piece of cartilage lying on top of the larynx. The "stem" of the epiglottis is attached to the thyroid cartilage, but the "leaf" portion is unattached and free to move up and down like a door on a hinge. In fact, the epiglottis is sometimes called the trap door. During swallowing, the free edge of the epiglottis forms a lid over the larynx (see Figure 19–3). In this way, the larynx is closed off and liquids and foods are routed into the esophagus and kept out of the trachea. If anything but air passes into the larynx, a cough reflex attempts to expel the material.

The *cricoid cartilage* is a ring of cartilage forming the lower walls of the larynx. It is attached to the first ring of cartilage of the trachea.

The paired *arytenoid cartilages* are pyramidal in shape and are located at the superior border of the cricoid cartilage. They attach to the vocal folds and pharyngeal muscles and by their action can move the vocal cords. The *corniculate cartilages* are paired, cone-shaped cartilages. Each is located at the apex of each arytenoid cartilage. The paired *cuneiform cartilages* are rod-shaped cartilages located in the mucous membrane fold that connects the epiglottis to the arytenoid cartilages.

Like the other respiratory passageways, the larynx is lined with a ciliated mucous membrane. Dust not removed in the upper passages can be trapped by the mucus and moved back up toward the throat, where the mucus can be swallowed or eliminated.

The mucous membrane of the larynx is arranged into two pairs of folds, an upper pair called the *ventricular folds* or *false vocal folds,* and a lower pair called simply the *vocal folds* or *true vocal cords* (Figure 19–4c). The air passageway between the folds is called the *glottis.* Underneath the mucous membrane of the true vocal cords lie bands of elastic ligaments that are stretched between pieces of rigid cartilage like the strings on a guitar. Skeletal muscles of the larynx, called intrinsic muscles, are attached internally to the pieces of rigid cartilage and to the vocal folds themselves. When the muscles contract, they pull the strings of elastic ligaments tight and stretch the cords out into the air passageways so that the glottis is narrowed (Figure 19–4c). If air is directed against the vocal folds, they vibrate and set up sound waves in the column of air in the pharynx, nose, and mouth. The greater the pressure of air, the louder the sound.

Pitch is controlled by the tension on the true vocal cords. If the cords are pulled taut by the muscles, they vibrate more rapidly and a higher pitch results. Lower sounds are produced by decreasing the muscular tension on the cords. Vocal cords are usually thicker and longer in males than they are in females, and they vibrate more slowly. This is why men have a lower range of pitch than women.

Sound originates from the vibration of the true vocal cords. But other structures are necessary for converting the sound into recognizable speech. For instance, the pharynx, mouth, nasal cavities, and paranasal sinuses all act as resonating chambers that give the voice its human and individual quality. By

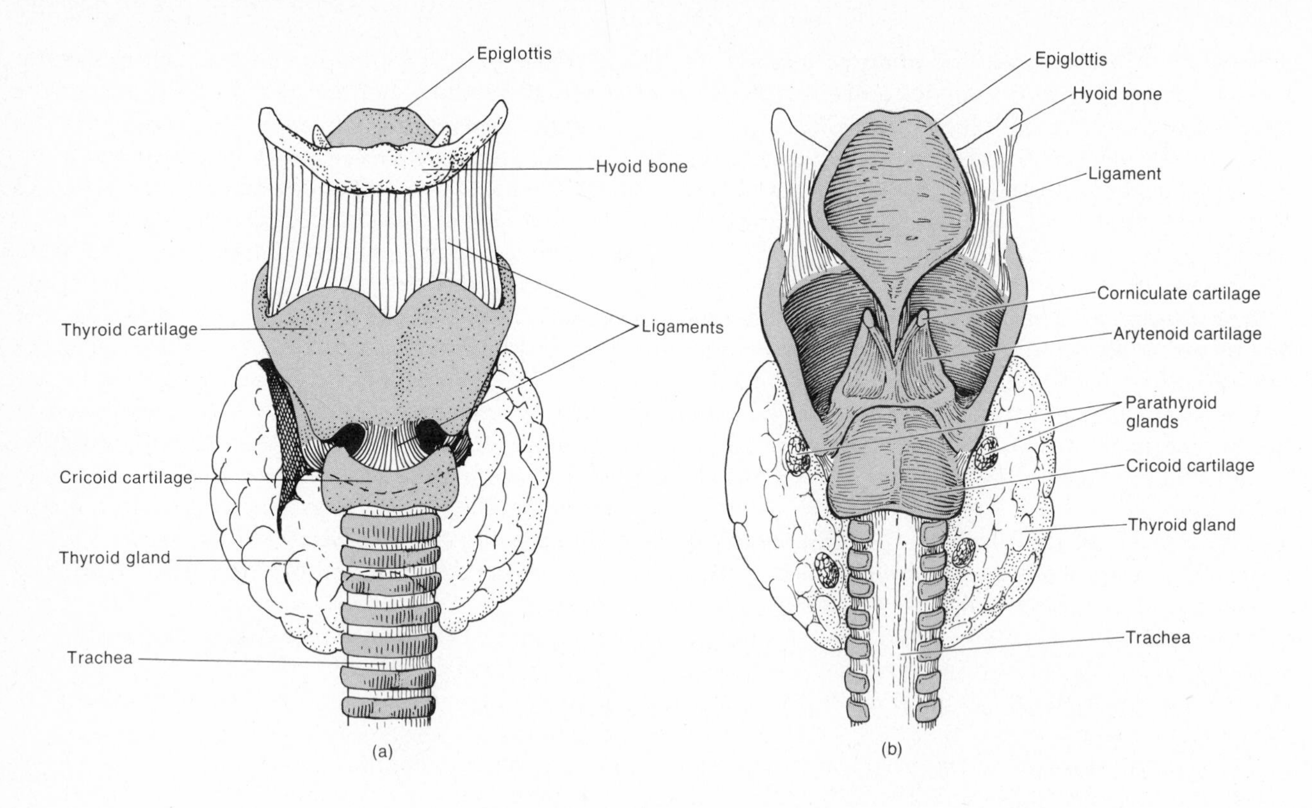

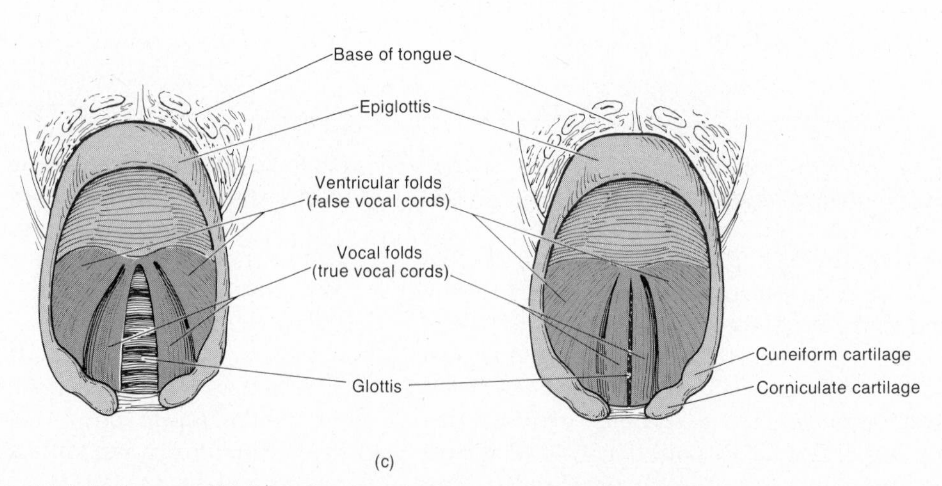

Figure 19–4. The larynx. (a) Anterior view. (b) Posterior view. (c) Viewed from above. In the figure on the left the true vocal cords are relaxed. In the figure on the right the true vocal cords are pulled taut.

constricting and relaxing the muscles in the walls of the pharynx, we produce the vowel sounds. Muscles of the face, tongue, and lips help us to enunciate words.

Laryngitis is an inflammation of the larynx that is most often caused by a respiratory infection or by irritants, such as cigarette smoke. Inflammation of the vocal folds themselves causes hoarseness or loss of voice by interfering with the contraction of the cords or by causing them to swell to the point where they cannot vibrate freely. Many long-term smokers acquire a permanent hoarseness from the damage done by chronic inflammation.

The arteries of the larynx are the superior laryngeal, inferior laryngeal, and cricothyroid. The superior and inferior laryngeal and cricothyroid veins accompany the arteries. The superior cricoid vein and cricothyroid vein empty into the superior thyroid vein, and the inferior vein empties into the inferior thyroid vein.

The nerves of the larynx are the superior and inferior laryngeal branches of the vagus.

Trachea

The **trachea,** or windpipe, is a tubular passageway for air, about 12 cm (4½ in.) in length and 2.5 cm (1 in.) in diameter. It is located in front of the esophagus, and it extends from the larynx to the fifth thoracic vertebra (see Figure 19–1), where it divides into right and left primary bronchi.

The trachea is lined with pseudostratified ciliated columnar epithelium providing the same protection against dust as the membrane lining the larynx (Figure 19–5). The walls of the trachea are composed of smooth muscle and elastic connective tissue. They are encircled by a series of horizontal rings of cartilage that look like a series of letter C's stacked one on top of the other. The open parts of the C's face the esophagus and permit the esophagus to expand into the trachea during the swallowing. The solid parts of the C's provide a rigid support so that the tracheal walls do not collapse inward and obstruct the air passageway.

Occasionally the respiratory passageways are unable to protect themselves from obstruction. For instance, the rings of cartilage may be accidentally crushed, or the mucous membrane may become inflamed and swell so much that it closes off the air space. Inflamed membranes also secrete a great deal of mucus that may clog the lower respiratory passageways. Or a large object may be breathed in (aspirated) while the epiglottis is open. In any case, the passageways must be cleared quickly. If the obstruction is above the level of the chest, a *tracheotomy* may be performed. The first step in a tracheotomy is to make an incision through the neck and

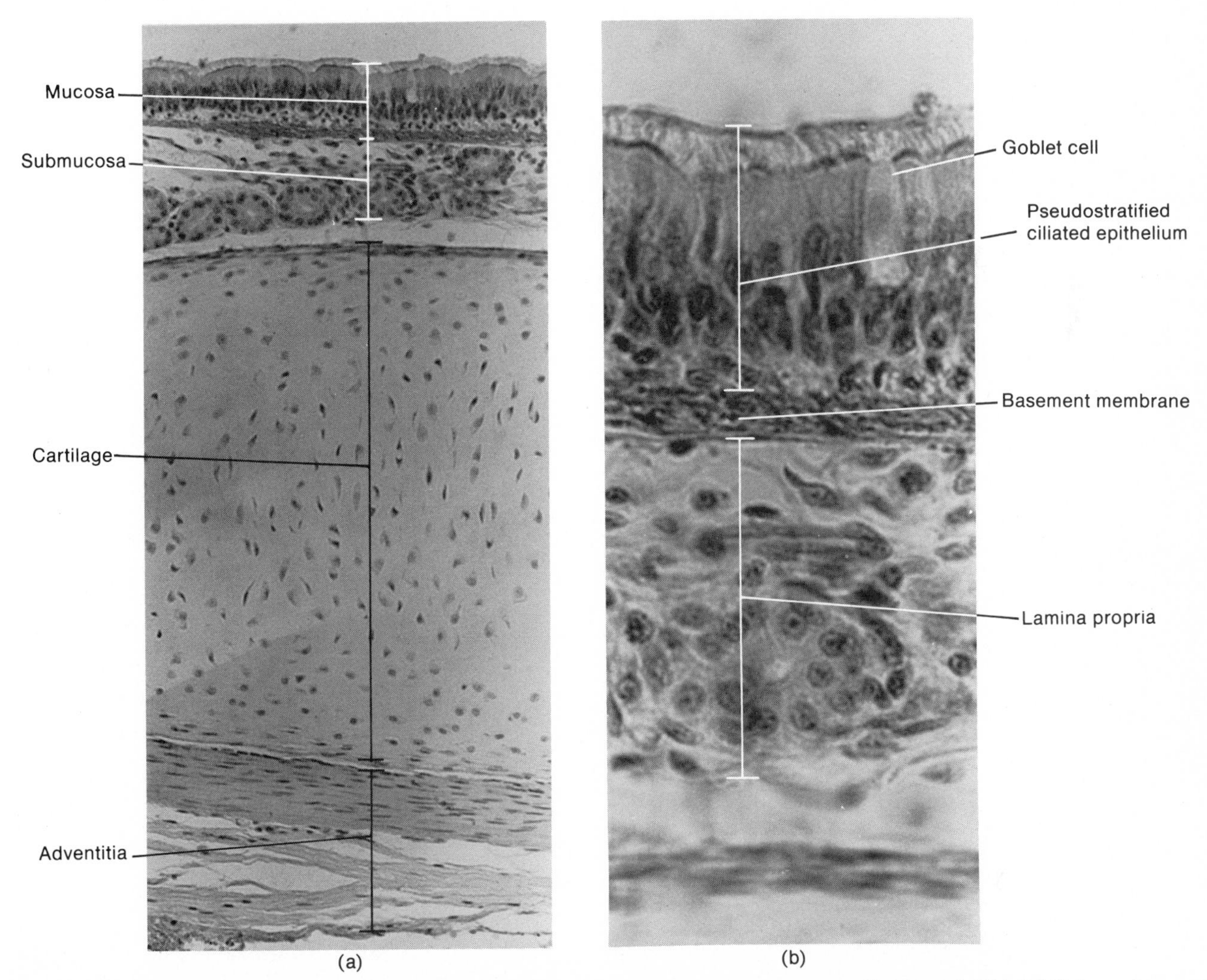

Figure 19–5. Histology of the trachea. (a) Section of the wall of the trachea at a magnification of 40X. (b) Enlarged aspect of the tracheal epithelium at a magnification of 80X. (Courtesy of Victor B. Eichler, Wichita State University.)

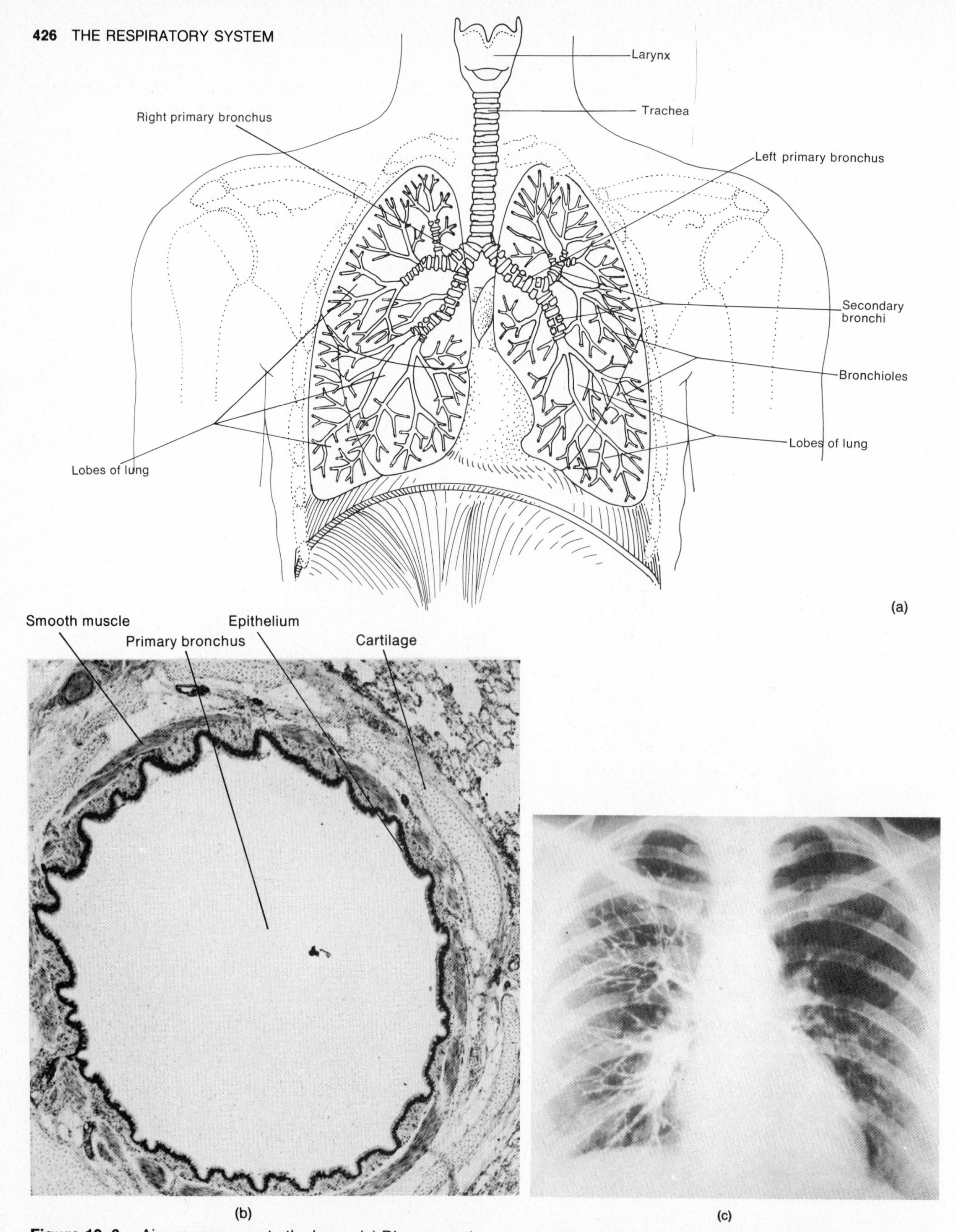

Figure 19–6. Air passageways to the lungs. (a) Diagrammatic representation of the bronchial tree in relationship to the lungs. (b) Histology of a primary bronchial tube seen in cross section at a magnification of 40X. (Courtesy of Edward J. Reith, from *Atlas of Descriptive Histology,* by Edward J. Reith and Michael H. Ross, Harper & Row, Publishers, Inc., New York, 1970.) (c) Anteroposterior bronchogram of the lungs. (Courtesy of Lester W. Paul and John H. Juhl, *The Essentials of Roentgen Interpretation,* 3d ed., Harper & Row, Publishers, Inc., New York, 1972.)

into the part of the trachea below the obstructed area. A tube is then inserted through the incision, and the patient breathes through the tube. Another method that may be employed is *intubation.* A tube is inserted into the mouth and passed down through the larynx and trachea. The firm walls of the tube push back any flexible obstruction, and the inside of the tube provides a passageway for air. If mucus is clogging the airways, it can be suctioned up through the tube.

The arteries of the trachea are branches of the inferior thyroid, internal thoracic, and bronchial arteries. The veins of the trachea terminate in the inferior thyroid veins.

The smooth muscle and glands of the trachea are innervated parasympathetically via the vagus nerve, directly, and by its recurrent laryngeal branches. Sympathetic innervation is through branches from the sympathetic trunk and its ganglia.

Bronchi

The trachea terminates in the chest by dividing into a **right primary bronchus,** which goes to the right lung, and a **left primary bronchus,** which goes to the left lung (Figure 19–6a). The right primary bronchus is more vertical, shorter, and wider than the left. As a result, foreign objects that enter the air passageways frequently lodge in it. Like the trachea, the primary bronchi contain incomplete rings of cartilage and are lined by a ciliated columnar epithelium.

The blood supply to the bronchi is via the left bronchial arteries and the right bronchial artery. The bronchial veins that drain the bronchi are the right bronchial vein which enters the azygous vein and the left bronchial vein that empties into the hemiazygous vein or the left superior intercostal vein.

Upon entering the lungs, the primary bronchi divide to form smaller bronchi, the *secondary* or *lobar bronchi,* one for each lobe of the lung. (The right lung has three lobes, the left lung has two.) The secondary bronchi continue to branch, forming still smaller bronchi, called *tertiary* or *segmental bronchi,* which divide into *bronchioles* (see Figure 19–8). Bronchioles, in turn, branch into even smaller tubes, called the *terminal bronchioles.* The continuous branching of the trachea into primary bronchi, secondary bronchi, bronchioles, and terminal bronchioles resembles a tree trunk with its branches and is commonly referred to as the *bronchial tree.* As the branching becomes more extensive in the bronchial tree, several structural changes may be noted. First, rings of cartilage are replaced by plates of cartilage that finally disappear in the bronchioles. Second, as the cartilage decreases, the amount of smooth muscle increases (Figure 19–6b). In addition, the epithelium changes from ciliated columnar to simple cuboidal in the terminal bronchioles. The fact that the walls of the bronchioles contain a great deal of muscle but no cartilage is clinically significant. During an asthma attack the muscles spasm. Because there is no supporting cartilage, the spasms tend to close off the air passageways.

Bronchography is a technique for examining the bronchial tree. The patient breathes in air that contains a safe dosage of a radioactive element. The element gives off rays that penetrate the chest walls and expose a film. The developed film, a *bronchogram,* provides a picture of the tree (see Figure 19–6c).

Lungs

The **lungs** are paired, cone-shaped organs lying in the thoracic cavity (Figure 19–7a). They are separated from each other by the heart and other structures in the mediastinum. Two layers of serous membrane, collectively called the *pleural membrane,* enclose and protect each lung. The outer layer is attached to the walls of the pleural cavity and is called the *parietal pleura.* The inner layer, the *visceral pleura,* covers the lungs themselves. Between the visceral and parietal pleura is a small potential space, the *pleural cavity,* which contains a lubricating fluid secreted by the membranes. This fluid prevents friction between the membranes and allows them to move easily on one another during breathing (see Figure 1–3). Inflammation of the pleural membrane, or *pleurisy,* causes friction during breathing that can be quite painful when the swollen membranes rub against each other.

The lungs extend from the diaphragm to a point just above the clavicles and lie against the ribs in front and back. The broad inferior portion of the lung, the *base,* is concave and fits over the convex area of the diaphragm. The narrow superior portion of the lung is referred to as the *apex.* The surface of the lung lying against the ribs, the *costal surface,* is rounded to match the curvature of the ribs. The *mediastinal (medial) surface* of each lung contains a vertical slit, the *hilum,* through which bronchi, pulmonary vessels, and nerves enter and exit. The blood vessels, bronchi, and nerves are held together by the pleura and connective tissue, and they constitute the *root* of the lung. Medially, the left lung also contains a concavity, the *cardiac notch,* in which the heart lies.

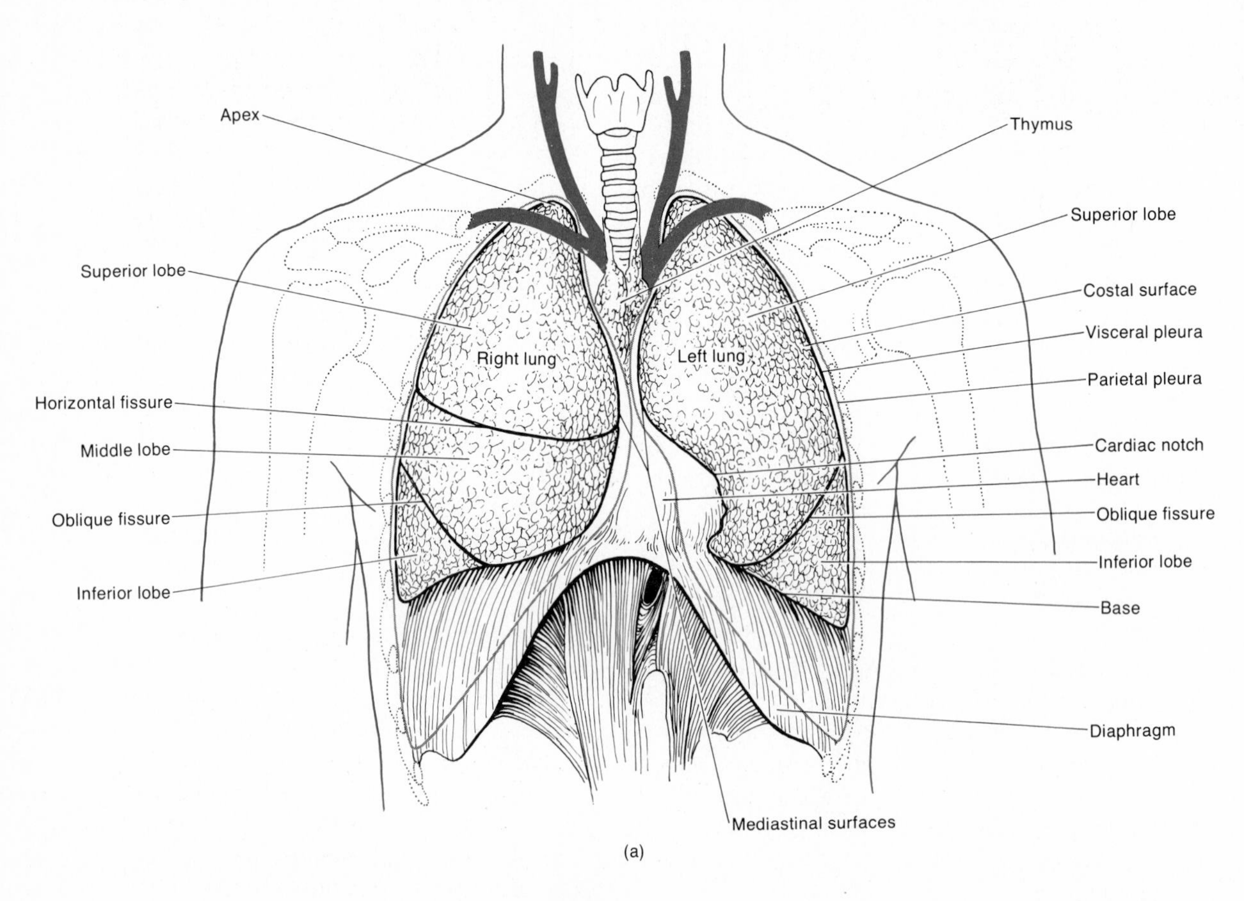

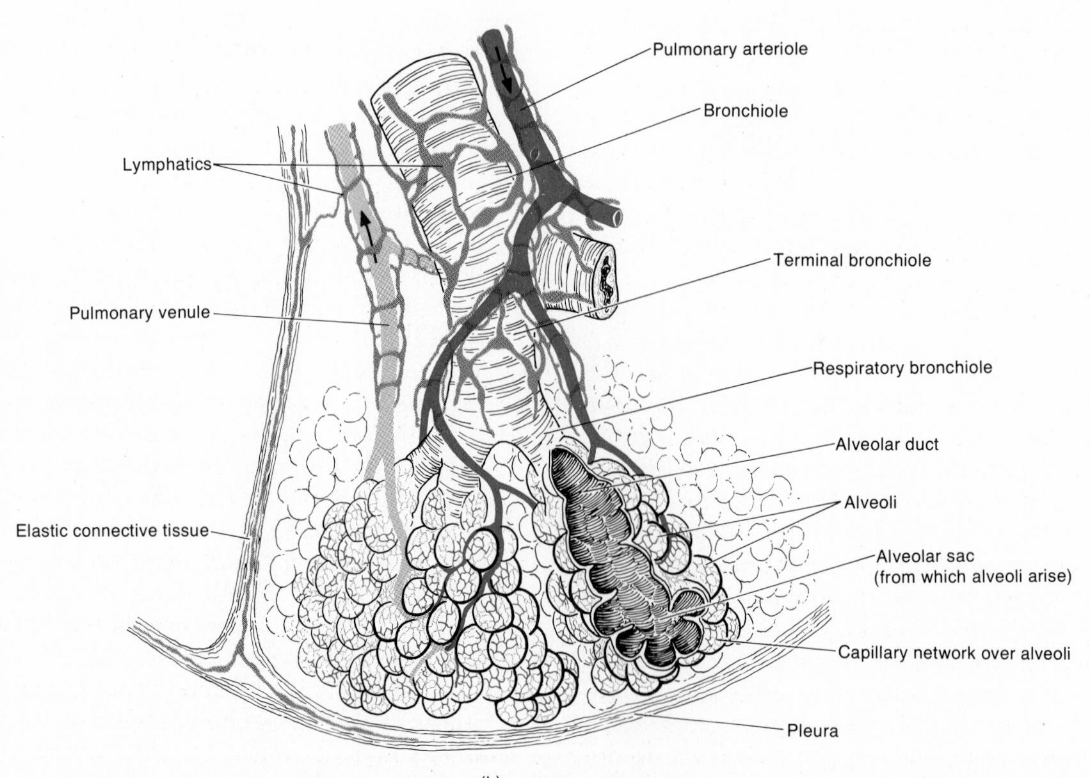

Figure 19–7. The lungs. (a) Coverings and external anatomy. (b) A lobule of the lung.

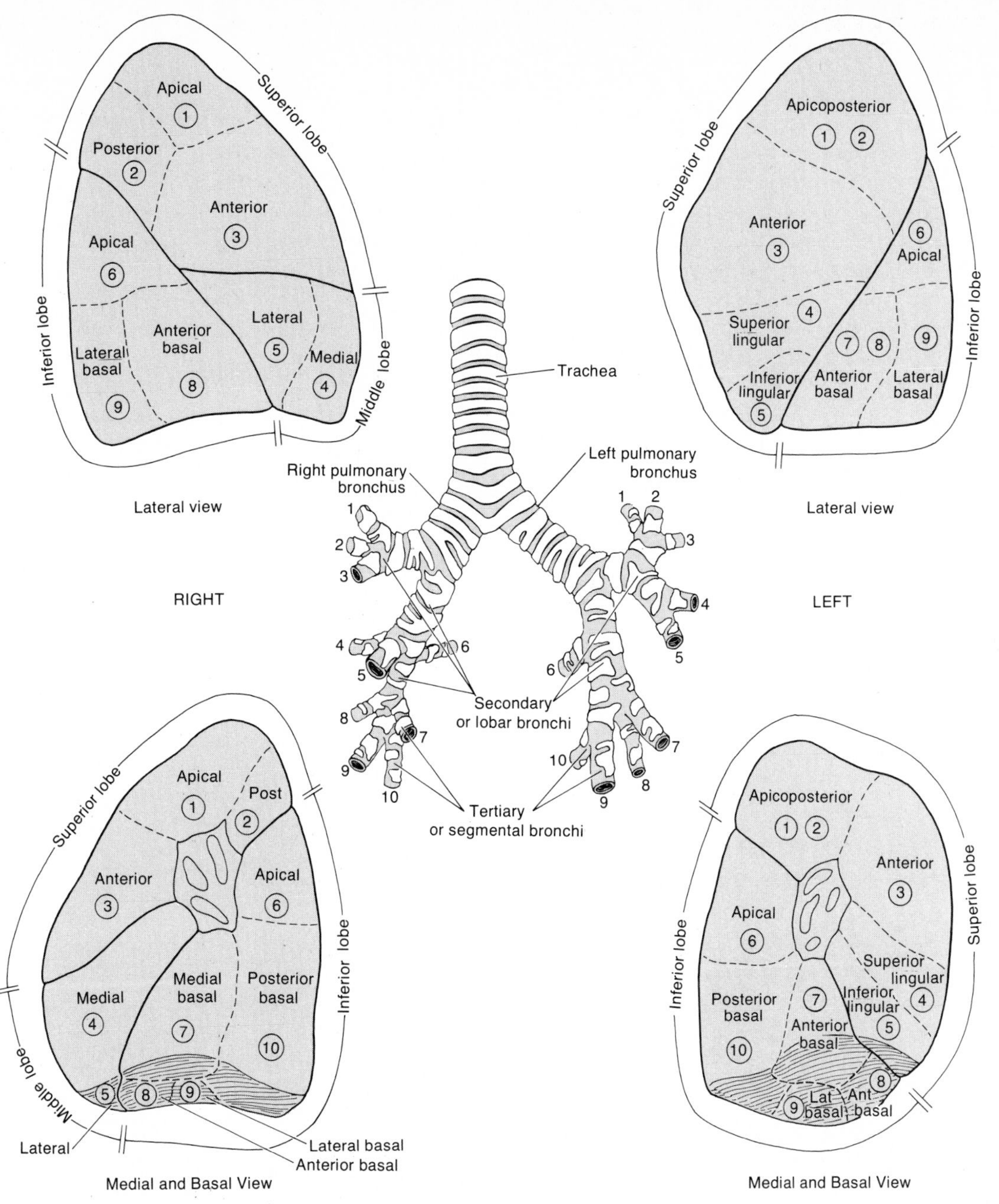

Figure 19–8. Bronchopulmonary segments of the lungs. The bronchial branches are shown in the center of the figure.

The right lung is thicker and broader than the left. It is also somewhat shorter than the left because the diaphragm is higher on the right side to accommodate the liver that lies below it. The left lung is thinner, narrower, and longer than the right.

Each lung is divided into lobes by one or more fissures. Both lungs have an *oblique fissure,* which extends downward and forward. The right lung also has a *horizontal fissure.* The oblique fissure in the left lung separates the superior from the inferior lobe. The upper part of the oblique fissure of the right lung separates the superior lobe from the inferior lobe, whereas the lower part of the oblique fissure separates the inferior lobe from the middle lobe. The horizontal fissure of the right lung separates the superior lobe from the middle lobe.

Each lobe receives its own secondary or lobar bronchus. Thus, the right primary bronchus gives rise to three lobar bronchi called the *superior, middle,* and *inferior lobar* or *secondary bronchi.* The left primary bronchus gives rise to a *superior* and an *inferior lobar* or *secondary bronchi.* Within the substance of the lung, the lobar bronchi give rise to branches that are relatively constant in both origin and distribution. Such branches are called *tertiary* or *segmental bronchi,* and the segment of lung tissue that each supplies is a *bronchopulmonary segment.* (Figure 19–8). The ability to recognize a segmental bronchus in a roentgenogram is important in diagnosing diseases of the lung. For example, by identifying a segmental bronchus that leads to a diseased segment of a lung, a surgeon may be able to remove only the diseased segment instead of removing the entire lobe or the entire lung.

Each bronchopulmonary segment of the lungs is broken up into many small compartments called *lobules* (Figure 19–7b). Every lobule is wrapped in elastic connective tissue and contains a lymphatic vessel, an arteriole, a venule, and a branch from a terminal bronchiole. Terminal bronchioles subdivide into microscopic branches called *respiratory bronchioles.* In the respiratory bronchioles, the epithelial lining changes from cuboidal to squamous as they become more distal. In addition, respiratory bronchioles contain some alveoli (described shortly) which appear as cup-shaped outpouchings of their walls. Respiratory bronchioles, in turn, subdivide into several (2 to 11) *alveolar ducts.* Around the circumference of the alveolar ducts are numerous alveoli and alveolar sacs. An **alveolus** is a cup-shaped outpouching lined by squamous epithelium and supported by a thin elastic membrane. An *alveolar sac* or air sac is a collection or cluster of alveoli that share a common opening (Figure 19–9). Over the alveoli, the arteriole and venule disperse into a network of capillaries. The exchange of gases between the lungs and blood takes place by diffusion across the alveoli and the walls of the capillaries. It has been estimated that each lung contains 150 million alveoli – a situation that provides an immense surface area for the exchange of gases.

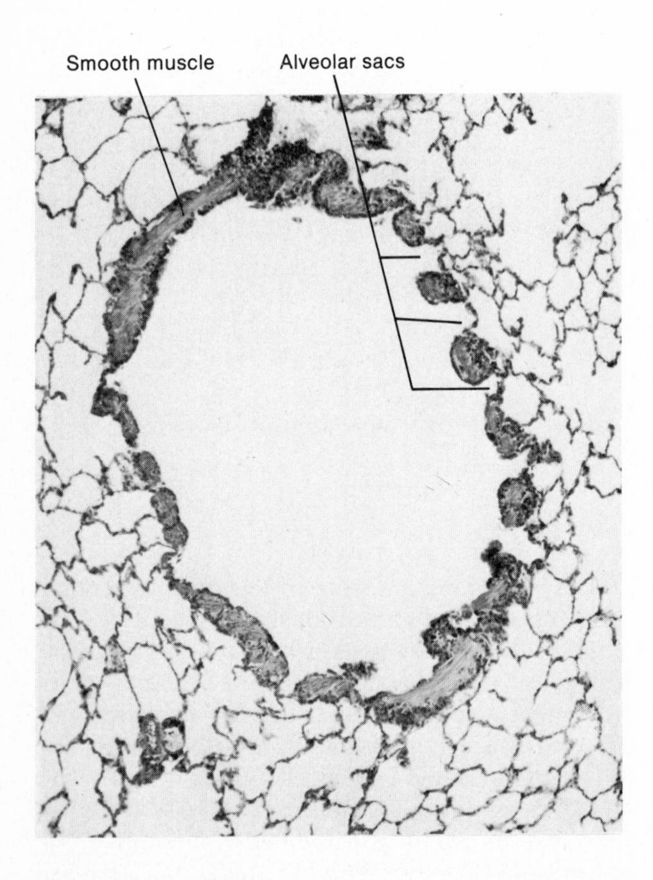

Figure 19–9. Histology of a respiratory bronchiole and associated structures at a magnification of 65X. (Courtesy of Edward J. Reith, from *Atlas of Descriptive Histology,* by Edward J. Reith and Michael H. Ross, Harper & Row, Publishers, Inc., New York, 1970.)

The arterial supply of the lungs is derived from the pulmonary trunk. It divides into a left pulmonary artery which enters the left lung and a right pulmonary artery which enters the right lung. The venous return of the oxygenated blood is by way of the pulmonary veins, typically two in number on each side – the right and left superior and inferior pulmonary veins. All four veins drain into the left atrium.

The nerve supply of the lungs is derived from the autonomic nervous system. Parasympathetic fibers come from the vagus nerve and sympathetic fibers are derived from the second, third, and fourth thoracic ganglia.

RESPIRATION

The principal purpose of **respiration** is to supply the cells of the body with oxygen and to remove the carbon dioxide produced by cellular activities. The path of oxygen from the atmosphere to the cells and that of carbon dioxide from the cells to the atmosphere involves three basic processes. The first process is *ventilation,* or breathing, which is the movement of air between the atmosphere and the lungs. The second and third processes involve the exchange of gases within the body. These processes are *external respiration,* which is the exchange of gases between the lungs and blood, and *internal respiration,* which is the exchange of gases between the blood and the cells of the body.

Ventilation

Ventilation or breathing is the process by which atmospheric gases are drawn down into the lungs and waste gases that have diffused into the lungs are expelled back up through the respiratory passageways. Air flows between the atmosphere and lungs for the same reason that blood flows through the body; that is, a pressure gradient exists. We breathe in when the pressure inside the lungs is less than the air pressure in the atmosphere, and we breathe out when the pressure inside the lungs is greater than the pressure in the atmosphere.

Inspiration

Breathing in is called **inspiration** or inhalation. Just before each inspiration the air pressure inside the lungs equals the pressure of the atmosphere, which is about 760 mm Hg at standard conditions. For air to flow into the lungs, the pressure inside the lungs must become lower than the pressure in the atmosphere. This is achieved by increasing the volume of the lungs. As you have observed, perhaps unknowingly, the pressure of a gas is inversely proportional to the volume of its container. If the size of a closed container is increased, the pressure of the air inside the container decreases. If the size of the container is decreased, then the pressure inside the container increases.

The first step toward increasing lung volume involves the contraction of the respiratory muscles, that is, the diaphragm and external intercostal muscles (Figure 19–10a). The diaphragm is the sheet of skeletal muscle that forms the floor of the thoracic cavity. As it contracts it moves downward, thereby increasing the depth of the thoracic cavity. At the same time, the external intercostal muscles contract, pulling the ribs upward and turning them slightly so the sternum is pushed forward. In this way the circumference of the thoracic cavity also is increased.

The overall increase in the size of the thoracic cavity causes its pressure, called *intrathoracic* or *intrapleural pressure,* to fall way below the pressure of the air inside the lungs (Figure 19–10b). Consequently, the walls of the lungs are sucked outward by the partial vacuum. Expansion of the lungs is aided by the pleural membranes. The parietal pleura lining the chest cavity tends to stick to the visceral pleura around the lungs and to pull the visceral pleura with it.

When the volume of the lungs increases, the pressure inside the lungs, called the *intrapulmonic* or *intraalveolar pressure,* drops from 760 to 758 mm Hg. A pressure gradient is thus established between the atmosphere and the alveoli. Air rushes from the atmosphere into the lungs, and an inspiration takes place. Inspiration is frequently referred to as an active process because it is initiated by muscle contraction.

Expiration

Breathing out, called **expiration** or exhalation, is also achieved by a pressure gradient. But this time the gradient is reversed so that the pressure in the lungs is greater than the pressure of the atmosphere. Expiration starts when the respiratory muscles relax and the size of the chest cavity decreases in depth and circumference (Figure 19–10a). As the intrathoracic pressure returns to its preinspiration level, the walls of the lungs are no longer sucked out. The highly elastic basement membranes of the alveoli snap back into their relaxed shape, and lung volume decreases (Figure 19–10b). Intrapulmonic pressure increases, and air moves from the area of higher pressure (the alveolar sacs) to the area of lower pressure (the atmosphere). Expiration is a passive process since no muscular contraction is required. However, the internal intercostals do aid in expiration.

If you look again at Figure 19–10b, you will notice that the intrathoracic pressure is always a little less than the pressure inside the lungs or in the atmosphere. The pleural cavities are sealed off from the outside environment and cannot equalize their pressure with that of the atmosphere. Nor can the diaphragm and rib cage move inward enough to bring the intrathoracic pressure up to atmospheric pressure. Actually, maintenance of a lower intrathoracic pressure is vital to the functioning of the lungs. The alveoli are so elastic that at the end of an expiration they attempt to snap inward and collapse on themselves like the walls of a deflated balloon. Such a collapse, which would obstruct the movement of air, is called *atelectasis.* It is prevented by the slightly lower pressure in the pleural cavities that keeps the alveoli slightly inflated.

Another factor related to preventing the collapse of alveoli is the presence of a phospholipid produced by the alveolar cells. This substance, called *surfactant,* decreases surface tension in the lungs. That is, it forms a thin lining of the alveoli and prevents them from sticking together following expiration. Thus, as alveoli become smaller, for example following expiration, the tendency of alveoli to collapse is minimized because the surface tension does not increase. You will see later that lack of surfactant results in a disorder called hyaline membrane disease.

Air volumes exchanged in respiration

In clinical practice the word respiration is used to mean one inspiration plus one expiration. The average healthy adult has 14 to 18 respirations a minute (that is, the individual inspires 14 to 18 times and expires 14 to 18 times). During each respiration the lungs exchange given volumes of air with the atmosphere. A lower than normal exchange volume is usually a sign of pulmonary malfunction. The apparatus commonly used to measure the amount of air exchanged during breathing is referred to as a *spirometer.* During normal quiet breathing, about 500 ml of air moves into the respiratory passageways with each inspiration, and the same amount moves out with each expiration. This

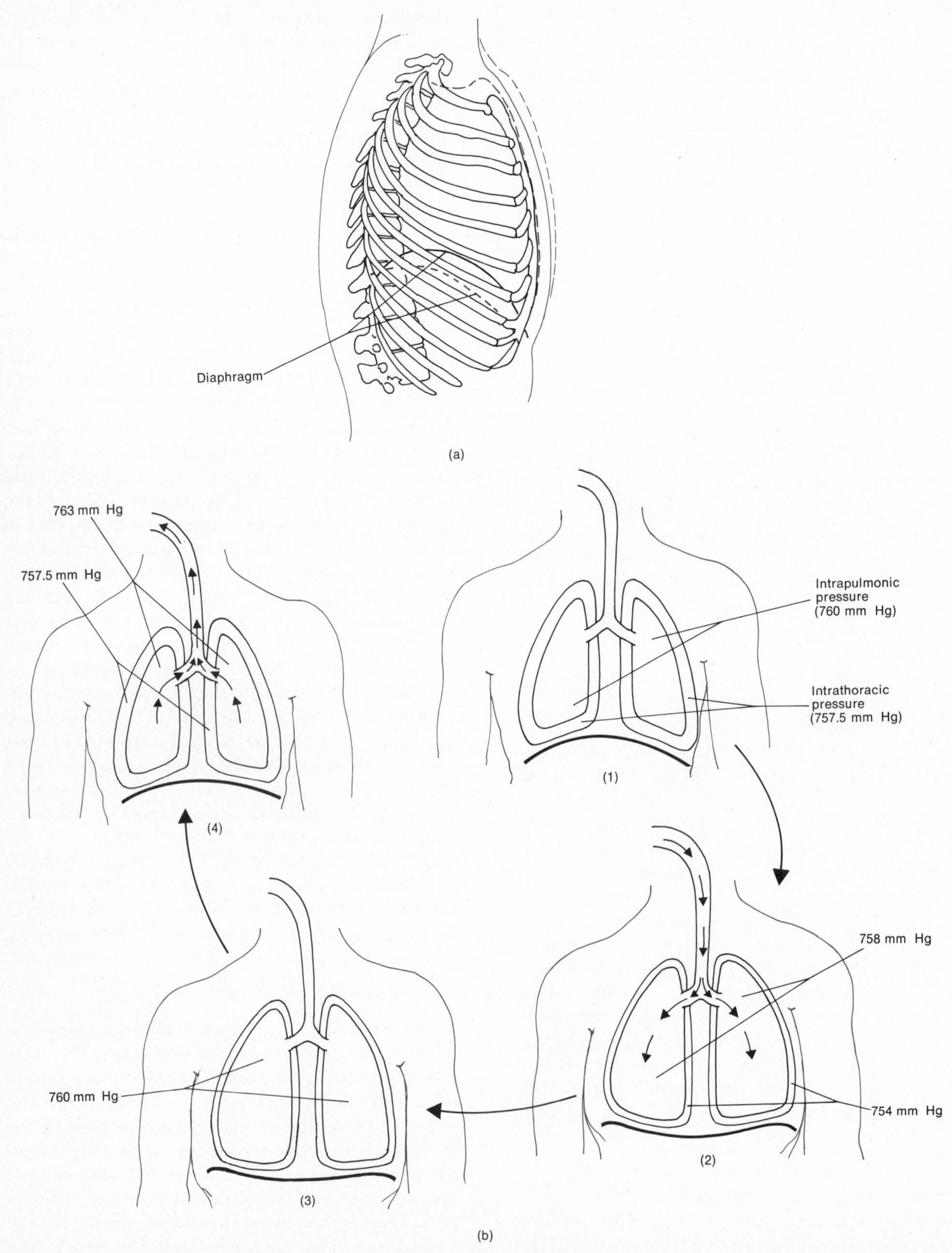

Figure 19–10. Breathing. (a) Changes in size of rib cage. Black indicates relaxed cage. Color indicates shape of rib cage during inspiration. (b) Pressure changes. (1) Lungs and pleural cavity just before inspiration. (2) Chest expanded and intrathoracic pressure decreased; lungs pulled outward and intrapulmonic pressure decreased. Air moves into lungs until intrapulmonic pressure equals atmospheric (3). (4) Chest relaxes, intrathoracic pressure rises, and lungs snap inward. Intrapulmonic pressure raised, forcing air out until intrapulmonic pressure equals atmospheric (1).

volume of air inspired (or expired) is called *tidal volume* (Figure 19–11). Actually, only about 350 ml of the tidal volume reaches the alveoli. The other 150 ml remains in the dead spaces of the nose, pharynx, larynx, trachea, and bronchi and is known as *dead air.*

By taking a very deep breath, we can inspire a good deal more than 500 ml. This excess inhaled air, called the *inspiratory reserve volume,* averages 3,000 ml above the 500 ml of tidal volume. Thus, the respiratory system can pull in as much as 3,500 ml of air. If we inhale normally and then exhale as forcibly as possible, we should be able to push out 1,100 ml of air in addition to the 500 ml tidal volume. This extra 1,100 ml is called the *expiratory reserve volume.* Even after the expiratory reserve volume is expelled, a good deal of air still remains in the lungs because the lower intrathoracic pressure keeps the alveoli slightly inflated. This air, the *residual volume,* amounts to about 1,200 ml. Opening the thoracic cavity allows the intrathoracic pressure to equal the atmospheric pressure, forcing out the residual volume. The air still remaining is called the *minimal volume.*

The presence of minimal volume can be demonstrated by placing a piece of lung in water and watching it float. Minimal volume provides a medical and legal tool for determining whether a baby was born dead or died after birth. Fetal lungs contain no air. If a baby is born dead, no minimal volume will be observed, but if the child died after he took his first breath, a minimal volume will be detected.

Lung capacity can be calculated by combining various lung volumes. *Inspiratory capacity,* the total inspiratory ability of the lungs, is the sum of tidal volume plus inspiratory reserve volume. *Functional residual capacity* is the sum of residual volume plus expiratory reserve volume. *Vital capacity* is the sum of inspiratory reserve volume, tidal volume, and expiratory reserve volume. Finally, *total lung capacity* is the sum of all volumes.

Gas exchange

As soon as the lungs fill with air, oxygen moves from the alveoli to the blood, through the interstitial fluid, and finally to the cells. Carbon dioxide moves in just the opposite direction—from the cells, through interstitial fluid to the blood, and to the alveoli.

Nervous control

As noted earlier, the size of the thorax is affected by the action of the respiratory muscles. These muscles contract and relax in turn as a result of nerve impulses transmitted to them from centers in the brain. The area from which nerve impulses are sent to respiratory muscles is located in the medulla oblongata of the brain and is referred to as the *respiratory center.* This center is functionally divided into two regions: the *inspiratory center,* which causes the inspiratory muscles to contract and thus brings on inspiration, and the *expiratory center,* which inhibits the inspiratory center and thereby allows the inspiratory muscles to relax. The respiratory center, along with its "pacemaker" centers in the pons, regulates the rhythm of respiration in the following manner.

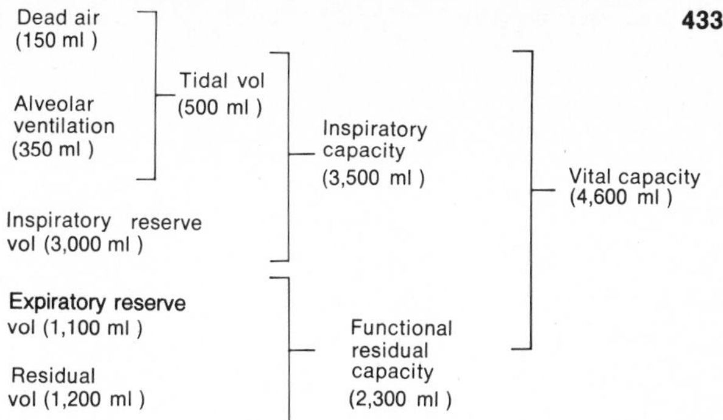

Figure 19–11. Air volumes exchanged in respiration.

An area of the pons, called the *apneustic center,* continually sends stimulatory impulses to the inspiratory center. The stimulated inspiratory center, in turn, sends impulses over motor neurons that stimulate the respiratory muscles to contract, and inspiration occurs. As soon as the lungs are filled, the inspiratory impulses are shut off by two controls.

The first set of controls involves the stretch receptors in the lung tissue. When the lungs expand to a critical point, the stretch receptors are stimulated and impulses are sent along the vagus nerves to the expiratory center. The expiratory center then sends out inhibitory impulses to the inspiratory center. This causes relaxation of the inspiratory muscles, and expiration follows. As air leaves the lungs during expiration, the lungs are deflated and the stretch receptors are no longer stimulated. Thus the inspiratory center is no longer inhibited, and a new respiration begins. These events are called the *Hering-Breuer reflex.* The Hering-Breuer reflex controls the depth and rhythm of respiration. It is also a protective reflex because it prevents the lungs from inflating to the point of bursting.

Should the Hering-Breuer reflex fail to operate, breathing will still continue. If the vagus nerves are severed, inspirations are longer and deeper than normal, but expiration is eventually initiated by a back-up control. This second control is provided by the *pneumotaxic center* of the pons. When the inspiratory center is stimulated, it sends impulses to the pneumotaxic center as well as to the respiratory muscles. After a delay, the pneumotaxic center sends impulses to the expiratory center, which inhibits the inspiratory center.

The term applied to normal quiet breathing is *eupnea.* Eupnea is a function of the Hering-Breuer reflex, and it involves shallow, deep, or combined shallow and deep breathing. Shallow, or chest, breathing is called *costal breathing.* It consists of an upward and outward movement of the chest as a result of contraction of the intercostal muscles. Deep, or abdominal, breathing is called *diaphragmatic breathing.* It consists of the outward movement of the abdomen as a result of the contraction and descent of the diaphragm.

The respiratory center has connections with the cerebral cortex, which means that we can voluntarily alter our pattern of breathing. Or we can refuse to breathe at all for a short time. Voluntary control is protective because it enables us to prevent water or irritating gases from entering the lungs. However, the ability to willfully stop breathing is limited by the buildup of carbon dioxide in the blood. When the amount of carbon dioxide increases to a certain critical level, the inspiratory center is stimulated, impulses are sent to inspiratory muscles, and breathing resumes whether or not the person wishes. Since the breath can be held for only a short period of time before the involuntary control centers take over, it is impossible for anyone to kill himself by holding his breath.

Some of the other factors that control respiration are:

1. A sudden cold stimulus such as plunging into cold water causes a temporary cessation of breathing, called *apnea.*
2. A sudden, severe pain brings about apnea, but a prolonged pain triggers off the general stress syndrome and increases respiration rate.
3. Stretching of the anal sphincter muscle increases the respiratory rate. Medical personnel sometimes employ this technique to stimulate respiration during emergencies.
4. Irritation of the pharynx or larynx by touch or by chemicals brings about an immediate cessation of breathing followed by coughing.

Exhibit 19–1. MODIFIED RESPIRATORY MOVEMENTS

MOVEMENT	COMMENT
Coughing	Preceded by a long-drawn and deep inspiration that is followed by a complete closure of the glottis – resulting strong expiration suddenly pushes glottis open and sends a blast of air through the upper respiratory passages. Stimulus for this reflex act could be a foreign body lodged in the larynx, trachea, or epiglottis.
Sneezing	Spasmodic contraction of muscles of expiration forcefully expels air through the nose and mouth. Stimulus may be an irritation of the nasal mucosa.
Sighing	A deep and long-drawn inspiration immediately followed by a shorter but forceful expiration.
Yawning	A deep inspiration through the widely opened mouth producing an exaggerated depression of the lower jaw. May be stimulated by drowsiness or fatigue, but precise stimulus-receptor cause is unknown.
Sobbing	Starts with a series of convulsive inspirations. Glottis closes earlier than normal after each inspiration so only a little air enters the lungs with each inspiration. Immediately followed by a single prolonged expiration.
Crying	An inspiration followed by many short convulsive expirations. Glottis remains open during the entire time, and the vocal cords vibrate. Accompanied by characteristic facial expressions.
Laughing	Involves the same basic movements as crying, but the rhythm of the movements and the facial expressions usually differ from those of crying. Laughing and crying are sometimes indistinguishable.
Hiccough	Spasmodic contraction of the diaphragm followed by a spasmodic closure of the glottis. Produces a sharp inspiratory sound. Stimulus is usually irritation of the sensory nerve endings of the digestive tract.

MODIFIED RESPIRATORY MOVEMENTS

Respirations provide human beings with methods for expressing emotions such as laughing, yawning, sighing, and sobbing. Moreover, respiratory air can be used to expel foreign matter from the upper air passages through actions such as sneezing and coughing. Some of the modified respiratory movements that express emotion or clear the air passageways are listed in Exhibit 19–1. All these movements are reflexes, but some of them also can be initiated voluntarily.

APPLICATIONS TO HEALTH

Hay fever

An allergic reaction to the proteins contained in foreign substances such as plant pollens, dust, and certain foods is called **hay fever.** Allergic reactions are a special type of antigen-antibody response that initiate either a localized or a systemic inflammatory response. In hay fever the response is localized in the respiratory membranes. The membranes become inflamed, and a watery fluid drains from the eyes and nose.

Bronchial asthma

Another disorder is **bronchial asthma.** This usually allergic reaction is characterized by attacks of wheezing and difficult breathing. Attacks are brought on by spasms of the smooth muscles that lie in the walls of the smaller bronchi and bronchioles, causing the passageways to close partially. The patient has trouble exhaling, and the alveoli may remain somewhat inflated during expiration. Usually the mucous membranes that line the respiratory passageways become irritated and secrete excessive amounts of mucus that may clog the bronchi and bronchioles and worsen the attack. About three out of four asthma victims are allergic to something they eat or to substances they breathe in, such as pollens, animal dander, house dust, or smog. Others are usually sensitive to the proteins of relatively harmless bacteria that inhabit the sinuses, nose, and throat. Asthma may also have a psychosomatic origin.

Emphysema

One lung disease that starts with the deterioration of some of the alveoli is **emphysema.** The alveolar walls lose their elasticity and remain filled with air during expiration. The name of the disease means "blown up" or "full of air." Reduced forced expiratory volume is the first symptom. Later, alveoli in other areas of the lungs are damaged. The lungs become permanently inflated. To adjust to the increased lung size, the size of the chest cage increases. The patient has to work to exhale. Oxygen diffusion does not occur as easily across the damaged alveoli, blood oxygen level is somewhat lowered, and any mild exercise that raises the oxygen requirements of the cells leaves the patient breathless. Carbon dioxide diffuses much more easily across the alveoli than does oxygen, so the level of carbon dioxide is not affected initially. But as the disease progresses, the alveoli degenerate and are replaced with thick fibrous connective tissue. Even carbon dioxide does not diffuse easily through this fibrous tissue. If the blood cannot buffer all the carbonic acid that accumulates, the blood acidity drops. Or, unusually high amounts of carbon dioxide may dissolve in the plasma. High carbon dioxide levels are toxic to the brain cells. Consequently, the inspiratory center becomes less active and the respiration rate slows down, further aggravating the problem. The capillaries that lie around the deteriorating alveoli are compressed and damaged and may no longer be able to receive blood. As a result, pressure increases in the pulmonary artery, and the right atrium overworks as it attempts to force blood through the remaining capillaries.

Emphysema is generally caused by any of a number of long-term irritations. Air pollution, occupational exposure to industrial dusts, and cigarette smoke are the most common irritants. Chronic bronchial asthma also may produce alveolar damage. Cases of emphysema are becoming more and more frequent in the United States. The irony is that the disease can be prevented and the progressive deterioration can be stopped by eliminating the harmful stimuli.

Pneumonia

The term **pneumonia** means an acute infection or inflammation of the alveoli. In this disease the alveolar sacs fill up with fluid and dead white blood cells, reducing the amount of air space in the lungs. (Remember that one of the cardinal signs of inflammation is edema.) Oxygen has difficulty diffusing through the inflamed alveoli, and the blood level of oxygen may be drastically reduced. Blood carbon dioxide usually remains normal because carbon dioxide always diffuses through the alveoli more easily than oxygen does. If all the alveoli of a lobe are inflamed, the pneumonia is called *lobar pneumonia.* If only parts of the lobe are involved, it is called *lobular,* or *segmental, pneumonia.* If both the alveoli and the bronchial tubes are included, it is called *bronchopneumonia.*

The most common cause of pneumonia is the pneumococcus bacterium, but other bacteria or a fungus may be a source of the trouble. Viral pneumonia is caused by any of several viruses, including the influenza virus.

Tuberculosis

The bacterium called *Mycobacterium tuberculosis* produces an inflammation called **tuberculosis.** Tuberculosis most often affects the lungs and the pleura. The bacteria destroy parts of the lung tissue, and the tissue is replaced by fibrous connective tissue. Because the connective tissue is inelastic and relatively thick, the affected areas of the lungs do not snap back during expiration, and larger amounts of air are retained. Gases no longer diffuse easily through the fibrous tissue.

Tuberculosis bacteria are spread by inhalation. Although they can withstand exposure to many disinfectants, they die quickly in sunlight. This is why tuberculosis is sometimes associated with crowded, poorly lit housing conditions. Many drugs are successful in treating tuberculosis. Rest, sunlight, and good diet are vital parts of treatment.

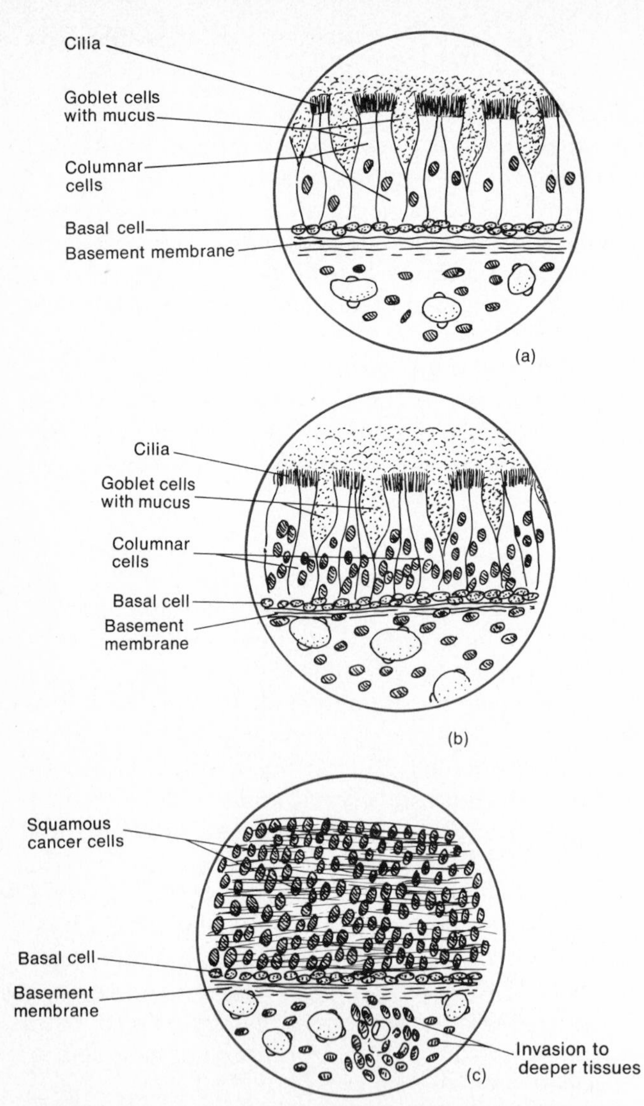

Figure 19–12. Effects of smoking on the respiratory epithelium. (a) Microscopic view of the normal epithelium of a bronchial tube. (b) Initial response of the bronchial epithelium to irritation by pollutants. (c) Advanced response of the bronchial epithelium.

Hyaline membrane disease (HMD)

Sometimes called glassy-lung disease, **hyaline membrane disease** is responsible for approximately 20,000 newborn infant deaths per year. It is more prevalent among premature infants. The disease is due to lack of surfactant, which is synthesized later in fetal life. As noted earlier, surfactant prevents the alveoli from sticking together following expiration, thus preventing collapse of the lungs. In the absence of surfactant, atelectasis occurs. Autopsies reveal a glassy appearance of the alveoli and alveolar ducts. Asphyxiation occurs for most infants within 72 hours of birth.

A new treatment currently being developed called PEEP–positive end expiratory pressure–could reverse the mortality rate from 90 percent deaths to 90 percent survival. This treatment consists of passing a tube through the air passage to the top of the lungs to provide needed oxygen-rich air at continuous pressures of up to 14 mm Hg. Continuous pressure keeps the baby's alveoli open and available for gas exchange.

SMOKING AND THE RESPIRATORY SYSTEM

As part of ordinary breathing, many irritating substances are inhaled. Almost all pollutants, including inhaled smoke, have an irritating effect on the bronchial tubes and lungs and may be regarded as stresses or irritating stimuli.

Close examination of the epithelium of a bronchial tube reveals that it consists of three kinds of cells (Figure 19–12a). The uppermost cells are columnar cells that contain the cilia on their surfaces. At intervals between the ciliated columnar cells are the mucus-secreting goblet cells. The bottom of the epithelium normally contains two rows of basal cells above the basement membrane. The bronchial epithelium is important clinically because researchers have learned that one of the most common types of lung cancer, **bronchogenic carcinoma,** starts in the walls of the bronchi.

The stress of constant irritation by inhaled smoke and pollutants causes an enlargement of the goblet cells of the bronchial epithelium (Figure 19–12b). They respond by secreting excessive amounts of mucus. The basal cells respond to the stress by undergoing cell division so fast that the basal cells push into the area occupied by the goblet and columnar cells. As many as 20 rows of basal cells may be produced. Many researchers believe that if the stress is removed at this point, the epithelium can return to normal.

If the stress persists, more and more mucus is secreted and the cilia become less effective. As a result, mucus is not carried toward the throat; instead, it remains trapped in the bronchial tubes. The individual then develops a "smoker's cough." Moreover, the constant irritation from the pollutant slowly destroys the alveoli, which are replaced with thick, inelastic connective tissue. Mucus that has accumulated becomes trapped in the air sacs. Millions of the sacs rupture. This results in a loss of diffusion surface for the exchange of oxygen and carbon dioxide. The individual has now developed emphysema. If the stress is removed at this point, there is little chance for improvement. Any alveolar tissue that has been destroyed cannot be repaired. But removal of the stress can stop further destruction of lung tissue.

Assuming that the stress continues, the emphysema gets progressively worse, and the basal cells of the bronchial tubes continue to divide and break through the basement membrane. At this point the stage is set for bronchogenic carcinoma. Columnar and goblet cells disappear and may be replaced with squamous cancer cells (Figure 19–12c). If this happens, the malignant growth spreads throughout the lung and may block a bronchial tube. If the obstruction occurs in a large bronchial tube, very little oxygen enters the lung, and disease-producing bacteria thrive on the mucoid secretions. In the end, the patient may develop emphysema, carcinoma, and a host of infectious diseases. Treatment involves surgical removal of the diseased lung. However, metastasis of the growth through the lymphatic or blood system may result in new growths in other parts of the body such as the brain and liver.

The processes that we have just described have been observed in some heavy smokers. One should realize, though, that there are other causes of emphysema, such as chronic asthma. There are also other factors that may be associated with lung cancer. For instance, breast, stomach, and prostate malignancies can metastasize to the lungs. People who apparently have not been exposed to pollutants do occasionally develop bronchogenic carcinoma. However, the occurrence of bronchogenic carcinoma is probably over 20 times higher in heavy cigarette smokers than it is in nonsmokers.

Key terms associated with the respiratory system

Apnea Absence of respirations.

Asphyxia Oxygen starvation due to low atmospheric oxygen or interference with ventilation, external respiration, or internal respiration.

Atelectasis (*asis* = state or condition) A collapsed lung or portion of a lung.

Bronchitis (*bronch* = bronchus, trachea) Inflammation of the bronchi and bronchioles.

Cheyne-Stokes respiration Irregular breathing beginning with shallow breaths that increase in depth and rapidity, then decrease and cease altogether for 15–20 seconds. The cycle repeats itself again and again. Cheyne-Stokes is normal in infants. It is also often seen just before death from pulmonary, cerebral, cardiac, and kidney disease and is referred to as the "death rattle."

Diphtheria An acute bacterial infection that causes the mucous membranes of the pharynx, nasopharynx, and larynx to enlarge and become leathery. Enlarged membranes may obstruct airways and cause death from asphyxiation.

Dyspnea (*dys* = painful, difficult) Labored or difficult breathing (short-winded).

Eupnea (*eu* = good, normal) Normal quiet breathing.

Hypoxia Reduction in oxygen supply to cells.

Influenza Viral infection that causes inflammation of respiratory mucous membranes as well as fever.

Orthopnea Inability to breathe in a horizontal position.

Pneumothorax (*pneumo* = lung) Air in pleural space causing collapse of the lung. Most common cause is surgical opening of chest during heart surgery, making intrathoracic pressure equal atmospheric pressure.

Pulmonary edema Excess amounts of interstitial fluid in the lungs producing cough and dyspnea. Common in failure of the left side of the heart.

Pulmonary embolism Presence of a blood clot or other foreign substance in a pulmonary arterial vessel stopping circulation to a part of the lungs.

Rales Sounds sometimes heard in the lungs that resemble bubbling or rattling. May be caused by air or an abnormal secretion in the lungs.

Respirator Metal chamber that entombs chest; also called "iron lung." Used to produce inspiration and expiration in patient with paralyzed respiratory muscles. Pressure inside chamber is rhythmically alternated to suck out and push in chest walls.

Chapter summary in outline

RESPIRATORY ORGANS

1. Respiratory organs include the nose, pharynx, larynx, trachea, bronchi, and lungs.
2. They act with the circulatory system to supply oxygen and remove carbon dioxide.

Nose

1. The external portion is made of cartilage and skin and lined with mucous membrane; openings to exterior are external nares.
2. The internal portion communicates with pharynx through internal nares and communicates with paranasal sinuses.
3. The nose is divided into cavities by a septum. Anterior portions of the cavities are called the vestibules.
4. The nose is adapted for the warming, moistening, and filtering of air; olfaction; and it assists in speech.

Pharynx

1. The pharynx, or throat, is a muscular tube lined by mucous membrane.
2. Anatomic regions are nasopharynx, oropharynx, and laryngopharynx.
3. The nasopharynx functions in respiration. The oropharynx and laryngopharynx function in digestion and respiration.

Larynx

1. The larynx is a passageway that connects the pharynx with the trachea.
2. Prominent cartilages are the thyroid, or Adam's apple, the epiglottis, which prevents food from entering the larynx, the cricoid, which connects the larynx and trachea, the arytenoid, the corniculate, and the cuneiform cartilages.
3. The larynx contains true vocal cords that produce sound. Taut cords produce high pitches, and relaxed cords produce low pitches.

Trachea

1. The trachea extends from the larynx to the primary bronchi.
2. It is composed of smooth muscle and C-shaped rings of cartilage and is lined with ciliated mucous membrane.

Bronchi

1. The bronchial tree consists of primary bronchi, secondary bronchi, bronchioles, and terminal bronchioles. Walls of bronchi contain rings of cartilage; walls of bronchioles do not.
2. A developed picture of the tree is called a bronchogram.

Lungs

1. Lungs are paired organs in the thoracic cavity. They are enclosed by the pleural membrane (parietal pleura is outer layer; visceral pleura is inner layer).
2. The right lung has three lobes separated by two fissures; the left lung has two lobes separated by one fissure and a depression, the cardiac notch.
3. The secondary bronchi give rise to branches called segmental bronchi which supply segments of lung tissue called bronchopulmonary segments.
4. Each bronchopulmonary segment consists of lobules, which contain lymphatics, arterioles, venules, terminal bronchioles, respiratory bronchioles, alveolar ducts, alveolar sacs, and alveoli.
5. Gas exchange occurs across alveolar-capillary membranes.

RESPIRATION

Ventilation

1. Inspiration occurs when intrapulmonic pressure falls below atmospheric pressure. Contraction of the diaphragm and intercostals increases the size of the thorax and decreases the intrathoracic pressure. Decreased intrathoracic pressure causes a decreased intrapulmonic pressure.
2. Expiration occurs when intrapulmonic pressure is higher than atmospheric pressure. Relaxation of diaphragm and intercostals increases intrathoracic pressure, which causes an increased intrapulmonic pressure.
3. Among the air volumes exchanged in ventilation are tidal, inspiratory reserve, expiratory reserve, residual, and minimal volumes.
4. Nervous control is regulated by the respiratory centers in medulla and pons, which control the rhythm of respiration. The Hering-Breuer reflex controls the depth and rhythm of respiration.
5. Coughing, sneezing, sighing, yawning, sobbing, crying, laughing, and hiccoughing involve modified respiratory movements.

APPLICATIONS TO HEALTH

1. Hay fever is an allergic reaction of respiratory membranes.
2. Bronchial asthma occurs when spasms of smooth muscle in bronchial tubes result in partial closure of air passageways, inflammation, inflated alveoli, and excess mucus production.
3. Emphysema is characterized by deterioration of alveoli leading to loss of their elasticity. Symptoms are reduced expiratory volume, inflated lungs, and enlarged chest.
4. Pneumonia is an acute inflammation or infection of alveoli.
5. Tuberculosis is an inflammation of pleura and lungs produced by a specific bacterium.
6. Hyaline membrane disease is an infant disorder in which surfactant is lacking and alveolar ducts and alveoli have a glassy appearance.
7. Smoking
   (a) Pollutants, including smoke, act as stresses on the epithelium of the bronchi and lungs. Constant irritation results in excessive secretion of mucus and rapid division of bronchial basal cells.
   (b) Additional irritation may cause retention of mucus in bronchioles, loss of elasticity of alveoli, and less surface area for gaseous exchange.
   (c) In the final stages, bronchial epithelial cells may be replaced by cancer cells. The growth may block a bronchial tube and spread throughout the lung and other body tissues.

Review questions and problems

1. What organs constitute the respiratory system? What function do the respiratory and circulatory systems have in common?
2. Describe the structures of the external and internal nose and describe their functions in filtering, warming, and moistening air.
3. What is the pharynx? Differentiate the three regions of the pharynx, and indicate their roles in respiration.
4. Where is the larynx located? Describe the positions and functions of the laryngeal cartilages. How does the larynx function in voice production?
5. Describe the location and structure of the trachea. What is a tracheotomy? Intubation?
6. What is the bronchial tree? Describe its structure. What is a bronchogram?
7. Where are the lungs located? Distinguish the parietal pleura from the visceral pleura. What is pleurisy?
8. Define each of the following parts of a lung: base, apex, costal surface, medial surface, hilum, root, cardiac notch, and bronchopulmonary segment.
9. What is a lobule of the lung? Describe its composition and function in respiration.
10. What are the basic differences among ventilation, external respiration, and internal respiration?
11. Discuss the basic steps involved in inspiration and expiration. Be sure to include values for all pressures involved.
12. What is a spirometer? Define the various lung volumes and capacities.
13. Discuss how the inspiratory and expiratory centers are related to the Hering-Breuer reflex. What is the role of the apneustic center and the pneumotaxic center in controlling respiration?
14. Define the various kinds of modified respiratory movements.
15. For each of the following, list the outstanding clinical symptoms: hay fever, bronchial asthma, emphysema, pneumonia, tuberculosis, hyaline membrane disease, and bronchogenic carcinoma.
16. Discuss the stages in the destruction of respiratory epithelium as a result of continued irritation by pollutants.
17. Refer to the glossary of key terms associated with respiratory system. Be sure that you can define each term.

Selected readings

Cunningham's Textbook of Anatomy. 11th ed., edited by G. J. Romanes. London: Oxford University Press, 1972. Pp. 469–503.

Dawson, Helen L. *Basic Human Anatomy.* 2d ed. New York: Appleton-Century-Crofts, 1974. Chap. 10.

Gray, Henry. *Anatomy of the Human Body.* 29th ed., edited by Charles Mayo Goss. Philadelphia: Lea and Febiger, 1973. Chap. 15.

Pace, W. R. *Pulmonary Physiology in Clinical Practice.* 2d ed. Philadelphia: F. A. Davis, 1970.

CHAPTER 20

THE DIGESTIVE SYSTEM

STUDENT OBJECTIVES

After you have read this chapter, you should be able to:

1. Define digestion as a chemical and mechanical process
2. Identify the organs of the gastrointestinal tract and the accessory organs of digestion
3. Describe the structure of the wall of the gastrointestinal tract
4. Define the mesentery, mesocolon, falciform ligament, lesser omentum, and greater omentum
5. Describe the structure of the mouth and its role in mechanical digestion
6. Identify the location and histology of the salivary glands and define the composition and function of saliva
7. Identify the parts of a typical tooth and compare deciduous and permanent dentitions
8. Discuss the sequence of events involved in swallowing
9. Describe the location, anatomy, and histology of the stomach and compare mechanical and chemical digestion
10. Describe the location, structure, and histology of the pancreas
11. Define the position, structure, and histology of the liver
12. Describe the structure and histology of the gallbladder
13. Describe those structural features of the small intestine that adapt it for digestion and absorption
14. Describe those digestive activities of the small intestine by which carbohydrates, proteins, and fats are reduced to their final products
15. Describe the mechanical movements of the small intestine and define absorption
16. Describe those structural features of the large intestine that adapt it for absorption and feces formation and elimination and describe the mechanical movements of the large intestine
17. Describe the processes involved in feces formation and discuss the mechanisms involved in defecation
18. List the causes and symptoms of dental caries and periodontal disease
19. Contrast between the location and effects of gastric and duodenal ulcers
20. Describe the causes and dangers of peritonitis
21. Compare cirrhosis and tumors as disorders of the accessory organs of digestion
22. Define key terms associated with the digestive system

We all know that food is vital to life. Food is required for the chemical reactions that occur in every cell–both those that synthesize new enzymes, cell structures, bone, and all the other components of the body and those that release the energy needed for the building processes. However, the vast majority of foods we eat are simply too large to pass through the plasma membranes of the cells. Therefore, chemical and mechanical **digestion** must occur first.

Chemical digestion is a series of catabolic reactions that break down the large carbohydrate, lipid, and protein molecules which we eat into monosaccharides, some glycerol and fatty acids, and amino acids, respectively. These products of digestion are small enough to pass through the walls of the digestive organs, into the blood and lymph capillaries, and eventually into the cells of the body. *Mechanical digestion* consists of various movements that aid chemical digestion. Food must be pulverized by the teeth before it is small and flexible enough to be swallowed. After it has been swallowed, the smooth muscles of the stomach and small intestine churn the food so it is thoroughly mixed with the enzymes that catalyze the reactions.

The function of the digestive system, then, is to prepare food for consumption by the cells. It does this through five basic activities:

1. Ingestion, or eating, which is taking food into the body
2. Movement of food along the digestive tract
3. Mechanical and chemical digestion
4. Absorption, the passage of digested food from the digestive tract into the circulatory and lymphatic systems for distribution to cells
5. Defecation, the elimination of indigestible substances from the body

We shall discuss each of these activities as we describe the organs that are involved in them. But first let us orient ourselves by looking at the general plan of the digestive system.

GENERAL ORGANIZATION

The organs of digestion are traditionally divided into two main groups. First is the **gastrointestinal tract** or **alimentary canal,** a continuous tube running through the ventral body cavity and extending from the mouth to the anus (Figure 20–1). The relationship of the digestive organs to the nine regions of the abdominopelvic cavity may be reviewed in Figure 1–4. The length of a tract taken from a cadaver is about 9 m (30 ft). In a living person it is somewhat shorter because the muscles lying in its walls are in a state of tone. Organs composing the gastrointestinal or GI tract include the mouth, pharynx, esophagus, stomach, small intestine, and large intestine. The GI tract contains the food while it is being eaten, digested, and prepared for elimination. Muscular contractions in the walls of the GI tract break down the food physically by churning it. Secretions produced by cells along the GI tract break down the food chemically.

The second group of organs composing the digestive system are the **accessory organs**–the teeth, tongue, salivary glands, liver, gallbladder, pancreas, and appendix. Teeth are cemented to bone, protrude into the GI tract, and aid in the physical breakdown of food. The other accessory organs lie totally outside the tract and produce or store secretions which aid in the chemical breakdown of the food. These secretions are secreted into the tract through ducts.

Since the organs composing the GI tract are structurally quite similar, let us first look at a generalized cross section of the tract. Later, as we follow the movement of food from the mouth to the anus, we shall describe the organs of the tract and their modifications, as well as the accessory organs, in more detail.

General histology of the gastrointestinal tract

The walls of the GI tract, especially from the esophagus to the anal canal, have the same basic arrangement of tissues. The four coats or tunics of the tract from the inside out are the mucosa, submucosa, muscularis, and serosa or adventitia (Figure 20–2).

The **tunica mucosa**, or inner lining of the tract, is a mucous membrane attached to a very thin layer of visceral muscle. Two layers compose the membrane: a lining epithelium, which is in direct contact with the food, and an underlying layer of connective tissue called the *lamina propria.* Under the lamina propria are two thin layers of visceral muscle called the *muscularis mucosa.*

The epithelial layer is composed of nonkeratinized cells that are stratified in the mouth and esophagus but are simple throughout the rest of the tract. The functions of the stratified epithelium are protection and secretion, and the functions of the simple epithelium are secretion and absorption. However, the lack of keratin allows some absorption to occur in all parts of the tract.

The lamina propria is made of connective tissue proper containing many blood and lymph vessels and scattered lymph nodules. This layer supports

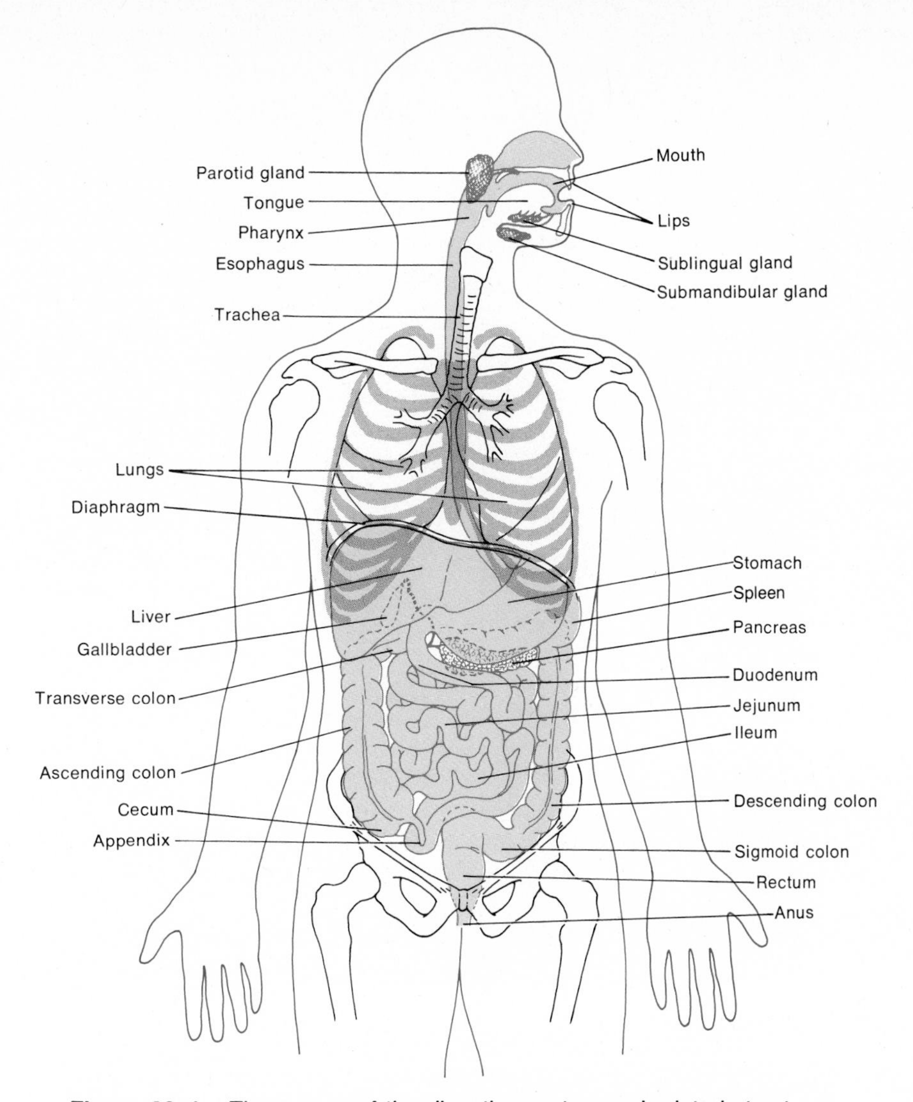

Figure 20–1. The organs of the digestive system and related structures.

the epithelium, binds it to the muscularis mucosa, and provides it with a blood and lymph supply. The blood and lymph vessels are the avenues by which nutrients in the tract reach the other tissues of the body. The lymph tissue also gives protection against disease. Remember that the GI tract is in contact with the outside environment and that it contains food which often carries harmful bacteria. Unlike the skin, the mucous membrane of the tract is not protected from bacterial entry by keratin.

The muscularis mucosa contains visceral muscle fibers that throw the mucous membrane of the intestine into small folds which increase the digestive and absorptive area. It also contains glandular epithelium that secretes products necessary for chemical digestion. With one exception, which will be described later, the other three coats of the intestine contain no glandular epithelium.

The **tunica submucosa** consists of dense connective tissue binding the mucosa to the muscularis. It is highly vascular and contains a portion of the *plexus of Meissner,* which is a part of the autonomic nerve supply to the muscularis mucosa.

The **tunica muscularis** of the mouth, pharynx, and esophagus consists in part of skeletal muscle that produces voluntary swallowing. Throughout the rest of the tract, the muscularis consists of smooth muscle that is generally found in two sheets: an inner ring of circular fibers and an outer sheet of longitudinal fibers. Contractions of the smooth muscles help to break down food physically, mix it with digestive secretions, and propel it through the tract. The muscularis also contains the major nerve supply to the alimentary tract. This is called the *plexus of Auerbach,* and it consists of fibers from both autonomic divisions.

The **tunica serosa,** the outermost layer of the canal, is a serous membrane composed of connective tissue and epithelium. This covering, also called the peritoneum, is worth discussing in more detail.

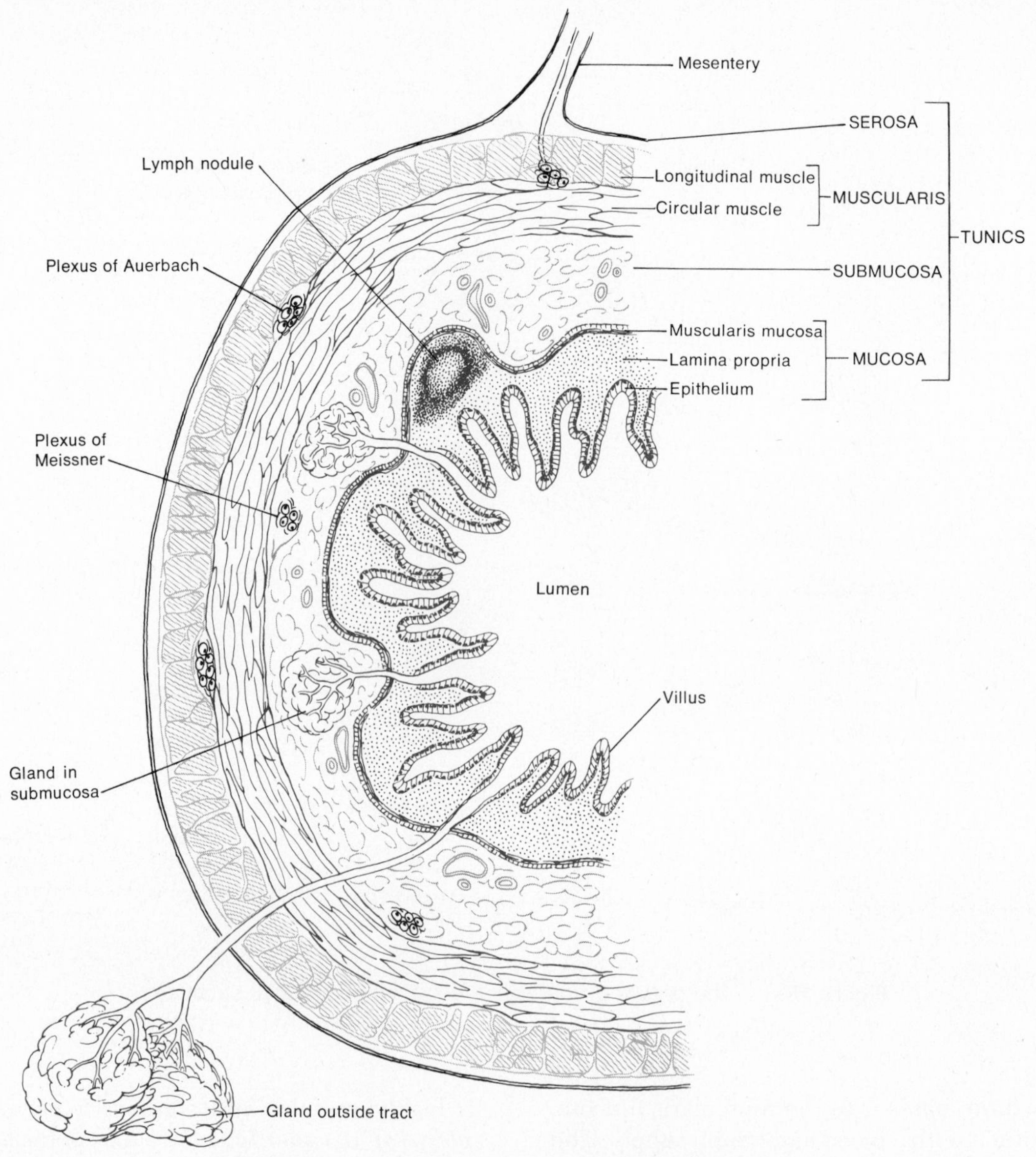

Figure 20–2. Cross section of the gastrointestinal tract.

The peritoneum

The **peritoneum** is the largest serous membrane of the body. It consists of a layer of simple squamous epithelium and an underlying supporting layer of connective tissue. The *visceral peritoneum* covers some of the organs and constitutes their serosa; the *parietal peritoneum* lines the walls of the abdominopelvic cavity. The space between the parietal and visceral portions of the peritoneum is called the *peritoneal cavity.*

Unlike the two other serous membranes of the body, the pericardium and pleura, the peritoneum contains large folds that weave in between the viscera. The folds bind the organs to each other and to the walls of the cavity and contain the blood and lymph vessels and the nerves that supply the abdominal organs. One extension of the peritoneum is called the **mesentery** and is an outward fold of the serous coat of the intestines (Figure 20–3c, d). Attached to the posterior abdominal wall is the tip of the fold. The mesentery binds the small intestine to the wall. A similar fold of parietal peritoneum, called the **mesocolon,** binds the large intestine to the posterior body wall (Figure 20–3c, d). It also carries blood vessels and lymphatics to the intestines.

Other important peritoneal folds are the falciform ligament, the lesser omentum, and the greater omentum. The **falciform ligament** attaches the liver to the anterior abdominal wall and diaphragm (Figure 20-3a, b, d). The **lesser omentum**

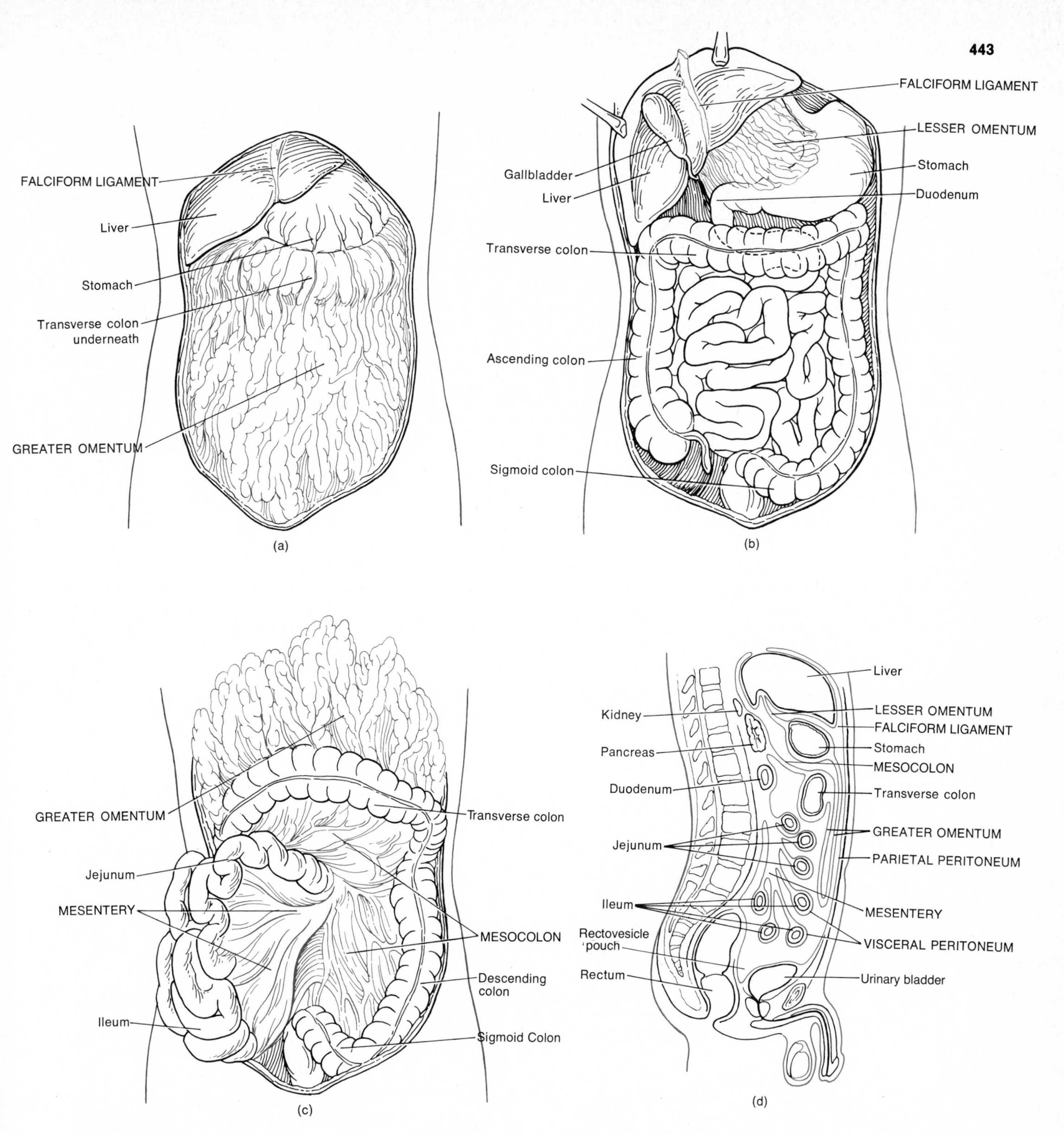

Figure 20–3. Extensions of the peritoneum. (a) Greater omentum. (b) Lesser omentum. The liver and gallbladder have been lifted. (c) Mesentery. The greater omentum has been lifted. (d) Sagittal section through the abdomen and pelvis indicating the relationship of the peritoneal extensions to each other.

arises as two folds in the serosa of the stomach and duodenum (Figure 20–3b). They extend anteriorly and connect with the visceral peritoneum of the liver. The lesser omentum suspends the stomach and duodenum from the liver (Figure 20–3d). An extension of the visceral peritoneum of the liver ties it, in turn, to the diaphragm and the upper abdominal wall. The **greater omentum** is a large fold in the serosa of the stomach that hangs down like an apron over the front of the intestines (Figure 20–3a, c, d). It then passes up to a part of the large intestine (the transverse colon), wraps itself around it, and finally attaches to the parietal peritoneum of the posterior wall of the abdominal cavity. Because the greater omentum contains large quantities of adipose tissue, it commonly is called the "fatty apron." The greater omentum contains numerous lymph nodes. If an infection occurs in the intestine, plasma cells formed in the nodes combat the infection and help to prevent it from spreading to the peritoneum. Inflammation of the peritoneum *(peritonitis)* is a serious condition because the peritoneal membranes are continuous with each other, enabling the infection to spread to all the organs in the cavity (Figure 20–3d).

THE DIGESTIVE ORGANS

The first portion of the GI tract we shall consider is the mouth – the first structure in the tract that is concerned with the physical and chemical digestion of food.

The mouth or oral cavity

The **mouth,** also referred to as the **oral** or **buccal cavity,** is formed by the cheeks, hard and soft palates, and tongue (Figure 20–4). Forming the lateral walls of the oral cavity are the cheeks, which are muscular structures lined by stratified squamous nonkeratinized epithelium. The anterior portions of the cheeks terminate in the superior and inferior lips. The lips are fleshy folds which surround the orifice of the mouth. They are covered on the outside by skin and on the inside by a mucous membrane. The transition zone where the two kinds of covering tissues meet is called the *vermilion* – this portion of the lips is not keratinized and the color of the blood in the underlying blood vessels is visible through the transparent surface layer of the vermilion. The inner surface of each lip is attached to its corresponding gum by a midline fold of mucous membrane called the *labial frenulum.* The orbicularis oris muscle and connective tissue lie between the external integumentary covering and the internal mucosal lining. During chewing the cheeks and lips help to keep food between the upper and lower teeth. They also assist in speech.

The *vestibule* of the oral cavity is bounded externally by the cheeks and lips and internally by the gums and teeth. The *oral cavity proper* extends

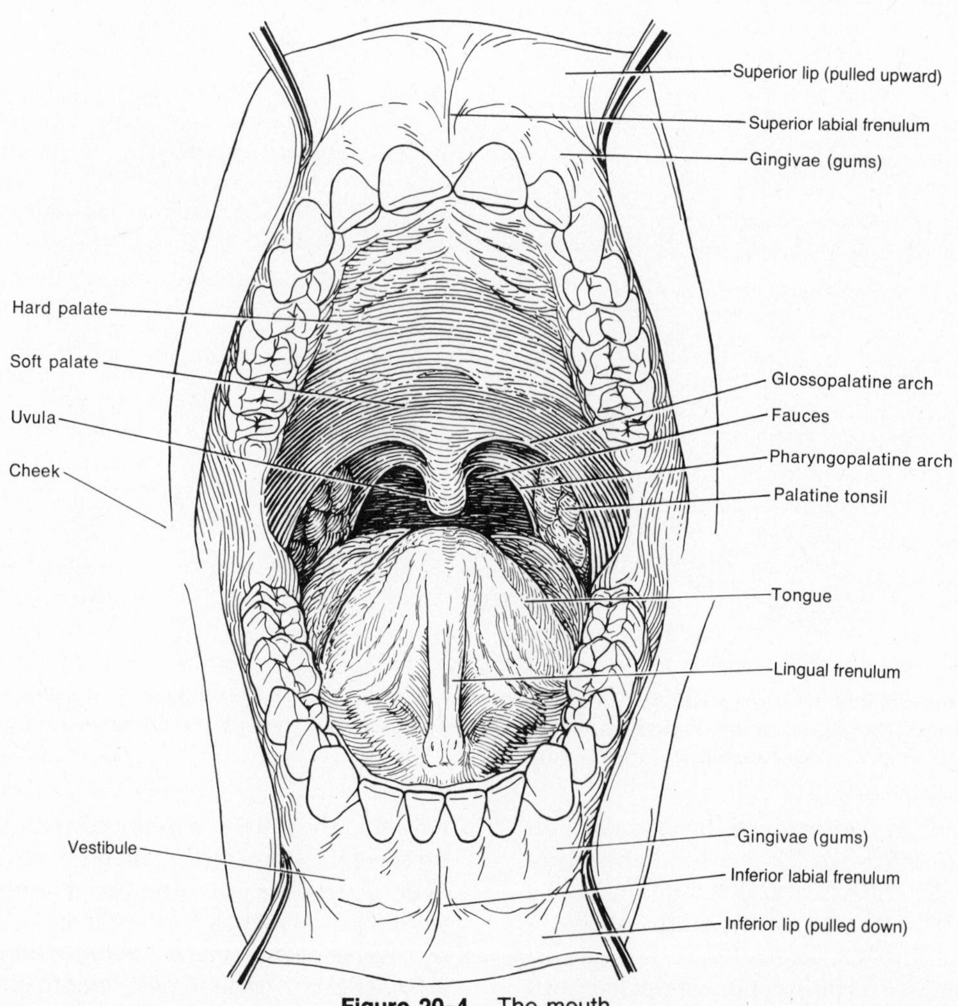

Figure 20–4. The mouth.

from the vestibule to the *fauces,* the opening between the oral cavity and the pharynx or throat.

The **hard palate,** which constitutes the anterior portion of the roof of the mouth, is formed by the maxillae and palatine bones and is lined by mucous membrane (see Figures 6–2 and 6–4). The **soft palate** forms the posterior portion of the roof of the mouth. It is an arch-shaped muscular partition between the oropharynx and nasopharynx and is lined by mucous membrane. Hanging from the middle of the lower border of the soft palate is a fingerlike muscular process called the *uvula.* On either side of the base of the uvula are two muscular folds that run down the lateral side of the soft palate. Anteriorly, the *glossopalatine arch* runs downward, laterally, and forward to the side of the base of the tongue. Posteriorly, the *pharyngopalatine arch* projects downward, laterally, and backward to the side of the pharynx. The palatine tonsils are situated between the arches, and the lingual tonsils are situated at the base of the tongue. At the posterior border of the soft palate, the mouth opens into the oropharynx through the *fauces.*

The **tongue,** together with its associated muscles, forms the floor of the oral cavity. It is composed of skeletal muscle covered with mucous membrane (Figure 20–5). The extrinsic muscles of the tongue originate outside the tongue and insert into it. They move the tongue from side to side and in and out and maneuver food for chewing and swallowing. The intrinsic muscles originate and insert within the tongue, and they alter the shape of the tongue for speech and swallowing. A fold of mucous membrane in the midline of the undersurface of the tongue, the *lingual frenulum,* attaches the tongue to the floor of the oral cavity (see Figure 20–4). In individuals whose frenulum is too short, tongue movements are restricted, speech is faulty, and the person is said to be "tongue-tied." These functional problems can be corrected very easily by cutting the frenulum surgically.

The upper surface and sides of the tongue contain projections of the lamina propria covered with epithelium called **papillae.** Taste buds are located within some papillae. *Filiform papillae* are conical projections distributed in parallel rows over the anterior two-thirds of the tongue and contain no taste buds. *Fungiform papillae* are mushroomlike elevations distributed among the filiform papillae and are more numerous near the tip of the tongue. They appear as red dots on the surface of the tongue, and most of them contain taste buds. *Circumvallate papillae* are arranged in the form of an inverted V on the posterior surface of the tongue, and all of them contain taste buds. Note the relative positions of the taste zones of the tongue in Figure 20–5.

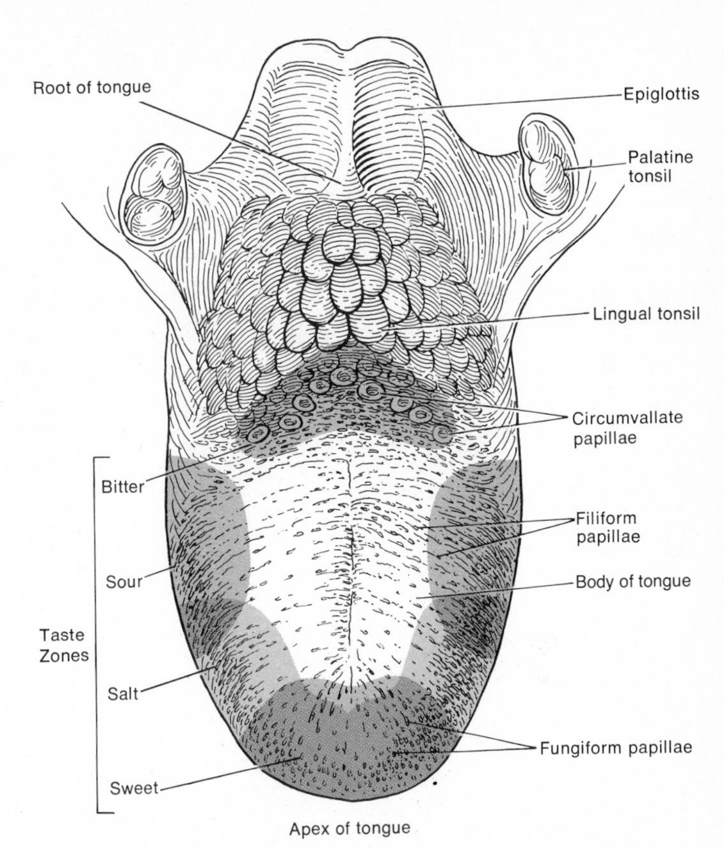

Figure 20–5. The dorsum of the tongue. Shown are the locations of the papillae and the four taste zones.

The tongue has a rich blood supply. The lingual artery is the main blood supply, and its veins eventually drain into the internal jugular vein. Most of the muscles of the tongue are supplied by the hypoglossal nerve. Taste sensations from the tongue are conveyed by the facial and glossopharyngeal nerves.

Salivary glands

Saliva is a fluid that is continuously secreted by glands lying in or near the mouth. Ordinarily, just enough saliva is secreted to keep the mucous membranes of the mouth moist. But when food enters the mouth, secretion increases so the saliva can lubricate, dissolve, and chemically break down the food. The mucous membrane lining the mouth contains many small glands, the *buccal glands,* that secrete small amounts of saliva. However, the major portion of saliva is secreted by the **salivary glands,** which lie outside the mouth and pour their contents into ducts that empty into the oral cavity. There are three pairs of salivary glands: the parotid, submandibular, and sublingual glands (Figure 20–6). The *parotid glands* are located under and in front of the ears. Each secretes into the oral cavity vestibule via a duct, called Stensen's duct, that opens into the

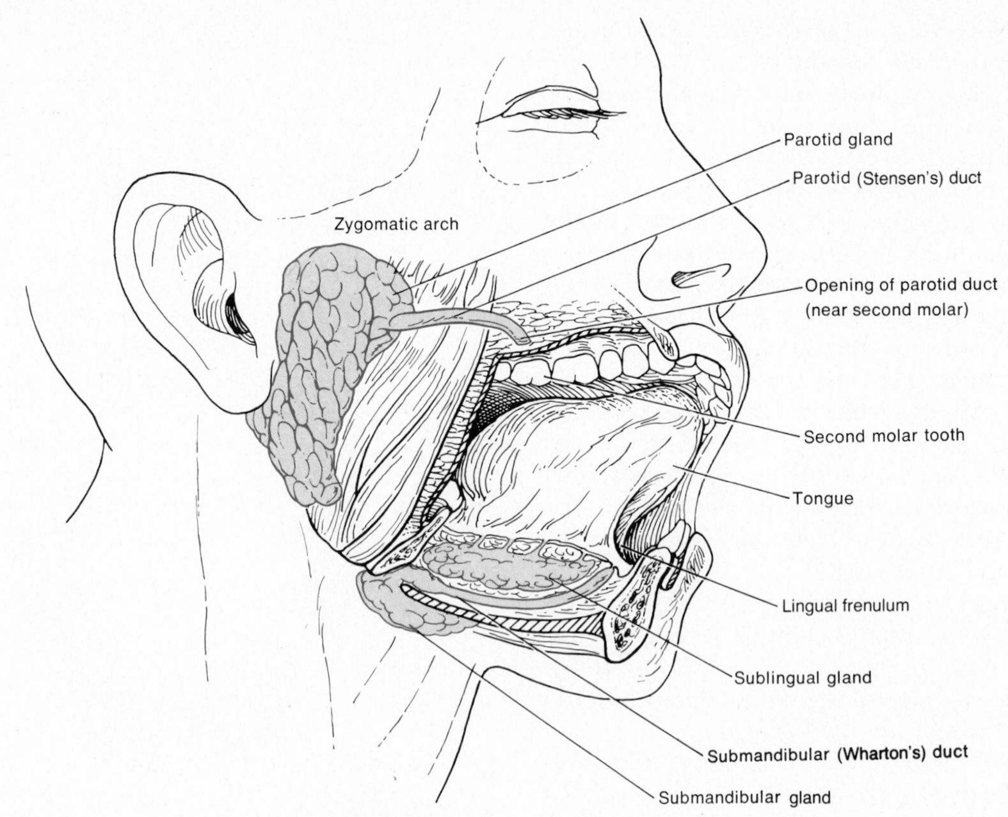

Figure 20–6. The salivary glands.

vestibule opposite the upper second molar tooth. The *submandibular glands* are found beneath the base of the tongue in the posterior part of the floor of the mouth, and their ducts (Wharton's ducts) are situated on either side of the lingual frenulum in the oral cavity proper. The *sublingual glands* are anterior to the submandibular glands, and their ducts open into the floor of the mouth in the oral cavity proper. The parotid glands are compound tubuloacinar glands, whereas the submandibular and sublingual glands are compound acinar glands (Figure 20–7).

Saliva

The fluids secreted by the buccal glands and the three pairs of salivary glands constitute **saliva.** Amounts of saliva secreted daily vary considerably but range from 1,000 to 1,500 ml. Chemically, saliva is 99.5 percent water and 0.5 percent solutes. Among the solutes are salts, such as chlorides, bicarbonates, and phosphates of sodium and potassium. Some dissolved gases and various organic substances including urea and uric acid, serum albumin and globulin, mucin, the bacteriolytic enzyme lysozyme, and the digestive enzyme amylase are also present.

The water in saliva provides a medium for dissolving foods so they can be tasted and in which digestive reactions can take place. The chlorides in the saliva activate the amylase. The bicarbonates and phosphates buffer chemicals that enter the mouth and keep the saliva at a slightly acidic pH of 6.35 to 6.85. Urea and uric acid are found in saliva because the saliva-producing glands like the sweat glands of the skin help the body to get rid of wastes. Mucin is a protein that forms mucus when it is dissolved in water. Mucus lubricates the food so it can be easily turned in the mouth and swallowed. The enzyme lysozyme destroys bacteria, thereby protecting the mucous membrane from infection and the teeth from decay.

Saliva and digestion

Depending on the kinds of cells the gland contains, each saliva-producing gland supplies different ingredients to saliva. The parotids contain cells that secrete a thin watery liquid containing the enzyme salivary amylase. The submandibular glands contain cells similar to those found in the parotids plus some mucous cells. Therefore, they secrete a fluid that is thickened with mucus but still contains quite a bit of enzyme. The sublingual glands contain mostly mucous cells, so they secrete a much thicker fluid that contributes only a small amount of enzyme to the saliva.

The enzyme salivary amylase initiates the breakdown of carbohydrates, which is the only chemical digestion that occurs in the mouth. Carbohydrates are starches and sugars and are classified as either monosaccharides, disaccharides, or polysaccharides. Monosaccharides are small molecules containing several carbon, hydrogen, and oxygen atoms. An example of a monosaccharide is glucose. Disaccharides consist of two monosaccharides linked together; polysaccharides are chains of three or more monosaccharides. The vast majority of carbohydrates that we eat are polysaccharides. Since only monosaccharides can be absorbed into the bloodstream, ingested disaccharides and polysaccharides must be broken down. The function of *salivary amylase* is to break the chemical bonds between some of the monosaccharides that make up the polysaccharides. In this way, the enzyme breaks the long-chain polysaccharides into shorter polysaccharides called *dextrins.* Given sufficient time, salivary amylase also can break down the dextrins into disaccharides. However, food usually is swallowed too quickly for more than 3 to 5 percent of the carbohydrates to be reduced to disaccharides in the mouth.

Teeth

The **teeth,** or **dentes,** are located in sockets of the alveolar processes of the mandible and maxillae. The alveolar processes are covered by the *gingivae* (gums), which extend slightly into each socket (Figure 20–8). Periosteum, the fibrous membrane that covers all bone, lines the sockets. The periosteum lining the sockets is called the *periodontal membrane.*

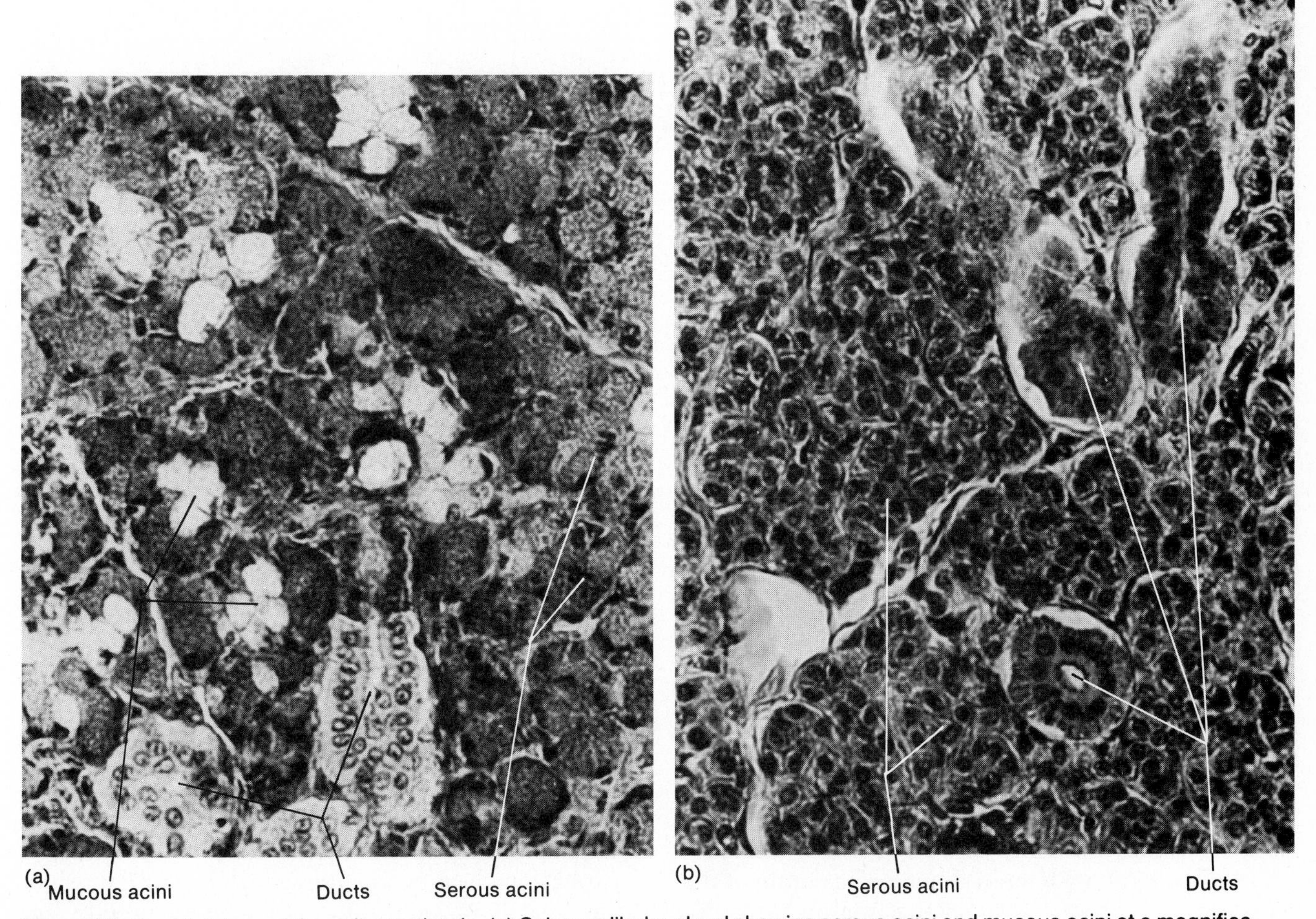

Figure 20–7. Histology of the salivary glands. (a) Submandibular gland showing serous acini and mucous acini at a magnification of 90X. (b) Parotid gland showing serous acini at a magnification of 90X. (Courtesy of Victor B. Eichler, Wichita State University.)

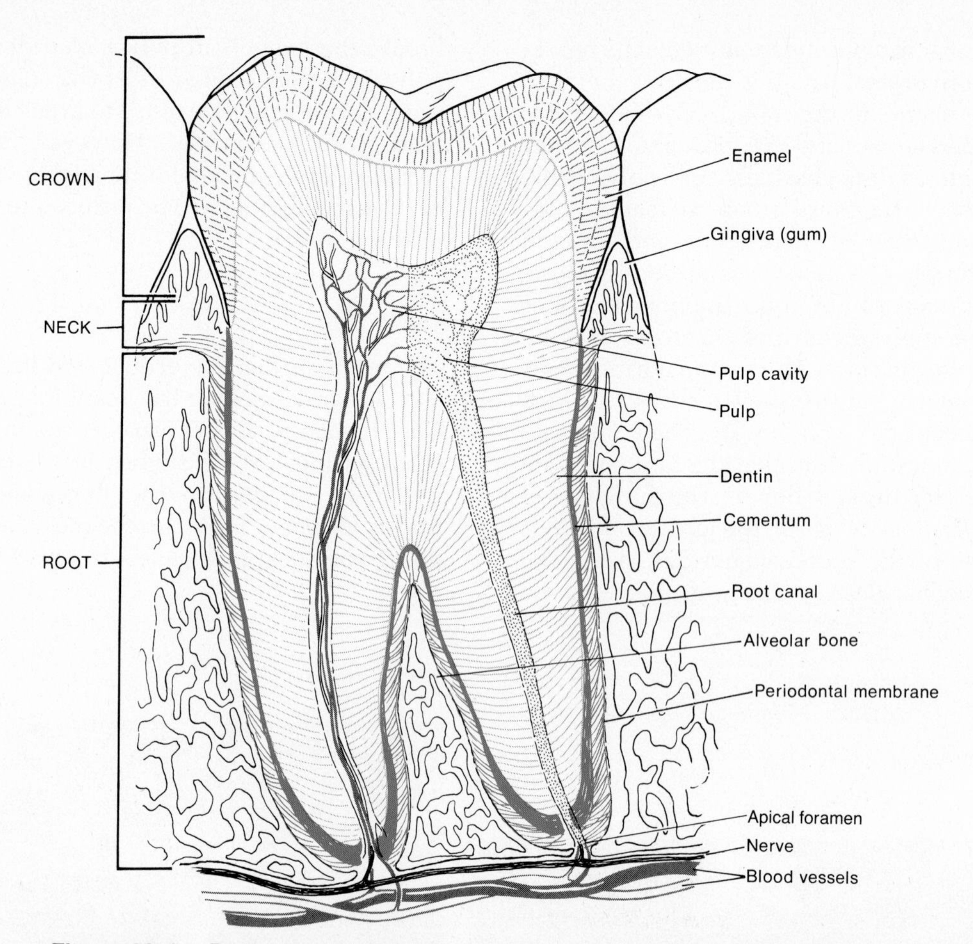

Figure 20–8. Parts of a typical tooth shown in a sagittal section through a molar.

A typical tooth consists of three principal portions (Figure 20–8). The *crown* is the portion above the level of the gums, the *root* consists of one to three projections embedded in the socket, and the *neck* is the constricted region between the crown and the root.

Teeth are composed primarily of *dentin,* a bonelike substance that gives the tooth its basic shape and rigidity. The dentin encloses a cavity. The enlarged part of the cavity, the *pulp cavity,* lies within the crown and neck and is filled with *pulp,* a connective tissue containing blood vessels, nerves, and lymphatics. Narrow extensions of the pulp cavity lying within the roots of the tooth are called the *root canals.* Each root canal has an opening at its base called the *apical foramen.* Blood and lymph vessels and nerves from the periodontal membrane run through the opening in the root and travel up the root canal to the pulp cavity, where they provide the tooth with nourishment. The dentin of the crown is covered by *enamel* that consists primarily of calcium phosphate. Enamel is the hardest substance in the body and protects the tooth from the wear and tear of chewing. It is also an effective barrier against acids that easily dissolve the dentin. The dentin of the root is covered by *cementum,* another bonelike substance. Cementum is bound to the underlying bone by the periodontal membrane. *Pyorrhea* is an inflammation of the periodontal membrane and adjacent gums. A prolonged, severe case of pyorrhea can weaken the periodontal membrane, erode the alveolar bone, and thereby cause the tooth to become loose.

Dentitions

Each individual has two *dentitions,* or sets of teeth. The first of these, the *deciduous* or *milk teeth,* begin to erupt at about 6 months of age, and one appears at about each month thereafter until all 20 are present. Figure 20–9a illustrates the deciduous teeth. The *incisors,* which are closest to the midline, are chisel-shaped and are adapted for cutting and biting food. Next to the incisors, moving posteriorly, are the *canines* or *cuspids,* which have a flat grinding surface called the cusp. Canines are used to tear and shred food. The incisors and canines have only one root apiece. Behind them lie the first and second *molars,* or *tricuspids,* which have three or four

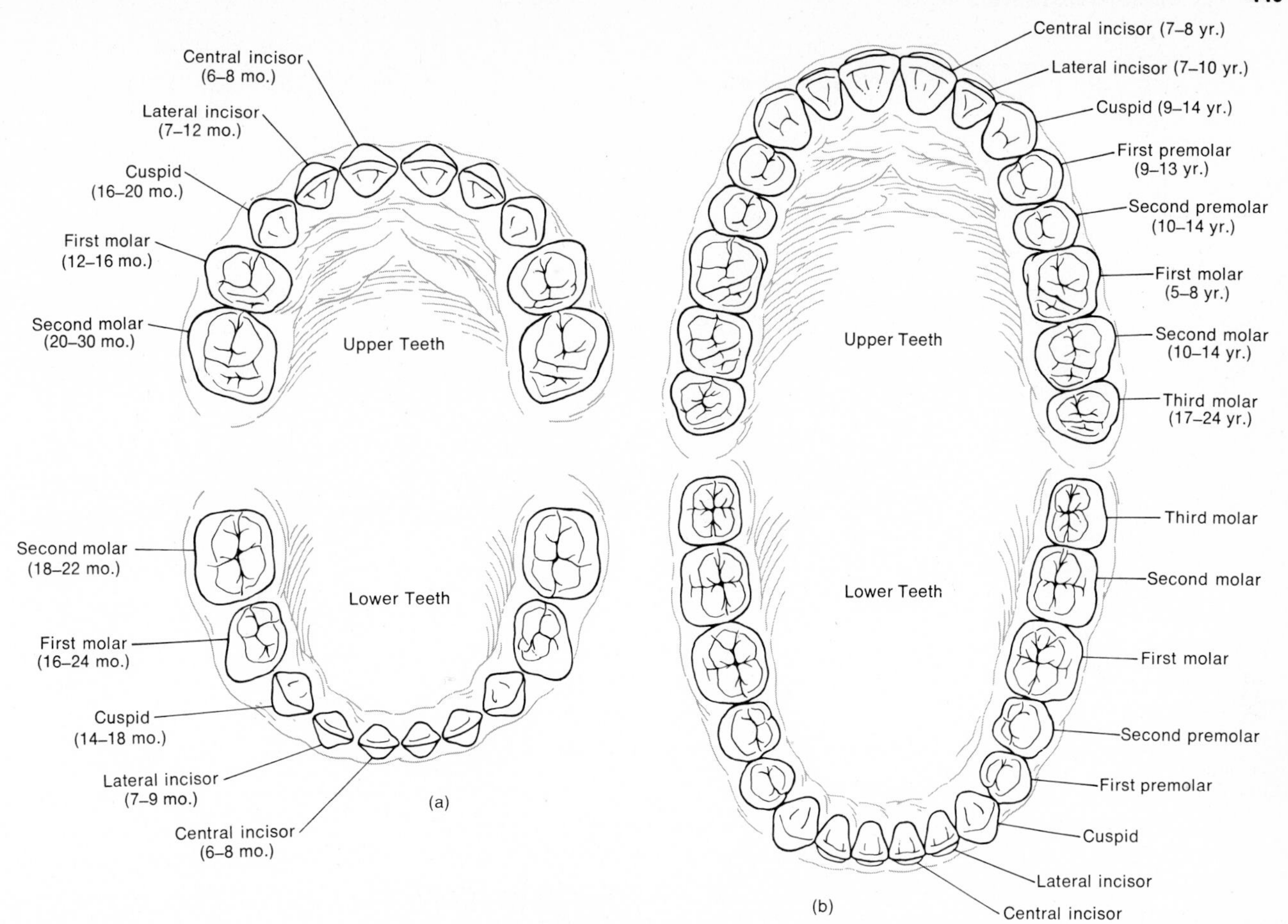

Figure 20–9. Dentitions and times of eruptions. The times of eruptions are indicated in parentheses. (a) Deciduous dentition. (b) Permanent dentition.

cusps and two roots. The molars crush and grind food.

All the deciduous teeth are lost—generally between 6 and 13 years of age—and are replaced by the *permanent dentition* (Figure 20–9b). The permanent dentition contains 32 teeth that appear between the age of 6 and adulthood. It resembles the deciduous dentition with the following exceptions: the deciduous first molars are replaced with *premolars* or *bicuspids,* which have two cusps and two roots and are used for crushing and grinding. Behind the premolars lie three sets of molars. The *third molars,* or *wisdom teeth,* do not appear before the age of 17. In fact, in some people they do not come through at all.

The arteries of the teeth receive their blood from the maxillary artery. The nerve supply is via the trigeminal nerve.

Teeth and digestion

Through chewing, or **mastication,** the teeth pulverize food and mix it with saliva. As a result, the food is reduced to a soft, flexible mass, called a *bolus,* that is easily swallowed.

Deglutition: mouth, pharynx, and esophagus

Swallowing, or **deglutition,** moves food from the mouth to the stomach. Swallowing starts with the bolus on the upper side of the tongue. Then the tip of the tongue rises and presses against the palate (Figure 20–10). The bolus slides to the back of the mouth and is pulled through the fauces by muscles that lie in the pharynx. During this period the respiratory passageways close, and breathing is temporarily interrupted. The soft palate and uvula move upward to close off the nasopharynx, and the larynx is pulled forward and upward under the tongue. As the larynx rises, it meets the epiglottis, which seals off the glottis. The movement of the larynx also pulls the vocal cords together, further sealing off the respiratory tract, and widens the opening between the pharynx and esophagus. The bolus passes through the pharynx and enters the esophagus in 1 second. The respiratory passageways then reopen, and breathing resumes.

The **esophagus,** the third organ involved in deglutition, is a muscular, collapsible tube that lies behind the trachea (see Figure 20–1). It is about 23

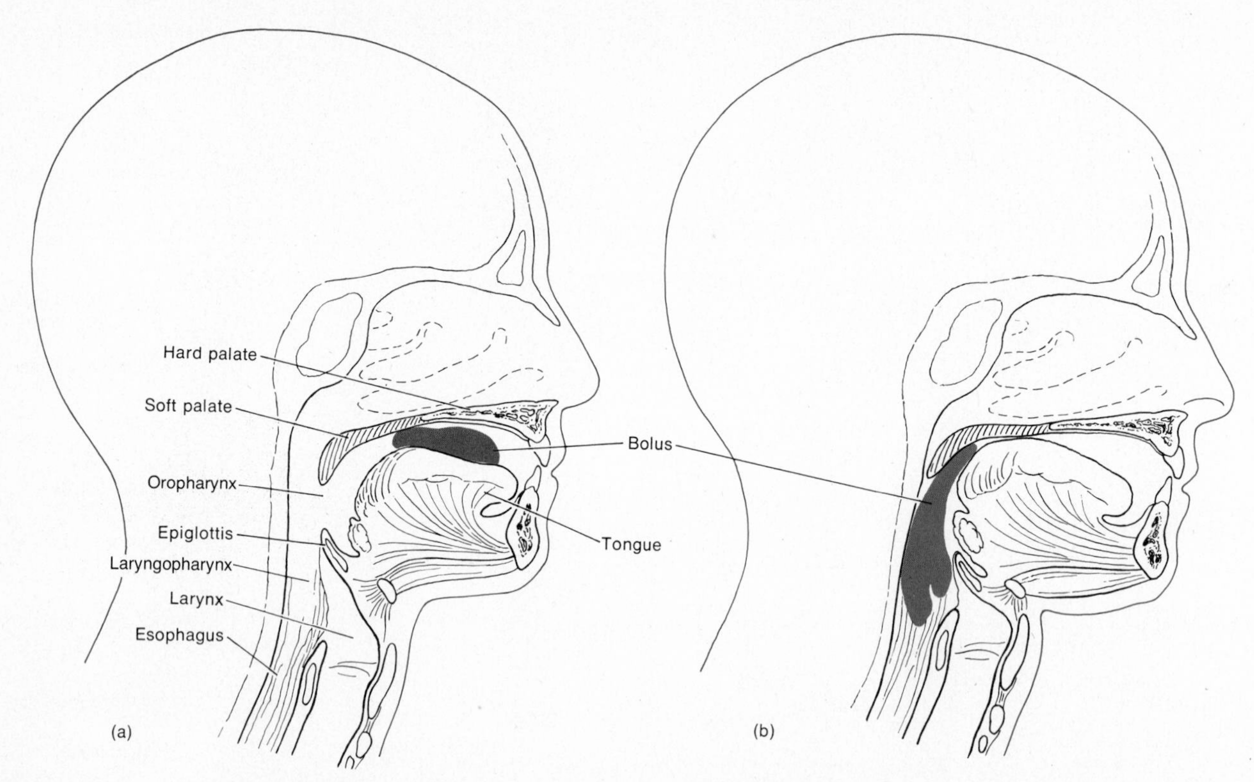

Figure 20–10. Deglutition. (a) Position of structures prior to swallowing. (b) During swallowing, the tongue rises against the palate, the nose is closed off, the larynx rises, the epiglottis seals off the larynx, and the bolus is passed into the esophagus.

to 25 cm (10 in.) long and begins at the end of the laryngopharynx, passes through the mediastinum in front of the vertebral column, pierces the diaphragm, and terminates in the upper portion of the stomach.

The mucosa of the esophagus consists of stratified squamous nonkeratinized epithelium, lamina propria, and a muscularis mucosa (Figure 20–11). The submucosa contains connective tissue and blood vessels. The muscularis of the upper third is striated, the middle third is striated and smooth, and the lower third is smooth. The outer layer is known as the *tunica adventitia* since it contains no serosa.

The esophagus does not produce digestive enzymes and does not carry on absorption. It does, however, secrete mucus and transport food to the stomach.

The arteries of the esophagus are derived from the inferior thyroid, thoracic aorta, intercostal arteries, phrenic, and left gastric arteries. The veins drain into the adjacent veins. The nerves of the esophagus are from laryngeal nerves, cervical sympathetic chain, and vagi.

Food is pushed through the esophagus by muscular movements called **peristalsis** (Figure 20–12). Peristalsis is a function of the tunica muscularis, and it occurs as follows: In the section of the esophagus lying just above and around the top of the bolus, the circular muscle fibers contract. The contraction constricts the esophageal wall and squeezes the bolus downward. Meanwhile, longitudinal fibers lying around the bottom of the bolus and just below it also contract. Contraction of the longitudinal fibers shortens this lower section, pushing its walls outward so it can receive the bolus. The contractions are repeated in a wave that moves down the esophagus, pushing the food toward the stomach. Passage of the bolus is further facilitated by glands secreting mucus. The passage of solid or semisolid food from the mouth to the stomach takes about 4 to 8 seconds. Very soft foods and liquids pass through in about 1 second.

Just above the level of the diaphragm, the esophagus is very slightly narrowed. This narrowing has been attributed to a physiological sphincter in the inferior part of the esophagus known as the *lower esophageal* or *gastroesophageal sphincter.* A **sphincter** is an opening that has a thick circle of muscle around it. The lower esophageal sphincter relaxes during swallowing and thus aids the passage of food from the esophagus into the stomach. The movement of the diaphragm against the stomach during ventilation presses on the stomach and helps to prevent the regurgitation of gastric contents from the stomach to the esophagus.

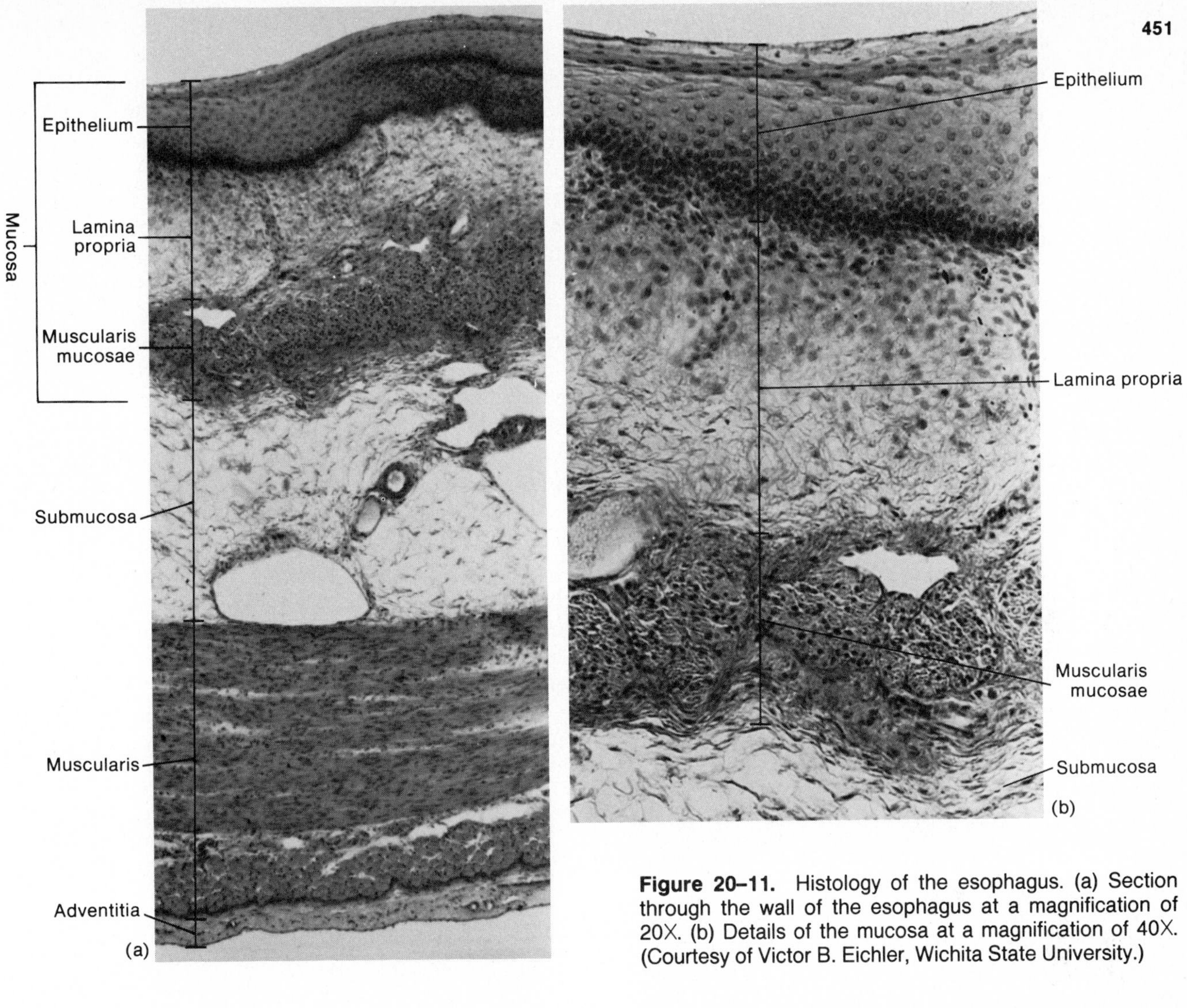

Figure 20–11. Histology of the esophagus. (a) Section through the wall of the esophagus at a magnification of 20X. (b) Details of the mucosa at a magnification of 40X. (Courtesy of Victor B. Eichler, Wichita State University.)

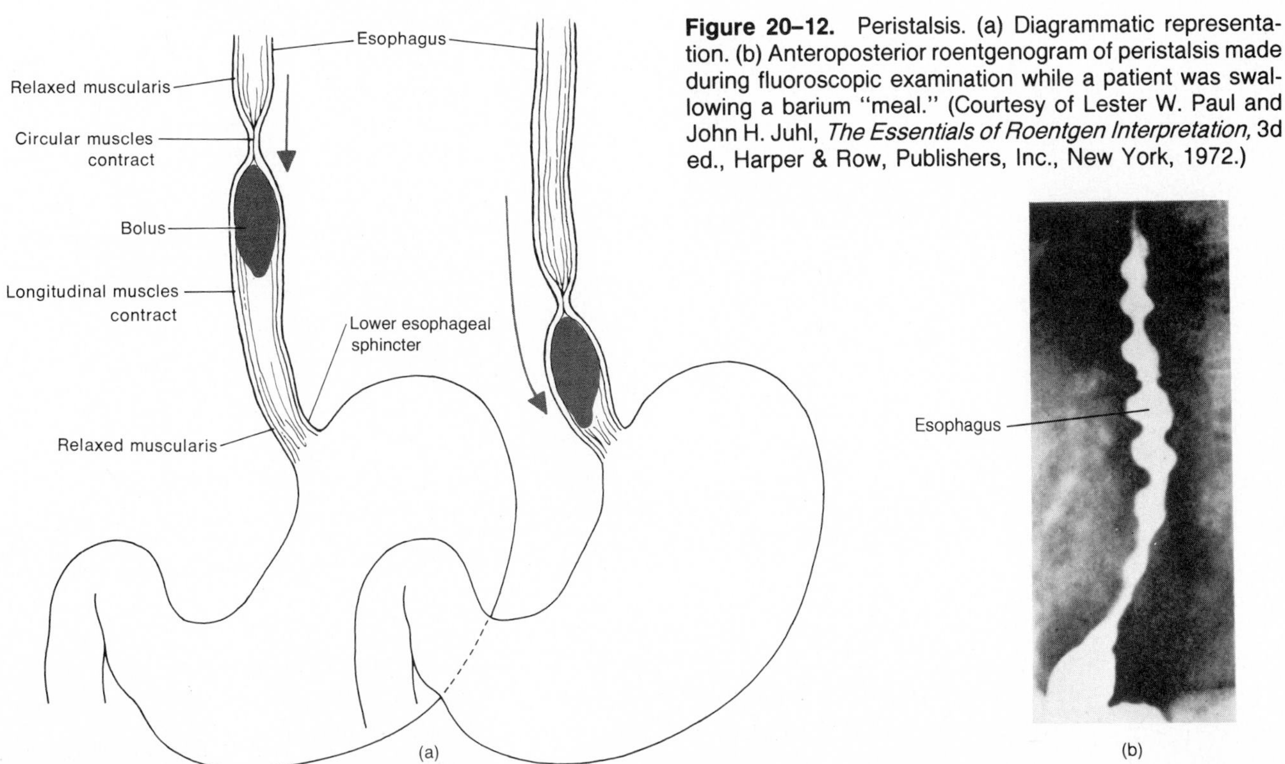

Figure 20–12. Peristalsis. (a) Diagrammatic representation. (b) Anteroposterior roentgenogram of peristalsis made during fluoroscopic examination while a patient was swallowing a barium "meal." (Courtesy of Lester W. Paul and John H. Juhl, *The Essentials of Roentgen Interpretation,* 3d ed., Harper & Row, Publishers, Inc., New York, 1972.)

Stomach

The **stomach** is a J-shaped enlargement of the GI tract directly under the diaphragm in the epigastric, umbilical, and left hypochondriac regions of the abdomen (see Figure 20–1). The superior portion of the stomach is a continuation of the esophagus, and the inferior portion empties into the duodenum, the first part of the small intestine. Within each individual, the position and size of the stomach vary continually. For instance, the diaphragm pushes the stomach downward with each inspiration and pulls it upward with each expiration. When the stomach is empty, it is about the size of a large sausage, but when food enters, the stomach can stretch itself immeasurably to accommodate its contents.

Anatomy

The stomach is divided into four areas: the cardia, fundus, body, and pylorus (Figure 20–13a). The *cardia* surrounds the lower esophageal sphincter, and the rounded portion above and to the left of the cardia is the *fundus.* Below the fundus, the large central portion of the stomach is called the *body,* and the more narrow, inferior region is the *pylorus.* The concave medial border of the stomach is called the *lesser curvature,* and the convex lateral border is referred to as the *greater curvature.* The pylorus communicates with the duodenum of the small intestine via a sphincter called the *pyloric valve.*

Two abnormalities of the pyloric valve sometimes are found in infants. One of these abnormalities, *pylorospasm,* is characterized by failure of the muscle fibers encircling the opening to relax normally. As a result, ingested food does not pass easily from the stomach to the small intestine. The stomach becomes overly full, and the infant vomits frequently to relieve the pressure. Pylorospasm is treated by adrenergic drugs that relax the muscle fibers of the valve. The other abnormality, called *pyloric stenosis,* is a narrowing of the pyloric valve caused by a tumorlike mass that apparently is formed by enlargement of the circular muscle fibers. The mass obstructs the passage of food and must be surgically corrected.

The wall of the stomach is composed of the same four basic layers as the rest of the alimentary canal,

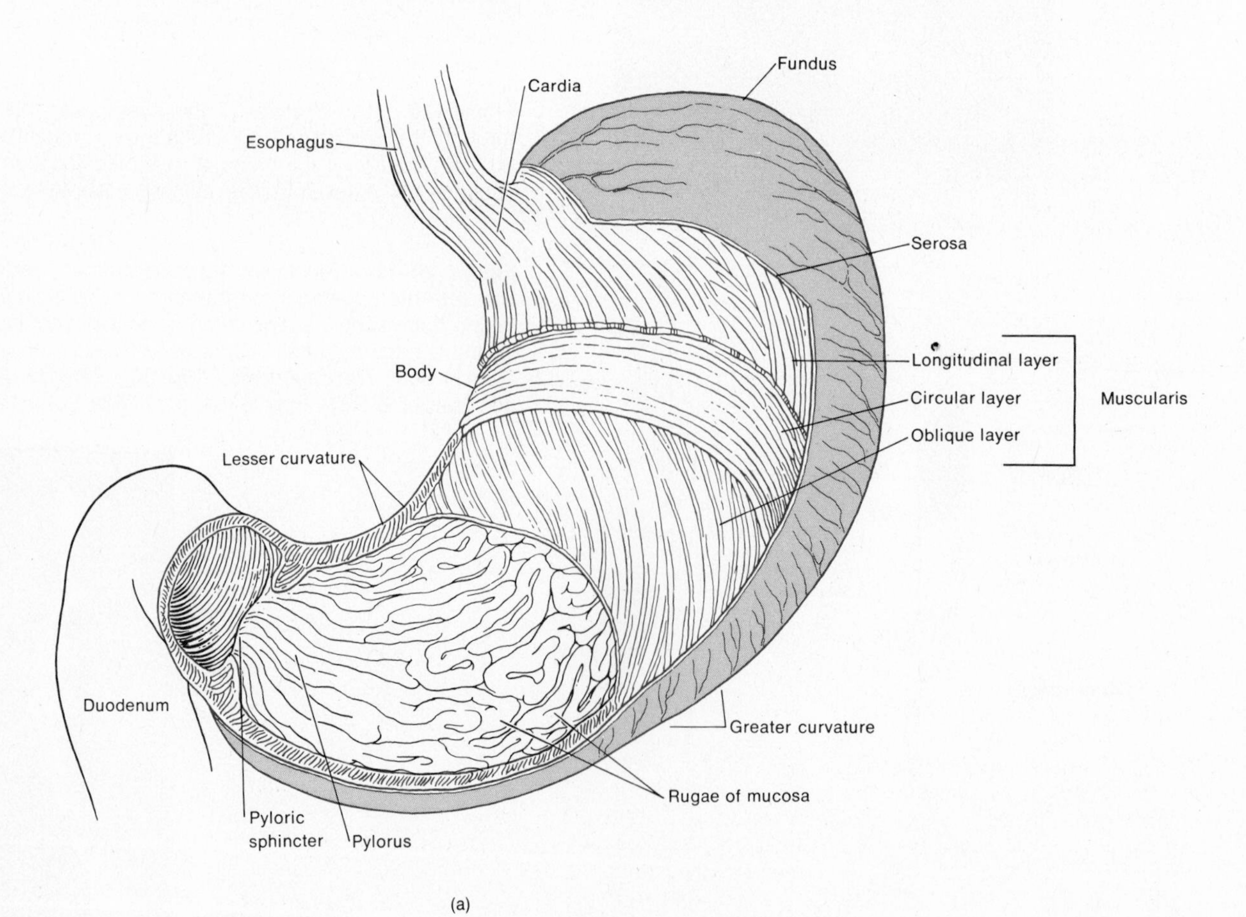

Figure 20–13. The stomach. (a) External and internal anatomy.

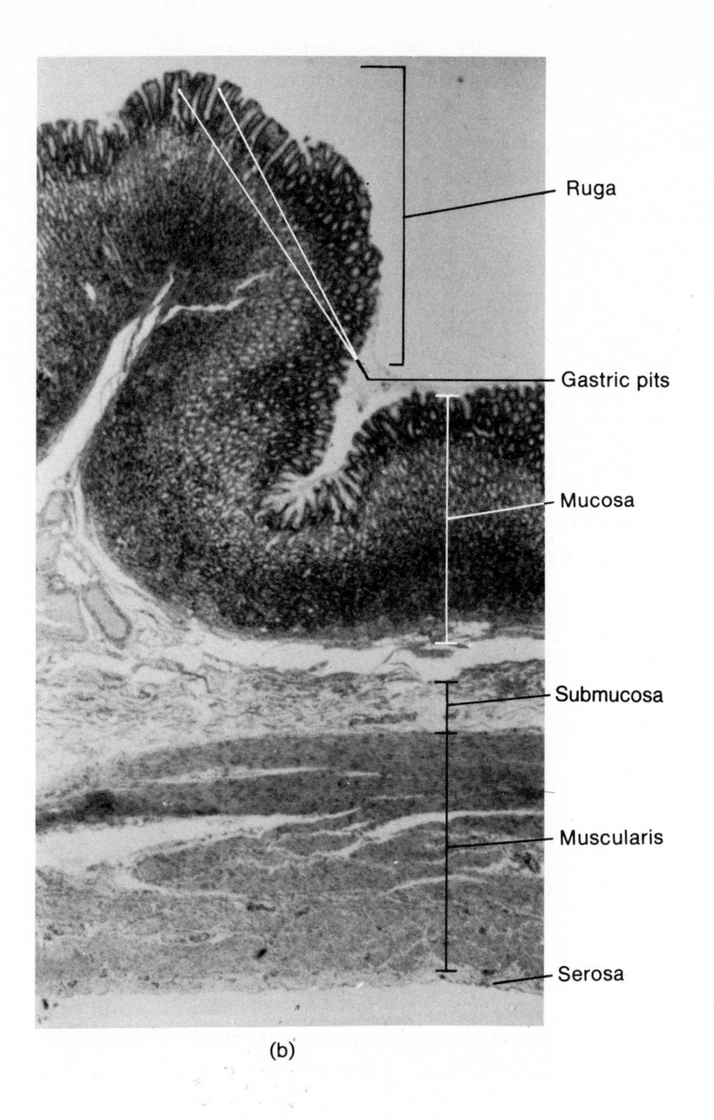

(b)

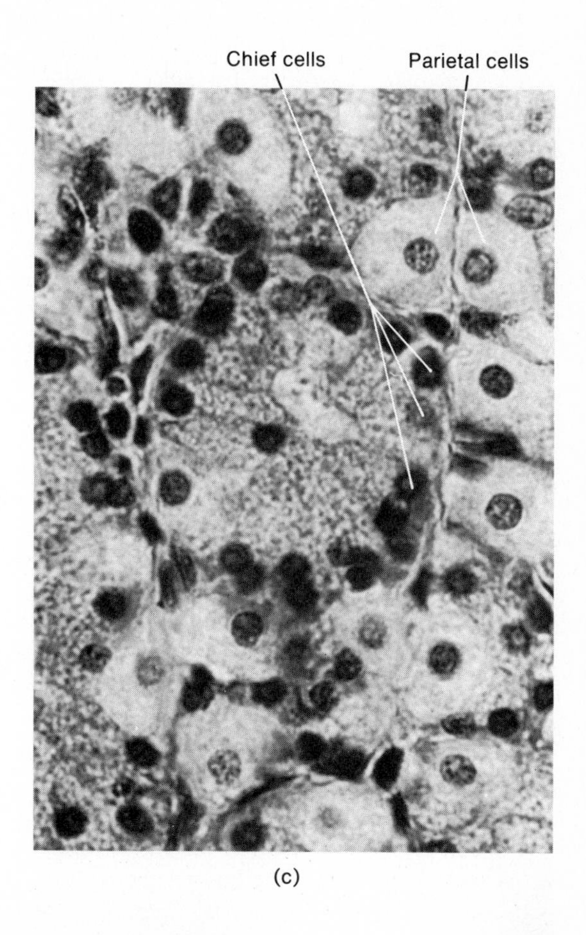

(c)

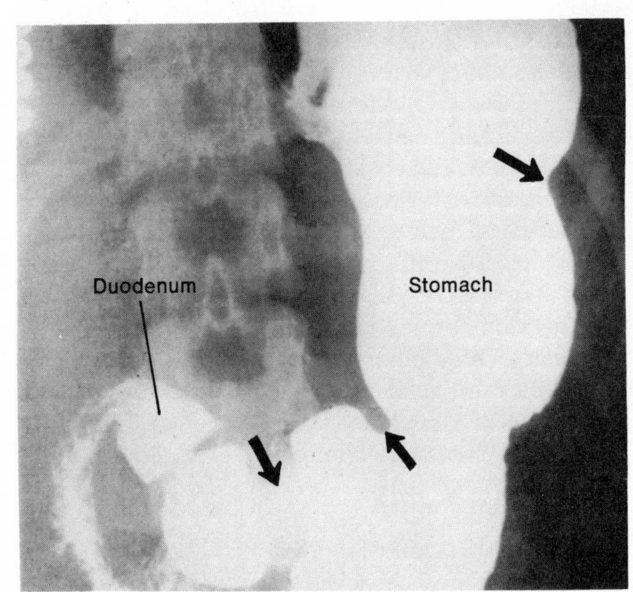

(d)

Figure 20–13 (cont.). The stomach. (b) Histology of the fundic wall at a magnification of 6X. (c) Enlarged aspect of parietal and chief cells at a magnification of 160X. (Courtesy of Victor B. Eichler, Wichita State University.) (d) Roentgenogram of a normal stomach. Note the peristaltic waves. (Courtesy of Lester W. Paul and John H. Juhl, *The Essentials of Roentgen Interpretation,* 3d ed., Harper & Row, Publishers, Inc., New York, 1972.)

with certain modifications. When the stomach is empty, the mucosa lies in large folds that can be seen with the naked eye. These folds are called *rugae* (Figure 20–13a). As the stomach fills and distends, the rugae gradually smooth out and disappear. Microscopic inspection of the mucosa reveals a layer of simple columnar epithelium containing many narrow openings that extend down into the lamina propria. These pits are called *gastric glands,* and they are lined with three kinds of secreting cells: zymogenic, parietal, and mucous (Figure 20–13b). The *zymogenic,* or *chief cells,* secrete the principal gastric enzyme called pepsinogen. Hydrochloric acid, which activates one of the digestive enzymes, is produced by the *parietal cells.* The *mucous cells* secrete mucus and the intrinsic factor, a substance involved in the absorption of vitamin B_{12}. Secretions of the gastric glands are collectively called *gastric juice.*

The submucosa of the stomach is composed of loose areolar connective tissue, and it connects the mucosa to the muscularis. The muscularis, unlike that in other areas of the alimentary canal, has three layers of smooth muscle: an outer longitudinal layer, a middle circular layer, and an inner oblique layer (Figure 20–13a). This arrangement of the fibers allows the stomach to contract in a variety of ways to churn food, break it into small particles, mix it with gastric juice, and pass it to the duodenum. The serosa covering the stomach is part of the peritoneum. At the lesser curvature the two layers of the peritoneum come together and extend upward to the liver as the lesser omentum. At the greater curvature, the peritoneum continues downward as the greater omentum hanging over the intestines.

The arterial supply of the stomach is derived from the celiac artery. The right and left gastric arteries form an anastomosing arch along the lesser curvature, and the right and left gastroepiploic arteries form a similar arch on the greater curvature. Short gastric arteries supply the fundus. The veins of the same name accompany the arteries and drain, directly or indirectly, into the hepatic portal vein.

The vagi convey parasympathetic fibers to the stomach. These fibers form synapses within the plexus of Auerbach between the muscular coats and the plexus of Meissner in the submucosa. The sympathetic nerves arise from the celiac ganglia.

Digestion in the stomach

Several minutes after food enters the stomach, gentle, rippling, peristaltic movements called *mixing waves* pass over the stomach. These waves occur about every 15 to 25 seconds and serve to macerate food, mix it with the secretions of the digestive glands, and reduce it to a thin liquid called *chyme.* Relatively few mixing waves are observed in the fundus, which is primarily a storage area. Foods may remain in the fundus for an hour or more without becoming mixed with gastric juice. During this time, salivary digestion continues.

The principal chemical activity of the stomach is to begin the digestion of proteins. In the adult, this is achieved primarily through the enzyme *pepsin.* Pepsin breaks some of the peptide bonds between the amino acids making up proteins. Thus a protein chain of many amino acids is broken down into fragments containing 4 to 12 amino acids—longer fragments are called *proteoses,* and the shorter ones are called *peptones.* Pepsin is most effective in the very acidic environment of the stomach, which has a very acid pH of 1. It becomes inactive in an alkaline environment. As you know, living cells are composed in part of proteins. What keeps pepsin from digesting the cells of the stomach along with the food? First of all, pepsin is secreted in an inactive form called *pepsinogen,* so it cannot digest the proteins in the zymogenic cells that produce it. When pepsinogen comes in contact with the hydrochloric acid secreted by the parietal cells, it is converted to active pepsin. Once pepsin has been activated, the cells of the stomach are protected by mucus. The mucus coats the mucosa and forms a barrier between the gastric juice and the cells. Sometimes the mucus fails to do its job, and the pepsin and hydrochloric acid eat a hole in the stomach wall known as a *gastric ulcer.*

The stomach wall is impermeable to the passage of most materials into the blood, so most substances are not absorbed until they reach the small intestine. However, the stomach does participate in the absorption of some water and salts, certain drugs, and alcohol.

The next step in the breakdown of food is digestion in the small intestine. However, chemical digestion in the small intestine is dependent not only on its own secretions but on those from three organs that lie outside the alimentary canal. These organs—the pancreas, liver, and gallbladder—will be discussed before we continue with the alimentary canal.

Pancreas

The **pancreas** is a soft, oblong-shaped tubuloacinar gland about 12.5 cm (6 in.) long and 2.5 cm (1 in.) thick. It lies along the greater curvature of the stomach and is connected by a duct (sometimes

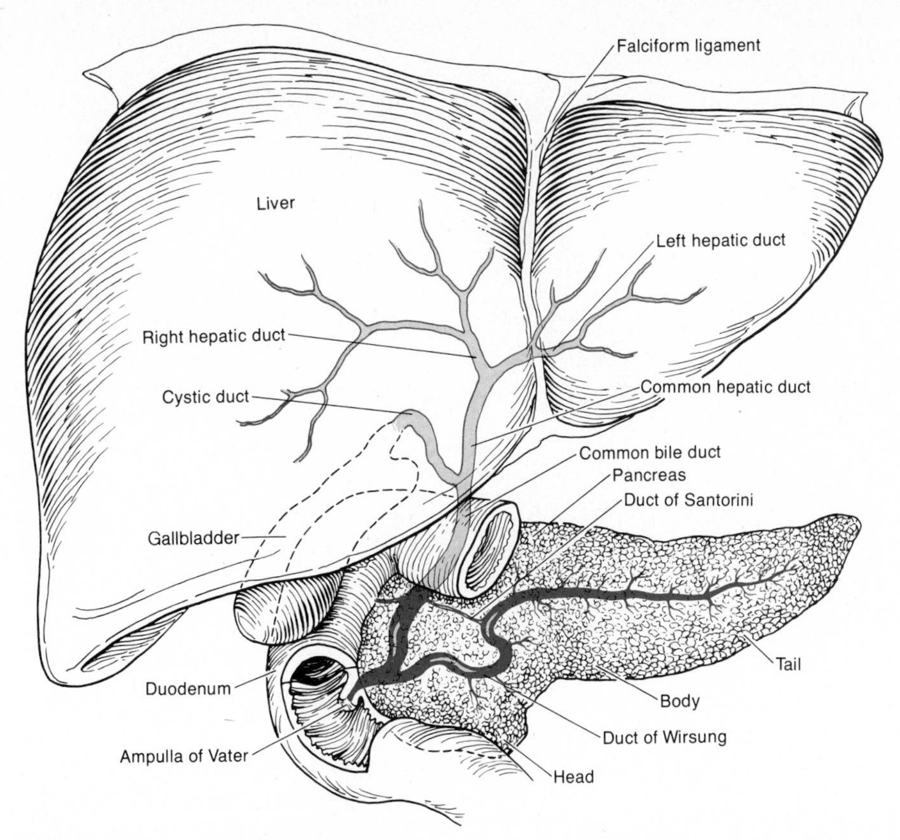

Figure 20–14. The pancreas in relation to the liver, gallbladder, and duodenum.

two) to the duodenum (Figure 20–14). The pancreas is divided into a head, body, and tail. The *head* is the more expanded portion near the C-shaped curve of the duodenum. Moving somewhat superiorly and to the left of the head are the centrally located *body* and the terminal tapering *tail.* The pancreas is made up of small clusters of glandular epithelial cells (see Figure 17–14). Some of the clusters, called islets of Langerhans, form the endocrine portions of the pancreas and consist of alpha and beta cells that secrete glucagon and insulin. The other masses of cells, called *acini,* are the exocrine portions of the organ. Secreting cells of the acini release a mixture of digestive enzymes called *pancreatic juice,* which is dumped into small ducts attached to the acini. Pancreatic juice eventually leaves the pancreas through a single large main tube called the *pancreatic duct,* or *duct of Wirsung.* In most people the pancreatic duct often unites with the common bile duct from the liver and gallbladder and enters the duodenum in a small, raised area called the *ampulla of Vater.* An accessory duct, the *duct of Santorini,* may also lead from the pancreas and empty into the duodenum about 2.5 cm (1 in.) above the ampulla of Vater.

The arterial supply of the pancreas is from the splenic artery through its pancreatic branches and from the superior mesenteric and common hepatic arteries by way of the inferior and superior pancreatoduodenal arteries. The veins, in general, correspond to the arteries. Venous blood reaches the hepatic portal vein by means of the splenic and superior mesenteric veins.

The nerves to the pancreas are branches of the celiac plexus that accompany the arteries entering the gland. The glandular innervation is from the parasympathetic division of the autonomic nervous system, and the innervation of the blood vessels is from the sympathetic division of the autonomic nervous system.

The functions of the pancreas are twofold. The acini secrete enzymes that digest food in the small intestine, and the alpha and beta cells secrete glucagon and insulin, which control the fate of digested and absorbed carbohydrates.

Liver

The **liver** performs so many vital functions that we cannot live long without it. The most important of these are listed below.

1. The liver manufactures the anticoagulant heparin and most of the other plasma proteins.
2. The reticuloendothelial cells of the liver phagocytoze worn-out red blood cells and some bacteria.

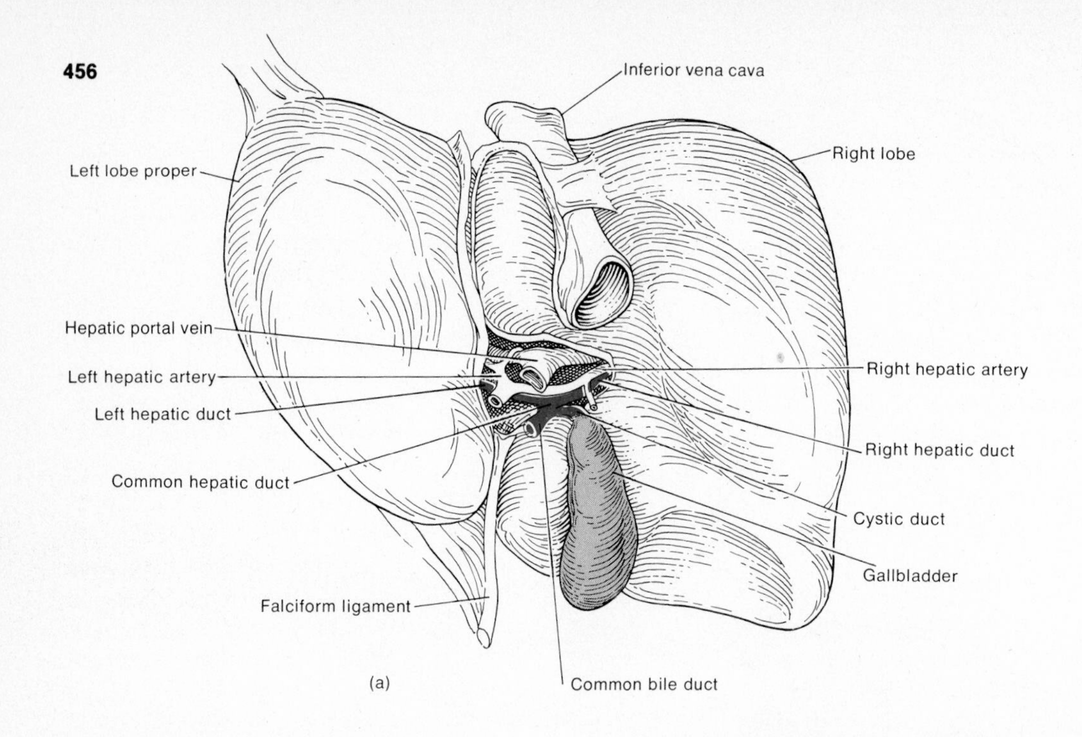

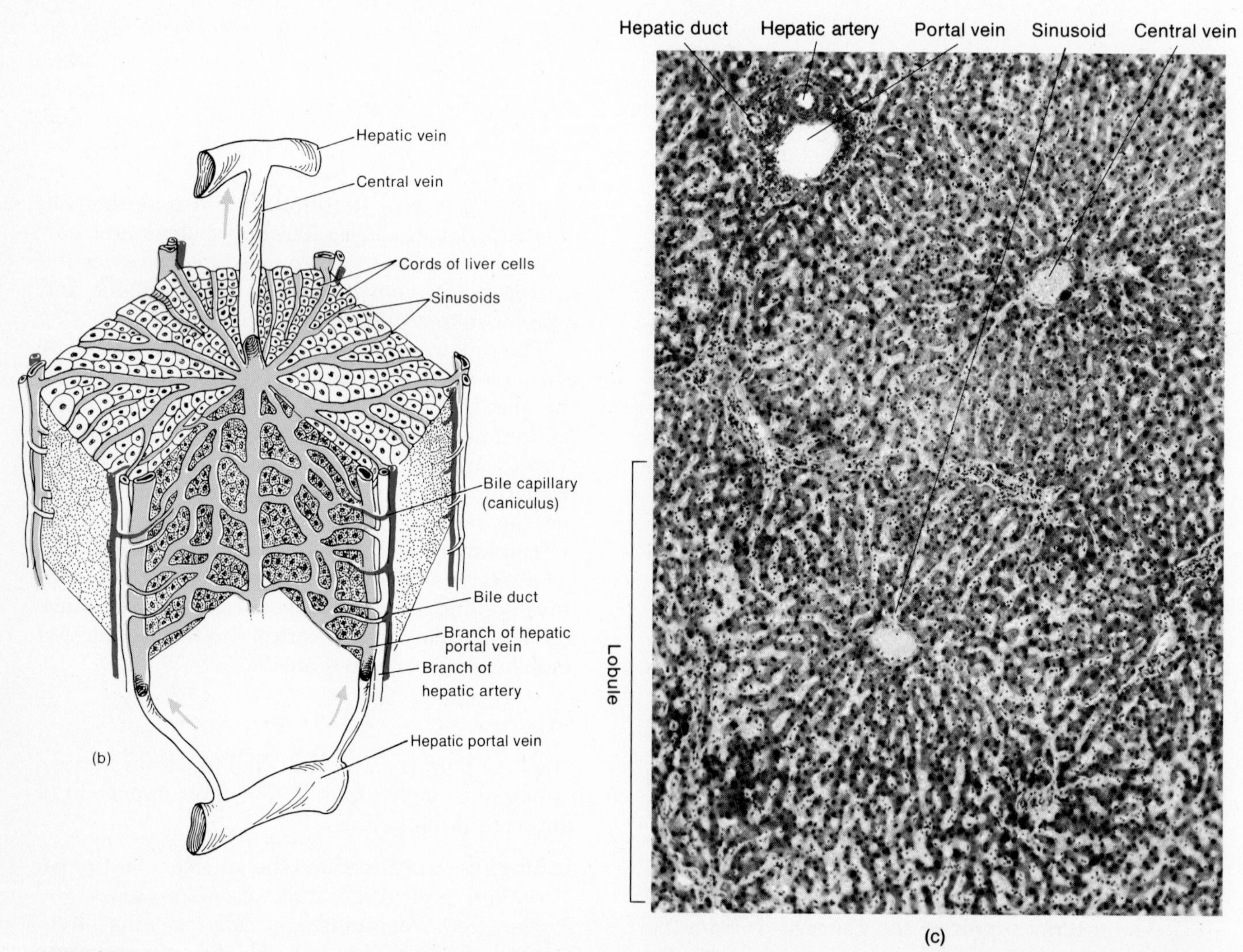

Figure 20–15. The liver. (a) External anatomy in posteroinferior view. (b) Diagrammatic representation of the microscopic appearance of a lobule. (c) Photomicrograph of a lobule at a magnification of 65X. (Courtesy of Edward J. Reith, from *Atlas of Descriptive Histology,* by Edward J. Reith and Michael H. Ross, Harper & Row, Publishers, Inc., New York, 1970.)

3. Liver cells contain enzymes that either break down poisons or transform them into less harmful compounds. For example, when amino acids are burned for energy, they leave behind toxic nitrogenous wastes that are converted to urea by the liver cells. Moderate amounts of urea are harmless to the body and are easily excreted by the kidneys and sweat glands.
4. Newly absorbed nutrients are collected in the liver. It can change any excess monosaccharides into glycogen or fat, both of which can be stored. In addition it can transform glycogen, fat, and protein into glucose and vice versa, depending on the body's needs.
5. The liver stores glycogen, copper, iron, and vitamins A, D, E, and K. It also stores some poisons that cannot be broken down and excreted. (This is why high levels of DDT are found in the livers of animals, including humans, who eat sprayed fruits and vegetables.)
6. Finally, the liver manufactures bile, which is used in the small intestine for the digestion and absorption of fats.

The liver is the largest single organ in the body, weighing about 1.4 kg (4 lb) in the average adult. It is located under the diaphragm and occupies most of the right hypochondrium and part of the epigastrium of the abdomen. The liver is covered largely by peritoneum and completely by a dense connective tissue layer that lies beneath the peritoneum. Anatomically, the liver is divided into two principal lobes – the **right lobe** and the **left lobe** – separated by the *falciform ligament* (Figure 20–15a). The right lobe, besides the main lobe, also has associated with it an inferior *quadrate lobe* and a posterior *caudate lobe.* The falciform ligament is a reflection of the parietal peritoneum which extends from the undersurface of the diaphragm to the superior surface of the liver where it separates the two principal lobes of the liver. In the free border of the falciform ligament is the *ligamentum teres (round ligament).* It extends from the liver to the umbilicus. The ligamentum teres is a fibrous cord homologous to the umbilical vein of the fetus (Chapter 11).

The lobes of the liver are made up of numerous functional units called *lobules,* which may be seen under a microscope (Figure 20–15b, c). A lobule consists of cords of *hepatic* (liver) *cells* arranged in a radial pattern around a *central vein.* Between the cords are endothelial-lined spaces called *sinusoids* through which blood passes. The sinusoids are also partly lined with phagocytic cells, termed *Kupffer cells,* that destroy worn-out white and red blood cells.

The liver receives a double supply of blood. From the hepatic artery it obtains oxygenated blood, and from the hepatic portal vein it receives deoxygenated blood containing newly absorbed nutrients (see Figure 11–22). Branches of both the hepatic artery and the hepatic portal vein carry the blood into the sinusoids of the lobules, where oxygen, most of the nutrients, and certain poisons are extracted by the hepatic cells. Nutrients are stored or used to make new materials, and the poisons are stored or detoxified. Products manufactured by the hepatic cells and nutrients needed by other cells are secreted back into the blood. The blood then drains into the central vein and eventually passes into a hepatic vein. Unlike the other products of the liver, bile normally is not secreted into the bloodstream.

The nerve supply to the liver consists of vagal preganglionic parasympathetic fibers and postganglionic sympathetic fibers from the celiac ganglia. Bile is manufactured by the hepatic cells and secreted into *bile capillaries* or *canaliculi* that empty into small ducts. These small ducts eventually merge to form the larger *right* and *left hepatic ducts,* which unite to leave the liver as the *common hepatic duct.* Further on, the common hepatic duct joins the *cystic duct* from the gallbladder, and the two tubes become the *common bile duct,* which empties into the duodenum (see Figure 20–14). The *sphincter of Oddi* is a valve in the common bile duct. When the small intestine is empty the sphincter closes, and the bile is forced up the cystic duct to the gallbladder, where it is stored.

Gallbladder

The **gallbladder** is a sac attached to the underside of the liver (see Figure 20–15a). Its inner walls consist of a mucous membrane arranged in rugae resembling those of the stomach (Figure 20–16). When the gallbladder fills with bile, the rugae allow it to expand to the size and shape of a pear. The middle, muscular coat of the wall consists of smooth muscle fibers. Contraction of these fibers ejects the bile into the cystic duct. The outer coat is the peritoneum.

The gallbladder receives its blood supply from the cystic artery, a branch of the right hepatic artery. The cystic veins pass directly into the hepatic portal vein. The nerve supply of the gallbladder is similar to that of the liver.

Before you begin to read the next section dealing with the small intestine, refer to Figure 20–17, a cross section of the abdomen at the level of the pancreas. Examine the figure carefully and note the relationship of the digestive organs to each other.

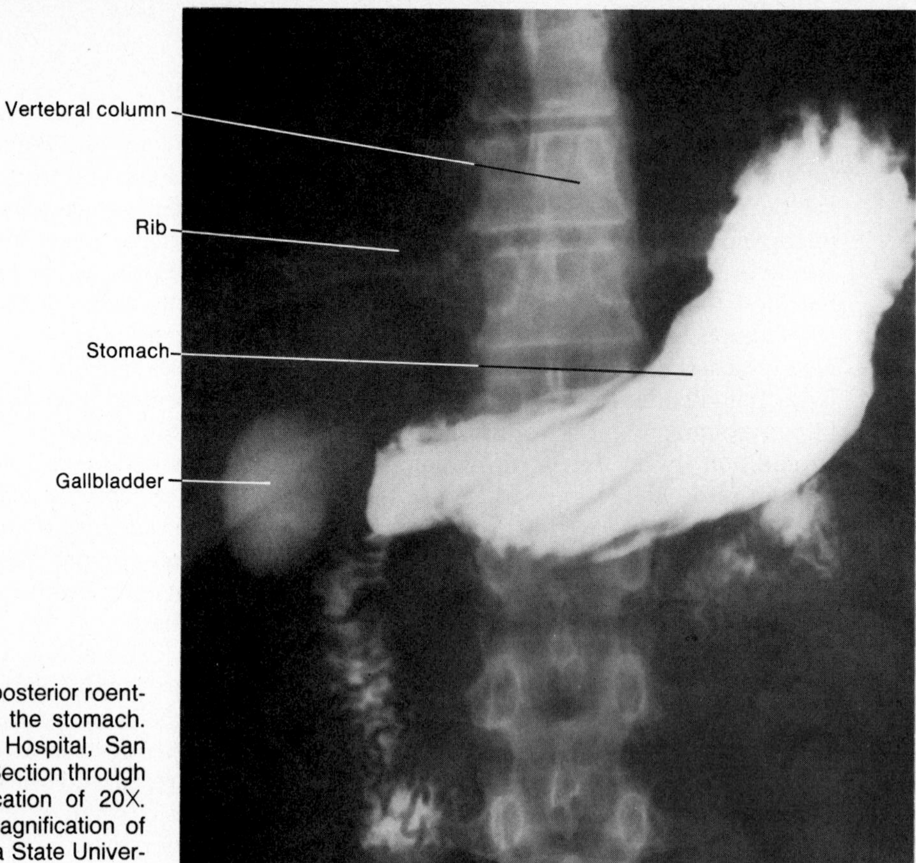

Figure 20–16. The gallbladder. (a) Anteroposterior roentgenogram of the gallbladder in relation to the stomach. (Courtesy of John C. Bennett, St. Mary's Hospital, San Francisco.) (b) Histology of the gallbladder. Section through the wall of the gallbladder at a magnification of 20X. (c) Enlarged aspect of the mucosa at a magnification of 100X. (Courtesy of Victor B. Eichler, Wichita State University.)

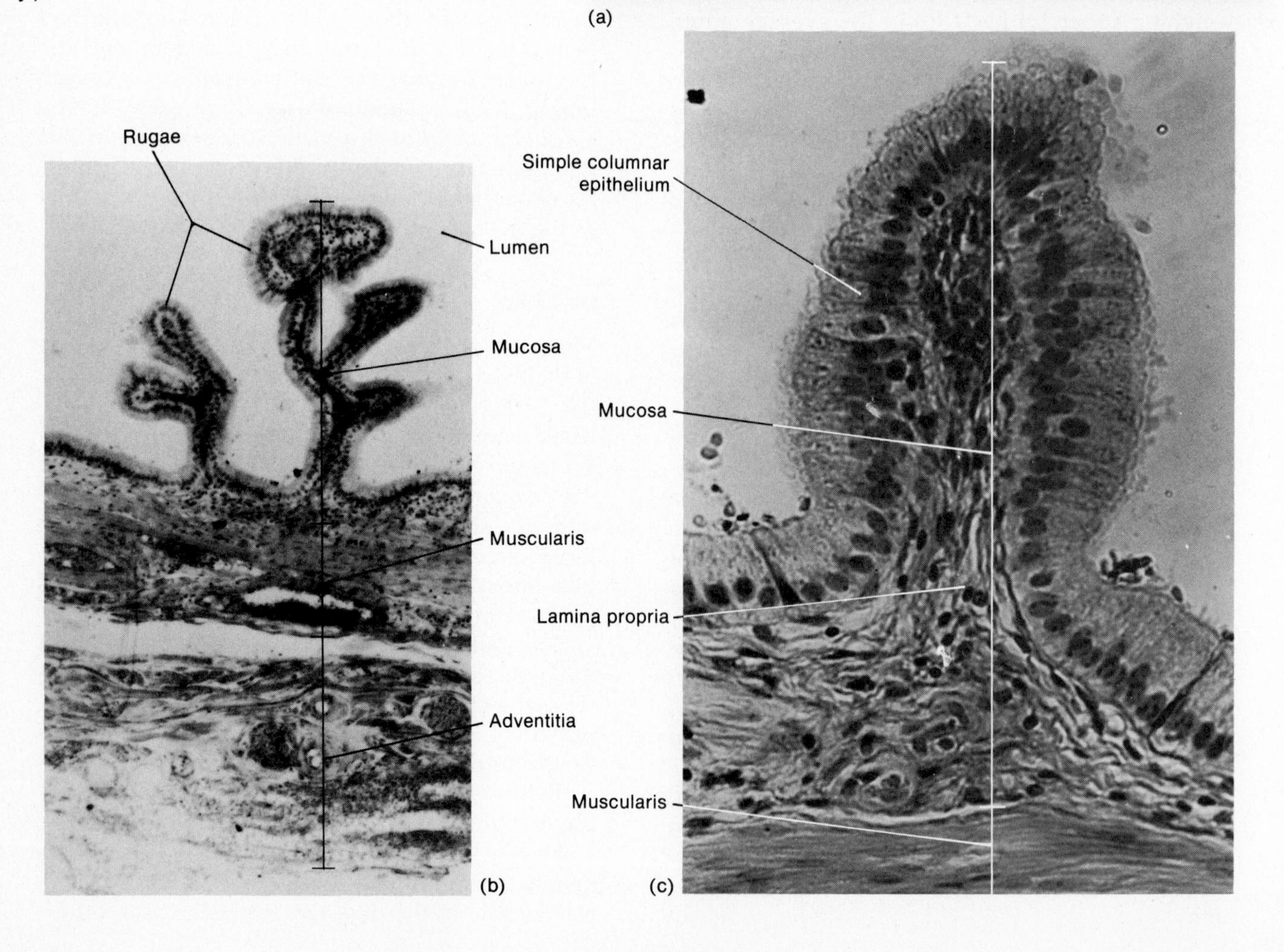

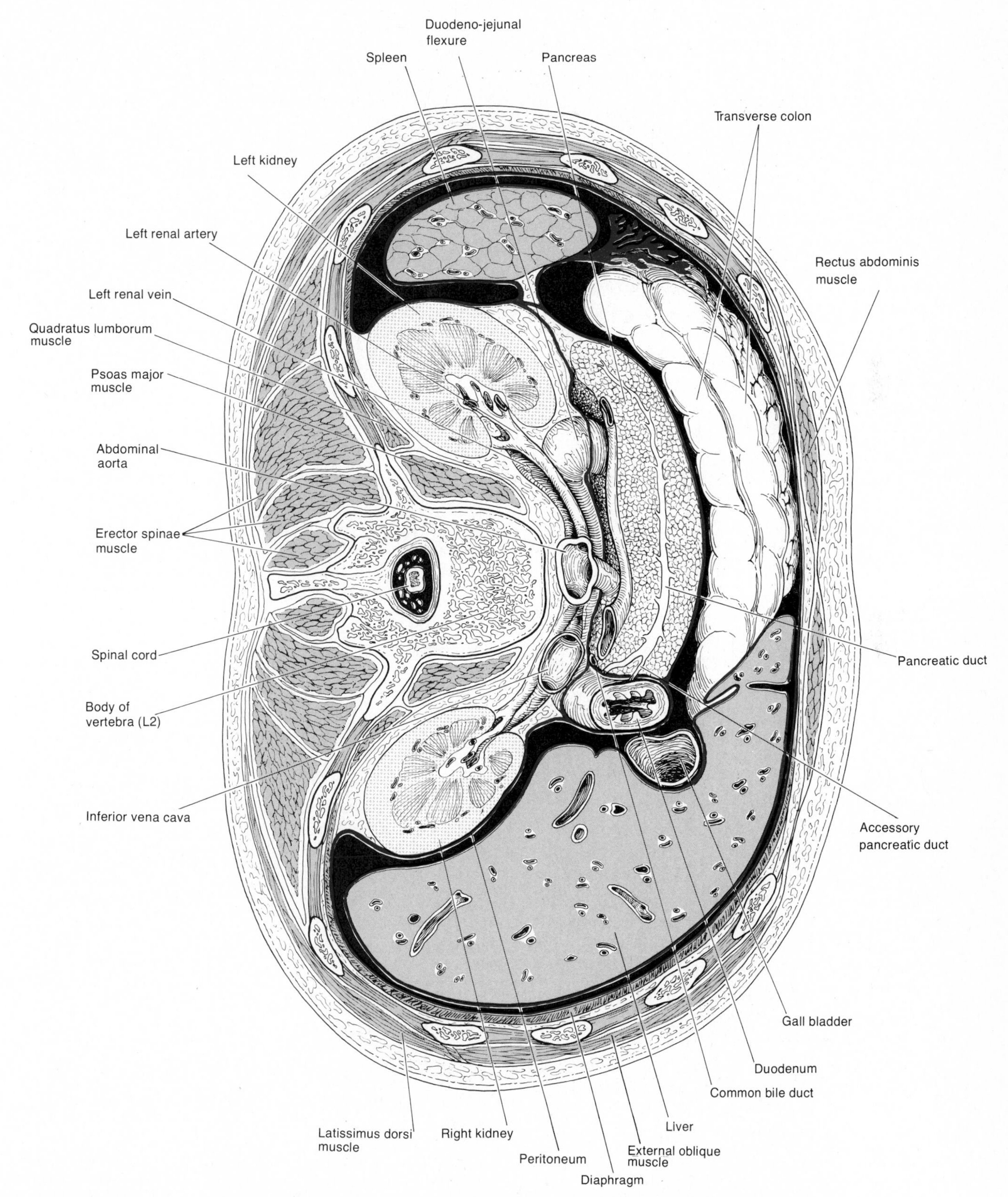

Figure 20–17. Cross section of the abdomen at the level of the pancreas.

Small intestine

The major portions of digestion and absorption occur within a long tube called the **small intestine.** The small intestine begins at the pyloric valve of the stomach, coils through the central and lower part of the abdominal cavity, and eventually opens into the large intestine (see Figure 20–1). In a living human it averages about 2.5 cm (1 in.) in diameter and 3 m (10 to 12 ft) in length.

Anatomy

The small intestine is divided into three segments: duodenum, jejunum, and ileum. The *duodenum,* the broadest part of the small intestine, originates at the pyloric valve of the stomach and extends about 25 cm (10 to 12 in.) until it merges with the jejunum. The *jejunum* is about 1 m (3 to 4 ft) long and extends to the ileum. The final portion of the small intestine, the *ileum,* measures about 2 m (6 to 7 ft) and joins the large intestine at the *ileocecal valve.* A roentgenogram of the normal small intestine is shown in Figure 20–18a.

The walls of the intestine are composed of the same four coats that make up most of the GI tract. However, both the mucosa and submucosa are modified to allow the small intestine to complete the processes of digestion and absorption. The mucosa contains many pits lined with glandular epithelium. These pits—the *intestinal glands,* or *crypts of Lieberkühn*—secrete the intestinal digestive enzymes (Figure 20–18d). The submucosa of the duodenum contains *Brunner's glands,* which secrete mucus to protect the walls of the small intestine from the action of the enzymes. *Succus entericus* is the name for the composite of all the intestinal secretions.

Since almost all the absorption of nutrients occurs in the small intestine, its walls need to be specially equipped to do this job. The epithelium covering and lining the mucosa consists of simple columnar epithelium. Some of the epithelial cells have been transformed to goblet cells, which secrete additional mucus. The rest contain microvilli—fingerlike projections of the plasma membrane (see Figure 2–7a). Digested nutrients diffuse more quickly into the intestinal wall because the microvilli increase the surface area of the plasma membrane.

The mucosa lies in a series of 0.5- to 1.5-mm high projections called *villi,* giving the intestinal mucosa its velvety appearance (Figure 20–18d). The enormous number of villi (4 to 5 million) vastly increases the surface area of the epithelium available for the epithelial cells specializing in absorption. Each villus is lined with the lamina propria, the connective tissue layer of the mucosa. Embedded in this connective tissue are an artery, a venule, a capillary, and a *lacteal* (lymphatic vessel). Nutrients that diffuse through the adjacent epithelial cells are able to pass through the walls of the capillary and lacteal and enter the blood.

A third set of projections called *plicae circulares* further increases the surface area for absorption. The plicae are deep folds in the mucosa and submucosa (Figure 20–18b). Some of the folds extend all the way around the circumference of the intestine, and others extend only partway around.

The muscularis of the small intestine consists of two layers of smooth muscle. The outer, thinner layer contains longitudinally arranged fibers, and the inner, thicker layer contains circularly arranged fibers. Except for a major portion of the duodenum, the serosa, or visceral peritoneum, completely covers the small intestine. The histological aspects of the small intestine are shown in Figure 20–19.

There is an abundance of lymphatic tissue in the walls of the small intestine. Single lymph nodules, called *solitary lymph nodules,* are most numerous in the lower part of the ileum. Aggregated lymph nodules, referred to as *Peyer's patches,* are also most numerous in the ileum.

The arterial blood supply of the small intestine is from the superior mesenteric artery and the gastroduodenal artery, coming indirectly from the common hepatic artery. Blood is returned by way of the superior mesenteric vein which, with the splenic vein, forms the hepatic portal vein.

The nerves to the small intestine are supplied by the superior mesenteric plexus. The branches of the plexus contain postganglionic sympathetic fibers, preganglionic parasympathetic fibers, and afferent fibers. The afferent fibers are both vagal and of spinal nerves. In the wall of the small intestine are two autonomic plexuses, the plexus of Auerbach between the muscular layers, and the plexus of Meissner in the submucosa. The nerve fibers are derived chiefly from the sympathetic division of the autonomic nervous system and partly from the vagus.

Chemical digestion

The digestion of carbohydrates, proteins, and lipids in the small intestine requires the combined actions of the secretions from the pancreas, liver, and intestinal glands.

SECRETIONS. Each day the liver secretes about 800 to 1,000 ml (almost 1 qt) of the yellow, brownish, or olive-green liquid called *bile.* Bile consists of wa-

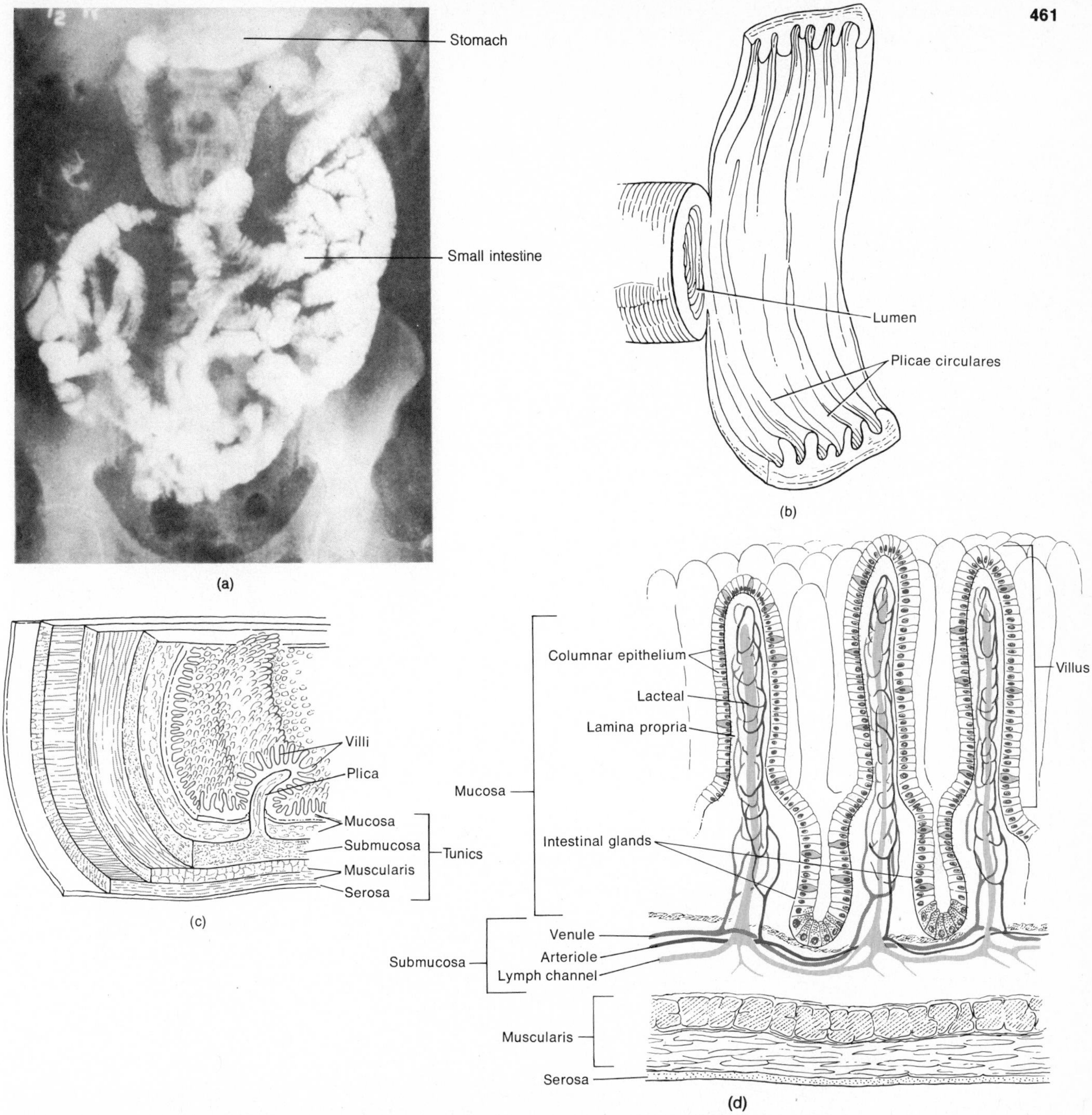

Figure 20–18. The small intestine. (a) Anteroposterior roentgenogram of the normal small intestine ½ hour after taking a barium "meal." (Courtesy of Lester W. Paul and John H. Juhl, *The Essentials of Roentgen Interpretation,* 3d ed., Harper & Row, Publishers, Inc., New York, 1972.) (b) Section of small intestine cut open to expose plicae circulares. (c) Villi in relation to the coats of the small intestine. (d) Enlarged aspect of several villi.

ter, bile salts, bile acids, a number of lipids, and two pigments called biliverdin and bilirubin. Bile is partially an excretory product and partially a digestive secretion. When red blood cells are broken down, iron, globin, and bilirubin are released. The iron and globin are recycled, but the bilirubin is excreted into the bile ducts. Bilirubin eventually is broken down in the intestines, and its breakdown products give feces their color. If the liver is unable to remove bilirubin from the blood or if the bile ducts are obstructed, large amounts of bilirubin circulate through the bloodstream and collect in other tissues, giving the skin and eyes a yellow color called *jaundice.* Other substances found in bile aid in the digestion of fats and are required for their absorption.

Each day the pancreas produces 1,200 to 1,500 ml (about 1 to 1½ qt) of a clear, colorless liquid called *pancreatic juice.* Pancreatic juice consists mostly of water, some salts, sodium bicarbonate,

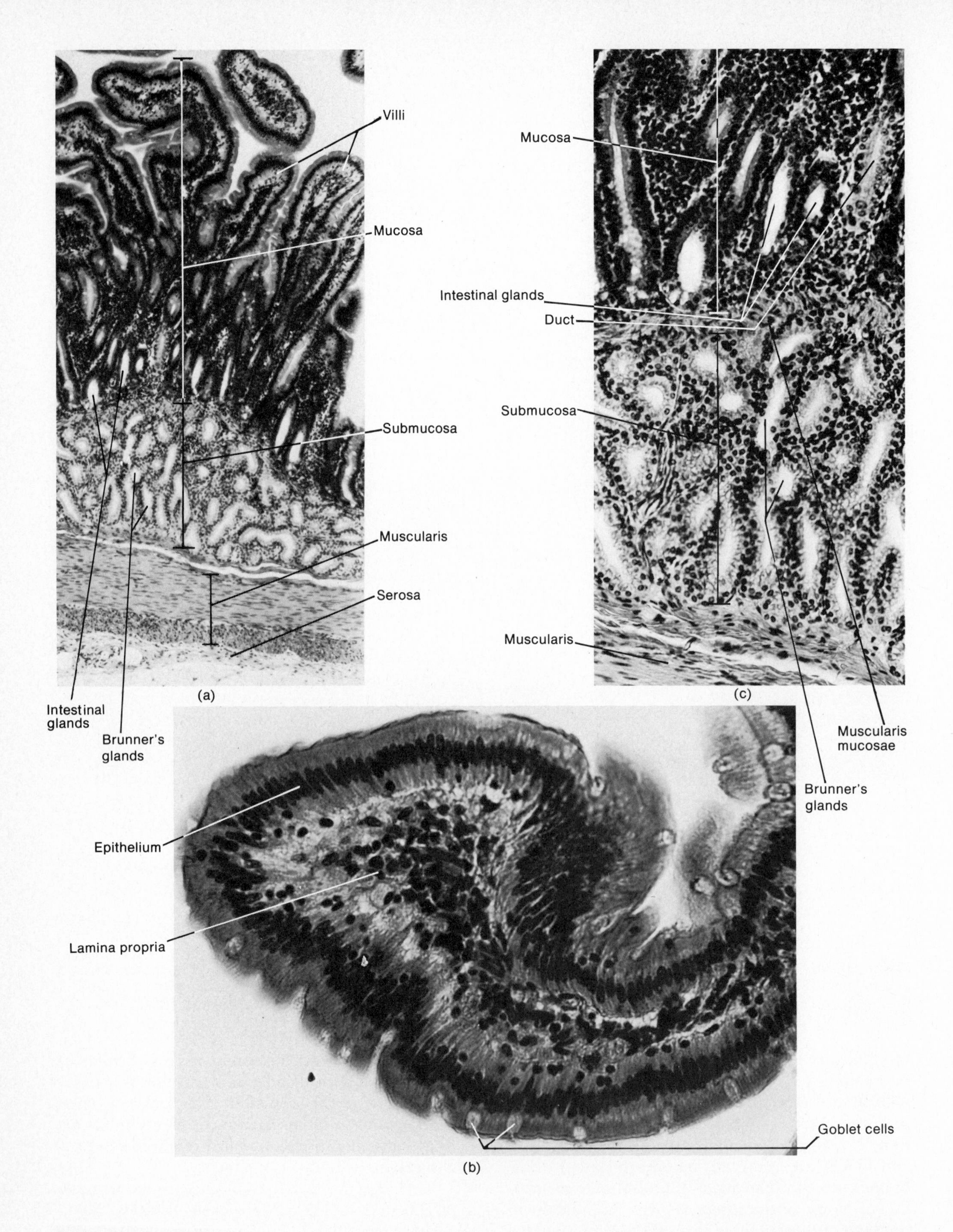

Figure 20–19. Histology of the small intestine. (a) Section through the duodenum showing various layers at a magnification of 15X. (b) Enlarged aspect of a villus at a magnification of 80X. (c) Enlarged aspect of intestinal glands at a magnification of 40X. (Courtesy of Victor B. Eichler, Wichita State University.)

and enzymes. The sodium bicarbonate gives pancreatic juice a slightly alkaline pH (7.1 to 8.2) that stops the action of pepsin from the stomach and creates the proper environment for the enzymes in the small intestine. The enzymes in pancreatic juice include a carbohydrate-digesting enzyme, several protein-digesting enzymes, and the only active fat-digesting enzyme in the adult body.

The intestinal juice, or *succus entericus,* is a clear yellow fluid secreted in amounts of about 2 to 3 liters (2 to 3 qt)/day. It has a pH of 7.6, which is slightly alkaline, and contains water, mucus, and enzymes that complete the digestion of carbohydrates and proteins.

THE DIGESTIVE PROCESS. When chyme reaches the small intestine, the carbohydrates and proteins have been only partially digested and are not ready for absorption. Lipid digestion has not even begun. Digestion in the small intestine continues as follows:

1. *Carbohydrates.* In the mouth, the carbohydrates are broken down into dextrins containing several monosaccharide units. Even though the action of salivary amylase may continue in the stomach, very few of the carbohydrates are reduced to disaccharides by the time chyme leaves the stomach. *Pancreatic amylase,* an enzyme in pancreatic juice, breaks dextrins into the disaccharides maltose, sucrose, and lactose. Next, three enzymes in the intestinal juice digest the disaccharides into monosaccharides. *Maltase* splits maltose into two molecules of glucose, *sucrase* breaks sucrose into a molecule of glucose and a molecule of fructose, and *lactase* digests lactose into a molecule of glucose and a molecule of galactose. This completes the digestion of carbohydrates.
2. *Proteins.* Protein digestion starts in the stomach, where most of the proteins are fragmented into short chains of amino acids called peptones and proteoses. Three enzymes found in pancreatic juice continue the digestion. *Trypsin* digests any intact proteins into peptones and proteoses, breaks the peptones and proteoses into dipeptides (containing only two amino acids), and breaks some of the dipeptides into single amino acids. *Chymotrypsin* duplicates trypsin's activities. *Carboxypeptidase,* the third enzyme, reduces whole or partially digested proteins to amino acids. To prevent these enzymes from digesting the proteins in the cells of the pancreas, they are secreted in inactive forms—trypsin as *trypsinogen,* activated by an intestinal enzyme called *enterokinase;* chymotrypsin as *chymotrypsinogen,* activated by trypsin; and carboxypeptidase as *procarboxypeptidase,* also activated by trypsin. Protein digestion is completed by several intestinal enzymes grouped together under the name *erepsin,* which converts all the remaining dipeptides into single amino acids. Single amino acids can be absorbed.
3. *Lipids.* In an adult, almost all lipid digestion occurs in the small intestine. The first step in the process is the *emulsification* of fats, which is a function of bile. Bile salts break the globules of fat into tiny droplets (emulsification) so the fat-splitting enzyme can get at the lipid molecules more easily. In the second step, *pancreatic lipase,* an enzyme found in pancreatic juice, hydrolyzes each fat molecule into glycerol and fatty acids, the end products of fat digestion.

Mechanical digestion

In the small intestine, three distinct types of movement occur as a result of contractions of the longitudinal and circular muscles. These movements are *rhythmic segmentation, pendular movements,* and *propulsive peristalsis.* Rhythmic segmentation and pendular movements are strictly localized contractions occurring in areas containing food. The two movements act to mix the chyme with the digestive juices and bring every particle of food into contact with the mucosa for absorption. They do not push the intestinal contents along the tract. Rhythmic segmentation starts with the contractions of some of the circular muscle fibers in a portion of the intestine, an action that constricts the intestine into segments. Next, muscle fibers that encircle the middle of each segment also contract, dividing each segment into two smaller segments. Finally, the fibers which contracted first relax, and each small segment unites with an adjoining small segment so that large segments are reformed. This sequence of events is repeated at a rate of 12 to 16 times a minute, sloshing the chyme back and forth and back and forth. Pendular movements consist of alternating contractions and relaxations of the longitudinal muscles. The contractions cause a portion of the intestine to shorten and lengthen, an action that also sends the chyme spilling back and forth.

The third kind of movement, propulsive peristalsis, propels the chyme onward through the intestinal tract. Peristaltic movement in the intestine is similar to that in the esophagus. In the intestine, these waves may be as slow as 5 cm (2 in.)/minute or as fast as 50 cm (20 in.)/second.

Absorption

All the chemical and mechanical phases of digestion occurring from the mouth down through the small intestine are directed toward changing foods into

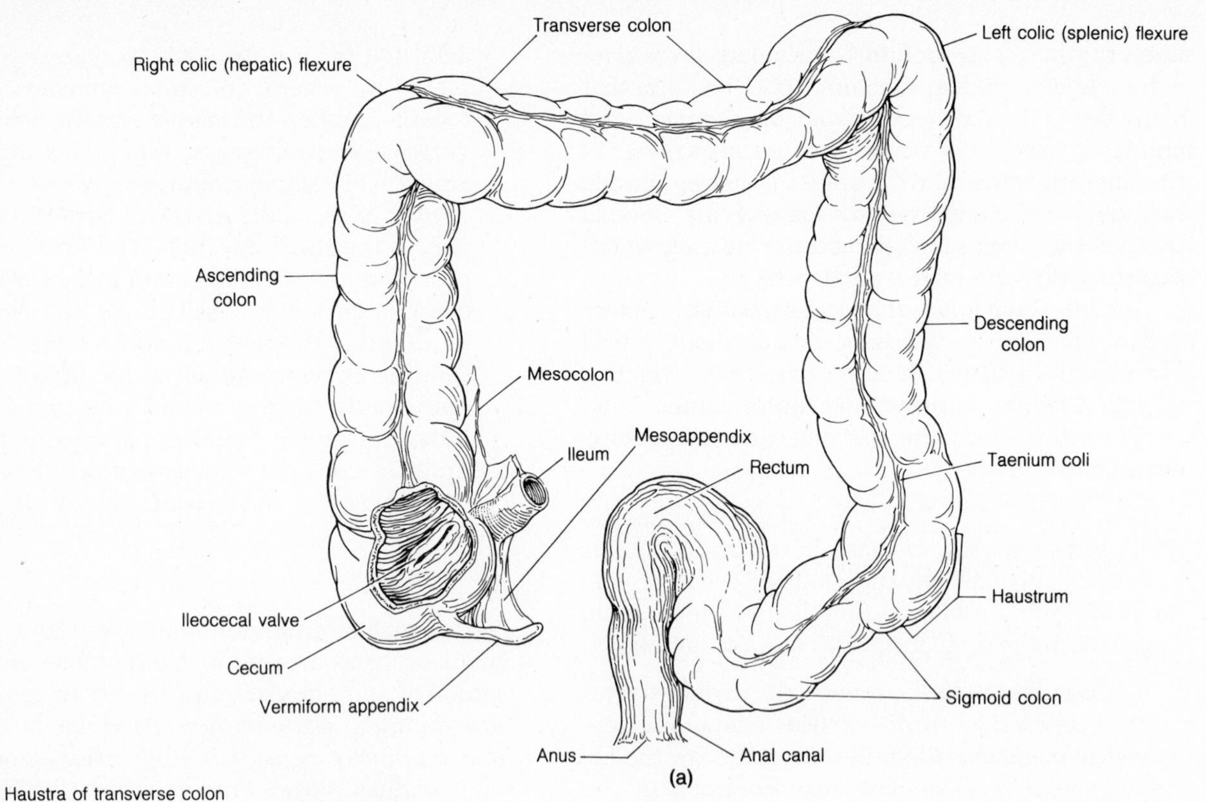

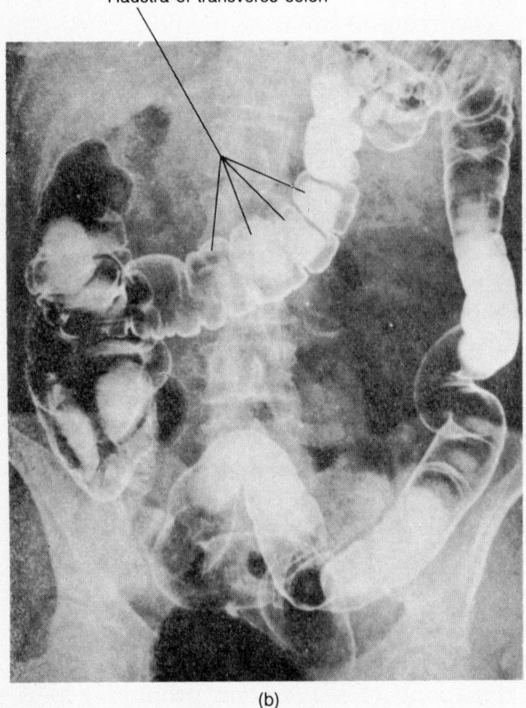

Figure 20–20. The large intestine. (a) Anatomy of the large intestine. (b) Anteroposterior roentgenogram of the large intestine in which several haustra are clearly visible. (Courtesy of Lester W. Paul and John H. Juhl, *The Essentials of Roentgen Interpretation,* 3d ed., Harper & Row, Publishers, Inc., New York, 1972.)

forms that can diffuse through the epithelial cells lining the mucosa into the underlying blood and lymph vessels. The diffusible forms are monosaccharides (glucose, fructose, and galactose), amino acids, fatty acids, and glycerol. Passage of these digested nutrients from the alimentary canal into the blood or lymph is called **absorption.**

About 90 percent of all absorption takes place throughout the length of the small intestine. The other 10 percent occurs in the stomach and large intestine. Absorption of materials in the small intestine occurs specifically through the villi (Figures 20–18 and 20–19) and depends on diffusion, osmosis, and active transport. Monosaccharides and amino acids are absorbed into the blood capillaries of the villi and are transported in the bloodstream to the liver via the hepatic portal system. Fatty acids and glycerol do not enter the bloodstream immediately. Glycerol is a water-soluble compound that passes rather easily into the lacteals of the villi. Fatty acids are not soluble and must combine with bile salts before they become soluble and pass into the lacteals. Once inside the lacteals, the fatty acids break apart from the bile salts and recombine with glycerol to form fats. The fat is carried as tiny droplets called *chylomicrons* to the thoracic duct, into the left subclavian vein, and eventually into the systemic circulation. It is finally delivered to the liver through the hepatic artery. Most of the products of carbohydrate, protein, and lipid digestion are processed by the liver before they are delivered to the other cells of the body. Large amounts of water, electrolytes, mineral salts, and some vitamins also are absorbed in the small intestine.

In summary, then, the principal chemical activity of the small intestine is to digest all foods into forms that are usable by body cells. Any undigested materials that are left behind are processed in the large intestine.

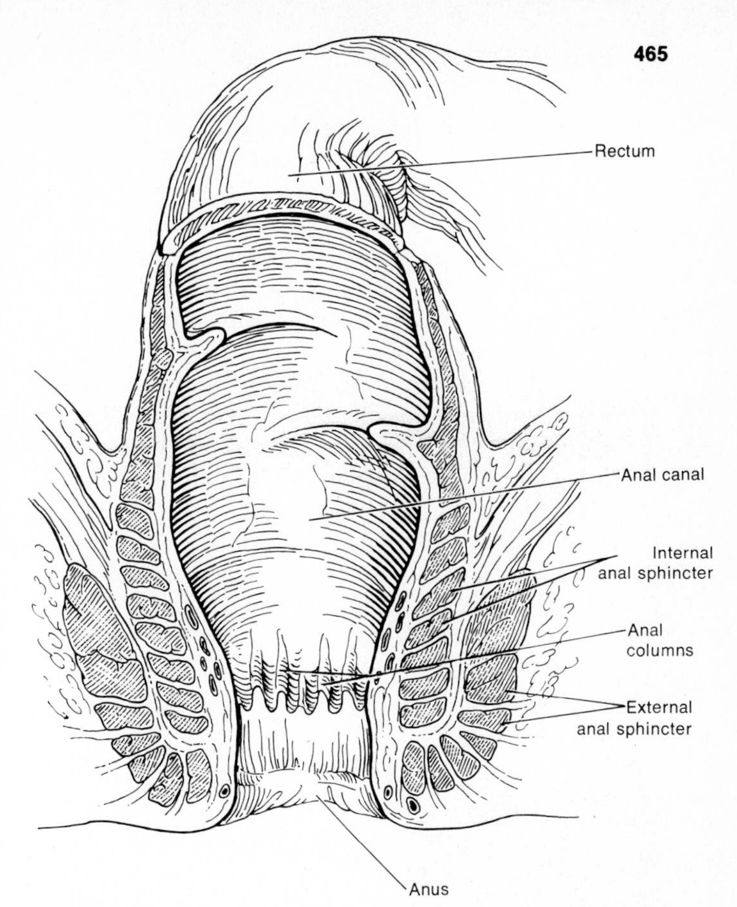

Figure 20–21. Longitudinal section through the anal canal.

Large intestine

The overall functions of the large intestine are the completion of absorption, the manufacture of some vitamins, the formation of feces, and the expulsion of feces from the body.

Anatomy

The **large intestine** is about 1.5 m (5 ft) in length and averages 6.5 cm (2.5 in.) in diameter. It extends from the ileum to the anus (see Figure 20–1) and is attached to the posterior abdominal wall by its mesocolon. Structurally, the large intestine is divided into four principal regions: the cecum, colon, rectum, and anal canal (Figure 20–20). Let us now look at the parts of the large intestine in the order in which food passes through them.

The opening from the ileum into the large intestine is guarded by a fold of mucous membrane called the *ileocecal valve.* This structure allows materials from the small intestine to pass into the large intestine but prevents them from moving in the opposite direction. Hanging below the ileocecal valve is the *cecum,* a blind pouch about 6 cm (2 to 3 in.) long. Attached to the cecum is a twisted, coiled tube, measuring about 8 cm (3 in.) in length on the average, that is called the *vermiform appendix* (*vermis* = worm). The mesentery of the appendix, called the mesoappendix, attaches the appendix to the inferior part of the ileum and adjacent part of the posterior abdominal wall. Inflammation of the appendix is called *appendicitis.*

The open end of the cecum merges with a long tube called the *colon.* Based on location, the colon is divided into ascending, transverse, descending, and sigmoid portions. The *ascending colon* ascends on the right side of the abdomen, reaches the undersurface of the liver, and turns abruptly to the left. Here it forms the *right colic (hepatic) flexure.* The colon continues across the abdomen to the left side as the *transverse colon.* It curves beneath the lower end of the spleen on the left side as the *left colic (splenic) flexure* and passes downward to the level of the iliac crest as the *descending colon.* The *sigmoid colon* is the S-shaped portion that begins at the iliac crest, projects inward to the midline, and terminates as the rectum at about the level of the third sacral vertebra.

The *rectum,* the last 20 cm (7 to 8 in.) of gastrointestinal tract, lies anterior to the sacrum and coccyx. The terminal 2 to 3 cm of the rectum is referred to as the *anal canal* (Figure 20–21). Internally, the mucous membrane of the anal canal is arranged in longitudinal folds called *anal columns* that contain a network of arteries and veins. Inflammation and enlargement of the anal veins is known as *hemorrhoids* or *piles.* The opening of the anal canal to the exterior is called the *anus.* It is guarded by an internal sphincter of smooth muscle and an external sphincter of skeletal muscle. Normally the anus is closed except during the elimination of the wastes of digestion.

The wall of the large intestine differs from that of the small intestine in several respects. No villi or permanent circular folds are found in the mucosa, which does, however, contain simple columnar epithelium with numerous goblet cells (Figure 20–22). These cells secrete mucus that lubricates the colonic contents as they pass through the colon. Solitary lymph nodes also are found in the mucosa. The submucosa of the large intestine is similar to that found in the rest of the alimentary canal. The muscularis consists of an external layer of longitudinal muscles and an internal layer of circular muscles. Unlike other parts of the digestive tract, however, the longitudinal muscles do not form a continuous sheet around the wall but are broken up into three flat bands called *taeniae coli.* Each band runs the length of the large intestine. Tonic contractions of the bands gather the colon into a series of pouches called *haustra,* which give the colon its puckered appearance. The serosa of the large intestine is part of the visceral peritoneum.

The arterial supply of the cecum and colon is derived from branches of the superior mesenteric and inferior mesenteric arteries. The venous return is by way of the superior and inferior mesenteric veins and ultimately to the hepatic portal vein and into the liver. The arterial supply of the rectum and anal canal is derived from the superior, middle, and inferior rectal arteries. The rectal veins correspond to the rectal arteries.

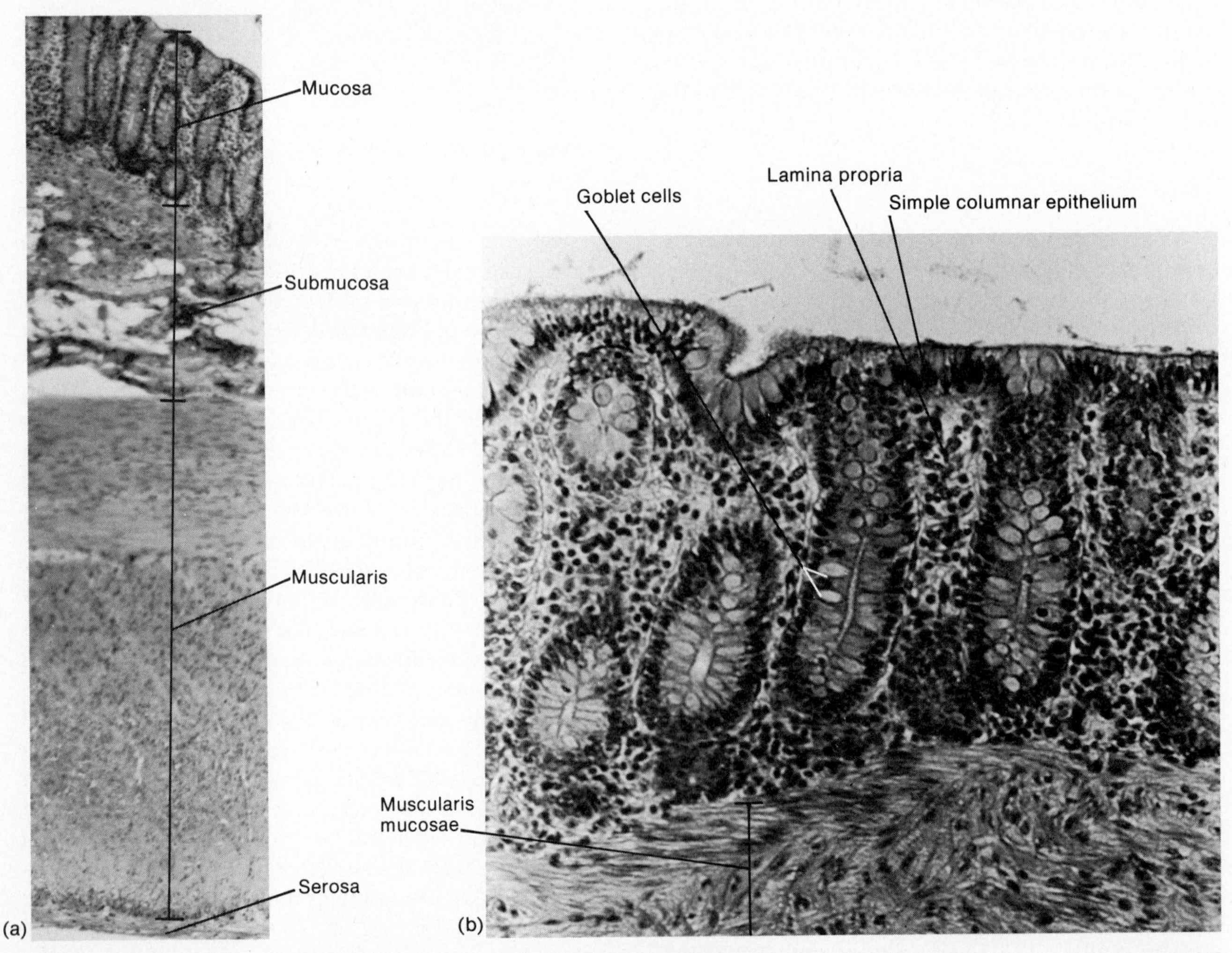

Figure 20–22. Histology of the large intestine. (a) Section through the wall of the large intestine at a magnification of 20X. (b) Enlarged aspect of the mucosa of the large intestine at a magnification of 50X. (Courtesy of Victor B. Eichler, Wichita State University.)

The nerves to the large intestine consist of sympathetic, parasympathetic, and afferent components. The sympathetic innervation is derived from the celiac, superior and inferior mesenteric, and internal iliac plexuses. The fibers reach the plexuses by way of the thoracic and lumbar splanchnic nerves. The parasympathetic innervation is derived from the vagus and pelvic splanchnic nerves.

Activities of the large intestine

The principal activities of the large intestine are concerned with mechanical movements, absorption, and the formation and elimination of feces.

MOVEMENTS. Movements of the colon begin when substances enter through the ileocecal valve. Since chyme moves through the small intestine at a fairly constant rate, the time required for a meal to pass into the colon is determined by gastric evacuation time. As food passes through the ileocecal valve, it fills the cecum and accumulates in the ascending colon.

One movement characteristic of the large intestine is *haustral churning.* In this process the haustra remain relaxed and distended while they fill up. When the distention reaches a certain point, the walls contract and squeeze the contents into the next haustrum. Peristalsis also occurs, although at a slower rate than in other portions of the tract (3 to 12 contractions per minute). A final type of movement is *mass peristalsis,* a very strong peristaltic wave that drives the colonic contents into the rectum. Food in the stomach initiates this reflex action in the colon. Thus, mass peristalsis usually takes place three or four times a day, during or immediately after a meal is eaten.

ABSORPTION AND FORMATION OF FECES. By the time the intestinal contents arrive at the large intestine, digestion and absorption are almost complete.

By the time the chyme has remained in the large intestine for about 3 to 10 hours, it has become solid or semisolid as a result of absorption and is now known as *feces.* Chemically, feces consist of water, inorganic salts, mucus, epithelial cells from the mucosa of the alimentary canal, bacteria, products of bacterial decomposition, and undigested parts of food not attacked by bacteria.

Mucus is secreted by the glands of the large intestine, but no enzymes are secreted. The mucus serves as a lubricant to aid the movement of the colonic materials and acts as a protective covering for the mucosa.

Chyme is prepared for elimination in the large intestine by the action of bacteria. These bacteria ferment any remaining carbohydrates and release hydrogen, carbon dioxide, and methane gas. They also convert remaining proteins to amino acids and break down the amino acids into simpler substances, such as indole, skatole, hydrogen sulfide, and fatty acids. Some of the indole and skatole is carried off in the feces and contributes to its smell, and the rest is absorbed. Bacteria also decompose bilirubin, the breakdown product of red blood cells that is excreted in bile, to simpler pigments which give feces its color. Intestinal bacteria also aid in the synthesis of several vitamins needed for normal metabolism, including some B vitamins (riboflavin, nicotinic acid, biotin, and folic acid) and vitamin K.

Although most water absorption occurs in the small intestine, the large intestine absorbs enough to make it an important organ in maintaining the water balance of the body. Intestinal water absorption is greatest in the cecum and ascending colon. The large intestine also absorbs inorganic solutes plus some products of bacterial action, including vitamins and large amounts of indole and skatole. The indole and skatole are transported to the liver, where they are converted to less toxic compounds that are excreted in the urine.

DEFECATION. Mass peristaltic movement pushes fecal material from the sigmoid colon into the rectum. The resulting distention of the rectal walls stimulates pressure-sensitive receptors, initiating a reflex for **defecation,** which is emptying of the rectum. Contraction of the longitudinal rectal muscles shortens the rectum, thereby increasing the pressure inside it. The pressure forces the sphincters open, and the feces are expelled through the anus. Voluntary contractions of the diaphragm and abdominal muscles aid defecation by increasing the pressure inside the abdomen, which pushes the walls of the sigmoid colon and rectum inward. If defecation does not occur, the feces remain in the rectum until the next wave of peristalsis again stimulates the pressure-sensitive receptors, creating an awareness of the desire to defecate.

APPLICATIONS TO HEALTH

Dental caries

Dental caries, or tooth decay, involve a gradual disintegration of the enamel and dentin (Figure 20–23a). If this condition remains untreated, various microorganisms may invade the pulp cavity, causing infection and inflammation of the living tissue. If the pulp is destroyed, the tooth is pronounced "dead."

No individual microbe is responsible for dental caries, but oral bacteria that create a pH of 5.5 or lower start the process. Acids can come directly from foods, such as the ascorbic acid of citrus fruits, or they may be breakdown products of carbohydrates. Microbes that digest carbohydrates include two bacteria, *Lactobacillus acidophilus* and streptococci, as well as some yeasts. Research suggests that the streptococci break down carbohydrates into *dental plaque,* a polysaccharide which adheres to the tooth surface. When other bacteria digest the plaque, acid is produced. Saliva cannot reach the tooth surface to buffer the acid because the plaque covers the teeth.

Certain measures can be taken to prevent dental caries. First, the diet of the mother during pregnancy is very important in forestalling tooth decay of the newborn. Simple, balanced meals are the best diet during pregnancy. Supplementation with multivitamins, with emphasis on vitamin D, and the minerals calcium and phosphorus is customary because they are responsible for normal bone and teeth development.

Other preventive measures have centered around fluoride treatment because teeth are less susceptible to acids when they are permeated with fluoride. Fluoride may be incorporated in the drinking water or applied topically to erupted teeth. Maximum benefit often occurs when fluoride is used in drinking water during the period when teeth are being calcified. Excessive fluoride may cause a light brown to brownish-black discoloration of the enamel of the permanent teeth called mottling.

Brushing the teeth immediately after eating removes the plaque from flat surfaces before the bacteria have a chance to go to work. Dentists also suggest that the plaque between the teeth be removed every 24 hours with dental floss.

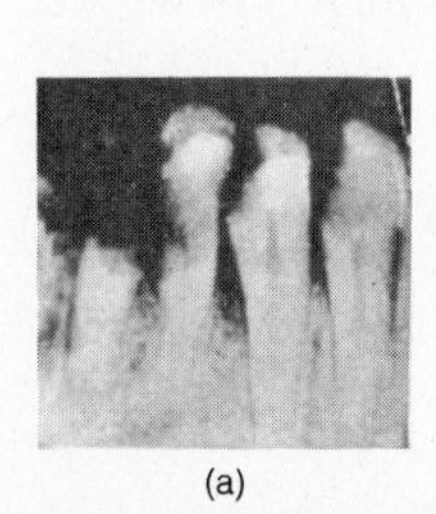

(a)

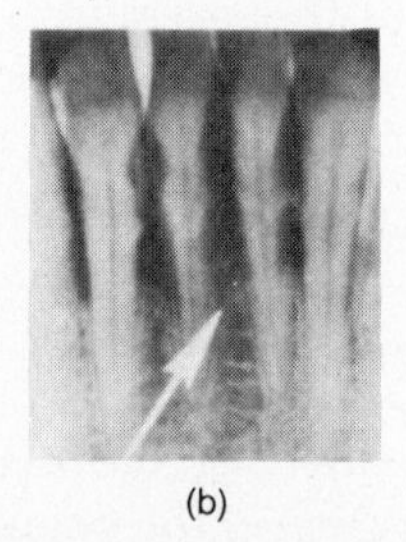

(b)

Figure 20–23. Diseases of the teeth. (a) Anteroposterior roentgenogram of dental caries. (b) Anteroposterior roentgenogram of periodontitis in which the alveolar bone (arrow) has been destroyed. (Courtesy of Lester W. Paul and John H. Juhl, *The Essentials of Roentgen Interpretation,* 3d ed., Harper & Row, Publishers, Inc., New York, 1972.)

Periodontal diseases

Periodontal disease is a collective term for a variety of conditions characterized by inflammation and/or degeneration of the gingivae, alveolar bone, periodontal membrane, and cementum. The initial symptoms are enlargement and inflammation of the soft tissue and bleeding gums. Without treatment, the soft tissue may deteriorate and the alveolar bone may be resorbed, causing loosening of the teeth and receding of the gums (Figure 20–23b).

Periodontal diseases are frequently caused by local irritants, such as bacteria, impacted food, cigarette smoke, or by a poor "bite." The latter may put a strain on the tissues supporting the teeth. Methods of prevention and treatment include good mouth care to remove plaque and other sources of irritation. Periodontal diseases may also be caused by allergies, vitamin deficiences, and a number of systemic disorders, especially those that affect bone, connective tissue, or circulation. In these cases, the systemic disorder must be treated as well.

Peptic ulcers

An **ulcer** is a craterlike lesion in a membrane. Ulcers that develop in areas of the alimentary canal exposed to acid gastric juice are called *peptic ulcers.* Peptic ulcers occasionally develop in the lower end of the esophagus. However, most of them occur on the lesser curvature of the stomach, in which case they are called *gastric ulcers,* or in the first part of the duodenum, where they are called *duodenal ulcers* (Figure 20–24).

The cause of ulcers is obscure. However, hypersecretion of acid gastric juice seems to be the immediate cause in the production of duodenal ulcers and in the reactivation of healed ulcers. Hypersecretion of acid gastric juice is not implicated as much in gastric ulcer patients because the stomach walls are highly adapted to resist gastric juice through their secretion of mucus. A possible cause of gastric ulcers is hyposecretion of mucus. Hypersecretion of pepsin also may contribute to ulcer formation.

Among the factors believed to stimulate an increase in acid secretion are certain foods or medications, such as alcohol, coffee, or aspirin, and overstimulation of the vagus nerve. Normally, the mucous membrane lining the stomach and duodenal walls resists the secretions of hydrochloric acid and pepsin, and no ulcer develops. In some people, however, this resistance breaks down, and an ulcer develops.

The danger inherent in ulcers is the erosion of the muscular portion of the wall of the stomach or

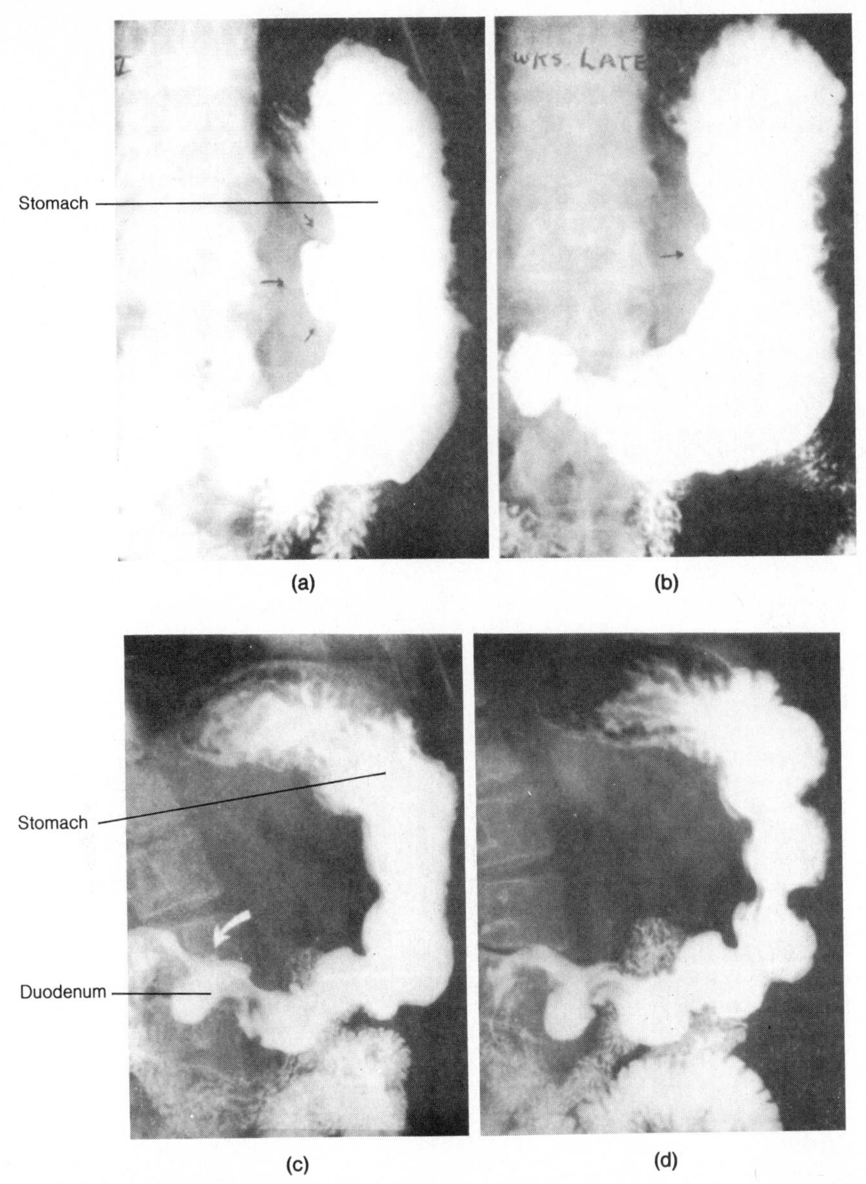

Figure 20–24. Peptic ulcers. (a) Anteroposterior roentgenogram of gastric ulcers (arrows) at the time of diagnosis. (b) The same ulcer 3 weeks after treatment. (c) Anteroposterior roentgenogram of a duodenal ulcer (arrow) prior to treatment. (d) The same ulcer after treatment. (Courtesy of Lester W. Paul and John H. Juhl, *The Essentials of Roentgen Interpretation,* 3d ed., Harper & Row, Publishers, Inc., New York, 1972.)

duodenum. This could damage blood vessels and possibly produce fatal hemorrhage. If an ulcer erodes all the way through the wall, the condition is called *perforation.* Perforation allows bacteria and partially digested food to pass into the peritoneal cavity, producing peritonitis.

Peritonitis

Peritonitis is an acute inflammation of the serous membrane lining the abdominal cavity and covering the abdominal viscera. One possible cause is contamination of the peritoneum by pathologic bacteria from the external environment. This contamination could result from accidental or surgical wounds in the abdominal wall or from perforation of organs exposed to the outside environment. Another possible cause is perforation of the walls of organs that contain bacteria or chemicals which are normally beneficial to the organ but are toxic to the peritoneum. For example, the large intestine contains colonies of bacteria that live on undigested nutrients and break them down so they can be eliminated more easily. But if the bacteria enter the peritoneal cavity, they attack the cells of the peritoneum for food and produce acute infection. As another example, the normal bacteria of the female reproductive tract protect the tract by giving off acid wastes that produce an acid environment unfavorable to many yeasts, protozoa, and bacteria which might otherwise attack the tract. However, these acid-producing bacteria are harmful to the peritoneum. A third cause may be chemical irritation. The peritoneum does not have any natural barriers that keep it from being irritated or digested by chemical substances such as bile and digestive enzymes. However, it does contain a great deal of lymphatic tissue and can fight infection fantastically well. The danger stems from the fact that the peritoneum is in contact with most of the abdominal organs. If the infection gets out of hand, it may destroy vital organs and bring on death. For these reasons, perforation of the alimentary canal from an ulcer or perforation of the uterus from an incompetent abortion are considered serious. If a surgeon plans to do extensive surgery on the colon, he may give the patient high doses of antibiotics for several days preceding surgery to kill intestinal bacteria and reduce the risk of peritoneal contamination.

Cirrhosis

Cirrhosis is a chronic disease of the liver in which the parenchymal (functional) liver cells are replaced by fibrous connective tissue, a process called *stromal repair.* Often there is a lot of replacement with adipose connective tissue as well. The liver has a high ability for parenchymal regeneration, so stromal repair occurs whenever any parenchymal cell is killed or when damage to the cells occurs continuously over a long time. These conditions could be caused by *hepatitis* (inflammation of the liver), certain chemicals that may destroy liver cells, parasites that sometimes infect the liver, and alcoholism.

Tumors

Both benign and malignant **tumors** occur in all parts of the gastrointestinal tract. The benign growths are much more common, but the malignant tumors are responsible for 30 percent of all deaths from cancer in the United States (Figure 20–25a). To achieve relatively early diagnosis, complete periodic routine examinations are necessary. Cancers of the mouth usually are detected through routine dental checkups.

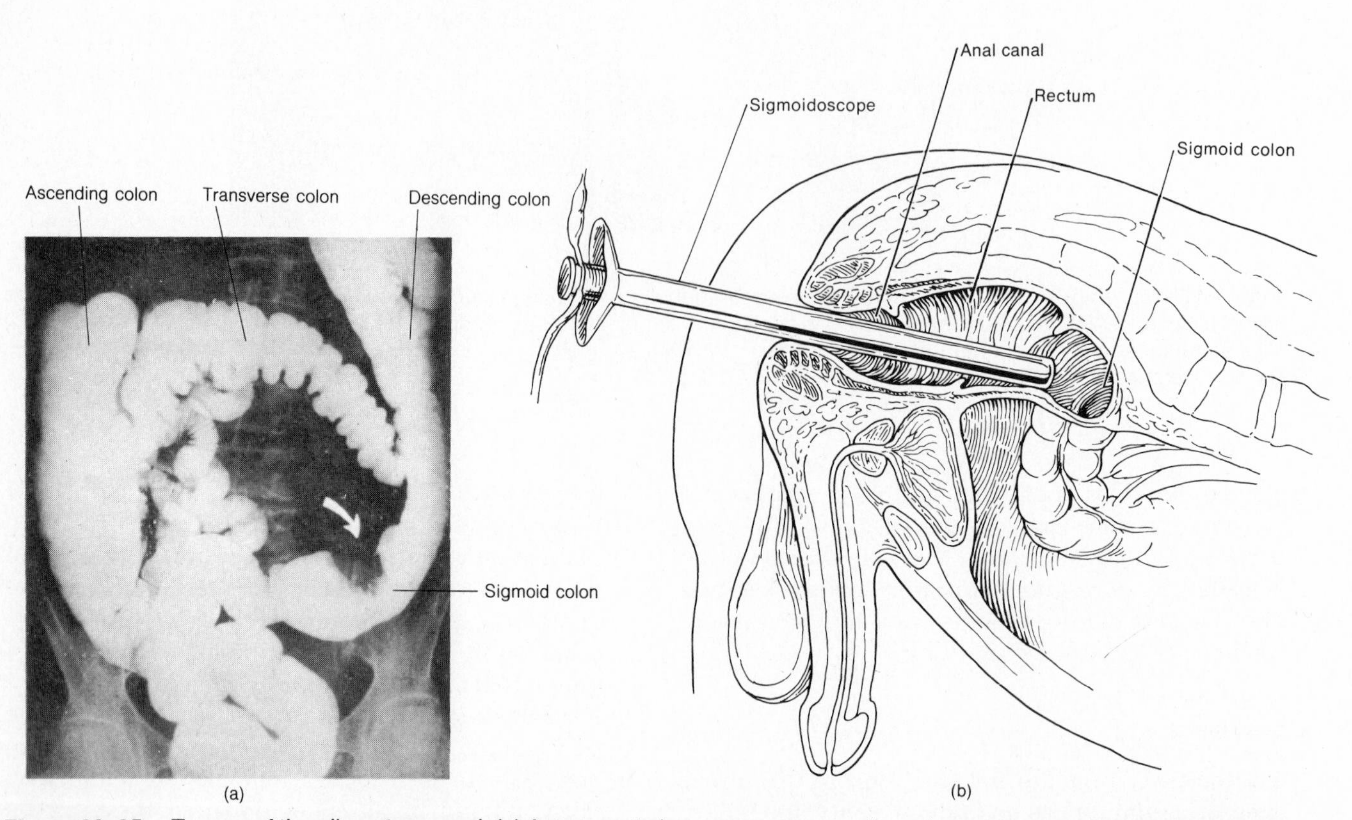

Figure 20–25. Tumors of the alimentary canal. (a) Anteroposterior roentgenogram of carcinoma (arrow) of the sigmoid colon. (Courtesy of Lester W. Paul and John H. Juhl, *The Essentials of Roentgen Interpretation,* 3d ed., Harper & Row, Publishers, Inc., New York, 1972.) (b) Detection of carcinomas by use of the sigmoidoscope.

A regular physical checkup should include rectal examination. Fifty percent of all rectal carcinomas are within reach of the finger, and 75 percent of all colonic carcinomas can be seen with the sigmoidoscope (Figure 20–25b). Both the fiberoptic sigmoidoscope and the more recent fiberoptic endoscope are flexible tubular instruments composed of a light and many tiny glass fibers. They allow visualization, magnification, and even photography of almost the entire length of the gastrointestinal tract and have been invaluable in the correct diagnosis of a wide range of gastrointestinal disorders without surgery. Another test in a routine examination for intestinal disorders is the filling of the gastrointestinal tract with barium, which is either swallowed or given in an enema. Barium, a mineral, shows up on x-rays the same way that calcium appears in bones. Tumors as well as ulcers can be diagnosed this way. The only definitive treatment of gastrointestinal carcinomas is surgery.

Key terms associated with the digestive system

Calculus (*calc* = stone) A stone in an organ. Enteroliths (*entero* = intestine; *lith* = stone, calcification) are intestinal stones or calculi.

Cholecystitis Inflammation of the gallbladder that often leads to infection. Some cases are caused by obstruction of the cystic duct with bile stones. Stagnating bile salts irritate the mucosa. Dead mucosal cells provide medium for bacteria.

Cholelithiases (*chole* = bile, gall) are calculi composed of bile salts, lecithin, and cholesterol that are formed in the gallbladder or its ducts and are also called *gallstones.*

Colitis An inflammation of the colon and rectum. Inflammation of the mucosa reduces absorption of water and salts, producing watery, bloody feces, and—in severe cases—dehydration and salt depletion. Irritated muscularis spasms produce cramps.

Colostomy The cutting of the colon and bringing the proximal end to the surface of the abdomen to serve as a substitute anus. Feces are eliminated through the upper end. A temporary colostomy may be done to allow a badly inflamed colon to rest and heal. Later the two halves may be rejoined and the abdominal opening closed. If the rectum is removed for malignancy, the colostomy provides a permanent outlet for feces.

Constipation Infrequent or difficult defecation.

Diarrhea Frequent defecation of liquid feces.

Flatus Excessive amounts of air(gas) in the stomach or intestine, usually expelled through the anus. If the gas is expelled through the mouth, it is called *belching* (burping). Flatus may result from gas released during the breakdown of foods in the stomach or from swallowing air or gas-containing substances such as carbonated drinks.

Heartburn A burning sensation in the region of the esophagus and stomach. It may result from regurgitation of gastric contents into the lower end of the esophagus or from distention stemming from causes such as the retention of regurgitated food and gastric contents in the lower esophagus.

Hepatitis (*hepato* = liver) A liver inflammation. It may be caused by organisms such as viruses, bacteria, and protozoa or by the absorption of materials, such as carbon tetrachloride and certain anesthetics and drugs, that are toxic to liver cells.

Hernia Protrusion of an organ or part of an organ through a membrane or through the wall of a cavity, usually the abdominal cavity. *Diaphragmatic hernia* is the protrusion of the lower esophagus, stomach, or intestine into the thoracic cavity through the hole in the diaphragm that allows passage of the esophagus. *Umbilical hernia* is the protrusion of abdominal organs through the naval area of the abdominal wall. *Inguinal hernia* is the protrusion of the hernial sac containing the intestine into the inguinal opening. It may extend into the scrotal compartment, causing strangulation of the herniated part.

Lesion Any pathological or traumatic discontinuity of tissue or loss of function of a part.

Mumps Viral disease causing painful inflammation and enlargement of the salivary glands, particularly the parotids. In adults, the sex glands and pancreas may be involved. Inflammation of testes may cause male sterility.

Nausea Discomfort preceding vomiting. Possibly, it is caused by distention or irritation of the gastrointestinal tract, most commonly the stomach.

Pancreatitis Inflammation of the pancreas. The pancreas secretes active trypsin instead of trypsinogen, and the trypsin digests the pancreatic cells and blood vessels.

Periodontal disease Diseases of the tissues surrounding the teeth. The following diseases are included: *Gingivitis* is an inflammation of the gingivae characterized by swelling, redness, and bleeding. *Periodontitis* is a progression of gingivitis to destruction of alveolar bone. *Gingivostomatitis,* better known as trench mouth, is an ulceration of the gingivae, the oral or pharyngeal mucosa, or the tonsils.

Vomiting Expulsion of stomach (and sometimes duodenal) contents through the mouth by reverse peristalsis. The abdominal muscle walls forcibly empty the stomach.

Chapter summary in outline

DIGESTION

1. Digestion is a series of chemical and mechanical processes by which foods are reduced to a form that the body can use.
2. Digestion occurs in the organs of the gastrointestinal tract and depends on the functioning of accessory organs as well.

3. The basic arrangement of tissues in the gastrointestinal tract from the inside outward is mucosa, submucosa, muscularis, and serosa.
4. Extensions of the peritoneum include the mesentery, mesocolon, falciform ligament, lesser omentum, and greater omentum.

MOUTH AND SALIVARY GLANDS
1. The mouth is formed by cheeks, palates, and tongue which aid mechanical digestion.
2. The teeth project into the mouth and are adapted for mechanical digestion.
3. The salivary glands produce saliva that lubricates foods and starts the chemical digestion of carbohydrates.

PHARYNX AND ESOPHAGUS
1. Both organs assume a role in deglutition, or swallowing.
2. When a bolus is swallowed, the respiratory tract is sealed off and the bolus moves into the esophagus.
3. Peristaltic movements of the esophagus pass the bolus into the stomach.

STOMACH
1. The stomach begins at the bottom of the esophagus and ends at the pyloric valve.
2. Adaptations of the stomach for digestion include rugae that permit distention; glands which produce mucus, hydrochloric acid, and enzymes; and a three-layered muscularis for efficient mechanical movements.
3. Proteins are chemically digested into peptones and proteoses through the action of pepsin in the stomach.
4. The stomach also stores food, produces the intrinsic factor, and carries on some absorption.

PANCREAS, LIVER, AND GALLBLADDER
1. Pancreatic acini produce enzymes that enter the duodenum via the pancreatic duct. Pancreatic enzymes digest proteins, carbohydrates, and fats.
2. Cells of the liver produce bile, which is needed to emulsify fats. Bile is stored in the gallbladder and passed into the duodenum via the common bile duct.

SMALL INTESTINE
1. This organ extends from the pyloric valve to the ileocecal valve.
2. It is very highly adapted for digestion and absorption. Its glands produce enzymes and mucus, and its wall contains microvilli, villi, and plicae circulares.
3. The enzymes of the small intestine digest carbohydrates, proteins, and fats into the end products of digestion: monosaccharides, amino acids, fatty acids, and glycerol.
4. Mechanical digestion in the small intestine involves rhythmic segmentation, pendular movements, and propulsive peristalsis.
5. Absorption is the passage of the end products of digestion from the alimentary canal into the blood or lymph.
6. Absorption in the small intestine occurs through the villi. Monosaccharides and amino acids pass into the blood capillary, and fatty acids and glycerol pass into the lacteal.

LARGE INTESTINE
1. This organ extends from the ileocecal valve to the anus.
2. Mechanical movements of the large intestine include haustral churning, mass peristalsis, and peristalsis.
3. The large intestine functions in the synthesis of several vitamins and in water absorption from chyme, leading to feces formation.
4. The elimination of feces from the rectum is called defecation. Defecation is a reflex action aided by voluntary contractions of the diaphragm and abdominal muscles.

APPLICATIONS TO HEALTH
1. Dental caries are started by acid-producing bacteria.
2. Periodontal diseases are characterized by inflammation and/or degeneration of gingivae, alveolar bone, peridontal membrane, and cementum.
3. Peptic ulcers are craterlike lesions that develop in the mucous membrane of the alimentary canal in areas exposed to gastric juice.
4. Peritonitis is inflammation of the peritoneum.
5. In cirrhosis, parenchymal cells of the liver are replaced by fibrous connective tissue.
6. Tumors may be detected by sigmoidoscope and barium x-rays.

Review questions and problems

1. Define digestion. Distinguish between chemical and mechanical digestion.
2. Identify the organs of the gastrointestinal tract in sequence. How does the gastrointestinal tract differ from the accessory organs of digestion?
3. Describe the structure of each of the four coats of the gastrointestinal tract.
4. What is the peritoneum? Describe the location and function of the mesentery, mesocolon, falciform ligament, lesser omentum, and greater omentum.
5. What structures form the oral cavity? How does each of the structures contribute to digestion?
6. Make a simple diagram of the tongue. Indicate the location of the papillae and the four taste zones.
7. Describe the location of the salivary glands and their ducts. What are buccal glands?
8. Contrast the histology of the salivary glands.
9. Describe the composition of saliva and the role of each of its components in digestion.
10. What are the principal portions of a typical tooth? What are the functions of each of the parts?
11. Compare deciduous and permanent dentitions with regard to numbers of teeth and times of eruption.
12. Contrast the functions of incisors, cuspids, premolars, and molars. What is pyorrhea?
13. What is a bolus? How is it formed?
14. Define deglutition. List the sequence of events involved in passing a bolus from the mouth to the stomach.
15. Describe the location of the stomach. List and briefly explain the anatomic features of the stomach.
16. What is the importance of rugae, zymogenic cells, parietal cells, and mucous cells in the stomach?
17. What is a gastric ulcer? How is it formed?
18. Where is the pancreas located? Describe the duct system by which the pancreas is connected to the duodenum.
19. What are pancreatic acini? Contrast their functions with those of the islets of Langerhans.
20. Where is the liver located? What are the principal functions of the liver?
21. Draw a labeled diagram of a liver lobule.
22. How is blood supplied to and drained from the liver?
23. Once bile has been formed by the liver, how is it collected and transported to the gallbladder for storage?
24. Where is the gallbladder located? How is the gallbladder connected to the duodenum?

25. What are the subdivisions of the small intestine? How are the coats of the small intestine adapted foı digestion and absorption?
26. Describe the movements that occur in the small intestine.
27. Define absorption. How are the end products of carbohydrate and protein digestion absorbed? How are the end products of fat digestion absorbed?
28. What routes are taken by absorbed nutrients to reach the liver?
29. What are the principal subdivisions of the large intestine? How does the muscularis of the large intestine differ from that of the rest of the digestive tract?
30. Describe the mechanical movements that occur in the large intestine.
31. Explain the activities of the large intestine that change chyme into feces.
32. Define defecation. How does defecation occur?
33. Define dental caries. How are they started? What is dental plaque? What are three preventive measures that can be taken against dental caries?
34. Define periodontal disease, and describe the best method of prevention.
35. What is a peptic ulcer? Distinguish between gastric and duodenal ulcers. Describe some of the suspected causes of ulcers. What is perforation?
36. Define peritonitis. Explain some possible causes. Why is peritonitis a potentially dangerous condition?
37. Define cirrhosis.
38. How are tumors of the alimentary canal detected?
39. Refer to the glossary of key terms associated with the digestive system and be sure that you can define each term.

Selected readings

Davenport, H. W. *Physiology of the Digestive Tract.* 3d ed. Chicago: Yearbook Medical Publishers, 1971.

Dawson, Helen L. *Basic Human Anatomy.* 2d ed. New York: Appleton-Century-Crofts, 1974. Chap. 9.

Gray, Henry. *Anatomy of the Human Body.* 29th ed., edited by Charles Mayo Goss. Philadelphia: Lea and Febiger, 1973. Chap. 16.

Netter, Frank H. *Upper Digestive Tract,* Vol. 3/I, 1959; *Lower Digestive Tract,* Vol. 3/II, 1962; *Liver, Biliary Tract and Pancreas,* Vol. 3/III, 1957. CIBA Collection of Medical Illustrations. Summit, NJ: CIBA Pharmaceutical Co.

CHAPTER 21

THE URINARY SYSTEM

STUDENT OBJECTIVES

After you have read this chapter, you should be able to:

1. Identify the external and internal gross anatomical features of the kidneys
2. Define the structural features of a nephron
3. Describe the blood and nerve supply to the kidneys
4. Describe the location, structure, and physiology of the ureters
5. Describe the location, structure, histology, and function of the urinary bladder
6. Describe the physiology of micturition
7. Compare the causes of incontinence, retention, and suppression
8. Describe the location, structure, and physiology of the urethra
9. Discuss the causes of ptosis, kidney stones, gout, glomerulonephritis, pyelitis, and cystitis
10. Discuss the operational principle of hemodialysis
11. Define key terms associated with the urinary system

The metabolism of nutrients results in the production of wastes by body cells, including carbon dioxide and excesses of water and heat. Protein catabolism produces toxic nitrogenous wastes, such as ammonia and urea. In addition, too many of the essential ions such as sodium, chloride, sulfate, phosphate, and hydrogen tend to be accumulated in the body. All the toxic materials and the excess essential materials must be eliminated by the body.

The primary function of the **urinary system** is to keep the body in homeostasis by controlling the concentration and volume of blood by removing and restoring selected amounts of water and solutes. It also excretes selected amounts of various wastes. Two kidneys, two ureters, one urinary bladder, and a single urethra make up the system (Figure 21–1). The kidneys control the concentration and volume of the blood and remove wastes from

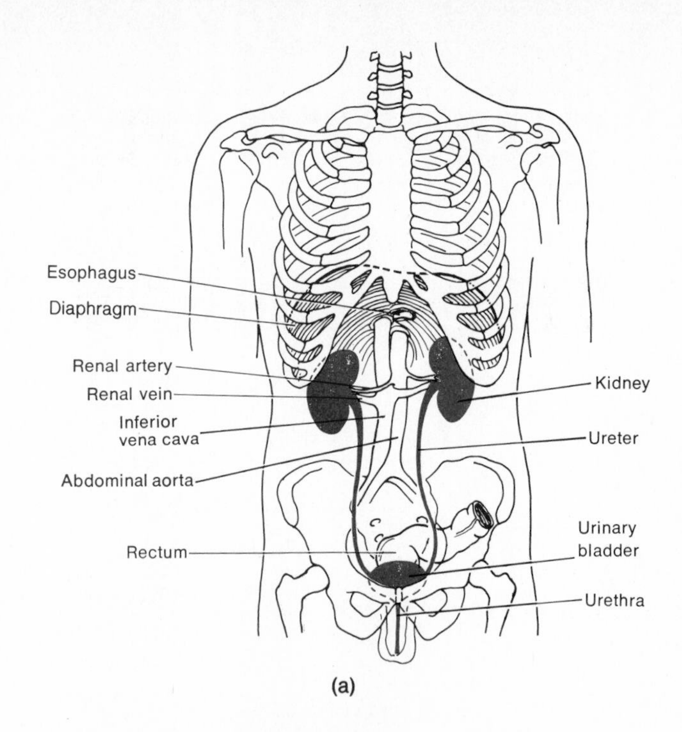

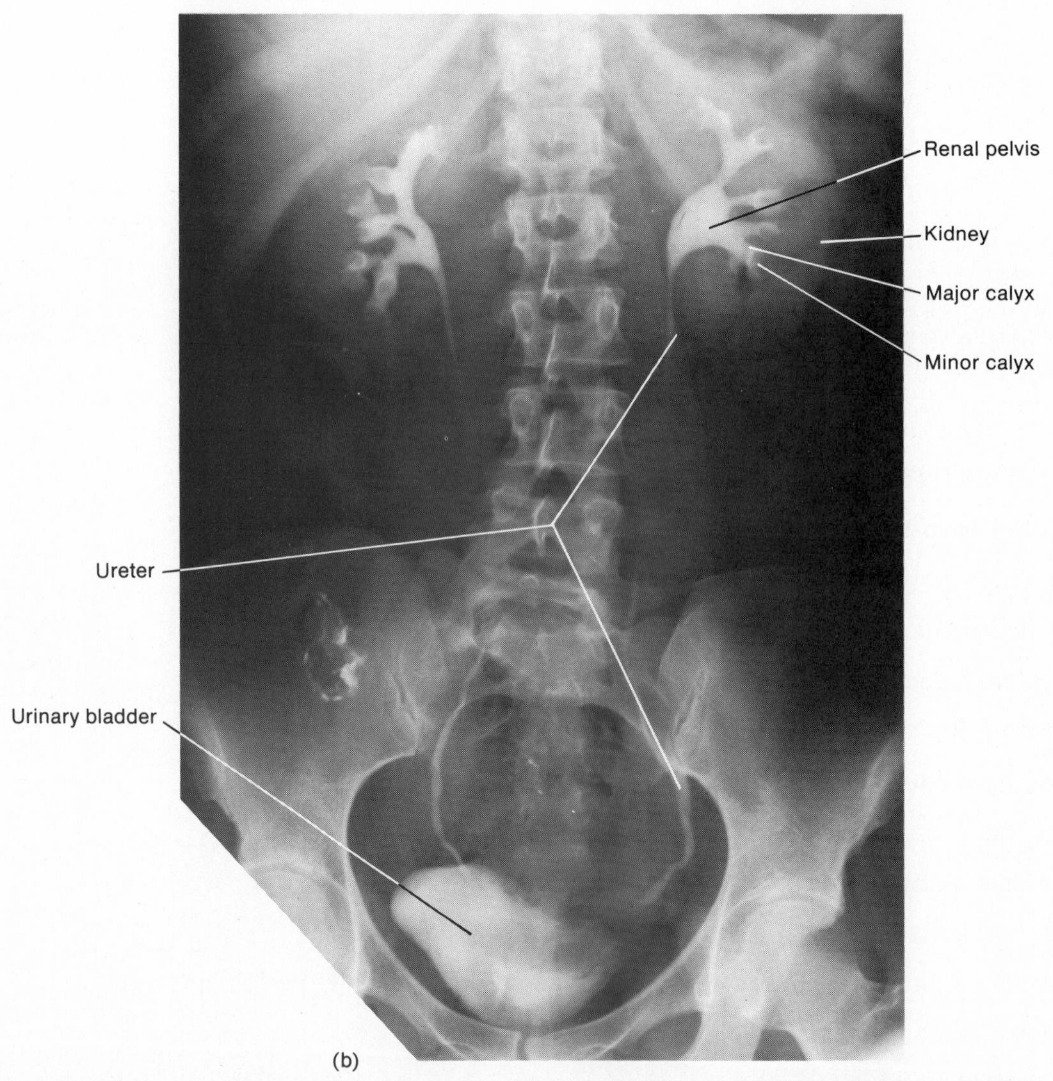

Figure 21–1. Organs of the urinary system. (a) Anterior view. (b) Anteroposterior roentgenogram. (Courtesy of John C. Bennett, St. Mary's Hospital, San Francisco.)

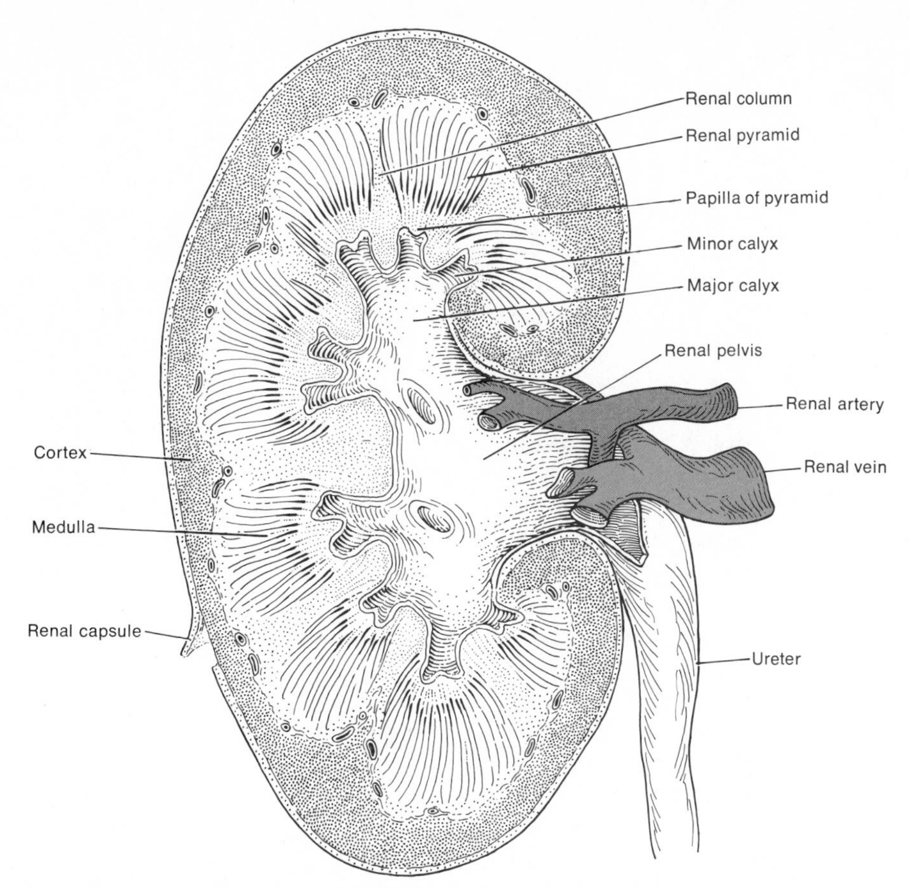

Figure 21–2. Coronal section through the right kidney illustrating gross internal anatomy.

the blood in the form of urine. Urine drains out of each kidney through its ureter and is stored in the urinary bladder until it is expelled from the body through the urethra. Other systems that aid in waste elimination are the respiratory, integumentary, and digestive systems.

THE KIDNEYS

The paired **kidneys** are reddish organs that resemble kidney beans in shape (see Figure 21–1). They are found just above the waist, between the parietal peritoneum and the posterior wall of the abdomen. Since they are external to the peritoneal lining of the abdominal cavity, their placement is described as *retroperitoneal* (see Figure 20–17). Relative to the vertebral column, the kidneys are located between the levels of the last thoracic and third lumbar vertebrae, with the right kidney slightly lower than the left because of the relatively large area occupied by the liver.

External anatomy

The average adult kidney measures about 11.25 cm (4 in.) long, 5.0 to 7.5 cm (2 to 3 in.) wide, and 2.5 cm (1 in.) thick. Its concave medial border faces the vertebral column. Near the center of the concave border is a notch called the *hilum,* through which the ureter leaves the kidney. Blood and lymph vessels and nerves also enter and exit the kidney through the hilum. The hilum is the entrance to a cavity in the kidney called the *renal sinus* (see Figure 21–5).

Three layers of tissue surround each kidney. The innermost layer, the *renal capsule,* is a smooth, transparent, fibrous membrane that can easily be stripped off the kidney and is continuous with the outer coat of the ureter at the hilum. It serves as a barrier against trauma and the spread of infection to the kidney. The second layer, the *adipose capsule,* is a mass of fatty tissue surrounding the renal capsule. It also protects the kidney from trauma and holds it firmly in place in the abdominal cavity. The outermost layer, the *renal fascia,* is a thin layer of fibrous connective tissue that anchors the kidneys to their surrounding structures and to the abdominal wall. Some individuals, especially thin ones in whom either the adipose capsule or renal fascia is deficient, may develop a condition called *ptosis* (dropping) of one or both kidneys. Ptosis is dangerous because it may cause kinking of the ureter with reflux of urine and back pressure. Ptosis of the kidneys below the rib cage also makes these organs more susceptible to blows and penetrating injuries.

Internal anatomy

If you make a coronal section through a kidney, you will see an outer, reddish area called the *cortex* and an inner, reddish-brown region called the *medulla* (Figure 21–2). The cortex is arbitrarily divided into an outer cortical zone and an inner juxtamedullary zone. Likewise, the medulla is also divided into an outer zone (one-third) and an inner zone (two-thirds). Within the medulla are 8 to 18 striated, triangular-shaped structures termed *renal,* or *medullary, pyramids.* The striated appearance is due to the presence of straight tubules and blood vessels (described later). The bases of the pyramids face the cortical area, and their apices, called *renal papillae,* are directed toward the center of the kidney. The cortex is the smooth-textured area extending from the renal capsule to the bases of the pyramids and into the spaces between the pyramids. The cortical substance between the renal pyramids is called the *renal column.* Together, the cortex and renal pyramids constitute the parenchyma of the kidney. Structurally, the parenchyma of each kidney consists of approximately 1 million microscopic units called nephrons, collecting ducts, and their associated vascular supply. Nephrons are the functional units of the kidney. They help form the urine and regulate the blood composition.

In the renal sinus of the kidney is a large cavity called the *renal pelvis* of the kidney. The edge of the pelvis is divided into cuplike extensions called the *major* and *minor calyces.* There are from 2 to 3 major calyces and 7 to 13 minor calyces. Each minor calyx collects urine from collecting ducts of the pyramids. From the major calyces, the urine drains into the pelvis and out through the ureter.

The nephron

The physiological unit of the kidney is referred to as a **nephron** (Figure 21–3). Essentially, each nephron is a *renal tubule.* The parts of a nephron are as follows: glomerular capsule, proximal convoluted tubule, descending limb of Henle, loop of Henle, ascending limb of Henle, and distal convoluted tubule. Let us examine each of these parts of a nephron in detail. It begins as a double-walled globe called the *glomerular* or *Bowman's capsule,* lying in the cortex of the kidney. The inner layer or wall of the capsule, known as the visceral layer, consists of simple squamous epithelium surrounding a capillary network called the *glomerulus.* A space separates the inner wall from the outer one, known as the parietal layer. It is also composed of simple squamous epithelium. Collectively, the Bowman's capsule and the enclosed glomerulus are called a *renal corpuscle.*

The different kinds of epithelium found in the nephron are adapted to perform specialized functions. Simple squamous epithelium provides a semipermeable membrane that offers minimal resistance to the passage of molecules. Water and solutes in the blood are filtered easily through the visceral layer of the Bowman's capsule and pass into the space between the visceral and parietal layers. From here the fluid drains into a tubule, which is subdivided into a number of sections.

The first section of the renal tubule, the *proximal convoluted tubule,* also lies in the cortex. Convoluted means that the tubule is highly coiled rather than straight, and the word proximal refers to the fact that the tubule is nearest its point of origin at the Bowman's capsule. The wall of the proximal convoluted tubule consists of cuboidal epithelium with microvilli. These cytoplasmic extensions, like those of the small intestine, increase the surface area for reabsorption and secretion.

The second section of the tubule, the *descending limb of Henle,* dips into the medulla. It consists of squamous epithelium. The tubule then bends into a U-shaped structure called the *loop of Henle.* As the tubule straightens out, it increases in diameter and ascends toward the cortex as the *ascending limb of Henle,* which consists of cuboidal and low columnar epithelium. In the cortex, the tubule again becomes convoluted. Because of its distance from the point of origin at the Bowman's capsule, this section is referred to as the *distal convoluted tubule.* Like those of the proximal tubule, the cells of the distal tubule are cuboidal. However, unlike the cells of the proximal tubule, the cells of the distal tubule have relatively few microvilli. The distal tubule terminates by merging with a straight *collecting duct.* In the medulla, the collecting ducts receive the distal tubules of several nephrons, pass through the renal pyramids, and open into the calyces of the pelvis through a series of larger ducts called *papillary ducts.* Cells of the collecting ducts are cuboidal, whereas those of the papillary ducts are columnar.

Nephrons are frequently classified into two kinds, based primarily upon the length of certain of their portions and the distribution of the efferent arteriole. This distribution is described in the next section. One type of nephron is called a *cortical nephron* (Figure 21–3b). These nephrons usually

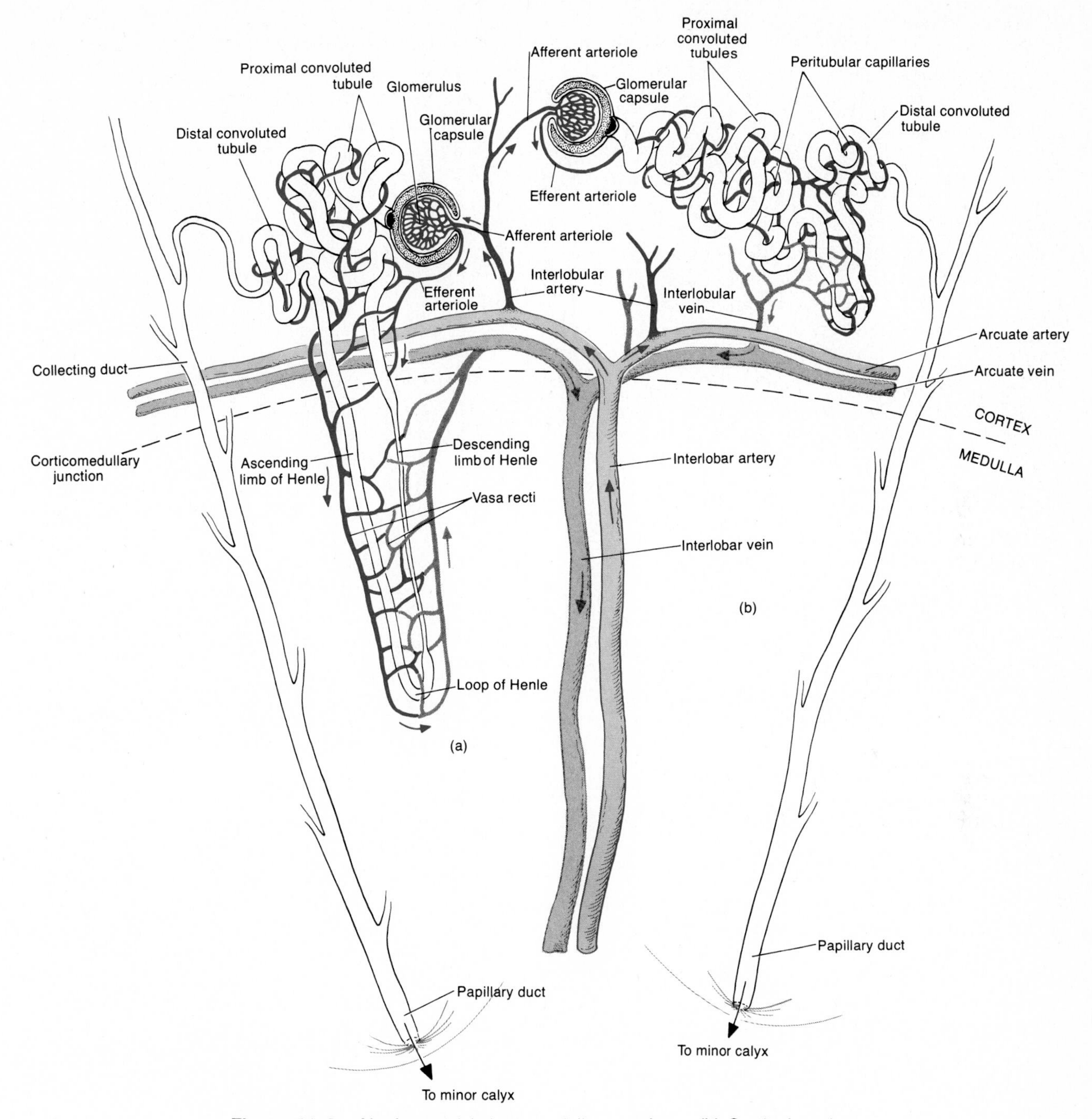

Figure 21–3. Nephrons. (a) Juxtamedullary nephron. (b) Cortical nephron.

have their glomeruli in the outer cortical zone, and the remainder of the nephrons rarely penetrate the medulla. A second type of nephron is called a *juxtamedullary nephron.* These nephrons usually have their glomeruli close to the corticomedullary junction, and other parts of them penetrate deeply into the medulla (Figure 21–3a).

The histology of the various components of a nephron and its glomerulus is shown in Figure 21–4.

Blood and nerve supply

Because the nephrons are responsible for removing wastes from the blood and regulating its fluid and electrolyte content, it should not seem surprising that they are abundantly supplied with blood vessels. The two *renal arteries* transport about one-fourth the total cardiac output to the kidneys (Figure 21–5). Thus, approximately 1,200 ml of blood passes through the kidneys each minute. Before or

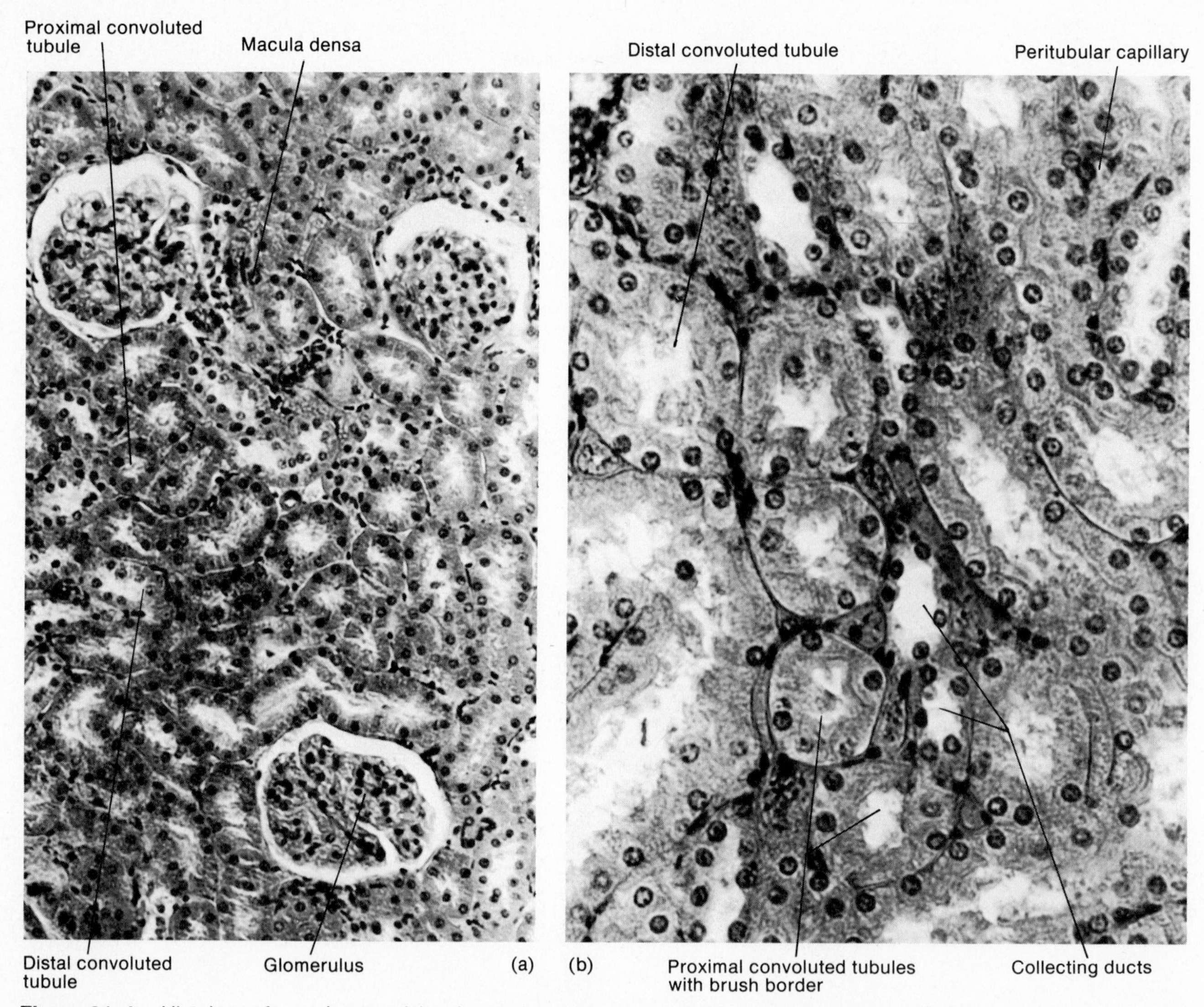

Figure 21–4. Histology of a nephron and the glomerulus. (a) Nephron and glomerulus at a magnification of 100X. (b) Enlarged aspect of tubules, collecting duct, and peritubular capillaries at a magnification of 100X. (Courtesy of Victor B. Eichler, Wichita State University.)

immediately after entering through the hilum, the renal artery divides into several branches that enter the parenchyma and pass as the *interlobar arteries* between the renal pyramids in the renal columns. At the base of the pyramids, the interlobar arteries arch between the medulla and cortex and become known as the *arcuate arteries.* Divisions of the arcuate arteries produce a series of *interlobular arteries* (see Figure 21–3). The interlobular arteries enter the cortex and divide into *afferent arterioles.* One afferent arteriole is distributed to each glomerular capsule, where the arteriole breaks up into the capillary network termed the *glomerulus.* The glomerular capillaries then reunite to form an *efferent arteriole,* leading away from the capsule, that is smaller in diameter than the afferent arteriole. This situation is unique because blood usually flows out of capillaries into venules and not into other arterioles. Each efferent arteriole of a cortical nephron divides to form a network of capillaries, called the *peritubular capillaries* around the convoluted tubules. The efferent arteriole of a juxtamedullary nephron also forms peritubular capillaries, and, in addition, it forms long loops of thin-walled vessels which dip down alongside the loop of Henle into the medullary region of the papillae. These blood vessels are called *vasa recta.* The peritubular capillaries eventually reunite to form *interlobular veins.* The blood then drains through the *arcuate veins* to the *interlobar veins* running between the pyramids and leaves the kidney through a single *renal vein* that exits at the hilum (Figure 21–5). The vasa recta pass blood into the arcuate vein. From here, it goes to the interlobar veins and then into the renal vein.

The nerve supply to the kidneys is derived from the *renal plexus* of the autonomic system. Nerves from the plexus accompany the renal arteries and their branches and are distributed to the vessels. Because the nerves are vasomotor, they regulate the circulation of blood in the kidney by regulating the diameters of the small blood vessels.

The juxtaglomerular apparatus

As the afferent arteriole approaches the renal corpuscle, the smooth muscle cells of the tunica media become modified. Their nuclei become rounded (instead of elongated), and their cytoplasm contains granules (instead of myofibrils). Such modified cells are called *juxtaglomerular cells.* Also, the cells of the distal convoluted tubule adjacent to the afferent arteriole are modified by becoming considerably narrower. These cells of the distal convoluted tubule are known as the *macula densa.* The modified cells of the afferent arteriole together with those of the distal convoluted tubule constitute the **juxtaglomerular apparatus** (Figure 21–6).

It is believed that when renal blood pressure falls below normal, the juxtaglomerular apparatus secretes an enzymelike substance called renin. Renin converts the plasma protein angiotensinogen into angiotensin I, which subsequently is transformed to angiotensin II. Angiotensin II raises blood pressure in two ways. It causes constriction of arteries throughout the body, and it stimulates the adrenal cortex to secrete aldosterone. Aldosterone, as you have learned, increases blood volume and decreases urine volume. The juxtaglomerular apparatus ensures that the pressure of blood entering the glomerulus is sufficient to filter the blood.

Physiology

The major work of the urinary system is done by the nephrons, whereas the other parts of the system are primarily passageways and storage areas. Nephrons carry out three important functions. They control the concentration and volume of the blood by removing selected amounts of water and solutes, help to regulate blood pH, and remove some types

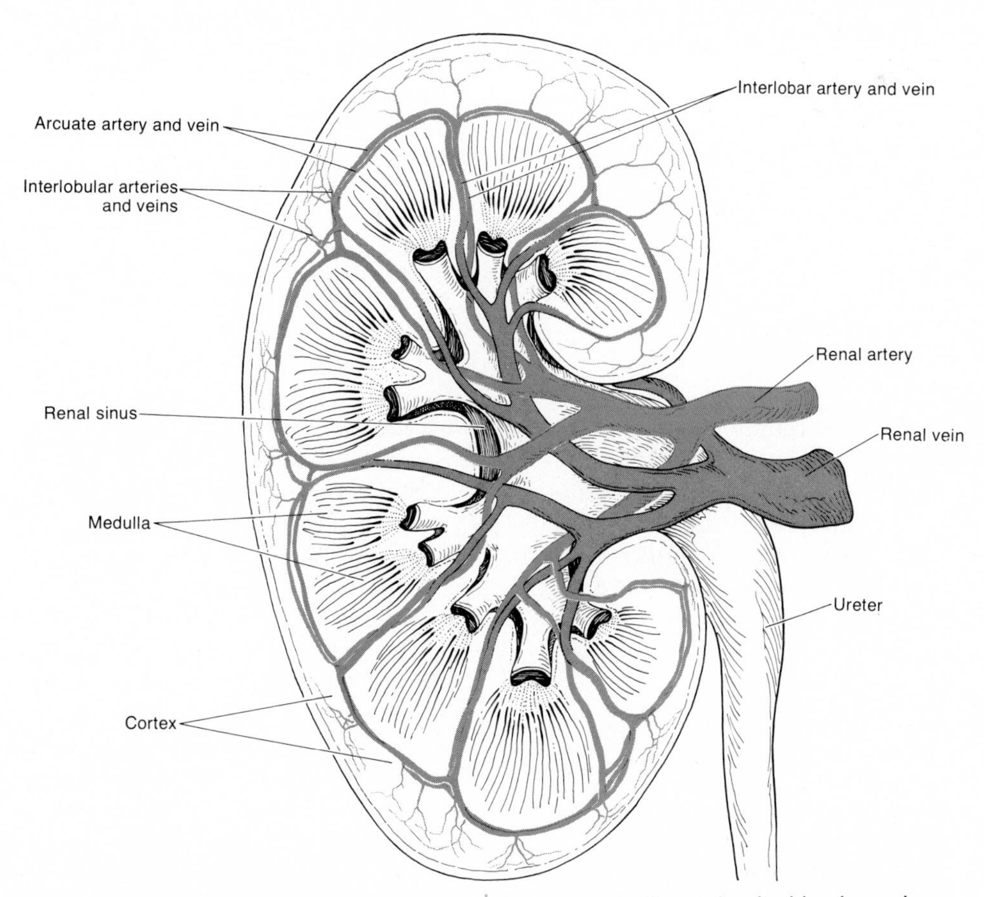

Figure 21–5. Coronal section through the right kidney illustrating its blood supply.

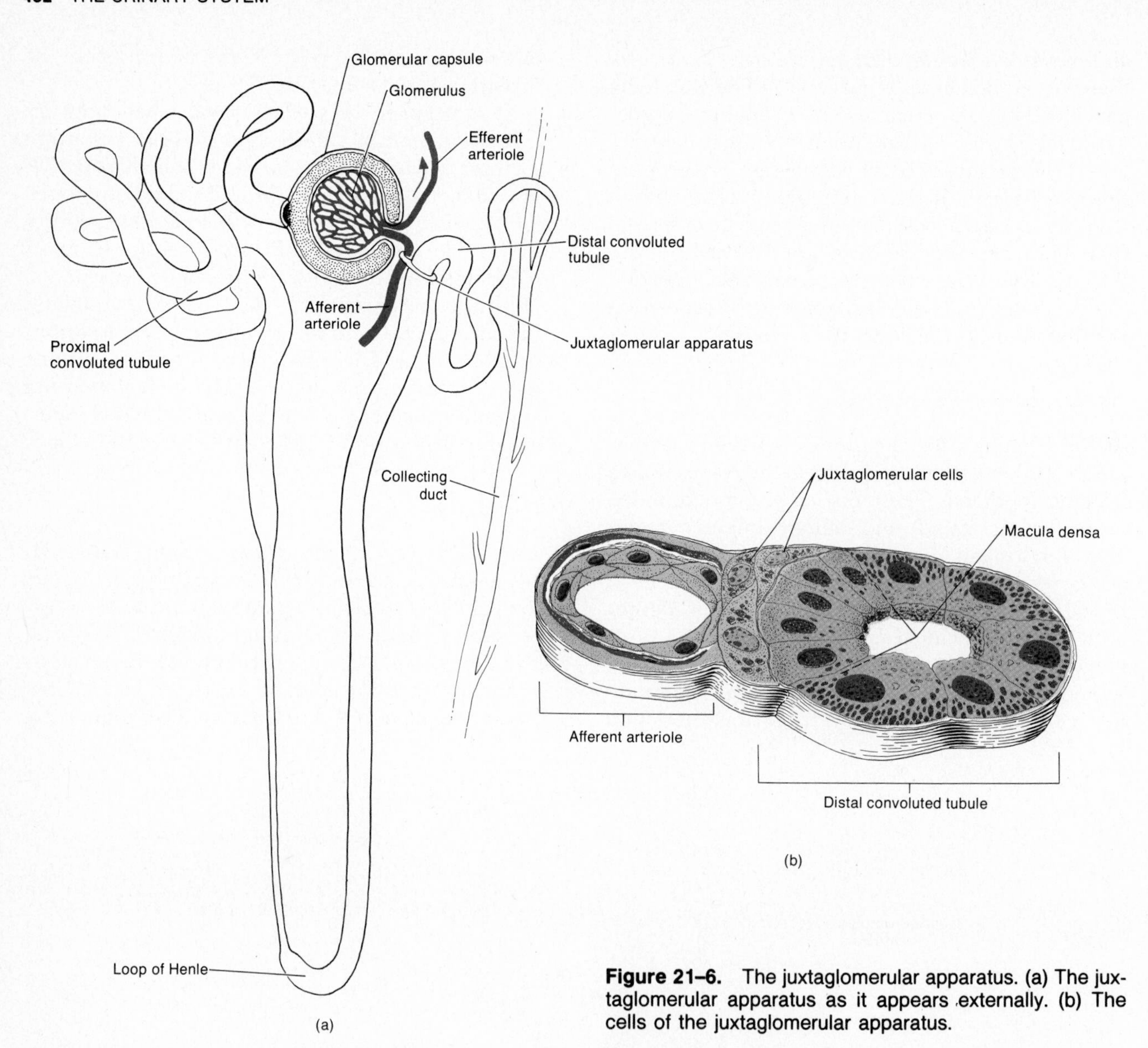

Figure 21–6. The juxtaglomerular apparatus. (a) The juxtaglomerular apparatus as it appears externally. (b) The cells of the juxtaglomerular apparatus.

of toxic wastes from the blood. As the nephrons go about these activities, they remove many materials from the blood, return the ones that the body requires, and eliminate the remainder. The eliminated materials are collectively called *urine.* The process of concentrating urine is called the countercurrent mechanism, and it depends on the special anatomical arrangement of the limbs of Henle and vasa recta.

THE URETERS

Once urine is formed by the nephrons, it drains through the collecting ducts and papillary ducts into the calyces surrounding the renal papillae. The minor calyces join with the major calyces that unite to become the renal pelvis (see Figure 21–2). From the pelvis, the urine drains into the ureters and is carried down by peristalsis to the urinary bladder. From the bladder, the urine is discharged from the body through the single urethra.

Structure

The body has two **ureters**—one for each kidney. Each ureter is an extension of the pelvis of the kidney and runs 25 to 30 cm (10 to 12 in.) to the bladder (see Figure 21–1). As the ureters descend, their thickened walls progressively increase in diameter, but at their widest point they measure less than 1.7 cm (½ in.) in diameter. Like the kidneys, the ureters are retroperitoneal in placement. The ureters enter the urinary bladder at the upper lateral angle of its base. Since there is no valve or sphincter at the openings of the ureters into the urinary bladder, it is quite possible for cystitis (bladder inflammation) to develop into kidney infection. Three coats of tissue form the walls of the ureters (Figure 21–7). A lining of mucous membrane, the *mucosa,* and transitional epithelium is the inner coat. The solute concentration and pH of urine differs drastically from the internal environment of the cells that form the walls of the ureters. Mucus secreted by the mucosa prevents the cells from coming in contact with the urine. Throughout most of the length of the ureters, the second or middle coat, the *muscularis,* is composed of inner longitudinal and outer circular layers of smooth muscle. The muscularis of the proximal one-third of the ureters also contains a layer of outer longitudinal muscle. Peristalsis is the major function of the muscularis. The third, or external, coat of the ureters is a *fibrous coat.* Extensions of the fibrous coat anchor the ureters in place.

The arterial supply of the ureters is from the renal, testicular (or ovarian), common iliac, and inferior vesical arteries. The veins terminate in the corresponding trunks. The ureters are innervated by the renal and vesical plexuses.

Physiology

The principal function of the ureters is to carry urine from the renal pelvis into the urinary bladder. Urine is carried through the ureters primarily by peristaltic contractions of the muscular walls of the

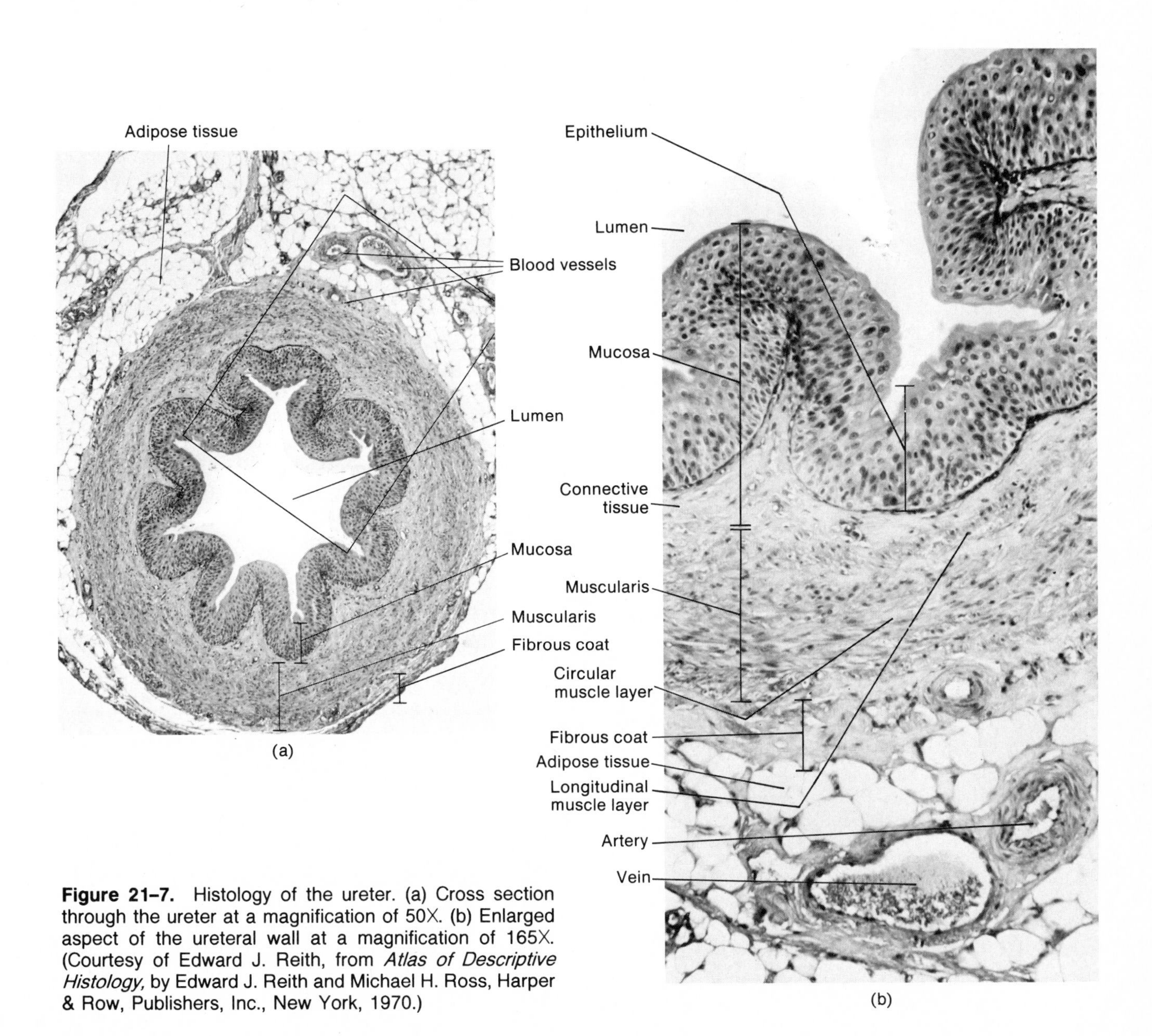

Figure 21–7. Histology of the ureter. (a) Cross section through the ureter at a magnification of 50X. (b) Enlarged aspect of the ureteral wall at a magnification of 165X. (Courtesy of Edward J. Reith, from *Atlas of Descriptive Histology,* by Edward J. Reith and Michael H. Ross, Harper & Row, Publishers, Inc., New York, 1970.)

ureters, but hydrostatic pressure and gravity also contribute. Peristaltic waves pass from the kidney to the urinary bladder, varying in rate from one to five per minute, depending on the amount of urine formation.

THE URINARY BLADDER

The **urinary bladder** (Figure 21–8) is a hollow muscular organ situated in the pelvic cavity posterior to the symphysis pubis. The urinary bladder is retroperitoneal. In the male it is directly in front of the rectum, whereas in the female it is also in front and under the uterus and in front of the vagina. It is a freely movable organ, but it is held in position by folds of the peritoneum. The shape of the urinary bladder depends on the volume of urine it contains. When empty it looks like a deflated balloon. It becomes more spherical when slightly distended, and as urine volume increases in the bladder, it becomes more pear-shaped and rises.

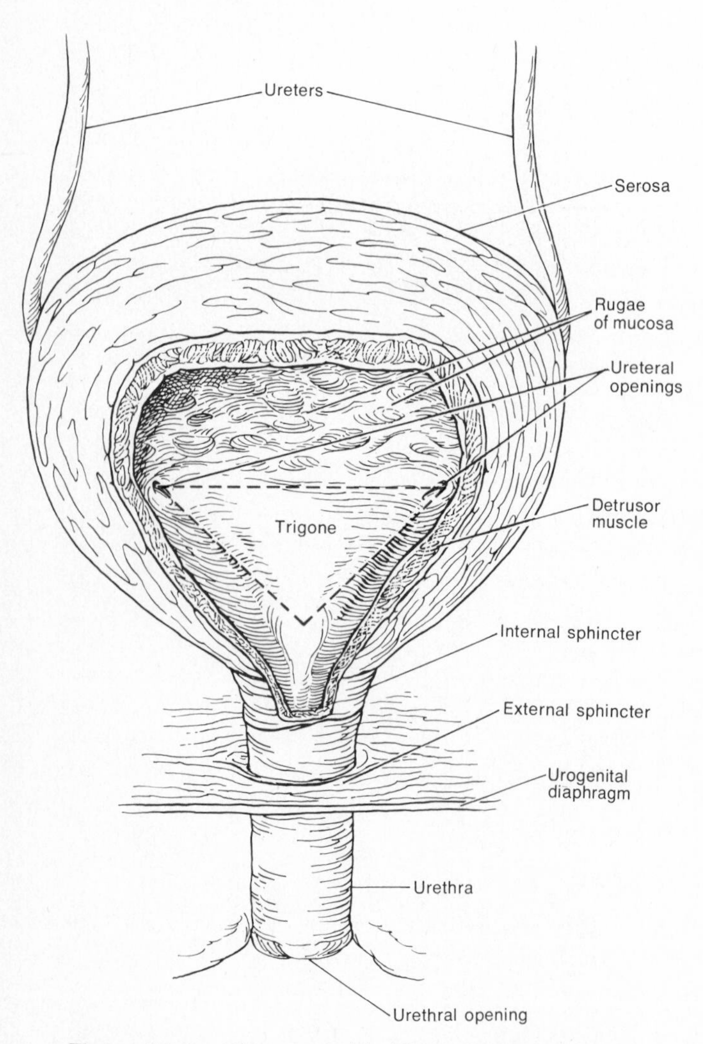

Figure 21–8. The urinary bladder and female urethra.

Structure

At the base of the bladder is a small triangular area with its apex pointing anteriorly. The opening to the urethra is found in the apex of the triangle. At the two points that form the base of the triangle, the ureters drain into the bladder. This triangular area is called the *trigone.*

Four coats make up the walls of the bladder (Figure 21–9). The mucosa, the innermost coat, is a mucous membrane containing transitional epithelium. Recall from Chapter 3 that transitional epithelium is able to stretch–a marked advantage for an organ that must continually inflate and deflate. Stretchability is further enhanced by the rugae, or folds in the mucosa, that appear when the bladder is empty. The second coat, the submucosa, is a layer of dense connective tissue that connects the mucous and muscular coats. The third coat–a muscular one called the *detrusor muscle*–consists of three layers: inner longitudinal, middle circular, and outer longitudinal muscles. In the area around the opening to the urethra, the circular fibers form an *internal sphincter* muscle. Below the internal sphincter is the *external sphincter* that is composed of skeletal muscle. The outermost coat, the serous coat, is formed by the peritoneum and covers only the superior surface of the organ.

The arteries of the urinary bladder are the superior vesical, the middle vesical, and the inferior vesical. The veins from the urinary bladder pass to the internal iliac trunk.

The nerves are derived partly from the hypogastric sympathetic plexus and partly from the second and third sacral nerves. The fibers from the sacral nerves constitute the nervi erigentes.

Physiology

Urine is expelled from the bladder by an act called *micturition,* commonly known as urination or voiding. This response is brought about by a combination of involuntary and voluntary nervous impulses. The average capacity of the bladder is 700 to 800 ml. When the amount of urine in the bladder exceeds about 200 to 400 ml, stretch receptors in the bladder wall transmit impulses to the lower portion of the spinal cord. These impulses initiate a conscious desire to expel urine and an unconscious reflex referred to as the *micturition reflex.* In the micturition reflex, parasympathetic impulses transmitted from the spinal cord reach the bladder wall and internal urethral sphincter, bringing about contraction of the bladder and relaxation of the internal sphincter. Then the conscious portion of the brain sends impulses to the external sphincter, the sphincter relaxes, and urination takes place. Al-

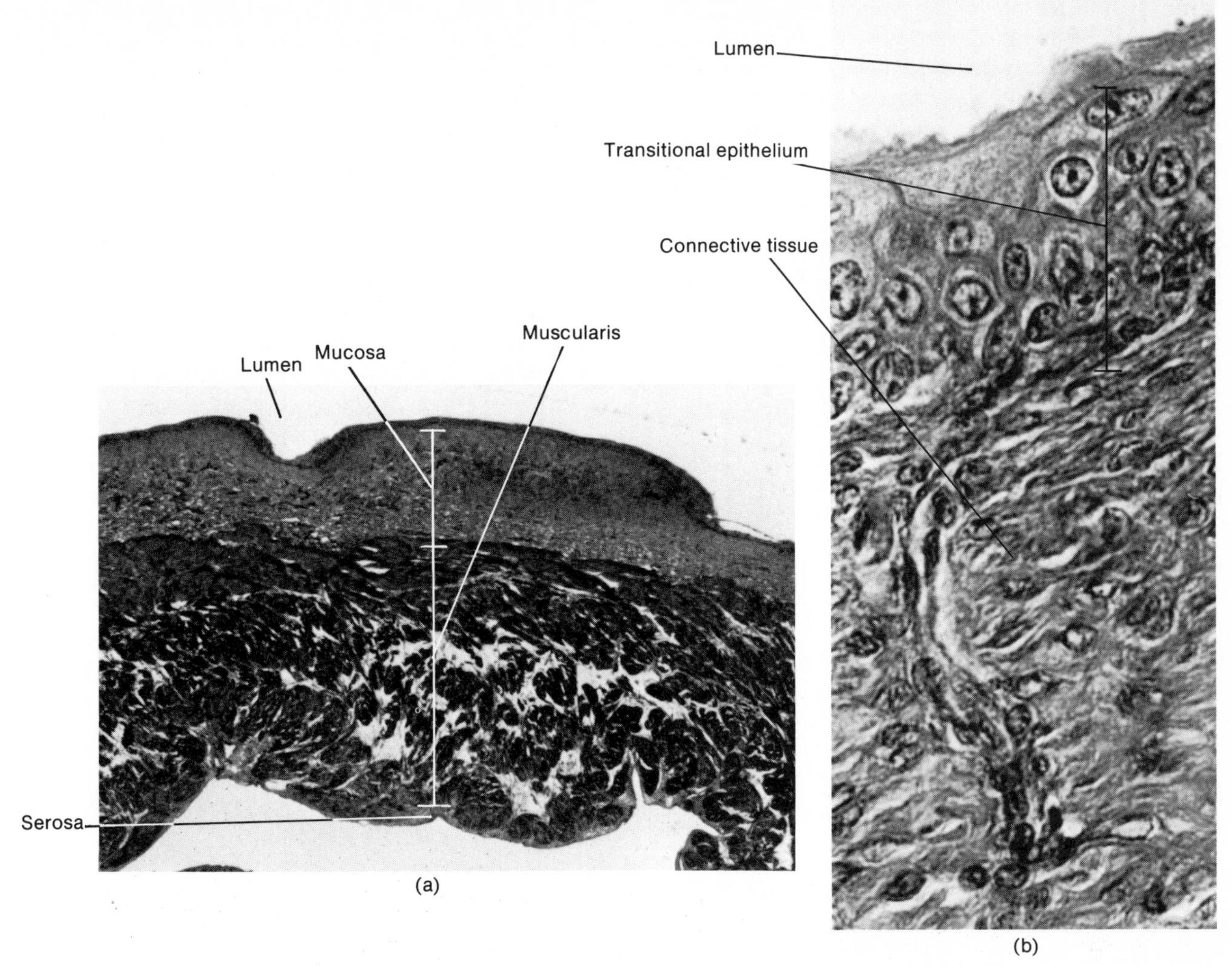

Figure 21–9. Histology of the urinary bladder. (a) Section through the wall of the urinary bladder at a magnification of 6X. (b) Enlarged aspect of the epithelium at a magnification of 100X. (Courtesy of Victor B. Eichler, Wichita State University.)

though emptying of the bladder is controlled by reflex, it may be initiated voluntarily and may be started or stopped at will because of cerebral control of the external sphincter.

A lack of voluntary control over micturition is referred to as *incontinence.* In infants about 2 years old and under, incontinence of urine is normal because they have not developed voluntary control over the external sphincter muscle. Infants void whenever the bladder is sufficiently distended to arouse a reflex stimulus. Proper training overcomes incontinence if the latter is not caused by emotional stress or irritation of the bladder.

Involuntary micturition in the adult may occur as a result of unconsciousness, injury to the spinal nerves controlling the bladder, irritation due to abnormal constituents in urine, disease of the bladder, and emotional stress due to failure of the detrusor muscle to relax.

Retention is a term used to describe a condition in which there is a failure to void urine. Retention may be due to an obstruction in the urethra or neck of the bladder, nervous contraction of the urethra, or lack of sensation to urinate.

A condition far more serious than retention is *suppression,* or *anuria*–the failure of the kidneys to secrete urine. It usually occurs when blood plasma is prevented from reaching the glomerulus as a result of inflammation of the glomeruli. Anuria also may be caused by a low filtration pressure.

THE URETHRA

The **urethra** is a small tube leading from the floor of the bladder to the exterior of the body (see Figure 21–8). In females, it lies directly behind the symphysis pubis and is embedded in the anterior wall of the vagina. Its undilated diameter is about

6 mm (1/4 in.), and its length is approximately 3.8 cm (1 1/2 in.). The female urethra is directed obliquely downward and forward, and the opening of the urethra to the exterior, the *urinary meatus*, is located between the clitoris and vaginal opening.

In males, the urethra is about 20 cm (8 in.) long, and it follows a different course from that of the female urethra. Immediately below the bladder it runs vertically through the prostate gland, then pierces the urogenital diaphragm, and finally penetrates the penis and takes a curved course through its body (see Figures 18–1 and 18–7). Unlike the female urethra, the male urethra serves as a common tube for urinary and reproductive systems.

Structure

The walls of the female urethra consist of three coats: an inner mucous coat that is continuous externally with that of the vulva, an intermediate thin layer of spongy tissue containing a plexus of veins, and an outer muscular coat which is continuous with that of the bladder and consists of circularly arranged fibers of smooth muscle. The male urethra is composed of an inner mucous membrane that is continuous with the mucous membrane of the bladder, and an outer submucous tissue which connects the urethra with the structures through which it passes.

Physiology

Since the urethra is the terminal portion of the urinary system, it serves as the passageway for discharging urine from the body. The male urethra also serves as the duct through which reproductive fluid (semen) is discharged from the body.

APPLICATIONS TO HEALTH

Disorders of the urinary system may be categorized as physical, chemical, or infectious. Let us now examine some of the more common disorders.

Physical

Floating kidney (ptosis) occurs when the kidney no longer is held in place securely by the adjacent organs or by its covering of fat and drifts or slips from its normal position. Pain occurs if the ureter is twisted or bent. Such an abnormal orientation also may obstruct the flow of urine.

Chemical

If urine becomes too concentrated, some of the chemicals that are normally dissolved in it may crystallize out, forming **kidney stones (renal calculi)**. Some common constituents of stones are uric acid, calcium oxalate, and calcium phosphate. The stones usually form in the pelvis of the kidney, where they cause pain, hematuria, and pyuria. Severe pain occurs when a stone passes through a ureter and stretches its walls. Ureteral stones are seldom completely obstructive because they are usually needle-shaped and urine can flow around them.

Gout, as you may recall, is a hereditary condition associated with a high level of uric acid in the blood. When the purine-type nucleic acids are catabolized, a certain amount of uric acid is produced as a waste. Some people, however, seem to produce excessive amounts of uric acid, and others seem to have trouble excreting normal amounts. In either case, uric acid accumulates in the body and tends to solidify into crystals that are deposited in the joints and kidney tissue. Gout is further aggravated by excessive use of diuretics, dehydration, and starvation.

Nephrons can be poisoned by various chemicals, such as mercury, gold, uranium, and other heavy metals, and blood transfusions can produce an agglutination reaction that blocks the nephrons. These common causes of toxicity of the kidneys have produced complete kidney shutdown. The body fluids become extremely acidic, and severe edema, coma, and death in 8 to 14 days follow.

Infectious

Glomerulonephritis is an inflammation of the kidney that involves the glomeruli. One of the most common causes of glomerulonephritis is an allergic reaction to the toxins given off by streptococci bacteria that have recently infected another part of the body, especially the throat. The glomeruli become so inflamed, swollen, and engorged with blood that the glomerular membranes become highly permeable and allow blood cells and proteins to enter the filtrate. Thus the urine contains many erythrocytes and much protein. The glomeruli may be permanently changed, leading to chronic renal disease and renal failure.

Pyelitis is an inflammation of the kidney pelvis and its calyces, and **pyelonephritis** is the interstitial inflammation of one or both kidneys. The latter usually involves both the parenchyma and the renal pelvis, due to bacterial invasion from the middle and lower urinary tracts or the bloodstream.

Cystitis is an inflammation of the urinary bladder involving principally the mucosa and submucosa. It may be caused by bacterial infection, chemicals, or mechanical injury.

HEMODIALYSIS

If the kidneys are impaired so severely by disease or injury that they are unable to excrete nitrogenous wastes and regulate pH and electrolyte concentration of the plasma, then the blood must be filtered by an artificial device. Such filtering of the blood is called *hemodialysis. Dialysis* means using a semipermeable membrane to separate large particles from smaller ones. One of the most well-known devices for accomplishing dialysis is the kidney machine (Figure 21–10). When the machine is in operation, a tube connects it with a much smaller tube implanted in the patient's radial artery. The blood is pumped from the artery and through the tubes to one side of a semiporous cellophane membrane. The other side of the membrane is continually washed with an artificial solution called the dialyzing solution.

All substances (including wastes) in the blood except protein molecules and blood cells can diffuse back and forth across the semipermeable membrane. The electrolyte level of the plasma is controlled by keeping the dialyzing solution electrolytes at the same concentration as that found in normal plasma. Any excess plasma electrolytes move down the concentration gradient and into the dialyzing solution. If the plasma electrolyte level is normal, it is in equilibrium with the dialyzing solution, and no electrolytes are gained or lost. Since the dialyzing solution contains no wastes, substances such as urea move down the concentration gradient and into the dialyzing solution. Thus, wastes are removed, and normal electrolyte balance is maintained.

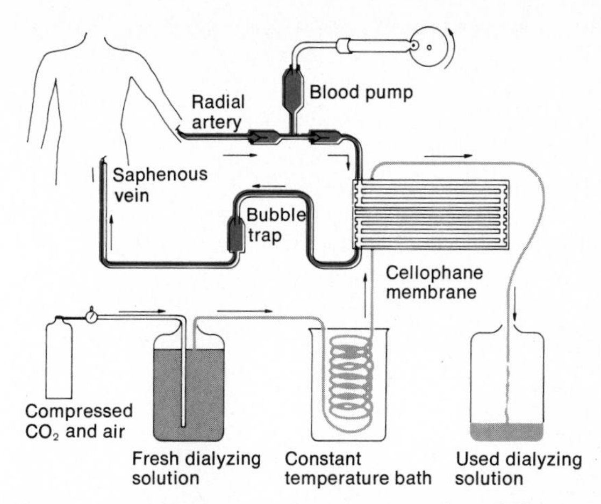

Figure 21–10. Diagrammatic representation of the operation of an artificial kidney. The blood route is indicated in color. The route of the dialyzing solution is indicated in gray.

One additional advantage of the kidney machine is that an individual's nutritional status can be bolstered by placing large quantities of glucose in the dialyzing solution. While the blood gives up its wastes, the glucose diffuses into the blood. Thus, the kidney machine beautifully accomplishes the principal function of the fundamental unit of the kidney – the nephron.

Obvious drawbacks to this artificial kidney system include the fact that the blood must be anticoagulated while dialysis is occurring, and a very large amount of blood must flow through this apparatus to make it work. To date, an artificial kidney that is capable of becoming a permanent implant has not yet been devised.

Key terms associated with the urinary system

Cystoscope (*cyst* = bladder, *scope* = to view) Instrument used for examination of the urinary bladder.

Dysuria (*dys* = painful) Painful urination.

Nephrosis (*neph* = kidney) Any disease of the kidney but usually one that is degenerative.

Oliguria (*olig* = scanty) Scanty urine.

Polyuria (*poly* = many, much) Excessive amounts of urine.

Stricture Narrowing of the lumen of a canal or hollow organ, as the ureter or urethra.

Uremia (*emia* = condition of blood) Toxic levels of urea in the blood resulting from severe malfunction of the kidneys.

Chapter summary in outline

KIDNEYS

1. The primary function of the urinary system is to regulate the concentration and volume of blood by removing and restoring selected amounts of water and solutes. It also excretes wastes.
2. The organs of the urinary system are the kidneys, ureters, urinary bladder, and urethra.
3. The kidneys are retroperitoneal and consist internally of a cortex, medulla, pyramids, papillae, columns, calyces, and a pelvis.
4. The functional unit of the kidney is the nephron.
5. The extensive blood flow through the kidney begins in the renal artery and terminates in the renal vein.

URETERS, URINARY BLADDER, AND URETHRA

1. The paired ureters convey urine from the kidneys to the urinary bladder, mostly by peristaltic contractions.
2. The urinary bladder stores urine. The expulsion of urine from the bladder is called micturition.
3. A lack of control over micturition is called incontinence, failure to void urine is referred to as retention, and inability of the kidneys to produce urine is called suppression.
4. The urethra extends from the floor of the bladder to the exterior and discharges urine from the body.

APPLICATIONS TO HEALTH

1. Ptosis or floating kidney occurs when the kidney slips from its normal position.
2. Renal calculi are kidney stones.
3. Gout is a high level of uric acid in the blood.
4. Glomerulonephritis is an inflammation of the glomeruli of the kidney.
5. Pyelitis is an inflammation of the kidney pelvis and calyces.
6. Cystitis is an inflammation of the urinary bladder.

HEMODIALYSIS

1. Filtering blood through an artificial device is called hemodialysis.
2. The kidney machine filters the blood of wastes and adds nutrients.

Review questions and problems

1. What are the functions of the urinary system? What organs constitute the system?
2. Describe the location of the kidneys. Why are they said to be retroperitoneal?
3. Prepare a labeled diagram that illustrates the principal external and internal gross features of the kidney.
4. What is a nephron? List and describe, in order, the parts of a nephron from the Bowman's capsule to the collecting duct.
5. How are nephrons supplied with blood?
6. Describe the structure, histology, and function of the ureters.
7. Describe the location, histology, and function of the urinary bladder.
8. How is the urinary bladder adapted to its storage function? What is micturition?
9. Contrast the causes of incontinence, retention, and suppression.
10. Compare the position of the urethra in the male and female. What is the function of the urethra?
11. Define each of the following: ptosis, kidney stones, gout, glomerulonephritis, pyelitis, and cystitis.
12. What is hemodialysis? Briefly describe the operation of an artificial kidney.
13. Refer to the glossary of key terms associated with the urinary system. Be sure that you can define each term.

Selected readings

Dawson, Helen L. *Basic Human Anatomy.* 2d ed. New York: Appleton-Century-Crofts, 1974. Chap. 11.

Gray, Henry. *Anatomy of the Human Body.* 29th ed., edited by Charles Mayo Goss. Philadelphia: Lea and Febiger, 1973. Pp. 1265–1299.

Netter, Frank H. *Kidneys, Ureters and Urinary Bladder.* CIBA Collection of Medical Illustrations, Vol. 6. Summit, NJ: CIBA Pharmaceutical Co., 1973.

Vander, Arthur J. *Renal Physiology.* New York: McGraw-Hill, 1975.

GLOSSARY

PRONUNCIATION KEY

Pronunciation appears in parentheses immediately following the word or phrase. The strongest accented syllable appears in capital letters; for example. ab-DO-men. If there is a secondary accent, it is noted by a double quotation mark ("); for example, ad"ĕ-no-hi-POF-ĭ-sis.

In a syllable ending with a consonant, the vowel preceding the consonant is pronounced with the short sound; for example, ab-DUK-shun.

ă as in *ăt*, ĕ as in *mĕt*, ĭ as in *bĭt*,
ŏ as in *nŏt*, ŭ as in *bŭd*

In a syllable ending with a vowel or if a vowel stands alone as a syllable, the vowel is pronounced with the long sound; for example, ab-DO-men and A-nal.

ā as in *māke*, ē as in *bē*, ī as in *īvy*,
ō as in *pōle*, ū as in *pūre*

If the pronunciation of a vowel does not follow these rules, then an accent mark appears above the vowel: ˘ for a short sound or ¯ for a long sound.

Other phonetics are as follows: *a* as in *father* is written *ah*, and the diphthong *oi* is written *oy*.

Abdomen (ab-DO-men) The area between the diaphragm and the pelvis.

Abduction (ab-DUK-shun) Movement away from the axis or midline of the body or one of its parts.

Absorption (ab-SORP-shun) The taking up of liquids by solids, or of gases by solids or liquids; especially, the movement of substances through epithelial cells into body fluids or tissues.

Accommodation A change in the shape of the eye lens so that vision is more acute; an adjustment of the eye lens for various distances; focusing.

Acetabulum (as"ĕ-TAB-yu-lum) "Little vinegar cup"; the rounded cavity on the lateral surface of the hipbone into which the head of the femur fits.

Acinus (AS-ĭ-nus) A small sac; a grapelike dilatation.

Actin (AK-tin) One of the contractile proteins in muscle fiber; another is myosin.

Active transport The movement of substances across cell membranes, against a concentration gradient, requiring the expenditure of energy.

Adduction (ad-DUK-shun) Movement toward the axis or midline of the body or one of its parts.

Adenohypophysis (ad"ĕ-no-hi-POF-ĭ-sis) The anterior, glandular portion of the pituitary gland or hypophysis.

Adenosine triphosphate or **ATP** (ah-DEN-o-sen tri-FOS-fāt) A compound containing sugar, adenine, nitrogen, and three phosphoric acids; the breakdown of ATP provides the energy for cellular work.

Adrenal cortex (ad-RE-nal KOR-teks) The outer portion of an adrenal gland; the cortex is divided into three zones, each of which has a different cellular arrangement and secretes different hormones.

Adrenal glands Two glands, one superior to each kidney; also called the suprarenal glands.

Adrenal medulla (mĕ-DUL-ah) The inner portion of an adrenal gland, consisting of cells which secrete epinephrine and norepinephrine in response to the stimulation of preganglionic sympathetic neurons.

Adrenergic fiber (ad"ren-ER-jik) A postganglionic sympathetic neuron which, when stimulated, releases epinephrine (adrenalin) or norepinephrine (noradrenalin) at its termination.

Afferent arteriole (AF-er-ent ar-TE-re-ōl) A blood vessel of a kidney that breaks up into the capillary network called a glomerulus; there is one afferent arteriole for each glomerulus.

Afferent neuron (NU-ron) A nerve cell that carries an impulse toward the central nervous system.

Agonist (AG-o-nist) Literally, "prime mover." The muscle directly engaged in the contraction that produces a desired movement; it is opposed by the antagonist or relaxed muscle.

All-or-none principle In muscle physiology: muscle fibers of a motor unit contract to their fullest extent or not at all. In neuron physiology: if a stimulus is strong enough to initiate an action potential, an impulse is transmitted along the entire neuron at a constant and maximum strength.

Alveolar sacs (al-VE-o-lar) A collection or cluster of alveoli opening into a central atrium or chamber.

Alveolus (al-VE-o-lus) A small hollow or cavity; a single air cell.

Amphiarthroses (am"fe-ar-THRO-sēs) Joints or articulations midway between diarthroses and synarthroses, in which the articulating bony surfaces are separated by an elastic substance to which both are attached, so that the mobility is slight, but may be exerted in all directions.

Ampulla (am-PU-lah) A saclike dilatation of a canal.

Ampulla of Vater (VAH-ter) A small, raised area in the duodenum where the combined common bile duct and main pancreatic duct empty into the duodenum.

Anal canal (A-nal) The terminal 2 or 3 cm of the rectum that open to the exterior at the anus.

Anal column A longitudinal fold in the mucous membrane of the anal canal that contains a network of arteries and veins.

Anal triangle The subdivision of the male or female perineum that contains the anus.

Anaphase (AN-ă-fāz) The third stage of mitosis when the chromatids that have separated at the centromeres move to opposite poles.

Anastomosis (ah-nas-to-MO-sis) A communication between either blood vessels, lymphatics, or nerves; an end-to-end union or joining together.

Anatomical position (an"ah-TOM-ĭ-kal) The body is erect, facing the observer, with the arms at the sides and the palms of the hands facing forward.

Anatomy (ah-NAT-o-me) The structure, or the study of the structure, of the body and the relationship of its parts to one another.

Anemia (ah-NEM-e-ah) A condition in which there is a decreased number of erythrocytes, or the erythrocytes contain a decreased amount of hemoglobin.

Aneurysm (AN-u-rizm) A saclike enlargement of a blood vessel, usually an artery, that occurs when the blood vessel wall becomes weakened.

Antagonist (an-TAG-o-nist) A muscle that exerts an action opposite to that of the prime mover or agonist.

Anterior or **ventral** (an-TE-re-or) Toward the front; opposite of posterior or dorsal.

Anterior root The structure composed of axons of motor or efferent fibers that emerges from the anterior aspect of the spinal cord and extends laterally to join a posterior root, forming a spinal nerve; there are 31 pairs of anterior or motor or ventral roots.

Anulus fibrosus (AN-u-lus fi-BRO-sus) A ring of fibrous tissue and fibrocartilage that encircles the pulpy substance (nucleus pulposus) of an intervertebral disc.

Anus (A-nus) The distal end and outlet of the rectum.

Aorta (a-OR-tah) The main systemic trunk of the arterial system of the body, emerging from the left ventricle.

Apex (A-peks) The pointed end of a conical structure.

Apocrine gland (AP-o-krin) A type of gland in which the secretory products gather at the free end of the secreting cell and are pinched off, along with some of the cytoplasm, to become the secretion, as in the mammary glands.

Aponeurosis (ap"o-nu-RO-sis) A white, flattened, sheetlike tendon that connects a muscle to the part that it moves, or functions as a sheath enclosing a muscle.

Aqueous humor (AK-we-us) The watery fluid that fills the anterior and posterior chambers of the anterior cavity of the eye.

Arachnoid (ar-AK-noyd) The middle of the three coverings (meninges) of the brain and spinal cord, so named because of its weblike or spidery structure.

Arachnoid villi (VIL-e) Berrylike tufts of arachnoid that protrude into the superior sagittal sinus and through which the cerebrospinal fluid enters the blood stream; arachnoid granulations.

Arbor vitae (AR-bor VI-te) The treelike appearance of the white-matter tracts of the cerebellum when seen in midsagittal section.

Arch of aorta (a-OR-tah) The most superior portion of the aorta, lying between the ascending and descending segments of the aorta; the brachiocephalic, left common carotid, and left subclavian arteries are the three branches of the aortic arch.

Areflexia (ah"re-FLEK-se-ah) Absence of reflexes.

Arm The portion of the upper extremity from the shoulder to the elbow.

Arrector pili (ah-REK-tor PI-le) Smooth muscles attached to hairs; contraction of the muscles pulls the hairs into a more vertical position, resulting in "goose bumps."

Arrhythmia (ah-RITH-me-ah) Irregular heart action causing absence of rhythm.

Arteriogram (ar-TE-re-o-gram) A roentgenogram of an artery, obtained by injecting radiopaque substances into the blood and then taking the roentgenogram.

Arteriole (ar-TE-re-ōl) A very small arterial branch.

Artery (AR-ter-e) A blood vessel carrying blood away from the heart.

Arthritis (ar-THRI-tis) Inflammation of a joint.

Arthrology (ar-THROL-o-je) The scientific study or description of joints.

Articular cartilage (ar-TIK-u-lar KAR-tĭ-lij) The gristle or white elastic substance attached to articular bone surfaces.

Articulation (ar-tik"u-LA-shun) A joint.

Arytenoid cartilages (ar-IT-en-oyd KAR-tĭ-lijs) A pair of small cartilages of the larynx that articulates with the cricoid cartilage.

Ascending colon (KO-lon) The portion of the large intestine that passes upward from the cecum to the lower edge of the liver where it bends at the hepatic flexure to become the transverse colon.

Association area A portion of the cerebral cortex connected by many motor and sensory fibers to other parts of the cortex; the association areas are concerned with motor patterns, memory, concepts of word-hearing and word-seeing, reasoning, will, judgment, and personality traits.

Association neuron (NU-ron) A nerve cell lying completely within the central nervous system that carries impulses from sensory neurons to motor neurons; internuncial or connecting or central neuron.

Astigmatism (ah-STIG-mah-tizm) An irregularity of the lens or cornea of the eye, causing the image to be out of focus, producing faulty vision.

Ataxia (ah-TAK-se-ah) The lack of muscular coordination; lack of precision.

Atherosclerosis (ath"er-o-sklĕ-RO-sis) A disease involving mostly the lining of large arteries, in which yellow patches of fat are deposited, forming plaques that decrease the size of the lumen.

Atrioventricular bundle (a"tre-o-ven-TRIK-u-lar) The portion of the conduction system of the heart that begins at the atrioventricular node and passes through the cardiac skeleton separating the atria and the ventricles, then runs a short distance down the interventricular septum before splitting into right and left bundle branches; the bundle of His or the AV bundle.

Atrioventricular node The portion of the conduction system of the heart made up of a compact mass of conducting cells located near the orifice of the coronary sinus in the right atrial wall; AV node or node of Tawara.

Atrioventricular valve A structure made up of membranous flaps or cusps that allows blood to flow in one direction only, from an atrium into a ventricle; AV valve.

Atrium (A-tre-um) Either of the two superior cavities of the heart that act as receiving chambers; a chamber at the end of an alveolar duct of the lungs that connects with a variable number of alveolar sacs.

Atrophy (AT-ro-fe) A wasting away or decrease in size of a part, due to a failure or abnormality of nutrition.

Auricle (AW-rĭ-kul) An ear-shaped appendage of each atrium of the heart; the trumpet or pinna of the ear.

Autonomic ganglion (aw"to-NOM-ik GANG-gle-on) Clusters of sympathetic or parasympathetic cell bodies located outside the central nervous system.

Autonomic nervous system Visceral efferent neurons, both sympathetic and parasympathetic, that transmit impulses from the central nervous system to smooth muscle, cardiac muscle, and glands; so named because this portion of the nervous system was thought to be self-governing or spontaneous.

Autonomic plexus (PLEK-sus) An extensive network of sympathetic and parasympathetic fibers; the cardiac, celiac, and pelvic plexuses are located in the thorax, abdomen, and pelvis, respectively.

Axilla (ak-SIL-ah) The small hollow beneath the arm where it joins the body at the shoulders; the armpit.

Axon (AK-son) The process of a nerve cell that carries an impulse away from the cell body.

Back The posterior part of the body; the dorsum.

Ball-and-socket joint A synovial joint in which the rounded surface of one bone moves within a cup-shaped depression or fossa of another bone; the shoulder or hip joint.

Basal ganglia (BA-sal GANG-gle-ah) Clusters of cell bodies that make up the central gray matter of the cerebral hemispheres; they include the caudate nucleus, the lentiform nucleus, the claustrum, and the amygdaloid body.

Base The broadest part of a pyramidal structure; a nonacid, or a proton acceptor, characterized by excess of OH^- ions and a pH greater than 7; a ring-shaped organic molecule, containing nitrogen, which is one of the components of a nucleotide, for example, adenine, guanine, cytosine, thymine, and uracil.

Basement membrane A sheet of extracellular material underlying the basal surface of epithelial cells; the basal lamina.

Basilar membrane (BAS-ĭ-lar) A membrane in the cochlea of the inner ear that separates the cochlear duct from the scala tympani; the organ of Corti rests on the basilar membrane.

Basophil (BA-so-fil) A polymorphonuclear leukocyte with a pale nucleus, and cytoplasm containing large granules that stain a deep blue, normally constituting about 0.5 percent of the white blood cells.

Belly The abdomen; the prominent, fleshy part of a skeletal muscle; gaster.

Benign (be-NĪN) Not malignant.

Blood The fluid that circulates through the heart, arteries, capillaries, and veins, and that constitutes the chief means of transport within the body.

Blood-brain barrier A special mechanism that prevents the passage of materials from the blood to the cerebrospinal fluid and brain; astrocytes, which in most parts of the brain are situated between neurons and capillaries, play a role in this barrier.

Body cavity A space within the body that contains various internal organs.

Bony labyrinth (LAB-ĭ-rinth) A series of cavities within the petrous portion of the temporal bone, or the vestibule, cochlea, and semicircular canals of the inner ear.

Bowman's capsule (KAP-sul) A double-walled globe at the proximal end of a nephron that encloses the glomerulus.

Brachial plexus (BRA-ke-al PLEK-sus) A network of nerve fibers of the anterior rami of spinal nerves C5, C6, C7, C8, and T1; the nerves that emerge from the brachial plexus supply the upper extremity.

Brain A mass of nerve tissue located in the cranial cavity.

Brain stem The portion of the brain immediately superior to the spinal cord, made up of the medulla oblongata, pons, and midbrain.

Broad ligament A double fold of parietal peritoneum attaching the uterus to the side of the pelvic cavity.

Bronchi (BRONG-ki) The two divisions of the trachea.

Bronchial tree (BRONG-ke-al) The bronchi and their branching structures.

Bronchioles (BRONG-ke-ōlz) The smaller divisions of the bronchi.

Bronchogenic carcinoma (brong"ko-JEN-ik kar"sĭ-NO-mah) Cancer originating in the bronchi.

Bronchogram (BRONG-ko-gram) A roentgenogram of the lungs and bronchi.

Bronchopulmonary segment (brong"ko-PUL-moner"e) One of the smaller divisions of a lobe of a lung supplied by its own branch of a bronchus.

Brunner's glands Glands in the submucosa of the duodenum that secrete intestinal juice.

Buccal cavity (BUK-al) The mouth or oral cavity.

Bulbourethral glands (BUL-bo-u-RE-thral) A pair of glands, also known as Cowper's glands, located inferior to the prostate on either side of the urethra that secrete an alkaline fluid into the cavernous urethra.

Bundle branch One of the two branches of the bundle of His or the atrioventricular bundle, made up of specialized muscle fibers that transmit electrical impulses to the ventricles.

Bunion (BUN-yun) Inflammation and thickening of the bursa of the joint at the base of the big toe.

Burn An injury caused by fire, steam, chemicals, electricity, lightning, or the ultraviolet rays of the sun.

Bursa (BUR-sah) A fluid-filled, saclike cavity in connective tissue situated at points of friction or pressure, usually about joints.

Bursitis (bur-SI-tis) Inflammation of a bursa.

Buttocks (BUT-oks) The two fleshy masses on the posterior aspect of the lower trunk, formed by the gluteal muscles.

Calcification (kal"sĭ-fĭ-KA-shun) The process by which calcium salts are deposited in the matrix of bone or cartilage.

Callus (KAL-lus) A growth of new bone tissue in and around a fractured area, ultimately replaced by mature bone; an acquired, localized thickening of the stratum corneum, as a result of continued physical trauma.

Calyx (KA-liks) A cup-shaped tube extending from the renal pelvis (a renal portion of the ureter) that encircles renal papillae.

Canal of Schlemm (shlem) A circular venous sinus located at the junction of the sclera and the cornea by which aqueous humor drains from the anterior chamber of the eyeball into the blood.

Canaliculus (kan"al-IK-u-lus) A small channel or canal; a microscopic channel in bone connecting lacunae; a short duct that conveys tears from the region of the medial canthus of the eye to the lacrimal sac.

Canthus (KAN-thus) The angle formed by the joining of the eyelids at either corner of the eye.

Capillary (KAP-ĭ-ler-e) A tiny blood vessel connecting an arteriole and a venule; the smallest of the lymphatic vessels.

Cardiac notch (KAR-de-ak) An angular notch in the anterior border of the left lung.

Cardinal ligament (KAR-dĭ-nal LIG-ah-ment) A ligament of the uterus, extending laterally from the cervix and vagina as a continuation of the broad ligament.

Carotene (KAR-o-teen) A yellow pigment in carrots, tomatoes, egg yolk, and other substances that can be converted by the body into vitamin A.

Carpus (KAR-pus) The wrist; a collective term for the eight bones that make up the wrist.

Cartilage (KAR-tĭ-lij) A specialized connective tissue forming part of the skeleton; gristle.

Cartilaginous joint (kar"tĭ-LAJ-ĭ-nus) A joint without a joint cavity where the articulating bones are held tightly together by cartilage, allowing little or no movement.

Cataract (KAT-ah-rakt) Loss of transparency of the crystalline lens of the eye, or its capsule, or both.

Cauda equina (KAW-dă e-KWI-nah) A taillike collection of roots of spinal nerves at the inferior end of the spinal canal.

Cecum (SE-kum) A blind pouch at the junction of

the ileum of the small intestine with the ascending colon, and to which the appendix is attached.

Celiac plexus (SE-le-ak PLEK-sus) A large mass of ganglia and nerve fibers located at the level of the upper part of the first lumbar vertebra; the solar plexus.

Cell The basic, living, structural and functional unit of all organisms; the smallest structure capable of performing all the activities vital to life.

Cell inclusion A lifeless, often temporary, constituent in the cytoplasm of a cell as opposed to an organelle.

Cementum (se-MEN-tum) Calcified tissue covering the root of a tooth.

Center An area in the brain where a particular function is localized.

Center of ossification (os"ĭ-fĭ-KA-shun) An area in the cartilage model of a future bone where the cartilage cells hypertrophy, secrete enzymes that result in the calcification of their matrix resulting in the death of the cartilage cells, followed by the invasion of the area by osteoblasts that then lay down bone.

Central canal A microscopic tube running the length of the spinal cord in the gray commissure.

Central fovea (FO-ve-ah) A cuplike depression in the center of the macula lutea of the retina, containing cones only; the area of clearest vision.

Central nervous system The brain and spinal cord.

Centrosome (SEN-tro-sōm) A rather dense area of cytoplasm, near the nucleus of a cell, containing a pair of centrioles.

Cerebellar peduncle (ser-ĕ-BEL-ar PE-dung-kl) A bundle of nerve fibers connecting the cerebellum with the brain stem.

Cerebellum (ser-ĕ-BEL-um) The portion of the hindbrain lying posterior to the medulla and pons, concerned with coordination of movements.

Cerebral aqueduct (SER-ĕ-bral AK-we-dukt) A channel through the midbrain connecting the third and fourth ventricles and containing cerebrospinal fluid; the aqueduct of Sylvius.

Cerebral peduncles (PE-dung-kls) A pair of nerve fiber bundles located on the ventral surface of the midbrain, conducting impulses between the pons and the cerebral hemispheres.

Cerebrospinal fluid (SER-ĕ-bro-SPI-nal) A fluid produced in the choroid plexuses of the ventricles of the brain.

Cerebrum (SER-ĕ-brum) The two hemispheres of the forebrain, making up the largest part of the brain.

Cervical (SER-vĭ-kal) Pertaining to the neck.

Cervical ganglion (GANG-glĭ-on) A cluster of nerve cell bodies of postganglionic sympathetic neurons, located in the neck, near the vertebral column.

Cervical plexus (PLEK-sus) A network of neuron fibers formed by the anterior rami of the first four cervical nerves.

Cervix (SER-viks) The neck or a necklike portion of an organ.

Cholinergic fiber (kol"in-ER-jik) An axon of a neuron that liberates acetylcholine at a synapse.

Chondrocyte (KON-dro-sīt) A cartilage cell.

Chordae tendineae (KOR-de ten-DIN-e-e) Fine strings of tendinous tissue connecting the papillary muscles of the ventricles of the heart with the atrioventricular valves.

Choroid (KO-royd) The vascular coat of the eye, located between the sclera and the retina.

Choroid plexus (PLEK-sus) A vascular structure located in the roof of each of the four ventricles of the brain that produces cerebrospinal fluid.

Chromaffin cells (kro-MAF-in) Cells that have an affinity for chrome salts, due in part to the presence of the precursors of the chemical transmitter epinephrine; also called pheochrome cells and are found, among other places, in the adrenal medulla.

Chromatin (KRO-mah-tin) The threadlike mass of the genetic material consisting principally of DNA, which is present in the nucleus of a nondividing or interphase cell.

Chromatolysis (kro"mah-TOL-ĭ-sis) Disappearance of the Nissl bodies of a neuron cell body as a result of trauma to the cell; disintegration of the chromatin of a cell nucleus.

Chromosome (KRO-mo-sōm) One of the 46 small, dark-staining bodies that appear in the nucleus of a human diploid cell during cell division.

Ciliary body (SIL-ĭ-ar-e) A part of the middle tunic of the eyeball that, along with the choroid and the iris, make up the vascular layer; the ciliary body includes the ciliary muscle and the ciliary processes.

Ciliary ganglion (GANG-glĭ-on) A very small parasympathetic ganglion whose preganglionic fibers come from the oculomotor nerve and whose postganglionic fibers carry impulses to the ciliary muscle and the sphincter muscle of the iris.

Cilium (SIL-ĭ-um) A hair or hairlike process; cilia projecting from cells can be used to move the entire cell or can move substances along the surface of the cell.

Circumduction (sir"kum-DUK-shun) A movement at a synovial joint in which the distal end of a bone moves in a circle while the proximal end remains relatively stable.

Circumvallate papilla (sir"kum-VAL-āt pah-PIL-ah) The largest of the elevations on the upper surface of the tongue; they are circular and arranged in an inverted V-shaped row at the posterior portion of the tongue.

Clitoris (KLIT-or-is) An erectile organ of the female homologous to the male penis.

Coccygeal (kok-SIJ-e-al) Pertaining to or located in the region of the coccyx, the caudal extremity of the vertebral column.

Coccygeal plexus (PLEK-sus) A network of nerves formed by the anterior rami of the coccygeal and the fourth and fifth sacral nerves; fibers from the plexus supply the skin in the region of the coccyx.

Cochlear duct (KOK-le-ar) The membranous cochlea consisting of a spirally arranged tube enclosed in the bony cochlea and lying along its outer wall.

Collagen (KOL-ah-jen) A protein that is the main organic constituent of connective tissue.

Common bile duct A tube formed by the union of the common hepatic duct and the cystic duct that empties bile into the duodenum at the ampulla of Vater.

Compact (dense) bone Bone tissue with no apparent spaces in which the layers or lamellae are fitted tightly together; compact bone is found immediately deep to the periosteum and external to spongy bone.

Conductivity The ability to transmit an impulse.

Condyle (KON-dīl) A rounded projection or process at the end of a bone that forms an articulation or joint.

Cone The light-sensitive receptor cell in the retina concerned with color vision.

Congenital (kon-JEN-ĭ-tal) Present at the time of birth.

Conjunctiva (kon"junk-TI-vah) The delicate membrane covering the eyeball and lining the eyelids.

Conjunctivitis (kon-junk"tiv-I-tis) Inflammation of the delicate membrane covering the eyeball and lining the eyelids.

Connective tissue The most abundant of the four tissue types in the body, performing the functions of binding and supporting; there are relatively few cells and a great deal of intercellular substance.

Contractility (kon"trak-TIL-ĭ-te) The ability to shorten.

Conus medullaris (KO-nus med"u-LAR-is) The terminal tapering portion of the spinal cord.

Convergence (kon-VER-jens) The coordinated movement of the eyes so that both focus on a single set of objects.

Convolution (kon"vo-LU-shun) An elevation of the gray matter of the cerebral cortex formed by the folding of the tissue upon itself.

Cornea (KOR-ne-ah) The clear, transparent, anterior covering of the eyeball; the anterior portion of the outermost tunic of the eyeball.

Coronary circulation (KOR-o-nă-re) The pathway followed by the blood from the ascending aorta through the blood vessels supplying the heart and returning to the right atrium.

Coronary sinus (SI-nus) A wide venous channel on the posterior surface of the heart that collects the blood from the coronary circulation and returns it to the right atrium.

Corpora quadrigemina (KOR-por-ah kwad-rĭ-JEM-ĭ-nah) Four small elevations on the dorsal region of the midbrain concerned with visual and auditory functions.

Corpus callosum (KOR-pus kal-LO-sum) An arch of white matter located at the bottom of the longitudinal fissure composed of myelinated fibers connecting the right and left hemispheres of the cerebrum.

Corpus striatum (stri-A-tum) An area in the interior of each cerebral hemisphere composed of the caudate and lentiform nuclei of the basal ganglia and white matter of the internal capsule, arranged in a striated manner.

Cortex (KOR-teks) An outer layer of an organ; the convoluted layer of gray matter covering each cerebral hemisphere.

Costal cartilage (KOS-tal KAR-tĭ-lij) A strip of hyaline cartilage by which each of the first through seventh ribs is connected directly to the sternum.

Cranial nerve (KRA-ne-al) One of twelve pairs of nerves that leaves the brain, passes through formina in the skull, to supply the head, neck, and part of the trunk; each is designated both with Roman numerals and with names.

Craniosacral outflow (kra"ne-o-SA-kral) The parasympathetic division of the autonomic system, with cell bodies of preganglionic neurons located in nuclei in the brain stem and in the lateral gray matter of the sacral portion of the spinal cord.

Cranium (KRA-ne-um) The bones of the skull that enclose and protect the brain and the organs of sight, hearing, and balance.

Cremaster muscle (kre-MAS-ter) A thin layer of skeletal muscle that arises from the internal oblique abdominal muscle and extends as a series of loops down the spermatic cord to attach to the tunica vaginalis of the scrotum; it draws the testis up toward the superficial inguinal ring.

Crest A projection, peak, or ridge—as the iliac crest, the thickened, expanded superior border of the ilium.

Cricoid cartilage (KRI-koyd KAR-tĭ-lij) A circle of cartilage shaped like a signet ring forming the inferior and posterior portion of the larynx.

Crista (KRIS-tah) A crest, as in the crista galli, a triangular process projecting superiorly from the cribriform plate of the ethmoid bone.

Crypts of Lieberkuhn (kripts of LI-ber-kun) Simple tubular glands that open onto the surface of the intestinal mucosa; these intestinal glands secrete digestive enzymes.

Cupula (KU-pu-lah) A mass of gelatinous material covering the hair cells of a crista, a receptor in the ampulla of a semicircular canal stimulated when the head moves.

Curvature (KUR-vah-tūr) A nonangular deviation of a straight line, as in the greater and lesser curvatures of the stomach; abnormal curvatures of the vertebral column include kyphosis, lordosis, and scoliosis.

Cutaneous membrane (ku-TA-ne-us) The skin.

Cystic duct (SIS-tik) The duct that transports bile from the gallbladder to the common bile duct.

Cytokinesis (si"to-kĭ-NE-sis) The division of the cytoplasm of a cell following mitosis (nuclear division).

Cytology (si-TOL-o-je) The study of cells.

Cytoplasm (SI-to-plazm) The protoplasm of a cell, excluding the nucleus.

Dartos (DAR-tos) Smooth muscle fibers in the scrotum.

Decussation of pyramids (de"kus-SA-shun) The crossing of the fibers of the pyramids of the medulla from one side to the other.

Deep fascia (FASH-e-ah) A sheet of connective tissue wrapped around a muscle to hold it in place.

Defecation (def-ĕ-KA-shun) The elimination of wastes and undigested food as feces from the rectum.

Deglutition (deg"lu-TĬ-shun) The act of swallowing.

Demineralization (de-min"er-al-ĭ-ZA-shun) The loss of calcium and phosphorus from bones.

Dendrite (DEN-drīt) A nerve cell process carrying an impulse toward the cell body.

Denticulate ligament (den-TIK-u-lāt LIG-ah-ment) A band of pia mater extending the entire length of the spinal cord on each side between the dorsal and ventral spinal nerve roots.

Dentin (DEN-tin) The main substance of a tooth, covered by enamel on the exposed part of the tooth and by cementum on the root of the tooth.

Dentition (den-TĬ-shun) The number, shape, and arrangement of teeth; the eruption of teeth (also called teething).

Deoxyribonucleic acid (de-ok"sĭ-ri"bo-nu-KLA-ik) A nucleic acid that transmits the genetic code from generation to generation and regulates protein synthesis; DNA is a double helix composed of units called nucleotides, each of which contains deoxyribose sugar, a phosphate group, and a nitrogen base.

Dermatome (DER-mah-tōm) An area of skin supplied by sensory fibers of a single dorsal root; an instrument for incising the skin or for cutting thin transplants of skin.

Dermis (DER-mis) A layer of dense connective tissue lying deep to the epidermis; the true skin or corium.

Descending colon (KO-lon) The part of the large intestine descending from the level of the lower end of the spleen on the left side to the level of the left iliac crest.

Detrusor muscle (de-TRU-sor) The smooth muscle coat or tunic of the urinary bladder.

Developmental anatomy The study of development from the fertilized egg to the adult form; the branch of anatomy called embryology is generally restricted to the study of development from the fertilized egg to the period just before birth.

Diabetes insipidus (di"ah-BE-tēz in-SIP-ĭ-dus) A metabolic disease resulting from a hyposecretion of antidiuretic hormone from the posterior pituitary gland, associated with excretion of large amounts of urine.

Diabetes mellitus (MEL-ĭ-tus) A metabolic disease resulting from a hyposecretion of insulin from the pancreas, associated with elevated blood sugar, sugar in the urine, excessive urination, and increased thirst.

Dialysis (di-AL-ĭ-sis) The process of separating crystalloids (smaller particles) from colloids (larger particles) by the difference in their rates of diffusion through a semipermeable membrane.

Diapedesis (di"ah-pĕ-DE-sis) The passage of blood cells through intact blood vessel walls.

Diaphysis (di-AF-ĭ-sis) The shaft of a long bone, or the portion of a long bone between the epiphyses.

Diarthrosis (di-ar-THRO-sis) An articulation in which opposing bones move freely, as in a hinge joint; a synovial joint.

Diencephalon (di"en-SEF-ah-lon) The "tween" brain, located between the cerebral hemispheres and the midbrain, consisting of the hypothalamus, thalamus, metathalamus, and epithalamus.

Diffusion (dĭ-FU-zhun) A passive process in which there is a net or greater movement of molecules or ions from a region of high concentration to a region of low concentration until equilibrium is reached.

Digestion (dĭ-JES-chun) The mechanical and chemical breakdown of food to simple molecules that can be absorbed by the body.

Distal (DIS-tal) Away from the attachment of a limb to the trunk; the opposite of proximal.

Dorsal ramus (DOR-sal RA-mus) A branch of a spinal nerve containing motor and sensory fibers supplying the muscles, skin, and bones of the posterior part of the head, neck, and trunk.

Dorsiflexion (dor"sĭ-FLEK-shun) Flexion of the foot at the ankle.

Duct of Santorini (san"tor-E-ne) An accessory duct of the pancreas that empties into the duodenum about 2.5 cm superior to the ampulla of Vater.

Ductus arteriosus (DUK-tus ar-ter-ĭ-O-sus) A fetal blood vessel connecting the pulmonary trunk with the aorta.

Ductus deferens (DEF-er-ens) The vas deferens or seminal duct that conducts spermatozoa from the epididymis to the ejaculatory duct.

Ductus venosus (ven-O-sus) A fetal blood vessel connecting the umbilical vein and the inferior vena cava.

Duodenum (du"o-DE-num) The first or proximal portion of the small intestine, originating at the pyloric valve of the stomach and extending about 25 cm until it merges with the jejunum.

Dura mater (DU-rah MA-ter) Literally, "hard or tough mother"; the outermost meninx.

Edema (ĕ-DE-mah) An abnormal accumulation of fluid in the body tissues.

Effector (ĕ-FEK-tor) The organ of the body, either a muscle or a gland, that responds to a motor neuron impulse.

Efferent arteriole (EF-er-ent ar-TE-re-ōl) A vessel of the renal vascular system that transports blood from the glomerulus to the peritubular capillary.

Efferent ducts A series of coiled tubes that transport spermatozoa from the rete testis to the epididymis.

Efferent neuron (NU-ron) A motor neuron that conveys impulses from the brain and spinal cord to effectors which may be either muscles or glands.

Ejaculatory duct (e-JAK-u-lah-tor"e) A tube that transports spermatozoa from the ductus deferens (seminal duct) to the prostatic urethra.

Elasticity (e-las-TIS-ĭ-te) The ability of muscle to return to its original shape after contraction or extension.

Elastin (e-LAS-tin) A substance found in elastic fibers of connective tissue that gives elasticity to the skin and to the tissues that form the walls of blood vessels.

Elbow The joint between the upper arm and the forearm.

Electrocardiogram (e-lek"tro-KAR-de-o-gram) A recording of the electrical changes that accompany the cardiac cycle.

Electroencephalogram (e-lek"tro-en-SEF-ah-lo-gram) A recording of the electrical impulses of the brain.

Ellipsoidal joint (e-lip-SOY-dal) A synovial or diarthrotic joint structured so that an oval-shaped condyle of one bone fits into an elliptical cavity of another bone, permitting side-to-side and back-and-forth movements, as at the joint at the wrist between the radius and carpals.

Embryology (em"bre-OL-o-je) The study of development from the fertilized egg through the eighth week in utero.

Emphysema (em"fĭ-SE-mah) A swelling or inflation of air passages with resulting stagnation of air in parts of the lungs; loss of elasticity in the alveoli.

Enamel (e-NAM-el) The very hard white substance covering the crown of a tooth.

End bulb of Krause (krows) The cutaneous receptor for the sensation of cold.

End organ of Ruffini (roo-FE-ne) A cutaneous receptor for heat.

Endocardium (en-do-KAR-dĭ-um) A thin layer of endothelium and smooth muscle that lines the inside of the heart and covers the valves and tendons that hold the valves open.

Endochondral ossification (en"do-KON-dral os"ĭ-fĭ-KA-shun) The replacement of cartilage by bone.

Endocrine gland (EN-do-krin) A gland that secretes a hormone directly into the blood.

Endometrium (en"do-ME-trĭ-um) The mucous membrane lining the uterus.

Endomysium (en"do-MIZ-ĭ-um) Connective tissue within a bundle or fascicle of muscle cells that separates the muscle cells from one another.

Endoneurium (en"do-NU-rĭ-um) Connective tissue within a bundle or fascicle of nerve fibers that separates and supports the nerve fibers.

Endoplasmic reticulum (en"do-PLAZ-mik re-TIK-u-lum) A network of canals running through the cytoplasm of a cell; in the areas where ribosomes are attached to the outer surface of the ER, it is referred to as granular or rough reticulum; portions of the ER that have no ribosomes are called agranular or smooth reticulum.

Endosteum (en-DOS-te-um) The membrane that lines the medullary cavity of bones.

Endothelium (en"do-THE-lĭ-um) The layer of simple squamous epithelium that lines the cavities of the heart and blood and lymphatic vessels.

Eosinophil (e"o-SIN-o-fil) A white blood cell readily stained by eosin, normally constituting about 2–4% of the white blood cells.

Epicardium (ep"ĭ-KAR-dĭ-um) The thin outer layer of the heart, also called the visceral pericardium.

Epicondyle (ep"ĭ-KON-dīl) A projection of a bone on or above a condyle.

Epidermis (ep"ĭ-DERM-is) The outermost layer of skin, composed of stratified squamous epithelium.

Epididymis (ep"ĭ-DID-ĭ-mus) A highly coiled tube located along the posterior border of the testis that receives spermatozoa from the efferent ducts and stores them until they are passed along to the ductus deferens.

Epidural space (ep"ĭ-DU-ral) A space between the spinal dura mater and the vertebral canal, containing loose connective tissue and a plexus of veins.

Epiglottis (ep"ĭ-GLOT-is) A large, leaf-shaped piece of cartilage lying on top of the larynx, with its "stem" attached to the thyroid cartilage and its "leaf" portion unattached and free to move up and down to cover the glottis.

Epimysium (ep"ĭ-MIZ-ĭ-um) A fibrous sheath of connective tissue wrapped around a muscle.

Epineurium (ep"ĭ-NU-rĭ-um) A sheath of connective tissue wrapped around a nerve.

Epiphyseal line (ep"ĭ-FIZ-e-al) The remnant of the epiphyseal plate in a long bone.

Epiphyseal plate A layer of cartilage between the epiphysis and diaphysis of a long bone that allows the bone to increase in length until early adulthood.

Epiphysis (ĕ-PIF-ĭ-sis) The end of a long bone, usually larger in diameter than the shaft (the diaphysis). The plural form is epiphyses.

Episiotomy (ĕ-piz"e-OT-o-me) Incision of the clinical perineum at the end of the second stage of labor to avoid tearing the perineum.

Epithelial tissue (ep"ĭ-THE-lĭ-al) The tissue that forms the outer part of the skin, lines blood vessels, hollow organs, and passages that lead to the outside of the body.

Eponychium (ep"o-NIK-ĭ-um) The thin layer of stratum corneum of the skin that overlaps the lunula of the nail.

Erythrocyte (ĕ-RITH-ro-sīt) A red blood cell or corpuscle.

Esophagus (ĕ-SOF-ah-gus) A hollow muscular tube connecting the pharynx and the stomach.

Etiology (e"te-OL-o-je) The study of the causes of disease, including theories of origin and what organisms, if any, are involved.

Eustachian tube (u-STA-she-an) A narrow channel connecting the middle ear with the nasopharynx; the auditory tube.

Eversion (e-VER-zhun) Turning outward.

Exocrine gland (EK-so-krin) One that secretes its product into a duct or tube that empties out at the surface of the covering and lining epithelium.

Exophthalmos (ek"sof-THAL-mos) An abnormal protrusion or bulging of the eyeball.

Expiration (ek"spĭ-RA-shun) Breathing out, or exhalation.

Extensibility (ek-sten"sĭ-BIL-ĭ-te) The ability to be stretched when pulled.

Extension (ek-STEN-shun) An increase in the anterior angle between two bones, except in extension of the knee and toes, in which the posterior angle is involved; restoring a body part to its anatomical position after flexion.

External or **superficial** (su"per-FISH-al) Located on or near the surface.

External auditory meatus (AW-dĭ-tor-e me-A-tus) A canal in the temporal bone that leads to the middle ear.

External ear The outer ear, consisting of the pinna, the external auditory canal, and the tympanic membrane or eardrum.

External nares (NA-rēs) The external nostrils, or the openings into the nasal cavity on the exterior of the body.

Exteroceptor (ek"ster-o-SEP-tor) A sense organ adapted for the reception of stimuli from outside the body.

Extrinsic (ek-STRIN-sik) Of external origin.

Eyebrow The hairy ridge above the eye.

Face The anterior aspect of the head.

Facet (FAS-et) A smooth surface for articulation.

Facilitated diffusion (fah-SIL-ĭ-ta-ted dĭ-FU-zuhn) A passive process that moves a substance down its own concentration gradient but that is carrier-mediated.

Falciform ligament (FAL-sĭ-form LIG-ah-ment) A sheet of parietal peritoneum that separates the two principal lobes of the liver; the ligamentum teres, or remnant of the umbilical vein, lies within its fold.

False vocal folds A pair of folds of the mucous membrane of the larynx superior to the true vocal cords.

Falx cerebelli (falks ser"ĕ-BEL-le) A small triangular process of the dura mater attached to the occipital bone in the posterior cranial fossa and projecting inward between the two cerebellar hemispheres.

Falx cerebri (SER-ĕ-bre) A fold of the dura mater extending down into the longitudinal fissure between the two cerebral hemispheres.

Fascia (FASH-e-ah) A fibrous membrane covering, supporting, and separating muscles.

Fasciculi (fah-SIK-u-le) Small bundles or clusters, especially of nerves or muscle fibers.

Fauces (FAW-sēz) The passageway from the mouth to the pharynx.

Feces (FE-sēz) Material discharged from the bowel that is made up of bacteria, secretions, and food residue; also called stool.

Fetal circulation (FE-tal) The circulatory system of the fetus, including the placenta and special blood vessels involved in the exchange of materials between fetus and mother.

Fibroblast (FI-bro-blast) A flat, long connective tissue cell that forms the fibroelastic tissues of the body.

Fibrocyte (FI-bro-sīt) A mature fibroblast, found in the fibroelastic connective tissues of the body.

Fibrous joint (FI-brus) A joint that allows little or no movement, such as a suture and syndesmosis.

Fibrous tunic (TU-nik) The outer coat of the eyeball, made up of the posterior sclera and the anterior cornea.

Fight-or-flight response The effect of the stimulation of the sympathetic division of the autonomic nervous system, particularly a marked increase in heart rate, blood pressure, oxygen consumption, respiration, and a feeling of tenseness; the defense-alarm reaction.

Filiform papilla (FIL-ĭ-form pah-PIL-ah) One of the many conical projections distributed in parallel rows over the anterior two-thirds of the tongue and containing no taste buds.

Filtration (fil-TRA-shun) A passive process involving the movement of solvents and dissolved substances across a semipermeable membrane by mechanical pressure, from an area of higher pressure to an area of lower pressure.

Filum terminale (FI-lum tur-min-AL-e) A thread of fibrous tissue continuous with the pia mater that extends from the terminal end of the spinal cord (conus medullaris) to the first segment of the coccyx.

Fimbriae (FIM-bre-e) A fringe of fingerlike projections surrounding the open end (infundibulum) of a uterine or Fallopian tube.

Fissure (FISH-er) A groove, fold, or slit that may be normal or abnormal.

Flagellum (fla-JEL-um) A hairlike, motile process on the extremity of a cell; in particular, the tail of a spermatozoan that propels it along its way.

Flatfoot A condition that results if the ligaments and tendons of the arches of the foot are weakened and the height in the longitudinal arch decreases or "falls."

Flexion (FLEK-shun) A folding movement in which there is a decrease in the angle between two bones anteriorly except in flexion of the knee and toes, in which the bones are approximated posteriorly.

Fontanel (fon-tah-NEL) A soft spot in a baby's skull; a membrane-covered spot where bone formation has not yet occurred.

Foot The terminal part of the lower extremity.

Foramen (for-A-men) A passage or opening; a communication between two cavities of an organ, or a hole in a bone for passage of vessels or nerves.

Foramen ovale (o-VAL-e) An opening in the fetal heart in the septum between the right and left atria; a hole in the greater wing of the sphenoid bone that transmits the mandibular branch of the trigeminal nerve (V).

Forearm (FŌR-arm) The part of the upper extremity between the elbow and the wrist.

Fornix (FŌR-niks) A tract in the brain made up of association fibers, connecting the hippocampus with the mammillary bodies; a recess around the cervix of the uterus where it protrudes into the vagina.

Fossa (FOS-ah) A depressed area or a shallow depression.

Fourth ventricle (VEN-trĭ-kl) A cavity within the brain lying between the cerebellum and the medulla and pons.

Fracture (FRAK-chūr) Any break in a bone.

Frontal or **coronal** (ko-RO-nal) A plane that runs vertical to the ground and divides the body into anterior and posterior portions.

Fungiform papilla (FUN-jĭ-form pah-PIL-ah) A mushroomlike elevation on the upper surface of the tongue appearing as a red dot; most fungiform papillae contain taste buds.

Gallbladder (GAWL-blad-er) A sac attached to the underside of the liver that serves as a storage place for bile.

Ganglion (GANG-glĭ-on) A cluster of nerve cell

bodies located outside the central nervous system.

Gastrointestinal tract (gas"tro-in-TES-tĭ-nal) The portion of the digestive tract from the cardia of the stomach to the anus.

Gingiva (jin-JIV-ah) The fleshy structure covering the tooth-bearing border of the maxilla and the mandible; the gum.

Gland A secretory structure.

Glans penis (glanz PE-nis) The slightly enlarged region at the distal end of the penis.

Gliding joint A synovial or diarthrotic joint having articulating surfaces that are usually flat, permitting only side-to-side and back-and-forth movements, as between carpal bones, tarsal bones, and the scapula and clavicle.

Glomerulus (glo-MER-u-lus) A rounded mass of nerves or blood vessels, especially the microscopic tuft of capillaries that is surrounded by the Bowman's capsule of each kidney tubule.

Glottis (GLOT-is) The air passageway between the vocal folds in the larynx.

Goblet cell A goblet-shaped unicellular gland that secretes mucus.

Goiter (GOY-ter) An enlargement of the thyroid gland.

Golgi complex (GOL-je) An organelle in the cytoplasm of cells consisting of four to eight flattened channels, stacked upon one another with expanded areas at their ends.

Golgi tendon organ A receptor found chiefly near the junction of tendons and muscles.

Gonadocorticoids (go-NAD-o-KOR-tĭ-koyds) Male and female sex hormones secreted by the adrenal cortex.

Gray commissure (KOM-ĭ-shur) A narrow strip of gray matter connecting the two lateral gray masses within the spinal cord.

Gray matter Areas in the central nervous system consisting of nonmyelinated nerve tissue.

Gray ramus communicans (RA-mus kom-U-nik-ans) A short nerve containing postganglionic sympathetic fibers; the cell bodies of the fibers are in a sympathetic chain ganglion, and the nonmyelinated axons run by way of the gray ramus to a spinal nerve and then to the periphery to supply smooth muscle in blood vessels, arrector pili muscles, and sweat glands.

Greater omentum (o-MEN-tum) A large fold in the serosa of the stomach that hangs down like an apron over the front of the intestines.

Greater vestibular glands (ves-TIB-u-lar) A pair of glands on either side of the vaginal orifice that open by a duct into the space between the hymen and the labia minora (also called Bartholin's glands); these glands are homologous to the male Cowper's or bulbourethral glands.

Groove or **sulcus** (SUL-kus) A shallow downfold of the gray matter of the cerebral cortex; a furrow on the surface of a bone that accommodates a soft structure such as a blood vessel, nerve, or tendon.

Gross anatomy The branch of anatomy that deals with structures that can be studied without using a microscope.

Hair A threadlike structure produced by the specialized epidermal structure developing from a papilla sunk in the dermis.

Hand The terminal portion of an upper extremity, including the carpus, metacarpus, and phalanges.

Hard palate (PAL-at) The anterior portion of the roof of the mouth, formed by the maxillae and palatine bones and lined by mucous membrane.

Haustra (HOS-trah) Pouches or sacculations of the colon.

Haversian canal (ha-VER-shan) A circular channel running longitudinally in the center of a Haversian system of mature compact bone, containing blood and lymph vessels and nerves.

Haversian system or **osteon** A Haversian canal with its concentrically arranged lamellae, lacunae, osteocytes, and canaliculi constituting the basic unit of structure in adult compact bone.

Head The superior part of a human being, cephalic to the neck; the superior or proximal part of a structure.

Heart A hollow muscular organ lying slightly to the left of the midline of the chest that pumps the blood through the cardiovascular system.

Heart block An arrhythmia of the heart in which the atria and ventricles contract independently because of a blocking of electrical impulses through the heart at a critical point in the conduction system.

Hematology (he"mah-TOL-o-je) The study of the blood.

Hematopoiesis (hem"ah-to-poy-E-sis) The process by which blood cells are formed.

Hemiballismus (hem"ĭ-bah-LIZ-mus) Violent muscular restlessness of half of the body, especially of the upper extremity.

Hemodialysis (he"mo-di-AL-ĭ-sis) Filtering of the blood by means of an artificial device so that certain substances are removed from the blood as a result of the difference in rates of their diffusion through a semipermeable membrane while the blood is being circulated outside the body.

Hemoglobin (he"mo-GLO-bin) A red pigment in erythrocytes constituting about 33% of the cell volume, involved in the transport of oxygen and carbon dioxide.

Hemorrhoids (HEM-o-royds) Dilated or varicosed blood vessels (usually veins) in the anal region; also called piles.

Hepatic duct (hē-PAT-ik) A duct that receives bile from the bile capillaries; small hepatic ducts merge to form the larger right and left hepatic ducts that unite to leave the liver as the common hepatic duct.

Hepatic portal circulation (POR-tal) The flow of blood from the digestive organs to the liver before returning to the heart.

Hilus (HI-lus) An area, depression, or pit where blood vessels and nerves enter or leave an organ; hilum.

Hinge joint A synovial or diarthrotic joint characterized by a convex surface of one bone that fits into a concave surface of another bone, such as the elbow, knee, ankle, and interphalangeal joints.

Histology (his-TOL-o-je) Microscopic study of the structure of tissues.

Holocrine gland (HOL-o-krin) The type of glandular secretion in which the entire secreting cell, along with its accumulated secretions, makes up the secretory product of the gland, as in the sebaceous glands.

Horizontal or **transverse** A plane that runs parallel to the ground and divides the body into superior and inferior portions.

Hormone (HŌR-mōn) A secretion of an endocrine or ductless gland secreted directly into the blood for transport.

Horn Principal area of gray matter in the spinal cord.

Hymen (HI-men) A thin fold of vascularized mucous membrane at the inferior end of the vaginal opening.

Hyperextension (hi"per-ek-STEN-shun) Continuation of extension beyond the anatomic position, as in bending the head backward.

Hypermetropia (hi"per-me-TRO-pe-ah) Farsightedness; hyperopia.

Hypersecretion (hi"per-se-KRE-shun) Overactivity of glands resulting in excessive secretion.

Hypertension (hi"per-TEN-shun) High blood pressure.

Hypertonic (hi"per-TON-ik) Having an osmotic pressure greater than that of the solution with which it is compared.

Hyponychium (hi"po-NIK-ĭ-um) The thickened epidermis beneath the free distal end of the nail of a digit.

Hyposecretion (hi"po-se-KRE-shun) Underactivity of glands resulting in diminished secretion.

Hypothalamic–hypophyseal tract (hi"po-thal-AM-ik–hi"po-FIZ-e-al) A bundle of nerve processes made up of fibers that have their cell bodies in the hypothalamus but that release their neurosecretions in the posterior pituitary gland or neurohypophysis.

Hypothalamus (hi"po-THAL-ȧ-mus) A portion of the forebrain, lying beneath the thalamus and forming the floor and part of the walls of the third ventricle.

Hypotonic (hi"po-TON-ik) Having an osmotic pressure lower than that of the solution with which it is compared.

Ileocecal valve (il"e-o-SE-kal) A fold of mucous membrane that guards the opening from the ileum into the large intestine.

Ileum (IL-e-um) The distal or terminal portion of the small intestine, extending between the jejunum of the small intestine and the large intestine.

Inclusion A secretion or storage area in the cytoplasm of a cell.

Incus (ING-kus) The middle of the three ossicles of the middle ear, so named because of its anvil shape; also called the anvil.

Infarction (in-FARK-shun) The presence of a localized area of necrotic tissue, produced by inadequate oxygenation of the tissue.

Inferior or **caudal** (KAW-dal) Away from the head, or toward the lower part of a structure.

Infundibulum (in"fun-DIB-u-lum) The stalklike structure that attaches the pituitary gland or hypophysis to the hypothalamus of the brain.

Insertion The manner or place of attachment of a muscle to the bone that it moves.

Inspiration (in"spī-RA-shun) The drawing of air into the lungs.

Insula (IN-su-lah) A triangular area of cerebral cortex that lies deep within the lateral cerebral fissure, under the parietal, frontal, and temporal lobes and cannot be seen in an external view of the brain; the island or isle of Reil.

Integument (in-TEG-u-ment) A covering, especially the skin.

Intercalated disc (in-TER-kah-la-ted) A transverse thickening of the sarcolemma that separates one cardiac cell from another.

Intercellular substance (in-ter-SEL-u-lar) The matrix of connective tissue that largely determines the quality of the tissue.

Intercostal nerve (in"ter-KOS-tal) A nerve supplying a muscle located between the ribs.

Internal capsule (KAP-sul) A thick sheet of white matter made up of myelinated fibers connecting various parts of the cerebral cortex and lying between the thalamus and the caudate and lentiform nuclei of the basal ganglia.

Internal or **deep** Away from the surface of the body.

Internal ear The inner ear or labyrinth, lying inside the temporal bone, containing the organs of hearing and balance.

Internal nares (NA-rēs) The two openings posterior to the nasal cavities opening into the nasopharynx; the choanae.

Interphase (IN-ter-fāz) The period during its life cycle when a cell is carrying on every life process except division; the stage between two mitotic divisions.

Interstitial cells of Leydig (in"ter-STĬ-shal . . . LI-dig) Secretory cells that secrete testosterone, located in the connective tissue between seminiferous tubules in a mature testis.

Interstitial fluid The fluid filling the microscopic spaces between the cells of tissues.

Interventricular foramen (in"ter-ven-TRIK-u-lar for-A-men) A narrow, oval opening through which the lateral ventricles of the brain communicate with the third ventricle; also called the foramen of Monro.

Intervertebral disc (in"ter-VER-tĕ-bral) A pad of fibrocartilage located between the bodies of two vertebrae.

Intramembranous ossification (in"tra-mem-BRA-nus os"sĭ-fĭ-KA-shun) The method of bone formation in which the bone is formed directly in membranous tissue.

Intrinsic (in-TRIN-sik) Of internal origin; for example, the intrinsic factor, a mucoprotein formed by the gastric mucosa that is necessary for the absorption of an extrinsic factor called vitamin B_{12}.

Inversion (in-VER-zhun) The movement of the sole inward at the ankle joint.

Irritability (ir"rĭ-tah-BIL-ĭ-te) The ability to receive and respond to stimuli.

Ischemia (is-KE-me-ah) A lack of sufficient blood to a part, because of obstruction of the circulation to it.

Islets of Langerhans (I-lets of LAHNG-er-hans) Clusters of endocrine gland cells in the pancreas that secrete insulin and glucagon.

Isotonic (i-so-TON-ik) Having equal tension or tone; the existence of equality of osmotic pressure between two different solutions or between two elements in a solution.

Isthmus (IS-mus) A narrow strip of tissue or a narrow passage connecting two larger parts; for example, the band of tissue connecting the two lobes of the thyroid gland.

Jejunum (je-JU-num) The second portion of the small intestine, located between the duodenum and the ileum.

Joint capsule (KAP-sūl) A sleevelike, fibrous cuff lined with a synovial membrane that encases the articulating bones of a diarthrotic or synovial joint.

Joint kinesthetic receptor (kin"es-THET-ik) A proprioceptive receptor located in a joint, stimulated by joint movement.

Keratin (KER-ă-tin) A special insoluble protein found in the hair, nails, and other horny tissues of the epidermis.

Kidney (KID-ne) One of the paired reddish organs located in the lumbar region that secrete urine.

Knee A hinge joint located between the thigh and the leg of the lower extremity.

Kupffer cells (KUP-fer) Phagocytic cells that partly line the sinusoids of the liver.

Kyphosis (ki-FO-sis) An increased curvature of the chest, giving a hunchback appearance.

Labia majora (LA-be-ah ma-JOR-ah) Two longitudinal folds of skin extending downward and backward from the mons pubis of the female.

Labia minora (min-OR-ah) Two small folds of skin lying between the labia majora of the female.

Labial frenulum (LA-be-al FREN-u-lum) A medial fold of mucous membrane between the inner surface of the lip and the gums.

Lacrimal canal (LAK-rĭ-mal) A duct, one in each eyelid, commencing at the punctum at the medial margin of an eyelid and conveying the tears medially into the nasolacrimal sac.

Lacrimal glands Secretory cells located at the superior lateral portion of both orbits that secrete tears into excretory ducts that open onto the surface of the conjunctiva.

Lacrimal sac The superior expanded portion of the nasolacrimal duct that receives the tears from a lacrimal canal.

Lacteal (LAK-te-al) Related to milk; one of the many intestinal lymph vessels that take up fat from digested food.

Lacuna (lă-KU-nah) A small, hollow space, such as that found in bones, in which lie the osteoblasts. The plural form is lacunae.

Lamella (lah-MEL-ah) One of the concentric rings surrounding the Haversian canal in a Haversian system of mature compact bone.

Lamina (LAM-in-ah) A thin, flat layer or membrane, as the flattened part of either side of the arch of a vertebra. The plural form is laminae.

Lamina propria (PRO-pre-ah) The connective tissue layer of a mucous membrane.

Large intestine The portion of the digestive tract extending from the ileum of the small intestine to the anus, divided structurally into the cecum, colon, rectum, and anal canal.

Laryngopharynx (lah-rin"go-FAR-ingks) The inferior portion of the pharynx, extending downward from the level of the hyoid bone to empty posteriorly into the esophagus and anteriorly into the larynx.

Larynx (LAR-ingks) The voicebox, a short passageway that connects the pharynx with the trachea.

Lateral (LAT-er-al) Farther from the midline of the body; toward the side.

Lateral apertures (AP-er-chūrz) Two of the three openings in the roof of the fourth ventricle through which cerebrospinal fluid enters the subarachnoid space of the brain and spinal cord; also known as the foramina of Luschka.

Lateral ventricle (VEN-trĭ-kl) A cavity within a cerebral hemisphere that communicates with the lateral ventricle in the other cerebral hemisphere and with the third ventricle by way of the interventricular foramen; cerebrospinal fluid flows through the ventricular system.

Leg The part of the lower extremity between the knee and the ankle.

Lens A transparent organ lying posterior to the pupil and iris of the eyeball and anterior to the vitreous humor.

Lesser omentum (o-MEN-tum) A fold of the peritoneum that extends from the liver to the lesser curvature of the stomach and the commencement of the duodenum.

Lesser vestibular glands (ves-TIB-u-lar) Mucous-secreting glands that have ducts that open on either side of the urethral orifice in the vestibule of the female; also known as Skene's glands; these glands are the female homologues of the male prostate gland.

Leucocyte (LU-ko-sīt) A white blood cell; also leukocyte.

Leukemia (lu-KE-me-ah) A malignant disease of the tissues in the bone marrow, spleen, and lymph nodes, characterized by an uncontrolled, greatly accelerated production of white blood cells; also known as cancer of the blood.

Ligament (LIG-ah-ment) Collagenous connective tissue, with numerous collagen molecules arranged parallel to one another, that attaches muscles to bones.

Limbic system A portion of the forebrain, sometimes termed the visceral brain, concerned with various aspects of emotion and behavior, that includes the limbic lobe (hippocampus and associated areas of gray matter plus the cingulate gyrus), certain parts of the temporal and frontal cortex, some thalamic and hypothalamic nuclei, and parts of the basal ganglia.

Line A mark, narrow ridge, stripe, or streak; often an imaginary line connecting different anatomical landmarks, as the imaginary line from the superior border of the pubic bone to the sacral promontory that separates the abdominopelvic cavity into the abdominal and pelvic subdivisions.

Lingual frenulum (LIN-gwal FREN-u-lum) A fold of mucous membrane that connects the tongue to the floor of the mouth.

Local Pertaining to or restricted to one spot or part.

Lordosis (lor-DO-sis) Abnormal anterior convexity of the lumbar spine; swayback.

Lower extremity An inferior limb, including the thigh, leg, and foot.

Lumbar (LUM-bar) The region of the back and side between the ribs and pelvis; loins.

Lumbar plexus (PLEK-sus) A network formed by the anterior branches of spinal nerves L1 through L4.

Lung One of the two main organs of respiration, lying on either side of the heart in the thoracic cavity.

Lunula (LU-nu-lah) The moon-shaped white area at the base of a nail.

Lymph (limf) Tissue fluid confined in lymphatic vessels and flowing through the lymphatic system to be returned to the blood.

Lymph node An oval- or bean-shaped structure located along the lymphatic vessels.

Lymphangiogram (lim-FAN-je-o-gram) A film produced by roentgenography in which lymphatic vessels and lymph organs are filled with an opaque substance in order to be filmed.

Lymphatic (lim-FAT-ik) Pertaining to lymph or referring to one of the vessels that collects lymph from the tissues and carries it to the blood.

Lymphocyte (LIM-fo-sīt) An agranular white blood cell that fights infection by producing antibodies, normally constituting about 20–25% of the white blood cells.

Lysosome (LI-so-sōm) An organelle in the cytoplasm of a cell, enclosed in a single membrane, and containing powerful digestive enzymes.

Macrophage or **histiocyte** (MAK-ro-fāj, HIS-te-o-sīt) A large phagocytic cell of the reticuloendothelial system common in loose connective tissue.

Macula lutea (MAK-u-lah LU-te-ah) A yellow spot

situated in the exact center of the posterior retina and containing the fovea centralis, the area of keenest vision.

Malignancy (mah-LIG-nan-se) A cancerous growth.

Malleus (MAL-e-us) The largest of the three ossicles of the middle ear; the one attached to the inner side of the eardrum; also called the hammer.

Mammary gland (MAM-ar-e) Gland of the female that secretes milk for the nourishment of the young.

Marrow (MAR-o) Soft, spongelike material in the cavities of bone; red marrow produces blood cells; yellow marrow, formed mainly of fatty tissue, has no blood-producing function.

Mast cell A cell found in loose connective tissue along blood vessels that produces heparin, an anticoagulant; the name given to a basophil after it has left the bloodstream and entered the tissues.

Mastication (mas"tĭ-KA-shun) Chewing.

Meatus (me-A-tus) A passageway or opening, especially the external portion of a canal.

Medial (ME-de-al) Nearer the midline of the body.

Medial lemniscus (lem-NIS-kus) A flat band of myelinated nerve fibers extending through the medulla, pons, and midbrain and terminating in the thalamus on the opposite side; second-order sensory neurons in this tract transmit impulses for discriminating touch and pressure sensations.

Median aperture (ME-de-an AP-er-chūr) One of the three openings in the roof of the fourth ventricle through which cerebrospinal fluid enters the subarachnoid space of the brain and cord; also called the foramen of Magendie.

Mediastinum (me"de-ah-STI-num) A space between the pleurae of the lungs that extends from the sternum to the backbone.

Medulla oblongata (mĕ-DUL-lah ob"lon-GAH-tah) A continuation of the upper portion of the spinal cord forming the most inferior portion of the brain stem; it lies just above the level of the foramen magnum and extends upward to the lower portion of the pons.

Medullary (marrow) cavity (MED-u-lar-rĭ) The space within the diaphysis of a long bone that contains fatty yellow marrow.

Meibomian gland (mi-BO-mĭ-an) A sebaceous gland embedded in the tarsal plates of each eyelid, with a duct that opens onto the edge of the eyelid; its secretion lubricates the margins of the lids and prevents the overflow of tears.

Meissner's corpuscle (MĪS-ners KOR-pusl) An encapsulated sensory receptor found in the papillae of the skin, sensitive to light touch; makes two-point discrimination possible.

Melanin (MEL-an-in) The dark pigment found in some parts of the body, such as the skin.

Melanocyte (MEL-ah-no-sīt") A pigmented cell located in the deepest layer of the epidermis that synthesizes melanin, a black pigment.

Membrane The combination of an epithelial layer and an underlying connective tissue layer.

Membranous labyrinth (mem-BRA-nus LAB-ĭ-rinth) The portion of the labyrinth of the inner ear that is located inside the bony labyrinth and separated from it by the perilymph; it is made up of the membranous semicircular canals, the saccule and utricle, and the cochlear duct.

Meninges (mĕ-NIN-jēz) Three membranes covering the brain and spinal cord, called the dura mater, arachnoid, and pia mater.

Menisci (mĕ-NIS-ke) Crescent-shaped fibrocartilages in the knee joint; also called the semilunar cartilages.

Merkel's disc (MER-kls) An encapsulated, cutaneous receptor for touch attached to deeper layers of epidermal cells.

Merocrine gland (MER-o-krin) A secretory cell that remains intact throughout the process of formation and discharge of the secretory product, as in the salivary and pancreatic glands.

Mesentery (MES-en-ter"e) A fold of peritoneum attaching the small intestine to the posterior abdominal wall.

Mesocolon (mez"o-KO-lon) A fold of peritoneum attaching the colon to the posterior abdominal wall.

Mesothelium (mez"o-THE-lĭ-um) The simple squamous epithelium that lines the serous cavities.

Mesovarium (mez"o-VAR-ĭ-um) A short fold of peritoneum that attaches an ovary to the broad ligament of the uterus.

Metacarpus (met"ah-KAR-pus) A collective term for the five bones that make up the palm of the hand.

Metastasis (mĕ-TAS-tah-sis) The transfer of disease from one organ or part of the body to another part that is not connected with it.

Metatarsus (met"ah-TAR-sus) A collective term for the five bones located in the foot between the tarsals and the phalanges.

Microvilli (mi"kro-VIL-e) Microscopic fingerlike projections of the cell membranes of certain cells; the brush border of light microscopy.

Micturition (mik-tu-RĬ-shun) The act of expelling urine from the bladder; urination or voiding.

Midbrain The part of the brain between the pons and the forebrain.

Middle ear A small, epithelial-lined cavity hollowed out of the temporal bone, separated from the external ear by the eardrum and from the internal ear by a thin bony partition containing the oval and round windows; extending across the middle ear are the three auditory ossicles. Also called the tympanic cavity.

Midsagittal (mid-SAJ-ĭ-tal) A plane through the midline of the body, running vertical to the ground and dividing the body into equal right and left sides; midline or median.

Mitochondrion (mi"to-KON-drĭ-on) The powerhouse of the cell; a double-membraned organelle that plays a central role in the production of ATP.

Mitosis (mi-TO-sis) The orderly division of the nucleus of a cell that ensures that each new daughter nucleus has the same number and kind of chromosomes as the original parent nucleus; the process of mitosis includes the replication of chromosomes and the distribution of the two sets of chromosomes into two separate and equal nuclei.

Modiolus (mo-DI-o-lus) The central pillar or column of the cochlea.

Monocyte (MON-o-sīt) The larger of the agranular white blood cells that combats inflammation and infection by phagocytosis, normally constitutes about 3–8% of the white blood cells.

Mons pubis (monz PU-bis) The rounded, fatty prominence over the symphysis pubis, covered by coarse pubic hair.

Motor area The region of the cerebral cortex that governs muscular movement, particularly the precentral gyrus of the frontal lobe.

Motor unit A motor neuron, together with the muscle cells it stimulates.

Mucosa (mu-KO-sah) A mucous membrane.

Mucous cell (MU-kus) A unicellular gland that secretes mucus; also called a goblet cell.

Mucous membrane A membrane that lines a body cavity that opens to the exterior; also called the mucosa.

Muscle Unless otherwise specified, the term "muscle" implies a skeletal muscle, an organ specialized for contraction, composed of striated muscle cells supported by connective tissue, attached to a bone by a tendon or an aponeurosis, and stimulated by a somatic efferent neuron.

Muscularis (mus"ku-LA-ris) A muscular layer or tunic of an organ.

Muscularis mucosa (mu-KO-sah) A thin layer of smooth muscle cells located in the outermost layer of the mucosa of the alimentary canal, underlying the lamina propria of the mucosa.

Myelin sheath (MI-ĕ-lin) A white, lipid, segmented covering, formed by Schwann cells, around the axons and dendrites of many peripheral neurons.

Myocardium (mi"o-KAR-dĭ-um) The layer of the heart made up of cardiac muscle that comprises the bulk of the heart and lies between the epicardium and the endocardium.

Myofibril (mi"o-FI-bril) A threadlike structure, running longitudinally through a muscle cell, consisting mainly of molecules of the protein actin and the protein myosin.

Myology (mi-OL-o-je) The science or study of the muscles and their parts.

Myometrium (mi"o-ME-trĭ-um) The smooth muscle coat or tunic of the uterus.

Myopia (mi-O-pe-ah) Defect in vision so that objects can be seen distinctly only when very close to the eyes; nearsightedness.

Myosin (MI-o-sin) The protein that makes up the thick filaments of myofibrils of muscle cells.

Nail A horny plate, composed largely of keratin, that develops from the epidermis of the skin to form a protective covering on the dorsal surface of the distal phalanges of the fingers and toes.

Nail matrix (MA-triks) The part of the nail beneath the body and root from which the nail is produced.

Nasal cavity (NA-zal) A mucosa-lined cavity on either side of the nasal septum that opens onto the face at an external nostril or naris and into the nasopharynx at an internal nostril or naris.

Nasal septum (SEP-tum) A vertical partition composed of bone and cartilage, covered with a mucous membrane, separating the right and left nasal cavities.

Nasolacrimal duct (na"zo-LAK-rĭ-mal) A canal that transports the lacrimal secretion from the nasolacrimal sac into the nose.

Nasopharynx (na"zo-FAR-ingks) The uppermost portion of the pharynx, lying posterior to the nose and extending down to the soft palate.

Neck The part of the body connecting the head and the trunk, or a constricted portion of an organ such as the neck of a femur or the neck of the uterus.

Nephron (NEF-ron) The physiological unit of a kidney, made up of a glomerular capsule, proximal convoluted tubule, descending limb of Henle, loop of Henle, ascending limb of Henle, and distal convoluted tubule.

Nerve A cordlike bundle of nerve fibers and their

associated connective tissue coursing together outside the central nervous system.

Nerve impulse An action potential or a wave of negativity that sweeps along the outside of the membrane of a neuron.

Neurilemma (nu"rĭ-LEM-ah) A delicate sheath around a nerve fiber composed of the cytoplasm and nucleus and enclosing cell membrane of a Schwann cell; there may or may not be a layer of myelin between the neurilemma and the axis cylinder (axon or dendrite) of the neuron.

Neuroeffector junction (nu"ro-ĕ-FEK-tor) A synapse between an autonomic fiber and a visceral effector.

Neurofibril (nu"ro-FI-bril) One of the delicate threads that form a complicated network in the cytoplasm of the cell body and processes of a neuron.

Neuroglia (nu-ro-GLE-ah) Cells of the nervous system that are specialized to perform the functions of connective tissue; the neuroglia of the central nervous system are the astrocytes, oligodendrocytes, and microglia; neuroglia of the peripheral nervous system include the Schwann cells and the ganglion satellite cells.

Neurohypophysis (nu"ro-hi-POF-ĭ-sis) The posterior lobe of the pituitary gland.

Neuromuscular junction (nu"ro-MUS-ku-lar) The area of contact or synapse between a neuron and a muscle fiber.

Neuromuscular spindle An encapsulated receptor in a skeletal muscle, consisting of specialized muscle cells and nerve endings, stimulated by change in length or tension of muscle cells; a proprioceptor.

Neuron (NU-ron) A nerve cell, consisting of a cell body, dendrites, and an axon.

Neurosyphilis (nu"ro-SIF-ĭ-lis) Syphilis of the nervous system; for example, tabes dorsalis.

Neutrophil (NU-tro-fil) A granular, polymorphonuclear white blood cell, actively phagocytic; normally constituting about 60–70% of the white blood cells.

Nipple A round or cone-shaped projection at the tip of the breast, or any similarly shaped structure.

Nissl bodies (NIS-l) Rough endoplasmic reticulum in the cell bodies of neurons.

Node of Ranvier (ron-ve-A) A space, along a myelinated nerve fiber, between the individual Schwann cells that form the myelin sheath and the neurilemma.

Nucleus (NU-kle-us) A spherical or oval organelle of a cell that contains the hereditary factors of the cell, called genes; a cluster of nerve cell bodies in the central nervous system.

Nucleus pulposus (pul-PO-sus) A soft, pulpy, highly elastic substance in the center of an intervertebral disc, a remnant of the notochord.

Olfactory bulb (ōl-FAK-to-re) A mass of gray matter at the termination of an olfactory nerve, lying beneath the frontal lobe of the cerebrum on either side of the crista galli of the ethmoid bone.

Olfactory cell A bipolar neuron with its cell body lying between supporting cells located in the mucous membrane lining the upper portion of each nasal cavity.

Olfactory tract A bundle of axons that extends from the olfactory bulb posteriorly to the olfactory portion of the cortex.

Olive A prominent oval mass on each lateral surface of the superior part of the medulla.

Optic chiasma (OP-tik ki-AZ-mah) A crossing point of the optic nerves, anterior to the pituitary gland.

Optic disc A small area of the retina containing openings through which the fibers of the ganglion neurons emerge as the optic nerve; the blind spot.

Optic tract A bundle of axons that transmits impulses from the retina of the eye between the optic chiasma and the thalamus.

Ora serrata (O-rah ser-AH-tah) The jagged, anterior margin of the retina, near the ciliary body, where the nervous portion of the retina ends.

Organ A group of two or more different kinds of tissue that performs a particular function.

Organ of Corti (KOR-te) The spiral organ or organ of hearing consisting of supporting cells and hair cells that rest on the basilar membrane and extend into the endolymph of the cochlear duct; the receptor for hearing.

Organelle (or-gan-EL) A tiny, specific particle of living material present in most cells and serving a specific function.

Organism (OR-gah-nizm) A total living form; one individual.

Origin (OR-ĭ-jin) The place of attachment of a muscle to the more stationary bone, or the end opposite the insertion.

Oropharynx (or"o-FAR-ingks) The second portion of the pharynx, lying posterior to the mouth and extending from the soft palate down to the hyoid bone.

Osmosis (os-MO-sis) The net movement of water molecules through a semipermeable membrane from an area of high water concentration to an area of lower water concentration.

Osseous tissue (OS-se-us) Bone tissue.

Ossification (os"ĭ-fĭ-KA-shun) Formation of bone substance.

Osteoblast (OS-te-o-blast") A bone-forming cell.

Osteoclast (OS-te-o-clast") A large, multinuclear cell that destroys or resorbs bone tissue.

Osteocyte (OS-te-o-sīt") A mature osteoblast that has lost its ability to produce new bone tissue.

Osteology (os"te-OL-o-je) The study of bones.

Osteomalacia (os"te-o-mah-LA-she-ah) Demineralization of bones due to vitamin D deficiency.

Osteomyelitis (os"te-o-mi-ĭ-LI-tis) Inflammation of bone marrow, or of the bone and the marrow.

Osteoporosis (os"te-o-pōr-O-sis) Increased porosity of bone.

Otic ganglion (O-tik GANG-glĭ-on) A ganglion of the parasympathetic subdivision of the autonomic nervous system, made up of cell bodies of postganglionic parasympathetics to the parotid gland.

Otolith (O-to-lith) A particle of calcium carbonate embedded in the gelatinous layer that coats the hair cells of the sensory receptor (the macula) in the saccule and utricle of the inner ear.

Oval window A small opening between the middle and inner ear into which the footplate of the stapes fits; also called the fenestra vestibuli.

Ovarian follicle (o-VAR-ĭ-an FOL-ĭ-kl) A general name for an ovum in any stage of development, along with its surrounding group of epithelial cells.

Ovarian ligament (LIG-ah-ment) A rounded cord of connective tissue that attaches the ovary to the uterus.

Ovary (O-var-e) The female gonad in which the ova are formed.

Pacinian corpuscle (pă-SIN-ĭ-an KOR-pusl) An encapsulated pressure receptor found in the deep subcutaneous tissues, in tissues that lie under mucous membranes, in serous membranes of the abdominal cavity, around joints and tendons, and in some viscera.

Paget's disease (PAJ-ets) A disorder characterized by an irregular thickening and softening of the bones; the cause is unknown, but the bone-producing osteoblasts and the bone-destroying osteoclasts apparently become uncoordinated.

Palpebra (PAL-pĕ-brah) An eyelid.

Pancreas (PAN-kre-as) A soft, oblong-shaped organ lying along the greater curvature of the stomach and connected by a duct to the duodenum; it is both exocrine (secreting pancreatic juice) and endocrine (secreting insulin and glucagon).

Pancreatic duct (pan"kre-AT-ik) A single large tube that drains pancreatic juice into the duodenum at the ampulla of Vater; also called the duct of Wirsung; sometimes an accessory duct, the duct of Santorini, is also present and empties into the duodenum above the ampulla of Vater.

Papillae (pah-PIL-e) Small projections or elevations.

Papillary muscle (PAP-ĭ-lar-e) Muscular columns located on the inner surface of the ventricles from which the chordae tendineae extend to attach to the cusps of the atrioventricular valves.

Paranasal sinus (par"ah-NA-zal SI-nus) A mucous-lined air cavity in a skull bone that communicates with the nasal cavity; paranasal sinuses are located in the frontal, maxillary, ethmoid, and sphenoid bones.

Parasympathetic (par"ah-sim-pah-THET-ik) One of the two subdivisions of the autonomic nervous system, having cell bodies of preganglionic neurons in nuclei in the brain stem and in the lateral gray matter of the sacral portion of the spinal cord and concerned with activities that restore and conserve body energy; the craniosacral division.

Parathyroids (par"ah-THI-roydz) The four small endocrine glands embedded on the posterior surfaces of the lateral lobes of the thyroid; their secretory product is parathormone, or parathyroid hormone.

Parietal (pah-RI-ĕ-tal) Pertaining to the walls of a cavity.

Parietal cell One of the three kinds of secreting cells of the gastric glands, the one that produces hydrochloric acid.

Parietal pleura (PLU-rah) The outer layer of the serous pleural membrane that encloses and protects the lung, the layer that is attached to the walls of the pleural cavity.

Parotid gland (pah-ROT-id) One of the paired salivary glands located inferior and anterior to the ears connected to the oral cavity via a duct that opens into the inside of the cheek opposite the upper second molar tooth.

Pathological (path"o-LOJ-ĭ-kal) Pertaining to or caused by disease.

Pathological anatomy The study of structural changes caused by disease.

Pedicle (PED-ĭ-kl) The part of the vertebral or neural arch of a vertebra that connects the body with a lamina.

Pelvic splanchnic nerves (PEL-vik SPLANGK-nik) Preganglionic parasympathetic fibers from the levels of S2, S3, and S4 that supply the urinary bladder, reproductive organs, and the descending and sigmoid colon and rectum.

Pelvis The basinlike structure formed by the two pelvic bones, the sacrum, and the coccyx; the expanded, proximal portion of the ureter, lying within the kidney and into which the major calyces open.

Penis (PE-nis) The male copulary organ, used to introduce spermatozoa into the female vagina.

Pericardium (per"ĭ-KAR-dĭ-um) A loose-fitting serous membrane that encloses the heart, consisting of an outer fibrous layer and an internal serous layer containing a parietal layer that lines the inside of the fibrous pericardium, and a visceral layer, also known as the epicardium.

Perichondrium (per"ĭ-KON-drĭ-um) A connective tissue that covers cartilage.

Perikaryon (per"ĭ-KAR-ĭ-on) The cell body of a neuron.

Perimysium (per"ĭ-MIZ-ĭ-um) Connective tissue around a bundle or fascicle of muscle cells.

Perineum (per"ĭ-NE-um) The pelvic floor; the space between the anus and the scrotum in the male, and the anus and the vulva in the female.

Perineurium (per"ĭ-NU-rĭ-um) Connective tissue around a bundle or fascicle of nerve fibers.

Periodontal membrane (per"ĭ-o-DON-tal) The periosteum lining the sockets for the teeth in the alveolar processes of the mandible and maxillae.

Periosteum (per"ĭ-OS-te-um) A dense, white, fibrous membrane covering the surface of a bone, except at the articular surface.

Peripheral nervous system (per-IF-er-al) The part of the nervous system that lies outside the central nervous system—nerves and ganglia.

Peristalsis (per"ĭ-STAL-sis) A wave of contraction along a hollow muscular tube, followed by relaxation.

Peritoneum (per"ĭ-to-NE-um) The serous membrane lining the abdominal cavity and covering the abdominal organs and some pelvic organs.

Peritonitis (per"ĭ-to-NI-tis) Inflammation of the peritoneum, the membranous coat lining the abdominal cavity and covering the viscera.

Peyer's patches (PI-urs) Clusters of lymph nodules on the walls of the ileum.

Phagocytosis (fag"o-si-TO-sis) Literally, "cell eating"; the engulfing of solid particles by living cells.

Phalanges (fah-LAN-jēz) Bones of a finger or toe.

Pharynx (FAR-ingks) The throat, a tube that starts at the internal nares and runs partway down the neck where it opens into the esophagus posteriorly and into the larynx anteriorly.

Physiology (fiz"e-OL-o-je) The study of the functions of body parts.

Pia mater (PE-a MA-ter) Literally, "tender mother"; the innermost meninx, the transparent layer of connective tissue that adheres to the surfaces of the brain and cord and contains blood vessels.

Pineal gland (PIN-e-al) The cone-shaped gland located in the roof of the third ventricle; also called the epiphysis cerebri.

Pinna (PIN-nah) The projecting part of the external ear; the auricle.

Pinocytosis (pin"o-si-TO-sis) Literally, "cell drinking"; the engulfing of liquid material by living cells.

Pituitary gland (pit-U-ĭ-tar-e) The hypophysis, nicknamed the "master gland," a small endocrine gland lying in the sella turcica of the sphenoid bone and attached to the hypothalamus by the infundibulum.

Pivot joint (PIV-ot) A synovial or diarthrotic joint in which a rounded, pointed, or conical surface of one bone articulates with a shallow depression of another bone, as in the joint between the atlas and axis and between the proximal ends of the radius and ulna.

Plantar flexion (PLAN-tar FLEK-shun) Extension of the foot at the ankle joint.

Plasma (PLAZ-mah) The extracellular fluid found in blood vessels.

Plasma cell A cell of loose connective tissue that gives rise to antibodies; the name given to a lymphocyte after it leaves the circulatory system and becomes a connective tissue cell.

Plexus (PLEK-sus) A network of nerves, veins, or lymphatic vessels.

Plexus of Auerbach (OW-ur-bok) A network of nerve fibers from both autonomic divisions located in the muscularis coat or tunic of the small intestine; also called the myenteric plexus.

Plexus of Meissner (MĪS-ner) A network of autonomic nerve fibers located in the outer portion of the submucous layer or tunic of the small intestine.

Plica circularis (PLI-ca sur-ku-LAR-is) A permanent, deep, transverse fold in the mucosa and submucosa of the small intestine that increases the surface area for absorption.

Pons varolii (ponz var-O-le-i) The portion of the brainstem that forms a "bridge" between the medulla and the midbrain, anterior to the cerebellum.

Posterior or **dorsal** (pos-TĒR-ĭ-or) Nearer to or at the back of the body.

Posterior root The structure composed of afferent (sensory) fibers lying between a spinal nerve and the dorsolateral aspect of the spinal cord; also called the dorsal root or sensory root.

Posterior root ganglion (GANG-glĭ-on) A group of cell bodies of sensory (afferent) neurons and their supporting cells located along the posterior root of a spinal nerve; also called a dorsal or sensory root ganglion.

Postganglionic neuron (pōst"gang-glĭ-ON-ik NU-ron) The second visceral efferent neuron in an autonomic pathway, having its cell body and dendrites located in an autonomic ganglion and its unmyelinated axon ending at cardiac muscle, smooth muscle, or a gland.

Preganglionic neuron (pre"gang-glĭ-ON-ik) The first visceral efferent neuron in an autonomic pathway, with its cell body and dendrites in the brain or spinal cord and its myelinated axon ending at an autonomic ganglion where it synapses with a postganglionic neuron.

Prepuce (PRE-pus) The foreskin; a fold of loosely fitting skin covering the glans of the penis or clitoris.

Prevertebral ganglion (pre-VER-tĕ-bral GANG-glĭ-on) A cluster of cell bodies of postganglionic sympathetic neurons lying anterior to the vertebral column and close to the large abdominal artery from which its name is derived; also called a collateral ganglion; examples are the superior mesenteric, inferior mesenteric, and celiac ganglia.

Pronation (pro-NA-shun) A movement of the flexed forearm in which the palm of the hand is turned backward (posteriorly).

Prophase (PRO-fāz) The first stage in mitosis.

Proprioceptor (pro"pre-o-SEP-tor) A receptor located in a muscle, tendon, or joint that provides information about body position and movements.

Prostate gland (PROS-tāt) A muscular and glandular doughnut-shaped organ inferior to the urinary bladder that surrounds the upper portion of the male urethra.

Protraction (pro-TRAK-shun) The movement of the mandible or clavicle forward on a plane parallel with the ground.

Proximal (PROK-sĭ-mal) Nearer the attachment of a limb to the trunk.

Pterygopalatine ganglion (ter"ĭ-go-PAL-ah-tĭn GANG-glĭ-on) A cluster of cell bodies of parasympathetic postganglionic neurons ending at the lacrimal and nasal glands.

Ptosis (TO-sis) A condition of dropping or downward displacement of an organ or body structure, for example, the kidney or the upper eyelid.

Pulmonary circulation (PUL-mo-ner"e) The flow of deoxygenated blood from the right ventricle to the lungs and the return of oxygenated blood from the lungs to the left atrium.

Pulp cavity A cavity within the crown and neck of a tooth, filled with pulp, a connective tissue containing blood vessels, nerves, and lymphatics.

Pupil The black hole in the center of the iris, the area through which light enters the posterior cavity of the eyeball.

Purkinje fibers (pur-KIN-je) Muscle fibers in the subendocardial tissue of the heart specialized for conducting an impulse to the myocardium; part of the conduction system of the heart.

Pyloric sphincter (pi-LOR-ik SFINGK-ter) A thickened ring of smooth muscle through which the pylorus of the stomach communicates with the duodenum; also called the pyloric valve.

Pyramids (PIR-ah-midz) Pointed or cone-shaped structures; two roughly triangular structures on the ventral side of the medulla composed of the largest motor tracts that run from the cerebral cortex to the spinal cord; triangular-shaped structures in the renal medulla composed of the straight segments of renal tubules.

Raphe (RA-fe) A seam or ridge.

Receptor (re-SEP-tor) A sense organ that receives stimuli from the environment; the distal end of a dendrite of a sensory neuron.

Rectouterine pouch (rek"to-U-ter-in) A pocket formed by the parietal peritoneum as it moves posteriorly from the surface of the uterus and is reflected onto the rectum; the lowest point in the pelvic cavity; also called the pouch or cul de sac of Douglas.

Rectum (REK-tum) The last 20 cm of the gastrointestinal tract, from the sigmoid colon to the anus.

Red nucleus A cluster of cell bodies in the midbrain, occupying a large portion of the tegmentum, and sending fibers into the rubroreticular and rubrospinal tracts.

Reflex arc The most basic conduction pathway through the nervous system, connecting a receptor and an effector and consisting of a receptor, a sensory neuron, a center in the central nervous system for a synapse, a motor neuron, and an effector.

Regional anatomy The division of anatomy dealing with a specific region of the body, such as the head, neck, chest, or abdomen.

Renal pelvis (RE-nal) A cavity in the center of the kidney formed by the expanded, proximal portion of the ureter, lying within the kidney, and into which the major calyces open.

Renal pyramid A triangular-shaped structure in the renal medulla composed of the straight segments of renal tubules.

Respiration (res-pĭ-RA-shun) The exchange of gases between the atmosphere and the cells.

Rete testis (RE-te TES-tis) A network of ducts in the center of the testis formed by the straight tubules.

Reticular formation (rĕ-TIK-u-lar) A network of small groups of nerve cells scattered among bundles of fibers beginning in the medulla as a continuation of the spinal cord and extending upward through the central part of the brain stem.

Reticulin (rĕ-TIK-u-lin) A fiber of small diameter in the matrix of connective tissue that branches to form a netlike supporting framework around fat cells, nerve fibers, muscle cells, and blood vessels; also called a reticular fiber.

Retina (RET-ĭ-nah) The inner coat of the eyeball, lying only in the posterior portion of the eye and consisting of nervous tissue and a pigmented layer comprised of epithelial cells lying in contact with the choroid.

Retraction (re-TRAK-shun) The movement of a protracted part of the body backward on a plane parallel to the ground, as in pulling the lower jaw back in line with the upper jaw.

Retroperitoneal (ret"ro-per-ĭ-to-NE-al) Behind the peritoneum.

Rhodopsin (ro-DOP-sin) A photo-sensitive, reddish-purple pigment in rod cells of the retina, consisting of the protein scotopsin plus retinene; visual purple.

Ribosomes (RI-bo-sōms) Organelles in the cytoplasm of cells, composed of ribosomal RNA, that synthesize proteins; nicknamed the "protein factories."

Rickets (RIK-ets) A disease of metabolism affecting children, caused by a vitamin D deficiency in which normal ossification does not take place, often resulting in deformities.

Right lymphatic duct (lim-FAT-ik) A vessel of the lymphatic system that drains lymph from the upper right side of the body and empties it into the right subclavian vein.

Rod A visual receptor in the retina of the eye that is specialized for vision in dim light.

Roentgenogram (RENT-gen-o-gram") A film exposed to X-rays.

Root canal A narrow extension of the pulp cavity lying within the root of a tooth.

Root hair plexus (PLEK-sus) A network of dendrites arranged around the root of a hair as free or naked nerve endings that are stimulated when a hair shaft is moved.

Rotation (ro-TA-shun) Moving a bone around its own axis, with no other movement permitted.

Round ligament (LIG-ah-ment) A band of fibrous connective tissue enclosed between the folds of the broad ligament of the uterus; it emerges from a point on the uterus just below the uterine tube, extends laterally along the pelvic wall, and penetrates the abdominal wall through the deep inguinal ring to end in the labia majora.

Round window A small opening between the middle and inner ear, directly below the oval window, covered by the second tympanic membrane; also called the fenestra cochleae.

Rugae (RU-je) Temporary large folds that appear in the mucosa of an empty hollow organ that gradually smooth out and disappear as the organ is distended; for example, in the mucosa of the stomach and vagina.

Saccule (SAK-ūl) The lower and smaller of the two chambers in the membranous labyrinth inside the vestibule of the inner ear containing a receptor organ for equilibrium.

Sacral (SA-kral) Pertaining to the sacrum.

Sacral plexus (PLEK-sus) A network formed by the anterior branches of spinal nerves L4 through S3.

Sacral promontory (PROM-on-tor-e) The superior surface of the body of the first sacral vertebra that projects anteriorly into the pelvic cavity; a line from the sacral promontory to the superior border of the symphysis pubis divides the abdominal and pelvic cavities.

Saddle joint A synovial or diarthrotic joint in which the articular surfaces of both of the bones is saddle-shaped or concave in one direction and convex in the other direction, as in the joint between the trapezium and the metacarpal of the thumb.

Sagittal (SAJ-ĭ-tal) A plane that runs vertically to the ground and divides the body into unequal left and right portions; also called parasagittal.

Salivary gland (SAL-ĭ-ver-e) The three pairs of glands that lie outside the mouth and pour their secretory product called saliva into ducts that empty into the oral cavity; the parotid, submandibular, and sublingual glands.

Sarcolemma (sar"ko-LEM-mah) The cell membrane of a muscle cell, especially of a skeletal muscle cell.

Sarcomere (SAR-ko-mēr) A contractile unit in a striated muscle cell extending from one Z-line to the next Z-line.

Sarcoplasm (SAR-ko-plazm) The cytoplasm of a striated muscle cell.

Scala tympani (SKA-lah TIM-pan-e) The lower spiral-shaped channel of the bony cochlea, filled with perilymph.

Scala vestibuli (ves-TIB-u-le) The upper spiral-shaped channel of the bony cochlea, filled with perilymph.

Schwann cell (shwon) A glial cell of the peripheral nervous system that forms the myelin sheath and neurilemma of a nerve fiber by spiraling around a nerve fiber in a jelly-roll fashion.

Sclera (SKLE-rah) The white coat of fibrous tissue that forms the outer protective covering over the eyeball except in the area of the anterior cornea; the posterior portion of the fibrous tunic.

Scoliosis (sko"le-O-sis) An abnormal curvature sideways (laterally) from the normal vertical line of the spine.

Scrotum (SKRO-tum) A skin-covered pouch that contains the testes and their accessory organs.

Sebaceous gland (se-BA-shus) An exocrine gland in the dermis of the skin, almost always associated with a hair follicle, that secretes sebum.

Segmental bronchi (seg-MEN-tal BRONG-ki) Branches of the secondary bronchi that lead into the segmental subdivisions of the lobes of the lungs.

Semen (SE-men) A fluid discharged at ejaculation by a male that consists of a mixture of spermatozoa and the secretions of the seminal vesicles, the prostate gland, and the bulbourethral glands.

Semicircular canals Three bony channels projecting upward and posteriorly from the vestibule of the inner ear, filled with perilymph, in which lie the membranous semicircular canals filled with endolymph; they contain receptors for equilibrium.

Semicircular ducts The membranous semicircular canals filled with endolymph and floating in the perilymph of the bony semicircular canals.

Semilunar valve (sem"i-LU-nar) A valve guarding the entrance into the aorta or the pulmonary trunk from a ventricle of the heart.

Seminal vesicles (SEM-ĭ-nal VES-ĭ-klz) A pair of convoluted pouchlike structures lying posterior and inferior to the urinary bladder, anterior to the rectum, that secrete their component of semen into the ejaculatory ducts.

Seminiferous tubule (sem"ĭ-NĬ-fer-us TU-būl) A tightly coiled duct, located in a lobule of the testis, where spermatozoa are produced.

Semipermeable Having pores of a size that will permit some but not all substances to diffuse through a membrane; differentially permeable.

Sensation A state of awareness of external or internal conditions of the body.

Sensory area A region of the cerebral cortex concerned with the interpretation of sensory impulses.

Septum (SEP-tum) A wall dividing two cavities.

Serosa (ser-O-sah) Any serous membrane; the outermost layer or tunic of an organ formed by a serous membrane, the membrane that lines the pleural, pericardial, and peritoneal cavities.

Serous membrane (SIR-us) An epithelial membrane (a combination of an epithelial and a connective tissue layer) that lines body cavities that do not open to the exterior and that covers the organs that lie within these cavities; each consists of two parts: a parietal portion and a visceral portion.

Sertoli cells (ser-TO-le) "Nurse" cells located in seminiferous tubules embedded between the developing spermatozoa that produce secretions for the supplying of nutrients to the spermatozoa.

Sesamoid bone (SES-ah-moyd) A bone that forms in a tendon and near a joint; for example, the patella.

Shoulder A synovial or diarthrotic joint where the humerus joins the scapula.

Sigmoid colon (SIG-moyd KO-lon) The S-shaped portion of the large intestine that begins at the level of the left iliac crest, projects inward to the midline, and terminates at the rectum at about the level of the third sacral vertebra.

Sinuatrial node (si"nu-A-tre-al) A compact mass of cardiac muscle cells specialized for conduction, located in the right atrium beneath the opening of the superior vena cava; also called the sinoatrial node, SA node, or pacemaker.

Sinusoid (SI-nus-oyd) A blood space in certain organs, as the liver or spleen.

Sliding-filament hypothesis The most commonly accepted explanation for muscle contraction in which actin and myosin move into interdigitation with each other, decreasing the length of the sarcomeres.

Slipped disc The popular name for a rupture of an intervertebral disc so that the nucleus pulposus protrudes into the vertebral cavity; also called a ruptured or herniated disc.

Small intestine A long tube of the gastrointestinal tract that begins at the pyloric valve of the stomach, coils through the central and lower part of the abdominal cavity, and ends at the large intestine; the small intestine is divided into three segments: duodenum, jejunum, and ileum.

Soft palate (PAL-at) The posterior portion of the roof of the mouth, extending posteriorly from the palatine bones and ending at the uvula; it is a muscular partition lined with mucous membrane.

Somatic nervous system (so-MAT-ik) The portion of the peripheral nervous system made up of the somatic efferent fibers that run between the central nervous system and the skeletal muscles and skin.

Somatic reflex (RE-fleks) A reflex arc in which the effector is a skeletal muscle.

Spermatic cord (sper-MAT-ik) A supporting structure of the male reproductive system, extending from a testis to the deep inguinal ring, that includes the ductus deferens, arteries, veins, lymphatics, nerves, cremaster muscle, and connective tissue.

Spermatogenesis (sper"mah-to-JEN-ĭ-sis) The formation and development of spermatozoa.

Spermatozoan (sper"mah-to-ZO-an) A mature male germ cell or sperm cell.

Sphincter of Oddi (SFINGK-ter of O-de) A circular muscle at the opening of the common bile and main pancreatic ducts in the duodenum that protrudes as a mass of tissue called the ampulla of Vater.

Spina bifida (SPI-nah BIF-ĭ-dah) A defect of the vertebral column in which the two halves of the neural arch of a vertebra fail to fuse in midline.

Spinal nerve (SPI-nal) One of the 31 pairs of nerves that originates on the spinal cord.

Spinal puncture Withdrawal of some of the cerebrospinal fluid from the subarachnoid space in the lumbar region.

Spinous process or **spine** (SPI-nus) A sharp or thornlike process or projection; a sharp ridge running diagonally across the posterior surface of the scapula.

Spongy or **cancellous bone** (SPUN-je or KAN-sel-lus) Bone tissue containing many large spaces filled with marrow.

Stapes (STA-pēz) The innermost of the three auditory ossicles of the middle ear, the one that fits into the oval window; also called the stirrup.

Stereocilia (ste"re-o-SIL-e-ah) Groups of extremely long, slender, nonmotile microvilli projecting from epithelial cells lining the epididymis.

Stomach The J-shaped enlargement of the gastrointestinal tract directly under the diaphragm in the epigastric, umbilical, and left hypochondriac regions of the abdomen, between the esophagus and the small intestine.

Straight tubules (TU-būlz) The ducts in a testis that lead from the convoluted seminiferous tubules to the rete testis.

Stratum basalis (STRA-tum ba-SAL-is) The outer layer of the endometrium, next to the myometrium, that is maintained during menstruation and gestation and produces a new functionalis following menstruation and parturition.

Stratum functionalis (funk"shun-AL-is) The inner layer of the endometrium, the layer next to the uterine cavity, that is shed during menstruation and that forms the maternal portion of the placenta during gestation.

Stroma (STRO-mah) The tissue that forms the ground substance, foundation, or framework of an organ, as opposed to its functional parts.

Subarachnoid space (sub"ah-RAK-noyd) A wide space between the arachnoid and the pia mater that surrounds the brain and spinal cord and through which cerebrospinal fluid circulates.

Subcutaneous layer or **superficial fascia** (sub"ku-TA-ne-us or su"per-FISH-al FASH-e-ah) A continuous sheet of fibrous connective tissue between the dermis of the skin and the deep fascia of the muscles.

Subdural space (sub-DU-ral) A space between the dura mater and the arachnoid of the brain and spinal cord that contains a small amount of fluid.

Sublingual gland (sub-LING-gwal) One of a pair of salivary glands situated in the floor of the mouth under the mucous membrane and to the side of the lingual frenulum, with a duct that opens into the floor of the mouth; the smallest of the three pairs of salivary glands.

Submandibular ganglion (sub-man-DIB-u-lar GANG-lĭ-on) A cluster of cell bodies of postganglionic parasympathetic neurons, located above the submandibular gland with fibers ending at the submandibular and sublingual salivary glands and other small salivary glands in the floor of the mouth.

Submandibular gland One of a pair of salivary glands found beneath the base of the tongue under the mucous membrane in the posterior part of the floor of the mouth, posterior to the sublingual glands, with a duct situated to the side of the lingual frenulum; also called the submaxillary gland.

Submucosa (sub-mu-KO-sah) A layer of connective tissue located beneath a mucous membrane, as in the alimentary canal where a submucosa binds the mucosa to the muscularis tunic, and in the urinary bladder where a submucosa connects the mucosa to the muscularis tunic.

Subserous fascia (sub-SE-rus FASH-e-ah) A layer

of connective tissue internal to the deep fascia, lying between the deep fascia and the serous membrane that lines the body cavities.

Sulcus (SUL-kus) A groove or depression between parts; especially a fissure between the convolutions of the brain; the plural form is sulci.

Superficial fascia (soo"per-FISH-al FASH-e-ah) A continuous sheet of fibrous connective tissue between the dermis of the skin and the deep fascia of the muscles; also called the subcutaneous layer.

Superior or **cephalic** (su-PĒR-e-or or sĕ-FAL-ik) Toward the head; toward the upper part of a structure.

Supination (su"pĭ-NA-shun) A movement of the forearm in which the palm of the hand is turned forward (anteriorly).

Surface anatomy The study of the structures that can be identified from the outside of the body.

Surfactant (sur-FAK-tant) A substance formed in the lungs that helps to keep the small air sacs expanded by reducing surface tension.

Suspensory ligament (sus-PEN-so-re LIG-ah-ment) A fold of peritoneum extending laterally from the surface of the ovary to the pelvic wall.

Suture (SU-chur) A type of fibrous joint found between bones of the skull where the bones are very slightly separated by a thin layer of fibrous tissue, with the result that the joints are immovable.

Sweat gland A gland widely distributed through the skin, particularly in the skin of the palms, soles, armpits, and forehead, with the secretory portion lying in the subcutaneous tissue and the excretory duct projecting upward through the dermis and epidermis to open in a pore at the surface.

Sympathetic (sim"pah-THET-ik) One of the two subdivisions of the autonomic nervous system, having cell bodies of preganglionic neurons in the lateral gray columns of the thoracic segment and first two or three lumbar segments of the spinal cord, primarily concerned with processes involving the expenditure of energy; the thoracolumbar division.

Sympathetic trunk ganglion (GANG-lĭ-on) A cluster of cell bodies of postganglionic sympathetic neurons lateral to the vertebral column, close to the body of a vertebra. These ganglia extend downward through the neck, thorax, and abdomen to the coccyx, on both sides of the vertebral column and are connected to one another to form a chain on each side of the vertebral column; also called lateral or sympathetic chain or vertebral chain ganglia.

Symphysis (SIM-fĭ-sis) A slightly movable cartilaginous joint in which the material connecting two bones is a broad, flat disc of fibrocartilage, as between the bodies of vertebrae and between the anterior surfaces of the pelvic bones.

Symphysis pubis (PU-bis) A slightly movable cartilaginous joint between the anterior surfaces of the pelvic bones.

Synapse (SIN-aps) A small gap that serves as the functional junction between two neurons, where a nerve impulse is conducted from one neuron to another by a neurohumor.

Synarthrosis (sin"ar-THRO-sis) An immovable joint, such as a suture, synchondrosis, and synostosis.

Synchondrosis (sin"kon-DRO-sis) An immovable cartilaginous joint in which the connecting material is hyaline cartilage; a synchondrosis may be temporary, as in the epiphyseal plate, or permanent, as between the ribs and the sternum.

Syndesmosis (sin"des-MO-sis) A fibrous joint in which two bones are united by dense fibrous tissue, producing a slightly movable joint such as the distal articulation between the tibia and the fibula.

Synergist (SIN-er-jist) A group of muscles that assists the prime mover or agonist by reducing undesired action or unnecessary movement; also called a fixator.

Synostosis (sin"os-TO-sis) An immovable joint in which the connecting tissue between two bones is bone, as in the epiphyseal line that replaces the cartilaginous epiphyseal plate of a growing bone.

Synovial cavity (sin-O-ve-al) The space between the articulating bones of a synovial or diarthrotic joint, filled with synovial fluid.

Synovial joint A fully movable or diarthrotic joint in which a joint or synovial cavity is present between the two articulating bones.

Synovial membrane The inner of the two layers of the articular capsule of a synovial joint, composed of loose connective tissue covered with epithelium that secretes synovial fluid into the joint cavity.

System An association of organs that have a common function.

Systemic (sis-TEM-ik) Affecting the whole body; generalized.

Systemic anatomy The study of particular systems of the body, such as the system of nerves, spinal cord, and brain, or the system of heart, blood vessels, and blood.

Systemic circulation All of the circulatory routes taken by oxygenated blood that leaves the left ventricle through the aorta and returns to the right atrium, carrying oxygenated blood to all of the organs of the body.

T tubule (TU-būl) In a muscle cell, an invagination of the sarcolemma that runs transversely through the fiber perpendicular to the sarcoplasmic reticulum.

Taeniae coli (TE-nĭ-e KO-li) Three flat bands of smooth muscle tissue, formed by concentrating the longitudinal muscle layer into bands instead of a sheet, that run the length of the large intestine and are easily seen in a gross specimen.

Target organ The organ or group of organs affected by a particular hormone.

Tarsal plate (TAR-sal) A thin, elongated sheet of connective tissue, one in each eyelid, giving the eyelid form and support; the aponeurosis of the levator palpebrae superioris is attached to the tarsal plate of the superior eyelid.

Tarsus (TAR-sus) A collective designation for the seven bones of the ankle.

Tectorial membrane (tek-TO-re-al) A gelatinous membrane projecting over and in contact with the hair cells of the organ of Corti in the cochlear duct.

Telophase (TEL-o-fāz) The final stage of mitosis in which the nuclei become established.

Tendinitis (ten"din-I-tis) A disorder involving the inflammation of a tendon and synovial membrane at a joint.

Tendon (TEN-don) A white fibrous cord of dense, regular connective tissue that attaches muscle to bone.

Tentorium cerebelli (ten-TO-re-um ser"ĕ-BEL-ē) A transverse shelf of dura mater that forms a partition between the occipital part of the cerebral hemispheres and the cerebellum and that covers the cerebellum.

Terminal ganglion (TER-min-al GANG-lĭ-on) A cluster of cell bodies of postganglionic parasympathetic neurons either lying very close to the visceral effectors or located within the walls of the visceral effectors supplied by the postganglionic fibers.

Testosterone (tes-TOS-ter-ōn) A hormone secreted by the interstitial cells (of Leydig) of a mature testis.

Thalamus (THAL-ah-mus) A large, oval structure, located above the midbrain, consisting of two masses of gray matter covered by a thin layer of white matter.

Thigh The portion of the lower extremity between the hip and the knee.

Third ventricle (VEN-trĭ-kl) A slitlike cavity between the right and left halves of the thalamus and between the lateral ventricles.

Thoracic (thor-Ă-sik) Pertaining to the chest.

Thoracic duct A lymphatic vessel that begins as a dilation called the cisterna chyli and receives lymph from the left side of the head, neck, and chest, and the left arm, and the entire body below the ribs, finally emptying the lymph into the left subclavian vein; also called the left lymphatic duct.

Thoracolumbar outflow (tho"rah-ko-LUM-bar) The sympathetic division of the autonomic nervous system, so called because the preganglionic sympathetic neurons have their cell bodies in the lateral gray columns of the thoracic segment and first two or three lumbar segments of the spinal cord.

Thorax (THO-raks) The chest.

Thrombocyte (THROM-bo-sīt) A blood platelet, a fragment of a megakaryocyte, that plays a role in the chain of reactions that results in blood clotting.

Thymus gland (THI-mus) A bilobed organ, located in the upper mediastinum posterior to the sternum and between the lungs, that plays a role in the immunity mechanism of the body.

Thyroid cartilage (THI-royd KAR-tĭ-lij) The largest single cartilage of the larynx, consisting of two fused plates that form the anterior wall of the larynx; also called the Adam's apple.

Tissue A group of similar cells, and their intercellular substance, joined together to perform a specific function.

Tongue A large skeletal muscle on the floor of the oral cavity.

Tonsil (TON-sil) A mass of lymphoid tissue embedded in mucous membrane.

Trabeculae (tra-BEK-u-le) Fibrous cords of connective tissue, serving as supporting fiber by forming a septum extending into an organ from its walls or capsule; an anastomosing network of lamellae of spongy bone, with the spaces filled with red marrow.

Trachea (TRA-ke-ah) The windpipe.

Tract A bundle of nerve fibers in the central nervous system.

Transection (tran-SEK-shun) A cross cut.

Transverse colon (trans-VERS KO-lon) The portion of the large intestine extending across the abdomen from the hepatic flexure to the splenic flexure.

Transverse fissure (FISH-er) The deep cleft that separates the cerebrum from the cerebellum.

Triad (TRI-ad) A complex of three units in a muscle cell composed of a T tubule and the segments of sarcoplasmic reticulum on both sides of it.

Trigone (TRI-gōn) A triangular area in the base of the urinary bladder between the ureteral and urethral orifices where the mucous membrane is firmly attached to the muscular coat and is always smooth.

Trochanter (tro-KAN-ter) A large projection on the femur that serves as a point of muscle attachment.

True vocal cords A pair of folds of the mucous membrane of the larynx enclosing two strong bands of fibrous tissue, the vocal ligaments, that vibrate to make sounds during speaking; the true vocal cords are inferior to the false vocal folds.

Trunk The part of the body to which the upper and lower extremities are attached.

Tubercle (TU-ber-kl) A small elevation.

Tuberosity (tu"bĕ-ROS-ĭ-te) An elevation or protuberance.

Tumor (TU-mor) A swelling or enlargement.

Tunica albuginea (TU-nĭ-kah al"bu-JIN-e-ah) A tough, whitish layer of fibrous tissue around an organ; for example, covering the testis.

Tunica externa (eks-TER-nah) The white, fibrous outer coat of an artery or vein.

Tunica interna (in-TER-nah) The inner coat of an artery or vein, consisting of a lining of endothelium and its supporting layer of connective tissue; also called the tunica intima.

Tunica media (ME-de-ah) The middle coat of an artery or vein composed of smooth muscle and elastic fibers.

Tympanic antrum (tim-PAN-ik AN-trum) An air space in the posterior wall of the middle ear that leads into the mastoid air cells or sinus.

Tympanic membrane The eardrum.

Ulcer (UL-ser) An open lesion upon the skin or mucous membrane of the body, with loss of substance and necrosis of the tissue.

Upper extremity The appendage attached at the shoulder, consisting of an arm, a forearm, and a hand.

Ureter (u-RE-ter) The duct that conveys urine from the kidney to the urinary bladder.

Urethra (u-RE-thrah) The duct that conveys urine from the urinary bladder to the exterior of the body.

Urinary bladder (U-rĭ-ner-e) A hollow, muscular organ situated in the pelvic cavity posterior to the symphysis pubis.

Urogenital triangle (u"ro-JEN-ĭ-tal) The region of the pelvic floor below the symphysis pubis, bounded by the symphysis pubis and the ischial tuberosities and containing the external genitalia.

Uterine tubes (U-ter-in) Ducts that transport ova from the ovary to the uterus; also called the Fallopian tubes or the oviducts.

Uterosacral ligament (u"ter-o-SA-kral LIG-ah-ment) A fibrous band of tissue extending from the cervix of the uterus laterally to attach to the sacrum.

Uterovesical pouch (u"ter-o-VES-ĭ-kal) A shallow pouch formed by the reflection of the peritoneum from the anterior surface of the uterus, at the junction of the cervix and the body, to the posterior surface of the urinary bladder.

Uterus (U-ter-us) An inverted, pear-shaped, hollow organ situated between the bladder and the rectum in the female pelvis; the womb.

Utricle (U-trĭ-kl) The larger of the two divisions of the membranous labyrinth located inside the vestibule of the inner ear.

Vagina (vah-JI-nah) A muscular, tubular organ that leads from the uterus to the vestibule, situated between the urinary bladder and the rectum of the female.

Varicose vein (VAR-ĭ-kōs) A swollen or distended vein, especially in the legs, due to the failure of the valves to prevent the backflow of blood.

Vasa vasorum (VA-sah va-SO-rum) Blood vessels supplying nutrients to the larger arteries and veins.

Vascular tunic (VAS-ku-lar TU-nik) The middle layer of the eyeball, composed of the choroid, the ciliary body, and the iris.

Vasectomy (vah-SEK-to-me) A means of sterilization of males in which a portion of each ductus deferens is removed.

Vein A blood vessel that conveys blood from the tissues back to the heart.

Venae cavae (VE-ne KA-ve) The two large veins that open into the right atrium, returning to the heart all of the deoxygenated blood from the systemic circulation except from the coronary circulation.

Ventilation (ven"tĭ-LA-shun) Breathing, or the process by which atmospheric gases are drawn down into the lungs and waste gases that have diffused into the lungs are expelled back up through the respiratory passageways.

Ventral ramus (VEN-tral RA-mus) The anterior branch of a spinal nerve, containing sensory and motor fibers to the muscles and skin of the anterior surface of the head, neck, and trunk and the extremities.

Ventricle (VEN-trĭ-kl) One of the two lower chambers of the heart; a cavity within the brain, containing cerebrospinal fluid.

Venule (VEN-ul) A small vein.
Vermiform appendix (VER-mĭ-form ah-PEN-diks) A twisted, coiled tube attached to the cecum.
Vermis (VER-mis) The central constricted area of the cerebellum that separates the two cerebellar hemispheres.
Vertebral or **spinal canal** (VER-tĕ-bral or SPI-nal) The cavity formed by the vertebral foramina of all vertebrae together; the inferior portion of the dorsal cavity, containing the spinal cord.
Vestibular membrane (ves-TIB-u-lar) The membrane that separates the cochlear duct from the scala vestibuli; also called Reissner's membrane.
Vestibule (VES-tib-ūl) A small space or cavity at the beginning of a canal; especially the inner ear, larynx, mouth, nose, vagina.
Villus (VIL-us) A minute, fingerlike projection of the intestinal mucosa into the lumen of the intestine that increases the surface area for absorption.
Viscera (VIS-er-ah) Organs within body cavities, especially within the abdomen; the singular form is viscus.
Visceral (VIS-er-al) Pertaining to the covering of an organ.
Visceral autonomic reflex (aw"to-NOM-ik RE-fleks) A quick, involuntary response in which the impulse travels over visceral efferent neurons to smooth muscle, cardiac muscle, or a gland.
Visceral effector (ĕ-FEK-tor) Cardiac muscle, smooth muscle, and glandular epithelium.
Visceral pleura (PLU-rah) The inner layer of the serous membrane that covers the lungs.
Visceroceptor (vis"er-o-SEP-tor) A sensory receptor, located in blood vessels and viscera, that picks up information about the internal environment.
Vitreous humor (VIT-re-us HU-mor) A soft, jelly-like substance that fills the posterior cavity of the eyeball, lying between the lens and the retina.
Volkmann's canal (FOLK-mahns) A minute passageway by means of which blood vessels and nerves from the periosteum penetrate into compact bone.
Vulva (VUL-vah) The external genitalia of the female; also called the pudendum.

Wallerian degeneration (wah-LER-ĭ-an) The changes that occur in a cut nerve cell proximal to the site of injury, first described by Waller, making it possible to follow peripheral nerve fibers into the central nervous system; the changes that occur in the proximal portion of the cut neuron include fragmentation of the fiber retrograde to the first node of Ranvier, phagocytosis of the myelin sheath, the Nissl reaction or chromatolysis in which the Nissl substance is dispersed in the cytoplasm, and the displacement of the nucleus to the periphery of the cell.
White matter Aggregations or bundles of myelinated axons located in the brain and spinal cord.
White pulp The portion of the spleen composed of ovoid masses of lymphoid tissue called lymph follicles that contain germinal centers and where lymphocytes are produced.
White ramus communicans (RA-mus com-MU-nĭ-kans) The portion of a preganglionic sympathetic nerve fiber that branches away from the anterior ramus of a spinal nerve to enter the nearest sympathetic trunk ganglion.
Wormian or **sutural bone** A small bone located within a suture or immovable joint of certain cranial bones.

Zona fasciculata (ZO-nah fah-sik"u-LAH-tah) The middle zone of the adrenal cortex that consists of cells arranged in long, straight cords and that secretes glucocorticoid hormones.
Zona glomerulosa (glo-mer"u-LO-sah) The outer zone of the adrenal cortex, directly under the connective tissue covering, that consists of cells arranged in arched loops or round balls and that secretes mineralocorticoids.
Zona reticularis (ret-ik"u-LAR-is) The inner zone of the adrenal cortex, consisting of cords of branching cells that secrete sex hormones, chiefly androgens.
Zymogenic cell (zi"mo-JEN-ik) A cell that secretes enzymes; for example, the chief cells of the gastric glands that secrete pepsinogen.

INDEX

Printer and Binder: Kingsport Press